Plate 2

Sufficiency of Boone's lemma

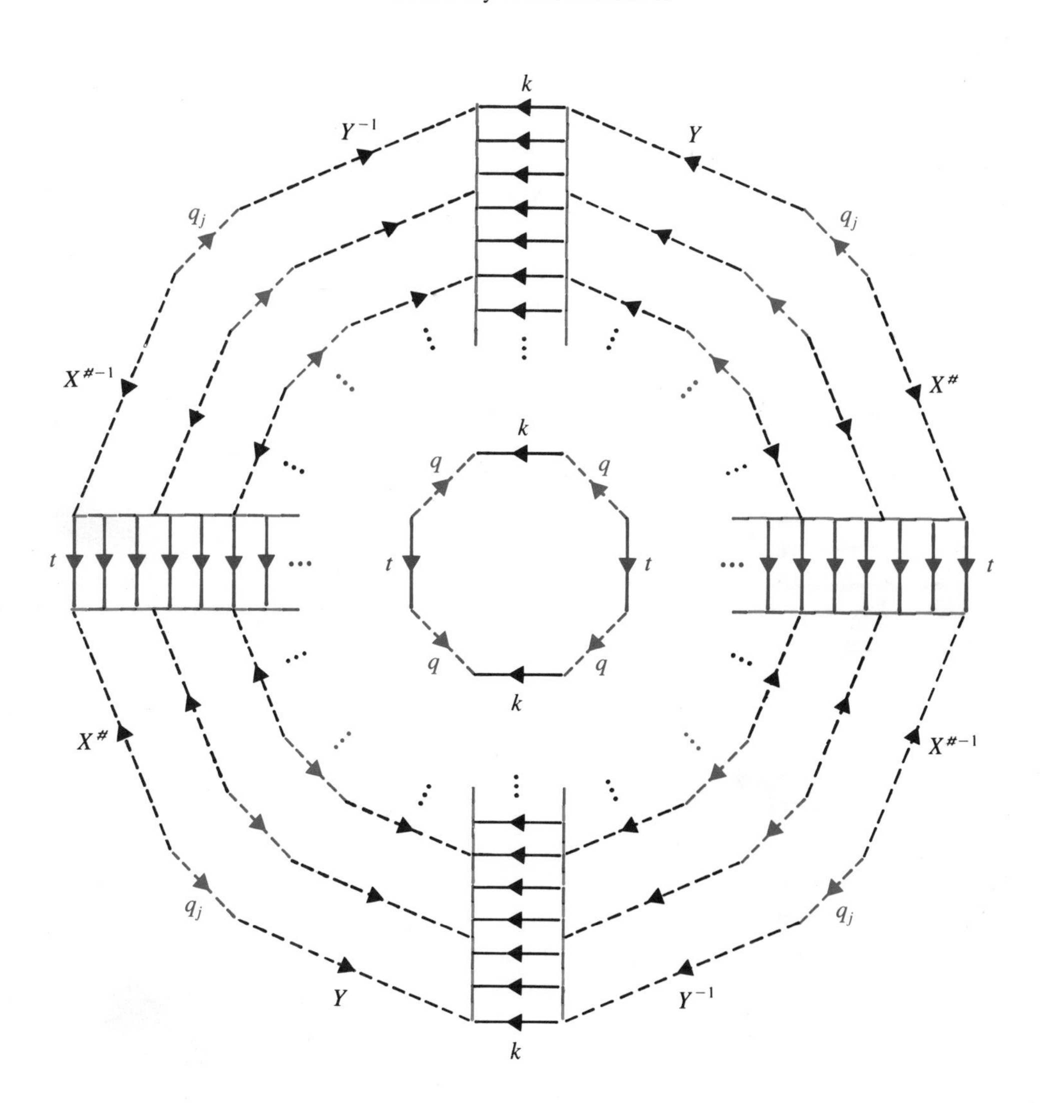

Graduate Texts in Mathematics 148

Graduate Texts in Mathematics

1. TAKEUTI/ZARING. Introduction to Axiomatic Set Theory. 2nd ed.
2. OXTOBY. Measure and Category. 2nd ed.
3. SCHAEFFER. Topological Vector Spaces.
4. HILTON/STAMMBACH. A Course in Homological Algebra.
5. MAC LANE. Categories for the Working Mathematician.
6. HUGHES/PIPER. Projective Planes.
7. SERRE. A Course in Arithmetic.
8. TAKEUTI/ZARING. Axiomatic Set Theory.
9. HUMPHREYS. Introduction to Lie Algebras and Representation Theory.
10. COHEN. A Course in Simple Homotopy Theory.
11. CONWAY. Functions of One Complex Variable. 2nd ed.
12. BEALS. Advanced Mathematical Analysis.
13. ANDERSON/FULLER. Rings and Categories of Modules. 2nd ed.
14. GOLUBITSKY/GUILEMIN. Stable Mappings and Their Singularities.
15. BERBERIAN. Lectures in Functional Analysis and Operator Theory.
16. WINTER. The Structure of Fields.
17. ROSENBLATT. Random Processes. 2nd ed.
18. HALMOS. Measure Theory.
19. HALMOS. A Hilbert Space Problem Book. 2nd ed.
20. HUSEMOLLER. Fibre Bundles. 3rd ed.
21. HUMPHREYS. Linear Algebraic Groups.
22. BARNES/MACK. An Algebraic Introduction to Mathematical Logic.
23. GREUB. Linear Algebra. 4th ed.
24. HOLMES. Geometric Functional Analysis and Its Applications.
25. HEWITT/STROMBERG. Real and Abstract Analysis.
26. MANES. Algebraic Theories.
27. KELLEY. General Topology.
28. ZARISKI/SAMUEL. Commutative Algebra. Vol.I.
29. ZARISKI/SAMUEL. Commutative Algebra. Vol.II.
30. JACOBSON. Lectures in Abstract Algebra I. Basic Concepts.
31. JACOBSON. Lectures in Abstract Algebra II. Linear Algebra.
32. JACOBSON. Lectures in Abstract Algebra III. Theory of Fields and Galois Theory.
33. HIRSCH. Differential Topology.
34. SPITZER. Principles of Random Walk. 2nd ed.
35. WERMER. Banach Algebras and Several Complex Variables. 2nd ed.
36. KELLEY/NAMIOKA et al. Linear Topological Spaces.
37. MONK. Mathematical Logic.
38. GRAUERT/FRITZSCHE. Several Complex Variables.
39. ARVESON. An Invitation to C^*-Algebras.
40. KEMENY/SNELL/KNAPP. Denumerable Markov Chains. 2nd ed.
41. APOSTOL. Modular Functions and Dirichlet Series in Number Theory. 2nd ed.
42. SERRE. Linear Representations of Finite Groups.
43. GILLMAN/JERISON. Rings of Continuous Functions.
44. KENDIG. Elementary Algebraic Geometry.
45. LOÈVE. Probability Theory I. 4th ed.
46. LOÈVE. Probability Theory II. 4th ed.
47. MOISE. Geometric Topology in Dimensions 2 and 3.
48. SACHS/WU. General Relativity for Mathematicians.
49. GRUENBERG/WEIR. Linear Geometry. 2nd ed.
50. EDWARDS. Fermat's Last Theorem.
51. KLINGENBERG. A Course in Differential Geometry.
52. HARTSHORNE. Algebraic Geometry.
53. MANIN. A Course in Mathematical Logic.
54. GRAVER/WATKINS. Combinatorics with Emphasis on the Theory of Graphs.
55. BROWN/PEARCY. Introduction to Operator Theory I: Elements of Functional Analysis.
56. MASSEY. Algebraic Topology: An Introduction.
57. CROWELL/FOX. Introduction to Knot Theory.
58. KOBLITZ. p-adic Numbers, p-adic Analysis, and Zeta-Functions. 2nd ed.
59. LANG. Cyclotomic Fields.
60. ARNOLD. Mathematical Methods in Classical Mechanics. 2nd ed.
61. WHITEHEAD. Elements of Homotopy Theory.
62. KARGAPOLOV/MERLZJAKOV. Fundamentals of the Theory of Groups.
63. BOLLOBAS. Graph Theory.
64. EDWARDS. Fourier Series. Vol. I. 2nd ed.

continued after index

Joseph J. Rotman

An Introduction to the Theory of Groups

Fourth Edition

With 37 Illustrations

Springer-Verlag
New York Berlin Heidelberg London Paris
Tokyo Hong Kong Barcelona Budapest

Joseph J. Rotman
Department of Mathematics
University of Illinois
at Urbana-Champaign
Urbana, IL 61801
USA

Mathematics Subject Classifications (1991): 20-01

Library of Congress Cataloging-in-Publication Data
Rotman, Joseph J., 1934–
An introduction to the theory of groups / Joseph Rotman. — 4th ed.
p. cm. — (Graduate texts in mathematics)
Includes bibliographical references and index.
ISBN 0-387-94285-8
1. Group theory. I. Title. II. Series.
QA174.2.R67 1994 94-6507
512′.2—dc20

Printed on acid-free paper.

Production coordinated by Brian Howe and managed by Bill Imbornoni; manufacturing supervised by Gail Simon.
Typeset by Asco Trade Typesetting Ltd., Hong Kong.
Printed and bound by R.R. Donnelley & Sons, Harrisonburg, VA.
Printed in the United States of America.

9 8 7 6 5 4 3 2 1

ISBN 0-387-94285-8 Springer-Verlag New York Berlin Heidelberg
ISBN 3-540-94285-8 Springer-Verlag Berlin Heidelberg New York

לזכרון נצח אבי מורי הנכבד
אליהו בן יוסף יונה הלוי

Preface to the Fourth Edition

Group Theory is a vast subject and, in this Introduction (as well as in the earlier editions), I have tried to select important and representative theorems and to organize them in a coherent way. Proofs must be clear, and examples should illustrate theorems and also explain the presence of restrictive hypotheses. I also believe that some history should be given so that one can understand the origin of problems and the context in which the subject developed.

Just as each of the earlier editions differs from the previous one in a significant way, the present (fourth) edition is genuinely different from the third. Indeed, this is already apparent in the Table of Contents. The book now begins with the unique factorization of permutations into disjoint cycles and the parity of permutations; only then is the idea of group introduced. This is consistent with the history of Group Theory, for these first results on permutations can be found in an 1815 paper by Cauchy, whereas groups of permutations were not introduced until 1831 (by Galois). But even if history were otherwise, I feel that it is usually good pedagogy to introduce a general notion only after becoming comfortable with an important special case. I have also added several new sections, and I have subtracted the chapter on Homological Algebra (although the section on Hom functors and character groups has been retained) and the section on Grothendieck groups.

The format of the book has been changed a bit: almost all exercises now occur at ends of sections, so as not to interrupt the exposition. There are several notational changes from earlier editions: I now write $H \leq G$ instead of $H \subset G$ to denote "H is a subgroup of G"; the dihedral group of order $2n$ is now denoted by D_{2n} instead of by D_n; the trivial group is denoted by 1 instead of by $\{1\}$; in the discussion of simple linear groups, I now distinguish *elementary transvections* from more general *transvections*; I speak of the *fundamental group* of an abstract simplicial complex instead of its *edgepath group*.

Here is a list of some other changes from earlier editions.

Chapter 3. The *cycle index* of a permutation group is given to facilitate use of Burnside's counting lemma in coloring problems; a brief account of motions in the plane introduces bilinear forms and symmetry groups; the affine group is introduced, and it is shown how affine invariants can be used to prove theorems in plane geometry.

Chapter 4. The number of subgroups of order p^s in a finite group is counted mod p; two proofs of the Sylow theorems are given, one due to Wielandt.

Chapter 5. Assuming Burnside's $p^\alpha q^\beta$ theorem, we prove P. Hall's theorem that groups having p-complements are solvable; we give Ornstein's proof of Schur's theorem that $G/Z(G)$ finite implies G' finite.

Chapter 6. There are several proofs of the basis theorem, one due to Schenkman; there is a new section on operator groups.

Chapter 7. An explicit formula is given for every outer automorphism of S_6; stabilizers of normal series are shown to be nilpotent; the discussion of the wreath product has been expanded, and it is motivated by computing the automorphism group of a certain graph; the theorem of Gaschütz on complements of normal p-subgroups is proved; a second proof of Schur's theorem on finiteness of G' is given, using the transfer; there is a section on projective representations, the Schur multiplier (as a cohomology group), and covers; there is a section on derivations and H^1, and derivations are used to give another proof (due to Gruenberg and Wehrfritz) of the Schur–Zassenhaus lemma. (Had I written a new chapter entitled Cohomology of Groups, I would have felt obliged to discuss more homological algebra than is appropriate here.)

Chapter 8. There is a new section on the classical groups.

Chapter 9. An imbedding of S_6 into the Mathieu group M_{12} is used to construct an outer automorphism of S_6.

Chapter 10. Finitely generated abelian groups are treated before divisible groups.

Chapter 11. There is a section on coset enumeration; the Schur multiplier is shown to be a homology group via Hopf's formula; the number of generators of the Schur multiplier is bounded in terms of presentations; universal central extensions of perfect groups are constructed; the proof of Britton's lemma has been redone, after Schupp, so that it is now derived from the normal form theorem for amalgams.

Chapter 12. Cancellation diagrams are presented before giving the difficult portion of the proof of the undecidability of the word problem.

In addition to my continuing gratitude to those who helped with the first three editions, I thank Karl Gruenberg, Bruce Reznick, Derek Robinson, Paul Schupp, Armond Spencer, John Walter, and Paul Gies for their help on this volume.

Urbana, Illinois
1994

Joseph J. Rotman

From Preface to the Third Edition

Quand j'ai voulu me restreindre, je suis tombé dans l'obscurité;
j'ai préféré passer pour un peu bavard.

H. POINCARÉ, Analysis situs,
Journal de l'École Polytechnique, 1895, pp. 1–121.

Although permutations had been studied earlier, the theory of groups really began with Galois (1811–1832) who demonstrated that polynomials are best understood by examining certain groups of permutations of their roots. Since that time, groups have arisen in almost every branch of mathematics. Even in this introductory text we shall see connections with number theory, combinatorics, geometry, topology, and logic.

By the end of the nineteenth century, there were two main streams of group theory: topological groups (especially Lie groups) and finite groups. In this century, a third stream has joined the other two: infinite (discrete) groups. It is customary, nowadays, to approach our subject by two paths: "pure" group theory (for want of a better name) and representation theory. This book is an introduction to "pure" (discrete) group theory, both finite and infinite.

We assume that the reader knows the rudiments of modern algebra, by which we mean that matrices and finite-dimensional vector spaces are friends, while groups, rings, fields, and their homomorphisms are only acquaintances. A familiarity with elementary set theory is also assumed, but some appendices are at the back of the book so that readers may see whether my notation agrees with theirs.

I am fortunate in having attended lectures on group theory given by I. Kaplansky, S. Mac Lane, and M. Suzuki. Their influence is evident throughout in many elegant ideas and proofs. I am happy to thank once again those who helped me (directly and indirectly) with the first two editions: K.I. Appel, M. Barr, W.W. Boone, J.L. Britton, G. Brown, D. Collins, C. Jockusch, T. McLaughlin, C.F. Miller, III. H. Paley, P. Schupp, F.D. Veldkamp, and C.R.B. Wright. It is a pleasure to thank the following who helped with the present edition: K.I. Appel, W.W. Boone, E.C. Dade, F. Haimo, L. McCulloh, P.M. Neumann, E. Rips, A. Spencer, and J. Walter. I particularly thank F. Hoffman, who read my manuscript, for his valuable comments and suggestions.

Contents

To the Reader

Exercises in a text generally have two functions: to reinforce the reader's grasp of the material and to provide puzzles whose solutions give a certain pleasure. Here, the exercises have a third function: to enable the reader to discover important facts, examples, and counterexamples. The serious reader should attempt all the exercises (many are not difficult), for subsequent proofs may depend on them; the casual reader should regard the exercises as part of the text proper.

CHAPTER 1

Groups and Homomorphisms

Generalizations of the quadratic formula for cubic and quartic polynomials were discovered in the sixteenth century, and one of the major mathematical problems thereafter was to find analogous formulas for the roots of polynomials of higher degree; all attempts failed. By the middle of the eighteenth century, it was realized that permutations of the roots of a polynomial $f(x)$ were important; for example, it was known that the coefficients of $f(x)$ are "symmetric functions" of its roots. In 1770, J.-L. Lagrange used permutations to analyze the formulas giving the roots of cubics and quartics,[1] but he could not fully develop this insight because he viewed permutations only as rearrangements, and not as bijections that can be composed (see below). Composition of permutations does appear in work of P. Ruffini and of P. Abbati about 1800; in 1815, A.L. Cauchy established the calculus of permutations, and this viewpoint was used by N.H. Abel in his proof (1824) that there exist quintic polynomials for which there is no generalization of the qua-

[1] One says that a polynomial (or a rational function) f of μ variables is ***r-valued*** if, by permuting the variables in all possible ways, one obtains exactly r distinct polynomials. For example, $f(x_1, x_2, x_3) = x_1 + x_2 + x_3$ is a 1-valued function, while $g(x_1, x_2, x_3) = x_1x_2 + x_3$ is a 3-valued function.

To each polynomial $f(x)$ of degree μ, Lagrange associated a polynomial, called its *resolvent*, and a rational function of μ variables. We quote Wussing (1984, English translation, p. 78): "This connection between the degree of the resolvent and the number of values of a rational function leads Lagrange ... to consider the number of values that can be taken on by a rational function of μ variables. His conclusion is that the number in question is always a divisor of $\mu!$. ... Lagrange saw the 'metaphysics' of the procedures for the solution of algebraic equations by radicals in this connection between the degree of the resolvent and the valuedness of rational functions. His discovery was the starting point of the subsequent development due to Ruffini, Abel, Cauchy, and Galois.... It is remarkable to see in Lagrange's work the germ, in admittedly rudimentary form, of the group concept." (See Examples 3.3 and 3.3′ as well as Exercise 3.38.)

dratic formula. In 1830, E. Galois (only 19 years old at the time) invented groups, associated to each polynomial a group of permutations of its roots, and proved that there is a formula for the roots if and only if the group of permutations has a special property. In one great theorem, Galois founded group theory and used it to solve one of the outstanding problems of his day.

Permutations

Definition. If X is a nonempty set, a ***permutation*** of X is a bijection $\alpha: X \to X$. We denote the set of all permutations of X by S_X.

In the important special case when $X = \{1, 2, \ldots, n\}$, we write S_n instead of S_X. Note that $|S_n| = n!$, where $|Y|$ denotes the number of elements in a set Y.

In Lagrange's day, a permutation of $X = \{1, 2, \ldots, n\}$ was viewed as a rearrangement; that is, as a list $i_1, i_2, \ldots, i_n$ with no repetitions of all the elements of X. Given a rearrangement $i_1, i_2, \ldots, i_n$, define a function $\alpha: X \to X$ by $\alpha(j) = i_j$ for all $j \in X$. This function α is an injection because the list has no repetitions; it is a surjection because all of the elements of X appear on the list. Thus, every rearrangement gives a bijection. Conversely, any bijection α can be denoted by two rows:

$$\alpha = \begin{pmatrix} 1 & 2 & \ldots & n \\ \alpha 1 & \alpha 2 & \ldots & \alpha n \end{pmatrix},$$

and the bottom row is a rearrangement of $\{1, 2, \ldots, n\}$. Thus, the two versions of permutation, rearrangement and bijection, are equivalent. The advantage of the new viewpoint is that two permutations in S_X can be "multiplied," for the composite of two bijections is again a bijection. For example, $\alpha = \begin{pmatrix} 1 & 2 & 3 \\ 3 & 2 & 1 \end{pmatrix}$ and $\beta = \begin{pmatrix} 1 & 2 & 3 \\ 2 & 3 & 1 \end{pmatrix}$ are permutations of $\{1, 2, 3\}$. The product $\alpha\beta$ is $\begin{pmatrix} 1 & 2 & 3 \\ 2 & 1 & 3 \end{pmatrix}$; we compute this product[2] by first applying β and then α:

$$\alpha\beta(1) = \alpha(\beta(1)) = \alpha(2) = 2,$$

$$\alpha\beta(2) = \alpha(\beta(2)) = \alpha(3) = 1,$$

$$\alpha\beta(3) = \alpha(\beta(3)) = \alpha(1) = 3.$$

Note that $\beta\alpha = \begin{pmatrix} 1 & 2 & 3 \\ 1 & 3 & 2 \end{pmatrix}$, so that $\alpha\beta \neq \beta\alpha$.

[2] We warn the reader that some authors compute this product in the reverse order: first α and then β. These authors will write functions on the right: instead of $f(x)$, they write $(x)f$ (see footnote 4 in this chapter).

EXERCISES

1.1. The identity function 1_X on a set X is a permutation, and we usually denote it by 1. Prove that $1\alpha = \alpha = \alpha 1$ for every permutation $\alpha \in S_X$.

1.2. For each $\alpha \in S_X$, prove that there is $\beta \in S_X$ with $\alpha\beta = 1 = \beta\alpha$ (*Hint*. Let β be the *inverse function* of the bijection α).

1.3. For all $\alpha, \beta, \gamma \in S_X$, prove that $\alpha(\beta\gamma) = (\alpha\beta)\gamma$. Indeed, if X, Y, Z, W are sets and $f: X \to Y, g: Y \to Z$, and $h: Z \to W$ are functions, then $h(gf) = (hg)f$. (*Hint*: Recall that two functions $f, g: A \to B$ are equal if and only if, for all $a \in A$, one has $f(a) = g(a)$.)

Cycles

The two-rowed notation for permutations is not only cumbersome but, as we shall see, it also disguises important features of special permutations. Therefore, we shall introduce a better notation.

Definition. If $x \in X$ and $\alpha \in S_X$, then α ***fixes*** x if $\alpha(x) = x$ and α ***moves*** x if $\alpha(x) \neq x$.

Definition. Let $i_1, i_2, \ldots, i_r$ be distinct integers between 1 and n. If $\alpha \in S_n$ fixes the remaining $n - r$ integers and if

$$\alpha(i_1) = i_2, \alpha(i_2) = i_3, \ldots, \alpha(i_{r-1}) = i_r, \alpha(i_r) = i_1,$$

then α is an ***r-cycle***; one also says that α is a cycle of ***length*** r. Denote α by $(i_1\ i_2\ \cdots\ i_r)$.

Every 1-cycle fixes every element of X, and so all 1-cycles are equal to the identity. A 2-cycle, which merely interchanges a pair of elements, is called a ***transposition.***

Draw a circle with $i_1, i_2, \ldots, i_r$ arranged at equal distances around the circumference; one may picture the r-cycle $\alpha = (i_1\ i_2\ \cdots\ i_r)$ as a rotation taking i_1 into i_2, i_2 into i_3, etc., and i_r into i_1. Indeed, this is the origin of the term *cycle*, from the Greek word *κύκλοσ* for circle; see Figure 1.1.

Here are some examples:

$$\begin{pmatrix} 1 & 2 & 3 & 4 \\ 2 & 3 & 4 & 1 \end{pmatrix} = (1\ 2\ 3\ 4);$$

$$\begin{pmatrix} 1 & 2 & 3 & 4 & 5 \\ 5 & 1 & 4 & 2 & 3 \end{pmatrix} = (1\ 5\ 3\ 4\ 2);$$

$$\begin{pmatrix} 1 & 2 & 3 & 4 & 5 \\ 2 & 3 & 1 & 4 & 5 \end{pmatrix} = (1\ 2\ 3)(4)(5) = (1\ 2\ 3).$$

Nous observerons d'abord que, si dans la substitution $\begin{pmatrix} A_s \\ A_t \end{pmatrix}$ formée par deux permutations prises à volonté dans la suite

$$A_1, \quad A_2, \quad A_3, \quad \ldots, \quad A_N,$$

les deux termes A_s, A_t renferment des indices correspondants qui soient respectivement égaux, on pourra, sans inconvénient, supprimer les mêmes indices pour ne conserver que ceux des indices correspondants qui sont respectivement inégaux. Ainsi, par exemple, si l'on fait $n = 5$, les deux substitutions

$$\begin{pmatrix} 1.2.3.4.5 \\ 2.3.1.4.5 \end{pmatrix} \quad \text{et} \quad \begin{pmatrix} 1.2.3 \\ 2.3.1 \end{pmatrix}$$

seront équivalentes entre elles. Je dirai qu'une substitution aura été réduite à sa plus simple expression lorsqu'on aura supprimé, dans les deux termes, tous les indices correspondants égaux.

Soient maintenant $\alpha, \beta, \gamma, \ldots, \zeta, \eta$ plusieurs des indices $1, 2, 3, \ldots, n$ en nombre égal à p, et supposons que la substitution $\begin{pmatrix} A_s \\ A_t \end{pmatrix}$ réduite à sa plus simple expression prenne la forme

$$\begin{pmatrix} \alpha & \beta & \gamma & \ldots & \zeta & \eta \\ \beta & \gamma & \delta & \ldots & \eta & \alpha \end{pmatrix},$$

en sorte que, pour déduire le second terme du premier, il suffise de ranger en cercle, ou plutôt en polygone régulier, les indices α, β, γ, δ, ..., ζ, η de la manière suivante :

et de remplacer ensuite chaque indice par celui qui, le premier, vient prendre sa place lorsqu'on fait tourner d'orient en occident le polygone

A. Cauchy, Mémoire sur le nombre des valeurs qu'une fonction peut acquérir, lorsqu'on y permute de toutes les manières possibles les quantités qu'elle renferme, *J. de l'École Poly XVII Cahier*, tome X (1815), pp. 1–28.
From: *Oeuvres Completes d'Augustin Cauchy*, II Serie, Tome I, Gauthier-Villars, Paris, 1905.

Figure 1.1

Multiplication is easy when one uses the cycle notation. For example, let us compute $\gamma = \alpha\beta$, where $\alpha = (1\ 2)$ and $\beta = (1\ 3\ 4\ 2\ 5)$. Since multiplication is composition of functions, $\gamma(1) = \alpha \circ \beta(1) = \alpha(\beta(1)) = \alpha(3) = 3$; Next, $\gamma(3) = \alpha(\beta(3)) = \alpha(4) = 4$, and $\gamma(4) = \alpha(\beta(4)) = \alpha(2) = 1$. Having returned to 1, we now seek $\gamma(2)$, because 2 is the smallest integer for which γ has not yet been evaluated. We end up with

$$(1\ 2)(1\ 3\ 4\ 2\ 5) = (1\ 3\ 4)(2\ 5).$$

The cycles on the right are *disjoint* as defined below.

Definition. Two permutations $\alpha, \beta \in S_X$ are ***disjoint*** if every x moved by one is fixed by the other. In symbols, if $\alpha(x) \neq x$, then $\beta(x) = x$ and if $\beta(y) \neq y$, then $\alpha(y) = y$ (of course, it is possible that there is $z \in X$ with $\alpha(z) = z = \beta(z)$). A family of permutations $\alpha_1, \alpha_2, \ldots, \alpha_m$ is ***disjoint*** if each pair of them is disjoint.

Exercises

1.4. Prove that $(1\ 2\ \cdots\ r-1\ r) = (2\ 3\ \cdots\ r\ 1) = (3\ 4\ \cdots\ 1\ 2) = \cdots = (r\ 1\ \cdots\ r-1)$. Conclude that there are exactly r such notations for this r-cycle.

1.5. If $1 \leq r \leq n$, then there are $(1/r)[n(n-1)\ldots(n-r+1)]$ r-cycles in S_n.

1.6. Prove the ***cancellation law*** for permutations: if either $\alpha\beta = \alpha\gamma$ or $\beta\alpha = \gamma\alpha$, then $\beta = \gamma$.

1.7. Let $\alpha = (i_1\ i_2\ \cdots\ i_r)$ and $\beta = (j_1\ j_2\ \cdots\ j_s)$. Prove that α and β are disjoint if and only if $\{i_1, i_2, \ldots, i_r\} \cap \{j_1, j_2, \ldots, j_s\} = \varnothing$.

1.8. If α and β are disjoint permutations, then $\alpha\beta = \beta\alpha$; that is, α and β ***commute***.

1.9. If $\alpha, \beta \in S_n$ are disjoint and $\alpha\beta = 1$, then $\alpha = 1 = \beta$.

1.10. If $\alpha, \beta \in S_n$ are disjoint, prove that $(\alpha\beta)^k = \alpha^k\beta^k$ for all $k \geq 0$. Is this true if α and β are not disjoint? (Define $\alpha^0 = 1$, $\alpha^1 = \alpha$, and, if $k \geq 2$, define α^k to be the composite of α with itself k times.)

1.11. Show that a power of a cycle need not be a cycle.

1.12. (i) Let $\alpha = (i_0\ i_1\ \ldots\ i_{r-1})$ be an r-cycle. For every $j, k \geq 0$, prove that $\alpha^k(i_j) = i_{k+j}$ if subscripts are read modulo r.
(ii) Prove that if α is an r-cycle, then $\alpha^r = 1$, but that $\alpha^k \neq 1$ for every positive integer $k < r$.
(iii) If $\alpha = \beta_1\beta_2\ldots\beta_m$ is a product of disjoint r_i-cycles β_i, then the smallest positive integer l with $\alpha^l = 1$ is the least common multiple of $\{r_1, r_2, \ldots, r_m\}$.

1.13. (i) A permutation $\alpha \in S_n$ is ***regular*** if either α has no fixed points and it is the product of disjoint cycles of the same length or $\alpha = 1$. Prove that α is regular if and only if α is a power of an n-cycle β; that is, $\alpha = \beta^m$ for some m. (*Hint*: if $\alpha = (a_1 a_2 \ldots a_k)(b_1 b_2 \ldots b_k)\ldots(z_1 z_2 \ldots z_k)$, where there are m letters $a, b, \ldots, z$, then let $\beta = (a_1 b_1 \ldots z_1 a_2 b_2 \ldots z_2 \ldots a_k b_k \ldots z_k)$.)
(ii) If α is an n-cycle, then α^k is a product of (n, k) disjoint cycles, each of length $n/(n, k)$. (Recall that (n, k) denotes the gcd of n and k.)
(iii) If p is a prime, then every power of a p-cycle is either a p-cycle or 1.

1.14. (i) Let $\alpha = \beta\gamma$ in S_n, where β and γ are disjoint. If β moves i, then $\alpha^k(i) = \beta^k(i)$ for all $k \geq 0$.

(ii) Let α and β be cycles in S_n (we do not assume that they have the same length). If there is i_1 moved by both α and β and if $\alpha^k(i_1) = \beta^k(i_1)$ for all positive integers k, then $\alpha = \beta$.

Factorization into Disjoint Cycles

Let us factor $\alpha = \begin{pmatrix} 1 & 2 & 3 & 4 & 5 & 6 & 7 & 8 & 9 \\ 6 & 4 & 1 & 2 & 5 & 3 & 8 & 9 & 7 \end{pmatrix}$ into a product of disjoint cycles. Now $\alpha(1) = 6$, and so α begins $(1\ \ 6$; as $\alpha(6) = 3$, α continues $(1\ \ 6\ \ 3$; since $\alpha(3) = 1$, the parentheses close, and α begins $(1\ \ 6\ \ 3)$. The smallest integer not having appeared is 2; write $(1\ \ 6\ \ 3)(2$, and then $(1\ \ 6\ \ 3)(2\ \ 4$; continuing in this way, we ultimately arrive at the factorization (which is a product of disjoint cycles)

$$\alpha = (1\ \ 6\ \ 3)(2\ \ 4)(5)(7\ \ 8\ \ 9).$$

Theorem 1.1. *Every permutation $\alpha \in S_n$ is either a cycle or a product of disjoint cycles.*

Proof. The proof is by induction on the number k of points moved by α. The base step $k = 0$ is true, for then α is the identity, which is a 1-cycle. If $k > 0$, let i_1 be a point moved by α. Define $i_2 = \alpha(i_1)$, $i_3 = \alpha(i_2)$, $\ldots$, $i_{r+1} = \alpha(i_r)$, where r is the smallest integer for which $i_{r+1} \in \{i_1, i_2, i_3, \ldots, i_r\}$ (the list i_1, i_2, $i_3, \ldots, i_k, \ldots$ cannot go on forever without a repetition because there are only n possible values). We claim that $\alpha(i_r) = i_1$. Otherwise, $\alpha(i_r) = i_j$ for some $j \geq 2$; but $\alpha(i_{j-1}) = i_j$, and this contradicts the hypothesis that α is an injection. Let σ be the r-cycle $(i_1\ \ i_2\ \ i_3\ \ \cdots\ \ i_r)$. If $r = n$, then α is the cycle σ. If $r < n$ and Y consists of the remaining $n - r$ points, then $\alpha(Y) = Y$ and σ fixes the points in Y. Now $\sigma|\{i_1, i_2, \ldots, i_r\} = \alpha|\{i_1, i_2, \ldots, i_r\}$. If α' is the permutation with $\alpha'|Y = \alpha|Y$ and which fixes $\{i_1, i_2, \ldots, i_r\}$, then σ and α' are disjoint and $\alpha = \sigma\alpha'$. Since α' moves fewer points than does α, the inductive hypothesis shows that α', and hence α, is a product of disjoint cycles. ■

One often suppresses all 1-cycles, if any, from this factorization of α, for 1-cycles equal the identity permutation. On the other hand, it is sometimes convenient to display all of them.

Definition. A ***complete factorization*** of a permutation α is a factorization of α as a product of disjoint cycles which contains one 1-cycle (i) for every i fixed by α.

In a complete factorization of a permutation α, every i between 1 and n occurs in exactly one of the cycles.

Theorem 1.2. *Let $\alpha \in S_n$ and let $\alpha = \beta_1 \dots \beta_t$ be a complete factorization into disjoint cycles. This factorization is unique except for the order in which the factors occur.*

Proof. Disjoint cycles commute, by Exercise 1.8, so that the order of the factors in a complete factorization is not uniquely determined; however, we shall see that the factors themselves are uniquely determined. Since there is exactly one 1-cycle (i) for every i fixed by α, it suffices to prove uniqueness of the cycles of length at least 2. Suppose $\alpha = \gamma_1 \dots \gamma_s$ is a second complete factorization into disjoint cycles. If β_t moves i_1, then $\beta_t^k(i_1) = \alpha^k(i_1)$ for all k, by Exercise 1.14(i). Now some γ_j must move i_1; since disjoint cycles commute, we may assume that $\gamma_j = \gamma_s$. But $\gamma_s^k(i_1) = \alpha^k(i_1)$ for all k, and so Exercise 1.14(ii) gives $\beta_t = \gamma_s$. The cancellation law, Exercise 1.6, gives $\beta_1 \dots \beta_{t-1} = \gamma_1 \dots \gamma_{s-1}$, and the proof is completed by an induction on $\max\{s, t\}$. ■

EXERCISES

1.15. Let α be the permutation of $\{1, 2, \dots, 9\}$ defined by $\alpha(i) = 10 - i$. Write α as a product of disjoint cycles.

1.16. Let p be a prime and let $\alpha \in S_n$. If $\alpha^p = 1$, then either $\alpha = 1$, α is a p-cycle, or α is a product of disjoint p-cycles. In particular, if $\alpha^2 = 1$, then either $\alpha = 1$, α is a transposition, or α is a product of disjoint transpositions.

1.17. How many $\alpha \in S_n$ are there with $\alpha^2 = 1$? (*Hint.* $(i\ j) = (j\ i)$ and $(i\ j)(k\ l) = (k\ l)(i\ j)$.)

1.18. Give an example of permutations α, β, and γ in S_5 with α commuting with β, with β commuting with γ, but with α not commuting with γ.

Even and Odd Permutations

There is another factorization of permutations that is useful.

Theorem 1.3. *Every permutation $\alpha \in S_n$ is a product of transpositions.*

Proof. By Theorem 1.1, it suffices to factor cycles, and

$$(1\ 2\ \dots\ r) = (1\ r)(1\ r-1)\dots(1\ 2). \quad ■$$

Every permutation can thus be realized as a sequence of interchanges. Such a factorization is not as nice as the factorization into disjoint cycles. First of all, the transpositions occurring need not commute: $(1\ 3)(1\ 2) = (1\ 2\ 3)$ and $(1\ 2)(1\ 3) = (1\ 3\ 2)$; second, neither the factors nor the number of factors are uniquely determined; for example,

$$\begin{aligned}(1\ 2\ 3) &= (1\ 3)(1\ 2) = (2\ 3)(1\ 3) \\ &= (1\ 3)(4\ 2)(1\ 2)(1\ 4) \\ &= (1\ 3)(4\ 2)(1\ 2)(1\ 4)(2\ 3)(2\ 3).\end{aligned}$$

Is there any uniqueness at all in such a factorization? We now prove that the *parity* of the number of factors is the same for all factorizations of a permutation α: that is, the number of transpositions is always even (as suggested by the above factorizations of $\alpha = (1\ 2\ 3)$) or is always odd.

Definition. A permutation $\alpha \in S$ is ***even*** if it is a product of an even number of transpositions; otherwise, α is ***odd***.

It is easy to see that $\alpha = (1\ 2\ 3)$ is even, for there is a factorization $\alpha = (1\ 3)(1\ 2)$ into two transpositions. On the other hand, we do not know whether there are any odd permutations α at all; if α is a product of an odd number of transpositions, perhaps it also has another factorization as a product of an even number of transpositions. The definition of odd permutation α, after all, says that there is no factorization of α into an even number of transpositions.

Lemma 1.4. *If* $k, l \geq 0$, *then*

$$(a\ b)(a\ c_1\ \ldots\ c_k\ b\ d_1\ \ldots\ d_l) = (a\ c_1\ \ldots\ c_k)(b\ d_1\ \ldots\ d_l)$$

and

$$(a\ b)(a\ c_1\ \ldots\ c_k)(b\ d_1\ \ldots\ d_l) = (a\ c_1\ \ldots\ c_k\ b\ d_1\ \ldots\ d_l).$$

Proof. The left side sends $a \mapsto c_1 \mapsto c_1$; $c_i \mapsto c_{i+1} \mapsto c_{i+1}$ if $i < k$; $c_k \mapsto b \mapsto a$; $b \mapsto d_1 \mapsto d_1$; $d_j \mapsto d_{j+1} \mapsto d_{j+1}$ if $j < l$; $d_l \mapsto a \mapsto b$. Similar evaluation of the right side shows that both permutations are equal. For the second equation, just multiply both sides of the first equation by $(a\ b)$ on the left. ■

Definition. If $\alpha \in S_n$ and $\alpha = \beta_1 \ldots \beta_t$ is a complete factorization into disjoint cycles, then ***signum*** α is defined by

$$\operatorname{sgn}(\alpha) = (-1)^{n-t}.$$

By Theorem 1.2, sgn is a *well defined* function (see Appendix III). If τ is a transposition, then it moves two numbers, say, i and j, and fixes each of the $n - 2$ other numbers; therefore, $t = (n - 2) + 1 = n - 1$, and so

$$\operatorname{sgn}(\tau) = (-1)^{n-(n-1)} = -1.$$

Lemma 1.5. *If* $\beta \in S_n$ *and* τ *is a transposition, then*

$$\operatorname{sgn}(\tau\beta) = -\operatorname{sgn}(\beta).$$

Proof. Let $\tau = (a \ b)$ and let $\beta = \gamma_1 \cdots \gamma_t$ be a complete factorization of β into disjoint cycles (there is one 1-cycle for each i fixed by β, and every number between 1 and n occurs in a unique γ). If a and b occur in the same γ, say, in γ_1, then $\gamma_1 = (a \ c_1 \ \ldots \ c_k \ b \ d_1 \ \ldots \ d_l)$, where $k \geq 0$ and $l \geq 0$. By Lemma 1.4,

$$\tau\gamma_1 = (a \ c_1 \ \cdots \ c_k)(b \ d_1 \ \cdots \ d_l),$$

and so $\tau\beta = (\tau\gamma_1)\gamma_2 \ldots \gamma_t$ is a complete factorization with an extra cycle ($\tau\gamma_1$ splits into two disjoint cycles). Therefore, $\operatorname{sgn}(\tau\beta) = (-1)^{n-(t+1)} = -\operatorname{sgn}(\beta)$. The other possibility is that a and b occur in different cycles, say, $\gamma_1 = (a \ c_1 \ \ldots \ c_k)$ and $\gamma_2 = (b \ d_1 \ \ldots \ d_l)$, where $k \geq 0$ and $l \geq 0$. But now $\tau\beta = (\tau\gamma_1\gamma_2)\gamma_3 \ldots \gamma_t$, and Lemma 1.4 gives

$$\tau\gamma_1\gamma_2 = (a \ c_1 \ \cdots \ c_k \ b \ d_1 \ \cdots \ d_l).$$

Therefore, the complete factorization of $\tau\beta$ has one fewer cycle than does β, and so $\operatorname{sgn}(\tau\beta) = (-1)^{n-(t-1)} = -\operatorname{sgn}(\beta)$. ■

Theorem 1.6. *For all* $\alpha, \beta \in S_n$,

$$\operatorname{sgn}(\alpha\beta) = \operatorname{sgn}(\alpha)\operatorname{sgn}(\beta).$$

Proof. Assume that $\alpha \in S_n$ is given and that $\alpha = \tau_1 \ldots \tau_m$ is a factorization of α into transpositions with m minimal. We prove, by induction on m, that $\operatorname{sgn}(\alpha\beta) = \operatorname{sgn}(\alpha)\operatorname{sgn}(\beta)$ for every $\beta \in S_n$. The base step is precisely Lemma 1.5. If $m > 1$, then the factorization $\tau_2 \ldots \tau_m$ is also minimal: if $\tau_2 \ldots \tau_m = \sigma_1 \ldots \sigma_q$ with each σ_j a transposition and $q < m - 1$, then the factorization $\alpha = \tau_1\sigma_1 \ldots \sigma_q$ violates the minimality of m. Therefore,

$$\begin{aligned}
\operatorname{sgn}(\alpha\beta) &= \operatorname{sgn}(\tau_1 \cdots \tau_m\beta) = -\operatorname{sgn}(\tau_2 \cdots \tau_m\beta) && \text{(Lemma 1.5)} \\
&= -\operatorname{sgn}(\tau_2 \cdots \tau_m)\operatorname{sgn}(\beta) && \text{(by induction)} \\
&= \operatorname{sgn}(\tau_1 \cdots \tau_m)\operatorname{sgn}(\beta) && \text{(by Lemma 1.5)} \\
&= \operatorname{sgn}(\alpha)\operatorname{sgn}(\beta).
\end{aligned}$$ ■

Theorem 1.7.

(i) *A permutation* $\alpha \in S_n$ *is even if and only if* $\operatorname{sgn}(\alpha) = 1$.
(ii) *A permutation* α *is odd if and only if it is a product of an odd number of transpositions.*

Proof. (i) We have seen that $\operatorname{sgn}(\tau) = -1$ for every transposition τ. Therefore, if $\alpha = \tau_1 \ldots \tau_q$ is a factorization of α into transpositions, then Theorem 1.6 gives $\operatorname{sgn}(\alpha) = \operatorname{sgn}(\tau_1) \ldots \operatorname{sgn}(\tau_q) = (-1)^q$. Thus, $\operatorname{sgn}(\alpha) = 1$ if and only if q is even. If α is even, then there exists a factorization with q even, and so $\operatorname{sgn}(\alpha) = 1$. Conversely, if $1 = \operatorname{sgn}(\alpha) = (-1)^q$, then q is even and hence α is even.

(ii) If α is odd, then it has no factorization into an even number of transpositions, and so it must be a product of an odd number of them. Conversely, if $\alpha = \tau_1 \ldots \tau_q$ with q odd, then $\operatorname{sgn}(\alpha) = (-1)^q = -1$; by (i), α is not even, and hence α is odd. ■

EXERCISES

1.19. Show that an r-cycle is an even permutation if and only if r is odd.

1.20. Compute $\operatorname{sgn}(\alpha)$ for $\alpha = \begin{pmatrix} 1 & 2 & 3 & 4 & 5 & 6 & 7 & 8 & 9 \\ 9 & 8 & 7 & 6 & 5 & 4 & 3 & 2 & 1 \end{pmatrix}$.

1.21. Show that S_n has the same number of even permutations as of odd permutations. (*Hint.* If $\tau = (1\ 2)$, consider the function $f: S_n \to S_n$ defined by $f(\alpha) = \tau\alpha$.)

1.22. Let $\alpha, \beta \in S_n$. If α and β have the same parity, then $\alpha\beta$ is even; if α and β have distinct parity, then $\alpha\beta$ is odd.

Semigroups

We are now going to abstract certain features of S_X.

Definition. A (binary) ***operation*** on a nonempty set G is a function $\mu: G \times G \to G$.

An operation μ assigns to each ordered pair (a, b) of elements of G a third element of G, namely, $\mu(a, b)$. In practice, μ is regarded as a "multiplication" of elements of G, and, instead of $\mu(a, b)$, more suggestive notations are used, such as ab, $a + b$, $a \circ b$, or $a * b$. In this first chapter, we shall use the star notation $a * b$.

It is quite possible that $a * b$ and $b * a$ are distinct elements of G. For example, we have already seen that $(1\ 2)(1\ 3) \neq (1\ 3)(1\ 2)$ in $G = S_3$.

The ***Law of Substitution*** (if $a = a'$ and $b = b'$, then $a * b = a' * b'$) is just the statement that μ is a well defined function: since $(a, b) = (a', b')$, it follows that $\mu(a, b) = \mu(a', b')$; that is, $a * b = a' * b'$.

One cannot develop a theory in this rarefied atmosphere; conditions on the operation are needed to obtain interesting results (and to capture the essence of composition in S_X). How can we multiply three elements of G? Given (not necessarily distinct) elements $a_1, a_2, a_3 \in G$, the expression $a_1 * a_2 * a_3$ is ambiguous. Since we can $*$ only two elements of G at a time, there is a choice: form $a_1 * a_2$ first, and then $*$ this new element of G with a_3 to get $(a_1 * a_2) * a_3$; or, form $a_1 * (a_2 * a_3)$. In general, these two elements of G may be different. For example, let $G = \mathbb{Z}$, the set of all integers (positive, negative, and zero), and let the operation be subtraction: $a * b = a - b$; any choice of integers a, b, c with $c \neq 0$ gives an example with $(a - b) - c \neq a - (b - c)$.

Definition. An operation $*$ on a set G is ***associative*** if

$$(a * b) * c = a * (b * c)$$

for every $a, b, c \in G$.

Exercise 1.3 shows that multiplication in S_X is associative. Associativity allows one to multiply every ordered triple of elements in G unambiguously; parentheses are unnecessary, and there is no confusion in writing $a * b * c$. If we are confronted by four elements of G, or, more generally, by a finite number of elements of G, must we postulate more intricate associativity axioms to avoid parentheses?

Consider the elements of G that can be obtained from an expression $a_1 * a_2 * \cdots * a_n$. Choose two adjacent a's, multiply them, and obtain an expression with only $n - 1$ factors in it: the product just formed and $n - 2$ original factors. In this new expression, choose two adjacent factors (either an original pair or an original a_i adjacent to the new product from the first step) and multiply them. Repeat this procedure until there is an expression with only two factors; multiply them and obtain an element of G. Let us illustrate this process with the expression $a * b * c * d$. We may first multiply $a * b$, arriving at $(a * b) * c * d$, an expression with three factors, namely, $a * b$, c, and d. Now choose either the pair c, d or the pair $a * b$, c; in either case, multiply the chosen pair, and obtain the shorter expressions $(a * b) * (c * d)$ or $[(a * b) * c] * d$. Each of these last two expressions involves only two factors which can be multiplied to give an element of G. Other ways to evaluate the original expression begin by forming $b * c$ or $c * d$ as the first step. It is not obvious whether all the elements arising from a given expression are equal.

Definition. An expression $a_1 * a_2 * \cdots * a_n$ ***needs no parentheses*** if, no matter what choices of multiplications of adjacent factors are made, the resulting elements of G are all equal.

Theorem 1.8 (Generalized Associativity). *If $*$ is an associative operation on a set G, then every expression $a_1 * a_2 * \cdots * a_n$ needs no parentheses.*

Proof. The proof is by induction on $n \geq 3$. The base step $n = 3$ holds because $*$ is associative . If $n \geq 3$, consider two elements obtained from an expression $a_1 * a_2 * \cdots * a_n$ after two series of choices:

(1) $(a_1 * \cdots * a_i) * (a_{i+1} * \cdots * a_n)$ and $(a_1 * \cdots * a_j) * (a_{j+1} * \cdots * a_n)$

(the choices yield a sequence of shorter expressions, and the parentheses indicate ultimate expressions of length 2). We may assume that $i \leq j$. Since each of the four expressions in parentheses has fewer than n factors, the inductive hypothesis says that each of them needs no parentheses. If $i = j$, it follows that the two products in (1) are equal. If $i < j$, then rewrite the first expression as

(2) $$(a_1 * \cdots * a_i) * ([a_{i+1} * \cdots * a_j] * [a_{j+1} * \cdots * a_n])$$

and rewrite the second expression as

(3) $$([a_1 * \cdots * a_i] * [a_{i+1} * \cdots * a_j]) * (a_{j+1} * \cdots * a_n).$$

By induction, each of the expressions $a_1 * \cdots * a_i$, $a_{i+1} * \cdots * a_j$, and $a_{j+1} * \cdots * a_n$ yield (uniquely defined) elements A, B, and C of G, respectively. Since (2) is the expression $A * (B * C)$ and (3) is the expression $(A * B) * C$, associativity says that both these expressions give the same element of G. ■

Definition. A ***semigroup*** $(G, *)$ is a nonempty set G equipped with an associative operation $*$.

Usually, one says "Let G be a semigroup ...," displaying the set G, but tacitly assuming that the operation $*$ is known. The reader must realize, however, that there are many possible operations on a set making it a semigroup. For example, the set of all positive integers is a semigroup under either of the operations of ordinary addition or ordinary multiplication.

Definition. Let G be a semigroup and let $a \in G$. Define $a^1 = a$ and, for $n \geq 1$, define $a^{n+1} = a * a^n$.

Corollary 1.9. *Let G be a semigroup, let $a \in G$, and let m and n be positive integers. Then $a^m * a^n = a^{m+n} = a^n * a^m$ and $(a^m)^n = a^{mn} = (a^n)^m$.*

Proof. Both sides of the first (or second) equations arise from an expression having $m + n$ (or mn) factors all equal to a. But these expressions need no parentheses, by Theorem 1.8. ■

The notation a^n obviously comes from the special case when $*$ is multiplication; $a^n = aa \ldots a$ (n times). When the operation is denoted by $+$, it is more natural to denote $a * a * \cdots * a = a + a + \cdots + a$ by na. In this additive notation, Corollary 1.9 becomes $ma + na = (m + n)a$ and $(mn)a = m(na)$.

Groups

The most important semigroups are groups.

Definition. A ***group*** is a semigroup G containing an element e such that:

(i) $e * a = a = a * e$ for all $a \in G$;
(ii) for every $a \in G$, there is an element $b \in G$ with

$$a * b = e = b * a.$$

Exercises 1.1, 1.2, and 1.3 show that S_X is a group with composition as operation; it is called the ***symmetric group*** on X. When $X = \{1, 2, \ldots, n\}$, then S_X is denoted by S_n and it is called the ***symmetric group on n letters***.

Definition. A pair of elements a and b in a semigroup ***commutes*** if $a * b = b * a$. A group (or a semigroup) is ***abelian*** if every pair of its elements commutes.

It is easy to see, for all $n \geq 3$, that S_n is not abelian.

There are many interesting examples of groups; we mention only a few of them now.

The set $\mathbb{Z}$ of all integers (positive, negative, and zero) is an abelian group with ordinary addition as operation: $a * b = a + b$; $e = 0$; $-a + a = 0$. Some other additive abelian groups are the rational numbers $\mathbb{Q}$, the real numbers $\mathbb{R}$, and the complex numbers $\mathbb{C}$. Indeed, every ring is an additive abelian group (it is only a semigroup with 1 under multiplication).

Recall that if $n \geq 2$ and a and b are integers, then $a \equiv b \bmod n$ (pronounced: a is congruent to b modulo n) means that n is a divisor of $a - b$. Denote the ***congruence class*** of an integer a mod n by $[a]$; that is,

$$\begin{aligned} [a] &= \{b \in \mathbb{Z}: b \equiv a \bmod n\} \\ &= \{a + kn: k \in \mathbb{Z}\}. \end{aligned}$$

The set $\mathbb{Z}_n$ of all the congruence classes mod n is called the ***integers modulo n***; it is an abelian group when equipped with the operation: $[a] + [b] = [a + b]$; here $e = [0]$ and $[-a] + [a] = [0]$ ($\mathbb{Z}_n$ is even a commutative ring when "one" is $[1]$ and multiplication is defined by $[a][b] = [ab]$). The reader should prove that these operations are well defined: if $[a'] = [a]$ and $[b'] = [b]$, that is, if $a' \equiv a \bmod n$ and $b' \equiv b \bmod n$, then $[a' + b'] = [a + b]$ and $[a'b'] = [ab]$.

If k is a field, then the set of all $n \times n$ nonsingular matrices with entries in k is a group, denoted by $\mathrm{GL}(n, k)$, called the ***general linear group***: here the operation is matrix multiplication, e is the identity matrix E, and if A^{-1} is the inverse of the matrix A, then $AA^{-1} = E = A^{-1}A$. If $n \geq 2$, then $\mathrm{GL}(n, k)$ is not abelian; if $n = 1$, then $\mathrm{GL}(1, k)$ is abelian: it is the multiplicative group $k^\times$ of all the nonzero elements in k.

If R is an associative ring (we insist that R has an element 1), then an element u is a ***unit*** in R if there exists $v \in R$ with $uv = 1 = vu$. If a is a unit, so that $ab = 1 = ba$ for some $b \in R$, then it is easy to see that ua is also a unit in R (with inverse bv) and that $U(R)$, the ***group of units*** in R, is a multiplicative group. If R is a field k, then $U(k) = k^\times$. If R is the ring of all $n \times n$ matrices over a field k, then $U(R) = \mathrm{GL}(n, k)$.

Theorem 1.10. *If G is a group, there is a unique element e with $e * a = a = a * e$ for all $a \in G$. Moreover, for each $a \in G$, there is a unique $b \in G$ with $a * b = e = b * a$.*

Proof. Suppose that $e' * a = a = a * e'$ for all $a \in G$. In particular, if $a = e$, then $e' * e = e$. On the other hand, the defining property of e gives $e' * e = e'$, and so $e' = e$.

Suppose that $a * c = e = c * a$. Then $c = c * e = c * (a * b) = (c * a) * b = e * b = b$, as desired. ■

As a result of the uniqueness assertions of the theorem, we may now give names to e and to b. We call e the ***identity*** of G and, if $a * b = e = b * a$, then we call b the ***inverse*** of a and denote it by a^{-1}.

Corollary 1.11. *If G is a group and $a \in G$, then*

$$(a^{-1})^{-1} = a.$$

Proof. By definition, $(a^{-1})^{-1}$ is that element $g \in G$ with $a^{-1} * g = e = g * a^{-1}$. But a is such an element, and so the uniqueness gives $g = a$. ■

Definition. If G is a group and $a \in G$, define the ***powers*** of a as follows: if n is a positive integer, then a^n is defined as in any semigroup; define $a^0 = e$; define $a^{-n} = (a^{-1})^n$.

Even though the list of axioms defining a group is short, it is worthwhile to make it even shorter so it will be as easy as possible to verify that a particular example is, in fact, a group.

Theorem 1.12. *If G is a semigroup with an element e such that:*

(i') *$e * a = a$ for all $a \in G$; and*
(ii') *for each $a \in G$ there is an element $b \in G$ with $b * a = e$, then G is a group.*

Proof. We claim that if $x * x = x$ in G, then $x = e$. There is an element $y \in G$ with $y * x = e$, and $y * (x * x) = y * x = e$. On the other hand, $y * (x * x) = (y * x) * x = e * x = x$. Therefore, $x = e$.

If $b * a = e$, let us show that $a * b = e$. Now $(a * b) * (a * b) = a * [(b * a) * b] = a * [e * b] = a * b$, and so our claim gives $a * b = e$. (Observe that we have used associativity for an expression having four factors.)

If $a \in G$, we must show that $a * e = a$. Choose $b \in G$ with $b * a = e = a * b$ (using our just finished calculation). Then $a * e = a * (b * a) = (a * b) * a = e * a = a$, as desired. ■

Exercises

1.23. If G is a group and $a_1, a_2, \ldots, a_n \in G$, then

$$(a_1 * a_2 * \cdots * a_n)^{-1} = a_n^{-1} * a_{n-1}^{-1} * \cdots * a_1^{-1}.$$

Conclude that if $n \geq 0$, then $(a^{-1})^n = (a^n)^{-1}$.

1.24. Let $a_1, a_2, \ldots, a_n$ be elements of an abelian semigroup. If $b_1, b_2, \ldots, b_n$ is a rearrangement of the a_i, then

$$a_1 * a_2 * \cdots * a_n = b_1 * b_2 * \cdots * b_n.$$

1.25. Let a and b lie in a semigroup G. If a and b commute, then $(a * b)^n = a^n * b^n$ for every $n \geq 1$; if G is a group, then this equation holds for every $n \in \mathbb{Z}$.

1.26. A group in which $x^2 = e$ for every x must be abelian.

1.27. (i) Let G be a finite abelian group containing no elements $a \neq e$ with $a^2 = e$. Evaluate $a_1 * a_2 * \cdots * a_n$, where $a_1, a_2, \ldots, a_n$ is a list with no repetitions, of all the elements of G.
(ii) Prove ***Wilson's theorem***: If p is prime, then

$$(p-1)! \equiv -1 \mod p.$$

(*Hint*. The nonzero elements of $\mathbb{Z}_p$ form a multiplicative group.)

1.28. (i) If $\alpha = (1\ 2\ \ldots\ r-1\ r)$, then $\alpha^{-1} = (r\ r-1\ \ldots\ 2\ 1)$.
(ii) Find the inverse of $\begin{pmatrix} 1 & 2 & 3 & 4 & 5 & 6 & 7 & 8 & 9 \\ 6 & 4 & 1 & 2 & 5 & 3 & 8 & 9 & 7 \end{pmatrix}$.

1.29. Show that $\alpha: \mathbb{Z}_{11} \to \mathbb{Z}_{11}$, defined by $\alpha(x) = 4x^2 - 3x^7$, is a permutation of $\mathbb{Z}_{11}$, and write it as a product of disjoint cycles. What is the parity of α? What is α^{-1}?

1.30. Let G be a group, let $a \in G$, and let $m, n \in \mathbb{Z}$ be (possibly negative) integers. Prove that $a^m * a^n = a^{m+n} = a^n * a^m$ and $(a^m)^n = a^{mn} = (a^n)^m$.

1.31. Let G be a group, let $a \in G$, and let m and n be relatively prime integers. If $a^m = e$, show that there exists $b \in G$ with $a = b^n$. (*Hint*. There are integers s and t with $1 = sm + tn$.)

1.32 **(Cancellation Laws).** In a group G, either of the equations $a * b = a * c$ and $b * a = c * a$ implies $b = c$.

1.33. Let G be a group and let $a \in G$.
(i) For each $a \in G$, prove that the functions $L_a: G \to G$, defined by $x \mapsto a * x$ (called ***left translation*** by a), and $R_a: G \to G$, defined by $x \mapsto x * a^{-1}$ (called ***right translation*** by a), are bijections.
(ii) For all $a, b \in G$, prove that $L_{a*b} = L_a \circ L_b$ and $R_{a*b} = R_a \circ R_b$.
(iii) For all a and b, prove that $L_a \circ R_b = R_b \circ L_a$.

1.34. Let G denote the multiplicative group of positive rationals. What is the identity of G? If $a \in G$, what is its inverse?

1.35. Let n be a positive integer and let G be the multiplicative group of all nth roots of unity; that is, G consists of all complex numbers of the form $e^{2\pi ik/n}$, where $k \in \mathbb{Z}$. What is the identity of G? If $a \in G$, what is its inverse? How many elements does G have?

1.36. Prove that the following four permutations form a group $\mathbf{V}$ (which is called the ***4-group***):

$$1; \qquad (1\ 2)(3\ 4); \qquad (1\ 3)(2\ 4); \qquad (1\ 4)(2\ 3).$$

1.37. Let $\hat{\mathbb{R}} = \mathbb{R} \cup \{\infty\}$, and define $1/0 = \infty$, $1/\infty = 0$, $\infty/\infty = 1$, and $1 - \infty = \infty = \infty - 1$. Show that the six functions $\hat{\mathbb{R}} \to \hat{\mathbb{R}}$, given by x, $1/x$, $1-x$, $1/(1-x)$, $x/(x-1)$, $(x-1)/x$, form a group with composition as operation.

Homomorphisms

Let G be a finite group with n elements $a_1, a_2, \ldots, a_n$. A ***multiplication table*** for G is the $n \times n$ matrix with i, j entry $a_i * a_j$:

G	a_1	a_2	$\cdots$	a_n
a_1	$a_1 * a_1$	$a_1 * a_2$	$\cdots$	$a_1 * a_n$
a_2	$a_2 * a_1$	$a_2 * a_2$	$\cdots$	$a_2 * a_n$
$\cdots$	$\cdots$	$\cdots$	$\cdots$	$\cdots$
a_n	$a_n * a_1$	$a_n * a_2$	$\cdots$	$a_n * a_n$

Informally, we say that we "know" a finite group G if we can write a multiplication table for it. Notice that we say "a" multiplication table and not "the" multiplication table, for a table depends on the particular ordering $a_1, a_2, \ldots, a_n$ of the elements of G. (One may also speak of multiplication tables of infinite groups, but in this case, of course, the matrices are infinite.) It is customary to list the identity e first so that the first row (and first column) display the elements in the order they occur on a chosen list.

Let us now consider two almost trivial examples of groups. Let G be the group whose elements are the numbers 1 and -1, with operation multiplication; let H be the additive group $\mathbb{Z}_2$. Compare multiplication tables of these two groups:

G	1	-1
1	1	-1
-1	-1	1

H	[0]	[1]
[0]	[0]	[1]
[1]	[1]	[0]

It is quite clear that G and H are distinct groups; on the other hand, it is equally clear that there is no significant difference between them. Let us make this idea precise.

Definition. Let $(G, *)$ and $(H, \circ)$ be groups.[3] A function $f: G \to H$ is a ***homomorphism*** if, for all $a, b \in G$,

$$f(a * b) = f(a) \circ f(b).$$

An ***isomorphism*** is a homomorphism that is also a bijection. We say that G is ***isomorphic*** to H, denoted by $G \cong H$, if there exists an isomorphism $f: G \to H$.

The two-element groups G and H, whose multiplication tables are given above, are isomorphic: define $f: G \to H$ by $f(1) = [0]$ and $f(-1) = [1]$.

[3] This definition also applies to semigroups.

Let $f: G \to H$ be an isomorphism, and let $a_1, a_2, \ldots, a_n$ be a list, with no repetitions, of all the elements of G. Since f is a bijection, every element of H occurs exactly once on the list $f(a_1), f(a_2), \ldots, f(a_n)$, and so this list can be used to form a multiplication table for H. That f is a homomorphism, that is, $f(a_i * a_j) = f(a_i) \circ f(a_j)$, says that if we superimpose the multiplication table of G onto that of H, then the tables "match." In this sense, isomorphic groups G and H have the "same" multiplication tables. Informally, one regards G and H as being essentially the same, the only distinction being that G is written in English and H is written in French; an isomorphism f is a dictionary which translates one to the other.

Two basic problems occurring in mathematics are: classification of all systems of a given type (e.g., groups, semigroups, vector spaces, topological spaces); classification of all the "maps" or transformations from one such system into another. By a classification of systems, we mean a way to distinguish different systems or, what is the same thing, a way to tell when two systems are essentially the same (isomorphic). For example, finite-dimensional vector spaces over a field k are classified by the theorem that two such are isomorphic if and only if they have the same dimension. One can even classify all the maps (linear transformations) between vector spaces; they give rise to similarity classes of matrices which are classified by canonical forms. The same two problems arise in Group Theory: when are two groups isomorphic; describe all the homomorphisms from one group to another. Both of these problems are impossibly hard, but partial answers are known and are very useful.

Theorem 1.13. *Let $f: (G, *) \to (G', \circ)$ be a homomorphism.*

(i) *$f(e) = e'$, where e' is the identity in G'.*
(ii) *If $a \in G$, then $f(a^{-1}) = f(a)^{-1}$.*
(iii) *If $a \in G$ and $n \in \mathbb{Z}$, then $f(a^n) = f(a)^n$.*

Proof. (i) Applying f to the equation $e = e * e$ gives $f(e) = f(e * e) = f(e) \circ f(e)$. Now multiply each side of the equation by $f(e)^{-1}$ to obtain $e' = f(e)$.

(ii) Applying f to the equations $a * a^{-1} = e = a^{-1} * a$ gives $f(a) * f(a^{-1}) = e' = f(a^{-1}) * f(a)$. It follows from Theorem 1.10, the uniqueness of the inverse, that $f(a^{-1}) = f(a)^{-1}$.

(iii) An easy induction proves $f(a^n) = f(a)^n$ for all $n \geq 0$, and then $f(a^{-n}) = f((a^{-1})^n) = f(a^{-1})^n = f(a)^{-n}$. ■

Here are some examples. Theorem 1.6 shows that sgn: $S_n \to \{\pm 1\}$ is a homomorphism; the function $v: \mathbb{Z} \to \mathbb{Z}_n$, defined by $v(a) = [a]$, is a homomorphism; if $k^\times$ denotes the multiplicative group of nonzero elements of a field k, then determinant is a homomorphism det: $\mathrm{GL}(n, k) \to k^\times$.

EXERCISES

1.38. (i) Write a multiplication table for S_3.
(ii) Show that S_3 is isomorphic to the group of Exercise 1.37. (*Hint.* The elements in the latter group permute $\{0, 1, \infty\}$.)

1.39. Let $f: X \to Y$ be a bijection between sets X and Y. Show that $\alpha \mapsto f \circ \alpha \circ f^{-1}$ is an isomorphism $S_X \to S_Y$.

1.40. Isomorphic groups have the same number of elements. Prove that the converse is false by showing that $\mathbb{Z}_4$ is not isomorphic to the 4-group $\mathbf{V}$ defined in Exercise 1.36.

1.41. If isomorphic groups are regarded as being the same, prove, for each positive integer n, that there are only finitely many distinct groups with exactly n elements.

1.42. Let $G = \{x_1, \ldots, x_n\}$ be a set equipped with an operation $*$, let $A = [a_{ij}]$ be its multiplication table (i.e., $a_{ij} = x_i * x_j$), and assume that G has a (two-sided) identity e (that is, $e * x = x = x * e$ for all $x \in G$).
(i) Show that $*$ is commutative if and only if A is a symmetric matrix.
(ii) Show that every element $x \in G$ has a (two-sided) inverse (i.e., there is $x' \in G$ with $x * x' = e = x' * x$) if and only if the multiplication table A is a ***Latin square***; that is, no $x \in G$ is repeated in any row or column (equivalently, every row and every column of A is a permutation of G.)
(iii) Assume that $e = x_1$, so that the first row of A has $a_{1i} = x_i$. Show that the first column of A has $a_{i1} = x_i^{-1}$ for all i if and only if $a_{ii} = e$ for all i.
(iv) With the multiplication table as in (iii), show that $*$ is associative if and only if $a_{ij}a_{jk} = a_{ik}$ for all i, j, k.

1.43. (i) If $f: G \to H$ and $g: H \to K$ are homomorphisms, then so is the composite $g \circ f: G \to K$.
(ii) If $f: G \to H$ is an isomorphism, then its inverse $f^{-1}: H \to G$ is also an isomorphism.
(iii) If $\mathscr{C}$ is a class of groups, show that the relation of isomorphism is an equivalence relation on $\mathscr{C}$.

1.44. Let G be a group, let X be a set, and let $f: G \to X$ be a bijection. Show that there is a unique operation on X so that X is a group and f is an isomorphism.

1.45. If k is a field, denote the columns of the $n \times n$ identity matrix E by $\varepsilon_1, \ldots, \varepsilon_n$. A ***permutation matrix*** P over k is a matrix obtained from E by permuting its columns; that is, the columns of P are $\varepsilon_{\alpha 1}, \ldots, \varepsilon_{\alpha n}$ for some $\alpha \in S_n$. Prove that the set of all permutation matrices over k is a group isomorphic to S_n. (*Hint.* The inverse of P is its transpose P^t, which is also a permutation matrix.)

1.46. Let $\mathbf{T}$ denote the ***circle group***: the multiplicative group of all complex numbers of absolute value 1. For a fixed real number y, show that $f_y: \mathbb{R} \to \mathbf{T}$, given by $f_y(x) = e^{iyx}$, is a homomorphism. (The functions f_y are the only *continuous* homomorphisms $\mathbb{R} \to \mathbf{T}$.)

1.47. If a is a fixed element of a group G, define $\gamma_a: G \to G$ by $\gamma_a(x) = a * x * a^{-1}$ (γ_a is called ***conjugation*** by a).

(i) Prove that γ_a is an isomorphism.
(ii) If $a, b \in G$, prove that $\gamma_a \gamma_b = \gamma_{a*b}$.[4]

1.48. If G denotes the multiplicative group of all complex nth roots of unity (see Exercise 1.35), then $G \cong \mathbb{Z}_n$.

1.49. Describe all the homomorphisms from $\mathbb{Z}_{12}$ to itself. Which of these are isomorphisms?

1.50. (i) Prove that a group G is abelian if and only if the function $f\colon G \to G$, defined by $f(a) = a^{-1}$, is a homomorphism.
(ii) Let $f\colon G \to G$ be an isomorphism from a finite group G to itself. If f has no nontrivial fixed points (i.e., $f(x) = x$ implies $x = e$) and if $f \circ f$ is the identity function, then $f(x) = x^{-1}$ for all $x \in G$ and G is abelian. (*Hint.* Prove that every element of G has the form $x * f(x)^{-1}$.)

1.51 **(Kaplansky).** An element a in a ring R has a ***left quasi-inverse*** if there exists an element $b \in R$ with $a + b - ba = 0$. Prove that if every element in a ring R except 1 has a left quasi-inverse, then R is a division ring. (*Hint.* Show that $R - \{1\}$ is a group under the operation $a \circ b = a + b - ba$.)

1.52. (i) If G is the multiplicative group of all positive real numbers, show that $\log\colon G \to (\mathbb{R}, +)$ is an isomorphism. (*Hint*: Find a function inverse to log.)
(ii) Let G be the additive group of $\mathbb{Z}[x]$ (all polynomials with integer coefficients) and let H be the multiplicative group of all positive rational numbers. Prove that $G \cong H$. (*Hint.* Use the Fundamental Theorem of Arithmetic.)

Having solved Exercise 1.52, the reader may wish to reconsider the question when one "knows" a group. It may seem reasonable that one knows a group if one knows its multiplication table. But addition tables of $\mathbb{Z}[x]$ and of H are certainly well known (as are those of the multiplicative group of positive reals and the additive group of all reals), and it was probably a surprise that these groups are essentially the same. As an alternative answer to the question, we suggest that a group G is "known" if it can be determined, given any other group H, whether or not G and H are isomorphic.

[4] It is easy to see that $\delta_a\colon G \to G$, defined by $\delta_a(x) = a^{-1} * x * a$, is also an isomorphism; however, $\delta_a \delta_b = \delta_{b*a}$. Since we denote the value of a function f by $f(x)$, that is, the symbol f is on the left, the isomorphisms γ_a are more natural for us than the δ_a. On the other hand, if one denotes $\delta_a(x)$ by x^a, then one has put the function symbol on the right, and the δ_a are more convenient: $x^{a*b} = (x^a)^b$. Indeed, many group theorists nowadays put all their function symbols on the right!

CHAPTER 2

The Isomorphism Theorems

We now drop the $*$ notation for the operation in a group. Henceforth, we shall write ab instead of $a * b$, and we shall denote the identity element by 1 instead of by e.

Subgroups

Definition. A nonempty subset S of a group G is a ***subgroup*** of G if $s \in G$ implies $s^{-1} \in G$ and $s, t \in G$ imply $st \in G$.

If X is a subset of a group G, we write $X \subset G$; if X is a subgroup of G, we write $X \leq G$.

Theorem 2.1. *If $S \leq G$ (i.e., if S is a subgroup of G), then S is a group in its own right.*

Proof. The hypothesis "$s, t \in S$ imply $st \in S$" shows that S is equipped with an operation (if μ: $G \times G \to G$ is the given multiplication in G, then its restriction $\mu | S \times S$ has its image contained in S). Since S is nonempty, it contains an element, say, s, and the definition of subgroup says that $s^{-1} \in S$; hence, $1 = ss^{-1} \in S$. Finally, the operation on S is associative because $a(bc) = (ab)c$ for every $a, b, c \in G$ implies, in particular, that $a(bc) = (ab)c$ for every $a, b, c \in S$. ■

Verifying associativity is the most tedious part of showing that a given set G equipped with a multiplication is actually a group. Therefore, if G is given as a subset of a group G^*, then it is much simpler to show that G is a subgroup of G^* than to verify all the group axioms for G. For example, the four permutations of the 4-group **V** form a group because they constitute a subgroup of S_4.

Theorem 2.2. *A subset S of a group G is a subgroup if and only if $1 \in S$ and $s, t \in S$ imply $st^{-1} \in S$.*

Proof. If $s \in S$, then $1s^{-1} = s^{-1} \in S$, and if $s, t \in S$, then $s(t^{-1})^{-1} = st \in S$. The converse is also easy. ∎

Definition. If G is a group and $a \in G$, then the ***cyclic subgroup generated by a***, denoted by $\langle a \rangle$, is the set of all the powers of a. A group G is called ***cyclic*** if there is $a \in G$ with $G = \langle a \rangle$; that is, G consists of all the powers of a.

It is plain that $\langle a \rangle$ is, indeed, a subgroup of G. Notice that different elements can generate the same cyclic subgroup. For example, $\langle a \rangle = \langle a^{-1} \rangle$.

Definition. If G is a group and $a \in G$, then the ***order*** of a is $|\langle a \rangle|$, the number of elements in $\langle a \rangle$.

Theorem 2.3. *If G is a group and $a \in G$ has finite order m, then m is the smallest positive integer such that $a^m = 1$.*

Proof. If $a = 1$, then $m = 1$. If $a \neq 1$, there is an integer $k > 1$ so that $1, a, a^2, \ldots, a^{k-1}$ are distinct elements of G while $a^k = a^i$ for some i with $0 \leq i \leq k - 1$. We claim that $a^k = 1 = a^0$. If $a^k = a^i$ for some $i \geq 1$, then $k - i \leq k - 1$ and $a^{k-i} = 1$, contradicting the original list $1, a, a^2, \ldots, a^{k-1}$ having no repetitions. It follows that k is the smallest positive integer with $a^k = 1$.

It now suffices to prove that $k = m$; that is, that $\langle a \rangle = \{1, a, a^2, \ldots, a^{k-1}\}$. Clearly $\langle a \rangle \supset \{1, a, a^2, \ldots, a^{k-1}\}$. For the reverse inclusion, let a^l be a power of a. By the division algorithm, $l = qk + r$, where $0 \leq r < k$. Hence, $a^l = a^{qk+r} = a^{qk}a^r = a^r$ (because $a^k = 1$), and so $a^l = a^r \in \{1, a, a^2, \ldots, a^{k-1}\}$. ∎

If $\alpha \in S_n$ is written as a product of disjoint cycles, say, $\alpha = \beta_1 \ldots \beta_t$, where β_i is an r_i-cycle for every i, then Exercise 1.12(iii) shows that the order of α is $\operatorname{lcm}\{r_1, \ldots, r_t\}$.

Corollary 2.4. *If G is a finite group, then a nonempty subset S of G is a subgroup if and only if $s, t \in S$ imply $st \in S$.*

Proof. Necessity is obvious. For sufficiency, we must show that $s \in S$ implies $s^{-1} \in S$. It follows easily by induction that S contains all the powers of s. Since G is finite, s has finite order, say, m. Therefore, $1 = s^m \in S$ and $s^{-1} = s^{m-1} \in S$. ∎

Example 2.1. If G is a group, then G itself and $\{1\}$ are always subgroups (we shall henceforth denote the subgroup $\{1\}$ by 1). Any subgroup H other than G is called ***proper***, and we denote this by $H < G$; the subgroup 1 is often called the ***trivial*** subgroup.

Example 2.2. Let $f: G \to H$ be a homomorphism, and define

$$\textbf{\textit{kernel}}\, f = \{a \in G: f(a) = 1\}$$

and

$$\textbf{\textit{image}}\, f = \{h \in H: h = f(a) \text{ for some } a \in G\}.$$

Then $K = \text{kernel } f$ is a subgroup of G and image f is a subgroup of H. To see that $K \leq G$, note first that $f(1) = 1$, so that $1 \in K$. Also, if $s, t \in K$, then $f(s) = 1 = f(t)$, and so $f(st^{-1}) = f(s)f(t)^{-1} = 1$; hence $st^{-1} \in K$, and so K is a subgroup of G. It is equally easy to see that image f is a subgroup of H.

Notation. We usually write ***ker f*** instead of kernel f and ***im f*** instead of image f.

We have been using multiplicative notation, but it is worth writing the definition of subgroup in additive notation as well. If G is an additive group, then a nonempty subset S of G is a subgroup of G if $s \in S$ implies $-s \in S$ and $s, t \in S$ imply $s + t \in S$. Theorem 2.2 says that S is a subgroup if and only if $0 \in S$ and $s, t \in S$ imply $s - t \in S$.

Theorem 2.5. *The intersection of any family of subgroups of a group G is again a subgroup of G.*

Proof. Let $\{S_i: i \in I\}$ be a family of subgroups of G. Now $1 \in S_i$ for every i, and so $1 \in \bigcap S_i$. If $a, b \in \bigcap S_i$, then $a, b \in S_i$ for every i, and so $ab^{-1} \in S_i$ for every i; hence, $ab^{-1} \in \bigcap S_i$, and $\bigcap S_i \leq G$. ∎

Corollary 2.6. *If X is a subset of a group G, then there is a* ***smallest*** *subgroup H of G containing X; that is, if $X \subset S$ and $S \leq G$, then $H \leq S$.*

Proof. There are subgroups of G containing X; for example, G itself contains X; define H as the intersection of all the subgroups of G which contain X. Note that H is a subgroup, by Theorem 2.5, and $X \subset H$. If $S \leq G$ and $X \subset S$, then S is one of the subgroups of G being intersected to form H; hence, $H \leq S$, and so H is the smallest such subgroup. ∎

Definition. If X is a subset of a group G, then the smallest subgroup of G containing X, denoted by $\langle X \rangle$, is called the ***subgroup generated by*** X. One also says that X ***generates*** $\langle X \rangle$.

In particular, if H and K are subgroups of G, then the subgroup $\langle H \cup K \rangle$ is denoted by $H \vee K$.

If X consists of a single element a, then $\langle X \rangle = \langle a \rangle$, the cyclic subgroup generated by a. If X is a finite set, say, $X = \{a_1, a_2, \ldots, a_n\}$ then we write $\langle X \rangle = \langle a_1, a_2, \ldots, a_n \rangle$ instead of $\langle X \rangle = \langle \{a_1, a_2, \ldots, a_n\} \rangle$.

Here is a description of the elements in $\langle X \rangle$.

Definition. If X is a nonempty subset of a group G, then a ***word*** on X is an element $w \in G$ of the form

$$w = x_1^{e_1} x_2^{e_2} \ldots x_n^{e_n},$$

where $x_i \in X$, $e_i = \pm 1$, and $n \geq 1$.

Theorem 2.7. *Let X be a subset of a group G. If $X = \varnothing$, then $\langle X \rangle = 1$; if X is nonempty, then $\langle X \rangle$ is the set of all the words on X.*

Proof. If $X = \varnothing$, then the subgroup $1 = \{1\}$ contains X, and so $\langle X \rangle = 1$. If X is nonempty, let W denote the set of all the words on X. It is easy to see that W is a subgroup of G containing X: $1 = x_1^{-1} x_1 \in W$; the inverse of a word is a word; the product of two words is a word. Since $\langle X \rangle$ is the smallest subgroup containing X, we have $\langle X \rangle \subset W$. The reverse inclusion also holds, for every subgroup H containing X must contain every word on X. Therefore, $W \leq H$, and W is the smallest subgroup containing X. ■

Exercises

2.1. Show that A_n, the set of all even permutations in S_n, is a subgroup with $n!/2$ elements. (A_n is called the ***alternating group*** on n letters.) (*Hint.* Exercise 1.21.)

2.2. If k is a field, show that $\mathrm{SL}(n, k)$, the set of all $n \times n$ matrices over k having determinant 1, is a subgroup of $\mathrm{GL}(n, k)$. ($\mathrm{SL}(n, k)$ is called the ***special linear group*** over k.)

2.3. The set theoretic union of two subgroups is a subgroup if and only if one is contained in the other. Is this true if we replace "two subgroups" by "three subgroups"?

2.4. Let S be a proper subgroup of G. If $G - S$ is the complement of S, prove that $\langle G - S \rangle = G$.

2.5. Let $f: G \to H$ and $g: G \to H$ be homomorphisms, and let

$$K = \{a \in G: f(a) = g(a)\}.$$

Must K be a subgroup of G?

2.6. Suppose that X is a nonempty subset of a set Y. Show that S_X can be ***imbedded*** in S_Y; that is, S_X is isomorphic to a subgroup of S_Y.

2.7. If $n > 2$, then A_n is generated by all the 3-cycles. (*Hint.* $(ij)(jk) = (ijk)$ and $(ij)(kl) = (ijk)(jkl)$.)

2.8. Imbed S_n as a subgroup of A_{n+2}, but show, for $n \geq 2$, that S_n cannot be imbedded in A_{n+1}.

2.9. (i) Prove that S_n can be generated by $(1\ 2), (1\ 3), \ldots, (1\ n)$.
(ii) Prove that S_n can be generated by $(1\ 2), (2\ 3), \ldots, (i\ i+1), \ldots, (n-1, n)$.
(iii) Prove that S_n can be generated by the two elements $(1\ 2)$ and $(1\ 2\ \ldots\ n)$.
(iv) Prove that S_4 cannot be generated by $(1\ 3)$ and $(1\ 2\ 3\ 4)$. (Thus, S_4 can be generated by a transposition and a 4-cycle, but not every choice of transposition and 4-cycle gives a generating set.)

Lagrange's Theorem

Definition. If S is a subgroup of G and if $t \in G$, then a ***right coset*** of S in G is the subset of G

$$St = \{st: s \in S\}$$

(a ***left coset*** is $tS = \{ts: s \in S\}$). One calls t a ***representative*** of St (and also of tS).

EXAMPLE 2.3. Let G be the additive group of the plane $\mathbb{R}^2$: the elements of G are vectors (x, y), and addition is given by the "parallelogram law": $(x, y) + (x', y') = (x + x', y + y')$. A line ℓ through the origin is the set of all scalar multiples of some nonzero vector $v = (x_0, y_0)$; that is, $\ell = \{rv: r \in \mathbb{R}\}$. It is easy to see that ℓ is a subgroup of G. If $u = (a, b)$ is a vector, then the coset $u + \ell$ is easily seen to be the line parallel to ℓ which contains u.

EXAMPLE 2.4. If G is the additive group $\mathbb{Z}$ of all integers, if S is the set of all multiples of an integer n ($S = \langle n \rangle$, the cyclic subgroup generated by n), and if $a \in \mathbb{Z}$, then the coset $a + S = \{a + qn: q \in \mathbb{Z}\} = \{k \in \mathbb{Z}: k \equiv a \bmod n\}$; that is, the coset $a + \langle n \rangle$ is precisely the congruence class $[a]$ of a mod n.

EXAMPLE 2.5. Let $G = S_3$ and let $H = \langle \tau \rangle = \{1, \tau\}$, where $\tau = (1\ 2)$. The right cosets of H in G are

$$H = \{1, \tau\}; \qquad H(1\ 2\ 3) = \{(1\ 2\ 3), (2\ 3)\};$$
$$H(1\ 3\ 2) = \{(1\ 3\ 2), (1\ 3)\}.$$

The left cosets of H in G are

$$H = \{1, \tau\}; \qquad (1\ 2\ 3)H = \{(1\ 2\ 3), (1\ 3)\};$$
$$(1\ 3\ 2)H = \{(1\ 3\ 2), (2\ 3)\}.$$

Notice that distinct right cosets are disjoint (as are distinct left cosets), just as in the example of parallel lines. Notice also that right cosets and left cosets can be distinct; for example, $(1\ 2\ 3)H \neq H(1\ 2\ 3)$; indeed, $(1\ 2\ 3)H$ is not equal to any right coset of H in G.

A right coset St has many representatives; every element of the form st for $s \in S$ is a representative of St. The next lemma gives a criterion for determining whether two right cosets of S are the same when a representative of each is known.

Lemma 2.8. *If $S \leq G$, then $Sa = Sb$ if and only if $ab^{-1} \in S$ ($aS = bS$ if and only if $b^{-1}a \in S$).*

Proof. If $Sa = Sb$, then $a = 1a \in Sa = Sb$, and so there is $s \in S$ with $a = sb$; hence, $ab^{-1} = s \in S$. Conversely, assume that $ab^{-1} = \sigma \in S$; hence, $a = \sigma b$. To prove that $Sa = Sb$, we prove two inclusions. If $x \in Sa$, then $x = sa$ for some $s \in S$, and so $x = s\sigma b \in Sb$; similarly, if $y \in Sb$, then $y = s'b$ for some $s' \in S$, and $y = s'\sigma^{-1}a \in Sa$. Therefore, $Sa = Sb$. ■

Theorem 2.9. *If $S \leq G$, then any two right (or any two left) cosets of S in G are either identical or disjoint.*

Proof. We show that if there exists an element $x \in Sa \cap Sb$, then $Sa = Sb$. Such an x has the form $sb = x = ta$, where $s, t \in S$. Hence, $ab^{-1} = t^{-1}s \in S$, and so the lemma gives $Sa = Sb$. ■

Theorem 2.9 may be paraphrased to say that the right cosets of a subgroup S comprise a partition of G (each such coset is nonempty, and G is their disjoint union). This being true, there must be an equivalence relation on G lurking somewhere in the background: it is given, for $a, b \in G$, by $a \equiv b$ if $ab^{-1} \in S$, and its equivalence classes are the right cosets of S.

Theorem 2.10. *If $S \leq G$, then the number of right cosets of S in G is equal to the number of left cosets of S in G.*

Proof. We give a bijection $f: \mathscr{R} \to \mathscr{L}$, where $\mathscr{R}$ is the family of right cosets of S in G and $\mathscr{L}$ is the family of left cosets. If $Sa \in \mathscr{R}$, your first guess is to define $f(Sa) = aS$, but this does not work. Your second guess is to define $f(Sa) = a^{-1}S$, and this does work. It must be verified that f is well defined; that is, if $Sa = Sb$, then $a^{-1}S = b^{-1}S$ (this is why the first guess is incorrect). It is routine to prove that f is a bijection. ■

Definition. If $S \leq G$, then the ***index*** of S in G, denoted by $[G : S]$, is the number of right cosets of S in G.

Theorem 2.10 shows that there is no need to define a right index and a left index, for the number of right cosets is equal to the number of left cosets.

It is a remarkable theorem of P. Hall (1935) that in a finite group G, one can always (as above) choose a ***common system of representatives*** for the right and left cosets of a subgroup S; if $[G : S] = n$, there exist elements $t_1, \ldots, t_n \in G$ so that $t_1S, \ldots, t_nS$ is the family of all left cosets and $St_1, \ldots, St_n$ is the family of all right cosets.

Definition. If G is a group, then the ***order*** of G, denoted by $|G|$, is the number of elements in G.

The next theorem was inspired by work of Lagrange (1770), but it was probably first proved by Galois.

Theorem 2.11 (Lagrange). *If G is a finite group and $S \leq G$, then $|S|$ divides $|G|$ and $[G : S] = |G|/|S|$.*

Proof. By Theorem 2.9, G is partitioned into its right cosets

$$G = St_1 \cup St_2 \cup \cdots \cup St_n,$$

and so $|G| = \sum_{i=1}^{n} |St_i|$. But it is easy to see that $f_i \colon S \to St_i$, defined by $f_i(s) = st_i$, is a bijection, and so $|St_i| = |S|$ for all i. Thus $|G| = n|S|$, where $n = [G : S]$. ∎

Corollary 2.12. *If G is a finite group and $a \in G$. Then the order of a divides $|G|$.*

Proof. By definition, the order of a is $|\langle a \rangle|$, and so the result follows at once from Lagrange's theorem. ∎

Definition. A group G has ***exponent*** n if $x^n = 1$ for all $x \in G$.

Remark. Some people use the term "exponent" to mean the smallest possible n such that $x^n = 1$ for all $x \in G$. For us, the 4-group $\mathbf{V}$ has exponent 4 as well as exponent 2.

Lagrange's theorem shows that a finite group G of order n has exponent n.

Corollary 2.13. *If p is a prime and $|G| = p$, then G is a cyclic group.*

Proof. Take $a \in G$ with $a \neq 1$. Then the cyclic subgroup $\langle a \rangle$ has more than one element (it contains a and 1), and its order $|\langle a \rangle| > 1$ is a divisor of p. Since p is prime, $|\langle a \rangle| = p = |G|$, and so $\langle a \rangle = G$. ∎

Corollary 2.14 (Fermat). *If p is a prime and a is an integer, then $a^p \equiv a \bmod p$.*

Proof. Let $G = U(\mathbb{Z}_p)$, the multiplicative group of nonzero elements of $\mathbb{Z}_p$; since p is prime, $\mathbb{Z}_p$ is a field and G is a group of order $p - 1$.

Recall that for integers a and b, one has $a \equiv b \bmod p$ if and only if $[a] = [b]$ in $\mathbb{Z}_p$. If $a \in \mathbb{Z}$ and $[a] = [0]$ in $\mathbb{Z}_p$, then it is clear that $[a]^p = [0] = [a]$. If $[a] \neq [0]$, then $[a] \in G$ and so $[a]^{p-1} = [1]$, by Corollary 2.12; multiplying by $[a]$ now gives the desired result. ∎

EXERCISES

2.10. If G is a finite group and $K \leq H \leq G$, then

$$[G:K] = [G:H][H:K].$$

2.11. Let $a \in G$ have order $n = mk$, where $m, k \geq 1$. Prove that a^k has order m.

2.12. (i) Prove that every group G of order 4 is isomorphic to either $\mathbb{Z}_4$ or the 4-group $\mathbf{V}$.
(ii) If G is a group with $|G| \leq 5$, then G is abelian.

2.13. If $a \in G$ has order n and k is an integer with $a^k = 1$, then n divides k. Indeed, $\{k \in \mathbb{Z}: a^k = 1\}$ consists of all the multiplies of n.

2.14. If $a \in G$ has finite order and $f: G \to H$ is a homomorphism, then the order of $f(a)$ divides the order of a.

2.15. Prove that a group G of even order has an odd number of elements of order 2 (in particular, it has at least one such element). (*Hint.* If $a \in G$ does not have order 2, then $a \neq a^{-1}$.)

2.16. If $H \leq G$ has index 2, then $a^2 \in H$ for every $a \in G$.

2.17. (i) If $a, b \in G$ commute and if $a^m = 1 = b^n$, then $(ab)^k = 1$, where $k = \operatorname{lcm}\{m, n\}$. (The order of ab may be smaller than k; for example, take $b = a^{-1}$.) Conclude that if a and b have finite order, then ab also has finite order.
(ii) Let $G = \mathrm{GL}(2, \mathbb{Q})$ and let $A, B \in G$ be given by

$$A = \begin{bmatrix} 0 & -1 \\ 1 & 0 \end{bmatrix} \quad \text{and} \quad B = \begin{bmatrix} 0 & 1 \\ -1 & -1 \end{bmatrix}.$$

Show that $A^4 = E = B^3$, but that AB has infinite order.

2.18. Prove that every subgroup of a cyclic group is cyclic. (*Hint.* Use the division algorithm.)

2.19. Prove that two cyclic groups are isomorphic if and only if they have the same order.

Definition. The ***Euler φ-function*** is defined as follows:

$$\varphi(1) = 1; \qquad \text{if } n > 1, \quad \text{then } \varphi(n) = |\{k: 1 \leq k < n \text{ and } (k, n) = 1\}|.$$

2.20. If $G = \langle a \rangle$ is cyclic of order n, then a^k is also a generator of G if and only if $(k, n) = 1$. Conclude that the number of generators of G is $\varphi(n)$.

2.21. (i) Let $G = \langle a \rangle$ have order rs, where $(r, s) = 1$. Show that there are unique $b, c \in G$ with b of order r, c of order s, and $a = bc$.
(ii) Use part (i) to prove that if $(r, s) = 1$, then $\varphi(rs) = \varphi(r)\varphi(s)$.

2.22. (i) If p is prime, then $\varphi(p^k) = p^k - p^{k-1} = p^k(1 - 1/p)$.
(ii) If the distinct prime divisors of n are $p_1, \ldots, p_t$, then

$$\varphi(n) = n(1 - 1/p_1)\ldots(1 - 1/p_t).$$

2.23 **(Euler).** If $(r, s) = 1$, then $s^{\varphi(r)} \equiv 1 \bmod r$. (*Hint.* The order of the group of units $U(\mathbb{Z}_n)$ is $\varphi(n)$.)

Cyclic Groups

Lemma 2.15. *If G is a cyclic group of order n, then there exists a unique subgroup of order d for every divisor d of n.*

Proof. If $G = \langle a \rangle$, then $\langle a^{n/d} \rangle$ is a subgroup of order d, by Exercise 2.11. Assume that $S = \langle b \rangle$ is a subgroup of order d (S must be cyclic, by Exercise 2.18). Now $b^d = 1$; moreover, $b = a^m$ for some m. By Exercise 2.13, $md = nk$ for some integer k, and $b = a^m = (a^{n/d})^k$. Therefore, $\langle b \rangle \leq \langle a^{n/d} \rangle$, and this inclusion is equality because both subgroups have order d. ■

Theorem 2.16. *If n is a positive integer, then*

$$n = \sum_{d|n} \varphi(d),$$

where the sum is over all divisors d of n with $1 \leq d \leq n$.

Proof. If C is a cyclic subgroup of a group G, let $\text{gen}(C)$ denote the set of all its generators. It is clear that G is the disjoint union

$$G = \bigcup \text{gen}(C),$$

where C ranges over all the cyclic subgroups of G. We have just seen, when G is cyclic of order n, that there is a unique cyclic subgroup C_d of order d for every divisor d of n. Therefore, $n = |G| = \sum_{d|n} |\text{gen}(C_d)|$. In Exercise 2.20, however, we saw that $|\text{gen}(C_d)| = \varphi(d)$; the result follows. ■

We now characterize finite cyclic groups.

Theorem 2.17. *A group G of order n is cyclic if and only if, for each divisor d of n, there is at most one cyclic subgroup of G having order d.*

Proof. If G is cyclic, then the result is Lemma 2.15. For the converse, recall from the previous proof that G is the disjoint union $\bigcup \text{gen}(C)$, where C ranges over all the cyclic subgroups of G. Hence, $n = |G| = \sum |\text{gen}(C)| \leq \sum_{d|n} \varphi(d) = n$, by Theorem 2.16. We conclude that G must have a cyclic subgroup of order d for every divisor d of n; in particular, G has a cyclic subgroup of order $d = n$, and so G is cyclic. ■

Observe that the condition in Theorem 2.17 is satisfied if, for every divisor d of n, there are at most d solutions $x \in G$ of the equation $x^d = 1$ (two cyclic subgroups of order d would contain more than d solutions).

Theorem 2.18.

(i) *If F is a field and if G is a finite subgroup of $F^\times$, the multiplicative group of nonzero elements of F, then G is cyclic.*
(ii) *If F is a finite field, then its multiplicative group $F^\times$ is cyclic.*

Proof. If $|G| = n$ and if $a \in G$ satisfies $a^d = 1$, where $d|n$, then a is a root in F of the polynomial $x^d - 1 \in F[x]$. Since a polynomial of degree d over a field has at most d roots, our observation above shows that the hypothesis of Theorem 2.17 is satisfied. Statement (ii) follows at once from (i). ■

When F is finite, the proof does not construct a generator of $F^\times$. Indeed, no algorithm is known which displays a generator of $\mathbb{Z}_p^\times$ for all primes p.

Theorem 2.19. *Let p be a prime. A group G of order p^n is cyclic if and only if it is an abelian group having a unique subgroup of order p.*

Proof. Necessity follows at once from Lemma 2.15. For the converse, let $a \in G$ have largest order, say p^k (it follows that $g^{p^k} = 1$ for all $g \in G$). Of course, the unique subgroup H of order p is a subgroup of $\langle a \rangle$. If $\langle a \rangle$ is a proper subgroup of G, then there is $x \in G$ with $x \notin \langle a \rangle$ but with $x^p \in \langle a \rangle$; let $x^p = a^l$. If $k = 1$, then $x^p = 1$ and $x \in H \leq \langle a \rangle$, a contradiction; we may, therefore, assume that $k > 1$. Now

$$1 = x^{p^k} = (x^p)^{p^{k-1}} = a^{lp^{k-1}},$$

so that $l = pm$ for some integer m, by Exercise 2.13. Hence, $x^p = a^{mp}$, and so $1 = x^{-p}a^{mp}$. Since G is abelian, $x^{-p}a^{mp} = (x^{-1}a^m)^p$, and so $x^{-1}a^m \in H \leq \langle a \rangle$. This gives $x \in \langle a \rangle$, a contradiction. Therefore, $G = \langle a \rangle$ and hence is cyclic. ■

EXERCISE

2.24. Let $G = \langle A, B \rangle \leq \mathrm{GL}(2, \mathbb{C})$, where

$$A = \begin{bmatrix} 0 & i \\ i & 0 \end{bmatrix} \quad \text{and} \quad B = \begin{bmatrix} 0 & 1 \\ -1 & 0 \end{bmatrix}.$$

Show that G is a nonabelian group (so G is not cyclic) of order 8 having a unique subgroup of order 2. (See Theorem 4.22.)

Normal Subgroups

This brief section introduces the fundamental notion of normal subgroups. We begin with a construction which generalizes that of cosets.

Definition. If S and T are nonempty subsets of a group G, then

$$ST = \{st: s \in S \text{ and } t \in T\}.$$

If $S \leq G$, $t \in G$, and $T = \{t\}$, then ST is the right coset St. Notice that the family of all the nonempty subsets of G is a semigroup under this operation:

if S, T, and U are nonempty subsets of G, then $(ST)U = S(TU)$, for either side consists of all the elements of G of the form $(st)u = s(tu)$ with $s \in S$, $t \in T$, and $u \in U$.

Theorem 2.20 (Product Formula). *If S and T are subgroups of a finite group G, then*

$$|ST||S \cap T| = |S||T|.$$

Remark. The subset ST need not be a subgroup.

Proof. Define a function $\varphi: S \times T \to ST$ by $(s, t) \mapsto st$. Since φ is a surjection, it suffices to show that if $x \in ST$, then $|\varphi^{-1}(x)| = |S \cap T|$. We show that $\varphi^{-1}(x) = \{(sd, d^{-1}t): d \in S \cap T\}$. It is clear that $\varphi^{-1}(x)$ contains the right side. For the reverse inclusion, let $(s, t), (\sigma, \tau) \in \varphi^{-1}(x)$; that is, $s, \sigma \in S$, $t, \tau \in T$, and $st = x = \sigma\tau$. Thus, $s^{-1}\sigma = t\tau^{-1} \in S \cap T$; let $d = s^{-1}\sigma = t\tau^{-1}$ denote their common value. Then $\sigma = s(s^{-1}\sigma) = sd$ and $d^{-1}t = \tau t^{-1}t = \tau$, as desired. ∎

There is one kind of subgroup that is especially interesting because it is intimately related to homomorphisms.

Definition. A subgroup $K \leq G$ is a ***normal subgroup***, denoted by $K \lhd G$, if $gKg^{-1} = K$ for every $g \in G$.

If $K \leq G$ and there are inclusions $gKg^{-1} \leq K$ for every $g \in G$, then $K \lhd G$: replacing g by g^{-1}, we have the inclusion $g^{-1}Kg \leq K$, and this gives the reverse inclusion $K \leq gKg^{-1}$.

The kernel K of a homomorphism $f: G \to H$ is a normal subgroup: if $a \in K$, then $f(a) = 1$; if $g \in G$, then $f(gag^{-1}) = f(g)f(a)f(g)^{-1} = f(g)f(g)^{-1} = 1$, and so $gag^{-1} \in K$. Hence, $gKg^{-1} \leq K$ for all $g \in G$, and so $K \lhd G$. Conversely, we shall see later that every normal subgroup is the kernel of some homomorphism.

In Example 2.5, we saw that if H is the cyclic subgroup of S_3 generated by the transposition $\tau = (1\ 2)$, then there are right cosets of H which are not left cosets. When K is normal, then every left coset of K in G is a right coset. Indeed, a subgroup K of G is normal in G if and only if $Kg = gK$ for every $g \in G$, for associativity of the multiplication of nonempty subsets gives $K = (Kg)g^{-1} = gKg^{-1}$. In terms of elements, this says that there is a *partial commutativity* when $K \lhd G$: if $g \in G$ and $k \in K$, then there exists $k' \in K$ with $ak = k'a$. It may not be true that g commutes with every element of K. For example, the reader should check that the cyclic subgroup K of S_3 generated by the 3-cycle $(1\ 2\ 3)$ is a normal subgroup. It follows that $(1\ 2)K = K(1\ 2)$ even though $(1\ 2)$ does not commute with $(1\ 2\ 3)$.

Normal subgroups are also related to conjugations $\gamma_a: G \to G$, where $\gamma_a(x) = axa^{-1}$ (see Exercise 2.34 below).

Definition. If $x \in G$, then a ***conjugate*** of x in G is an element of the form axa^{-1} for some $a \in G$; equivalently, x and y are conjugate if $y = \gamma_a(x)$ for some $a \in G$.

For example, if k is a field, then matrices A and B in $\mathrm{GL}(n, k)$ are conjugate if and only if they are similar.

EXERCISES

2.25. If S is a subgroup of G, then $SS = S$; conversely, show that if S is a *finite* nonempty subset of G with $SS = S$, then S is a subgroup. Give an example to show that the converse may be false when S is infinite.

2.26. Let $\{S_i: i \in I\}$ be a family of subgroups of a group G, let $\{S_i t_i: i \in I\}$ be a family of right cosets, and let $D = \bigcap S_i$. Prove that either $\bigcap S_i t_i = \varnothing$ or $\bigcap S_i t_i = Dg$ for some g.

2.27. If S and T are (not necessarily distinct) subgroups of G, then an **(S-T)-double coset** is a subset of G of the form SgT, where $g \in G$. Prove that the family of all (S-T)-double cosets partitions G. (*Hint.* Define an equivalence relation on G by $a \equiv b$ if $b = sat$ for some $s \in S$ and $t \in T$.)

2.28. Let $S, T \leq G$, where G is a finite group, and suppose that G is the disjoint union

$$G = \bigcup_{i=1}^{n} Sg_i T.$$

Prove that $[G:T] = \sum_{i=1}^{n} [S: S \cap g_i T g_i^{-1}]$. (Note that Lagrange's theorem is the special case of this when $T = 1$.)

2.29 (i) **(H. B. Mann).** Let G be a finite group, and let S and T be (not necessarily distinct) nonempty subsets. Prove that either $G = ST$ or $|G| \geq |S| + |T|$.
(ii) Prove that every element in a finite field F is a sum of two squares.

2.30. If $S \leq G$ and $[G:S] = 2$, then $S \lhd G$.

2.31. If G is abelian, then every subgroup of G is normal. The converse is false: show that the group of order 8 in Exercise 2.24 (the *quaternions*) is a counterexample.

2.32. If $H \leq G$, then $H \lhd G$ if and only if, for all $x, y \in G$, $xy \in H$ if and only if $yx \in H$.

2.33. If $K \leq H \leq G$ and $K \lhd G$, then $K \lhd H$.

2.34. A subgroup S of G is normal if and only if $s \in S$ implies that every conjugate of s is also in S. Conclude that if $S \leq G$, then $S \lhd G$ if and only if $\gamma(S) \leq S$ for every conjugation γ.

2.35. Prove that $\mathrm{SL}(n, k) \lhd \mathrm{GL}(n, k)$ for every $n \geq 1$ and every field k.

2.36. Prove that $A_n \lhd S_n$ for every n.

2.37. (i) The intersection of any family of normal subgroups of a group G is itself a normal subgroup of G. Conclude that if X is a subset of G, then there is a smallest normal subgroup of G which contains X; it is called the ***normal subgroup generated by*** X (or the *normal closure* of X; it is often denoted by $\langle X \rangle^G$).

(ii) If $X = \varnothing$, then $\langle X \rangle^G = 1$. If $X \neq \varnothing$, then $\langle X \rangle^G$ is the set of all words on the conjugates of elements in X.

(iii) If $gxg^{-1} \in X$ for all $x \in X$ and $g \in G$, then $\langle X \rangle = \langle X \rangle^G \lhd G$.

2.38. If H and K are normal subgroups of G, then $H \vee K \lhd G$.

2.39. Prove that if a normal subgroup H of G has index n, then $g^n \in H$ for all $g \in G$. Give an example to show this may be false when H is not normal.

Quotient Groups

The construction of the *quotient group* (or *factor group*) G/N in the next theorem is of fundamental importance.

We have already seen that if X and Y are nonempty subsets of a group G, then their product

$$XY = \{xy: x \in X \text{ and } y \in Y\}$$

defines an associative operation on the family of all nonempty subsets of G. If H is a subgroup of G, then the family of all right cosets of H in G need not be closed under this operation. In Example 2.5, we looked at the right cosets of $H = \langle(1\ 2)\rangle$ in S_3. The product of right cosets

$$H(1\ 2\ 3)H(1\ 3\ 2) = \{(1\ 2\ 3), (1\ 3\ 2), (2\ 3), 1\ 3)\}$$

is not a right coset of H, for it has four elements while right cosets of H have two elements. In the proof of the next theorem, we shall see that if H is a *normal* subgroup, then the product of two right cosets of H is also a right coset of H.

Theorem 2.21. *If $N \lhd G$, then the cosets of N in G form a group, denoted by G/N, of order $[G:N]$.*

Proof. To define a group, one needs a set and an operation. The set here is the family of all cosets of N in G (notice that we need not bother with the adjectives "left" and "right" because N is a normal subgroup). As operation, we propose the multiplication of nonempty subsets of G defined earlier. We have already observed that this operation is associative. Now

$$\begin{aligned} NaNb &= Na(a^{-1}Na)b \quad \text{(because } N \text{ is normal)} \\ &= N(aa^{-1})Nab = NNab = Nab \quad \text{(because } N \leq G\text{)}. \end{aligned}$$

Thus, $NaNb = Nab$, and so the product of two cosets is a coset. We let the reader prove that the identity is the coset $N = N1$ and that the inverse of Na is $N(a^{-1})$. This group is denoted by G/N, and the definition of index gives $|G/N| = [G:N]$. ■

Corollary 2.22. *If $N \lhd G$, then the* ***natural map*** *(i.e., the function $\nu: G \to G/N$ defined by $\nu(a) = Na$) is a surjective homomorphism with kernel N.*

Proof. The equation $\nu(a)\nu(b) = \nu(ab)$ is just the formula $NaNb = Nab$; hence, ν is a homomorphism. If $Na \in G/N$, then $Na = \nu(a)$, and so ν is surjective. Finally, $\nu(a) = Na = N$ if and only if $a \in N$, by Lemma 2.8, so that $N = \ker \nu$. ∎

We have now shown that every normal subgroup is the kernel of some homomorphism. Different homomorphisms can have the same kernel. For example, if $a = (1\ 2)$ and $b = (1\ 3)$, then $\gamma_a, \gamma_b: S_3 \to S_3$ are distinct and $\ker \gamma_a = 1 = \ker \gamma_b$.

The quotient group construction is a generalization of the construction of $\mathbb{Z}_n$ from $\mathbb{Z}$. Recall that if n is a fixed integer, then $[a]$, the congruence class of $a \bmod n$, is the coset $a + \langle n \rangle$. Now $\langle n \rangle \lhd \mathbb{Z}$, because $\mathbb{Z}$ is abelian, and the quotient group $\mathbb{Z}/\langle n \rangle$ has elements all cosets $a + \langle n \rangle$, where $a \in \mathbb{Z}$, and operation $(a + \langle n \rangle) + (b + \langle n \rangle) = a + b + \langle n \rangle$; in congruence class notation, $[a] + [b] = [a + b]$. Therefore, the quotient group $\mathbb{Z}/\langle n \rangle$ is equal to $\mathbb{Z}_n$, the group of integers modulo n. An arbitrary quotient group G/N is often called $G \bmod N$ because of this example.

Definition. If $a, b \in G$, the ***commutator***[1] of a and b, denoted by $[a, b]$, is

$$[a, b] = aba^{-1}b^{-1}.$$

The ***commutator subgroup*** (or *derived subgroup*) of G, denoted by G', is the subgroup of G generated by all the commutators.

We shall see, in Exercise 2.43 below, that the subset of all commutators need not be a subgroup (the product of two commutators need not be a commutator).

Theorem 2.23. *The commutator subgroup G' is a normal subgroup of G. Moreover, if $H \lhd G$, then G/H is abelian if and only if $G' \leq H$.*

Proof. If $f: G \to G$ is a homomorphism, then $f(G') \leq G'$ because $f([a, b]) = [fa, fb]$. It follows from Exercise 2.34 that $G' \lhd G$.

Let $H \lhd G$. If G/H is abelian, then $HaHb = HbHa$ for all $a, b \in G$; that is, $Hab = Hba$, and so $ab(ba)^{-1} = aba^{-1}b^{-1} = [a, b] \in H$. By Corollary 2.6, $G' \leq H$. Conversely, suppose that $G' \leq H$. For every $a, b \in G$, $ab(ba)^{-1} = [a, b] \in G' \leq H$, and so $Hab = Hba$; that is, G/H is abelian. ∎

[1] Those who write conjugates as $b^{-1}ab$ write commutators as $a^{-1}b^{-1}ab$.

EXERCISES

2.40. Let $H \lhd G$, let $\nu: G \to G/H$ be the natural map, and let $X \subset G$ be a subset such that $\nu(X)$ generates G/H. Prove that $G = \langle H \cup X \rangle$.

2.41. Let G be a finite group of odd order, and let x be the product of all the elements of G in some order. Prove that $x \in G'$.

2.42 **(P. Yff).** For any group G, show that G' is the subset of all "long commutators":

$$G' = \{a_1 a_2 \dots a_n a_1^{-1} a_2^{-1} \dots a_n^{-1}: a_i \in G \text{ and } n \geq 2\}.$$

(*Hint* **(P.M. Weichsel).**

$$(aba^{-1}b^{-1})(cdc^{-1}d^{-1}) = a(ba^{-1})b^{-1}c(dc^{-1})d^{-1}a^{-1}(ab^{-1})bc^{-1}(cd^{-1})d.)$$

2.43. The fact that the set of all commutators in a group need not be a subgroup is an old result; the following example is due to P.J. Cassidy (1979).

(i) Let $k[x, y]$ denote the ring of all polynomials in two variables over a field k, and let $k[x]$ and $k[y]$ denote the subrings of all polynomials in x and in y, respectively. Define G to be the set of all matrices of the form

$$A = \begin{bmatrix} 1 & f(x) & h(x, y) \\ 0 & 1 & g(y) \\ 0 & 0 & 1 \end{bmatrix},$$

where $f(x) \in k[x]$, $g(y) \in k[y]$, and $h(x, y) \in k[x, y]$. Prove that G is a multiplicative group and that G' consists of all those matrices for which $f(x) = 0 = g(y)$. (*Hint.* If A is denoted by (f, g, h), then $(f, g, h)(f', g', h') = (f + f', g + g', h + h' + fg')$. If $h = h(x, y) = \sum a_{ij}x^i y^j$, then

$$(0, 0, h) = \prod_{i,j} [(a_{ij}x^i, 0, 0), (0, y^j, 0)].)$$

(ii) If $(0, 0, h)$ is a commutator, then there are polynomials $f(x), f'(x) \in k[x]$ and $g(y), g'(y) \in k[y]$ with $h(x, y) = f(x)g'(y) - f'(x)g(y)$.

(iii) Show that $h(x, y) = x^2 + xy + y^2$ does not possess a decomposition as in part (ii), and conclude that $(0, 0, h) \in G'$ is not a commutator. (*Hint.* If $f(x) = \sum b_i x^i$ and $f'(x) = \sum c_i x^i$, then there are equations

$$b_0 g'(y) - c_0 g(y) = y^2,$$
$$b_1 g'(y) - c_1 g(y) = y,$$
$$b_2 g'(y) - c_2 g(y) = 1.$$

Considering $k[x, y]$ as a vector space over k, one obtains the contradiction that the independent set $\{1, y, y^2\}$ is in the subspace spanned by $\{g, g'\}$.)

Remark. With a little ring theory, one can modify this construction to give a finite example. If $k = \mathbb{Z}_p$ and $k[x, y]$ is replaced by its quotient ring $k[x, y]/I$, where I is the ideal generated by $\{x^3, y^3, x^2y, xy^2\}$, then the corresponding group G has order p^{12}. Using the computer language CAYLEY (now called MAGMA), I found that the smallest group in which the product of two commutators is not a commutator has order 96. There are exactly two such groups: in CAYLEY notation, they are library g96n197 and library g96n201; in each of these groups, the commutator subgroup has order 32 while there are only 29 commutators.

There is an explicit example in [Carmichael, p. 39] of a group $G \leq S_{16}$ (generated by eight permutations) with $|G| = 256$, $|G'| = 16$, and with a specific element of G' which is not a commutator.

The Isomorphism Theorems

There are three theorems, formulated by E. Noether, describing the relationship between quotient groups, normal subgroups, and homomorphisms. A testimony to the elementary character of these theorems is that analogues of them are true for most types of algebraic systems, e.g., groups, semigroups, rings, vector spaces, modules, operator groups.

Theorem 2.24 (First Isomorphism Theorem). *Let $f: G \to H$ be a homomorphism with kernel K. Then K is a normal subgroup of G and $G/K \cong \operatorname{im} f$.*

Proof. We have already noted that $K \lhd G$. Define $\varphi: G/K \to H$ by

$$\varphi(Ka) = f(a).$$

To see that φ is well defined, assume that $Ka = Kb$; that is, $ab^{-1} \in K$. Then $1 = f(ab^{-1}) = f(a)f(b)^{-1}$, and $f(a) = f(b)$; it follows that $\varphi(Ka) = \varphi(Kb)$, as desired. Now φ is a homomorphism:

$$\varphi(KaKb) = \varphi(Kab) = f(ab) = f(a)f(b) = \varphi(Ka)\varphi(Kb).$$

It is plain that $\operatorname{im} \varphi = \operatorname{im} f$. Finally, we show that φ is an injection. If $\varphi(Ka) = \varphi(Kb)$, then $f(a) = f(b)$; hence $f(ab^{-1}) = 1$, $ab^{-1} \in K$, and $Ka = Kb$ (note that φ being an injection is the converse of φ being well defined). We have shown that φ is an isomorphism. ■

It follows that there is no significant difference between a quotient group and a homomorphic image.

If $v: G \to G/K$ is the natural map, then the following "commutative diagram" (i.e., $f = \varphi \circ v$) with surjection v and injection φ describes the theorem:

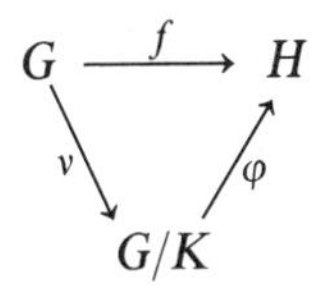

It is easy to describe $\varphi^{-1}: \operatorname{im} f \to G/K$: if $x \in \operatorname{im} f$, then there exists $a \in G$ with $f(a) = x$, and $\varphi^{-1}(x) = Ka$. The reader should check that φ^{-1} is well defined; that is, if $f(b) = x$, then $Ka = Kb$.

Given a homomorphism f, one must always salivate, like Pavlov's dog, by asking for its kernel and image; once these are known, there is a normal subgroup and f can be converted into an isomorphism. Let us illustrate this by solving Exercise 2.19: If $G = \langle g \rangle$ and $H = \langle h \rangle$ are cyclic groups of order n, then $G \cong H$. Define a homomorphism $f: \mathbb{Z} \to G$ by $f(k) = g^k$. It is easy to see that f is a surjection, while Exercise 2.13 shows that $\ker f = \langle n \rangle$. The first isomorphism theorem gives $\mathbb{Z}/\langle n \rangle \cong G$. Similarly, $\mathbb{Z}/\langle n \rangle \cong H$, and so $G \cong H$. Since $\mathbb{Z}/\langle n \rangle = \mathbb{Z}_n$, every cyclic group of order n is isomorphic to $\mathbb{Z}_n$.

Lemma 2.25. *If S and T are subgroups of G and if one of them is normal, then $ST = S \vee T = TS$.*

Proof. Recall that ST is just the set of all products of the form st, where $s \in S$ and $t \in T$; hence ST and TS are subsets of $S \vee T$ containing $S \cup T$. If ST and TS are subgroups, then the reverse inclusion will follow from Corollary 2.6. Assume that $T \lhd G$. If $s_1 t_1$ and $s_2 t_2 \in ST$, then

$$\begin{aligned}(s_1 t_1)(s_2 t_2)^{-1} &= s_1 t_1 t_2^{-1} s_2^{-1} \\ &= s_1 (s_2^{-1} s_2) t_1 t_2^{-1} s_2^{-1} \\ &= s_1 s_2^{-1} t_3 \\ &= (s_1 s_2^{-1}) t_3 \in ST,\end{aligned}$$

where $t_3 = s_2 (t_1 t_2^{-1}) s_2^{-1} \in T$ because $T \lhd G$. Therefore, $ST = S \vee T$. A similar proof shows that TS is a subgroup, and so $TS = S \vee T = ST$. ■

Suppose that $S \leq H \leq G$ are subgroups with $S \lhd G$. Then $S \lhd H$ and the quotient H/S is defined; it is the subgroup of G/S consisting of all those cosets Sh with $h \in H$. In particular, if $S \lhd G$ and T is any subgroup of G, then $S \leq ST \leq G$ and ST/S is the subgroup of G/S consisting of all those cosets Sst, where $st \in ST$. Since $Sst = St$, it follows that ST/S consists precisely of all those cosets of S having a representative in T.

Recall the product formula (Theorem 2.20): If $S, T \leq G$, then $|ST||S \cap T| = |S||T|$; equivalently, $|T|/|S \cap T| = |ST|/|S|$. This suggests the following theorem.

Theorem 2.26 (Second Isomorphism Theorem). *Let N and T be subgroups of G with N normal. Then $N \cap T$ is normal in T and $T/(N \cap T) \cong NT/N$.*

Remark. The following diagram is a mnemonic for this theorem:

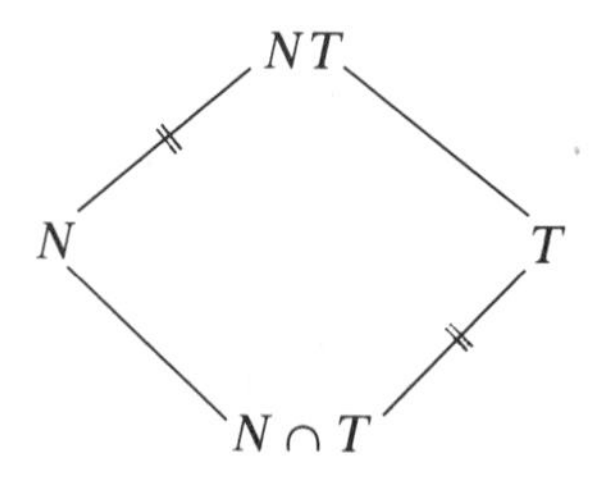

Proof. Let $\nu: G \to G/N$ be the natural map, and let $\nu' = \nu|T$, the restriction of ν to T. Since ν' is a homomorphism whose kernel is $N \cap T$, Theorem 2.24 gives $N \cap T \lhd T$ and $T/(N \cap T) \cong \operatorname{im} \nu'$. Our remarks above show that im ν' is just the family of all those cosets of N having a representative in T; that is, im ν' consists of all the cosets in NT/N. ■

Theorem 2.27 (Third Isomorphism Theorem). *Let $K \leq H \leq G$, where both K and H are normal subgroups of G. Then H/K is a normal subgroup of G/K and*

$$(G/K)/(H/K) \cong G/H.$$

Proof. Again we let the first isomorphism theorem do the dirty work. Define $f: G/K \to G/H$ by $f(Ka) = Ha$ (this "enlargement of coset" map f is well defined because $K \leq H$). The reader may check easily that f is a surjection with kernel H/K. ■

Imagine trying to prove the third isomorphism theorem directly; the elements of $(G/K)/(H/K)$ are cosets whose representatives are cosets!

EXERCISES

2.44. Prove that a homomorphism $f: G \to H$ is an injection if and only if $\ker f = 1$.

2.45. (i) Show that the 4-group $\mathbf{V}$ is a normal subgroup of S_4. (We shall do this more efficiently in the next chapter.)
(ii) If $K = \langle(1\ 2)(3\ 4)\rangle$, show that $K \lhd \mathbf{V}$ but that K is not a normal subgroup of S_4. Conclude that normality need not be transitive; that is, $K \lhd H$ and $H \lhd G$ need not imply $K \lhd G$.

2.46. Let $N \lhd G$ and let $f: G \to H$ be a homomorphism whose kernel contains N. Show that f induces a homomorphism $f_*: G/N \to H$ by $f_*(Na) = f(a)$.

2.47. If $S, T \leq G$, then ST is a subgroup of G if and only if $ST = TS$.

2.48 **(Modular Law).** Let A, B, and C be subgroups of G with $A \leq B$. If $A \cap C = B \cap C$ and $AC = BC$ (we do not assume that either AC or BC is a subgroup), then $A = B$.

2.49 **(Dedekind Law).** Let H, K, and L be subgroups of G with $H \leq L$. Then $HK \cap L = H(K \cap L)$ (we do not assume that either HK or $H(K \cap L)$ is a subgroup).

2.50. Let $f: G \to G^*$ be a homomorphism and let S^* be a subgroup of G^*. Then $f^{-1}(S^*) = \{x \in G: f(x) \in S^*\}$ is a subgroup of G containing $\ker f$.

Correspondence Theorem

The theorem in this section should be called the fourth isomorphism theorem. Let X and X^* be sets. A function $f: X \to X^*$ induces a "forward motion" and a "backward motion" between subsets of X and subsets of X^*.

The forward motion assigns to each subset $S \subset X$ the subset $f(S) = \{f(s): s \in S\}$ of X^*; the backward motion assigns to each subset S^* of X^* the subset $f^{-1}(S^*) = \{x \in X: f(x) \in S^*\}$ of X. Moreover, if f is a surjection, these motions define a bijection between all the subsets of X^* and certain subsets of X. The following theorem is the group-theoretic version of this.

Theorem 2.28 (Correspondence Theorem). *Let $K \lhd G$ and let $\nu: G \to G/K$ be the natural map. Then $S \mapsto \nu(S) = S/K$ is a bijection from the family of all those subgroups S of G which contain K to the family of all the subgroups of G/K.*

Moreover, if we denote S/K by S^*, then:

(i) *$T \leq S$ if and only if $T^* \leq S^*$, and then $[S : T] = [S^* : T^*]$; and*
(ii) *$T \lhd S$ if and only if $T^* \lhd S^*$, and then $S/T \cong S^*/T^*$.*

Remark. A mnemonic diagram for this theorem is:

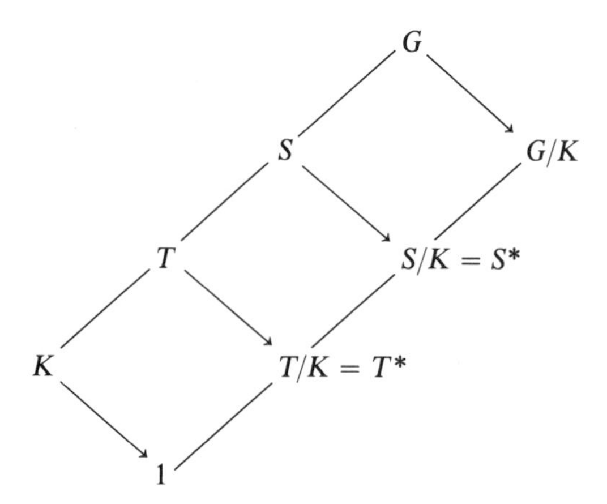

Proof. We show first that $S \mapsto S/K$ is an injection: if S and T are subgroups containing K, and if $S/K = T/K$, then $S = T$. To see this, let $s \in S$; since $S/K = T/K$, there exists $t \in T$ with $Ks = Kt$. Hence, $s = kt$ for some $k \in K \leq T$ and $s \in T$. The reverse inclusion is proved similarly. To see that the correspondence $S \mapsto S/K$ is a surjection, we must show that if $A \leq G/K$, then there is a subgroup S of G containing K with $A = S/K$. By Exercise 2.50, $S = \nu^{-1}(A)$ is a subgroup of G containing K; moreover, that ν is a surjection[2] implies that $S/K = \nu(S) = \nu\nu^{-1}(A) = A$.

It is plain that if $K \leq T \leq S$, then $T/K \leq S/K$. To prove that $[S : T] = [S^* : T^*]$, it suffices to show that there is a bijection from the family of all cosets of the form Ts, where $s \in S$, to the family of all cosets T^*s^*, where $s^* \in S^*$. The reader may check that α, defined by $\alpha: Ts \mapsto T^*\nu(s)$, is such a

[2] If $f: X \to X^*$ is a function and $A \subset X^*$, then $ff^{-1}(A) \subset A$; if f is a surjection, then $ff^{-1}(A) = A$.

bijection. (If G is finite, then we may prove $[S:T] = [S^*:T^*]$ as follows:

$$[S^*:T^*] = |S^*|/|T^*| = |S/K|/|T/K|$$

$$= (|S|/|K|)/(|T|/|K|) = |S|/|T| = [S:T].)$$

If $T \lhd S$, then the third isomorphism theorem gives $T/K \lhd S/K$ and $(S/K)/(T/K) \cong S/K$; that is, $T^* \lhd S^*$ and $S^*/T^* \cong S/T$. It remains to show that if $T^* \lhd S^*$, then $T \lhd S$. The reader may verify that $T = \ker \mu\nu_0\colon S \to S^*/T^*$, where $\nu_0 = \nu|S$ and $\mu\colon S^* \to S^*/T^*$ is the natural map. ■

EXERCISES

2.51. If $G' \leq H \leq G$, where G' is the commutator subgroup of G, then $H \lhd G$ and G/H is abelian.

2.52. Give an example to show that if $H \lhd G$, then G need not contain a subgroup isomorphic to G/H.

2.53. Prove that the circle group T is isomorphic to $\mathbb{R}/\mathbb{Z}$.

2.54. (i) Let $H, K \leq G$. If $(|H|, |K|) = 1$, then $H \cap K = 1$.
(ii) Let G be a finite group, and let H be a normal subgroup with $(|H|, [G:H]) = 1$. Prove that H is the unique such subgroup in G. (*Hint*: If K is another such subgroup, what happens to K in G/H?)

2.55 **(Zassenhaus).** Let G be a finite group such that, for some fixed integer $n > 1$, $(xy)^n = x^n y^n$ for all $x\ y \in G$. If $G[n] = \{z \in G: z^n = 1\}$ and $G^n = \{x^n: x \in G\}$, then both $G[n]$ and G^n are normal subgroups of G and $|G^n| = [G:G[n]]$.

2.56. A subgroup $H \leq G$ is a ***maximal normal subgroup*** of G if there is no normal subgroup N of G with $H < N < G$. Prove that H is a maximal normal subgroup of G if and only if G/H has no normal subgroups (other than itself and 1).

Definition. A group $G \neq 1$ is ***simple*** if it has no normal subgroups other than G and 1.

We may restate Exercise 2.56: H is a maximal normal subgroup of G if and only if G/H is simple.

2.57. An abelian group is simple if and only if it is finite and of prime order.

2.58. Let M be a maximal subgroup of G; that is, there is no subgroup S with $M < S < G$. Prove that if $M \lhd G$, then $[G:M]$ is finite and equal to a prime

2.59 **(Schur).** Let $f\colon G \to H$ be a homomorphism that does not send every element of G into 1. If G is simple, then f must be an injection.

Direct Products

Definition. If H and K are groups, then their ***direct product***, denoted by $H \times K$, is the group with elements all ordered pairs (h, k), where $h \in H$ and $k \in K$, and with operation

$$(h, k)(h', k') = (hh', kk').$$

It is easy to check that $H \times K$ is a group: the identity is $(1, 1)$; the inverse $(h, k)^{-1}$ is (h^{-1}, k^{-1}). Notice that neither H nor K is a subgroup of $H \times K$, but $H \times K$ does contain isomorphic replicas of each, namely, $H \times 1 = \{(h, 1): h \in H\}$ and $1 \times K = \{(1, k): k \in K\}$.

EXERCISES

2.60. (i) Show that $(h, 1) \in H \times 1$ and $(1, k) \in 1 \times K$ commute.
(ii) $H \times 1$ and $1 \times K$ are normal subgroups of $H \times K$.
(iii) $(H \times 1) \cap (1 \times K) = 1$ and $(H \times 1)(1 \times K) = H \times K$.

2.61. $H \times K$ is abelian if and only if both H and K are abelian.

2.62. (i) Prove that $\mathbb{Z}_6 \cong \mathbb{Z}_2 \times \mathbb{Z}_3$.
(ii) If $(m, n) = 1$, then $\mathbb{Z}_{mn} \cong \mathbb{Z}_m \times \mathbb{Z}_n$. (*Hint.* Use the Chinese Remainder Theorem).

2.63. If p is a prime, prove that $\mathbb{Z}_{p^2} \not\cong \mathbb{Z}_p \times \mathbb{Z}_p$.

2.64. Let $\mu: G \times G \to G$ be the operation on a group G; that is, $\mu(a, b) = ab$. If $G \times G$ is regarded as the direct product, prove that μ is a homomorphism if and only if G is abelian.

2.65. Let A be an abelian group, and let $\alpha: H \to A$ and $\beta: K \to A$ be homomorphisms. Prove that there exists a unique homomorphism $\gamma: H \times K \to A$ with $\gamma(h, 1) = \alpha(h)$ for all $h \in H$ and $\gamma(1, k) = \beta(k)$ for all $k \in K$. Show that this may be false if A is not abelian.

We now take another point of view. It is easy to multiply two polynomials together; it is harder to factor a given polynomial. We have just seen how to multiply two groups together; can one factor a given group?

Theorem 2.29. *Let G be a group with normal subgroups H and K. If $HK = G$ and $H \cap K = 1$, then $G \cong H \times K$.*

Proof. If $a \in G$, then $a = hk$ for some $h \in H$ and $k \in K$ (because $G = HK$). We claim that h and k are uniquely determined by a. If $a = h_1 k_1$ for $h_1 \in H$ and $k_1 \in K$, then $hk = h_1 k_1$ and $h^{-1}h_1 = kk_1^{-1} \in H \cap K = 1$; hence $h = h_1$ and $k = k_1$.

Define $f: G \to H \times K$ by $f(a) = (h, k)$, where $a = hk$. Is f a homomorphism? If $a = hk$ and $a' = h'k'$, then $aa' = hkh'k'$ which is not in the proper form for evaluating f. Were it true that $kh' = h'k$, however, then we could evaluate $f(aa')$. Consider the commutator $h'kh'^{-1}k^{-1}$. Now $(h'kh'^{-1})k^{-1} \in K$ (for $h'kh'^{-1} \in K$ because K is normal), and, similarly, $h'(kh'^{-1}k^{-1}) \in H$ (because H is normal); therefore, $h'kh'^{-1}k^{-1} \in H \cap K = 1$ and h' and k commute. The reader can now check that f is a homomorphism and a bijection; that is, f is an isomorphism. ■

We pause to give an example showing that all the hypotheses in Theorem 2.29 are necessary. Let $G = S_3$, $H = \langle(1\ 2\ 3)\rangle$, and $K = \langle(1\ 2)\rangle$. It is easy to see that $HK = G$ and $H \cap K = 1$; moreover, $H \lhd G$ but K is not a normal subgroup. The direct product $H \times K \cong \mathbb{Z}_3 \times \mathbb{Z}_2$ is abelian, and so the nonabelian group $G = S_3$ is not isomorphic to $H \times K$.

Theorem 2.30. *If $A \lhd H$ and $B \lhd K$, then $A \times B \lhd H \times K$ and*

$$(H \times K)/(A \times B) \cong (H/A) \times (K/B).$$

Proof. The homomorphism $\varphi\colon H \times K \to (H/A) \times (K/B)$, defined by $\varphi(h, k) = (Ah, Bk)$, is surjective and $\ker \varphi = A \times B$. The first isomorphism theorem now gives the result. ■

It follows, in particular, that if $N \lhd H$, then $N \times 1 \lhd H \times K$.

Corollary 2.31. *If $G = H \times K$, then $G/(H \times 1) \cong K$.*

There are two versions of the direct product $H \times K$: the ***external*** version, whose elements are ordered pairs and which contains isomorphic copies of H and K (namely, $H \times 1$ and $1 \times K$); the ***internal*** version which does contain H and K as normal subgroups and in which $HK = G$ and $H \cap K = 1$. By Theorem 2.29, the two versions are isomorphic. In the future, we shall not distinguish between external and internal; in almost all cases, however, our point of view is internal. For example, we shall write Corollary 2.31 as $(H \times K)/H \cong K$.

Exercises

2.66. Prove that $\mathbf{V} \cong \mathbb{Z}_2 \times \mathbb{Z}_2$.

2.67. Show that it is possible for a group G to contain three distinct normal subgroups H, K, and L such that $G = H \times L = K \times L$; that is, $HL = G = KL$ and $H \cap L = 1 = K \cap L$. (*Hint*: Try $G = \mathbf{V}$).

2.68. Prove that an abelian group G of order p^2, where p is a prime, is either cyclic or isomorphic to $\mathbb{Z}_p \times \mathbb{Z}_p$. (We shall see in Corollary 4.5 that every group of order p^2 must be abelian).

2.69. Let G be a group with normal subgroups H and K. Prove that $HK = G$ and $H \cap K = 1$ if and only if each $a \in G$ has a unique expression of the form $a = hk$, where $h \in H$ and $k \in K$.

2.70. If $N \lhd H \times K$, then either N is abelian or N intersects one of the factors H or K nontrivially.

2.71. Give an example of an abelian group $H \times K$ which contains a nontrivial subgroup N such that $N \cap H = 1 = N \cap K$. Conclude that it is possible that $N \leq H \times K$ and $N \neq (N \cap H) \times (N \cap K)$.

2.72. Let G be a group having a simple subgroup H of index 2. Prove that either H is the unique proper normal subgroup of G or that G contains a normal subgroup K of order 2 with $G = H \times K$. (*Hint*. Use the second isomorphism theorem.)

2.73. Let 0 denote the trivial homomorphism which sends every element to the identity. Prove that $G \cong H \times K$ if and only if there exist homomorphisms

$$H \xrightarrow{i} G \xrightarrow{p} K \quad \text{and} \quad K \xrightarrow{j} G \xrightarrow{q} H$$

with $qi = 1_H$ (the identity function on H), $pj = 1_K$, $pi = 0$, $qj = 0$, and $(i \circ q)(x)(j \circ p)(x) = x$ for all $x \in G$.

2.74. The operation of direct product is commutative and associative in the following sense: for groups H, K, and L,

$$H \times K \cong K \times H \quad \text{and} \quad H \times (K \times L) \cong (H \times K) \times L.$$

Conclude that the notations $H_1 \times \cdots \times H_n$ and $\prod_{i=1}^n H_i$ are unambiguous.

2.75. Let G be a group having normal subgroups $H_1, \ldots, H_n$.
(i) If $G = \langle \bigcup_{i=1}^n H_i \rangle$ and, for all j, $1 = H_j \cap \langle \bigcup_{i \neq j} H_i \rangle$, then $G \cong H_1 \times \cdots \times H_n$.
(ii) If each $a \in G$ has a unique expression of the form $a = h_1 \ldots h_n$, where each $h_i \in H_i$, then $G \cong H_1 \times \cdots \times H_n$.

2.76. Let $H_1, \ldots, H_n$ be normal subgroups of a group G, and define $\varphi: G \to G/H_1 \times \cdots \times G/H_n$ by $\varphi(x) = (H_1 x, \ldots, H_n x)$.
(i) Prove that $\ker \varphi = H_1 \cap \cdots \cap H_n$.
(ii) If each H_i has finite index in G and if $(|G/H_i|, |G/H_j|) = 1$ for all $i \neq j$, then φ is a surjection and

$$[G: H_1 \cap \cdots \cap H_n] = \prod_{i=1}^n |G/H_i|.$$

2.77. Let V be an n-dimensional vector space over a field F. Prove that, as abelian groups, $V \cong F_1 \times \cdots \times F_n$, where $F_i \cong F$ for all i.

Definition. If p is a prime, then an ***elementary abelian p-group*** is a finite group G isomorphic to $\mathbb{Z}_p \times \cdots \times \mathbb{Z}_p$.

2.78. Prove that every abelian group G of prime exponent p is elementary abelian, that G is a vector space over $\mathbb{Z}_p$, and that every homomorphism $\varphi: G \to G$ is a linear transformation.

CHAPTER 3

Symmetric Groups and *G*-Sets

The definition of group arose from fundamental properties of the symmetric group S_n. But there is another important feature of S_n: its elements are functions acting on some underlying set, and this aspect is not explicit in our presentation so far. The notion of *G-set* is the appropriate abstraction of this idea.

Conjugates

In this section we study conjugates and conjugacy classes for arbitrary groups; in the next section, we consider the special case of symmetric groups.

Lemma 3.1. *If G is a group, then the relation "y is a* ***conjugate*** *of x in G," that is, $y = gxg^{-1}$ for some $g \in G$, is an equivalence relation.*

Proof. Routine. ■

Definition. If G is a group, then the equivalence class of $a \in G$ under the relation "y is a conjugate of x in G" is called the ***conjugacy class*** of a; it is denoted by a^G.

Of course, the conjugacy class a^G is the set of all the conjugates of a in G. Exercise 2.45(i) can be rephrased: a subgroup is normal if and only if it is a (disjoint) union of conjugacy classes. If a and b are conjugate in G, say, $b = gag^{-1}$, then there is an isomorphism $\gamma: G \to G$, namely, conjugation by g, with $\gamma(a) = b$. It follows that all the elements in the same conjugacy class have the same order. In particular, for any two elements $x, y \in G$, the elements xy and yx have the same order.

If $a \in G$ is the sole resident of its conjugacy class, then $a = gag^{-1}$ for all $g \in G$; that is, a commutes with every element of G.

Definition. The ***center*** of a group G, denoted by $Z(G)$, is the set of all $a \in G$ that commute with every element of G.

It is easy to check that $Z(G)$ is a normal abelian subgroup of G.

The following subgroup is introduced to count the number of elements in a conjugacy class.

Definition. If $a \in G$, then the ***centralizer*** of a in G, denoted by $C_G(a)$, is the set of all $x \in G$ which commute with a.

It is immediate that $C_G(a)$ is a subgroup of G.

Theorem 3.2. *If $a \in G$, the number of conjugates of a is equal to the index of its centralizer*:

$$|a^G| = [G : C_G(a)],$$

and this number is a divisor of $|G|$ when G is finite.

Proof. Denote the family of all left cosets of $C = C_G(a)$ in G by G/C, and define $f: a^G \to G/C$ by $f(gag^{-1}) = gC$. Now f is well defined: if $gag^{-1} = hah^{-1}$ for some $h \in G$, then $h^{-1}gag^{-1}h = a$ and $h^{-1}g$ commutes with a; that is, $h^{-1}g \in C$, and so $hC = gC$. The function f is an injection: if $gC = f(gag^{-1}) = f(kak^{-1}) = kC$ for some $k \in G$, then $k^{-1}g \in C$, $k^{-1}g$ commutes with a, $k^{-1}gag^{-1}k = a$, and $gag^{-1} = kak^{-1}$; the function f is a surjection: if $g \in G$, then $gC = f(gag^{-1})$. Therefore, f is a bijection and $|a^G| = |G/C| = [G : C_G(a)]$. When G is finite, Lagrange's theorem applies. ■

One may conjugate subgroups as well as elements.

Definition. If $H \leq G$ and $g \in G$, then the ***conjugate*** gHg^{-1} is $\{ghg^{-1}: h \in H\}$. The conjugate gHg^{-1} is often denoted by H^g.

The conjugate gHg^{-1} is a subgroup of G isomorphic to H: if $\gamma_g: G \to G$ is conjugation by g, then $\gamma_g|H$ is an isomorphism from H to gHg^{-1}.

Note that a subgroup H is a normal subgroup if and only if it has only one conjugate.

Definition. If $H \leq G$, then the ***normalizer*** of H in G, denoted by $N_G(H)$, is

$$N_G(H) = \{a \in G: aHa^{-1} = H\}.$$

It is immediate that $N_G(H)$ is a subgroup of G. Notice that $H \lhd N_G(H)$; indeed, $N_G(H)$ is the largest subgroup of G in which H is normal.

Theorem 3.3. *If $H \leq G$, then the number c of conjugates of H in G is equal to the index of its normalizer: $c = [G : N_G(H)]$, and c divides $|G|$ when G is finite. Moreover, $aHa^{-1} = bHb^{-1}$ if and only if $b^{-1}a \in N_G(H)$.*

Proof. Let $[H]$ denote the family of all the conjugates of H, and let G/N denote the family of all left cosets of $N = N_G(H)$ in G. Define $f: [H] \to G/N$ by $f(aHa^{-1}) = aN$. Now f is well defined: if $aHa^{-1} = bHb^{-1}$ for some $b \in G$, then $b^{-1}aHa^{-1}b = H$ and $b^{-1}a$ normalizes H; that is, $b^{-1}a \in N$, and so $bN = aN$. The function f is an injection: if $aN = f(aHa^{-1}) = f(cHc^{-1}) = cN$ for some $c \in G$, then $c^{-1}a \in N$, $c^{-1}a$ normalizes H, $c^{-1}aHa^{-1}c = H$, and $aHa^{-1} = cHc^{-1}$; the function f is a surjection: if $a \in G$, then $aN = f(aHa^{-1})$. Therefore, f is a bijection and $|[H]| = |G/N| = [G : N_G(H)]$. When G is finite, Lagrange's theorem applies. ■

The strong similarity of Theorems 3.2 and 3.3 will be explained when we introduce G-sets.

Exercises

3.1. (i) A group G is ***centerless*** if $Z(G) = 1$. Prove that S_n is centerless if $n \geq 3$.
 (ii) Prove that A_4 is centerless.

3.2. If $\alpha \in S_n$ is an n-cycle, then its centralizer is $\langle \alpha \rangle$.

3.3. Prove that if G is not abelian, then $G/Z(G)$ is not cyclic.

3.4. (i) A finite group G with exactly two conjugacy classes has order 2.
 (ii) Let G be a group containing an element of finite order $n > 1$ and exactly two conjugacy classes. Prove that $|G| = 2$. (*Hint.* There is a prime p with $a^p = 1$ for all $a \in G$. If p is odd and $a \in G$, then $a^2 = xax^{-1}$ for some x, and so $a^{2^k} = x^k a x^{-k}$ for all $k \geq 1$.)
 (There are examples of infinite groups G with no elements of finite order which do have exactly two conjugacy classes.)

3.5. Prove that $Z(G_1 \times \cdots \times G_n) = Z(G_1) \times \cdots \times Z(G_n)$.

3.6. (i) Prove, for every $a, x \in G$, that $C_G(axa^{-1}) = aC_G(x)a^{-1}$.
 (ii) Prove that if $H \leq G$ and $h \in H$, then $C_H(h) = C_G(h) \cap H$.

3.7. Let G be a finite group, let H be a normal subgroup of prime index, and let $x \in H$ satisfy $C_H(x) < C_G(x)$. If $y \in H$ is conjugate to x in G, then y is conjugate to x in H.

3.8. If $a_1, \ldots, a_n$ is a list of (not necessarily distinct) elements of a group G, then, for all i, $a_i \ldots a_n a_1 \ldots a_{i-1}$ is conjugate to $a_1 \ldots a_n$.

3.9. (i) Prove that $N_G(aHa^{-1}) = aN_G(H)a^{-1}$.
 (ii) If $H \leq K \leq G$, then $N_K(H) = N_G(H) \cap K$.

(iii) If $H, K \leq G$, prove that $N_G(H) \cap N_G(K) \leq N_G(H \cap K)$. Give an example in which the inclusion is proper.

3.10. If $f: G \to H$ is surjective and $A \leq Z(G)$, then $f(A) \leq Z(H)$.

3.11. If $H \leq G$, then $N_G(H) \leq \{a \in G: aHa^{-1} \leq H\}$; when H is finite, then there is equality. (There are examples of infinite subgroups $H \leq G$ with $aHa^{-1} < H$ for some $a \in G$).

Definition. An $n \times n$ matrix $M = [m_{ij}]$ over a field K is ***monomial*** if there is $\alpha \in S_n$ and (not necessarily distinct) nonzero elements $x_1, \ldots, x_n \in K$ such that

$$m_{ij} = \begin{cases} x_i & \text{if } j = \alpha(i), \\ 0 & \text{otherwise.} \end{cases}$$

Monomial matrices thus have only one nonzero entry in any row or column. Of course, a monomial matrix in which each $x_i = 1$ is a permutation matrix over K. (This definition will be generalized when we discuss wreath products.)

3.12. (i) Let k be a field. If $G = \mathrm{GL}(n, k)$ and T is the subgroup of G of all diagonal matrices, then $N_G(T)$ consists of all the monomial matrices over k.
(ii) Prove that $N_G(T)/T \cong S_n$.

3.13. (i) If H is a proper subgroup of G, then G is not the union of all the conjugates of H.
(ii) If G is a finite group with conjugacy classes $C_1, \ldots, C_m$, and if $g_i \in C_i$, then $G = \langle g_1, \ldots, g_m \rangle$.

Symmetric Groups

Definition. Two permutations $\alpha, \beta \in S_n$ have the ***same cycle structure*** if their complete factorizations into disjoint cycles have the same number of r-cycles for each r.

Lemma 3.4. *If $\alpha, \beta \in S_n$, then $\alpha\beta\alpha^{-1}$ is the permutation with the same cycle structure as β which is obtained by applying α to the symbols in β.*

EXAMPLE 3.1. If $\beta = (1\ 3)(2\ 4\ 7)$ and $\alpha = (2\ 5\ 6)(1\ 4\ 3)$, then $\alpha\beta\alpha^{-1} = (\alpha 1\ \alpha 3)(\alpha 2\ \alpha 4\ \alpha 7) = (4\ 1)(5\ 3\ 7)$.

Proof. Let π be the permutation defined in the lemma. If β fixes a symbol i, then π fixes $\alpha(i)$, for $\alpha(i)$ resides in a 1-cycle; but $\alpha\beta\alpha^{-1}(\alpha(i)) = \alpha\beta(i) = \alpha(i)$, and so $\alpha\beta\alpha^{-1}$ fixes $\alpha(i)$ as well. Assume that β moves i; say, $\beta(i) = j$. Let the complete factorization of β be

$$\beta = \gamma_1\gamma_2\cdots(\cdots\ i\ j\ \cdots)\cdots\gamma_t.$$

If $\alpha(i) = k$ and $\alpha(j) = l$, then $\pi: k \mapsto l$. But $\alpha\beta\alpha^{-1}: k \mapsto i \mapsto j \mapsto l$, and so $\alpha\beta\alpha^{-1}(k) = \pi(k)$. Therefore, π and $\alpha\beta\alpha^{-1}$ agree on all symbols of the form $k = \alpha(i)$; since α is a surjection, it follows that $\pi = \alpha\beta\alpha^{-1}$. ■

Theorem 3.5. *Permutations $\alpha, \beta \in S_n$ are conjugate if and only if they have the same cycle structure.*

Proof. The lemma shows that conjugate permutations do have the same cycle structure. For the converse, define $\gamma \in S_n$ as follows: place the complete factorization of α over that of β so that cycles of the same length correspond, and let γ be the function sending the top to the bottom. For example, if

$$\alpha = \gamma_1\gamma_2\cdots(\cdots\ i\ j\ \cdots)\cdots\gamma_t,$$

$$\beta = \delta_1\delta_2\cdots(\cdots\ k\ l\ \cdots)\cdots\delta_t,$$

then $\gamma(i) = k$, $\gamma(j) = l$, etc. Notice that γ is a permutation, for every i between 1 and n occurs exactly once in a complete factorization. The lemma gives $\gamma\alpha\gamma^{-1} = \beta$, and so α and β are conjugate. ■

EXAMPLE 3.2. If

$$\alpha = (2\ \ 3\ \ 1)(4\ \ 5)(6),$$

$$\beta = (5\ \ 6\ \ 2)(3\ \ 1)(4),$$

then $\gamma = \begin{pmatrix} 1 & 2 & 3 & 4 & 5 & 6 \\ 2 & 5 & 6 & 3 & 1 & 4 \end{pmatrix} = (1\ \ 2\ \ 5)(3\ \ 6\ \ 4)$. Notice that γ is not unique; for example, the 3-cycle in α could also be written $(1\ \ 2\ \ 3)$, and the "downward" permutation is now $\gamma' = (1\ \ 5)(2\ \ 6\ \ 4\ \ 3)$. The multiplicity of choices for γ is explained by Theorem 3.2.

Corollary 3.6. *A subgroup H of S_n is a normal subgroup if and only if, whenever $\alpha \in H$, then every β having the same cycle structure as α also lies in H.*

Proof. By Exercise 2.34, $H \lhd S_n$ if and only if H contains every conjugate of its elements. ■

The solution of Exercise 2.45(i), which states that $\mathbf{V} \lhd S_4$, follows from the fact that $\mathbf{V}$ contains all products of disjoint transpositions.

If $1 \leq r \leq n$, then Exercise 1.5 shows that there are exactly $(1/r)[n(n-1)\cdots(n-r+1)]$ distinct r-cycles in S_n. This formula can be used to compute the number of permutations having any given cycle structure if one is careful about factorizations with several factors of the same length. For example, the number of permutations in S_4 of the form $(a\ \ b)(c\ \ d)$ is

$$\tfrac{1}{2}[(4 \times 3)/2 \times (2 \times 1)/2] = 3,$$

the factor $\frac{1}{2}$ occurring so that we do not count $(a\ \ b)(c\ \ d) = (c\ \ d)(a\ \ b)$ twice.

S_4

Cycle Structure	Number	Order	Parity
(1)	1	1	Even
(12)	$6 = (4 \times 3)/2$	2	Odd
(123)	$8 = (4 \times 3 \times 2)/3$	3	Even
(1234)	$6 = 4!/4$	4	Odd
(12)(34)	$3 = \frac{1}{2}\left(\frac{4 \times 3}{2} \times \frac{2 \times 1}{2}\right)$	2	Even
	$24 = 4!$		

Table 3.1

Let us now examine S_4 using Table 3.1. The 12 elements of A_4 are eight 3-cycles, three products of disjoint transpositions, and the identity. These elements are the 4-group $\mathbf{V}$ together with

$$(1\ 2\ 3); \quad (1\ 3\ 2); \quad (2\ 3\ 4); \quad (2\ 4\ 3);$$

$$(3\ 4\ 1); \quad (3\ 1\ 4); \quad (4\ 1\ 2); \quad (4\ 2\ 1).$$

We can now see that the converse of Lagrange's theorem is false.

Theorem 3.7. *A_4 is a group of order 12 having no subgroup of order 6.*

Proof **(T.-L. Sheu).** If such a subgroup H exists, then it has index 2, and so Exercise 2.16 gives $\alpha^2 \in H$ for every $\alpha \in A_4$. If α is a 3-cycle, however, then $\alpha = \alpha^4 = (\alpha^2)^2$, and this gives 8 elements in H, a contradiction. ■

EXERCISES

3.14. (i) If the conjugacy class of $x \in G$ is $\{a_1, \ldots, a_k\}$, then the conjugacy class of x^{-1} is $\{a_1^{-1}, \ldots, a_k^{-1}\}$.
(ii) If $\alpha \in S_n$, then α is conjugate to α^{-1}.

3.15. A_4 is the only subgroup of S_4 having order 12.

Definition. If n is a positive integer, then a ***partition of n*** is a sequence of integers $1 \leq i_1 \leq i_2 \leq \cdots \leq i_r$ with $\sum i_j = n$.

3.16. Show that the number of conjugacy classes in S_n is the number of partitions of n.

3.17. If $n < m$, then A_n can be imbedded in A_m (as all even permutations fixing $\{n + 1, \ldots, m\}$).

3.18. Verify the entries in Table 3.2.

3.19. Verify the entries in Table 3.3.

S_5

Cycle Structure	Number	Order	Parity
(1)	1	1	Even
(12)	$10 = (5 \times 4)/2$	2	Odd
(123)	$20 = (5 \times 4 \times 3)/3$	3	Even
(1234)	$30 = (5 \times 4 \times 3 \times 2)/4$	4	Odd
(12345)	$24 = 5!/5$	5	Even
(12)(34)	$15 = \frac{1}{2}\left(\frac{5 \times 4}{2} \times \frac{3 \times 2}{2}\right)$	2	Even
(123)(45)	$20 = \frac{5 \times 4 \times 3}{3} \times \frac{2 \times 1}{2}$	6	Odd
	$120 = 5!$		

A_5

Cycle Structure	Number	Order	Parity
(1)	1	1	Even
(123)	20	3	Even
(12345)	24	5	Even
(12)(34)	15	2	Even
	60		

Table 3.2

S_6

	Cycle Structure	Number	Order	Parity
C_1	(1)	1	1	Even
C_2	(12)	15	2	Odd
C_3	(123)	40	3	Even
C_4	(1234)	90	4	Odd
C_5	(12345)	144	5	Even
C_6	(123456)	120	6	Odd
C_7	(12)(34)	45	2	Even
C_8	(12)(345)	120	6	Odd
C_9	(12)(3456)	90	4	Even
C_{10}	(12)(34)(56)	15	2	Odd
C_{11}	(123)(456)	40	3	Even
		$720 = 6!$		

A_6

Cycle Structure	Number	Order	Parity
(1)	1	1	Even
(123)	40	3	Even
(12345)	144	5	Even
(12)(34)	45	2	Even
(12)(3456)	90	4	Even
(123)(456)	40	3	Even
	360		

Table 3.3

The Simplicity of A_n

We are going to prove that A_n is simple for all $n \geq 5$. The alternating group A_4 is not simple, for it contains a normal subgroup, namely, $\mathbf{V}$.

Lemma 3.8. *A_5 is simple.*

Proof. (i) *All 3-cycles are conjugate in A_5.* (We know that this is true in S_5, but now we are allowed to conjugate only by even permutations.)

If, for example, $\alpha = (1\ 2\ 3)$, then the odd permutation $(4\ 5)$ commutes with α. Since A_5 has index 2 in S_5, it is a normal subgroup of prime index, and so Exercise 3.7 says that α has the same number of conjugates in A_5 as it does in S_5 because $C_{A_5}(\alpha) < C_{S_5}(\alpha)$.

(ii) *All products of disjoint transpositions are conjugate in A_5.*

If, for example, $\alpha = (1\ 2)(3\ 4)$, then the odd permutation $(1\ 2)$ commutes with α. Since A_5 has index 2 in S_5, Exercise 3.7 says that α has the same number of conjugates in A_5 as it does in S_5.

(iii) *There are two conjugacy classes of 5-cycles in A_5, each of which has 12 elements.*

In S_5, $\alpha = (1\ 2\ 3\ 4\ 5)$ has 24 conjugates, so that $C_{S_5}(\alpha)$ has 5 elements; these must be the powers of α. By Exercise 3.2, $C_{A_5}(\alpha)$ has order 5, hence, index $60/5 = 12$.

We have now surveyed all the conjugacy classes occurring in A_5. Since every normal subgroup H is a union of conjugacy classes, $|H|$ is a sum of 1 and certain of the numbers: 12, 12, 15, and 20. It is easily checked that no such sum is a proper divisor of 60, so that $|H| = 60$ and A_5 is simple. ■

Lemma 3.9. *Let $H \lhd A_n$, where $n \geq 5$. If H contains a 3-cycle, then $H = A_n$.*

Proof. We show that $(1\ 2\ 3)$ and $(i\ j\ k)$ are conjugate in A_n (and thus that all 3-cycles are conjugate in A_n). If these cycles are not disjoint, then each fixes all the symbols outside of $\{1, 2, 3, i, j\}$, say, and the two 3-cycles lie in A^*, the group of all even permutations on these 5 symbols. Of course, $A^* \cong A_5$, and, as in part (i) of the previous proof, $(1\ 2\ 3)$ and $(i\ j\ k)$ are conjugate in A^*; *a fortiori*, they are conjugate in A_n. If the cycles are disjoint, then we have just seen that $(1\ 2\ 3)$ is conjugate to $(3\ j\ k)$ and that $(3\ j\ k)$ is conjugate to $(i\ j\ k)$, so that $(1\ 2\ 3)$ is conjugate to $(i\ j\ k)$ in this case as well.

A normal subgroup H containing a 3-cycle α must contain every conjugate of α; as all 3-cycles are conjugate, H contains every 3-cycle. But Exercise 2.7 shows that A_n is generated by the 3-cycles, and so $H = A_n$. ■

Lemma 3.10. *A_6 is simple.*

Proof. Let $H \neq 1$ be a normal subgroup of A_6, and let $\alpha \in H$ be distinct from 1. If α fixes some i, define

$$F = \{\beta \in A_6: \beta(i) = i\}.$$

Now $F \cong A_5$ and $\alpha \in H \cap F$. But $H \cap F \lhd F$, by the second isomorphism theorem, so that F simple and $H \cap F \neq 1$ give $H \cap F = F$; that is, $F \leq H$. Therefore, H contains a 3-cycle, $H = A_6$ (by the lemma), and we are done.

We may now assume that no $\alpha \in H$ with $\alpha \neq 1$ fixes any i, for $1 \leq i \leq 6$. A glance at Table 3.3 shows that the cycle structure of α is either $(1\ 2)(3\ 4\ 5\ 6)$ or $(1\ 2\ 3)(4\ 5\ 6)$. In the first case, $\alpha^2 \in H$, $\alpha^2 \neq 1$, and α^2 fixes 1 (and 2), a contradiction. In the second case, H contains $\alpha(\beta\alpha^{-1}\beta^{-1})$, where $\beta = (2\ 3\ 4)$, and it is easily checked that this element is not the identity and it fixes 1, a contradiction. Therefore, no such normal subgroup H can exist. ■

Theorem 3.11. *A_n is simple for all $n \geq 5$.*

Proof. Let $n \geq 5$ and let $H \neq 1$ be a normal subgroup of A_n. If $\beta \in H$ and $\beta \neq 1$, then there is an i with $\beta(i) = j \neq i$. If α is a 3-cycle fixing i and moving j, then α and β do not commute: $\beta\alpha(i) = \beta(i) = j$ and $\alpha\beta(i) = \alpha(j) \neq j$; therefore, their commutator is not the identity. Furthermore, $\alpha(\beta\alpha^{-1}\beta^{-1})$ lies in the normal subgroup H, and, by Lemma 3.4, it is a product of two 3-cycles $(\alpha\beta\alpha^{-1})\beta^{-1}$; thus it moves at most 6 symbols, say, $i_1, \ldots, i_6$. If $F = \{\gamma \in A_m$: γ fixes the other symbols$\}$, then $F \cong A_6$ and $\alpha\beta\alpha^{-1}\beta^{-1} \in H \cap F \lhd F$. Since A_6 is simple, $H \cap F = F$ and $F \leq H$. Therefore H contains a 3-cycle, $H = A_n$ (by Lemma 3.9), and the proof is complete. ■

EXERCISES

3.20. Show that A_5, a group of order 60, has no subgroup of order 30.

3.21. If $n \neq 4$, prove that A_n is the only proper nontrivial normal subgroup of S_n.

3.22. If $G \leq S_n$ contains an odd permutation, then $|G|$ is even and exactly half the elements of G are odd permutations.

3.23. If $X = \{1, 2, \ldots\}$ is the set of all positive integers, then the ***infinite alternating group*** A_∞ is the subgroup of S_X generated by all the 3-cycles. Prove that A_∞ is an infinite simple group. (*Hint.* Adapt the proof of Theorem 3.11.)

Some Representation Theorems

A valuable technique in studying a group is to represent it in terms of something familiar and concrete. After all, an abstract group is a cloud; it is a capital letter G. If the elements of G happen to be permutations or matrices, however, we may be able to obtain results by using this extra information. In this section we give some elementary theorems on *representations*; that is, on homomorphisms into familiar groups.

The first such theorem was proved by Cayley; it shows that the study of subgroups of symmetric groups is no less general than the study of all groups.

Theorem 3.12 (Cayley, 1878). *Every group G can be imbedded as a subgroup of S_G. In particular, if $|G| = n$, then G can be imbedded in S_n.*

Proof. Recall Exercise 1.33: for each $a \in G$, left translation $L_a\colon G \to G$, defined by $x \mapsto ax$, is a bijection; that is, $L_a \in S_G$. The theorem is proved if the function $L\colon G \to S_G$, given by $a \mapsto L_a$, is an injection and a homomorphism, for then $G \cong \operatorname{im} L$. If $a \neq b$, then $L_a(1) = a \neq b = L_b(1)$, and so $L_a \neq L_b$. Finally, we show that $L_{ab} = L_a \circ L_b$. If $x \in G$, then $L_{ab}(x) = (ab)x$, while $(L_a \circ L_b)(x) = L_a(L_b(x)) = L_a(bx) = a(bx)$; associativity shows that these are the same. ■

Definition. The homomorphism $L\colon G \to S_G$, given by $a \mapsto L_a$, is called the ***left regular representation*** of G.

The reason for this name is that each L_a is a regular permutation, as we shall see in Exercise 3.29 below.

Corollary 3.13. *If k is a field and G is a finite group of order n, then G can be imbedded in* $\mathrm{GL}(n, k)$.

Proof. The group $P(n, k)$ of all $n \times n$ permutation matrices is a subgroup of $\mathrm{GL}(n, k)$ that is isomorphic to S_n (see Exercise 1.45). Now apply Cayley's theorem to imbed G into $P(n, k)$. ■

The left regular representation gives another way to view associativity. Assume that G is a set equipped with an operation $*$ such that there is an identity e (that is, $e * a = a = a * e$ for all $a \in G$) and each element $a \in G$ has a (two-sided) inverse a' (i.e., $a * a' = e = a' * a$). Then, for each $a \in G$, the function $L_a\colon G \to G$, defined by $L_a(x) = a * x$, is a permutation of G with inverse $L_{a'}$. If $*$ is associative, then G is a group and Cayley's theorem shows that the function $L\colon G \to S_G$, defined by $a \mapsto L_a$, is a homomorphism; hence, $\operatorname{im} L$ is a subgroup of S_G. Conversely, if $\operatorname{im} L$ is a subgroup of S_G, then $*$ is associative. For if $L_a \circ L_b \in \operatorname{im} L$, there is $c \in G$ with $L_a \circ L_b = L_c$. Thus, $L_a \circ L_b(x) = L_c(x)$ for all $x \in G$; that is, $a * (b * x) = c * x$ for all $x \in G$. But if $x = e$, then $a * b = c$, and so $c * x = (a * b) * x$. This observation can be used as follows. Assume that $G = \{x_1, \ldots, x_n\}$ is a set equipped with an operation $*$, and assume that its multiplication table $[a_{ij}]$ is a Latin square, where $a_{ij} = x_i * x_j$. Each row of the table is a permutation of G, and so $*$ is associative (and G is a group) if the composite of every two rows of the table is again a row of the table. This test for associativity, however, is roughly as complicated as that in Exercise 1.42; both require about n^3 computations.

We now generalize Cayley's theorem.

Theorem 3.14. *If $H \leq G$ and $[G:H] = n$, then there is a homomorphism $\rho: G \to S_n$ with $\ker \rho \leq H$.*

Proof. If $a \in G$ and X is the family of all the left cosets of H in G, define a function $\rho_a: X \to X$ by $gH \mapsto agH$ for all $g \in G$. It is easy to check that each ρ_a is a permutation of X (its inverse is $\rho_{a^{-1}}$) and that $a \mapsto \rho_a$ is a homomorphism $\rho: G \to S_X \cong S_n$. If $a \in \ker \rho$, then $agH = gH$ for all $g \in G$; in particular, $aH = H$, and so $a \in H$; therefore, $\ker \rho \leq H$. ■

Definition. The homomorphism ρ in Theorem 3.14 is called the ***representation of G on the cosets of H***.

When $H = 1$, Theorem 3.14 specializes to Cayley's theorem.

Corollary 3.15. *A simple group G which contains a subgroup H of index n can be imbedded in S_n.*

Proof. There is a homomorphism $\rho: G \to S_n$ with $\ker \rho \leq H < G$. Since G is simple, $\ker \rho = 1$, and so ρ is an injection. ■

Corollary 3.16. *An infinite simple group G has no proper subgroups of finite index.*

Corollary 3.15 provides a substantial improvement over Cayley's theorem, at least for simple groups. For example, if $G \cong A_5$, then Cayley's theorem imbeds G in S_{60}. But G has a subgroup $H \cong A_4$ of order 12 and index $60/12 = 5$, and so Corollary 3.15 says that G can be imbedded in S_5.

Theorem 3.17. *Let $H \leq G$ and let X be the family of all the conjugates of H in G. There is a homomorphism $\psi: G \to S_X$ with $\ker \psi \leq N_G(H)$.*

Proof. If $a \in G$, define $\psi_a: X \to X$ by $\psi_a(gHg^{-1}) = agHg^{-1}a^{-1}$. If $b \in G$, then

$$\psi_a\psi_b(gHg^{-1}) = \psi_a(bgHg^{-1}b^{-1}) = abgHg^{-1}b^{-1}a^{-1} = \psi_{ab}(gHg^{-1}).$$

We conclude that ψ_a has inverse $\psi_{a^{-1}}$, so that $\psi_a \in S_X$ and $\psi: G \to S_X$ is a homomorphism.

If $a \in \ker \psi$, then $agHg^{-1}a^{-1} = gHg^{-1}$ for all $g \in G$. In particular, $aHa^{-1} = H$, and so $a \in N_G(H)$; hence $\ker \psi \leq N_G(H)$. ■

Definition. The homomorphism ψ of Theorem 3.17 is called the ***representation of G on the conjugates of H***.

Exercises

3.24. Let $a \in G$, where G is finite. If a^n has m conjugates and a has k conjugates, then $m|k$. (*Hint.* $C_G(a) \leq C_G(a^n)$.)

3.25. Show that S_n has no subgroup of index t for $2 < t < n$.

3.26. (i) If ρ is the representation of a group G on the cosets of a subgroup H, then $\ker \rho = \bigcap_{x \in G} xHx^{-1}$. Conclude that if $H \lhd G$, then $\ker \rho = H$.
(ii) If ψ is the representation of a group G on the conjugates of a subgroup H, then $\ker \psi = \bigcap_{x \in G} xN_G(H)x^{-1}$.

3.27. The ***right regular representation*** of a group G is the function $R: G \to S_G$ defined by $a \mapsto R_a$, where $R_a(x) = xa^{-1}$.
(i) Show that R is an injective homomorphism. (*Hint.* See Exercises 1.33 and 1.47.) (This exercise is the reason why R_a is defined as right multiplication by a^{-1} and not by a.)
(ii) If L and R are, respectively, the left and right regular representations of S_3, prove that im L and im R are conjugate subgroups of S_6.

3.28. If p is prime and $\tau, \alpha \in S_p$ are a transposition and a p-cycle, respectively, show that $S_p = \langle \tau, \alpha \rangle$. (See Exercise 2.9(iii).)

3.29. If G is a finite group and $a \in G$, then L_a is a regular permutation of G. (*Hint.* If $L_a = \beta_1 \ldots \beta_t$ is the complete factorization of L_a and if g is a symbol occurring in some β_i, then the set of all symbols in β_i is the right coset $\langle a \rangle g$.)

3.30. (i) Let G be a group of order $2^m k$, where k is odd. Prove that if G contains an element of order 2^m, then the set of all elements of odd order in G is a (normal) subgroup of G. (*Hint.* Consider G as permutations via Cayley's theorem, and show that it contains an odd permutation.)
(ii) Show that a finite simple group of even order must have order divisible by 4.

3.31 (i) **(Poincaré).** If H and K are subgroups of G having finite index, then $H \cap K$ also has finite index in G. (*Hint.* Show that $[G: H \cap K] \leq [G:H][G:K]$.)
(ii) If H has finite index in G, then the intersection of all the conjugates of H is a normal subgroup of G having finite index.
(iii) If $([G:H], [G:K]) = 1$, then $[G: H \cap K] = [G:H][G:K]$.

3.32. Prove that A_6 has no subgroup of prime index.

3.33. Let G be a finite group containing a subgroup H of index p, where p is the smallest prime divisor of $|G|$. Prove that H is a normal subgroup of G.

3.34. Let G be an infinite simple group.
(i) Every $x \in G$ with $x \neq 1$ has infinitely many conjugates.
(ii) Every proper subgroup $H \neq 1$ has infinitely many conjugates.

3.35 **(Eilenberg–Moore).** (i) If $H < G$, then there exists a group L and distinct homomorphisms $f, g: G \to L$ with $f|H \neq g|H$. (*Hint.* Let $L = S_X$, where X denotes the family of all the left cosets of H in G together with an additional element ∞. If $a \in G$, define $f_a \in S_X$ by $f_a(\infty) = \infty$ and $f_a(bH) = abH$; define $g: G \to S_X$ by $g = \gamma \circ f$, where $\gamma: S_X \to S_X$ is conjugation by the transposition which interchanges H and ∞.)

(ii) If A and G are groups, then a homomorphism $h: A \to G$ is a surjection if and only if it is *right cancellable*: for every group L and every pair of homomorphisms $f, g: G \to L$, the equation $f \circ h = g \circ h$ implies $f = g$.

G-Sets

The elements of symmetric groups are functions; here is the appropriate abstraction of this property.

Definition. If X is a set and G is a group, then X is a ***G-set*** if there is a function $\alpha: G \times X \to X$ (called an ***action***),[1] denoted by $\alpha: (g, x) \mapsto gx$, such that:

(i) $1x = x$ for all $x \in X$; and
(ii) $g(hx) = (gh)x$ for all $g, h \in G$ and $x \in X$.

One also says that G ***acts*** on X. If $|X| = n$, then n is called the ***degree*** of the G-set X.

For example, if $G \leq S_X$, then $\alpha: G \times X \to X$ is evaluation: $\alpha(\sigma, x) = \sigma(x)$; using the notation of the definition, one often writes σx instead of $\sigma(x)$.

The first result is that G-sets are just another way of looking at permutation representations.

Theorem 3.18. *If X is a G-set with action α, then there is a homomorphism $\tilde{\alpha}: G \to S_X$ given by $\tilde{\alpha}(g): x \mapsto gx = \alpha(g, x)$. Conversely, every homomorphism $\varphi: G \to S_X$ defines an action, namely, $gx = \varphi(g)x$, which makes X into a G-set.*

Proof. If X is a G-set, $g \in G$, and $x \in X$, then

$$\tilde{\alpha}(g^{-1})\tilde{\alpha}(g): x \mapsto \tilde{\alpha}(g^{-1})(gx) = g^{-1}(gx) = (g^{-1}g)x = 1x = x;$$

it follows that each $\tilde{\alpha}(g)$ is a permutation of X with inverse $\tilde{\alpha}(g^{-1})$. That $\tilde{\alpha}$ is a homomorphism is immediate from (ii) of the definition of G-set. The converse is also routine. ∎

The first mathematicians who studied group-theoretic problems, e.g., Lagrange, were concerned with the question: What happens to the polynomial $g(x_1, \ldots, x_n)$ if one permutes the variables? More precisely, if $\sigma \in S_n$,

[1] In this definition, the elements of G act on the left. There is a "right" version of G-set that is sometimes convenient. Define a ***right action*** $\alpha': G \times X \to X$, denoted by $(g, x) \mapsto xg$, to be a function such that:

(i) $x1 = x$ for all $x \in X$; and
(ii) $x(gh) = (xg)h$ for all $g, h \in G$ and $x \in X$.

It is easy to see that every right G-set gives rise to a (left) G-set if one defines $\alpha: G \times X \to X$ by $\alpha(g, x) = xg^{-1} = \alpha'(g^{-1}, x)$.

define

$$g^\sigma(x_1, \ldots, x_n) = g(x_{\sigma 1}, \ldots, x_{\sigma n});$$

given g, how many distinct polynomials g^σ are there? If $g^\sigma = g$ for all $\sigma \in S_n$, then g is called a ***symmetric function***. If a polynomial $f(x) = \sum_{i=0}^n a_i x^i$ has roots $r_1, \ldots, r_n$, then each of the coefficients a_i of $f(x) = a_n \prod_{i=0}^n (x - r_i)$ is a symmetric function of $r_1, \ldots, r_n$. Other interesting functions of the roots may not be symmetric. For example, the ***discriminant*** of $f(x)$ is defined to be the number d^2, where $d = \prod_{i<j} (r_i - r_j)$. If $D(x_1, \ldots, x_n) = \prod_{i<j} (x_i - x_j)$, then it is easy to see, for every $\sigma \in S_n$, that $D^\sigma = \pm D$ (for all $i < j$, either $x_i - x_j$ or $x_j - x_i = -(x_i - x_j)$ occurs as a factor of D^σ). Indeed, D is an ***alternating function*** of the roots: $D^\sigma = D$ if and only if $\sigma \in A_n$. This suggests a slight change in viewpoint. Given $g(x_1, \ldots, x_n)$, find

$$\mathscr{S}(g) = \{\sigma \in S_n : g^\sigma = g\};$$

this is precisely what Lagrange did (see Examples 3.3 and 3.3′ below). It is easy to see that $\mathscr{S}(g) \leq S_n$; moreover, g is symmetric if and only if $\mathscr{S}(g) = S_n$, while $\mathscr{S}(D) = A_n$. Modern mathematicians are concerned with this same type of problem. If X is a G-set, then the set of all $f: X \to X$ such that $f(\sigma x) = f(x)$ for all $x \in X$ and all $\sigma \in G$ is usually valuable in analyzing X.

EXAMPLE 3.3. If k is a field, then S_n acts on $k[x_1, \ldots, x_n]$ by $\sigma g = g^\sigma$, where $g^\sigma(x_1, \ldots, x_n) = g(x_{\sigma 1}, \ldots, x_{\sigma n})$.

EXAMPLE 3.4. Every group G acts on itself by conjugation.

EXAMPLE 3.5. Every group G acts on the family of all its subgroups by conjugation.

There are two fundamental aspects of a G-set.

Definition. If X is a G-set and $x \in X$, then the ***G-orbit*** of x is

$$\mathcal{O}(x) = \{gx : g \in G\} \subset X.$$

One often denotes the orbit $\mathcal{O}(x)$ by Gx. Usually, we will say *orbit* instead of G-orbit. The orbits of X form a partition; indeed, the relation $x \equiv y$ defined by "$y = gx$ for some $g \in G$" is an equivalence relation whose equivalence classes are the orbits.

Definition. If X is a G-set and $x \in X$, then the ***stabilizer*** of x, denoted by G_x, is the subgroup

$$G_x = \{g \in G : gx = x\} \leq G.$$

Let us see the orbits and stabilizers in the G-sets above.

EXAMPLE 3.3′. Let $X = k[x_1, \ldots, x_n]$ and $G = S_n$. If $g \in k[x_1, \ldots, x_n]$, then

$\mathcal{O}(g)$ is the set of distinct polynomials of the form g^σ, and $G_g = \mathcal{S}(g) = \{\sigma \in G = S_n: g^\sigma = g\}$.

Given g, Lagrange defined

$$g^*(x_1, \ldots, x_n) = \prod_{\sigma \in S_n} (x - g^\sigma(x_1, \ldots, x_n));$$

he then defined the *resolvent* $\lambda(g)$ of g to be the polynomial obtained from g^* by removing redundant factors. If r is the degree of $\lambda(g)$, then Lagrange claimed that $r = n!/|\mathcal{S}(g)|$ (Abbati (1803) proved this claim). This formula is the reason Lagrange's theorem is so-called; Lagrange's theorem for subgroups of arbitrary finite groups was probably first proved by Galois.

EXAMPLE 3.4′. If G acts on itself by conjugation and $x \in G$, then $\mathcal{O}(x)$ is the conjugacy class of x and $G_x = C_G(x)$.

EXAMPLE 3.5′. If G acts by conjugation on the family of all its subgroups and it $H \le G$, then $\mathcal{O}(H) = \{$all the conjugates of $H\}$ and $G_H = N_G(H)$.

Theorem 3.19. *If X is a G-set and $x \in X$, then*

$$|\mathcal{O}(x)| = [G : G_x].$$

Proof. If $x \in X$, let G/G_x denote the family of all left cosets of G_x in G. Define $f: \mathcal{O}(x) \to G/G_x$ by $f(ax) = aG_x$. Now f is well defined: if $ax = bx$ for some $b \in G$, then $b^{-1}ax = x$, $b^{-1}a \in G_x$, and $aG_x = bG_x$. The function f is an injection: if $aG_x = f(ax) = f(cx) = cG_x$ for some $c \in G$, then $c^{-1}a \in G_x$, $c^{-1}ax = x$, and $ax = cx$; the function f is a surjection: if $a \in G$, then $aG_x = f(ax)$. Therefore, f is a bijection and $|\mathcal{O}(x)| = |G/G_x| = [G : G_x]$. ■

Corollary 3.20. *If a finite group G acts on a set X, then the number of elements in any orbit is a divisor of $|G|$.*

Corollary 3.21.

(i) *If G is a finite group and $x \in G$, then the number of conjugates of x in G is $[G : C_G(x)]$.*
(ii) *If G is a finite group and $H \le G$, then the number of conjugates of H in G is $[G : N_G(H)]$.*

Proof. Use Examples 3.4′ and 3.5′. ■

We have now explained the similarity of the proofs of Theorems 3.2 and 3.3.

EXERCISES

3.36. If $D(x_1, \ldots, x_n) = \prod_{i<j} (x_i - x_j)$, prove that $\mathcal{S}(D) = \{\sigma \in S_n: D^\sigma = D\} = A_n$.

3.37. Let X be a G-set, let $x, y \in X$, and let $y = gx$ for some $g \in G$. Prove that $G_y = gG_xg^{-1}$; conclude that $|G_y| = |G_x|$.

3.38 **(Abbati).** If k is a field, $g \in k[x_1, \dots, x_n]$, and $\sigma \in S_n$, write $g^\sigma(x_1, \dots, x_n) = g(x_{\sigma 1}, \dots, x_{\sigma n})$, as in Example 3.3. Show that, for any given g, the number of distinct polynomials of the form g^σ is a divisor of $n!$. (*Hint.* Theorem 3.19.)

3.39. If $G \leq S_n$, then G acts on $X = \{1, \dots, n\}$. In particular, $\langle \alpha \rangle$ acts on X for every $\alpha \in S_n$. If the complete factorization of α into disjoint cycles is $\alpha = \beta_1 \dots \beta_t$ and if i is a symbol appearing in β_j, then $\mathcal{O}(i) = \{\alpha^k(i): k \in \mathbb{Z}\}$ consists of all the symbols appearing in β_j. (Compare Exercise 3.29.)

3.40. Cayley's theorem shows that every group G acts on itself via left translations. Show that there is just one orbit (if $x \in G$, then $G = \{gx: g \in G\}$) and that $G_x = 1$ for every $x \in G$.

Definition. A G-set X is ***transitive*** if it has only one orbit; that is, for every $x, y \in X$, there exists $\sigma \in G$ with $y = \sigma x$.

3.41. If X is a G-set, then each of its orbits is a transitive G-set.

3.42. If $H \leq G$, then G acts transitively on the set of all left cosets of H (Theorem 3.14) and G acts transitively on the set of all conjugates of H (Theorem 3.17).

3.43. (i) If $X = \{x_1, \dots, x_n\}$ is a transitive G-set and $H = G_{x_1}$, then there are elements $g_1, \dots, g_n$ in G with $g_i x_1 = x_i$ such that $g_1 H, \dots, g_n H$ are the distinct left cosets of H in G.
(ii) The stabilizer H acts on X, and the number of H-orbits of X is the number of $(H\text{-}H)$-double cosets in G.

3.44. Let X be a G-set with action $\alpha: G \times X \to X$, and let $\tilde{\alpha}: G \to S_X$ send $g \in G$ into the permutation $x \mapsto gx$.
(i) If $K = \ker \tilde{\alpha}$, then X is a (G/K)-set if one defines

$$(gK)x = gx.$$

(ii) If X is a transitive G-set, then X is a transitive (G/K)-set.
(iii) If X is a transitive G-set, then $|\ker \tilde{\alpha}| \leq |G|/|X|$. (*Hint.* If $x \in X$, then $|\mathcal{O}(x)| = [G : G_x] \leq [G : \ker \tilde{\alpha}]$.)

Counting Orbits

Let us call a G-set X ***finite*** if both X and G are finite.

Theorem 3.22 (Burnside's Lemma[2]). *If X is a finite G-set and N is the number*

[2] What is nowadays called Burnside's lemma was proved by Frobenius (1887), as Burnside himself wrote in the first edition (1897) of his book. This is another example (we have already mentioned Lagrange's theorem) of a name of a theorem that is only a name; usually "Smith's theorem" was discovered by Smith, but this is not always the case. It is futile to try to set things right, however, for trying to change common usage would be as successful as spelling reform.

of G-orbits of X, then

$$N = (1/|G|) \sum_{\tau \in G} F(\tau),$$

where, for $\tau \in G$, $F(\tau)$ is the number of $x \in X$ fixed by τ.

Proof. In the sum $\sum_{\tau \in G} F(\tau)$, each $x \in X$ is counted $|G_x|$ times (for G_x consists of all those $\tau \in G$ which fix x). If x and y lie in the same orbit, then Exercise 3.37 gives $|G_y| = |G_x|$, and so the $[G : G_x]$ elements constituting the orbit of x are, in the above sum, collectively counted $[G : G_x]|G_x| = |G|$ times. Each orbit thus contributes $|G|$ to the sum, and so $\sum_{\tau \in G} F(\tau) = N|G|$. ■

Corollary 3.23. *If X is a finite transitive G-set with $|X| > 1$, then there exists $\tau \in G$ having no fixed points.*

Proof. Since X is transitive, the number N of orbits of X is 1, and so Burnside's lemma gives

$$1 = (1/|G|) \sum_{\tau \in G} F(\tau).$$

Now $F(1) = |X| > 1$; if $F(\tau) > 0$ for every $\tau \in G$, then the right hand side is too large. ■

Burnside's lemma is quite useful in solving certain combinatorial problems. Given q distinct colors, how many striped flags are there having n stripes (of equal width)? Clearly the two flags below are the same (just turn over the top flag and put its right end at the left).

c_1	c_2	$\dots$	c_{n-1}	c_n

c_n	c_{n-1}	$\dots$	c_2	c_1

Let $\tau \in S_n$ be the permutation $\begin{pmatrix} 1 & 2 & \dots & n-1 & n \\ n & n-1 & \dots & 2 & 1 \end{pmatrix}$. If $\mathscr{C}^n$ is the set of all n-tuples $\mathbf{c} = (c_1, \dots, c_n)$, where each c_i is any of the q colors, then the cyclic group $G = \langle \tau \rangle$ acts on $\mathscr{C}^n$ if we define $\tau\mathbf{c} = \tau(c_1, \dots, c_n) = (c_n, \dots, c_1)$. Since both $\mathbf{c}$ and $\tau\mathbf{c}$ give the same flag, a flag corresponds to a G-orbit, and so the number of flags is the number N of orbits. By Burnside's lemma, it suffices to compute $F(1)$ and $F(\tau)$. Now $F(1) = |\mathscr{C}^n| = q^n$. An n-tuple $(c_1, \dots, c_n)$ is fixed by τ if and only if it is a "palindrome": $c_1 = c_n$; $c_2 = c_{n-1}$; etc. If $n = 2k$, then $\tau = (1\ n)(2\ n-1)\dots(k\ k+1)$; if $n = 2k+1$, then $\tau = (1\ n)(2\ n-1)\dots(k\ k+2)$. It follows that $F(\tau) = q^{[(n+1)/2]}$, where $[(n+1)/2]$ denotes the greatest integer in $(n+1)/2$. The number of flags is thus

$$N = \tfrac{1}{2}(q^n + q^{[(n+1)/2]}).$$

Let us make the notion of coloring more precise.

Definition. If $G \leq S_X$, where $X = \{1, \ldots, n\}$, and if $\mathscr{C}$ is a set of *colors*, then $\mathscr{C}^n$ is a G-set if we define $\tau(c_1, \ldots, c_n) = (c_{\tau 1}, \ldots, c_{\tau n})$ for all $\tau \in G$. If $|\mathscr{C}| = q$, then an orbit of $\mathscr{C}^n$ is called a ***(q, G)-coloring*** of X.

Lemma 3.24. *Let $\mathscr{C}$ be a set of q colors, and let $G \leq S_X \cong S_n$. If $\tau \in G$, then $F(\tau) = q^{t(\tau)}$, where $t(\tau)$ is the number of cycles occurring in the complete factorization of τ.*

Proof. Since $\tau(c_1, \ldots, c_n) = (c_{\tau 1}, \ldots, c_{\tau n}) = (c_1, \ldots, c_n)$, we see that $c_{\tau i} = c_i$ for all i, and so τi has the same color as i. It follows that $\tau^k i$ has the same color as i, for all k; that is, all i in the $\langle \tau \rangle$-orbit of X have the same color. But Exercise 3.39 shows that if the complete factorization of τ is $\tau = \beta_1 \ldots \beta_{t(\tau)}$, and if i occurs in β_j, then the set of symbols occurring in β_j is the $\langle \tau \rangle$-orbit containing i. Since there are $t(\tau)$ orbits and q colors, there are $q^{t(\tau)}$ n-tuples fixed by τ in its action on $\mathscr{C}^n$. ∎

Definition. If the complete factorization of $\tau \in S_n$ has $e_r(\tau) \geq 0$ r-cycles, then the ***index*** of τ is

$$\operatorname{ind}(\tau) = x_1^{e_1(\tau)} x_2^{e_2(\tau)} \ldots x_n^{e_n(\tau)}.$$

If $G \leq S_n$, then the ***cycle index*** of G is the polynomial

$$P_G(x_1, \ldots, x_n) = (1/|G|) \sum_{\tau \in G} \operatorname{ind}(\tau) \in \mathbb{Q}[x_1, \ldots, x_n].$$

For example, let us consider all possible blue and white flags having nine stripes. Here $|X| = 9$ and $G = \langle \tau \rangle \leq S_9$, where $\tau = (1\ 9)(2\ 8)(3\ 7)(4\ 6)(5)$. Now, $\operatorname{ind}(1) = x_1^9$, $\operatorname{ind}(\tau) = x_1 x_2^4$, and the cycle index of $G = \langle \tau \rangle = \{1, \tau\}$ is

$$P_G(x_1, \ldots, x_9) = \tfrac{1}{2}(x_1^9 + x_1 x_2^4);$$

Corollary 3.25. *If $|X| = n$ and $G \leq S_n$, then the number of (q, G)-colorings of X is $P_G(q, \ldots, q)$.*

Proof. By Burnside's lemma for the G-set $\mathscr{C}^n$, the number of (q, G)-colorings of X is

$$(1/|G|) \sum_{\tau \in G} F(\tau).$$

By Lemma 3.24, this number is

$$(1/|G|) \sum_{\tau \in G} q^{t(\tau)},$$

where $t(\tau)$ is the number of cycles in the complete factorization of τ. On the other hand,

$$P_G(x_1, \dots, x_n) = (1/|G|) \sum_{\tau \in G} \text{ind}(\tau)$$
$$= (1/|G|) \sum_{\tau \in G} x_1^{e_1(\tau)} x_2^{e_2(\tau)} \dots x_n^{e_n(\tau)},$$

and so

$$P_G(q, \dots, q) = (1/|G|) \sum_{\tau \in G} q^{e_1(\tau)+e_2(\tau)+\dots+e_n(\tau)}$$
$$= (1/|G|) \sum_{\tau \in G} q^{t(\tau)}. \quad \blacksquare$$

In 1937, Polyá pushed this technique further. Burnside's lemma allows one to compute the number of blue and white flags having nine stripes; there are 264 of them. How many of these flags have four blue stripes and five white stripes?

Theorem (Polyá, 1937). *Let $G \leq S_X$, where $|X| = n$, let $|\mathscr{C}| = q$, and, for each $i \geq 1$, define $\sigma_i = c_1^i + \dots + c_q^i$. Then the number of (q, G)-colorings of X with f_r elements of color c_r, for every r, is the coefficient of $c_1^{f_1} c_2^{f_2} \dots c_q^{f_q}$ in $P_G(\sigma_1, \dots, \sigma_n)$.*

The proof of Polyá's theorem can be found in combinatorics books (e.g., see Biggs (1989), *Discrete Mathematics*). Let us solve the flag problem posed above; we seek the coefficient of $b^4 w^5$ in

$$P_G(\sigma_1, \dots, \sigma_9) = \tfrac{1}{2}((b + w)^9 + (b + w)(b^2 + w^2)^4).$$

A short exercise with the binomial theorem shows that the coefficient of $b^4 w^5$ is 66.

Exercises

3.45. If G is a finite group and c is the number of conjugacy classes in G, then

$$c = (1/|G|) \sum_{\tau \in G} |C_G(\tau)|.$$

3.46. (i) Let p be a prime and let X be a finite G-set, where $|G| = p^n$ and $|X|$ is not divisible by p. Prove that there exists $x \in X$ with $\tau x = x$ for all $\tau \in G$.
(ii) Let V be a d-dimensional vector space over $\mathbb{Z}_p$, and let $G \leq \mathrm{GL}(d, \mathbb{Z}_p)$ have order p^n. Prove that there is a nonzero vector $v \in V$ with $\tau v = v$ for all $\tau \in G$.

3.47. If there are q colors available, prove that there are

$$\tfrac{1}{4}(q^{n^2} + 2q^{[(n^2+3)/4]} + q^{[(n^2+1)/2]})$$

distinct $n \times n$ colored chessboards. (*Hint.* The set X consists of all $n \times n$ arrays, and the group G is a cyclic group $\langle \tau \rangle$, where τ is a rotation by 90°. Show that τ is product of disjoint 4-cycles.)

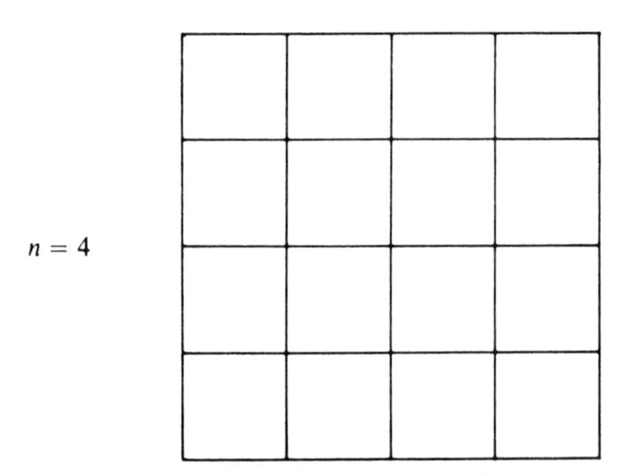

Figure 3.1

3.48. If there are q colors available, prove that there are

$$(1/n) \sum_{d|n} \varphi(n/d) q^d$$

colored roulette wheels having n congruent compartments, each a circular sector (in the formula, φ is the Euler φ-function and the summation ranges over all divisors d of n with $1 \leq d \leq n$). (*Hint*. The group $G = \langle \tau \rangle$ acts on n-tuples, where $\tau(c_1, c_2, \ldots, c_n) = (c_n, c_1, c_2, \ldots, c_{n-1})$. Use Corollary 3.25 and Theorem 2.15.)

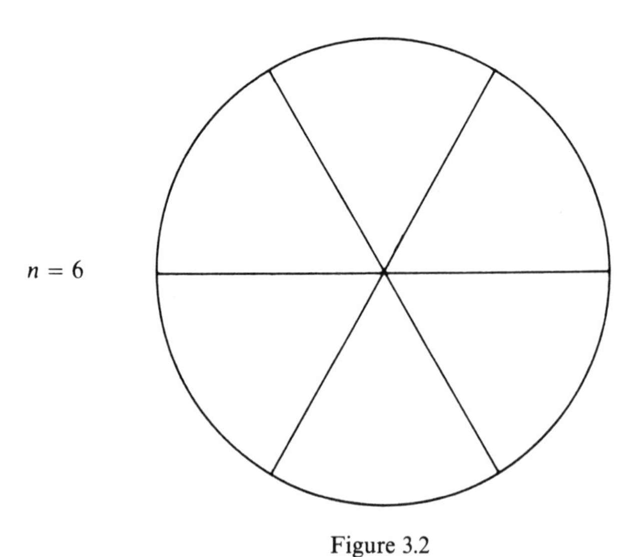

Figure 3.2

Some Geometry

The familiar ***euclidean n-space*** is the vector space $\mathbb{R}^n$ consisting of all n-tuples of real numbers together with an *inner* (or *dot*) *product*. If ε_i is the vector having ith coordinate 1 and all other coordinates 0, then the ***standard basis*** is $\{\varepsilon_1, \ldots, \varepsilon_n\}$. A vector $x = (x_1, \ldots, x_n)$ has the unique expression $x = \sum_i x_i\varepsilon_i$, and if $y = (y_1, \ldots, y_n)$, then the ***inner product*** (x, y) is defined to be the number $\sum_i x_i y_i$.

A subset $\{u_1, \ldots, u_n\}$ of $\mathbb{R}^n$ is called an ***orthonormal basis*** if $(u_i, u_j) = \delta_{ij}$; that is, $(u_i, u_i) = 1$ for all i and $(u_i, u_j) = 0$ when $i \neq j$ (it is easy to see that an orthonormal basis is a basis, for it is a linearly independent subset of $\mathbb{R}^n$ having n elements). The standard basis is an orthonormal basis. If $\{u_1, \ldots, u_n\}$ is an orthonormal basis and if $x = \sum_i x_i u_i$, then $(x, x) = (\sum_i x_i u_i, \sum_j x_j u_j) = \sum_{i,j} x_i x_j (u_i, u_j) = \sum_i x_i^2$. If $x = (x_1, \ldots, x_n) \in \mathbb{R}^n$, define $\|x\| = \sqrt{\sum_i x_i^2}$ (thus, $\|x\|^2 = (x, x)$), and define the ***distance*** between x and y to be $\|x - y\|$.

Definition. A ***motion*** is a distance-preserving function $T\colon \mathbb{R}^n \to \mathbb{R}^n$; that is, $\|Tx - Ty\| = \|x - y\|$ for all $x, y \in \mathbb{R}^n$.

It is plain that if $w \in \mathbb{R}^n$, then the function $T_w\colon \mathbb{R}^n \to \mathbb{R}^n$, defined by $T_w(x) = x + w$ for all $x \in \mathbb{R}^n$, is a motion (T_w is called ***translation*** by w). Of course, $T_w(0) = w$, so that T_w is not a linear transformation if $w \neq 0$.

Definition. A linear transformation $S\colon \mathbb{R}^n \to \mathbb{R}^n$ is ***orthogonal*** if $\|Sx\| = \|x\|$ for all $x \in \mathbb{R}^n$.

Lemma 3.26.

(i) *A linear transformation $S\colon \mathbb{R}^n \to \mathbb{R}^n$ is orthogonal if and only if $\{S\varepsilon_1, \ldots, S\varepsilon_n\}$ is an orthonormal basis (where $\{\varepsilon_1, \ldots, \varepsilon_n\}$ is the standard basis).*

(ii) *Every orthogonal transformation S is a motion.*

Proof. (i) Assume that S is orthogonal. If $x = \sum_i x_i\varepsilon_i$ and $y = \sum_i y_i\varepsilon_i$, then

$$\|x + y\|^2 - \|x\|^2 - \|y\|^2 = 2\sum_i x_i y_i = 2(x, y).$$

In particular,

$$\|Sx + Sy\|^2 - \|Sx\|^2 - \|Sy\|^2 = 2(Sx, Sy).$$

Since $\|x + y\|^2 = \|S(x + y)\|^2 = \|Sx + Sy\|^2$, we have $(Sx, Sy) = (x, y)$ for all x, y. In particular,

$$\delta_{ij} = (\varepsilon_i, \varepsilon_j) = (S\varepsilon_i, S\varepsilon_j),$$

so that $\{S\varepsilon_1, \ldots, S\varepsilon_n\}$ is an orthonormal basis.

Conversely, if $x = \sum_i x_i\varepsilon_i$, then $Sx = \sum_i x_i S\varepsilon_i$,

$$\|Sx\|^2 = (Sx, Sx) = \sum_{i,j} x_i x_j (S\varepsilon_i, S\varepsilon_j) = \sum_i x_i^2 = \|x\|^2,$$

and S is orthogonal.

(ii) If $x = \sum_i x_i\varepsilon_i$ and $y = \sum_i y_i\varepsilon_i$, then $\|Sx - Sy\|^2 = \|S(x - y)\|^2 = \|\sum_i (x_i - y_i)S\varepsilon_i\|^2 = \sum_i (x_i - y_i)^2 = \|x - y\|^2$. ■

Notice that every orthogonal transformation (as is every motion) is an injection, for if $x \neq y$, then $0 \neq \|x - y\| = \|Sx - Sy\|$ and $Sx \neq Sy$; since S is a linear transformation on $\mathbb{R}^n$ with nullity 0, it follows that S is invertible. It is easy to see that if S is orthogonal, then so is S^{-1}.

Lemma 3.27. *Every motion $S: \mathbb{R}^n \to \mathbb{R}^n$ fixing the origin is a linear transformation, hence is orthogonal.*

Proof. We begin by showing that a motion T fixing 0 and each of the elements in the standard basis must be the identity. If $x = (x_1, \ldots, x_n)$, denote Tx by $(y_1, \ldots, y_n)$. Since $T0 = 0$, $\|Tx\| = \|Tx - T0\| = \|x - 0\| = \|x\|$ gives

$$y_1^2 + \cdots + y_n^2 = x_1^2 + \cdots + x_n^2.$$

But also $\|Tx - \varepsilon_1\| = \|Tx - T\varepsilon_1\| = \|x - \varepsilon_1\|$ gives

$$(y_1 - 1)^2 + y_2^2 + \cdots + y_n^2 = (x_1 - 1)^2 + x_2^2 + \cdots + x_n^2.$$

Subtracting gives $2y_1 - 1 = 2x_1 - 1$, and $y_1 = x_1$. A similar argument gives $y_i = x_i$ for all i, so that $Tx = x$.

Assume now that $T\varepsilon_i = u_i$ for $i = 1, \ldots, n$, and let $S: \mathbb{R}^n \to \mathbb{R}^n$ be the linear transformation with $S\varepsilon_i = u_i$ for all i. Now TS^{-1} is a motion (being the composite of two motions) that fixes the standard basis and 0, and so $T = S$. ■

Theorem 3.28.

(i) *The set $O(n, \mathbb{R})$ of all motions $S: \mathbb{R}^n \to \mathbb{R}^n$ fixing the origin is a subgroup of $GL(n, \mathbb{R})$ (called the real **orthogonal group**).*

(ii) *Every motion $T: \mathbb{R}^n \to \mathbb{R}^n$ is the composite of a translation and an orthogonal transformation, and the set $M(n, \mathbb{R})$ of all motions is a group (called the real **group of motions**).*

Proof. (i) Routine, using the lemma.

(ii) Let T be a motion, and let $T(0) = w$. If S is translation by $-w$ (i.e., $Sx = x - w$ for all x), then ST is a motion fixing 0, hence is orthogonal, hence is a bijection; therefore, $T = S^{-1}(ST)$ is a bijection. The reader may now show that the inverse of a motion is a motion, and that the composite of motions is a motion. ■

Theorem 3.29. *A function $T: \mathbb{R}^n \to \mathbb{R}^n$ fixing the origin is a motion if and only* *if it preserves inner products*:

$$(Tx, Ty) = (x, y) \qquad \textit{for all} \quad x, y \in \mathbb{R}^n.$$

Proof. If $x, y \in \mathbb{R}^n$,

$$\|x + y\|^2 = (x + y, x + y) = \|x\|^2 + 2(x, y) + \|y\|^2.$$

Similarly, since T is linear (by Lemma 3.27),

$$\begin{aligned}\|T(x + y)\|^2 &= (T(x + y), T(x + y)) \\ &= (Tx + Ty, Tx + Ty) \\ &= \|Tx\|^2 + 2(Tx, Ty) + \|Ty\|^2.\end{aligned}$$

By hypothesis, $\|x + y\|^2 = \|T(x + y)\|^2$, $\|x\|^2 = \|Tx\|^2$, and $\|y\|^2 = \|Ty\|^2$, so that $2(x, y) = 2(Tx, Ty)$ and $(x, y) = (Tx, Ty)$.

Conversely, if T preserves inner products, then

$$\|x - y\|^2 = (x - y, x - y) = (Tx - Ty, Tx - Ty) = \|Tx - Ty\|^2$$

for all $x, y \in \mathbb{R}^n$. Therefore, T preserves distance, hence is a motion. ∎

The geometric interpretation of this theorem is that every motion fixing the origin preserves angles, for $(x, y) = \|x\| \|y\| \cos \theta$, where θ is the angle between x and y. Of course, all motions preserve lines and planes; they are, after all, linear. For example, given a line $\ell = \{y + rx : r \in \mathbb{R}\}$ (where x and y are fixed vectors) and a motion $T_w S$ (where T_w is translation by w and S is orthogonal), then $T_w S(\ell) = \{T_w(Sy + rSx) : r \in \mathbb{R}\} = \{(w + Sy) + rSx : r \in \mathbb{R}\}$ is also a line.

Definition. A matrix $A \in \mathrm{GL}(n, \mathbb{R})$ is ***orthogonal*** if $AA^t = E$, where A^t denotes the transpose of A.

Denote the ith row of A by a_i. Since the i, j entry of AA^t is (a_i, a_j), it follows that $\{a_1, \ldots, a_n\}$ is an orthonormal basis of $\mathbb{R}^n$. If T is an orthogonal transformation with $T\varepsilon_i = a_i$ for all i, then the matrix of T relative to the standard basis is an orthogonal matrix. It follows that $\mathrm{O}(n, \mathbb{R})$ is isomorphic to the multiplicative group of all $n \times n$ orthogonal matrices.

Since $\det A^t = \det A$, it follows that if A is orthogonal, then $(\det A)^2 = 1$, and so $\det A = \pm 1$.

Definition. A motion T fixing the origin is called a ***rotation*** (or is ***orientation-preserving***) if $\det T = 1$. The set of all rotations form a subgroup $\mathrm{SO}(n, \mathbb{R}) \leq \mathrm{O}(n, \mathbb{R})$, called the ***rotation group***. A motion fixing the origin is called ***orientation-reversing*** if $\det T = -1$.

Of course, $[\mathrm{O}(n, \mathbb{R}) : \mathrm{SO}(n, \mathbb{R})] = 2$.

Here are some examples of orientation-reversing motions. It is a standard result of linear algebra that if W is any subspace of $\mathbb{R}^n$ and $W^\perp = \{v \in V: (v, w) = 0 \text{ for all } w \in W\}$, then $\dim W^\perp = n - \dim W$. A ***hyperplane*** H in $\mathbb{R}^n$ is a translate of a subspace W of dimension $n - 1$: $H = W + v_0$ for some vector v_0. If H is a hyperplane through the origin (that is, $H = W$ is a subspace of dimension $n - 1$), then $\dim H^\perp = 1$, and so there is a nonzero vector a with $(a, h) = 0$ for all $h \in H$; multiplying by a scalar if necessary, we may assume that a is a unit vector.

If ℓ is a line in the plane, then the ***reflection*** in ℓ is the motion $\rho: \mathbb{R}^2 \to \mathbb{R}^2$ which fixes every point on ℓ and which interchanges all points x and x' equidistant from ℓ (as illustrated in Figure 3.3; thus, ℓ behaves as a mirror). More generally, define the ***reflection*** in a hyperplane H as the motion that fixes every point of H and that interchanges points equidistant from H. If ρ is to be a linear transformation, then H must be a line through the origin, for the only points fixed by ρ lie on H.

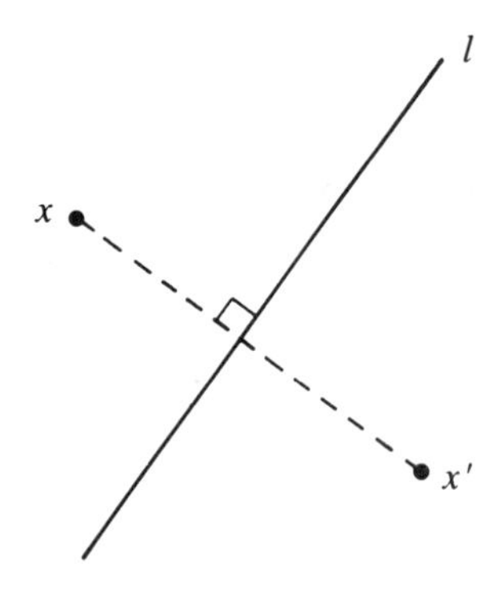

Figure 3.3

Theorem 3.30. *Every reflection ρ in a hyperplane H through the origin is orientation-reversing.*

Proof. Choose a unit vector $a \in \mathbb{R}^n$ with $(h, a) = 0$ for all $h \in H$. Define $\rho': \mathbb{R}^n \to \mathbb{R}^n$ by $\rho'(x) = x - 2(x, a)a$ for all $x \in \mathbb{R}^n$. If $x \in H$, then $(x, a) = 0$, $x - 2(x, a)a = x$, and ρ' fixes x; if $x \notin H$, then $x = h + ra$, where $h \in H$ and $r \in \mathbb{R}$. Now $(x, a) = (h + ra, a) = r$ and $x - 2(x, a)a = h - ra$; hence, $\rho'(h + ra) = h - ra$, so that ρ' interchanges pairs of vectors equidistant from H and fixes H pointwise. Hence, $\rho' = \rho$.

If $\{h_1, \ldots, h_{n-1}\}$ is a basis of H, then $\{h_1, \ldots, h_{n-1}, a\}$ is a basis of $\mathbb{R}^n$. Relative to the latter basis, the matrix of ρ is diagonal with diagonal entries $1, 1, \ldots, 1, -1$; therefore, $\det \rho = -1$ and ρ is orientation-reversing. ■

Let us now consider the case $n = 2$. If we identify $\mathbb{R}^2$ with the complex numbers $\mathbb{C}$, then perpendicular unit vectors u_1 and u_2 have the form $u_1 = e^{i\theta}$ and $u_2 = e^{i\varphi}$, where $\varphi = \theta \pm \pi/2$. It follows that if

$$A = \begin{bmatrix} a & b \\ c & d \end{bmatrix}$$

is an orthogonal matrix, then its columns are the real and imaginary parts of $A\varepsilon_1$ and $A\varepsilon_2$. Therefore, either

$$A = \begin{bmatrix} \cos\theta & -\sin(\theta + \pi/2) \\ \sin\theta & \cos(\theta + \pi/2) \end{bmatrix} = \begin{bmatrix} \cos\theta & -\sin\theta \\ \sin\theta & \cos\theta \end{bmatrix}$$

and $\det A = 1$ (so that A corresponds to rotation about the origin by the angle θ) or

$$A = \begin{bmatrix} \cos\theta & -\sin(\theta - \pi/2) \\ \sin\theta & \cos(\theta - \pi/2) \end{bmatrix} = \begin{bmatrix} \cos\theta & \sin\theta \\ \sin\theta & -\cos\theta \end{bmatrix}$$

and $\det A = -1$ (so that A corresponds to reflection in the line ℓ through the origin having slope $\tan\theta$). In particular, the matrix

$$B = \begin{bmatrix} 1 & 0 \\ 0 & -1 \end{bmatrix}$$

corresponds to the motion $(x, y) \mapsto (x, -y)$ which is the reflection in the x-axis.

With this background, we now pose the following problem. Let Δ be a figure in the plane having its center of gravity at the origin. Define

$$\mathscr{S}(\Delta) = \{S \in \mathrm{O}(2, \mathbb{R}) \colon S(\Delta) = \Delta\}.$$

Of course, $S(\Delta) = \{x \in \mathbb{R}^2 \colon x = S(y) \text{ for some } y \in \Delta\}$. If Δ is a triangle with vertices a, b, and c and if S is a motion, then $S(\Delta)$ is also a triangle, say, with vertices Sa, Sb, and Sc; if $S \in \mathscr{S}(\Delta)$, then S permutes $X = \{a, b, c\}$. It follows that $\mathscr{S}(\Delta)$ acts on X: there is a homomorphism $\psi \colon \mathscr{S}(\Delta) \to S_X$, namely, $S \mapsto S|X$, the restriction of S to X. Now ψ is an injection, for a linear transformation on $\mathbb{R}^2$ is determined by its values on an independent set of two vectors. It follows that $\mathscr{S}(\Delta)$ is a finite group; indeed, $\mathscr{S}(\Delta)$ is isomorphic to a subgroup of S_3. If Δ is an equilateral triangle, then $\mathscr{S}(\Delta) \cong S_3$ (see Exercise 3.58 below); if Δ is only an isosceles triangle, then $\mathscr{S}(\Delta) \cong \mathbb{Z}_2$; if Δ is not even isosceles, then $\mathscr{S}(\Delta) = 1$. The group $\mathscr{S}(\Delta)$ thus "measures" the amount of symmetry present in Δ: bigger groups arise from "more symmetric" triangles. A circle Δ with center at the origin is very symmetric, for $\mathscr{S}(\Delta)$ is an infinite group (for every θ, it contains rotation about the origin by the angle θ). One calls $\mathscr{S}(\Delta)$ the ***symmetry group*** of the figure Δ.

Theorem 3.31. *If Δ is a regular polygon with n vertices, then $\mathscr{S}(\Delta)$ is a group of* *order* $2n$ *which is generated by two elements* S *and* T *such that*

$$S^n = 1, \qquad T^2 = 1, \qquad \textit{and} \qquad TST = S^{-1}.$$

Proof. We may assume that the origin is the center of gravity of Δ and that one vertex of Δ lies on the x-axis. Observe that $\mathscr{S}(\Delta)$ is finite: as in the example of triangles, it can be imbedded in the group of all permutatons of the vertices. Now each $S \in \mathscr{S}(\Delta)$ also permutes the edges of Δ; indeed, regularity (that is, all edges having the same length) implies that $\mathscr{S}(\Delta)$ acts transitively on the n edges. Since the stabilizer of an edge has order 2 (the endpoints can be interchanged), Theorem 3.19 gives $|\mathscr{S}(\Delta)| = 2n$. If S is rotation by $(360/n)^\circ$ and T is reflection in the x-axis, then it is easy to check that $\mathscr{S}(\Delta) = \langle S, T \rangle$ and that S and T satisfy the displayed relations. ■

Definition. The ***dihedral group***[3] D_{2n}, for $2n \geq 4$, is a group of order $2n$ which is generated by two elements s and t such that

$$s^n = 1, \qquad t^2 = 1, \qquad \text{and} \qquad tst = s^{-1}.$$

Note that D_{2n} is not abelian for all $n \geq 3$, while D_4 is the 4-group $\mathbf{V}$.
The next result explains the ubiquity of dihedral groups.

Theorem 3.32. *If* G *is a finite group and if* $a, b \in G$ *have order* 2, *then* $\langle a, b \rangle \cong D_{2n}$ *for some* n.

Proof. Since G is finite, the element ab has finite order n, say. If $s = ab$, then $asa = a(ab)a = ba = (ab)^{-1} = s^{-1}$, because both a and b have order 2.

It remains to show that $|\langle a, b \rangle| = 2n$ (of course, $|\langle a, b \rangle| = 2m$ for some m, but it is not obvious that $m = n$). We claim that $as^i \neq 1$ for all $i \geq 0$. Otherwise, choose $i \geq 0$ minimal with $as^i = 1$. Now $i \neq 0$ (for $a \neq 1$) and $i \neq 1$ (lest $1 = as = aab = b$), so that $i \geq 2$. But $1 = as^i = aabs^{i-1} = bs^{i-1}$; conjugating by b gives $1 = s^{i-1}b = s^{i-2}abb = s^{i-2}a$, and conjugating by a now gives $as^{i-2} = 1$, contradicting the minimal choice of i. It follows that $as^i \neq s^j$ for all i, j; hence $\langle a, b \rangle$ contains the disjoint union $\langle s \rangle \cup a\langle s \rangle$, and so $|\langle a, b \rangle| = |\langle a, s \rangle| \geq 2n$. For the reverse inequality, it suffices to show that

$$H = \{a^j s^i \colon 0 \leq j < 2, 0 \leq i < n\}$$

is a subgroup. Using Corollary 2.4, one need check only four cases: $as^i as^k = s^{-i}s^k = s^{k-i} \in H$; $as^i s^k = as^{i+k} \in H$; $s^i s^k = s^{i+k} \in H$; $s^i as^k = a(as^i a)s^k = s^{-i}s^k = s^{k-i} \in H$. ■

Elements of order 2 arise often enough to merit a name; they are called ***involutions***.

[3] In earlier editions, I denoted D_{2n} by D_n.

Let Ω be a figure in $\mathbb{R}^3$ with its center of gravity at the origin, and define

$$\mathscr{S}(\Omega) = \{S \in O(3, \mathbb{R}): S(\Omega) = \Omega\}.$$

If $\Omega = \Sigma_n$ is the regular solid having n (congruent) faces, each of which has k edges, then $\mathscr{S}(\Sigma_n)$ acts transitively on the set of n faces of Σ_n (this is essentially the definition of regularity), while the stabilizer of a face f (which is a regular k-gon) consists of the k rotations of f about its center. By Theorem 3.19, $\mathscr{S}(\Sigma_n)$ has order nk.

It is a classical result that there are only five regular solids: the *tetrahedron* Σ_4 with 4 triangular faces; the *cube* Σ_6 with 6 square faces; the *octahedron* Σ_8 with 8 triangular faces; the *dodecahedron* Σ_{12} with 12 pentagonal faces; the *icosahedron* Σ_{20} with 20 triangular faces. The rotation groups of these solids thus have orders 12, 24, 24, 60, and 60, respectively.

These considerations suggest investigation of the finite subgroups of the orthogonal groups $O(n, \mathbb{R})$. It can be shown that the finite subgroups of $O(2, \mathbb{R})$ are isomorphic to either D_{2n} or $\mathbb{Z}_n$, and that the finite subgroups of $O(3, \mathbb{R})$ are isomorphic to either D_{2n}, $\mathbb{Z}_n$, A_4, S_4, or A_5.

EXERCISES

3.49. Prove that $D_4 \cong \mathbf{V}$ and $D_6 \cong S_3$.

3.50. Prove that $D_{12} \cong S_3 \times \mathbb{Z}_2$.

3.51. Let G be a transitive subgroup of S_4.
(i) If $m = [G: G \cap \mathbf{V}]$, then $m | 6$.
(ii) If $m = 6$, then $G = S_4$; if $m = 3$, then $G = A_4$; if $m = 1$, then $G = \mathbf{V}$; if $m = 2$, then either $G \cong \mathbb{Z}_4$ or $G \cong D_8$.

3.52. (i) **(von Dyck (1882)).** Prove that $\mathscr{S}(\Sigma_4) \cong A_4$ and that $\mathscr{S}(\Sigma_6) \cong S_4 \cong \mathscr{S}(\Sigma_8)$. (*Hint.* $A_4 = \langle s, t \rangle$, where $s^2 = t^3 = (st)^3 = 1$; $S_4 = \langle s, t \rangle$, where $s^2 = t^3 = (st)^4 = 1$.)
(ii) **(Hamilton (1856)).** Prove that $\mathscr{S}(\Sigma_{12}) \cong A_5 \cong \mathscr{S}(\Sigma_{20})$. (*Hint.* $A_5 = \langle s, t \rangle$, where $s^2 = t^3 = (st)^5 = 1$.)
(Because of this exercise, A_4 is also called the ***tetrahedral group***, S_4 is also called the ***octahedral group***, and A_5 is also called the ***icosahedral group***.)

3.53. Let $\mathrm{Tr}(n, \mathbb{R})$ denote the set of all the translations of $\mathbb{R}^n$. Show that $\mathrm{Tr}(n, \mathbb{R})$ is an abelian normal subgroup of the group of motions $M(n, \mathbb{R})$, and $M(n, \mathbb{R})/\mathrm{Tr}(n, \mathbb{R}) \cong O(n, \mathbb{R})$.

3.54. It can be shown that every $S \in SO(3, \mathbb{R})$ has 1 as an eigenvalue (there is thus a nonzero vector v with $Sv = v$). Using this, show that the matrix of S (relative to a suitable basis of $\mathbb{R}^3$) is

$$\begin{bmatrix} 1 & 0 & 0 \\ 0 & \cos\theta & -\sin\theta \\ 0 & \sin\theta & \cos\theta \end{bmatrix}.$$

3.55. Prove that the circle group $\mathbf{T}$ is isomorphic to $SO(2, \mathbb{R})$.

3.56. Let $\omega = e^{2\pi i/n}$ be a primitive nth root of unity. Show that the matrices

$$A = \begin{bmatrix} \omega & 0 \\ 0 & \omega^{-1} \end{bmatrix} \quad \text{and} \quad B = \begin{bmatrix} 0 & 1 \\ 1 & 0 \end{bmatrix}$$

generate a subgroup of $GL(2, \mathbb{C})$ isomorphic to D_{2n}.

3.57. What is the center $Z(D_{2n})$? (*Hint.* Every element of D has a factorization $s^i t^j$.)

3.58. If Δ is an equilateral triangle in $\mathbb{R}^2$ with its center of gravity at the origin, show that $\mathscr{S}(\Delta)$ is generated by

$$A = \begin{bmatrix} -\frac{1}{2} & \sqrt{3}/2 \\ \sqrt{3}/2 & -\frac{1}{2} \end{bmatrix} \quad \text{and} \quad B = \begin{bmatrix} 1 & 0 \\ 0 & -1 \end{bmatrix}.$$

3.59. How many bracelets are there having n beads each of which can be painted any one of q colors? (*Hint.* Use Corollary 3.25; D_{2n} is the group that is acting.)

We now show how one can use groups to prove geometric theorems.

Recall that if $u, v \in \mathbb{R}^2$, then the line segment with endpoints u and v, denoted by $[u, v]$, consists of all vectors $tu + (1 - t)v$, where $0 \le t \le 1$. If u, v, w are the vertices of a triangle Δ, then we will denote Δ by $[u, v, w]$.

Definition. If $v_1, \ldots, v_n \in \mathbb{R}^2$, then a ***convex combination*** of $v_1, \ldots, v_n$ is a linear combination $\sum t_i v_i$, where all $t_i \ge 0$ and $\sum t_i = 1$.

Lemma 3.33. *If $\Delta = [v_1, v_2, v_3]$ is a triangle, then Δ consists of all the convex combinations of v_1, v_2, v_3.*

Proof. Denote the set of all convex combinations of v_1, v_2, v_3 by C. We first show that $C \subset \Delta$. Let $c = t_1 v_1 + t_2 v_2 + t_3 v_3$ belong to C. If $t_3 = 1$, then $c = v_3 \in \Delta$. If $t_3 \ne 1$, then $q = t_1/(1 - t_3)v_1 + t_2/(1 - t_3)v_2$ is a convex combination of v_1 and v_2, hence lies on the line segment $[v_1, v_2] \subset \Delta$. Finally, $c = (1 - t_3)q + t_3 v_3 \in \Delta$, for it is a convex combination of q and v_3, and hence it lies on the line segment joining these two points (which is wholly inside of Δ).

For the reverse inclusion, take $\delta \in \Delta$. It is clear that C contains the perimeter of Δ (such points lie on line segments, which consist of convex combinations of two vertices). If δ is an interior point, then it lies on a line segment $[u, w]$, where u and w lie on the perimeter (indeed, it lies on many such segments). Thus, $\delta = tu + (1 - t)w$ for some $0 \le t \le 1$. Write $u = t_1 v_1 + t_2 v_2 + t_3 v_3$ and $w = s_1 v_1 + s_2 v_2 + s_3 v_3$, where $t_i \ge 0$, $s_i \ge 0$ and $\sum_{i=1}^{3} t_i = 1 = \sum_{i=1}^{3} s_i$. It suffices to show that $\delta = t(\sum t_i v_i) + (1 - t)(\sum s_i v_i) = \sum [t t_i v_i + (1 - t) s_i] v_i$ is a convex combination of v_1, v_2, v_3. But $t t_i + (1 - t) s_i \ge 0$, because each of its terms is nonnegative, while $\sum [t t_i + (1 - t) s_i] = t(\sum t_i) + (1 - t)(\sum s_i) = t + (1 - t) = 1$. ■

Definition. A function $\varphi: \mathbb{R}^2 \to \mathbb{R}^2$ is an ***affine map*** if there is a nonsingular linear transformation $\lambda: \mathbb{R}^2 \to \mathbb{R}^2$ and a vector $z \in \mathbb{R}^2$ such that, for all $v \in \mathbb{R}^2$,

$$\varphi(v) = \lambda(v) + z.$$

The set of all affine maps under composition, denoted by Aff(2, $\mathbb{R}$), is called the ***affine group of the plane***.

Lemma 3.34. *Let φ be an affine map.*

(i) *φ preserves all convex combinations.*
(ii) *φ preserves line segments: for all u, v, $\varphi([u, v]) = [\varphi u, \varphi v]$.*
(iii) *The point $tu + (1 - t)v$, for $0 \le t \le 1$, is called the **t-point** of $[u, v]$. If z is the t-point of $[u, v]$, where $0 \le t \le 1$, then φz is the t-point of $[\varphi u, \varphi v]$. In particular, φ preserves midpoints of line segments.*
(iv) *φ preserves triangles: if $\Delta = [u, v, w]$ is a triangle, then $\varphi(\Delta)$ is the triangle $[\varphi u, \varphi v, \varphi w]$.*

Proof. (i) Let $\varphi x = \lambda x + z$, where λ is a nonsingular linear transformation and $z \in \mathbb{R}^2$. If $\sum_i t_i v_i$ is a convex combination, then

$$\begin{aligned}\varphi\left(\sum_i t_i v_i\right) &= \lambda \sum_i t_i v_i + z \\ &= \lambda \sum_i t_i v_i + \left(\sum_i t_i\right) z \\ &= \sum_i t_i(\lambda v_i) + \left(\sum_i t_i\right) z \\ &= \sum_i t_i(\lambda v_i + z) = \sum_i t_i \varphi(v_i).\end{aligned}$$

(ii) Immediate from (i), for $[u, v]$ is the set of all convex combinations of u and v.

(iii) Immediate from (i).

(iv) Immediate from (i) and Lemma 3.33. ■

Lemma 3.35. *Points u, v, w in $\mathbb{R}^2$ are collinear if and only if $\{u - w, v - w\}$ is a linearly dependent set.*

Proof. Suppose that u, v, w lie on a line ℓ, and let ℓ consist of all vectors of the form $ry + z$, where $r \in \mathbb{R}$ and $y, z \in \mathbb{R}^2$. There are thus numbers r_i with $u = r_1 y + z$, $y = r_2 y + z$, and $w = r_3 y + z$. Therefore, $u - w = (r_1 - r_3)y$ and $v - w = (r_2 - r_3)y$ form a linearly dependent set.

Conversely, suppose that $u - w = r(v - w)$, where $r \in \mathbb{R}$. It is easily seen that u, v, w all lie on the line ℓ consisting of all vectors of the form $t(v - w) + w$, where $t \in \mathbb{R}$. ■

Lemma 3.36. *If $\Delta = [u, v, w]$ and $\Delta' = [u', v', w']$, are triangles, then there is* *an affine map* φ *with* $\varphi u = u'$, $\varphi v = v'$, *and* $\varphi w = w'$. *Thus,* $\mathrm{Aff}(2, \mathbb{R})$ *acts transitively on the family of all triangles in* $\mathbb{R}^2$.

Proof. The vertices u, v, w of a triangle are not collinear, so that the vectors $u - w$ and $v - w$ form a linearly independent set, hence comprise a basis of $\mathbb{R}^2$; similarly, $\{u' - w', v' - w'\}$ is also a basis of $\mathbb{R}^2$. There thus exists a nonsingular linear transformation λ with $\lambda(u - w) = u' - w'$ and $\lambda(v - w) = v' - w'$. If $z = w' - \lambda(w)$, then define φ by

$$\varphi(x) = \lambda(x) + w' - \lambda(w) = \lambda(x - w) + w'.$$

It is easy to see that φ is an affine map which carries u, v, w to u', v', w', respectively. It follows from Lemma 3.33 that $\varphi(\Delta) = \Delta'$. ■

Theorem 3.37. *For every triangle* Δ, *the medians meet in a common point which is a* $\frac{2}{3}$*-point on each of the medians.*

Proof. It is easy to see that the theorem is true in the special case of an equilateral triangle. E. By Lemma 3.36, there exists an affine map φ with $\varphi(E) = \Delta$. By Lemma 3.34, φ preserves collinearity, medians and $\frac{2}{3}$-points. ■

The reader is invited to prove other geometric theorems in this spirit, using the (easily established) fact that affine maps preserve parallel lines as well as conic sections (in particular, every ellipse is of the form $\varphi(\Delta)$, where Δ is the unit circle, for these are the only bounded (compact) conic sections).

F. Klein's ***Erlangen Program*** (1872) uses groups to classify different geometries on the plane (or more general spaces). If $G \leq S_{\mathbb{R}^2}$, then a property P of a figure Δ in $\mathbb{R}^2$ is an ***invariant*** of G if $\varphi(\Delta)$ has property P for all $\varphi \in G$. For example, invariants of the group $M(2, \mathbb{R})$ of all motions include collinearity, length, angle, and area; the corresponding geometry is the usual geometry of Euclid. Invariants of $\mathrm{Aff}(2, \mathbb{R})$ include collinearity, triangles, line segments, and t-points of line segments, parallelism, conic sections; the corresponding geometry is called *affine geometry*. Other groups may give other geometries. For example, if G is the group of all homeomorphisms of the plane, then invariants include connectedness, compactness, and dimensionality; the corresponding geometry is called *topology*.

CHAPTER 4

The Sylow Theorems

p-Groups

The order of a group G has consequences for its structure. A rough rule of thumb is that the more complicated the prime factorization of $|G|$, the more complicated the group. In particular, the fewer the number of distinct prime factors in $|G|$, the more tractible it is. We now study the "local" case when only one prime divides $|G|$.

Definition. If p is a prime, then a ***p-group*** is a group in which every element has order a power of p.

Corollary 4.3 below gives a simple characterization of finite p-groups.

Lemma 4.1. *If G is a finite abelian group whose order is divisible by a prime p, then G contains an element of order p.*

Proof. Write $|G| = pm$, where $m \geq 1$. We proceed by induction on m after noting that the base step is clearly true. For the inductive step, choose $x \in G$ of order $t > 1$. If $p \mid t$, then Exercise 2.11 shows that $x^{t/p}$ has order p, and the lemma is proved. We may, therefore, assume that the order of x is not divisible by p. Since G is abelian, $\langle x \rangle$ is a normal subgroup of G, and $G/\langle x \rangle$ is an abelian group of order $|G|/t = pm/t$. Since $p \nmid t$, we must have $m/t < m$ an integer. By induction, $G/\langle x \rangle$ contains an element y^* of order p. But the natural map $\nu\colon G \to G/\langle x \rangle$ is a surjection, and so there is $y \in G$ with $\nu(y) = y^*$. By Exercise 2.14, the order of y is a multiple of p, and we have returned to the first case. ■

We now remove the hypothesis that G is abelian.

Theorem 4.2 (Cauchy, 1845). *If G is a finite group whose order is divisible by a prime p, then G contains an element of order p.*

Proof. Recall Theorem 3.2. If $x \in G$, then the number of conjugates of x is $[G : C_G(x)]$, where $C_G(x)$ is the centralizer of x in G. If $x \notin Z(G)$, then its conjugacy class has more than one element, and so $|C_G(x)| < |G|$. If $p \,|\, |C_G(x)|$ for such a noncentral x, we are done, by induction. Therefore, we may assume that $p \nmid |C_G(x)|$ for all noncentral x in G. Better, since $|G| = [G : C_G(x)]|C_G(x)|$, we may assume that $p | [G : C_G(x)]$ (using Euclid's lemma, which applies because p is prime).

Partition G into its conjugacy classes and count (recall that $Z(G)$ consists of all the elements of G whose conjugacy class has just one element):

$$(*) \qquad |G| = |Z(G)| + \sum_i [G : C_G(x_i)],$$

where one x_i is selected from each conjugacy class with more than one element. Since $|G|$ and all $[G : C_G(x_i)|$ are divisible by p, it follows that $|Z(G)|$ is divisible by p. But $Z(G)$ is abelian, and so it contains an element of order p, by the lemma. ■

Definition. Equation $(*)$ above is called the ***class equation*** of the finite group G.

Here is a second proof of Cauchy's theorem, due to J.H. McKay, which avoids the class equation. Assume that p is a prime and that G is a finite group. Define

$$X = \{(a_1, \ldots, a_p) \in G \times \cdots \times G: a_1 a_2 \ldots a_p = 1\}.$$

Note that $|X| = |G|^{p-1}$, for having chosen the first $p - 1$ coordinates arbitrarily, we must set $a_p = (a_1 a_2 \ldots a_{p-1})^{-1}$. Now X is a $\mathbb{Z}_p$-set, where $g \in \mathbb{Z}_p$ acts by cyclically permuting the coordinates (since $a_i \ldots a_p a_1 \ldots a_{i-1}$ is a conjugate of $a_1 a_2 \ldots a_p$, the product of the permuted coordinates is also equal to 1). By Corollary 3.21, each orbit of X has either 1 or p elements. An orbit with just one element is a p-tuple having all its coordinates equal, say, $a_i = a$ for all i; in other words, such orbits correspond to elements $a \in G$ with $a^p = 1$. Clearly $(1, \ldots, 1)$ is such an orbit; were this the only such orbit, then we would have

$$|X| = |G|^{p-1} = 1 + kp$$

for some integer $k \geq 0$; that is, $|G|^{p-1} \equiv 1 \bmod p$. If p divides $|G|$, however, this is a contradiction, and so we conclude that G must have an element of order p. (As A. Mann remarked to me, if $|G|$ is not divisible by p, then we have proved Fermat's theorem.)

Corollary 4.3. *A finite group G is a p-group if and only if $|G|$ is a power of p.*

Proof. If $|G| = p^m$, then Lagrange's theorem shows that G is a p-group. Conversely, assume that there is a prime $q \neq p$ which divides $|G|$. By Cauchy's theorem, G contains an element of order q, and this contradicts G being a p-group. ■

Theorem 4.4. *If $G \neq 1$ is a finite p-group, then its center $Z(G) \neq 1$.*

Proof. Consider the class equation

$$|G| = |Z(G)| + \sum [G : C_G(x_i)].$$

Each $C_G(x_i)$ is a proper subgroup of G, for $x_i \notin Z(G)$. By Corollary 4.3, $[G : C_G(x_i)]$ is a power of p (since $|G|$ is). Thus, p divides each $[G : C_G(x_i)]$, and so p divides $|Z(G)|$. ■

If G is a finite simple p-group, then $G = Z(G)$ and G is abelian; therefore, G must be cyclic of order p. Theorem 4.4 is false for infinite p-groups.

Corollary 4.5. *If p is a prime, then every group G of order p^2 is abelian.*

Proof. If G is not abelian, then $Z(G) < G$; since $1 \neq Z(G)$, we must have $|Z(G)| = p$. The quotient group $G/Z(G)$ is defined, since $Z(G) \lhd G$, and it is cyclic, because $|G/Z(G)| = p$; this contradicts Exercise 3.3. ■

Theorem 4.6. *Let G be a finite p-group.*

(i) *If H is a proper subgroup of G, then $H < N_G(H)$.*
(ii) *Every maximal subgroup of G is normal and has index p.*

Proof. (i) If $H \lhd G$, then $N_G(H) = G$ and the theorem is true. If X is the set of all the conjugates of H, then we may assume that $|X| = [G : N_G(H)] \neq 1$. Now G acts on X by conjugation and, since G is a p-group, every orbit of X has size a power of p. As $\{H\}$ is an orbit of size 1, there must be at least $p - 1$ other orbits of size 1. Thus there is at least one conjugate $gHg^{-1} \neq H$ with $\{gHg^{-1}\}$ also an orbit of size 1. Now $agHg^{-1}a^{-1} = gHg^{-1}$ for all $a \in H$, and so $g^{-1}ag \in N_G(H)$ for all $a \in H$. But $gHg^{-1} \neq H$ gives at least one $a \in H$ with $g^{-1}ag \notin H$, and so $H < N_G(H)$.

(ii) If H is a maximal subgroup of G, then $H < N_G(H)$ implies that $N_G(H) = G$; that is, $H \lhd G$. By Exercise 2.58, $[G : H] = p$. ■

Lemma 4.7. *If G is a finite p-group and r_1 is the number of subgroups of G having order p, then $r_1 \equiv 1 \bmod p$.*

Proof. Let us first count the number of elements of order p. Since $Z(G)$ is abelian, all its elements of order p together with 1 form a subgroup H whose order is a power of p; hence the number of central elements of order p is $|H| - 1 \equiv -1 \bmod p$. If $x \in G$ is of order p and not central, then its conjugacy class x^G consists of several elements of order p; that is, $|x^G| > 1$ is an "honest" power of p, by Theorem 3.2. It follows that the number of elements in G of order p is congruent to $-1 \bmod p$; say, there are $mp - 1$ such elements. Since the intersection of any distinct pair of subgroups of order p is trivial, the number of elements of order p is $r_1(p-1)$. But $r_1(p-1) = mp - 1$ implies $r_1 \equiv 1 \bmod p$. ■

Theorem 4.8. *If G is a finite p-group and r_s is the number of subgroups of G having order p^s, then $r_s \equiv 1 \bmod p$.*

Proof. Let H be a subgroup of order p^s, and let $K_1, \ldots, K_a$ be the subgroups of G of order p^{s+1} which contain it; we claim that $a \equiv 1 \bmod p$. Every subgroup of G which normalizes H is contained in $N = N_G(H)$; in particular, each K_j lies in N, for Lemma 4.6(ii) shows that $H \lhd K_j$ for all j. By the Correspondence Theorem, the number of subgroups of order p in N/H is equal to the number of subgroups of N containing H which have order p^{s+1}. By the lemma, $a \equiv 1 \bmod p$.

Now let K be a subgroup of order p^{s+1}, and let $H_1, \ldots, H_b$ be its subgroups of order p^s; we claim that $b \equiv 1 \bmod p$. By the lemma, $H_i \lhd K$ for all i. Since $H_1 H_2 = K$ (for the H_i are maximal subgroups of K), the product formula (Theorem 2.20) gives $|D| = p^{s-1}$, where $D = H_1 \cap H_2$, and $[K : D] = p^2$. By Corollary 4.5, the group K/D is abelian; moreover, K/D is generated by two subgroups of order p, namely, H_i/D for $i = 1, 2$, and so it is not cyclic. By Exercise 2.68, $K/D \cong \mathbb{Z}_p \times \mathbb{Z}_p$. Therefore, K/D has $p^2 - 1$ elements of order p and hence has $p + 1 = (p^2 - 1)/(p - 1)$ subgroups of order p. The Correspondence Theorem gives $p + 1$ subgroups of K of order p^s containing D. Suppose there is some H_j with $D \not\leq H_j$. Let $E = H_1 \cap H_j$; as above, there is a new list of $p + 1$ subgroups of K of order p^s containing E, one of which is H_1. Indeed, $H_1 = ED$ is the only subgroup on both lists. Therefore, there are p new subgroups and $1 + 2p$ subgroups counted so far. If some H_l has not yet been listed, repeat this procedure beginning with $H_1 \cap H_l$ to obtain p new subgroups. Eventually, all the H_i will be listed, and so the number of them is $b = 1 + mp$ for some m. Hence, $b \equiv 1 \bmod p$.

Let $H_1, \ldots, H_{r_s}$ be all the subgroups of G of order p^s, and let $K_1, \ldots, K_{r_{s+1}}$ be all the subgroups of order p^{s+1}. For each H_i, let there be a_i subgroups of order p^{s+1} containing H_i; for each K_j, let there be b_j subgroups of order p^s contained in K_j.

Now

$$\sum_{i=1}^{r_s} a_i = \sum_{j=1}^{r_{s+1}} b_j,$$

for either sum counts each K_j with multiplicity the number of H's it contains.

Since $a_i \equiv 1 \bmod p$ for all i and $b_j \equiv 1 \bmod p$ for all j, it follows that $r_s \equiv r_{s+1} \bmod p$. Lemma 4.7 now gives the result, for $r_1 \equiv 1 \bmod p$. ■

Each term in the class equation of a finite group G is a divisor of $|G|$, so that multiplying by $|G|^{-1}$ gives an equation of the form $1 = \sum_j (1/i_j)$ with each i_j a positive integer; moreover, $|G|$ is the largest i_j occurring in this expression.

Lemma 4.9 (Landau, 1903). *Given $n > 0$ and $q \in \mathbb{Q}$, there are only finitely many n-tuples $(i_1, \ldots, i_n)$ of positive integers such that $q = \sum_{j=1}^n (1/i_j)$.*

Proof. We do an induction on n; the base step $n = 1$ is obviously true. Since there are only $n!$ permutations of n objects, it suffices to prove that there are only finitely many n-tuples $(i_1, \ldots, i_n)$ with $i_1 \leq i_2 \leq \cdots \leq i_n$ which satisfy the equation $q = \sum_{j=1}^n (1/i_j)$. For any such n-tuple, we have $i_1 \leq n/q$, for

$$q = 1/i_1 + \cdots + 1/i_n \leq 1/i_1 + \cdots + 1/i_1 = n/i_1.$$

But for each positive integer $k \leq n/q$, induction gives only finitely many $(n-1)$-typles $(i_2, \ldots, i_n)$ of positive integers with $q - (1/k) = \sum_{j=2}^n (1/i_j)$. This completes the proof, for there are only finitely many such k. ■

Theorem 4.10. *For every $n \geq 1$, there are only finitely many finite groups having exactly n conjugacy classes.*

Proof. Assume that G is a finite group having exactly n conjugacy classes. If $|Z(G)| = m$, then the class equation is

$$|G| = |Z(G)| + \sum_{j=m+1}^{n} [G : C_G(x_j)].$$

If $i_j = |G|$ for $1 \leq j \leq m$ and $i_j = |G|/[G : C_G(x_j)] = |C_G(x_j)|$ for $m + 1 \leq j \leq n$, then $1 = \sum_{j=1}^n (1/i_j)$. By the lemma, there are only finitely many such n-tuples, and so there is a maximum value for all possible i_j's occurring therein, say, M. It follows that a finite group G having exactly n conjugacy classes has order at most M. But Exercise 1.41 shows that there are only finitely many (nonisomorphic) groups of any given order. ■

Exercises

4.1. Let $H \lhd G$. If both H and G/H are p-groups, then G is a p-group.

4.2. If $|G| = p^n$, where p is prime, and if $0 \leq k \leq n$, then G contains a normal subgroup of order p^k.

4.3. Let G be a finite p-group, and let H be a nontrivial normal subgroup of G. Prove that $H \cap Z(G) \neq 1$.

4.4. Let G be a finite p-group; show that if H is a normal subgroup of G having order p, then $H \leq Z(G)$.

4.5. Let H be a proper subgroup of a finite p-group G. If $|H| = p^s$, then there is a subgroup of order p^{s+1} containing H.

4.6. Let p be a prime, let G be a finite group whose order is divisible by p, and assume that $P \leq G$ is a *maximal p-subgroup* (if $Q \leq G$ is a p-subgroup and $P \leq Q$, then $P = Q$).
(i) Every conjugate of P is also a maximal p-subgroup.
(ii) If P is the only maximal p-subgroup of G, then $P \lhd G$.

4.7. If p is a prime and G is a nonabelian group of order p^3, then $|Z(G)| = p$, $G/Z(G) \cong \mathbb{Z}_p \times \mathbb{Z}_p$, and $Z(G) = G'$, the commutator subgroup.

4.8. Prove that the number of normal subgroups of order p^s of a finite p-group G is congruent to 1 mod p.

The Sylow Theorems

The main results of this section are fundamental for understanding the structure of a finite group. If p^e is the largest power of p dividing $|G|$, then we shall see that G contains a subgroup of order p^e. Any two such subgroups are isomorphic (indeed, they are conjugate), and the number of them can be counted within a congruence.

Definition. If p is a prime, then a ***Sylow p-subgroup*** P of a group G is a maximal p-subgroup.

Observe that every p-subgroup of G is contained in some Sylow p-subgroup; this is obvious when G is finite, and it follows from Zorn's lemma when G is infinite. (Although we have allowed infinite groups G, the most important groups in this context are finite.)

Lemma 4.11. *Let P be a Sylow p-subgroup of a finite group G.*

(i) *$|N_G(P)/P|$ is prime to p.*
(ii) *If $a \in G$ has order some power of p and $aPa^{-1} = P$, then $a \in P$.*

Proof. (i) If p divides $|N_G(P)/P|$, then Cauchy's theorem (Theorem 4.2) shows that $N_G(P)/P$ contains some element Pa of order p; hence, $S^* = \langle Pa \rangle$ has order p. By the Correspondence Theorem, there is a subgroup $S \leq N_G(P) \leq G$ containing P with $S/P \cong S^*$. Since both P and S^* are p-groups, Exercise 4.1 shows that S is a p-group, contradicting the maximality of P.

(ii) Replacing a by a suitable power of a if necessary, we may assume that a has order p. Since a normalizes P, we have $a \in N_G(P)$. If $a \notin P$, then the coset $Pa \in N_G(P)/P$ has order p, and this contradicts (i). ■

The observation suggesting the coming proof is that every conjugate of a Sylow p-subgroup is itself a Sylow p-subgroup.

Theorem 4.12 (Sylow, 1872).

(i) *If P is a Sylow p-subgroup of a finite group G, then all Sylow p-subgroups of G are conjugate to P.*
(ii) *If there are r Sylow p-subgroups, then r is a divisor of $|G|$ and $r \equiv 1 \bmod p$.*[1]

Proof. Let $X = \{P_1, \ldots, P_r\}$ be the family of all the Sylow p-subgroups of G, where we have denoted P by P_1. In Theorem 3.17, we saw that G acts on X by conjugation: there is a homomorphism $\psi: G \to S_X$ sending $a \mapsto \psi_a$, where $\psi_a(P_i) = aP_ia^{-1}$. Let Q be a Sylow p-subgroup of G. Restricting ψ to Q shows that Q acts on X; by Corollary 3.21, every orbit of X under this action has size dividing $|Q|$; that is, every orbit has size some power of p. What does it mean to say that one of these orbits has size 1? There would be an i with $\psi_a(P_i) = P_i$ for all $a \in Q$; that is, $aP_ia^{-1} = P_i$ for all $a \in Q$. By Lemma 4.11(ii), if $a \in Q$, then $a \in P_i$; that is, $Q \leq P_i$; since Q is a Sylow p-subgroup, $Q = P_i$. If $Q = P = P_1$, we conclude that every P-orbit of X has size an "honest" power of p save $\{P_1\}$ which has size 1. Therefore, $|X| = r \equiv 1 \bmod p$.

Suppose there were a Sylow p-subgroup Q that is not a conjugate of P; that is, $Q \notin X$. If $\{P_i\}$ is a Q-orbit of size 1, then we have seen that $Q = P_i$, contradicting $Q \notin X$. Thus, every Q-orbit of X has size an honest power of p, and so p divides $|X|$; that is, $r \equiv 0 \bmod p$. The previous congruence is contradicted, and so no such subgroup Q exists. Therefore, every Sylow p-subgroup Q is conjugate to P.

Finally, the number r of conjugates of P is the index of its normalizer, and so it is a divisor of $|G|$. ■

For example, $|S_4| = 24 = 2^3 \cdot 3$, and so a Sylow 2-subgroup of S_4 has order 8. It is easily seen that $D_8 \leq S_4$ if one recalls the symmetries of a square. The Sylow theorem says that all the subgroups of S_4 of order 8 are conjugate (hence isomorphic) and that the number r of them is an odd divisor of 24. Since $r \neq 1$, there are 3 such subgroups.

We have seen, in Exercise 4.6, that if a finite group G has only one Sylow p-subgroup P, for some prime p, then $P \lhd G$. We can now see that the converse is also true.

Corollary 4.13. *A finite group G has a unique Sylow p-subgroup P, for some prime p, if and only if $P \lhd G$.*

Proof. If G has only one Sylow p-subgroup P, then $P \lhd G$, for any conjugate of P is also a Sylow p-subgroup. Conversely, if P is a normal Sylow p-subgroup of G, then it is unique, for all Sylow p-subgroups of G are conjugate. ■

[1] Since all Sylow p-subgroups have the same order, this congruence also follows from Theorem 4.8.

Theorem 4.14. *If G is a finite group of order $p^e m$, where $(p, m) = 1$, then every Sylow p-subgroup P of G has order p^e.*

Proof. We claim that $[G:P]$ is prime to p. Now $[G:P] = [G:N][N:P]$, where $N = N_G(P)$, and so it suffices to prove that each of the factors is prime to p. But $[G:N] = r$, the number of conjugates of p, so that $[G:N] \equiv 1 \bmod p$, while $[N:P] = |N/P|$ is prime to p, by Lemma 4.11(i).

By Lagrange's theorem, $|P| = p^k$, where $k \leq e$, and so $[G:P] = |G|/|P| = p^{e-k}m$. Since $[G:P]$ is prime to p, however, we must have $k = e$. ■

Corollary 4.15. *Let G be a finite group and let p be a prime. If p^k divides $|G|$, then G contains a subgroup of order p^k.*

Proof. If P is a Sylow p-subgroup of G, then p^k divides $|P|$, and the result now follows from Exercise 4.2. ■

We have now seen how much of the converse of Lagrange's theorem (if m divides $|G|$, then G has a subgroup of order m) can be salvaged. If m is a prime power, then G contains a subgroup of order m; if m has two distinct prime factors, however, we have already seen an example (Theorem 3.7) in which G has no subgroup of order m (namely, $m = 6$ and $G = A_4$, a group of order 12).

Since, for each prime p, any two Sylow p-subgroups of a finite group G are isomorphic (they are even conjugate), we may list the Sylow subgroups of G, one for each prime. It is plain that isomorphic groups G give the same list, but the converse is false. For example, both S_3 and $\mathbb{Z}_6$ give the same list.

Here is another proof of Sylow's theorem, due to Wielandt, that does not use Cauchy's theorem (and so it gives a third proof of Cauchy's theorem).

Lemma 4.16. *If p is a prime not dividing an integer m, then for all $n \geq 1$, the binomial coefficient $\binom{p^n m}{p^n}$ is not divisible by p.*

Proof. Write the binomial coefficient as follows:

$$\frac{p^n m(p^n m - 1)\cdots(p^n m - i)\cdots(p^n m - p^n + 1)}{p^n(p^n - 1)\cdots(p^n - i)\cdots(p^n - p^n + 1)}.$$

Since p is prime, each factor equal to p of the numerator (or of the denominator) arises from a factor of $p^n m - i$ (or of $p^n - i$). If $i = 0$, then the multiplicity of p in $p^n m$ and in p^n are the same because $p \nmid m$. If $1 \leq i \leq p^n$, then $i = p^k j$, where $0 \leq k < n$ and $p \nmid j$. Now p^k is the highest power of p dividing $p^n - i$, for $p^n - i = p^n - p^k j = p^k(p^{n-k} - j)$ and $p \nmid p^{n-k} - j$ (because $n - k > 0$). A similar argument shows that the highest power of p dividing $p^n m - i$ is also p^k. Therefore, every factor of p upstairs is canceled by a factor of p downstairs, and hence the binomial coefficient has no factor equal to p. ■

Theorem 4.17 (Wielandt's Proof). *If G is a finite group of order $p^n m$, where $(p, m) = 1$, then G has a subgroup of order p^n.*

Proof. If X is the family of all subsets of G of cardinal p^n, then $|X|$ is the binomial coefficient in the lemma, and so $p \nmid |X|$. Let G act on X by left translation: if B is a subset of G with p^n elements, then for each $g \in G$, define

$$gB = \{gb: b \in B\}.$$

Now p cannot divide the size of every orbit of X lest $p \mid |X|$; therefore, there is some $B \in X$ with $|\mathcal{O}(B)|$ not divisible by p, where $\mathcal{O}(B)$ is the orbit of B. If G_B is the stabilizer of B, then $|G|/|G_B| = [G : G_B] = |\mathcal{O}(B)|$ is prime to p. Hence, $|G_B| = p^n m' \geq p^n$ (for some m' dividing m). On the other hand, if $b_0 \in B$ and $g \in G_B$, then $gb_0 \in gB = B$ (definition of stabilizer); moreover, if g and h are distinct elements of G_B, then gb_0 and hb_0 are distinct elements of B. Therefore, $|G_B| \leq |B| = p^n$, and so $|G_B| = p^n$. ∎

The next technical result is useful.

Theorem 4.18 (Frattini Argument). *Let K be a normal subgroup of a finite group G. If P is a Sylow p-subgroup of K (for some prime p), then*

$$G = KN_G(P).$$

Proof. If $g \in G$, then $gPg^{-1} \leq gKg^{-1} = K$, because $K \lhd G$. If follows that gPg^{-1} is a Sylow p-subgroup of K, and so there exists $k \in K$ with $kPk^{-1} = gPg^{-1}$. Hence, $P = (k^{-1}g)P(k^{-1}g)^{-1}$, so that $k^{-1}g \in N_G(P)$. The required factorization is thus $g = k(k^{-1}g)$. ∎

Exercises

4.9. (i) Let X be a finite G-set, and let $H \leq G$ act transitively on X. Then $G = HG_x$ for each $x \in X$.
(ii) Show that the Frattini argument follows from (i).

4.10. Let $\{P_i: i \in I\}$ be a set of Sylow subgroups of a finite group G, one for each prime divisor of $|G|$. Show that G is generated by $\bigcup P_i$.

4.11. Let $P \leq G$ be a Sylow subgroup. If $N_G(P) \leq H \leq G$, then H is equal to its own normalizer; that is $H = N_G(H)$.

4.12. If a finite group G has a unique Sylow p-subgroup for each prime divisor p of $|G|$, then G is the direct product of its Sylow subgroups.

4.13. (i) Let G be a finite group and let $P \leq G$ be a Sylow subgroup. If $H \lhd G$, then $H \cap P$ is a Sylow subgroup of H and HP/H is a Sylow subgroup of G/H. (*Hint.* Compare orders.)
(ii) Let G be a finite group and let $P \leq G$ be a Sylow subgroup. Give an example of a (necessarily non-normal) subgroup H of G with $H \cap P$ not a Sylow subgroup of H.

4.14. Prove that a Sylow 2-subgroup of A_5 has exactly 5 conjugates.

4.15. (i) Prove that a Sylow 2-subgroup of S_5 is isomorphic to D_8.
(ii) Prove that $D_8 \times \mathbb{Z}_2$ is isomorphic to a Sylow 2-subgroup of S_6.

4.16. If Q is a normal p-subgroup of a finite group G, then $Q \leq P$ for every Sylow p-subgroup P.

Definition. An $n \times n$ matrix A over a commutative ring R is ***unitriangular*** if it has 0's below the diagonal and 1's on the diagonal. The set of all unitriangular 3×3 matrices over $\mathbb{Z}_p$ is denoted by $\mathrm{UT}(3, \mathbb{Z}_p)$.

4.17. (i) Show that $|\mathrm{GL}(3, \mathbb{Z}_p)| = (p^3 - 1)(p^3 - p)(p^3 - p^2)$.
(ii) If p is a prime, then $\mathrm{UT}(3, \mathbb{Z}_p)$ is a Sylow p-subgroup of $\mathrm{GL}(3, \mathbb{Z}_p)$.
(iii) Show that the center of $\mathrm{UT}(3, \mathbb{Z}_p)$ consists of all matrices of the form

$$\begin{bmatrix} 1 & 0 & x \\ 0 & 1 & 0 \\ 0 & 0 & 1 \end{bmatrix}.$$

4.18. Show that a finite group G can have three Sylow p-subgroups A, B, and C such that $A \cap B = 1$ and $A \cap C \neq 1$. (*Hint.* Take $G = S_3 \times S_3$.)

4.19. Let $|G| = p^n m$, where $p \nmid m$. If $s \leq n$ and r_s is the number of subgroups of G of order p^s, then $r_s \equiv 1 \bmod p$.

4.20. (i) Let $\sigma = (1\ 2\ 3\ 4\ 5)$, let $P = \langle \sigma \rangle \leq S_5$, and let $N = N_{S_5}(P)$. Show that $|N| = 20$ and $N = \langle \sigma, \alpha \rangle$, where $\alpha = (2\ 3\ 5\ 4)$.
(ii) If A is the group (under composition),

$$A = \{\varphi\colon \mathbb{Z}_5 \to \mathbb{Z}_5\colon \varphi(x) = \alpha x + \beta,\ \alpha, \beta \in \mathbb{Z}_5, \alpha \neq 0\},$$

then $N_{S_5}(P) \cong A$. (*Hint.* Show that $A = \langle s, t \rangle$, where $s\colon x \mapsto x + 1$, and $t\colon x \mapsto 2x$.)

Groups of Small Order

We illustrate the power of the Sylow theorems by classifying the groups of small order.

Theorem 4.19. *If p is a prime, then every group G of order $2p$ is either cyclic or dihedral.*

Proof. If $p = 2$, then $|G| = 4$, and the result is Exercise 2.12. If p is an odd prime, then Cauchy's theorem shows that G contains an element s of order p and an element t of order 2. If $H = \langle s \rangle$, then H has index 2 in G, and so $H \lhd G$. Therefore, $tst = s^i$ for some i. Now $s = t^2 s t^2 = t(tst)t = ts^i t = s^{i^2}$; hence, $i^2 \equiv 1 \bmod p$ and, because p is prime, Euclid's lemma gives $i \equiv \pm 1 \bmod p$.

Thus, either $tst = s$ or $tst = s^{-1}$. In the first case, s and t commute, G is abelian, and Exercises 4.12 and 2.62(ii) give $G \cong \mathbb{Z}_p \times \mathbb{Z}_2 \cong \mathbb{Z}_{2p}$; in the second case, $G \cong D_{2p}$. ■

We now generalize this result by replacing 2 by q.

Theorem 4.20. *Let $|G| = pq$, where $p > q$ are primes. Then either G is cyclic or $G = \langle a, b \rangle$, where*

$$b^p = 1, \qquad a^q = 1, \qquad aba^{-1} = b^m,$$

and $m^q \equiv 1 \bmod p$ but $m \not\equiv 1 \bmod p$. If $q \nmid p - 1$, then the second case cannot occur.

Proof. By Cauchy's theorem, G contains an element b of order p; let $S = \langle b \rangle$. Since S has order p, it has index q. It follows from Exercise 3.33 that $S \lhd G$.

Cauchy's theorem shows that G contains an element a of order q; let $T = \langle a \rangle$. Now T is a Sylow q-subgroup of G, so that the number c of its conjugates is $1 + kq$ for some $k \geq 0$. As above, either $c = 1$ or $c = p$. If $c = 1$, then $T \lhd G$ and $G \cong S \times T$ (by Exercise 4.12), and so $G \cong \mathbb{Z}_p \times \mathbb{Z}_q \cong \mathbb{Z}_{pq}$, by Exercise 2.62(ii). In case $c = kq + 1 = p$, then $q | p - 1$, and T is not a normal subgroup of G. Since $S \lhd G$, $aba^{-1} = b^m$ for some m; furthermore, we may assume that $m \not\equiv 1 \bmod p$ lest we return to the abelian case. The reader may prove, by induction on j, that $a^j b a^{-j} = b^{m^j}$. In particular, if $j = q$, then $m^q \equiv 1 \bmod p$. ■

Corollary 4.21. *If $p > q$ are primes, then every group G of order pq contains a normal subgroup of order p. Moreover, if q does not divide $p - 1$, then G must be cyclic.*

For example, the composite numbers $n \leq 100$ for which every group of order n is cyclic are:

$$15, 33, 35, 51, 65, 69, 77, 85, 87, 91, 95.$$

Definition. The ***quaternions*** is a group $\mathbf{Q} = \langle a, b \rangle$ of order 8 with $a^4 = 1$, $b^2 = a^2$, and $bab^{-1} = a^{-1}$.

We continue describing groups of small order.

Theorem 4.22. $\mathbf{Q}$ *and D_8 are the only nonabelian groups of order 8.*

Proof. A nonabelian group G of order 8 has no element of order 8 (lest it be cyclic), and not every nonidentity element has order 2 (Exercise 1.26); thus, G has an element a of order 4. Now $\langle a \rangle \lhd G$, for it has index 2, and $G/\langle a \rangle \cong \mathbb{Z}_2$. If $b \in G$ and $b \notin \langle a \rangle$, then $b^2 \in \langle a \rangle$ (Exercise 2.16). If $b^2 = a$ or $b^2 = a^3$, then

b has order 8, a contradiction; therefore, either

$$b^2 = a^2 \qquad \text{or} \qquad b^2 = 1.$$

Furthermore, $bab^{-1} \in \langle a \rangle$, because $\langle a \rangle$ is normal, so that

$$bab^{-1} = a \qquad \text{or} \qquad bab^{-1} = a^3$$

(these are the only possibilities because a and bab^{-1} have the same order). The first case is ruled out, for $G = \langle a, b \rangle$ and G is abelian if a and b commute. The following case remain:

(i) $a^4 = 1$, $b^2 = a^2$, and $bab^{-1} = a^3$; and
(ii) $a^4 = 1$, $b^2 = 1$, and $bab^{-1} = a^3$.

Since $a^3 = a^{-1}$, (i) describes $\mathbf{Q}$ and (ii) describes D_8. ■

Lemma 4.23. *If G has order 12 and $G \not\cong A_4$, then G contains an element of order 6; moreover, G has a normal Sylow 3-subgroup, hence has exactly two elements of order 3.*

Proof. If P is a Sylow 3-subgroup of G, then $|P| = 3$ and so $P = \langle b \rangle$ for some b of order 3. Since $[G:P] = 4$, there is a homomorphism $\psi: G \to S_4$ whose kernel K is a subgroup of P; as $|P| = 3$, either $K = 1$ or $K = P$. If $K = 1$, then ψ is an injection and G is isomorphic to a subgroup of S_4 of order 12; by Exercise 3.15, $G \cong A_4$. Therefore, $K = P$ and so $P \lhd G$; it follows that P is the unique Sylow 3-subgroup, and so the only elements in G of order 3 are b and b^2. Now $[G : C_G(b)]$ is the number of conjugates of b. Since every conjugate of b has order 3, $[G : C_G(b)] \leq 2$ and $|C_G(b)| = 6$ or 12; in either case, $C_G(b)$ contains an element a of order 2. But a commutes with b, and so ab has order 6. ■

We now define a new group.

Definition. T is a group of order 12 which is generated by two elements a and b such that $a^6 = 1$ and $b^2 = a^3 = (ab)^2$.

It is not obvious that such a group T exists; we shall construct such a group later, in Example 7.14.

Theorem 4.24. *Every nonabelian group G of order 12 is isomorphic to either A_4, D_{12}, or T.*

Remark. Exercise 3.50 shows that $S_3 \times \mathbb{Z}_2 \cong D_{12}$.

Proof. It suffices to show that if G is a nonabelian group of order 12 which is not isomorphic to A_4, then either $G \cong D_{12}$ or $G \cong T$.

Let K be a Sylow 3-subgroup of G (so that $K = \langle k \rangle$ is cyclic of order 3 and, by Lemma 4.23, $K \lhd G$), and let P be a Sylow 2-subgroup of G (so that

P has order 4). Notice that $G = KP$, for P is a maximal subgroup because it has prime index. Now either $P \cong \mathbf{V}$ or $P \cong \mathbb{Z}_4$.

In the first case, $P = \{1, x, y, z\}$, where x, y, and z are involutions. Not every element of P commutes with k lest G be abelian; therefore, there exists an involution in P, say x, with $xkx \neq k$. But xkx is a conjugate of k, hence has order 3; as G has only two elements of order 3, $xkx = k^{-1}$, and $\langle x, k\rangle \cong D_6$ ($\cong S_3$). We claim that either y or z commutes with k. If y does not commute with k, then, as above, $yky = k^{-1}$, so that $xkx = yky$ and $z = xy$ commutes with k. If $a = zk$, then $G = \langle a, x\rangle$, a has order 6, and $xax = xzkx = zxkx = zk^{-1} = a^{-1}$, so that $G \cong D_{12}$.

In the second case, $P = \langle x\rangle \cong \mathbb{Z}_4$. Now x and k do not commute, because $G = \langle x, k\rangle$ is not abelian, and so $xkx^{-1} = k^{-1}$. But x^2 does commute with k, for $x^2kx^{-2} = x(xkx^{-1})x^{-1} = xk^{-1}x^{-1} = k$; hence, $a = x^2k$ has order 6. Since x^2 and k commute, $a^3 = (x^2k)^3 = x^6k^3 = x^2$. Finally, $(ax)^2 = axax = kx^3kx^3 = (kx^{-1})kx^{-1} = (x^{-1}k^{-1})kx^{-1}$ (because $xkx^{-1} = k^{-1}$) $= x^{-2} = x^2$. We have shown that $G \cong T$ in this case. ■

One can prove this last result in the style of the proof of Theorem 4.22, beginning by choosing an element of order 6. The reader may see that the proof just given is more efficient.

Notice that T is almost a direct product: it contains subgroups $P = \langle x\rangle$ and $K = \langle k\rangle$ with $PK = G$, $P \cap K = 1$, but only P is normal.

If $n \geq 1$, define $\mathfrak{G}(n)$ to be the number of nonisomorphic groups of order n. No one knows a formula for $\mathfrak{G}(n)$, although it has been computed for $n \leq 100$ and beyond. Here is the classification of all groups of order ≤ 15 followed by a table of values of $\mathfrak{G}(n)$ for small n. Of course, $\mathfrak{G}(n) = 1$ whenever n is prime.

Table of Groups of Small Order

Order n	$\mathfrak{G}(n)$	Groups[2]
4	2	$\mathbb{Z}_4$, $\mathbf{V}$
6	2	$\mathbb{Z}_6 \cong \mathbb{Z}_2 \times \mathbb{Z}_3$, S_3
8	5	$\mathbb{Z}_8$, $\mathbb{Z}_4 \times \mathbb{Z}_2$, $\mathbb{Z}_2 \times \mathbb{Z}_2 \times \mathbb{Z}_2$, D_8, $\mathbf{Q}$
9	2	$\mathbb{Z}_9$, $\mathbb{Z}_3 \times \mathbb{Z}_3$
10	2	$\mathbb{Z}_{10}$, D_{10}
12	5	$\mathbb{Z}_{12}$, $\mathbb{Z}_6 \times \mathbb{Z}_2$, A_4, D_{12}, T
14	2	$\mathbb{Z}_{14}$, D_{14}
15	1	$\mathbb{Z}_{15}$

Some Other Values of $\mathfrak{G}(n)$

n	16	18	20	21	22	24	25	26	27	28	30	32	64
$\mathfrak{G}(n)$	14	5	5	2	2	15	2	2	5	4	4	51	267

[2] The abelian groups in the table will be discussed in Chapter 6.

R. James, M.F. Newman, and E.A. O'Brien (1990) have shown, using computers, that $\mathfrak{G}(128) = 2328$; E.A. O'Brien (1991) has shown that $\mathfrak{G}(256) = 56{,}092$, and that $\mathfrak{G}(512) > 8{,}000{,}000$ (counting only those groups of "class 2"). G. Higman (1960) and C. Sims (1965) showed that $\mathfrak{G}(p^n)$ is about $p^{2n^3/27}$. If the prime factorization of an integer n is $p_1^{e_1}p_2^{e_2}\ldots p_n^{e_n}$, define $\mu = \mu(n)$ to be the largest e_i. Using the fact that there are at most two nonisomorphic simple groups of the same finite order (which follows from the classification of the finite simple groups), A. McIver and P.M. Neumann (1987) have shown that $\mathfrak{G}(n) \leq n^{\mu^2+\mu+2}$.

Here is another type of application of the Sylow theorems.

EXAMPLE 4.1. There is no simple group of order 30.

Such a group would have r Sylow 5-subgroups, where $r \equiv 1 \bmod 5$ and $r|30$. Now $r \neq 1$, lest G have a normal subgroup, and so $r = 6$. Aside from the identity, this accounts for 24 elements (for distinct subgroups of order 5 intersect in 1). Similarly, there must be 10 Sylow 3-subgroups, accounting for 20 elements, and we have exceeded 30.

EXAMPLE 4.2. There is no simple group of order 36.

A Sylow 3-subgroup P of such a group G has 4 conjugates, and so $[G:P] = 4$. Representing G on the cosets of P gives a homomorphism $\psi: G \to S_4$ with $\ker \psi \leq P$; since G is simple, ψ must be an injection and G is imbedded in S_4. This contradicts $36 > 24$, and so no such group G can exist.

EXERCISES

4.21. A_5 is a group of order 60 containing no subgroups of order 15 or of order 30.

4.22. If G is the subgroup of $\mathrm{GL}(2, \mathbb{C})$ generated by

$$A = \begin{bmatrix} 0 & i \\ i & 0 \end{bmatrix} \quad \text{and} \quad B = \begin{bmatrix} 0 & 1 \\ -1 & 0 \end{bmatrix},$$

then $G \cong \mathbf{Q}$. (See Exercise 2.24.)

4.23. The division ring $\mathbb{H}$ of ***real quaternions*** is a four-dimensional vector space over $\mathbb{R}$ having a basis $\{1, i, j, k\}$ with multiplication satisfying

$$i^2 = j^2 = k^2 = -1,$$

$$ij = k, \quad jk = i, \quad ki = j, \quad ji = -k, \quad kj = -i, \quad ik = -j$$

(these equations eight determine the multiplication on all of $\mathbb{H}$.) Show that the eight elements $\{\pm 1, \pm i, \pm j, \pm k\}$ form a multiplicative group isomorphic to $\mathbf{Q}$.

4.24. Show that $\mathbf{Q}$ has a unique element of order 2, and that this element generates $Z(\mathbf{Q})$.

4.25. Prove that $D_8 \not\cong \mathbf{Q}$.

4.26. Prove that $\mathbf{Q}$ contains no subgroup isomorphic to $\mathbf{Q}/Z(\mathbf{Q})$.

4.27. Prove that every subgroup of $\mathbf{Q}$ is normal.

4.28. Let $G = \mathbf{Q} \times A \times B$, where A is a (necessarily abelian) group of exponent 2 and B is an abelian group in which every element has odd order. Prove that every subgroup of G is normal. (Dedekind proved the converse: if G is a finite group in which every subgroup is normal, then $G = \mathbf{Q} \times A \times B$ as above.) Groups with this property are called ***hamiltonian***, after W.R. Hamilton who discovered $\mathbf{Q}$.

4.29. Let SL(2, 5) denote the multiplicative group of all 2×2 matrices over $\mathbb{Z}_5$ which have determinant 1.
 (i) Show that $|\mathrm{SL}(2, 5)| = 120$.
 (ii) Show that $\mathbf{Q}$ is isomorphic to a Sylow 2-subgroup of SL(2, 5). (*Hint*. Show that SL(2, 5) has a unique involution.)
 (iii) Show that a Sylow 2-subgroup of S_5 is D_8, and conclude that $\mathrm{SL}(2, 5) \not\cong S_5$.
 (iv) Show that A_5 cannot be imbedded in SL(2, 5).

4.30. For every divisor d of 24, show that there is a subgroup of S_4 of order d. Moreover, if $d \neq 4$, then any two subgroups of order d are isomorphic.

4.31. Exhibit all the subgroups of S_4; aside from S_4 and 1, there are 26 of them.

4.32. (i) Let p be a prime. By Exercise 4.17, $P = \mathrm{UT}(3, \mathbb{Z}_p)$ is a Sylow p-subgroup of $\mathrm{GL}(3, \mathbb{Z}_p)$. If p is odd, prove that P is a nonabelian group of order p^3 of exponent p. (Compare Exercise 1.26.) If $p = 2$, show that $\mathrm{UT}(3, \mathbb{Z}_2) \cong \mathbf{Q}$.
 (ii) Let p be an odd prime. Prove that there are at most two nonabelian groups of order p^3. One is given in part (i) and has exponent p; the other has generators a and b satisfying the relations $a^{p^2} = 1$, $b^p = 1$, and $bab^{-1} = a^{1+p}$. (This group will be shown to exist in Example 7.16.)

4.33. Give an example of two nonisomorphic groups G and H such that, for each positive integer d, the number of elements in G of order d is equal to the number of elements in H of order d.

4.34. If G is a group of order 8 having only one involution, then either $G \cong \mathbb{Z}_8$ or $G \cong \mathbf{Q}$.

4.35. If p and q are primes, then there is no simple group of order p^2q.

4.36. Prove that there is no nonabelian simple group of order less than 60.

4.37. Prove that any simple group G of order 60 is isomorphic to A_5. (*Hint*. If P and Q are distinct Sylow 2-subgroups having a nontrivial element x in their intersection, then $C_G(x)$ has index 5; otherwise, every two Sylow 2-subgroups intersect trivially and $N_G(P)$ has index 5.)

4.38. In Corollary 7.55, we shall prove that a group of squarefree order cannot be simple. Use this result to prove that if $|G| = p_1 \dots p_s$, where $p_1 < p_2 < \cdots < p_s$ are primes, then G contains a normal Sylow p_s-subgroup.

Definition. If $n \geq 3$, a ***generalized quaternion group*** (or *dicyclic group*) is a group $\mathbf{Q}_n$ of order 2^n generated by elements a and b such that

$$a^{2^{n-1}} = 1, \quad bab^{-1} = a^{-1}, \qquad \text{and} \qquad b^2 = a^{2^{n-2}}.$$

4.39. Any two groups of order 2^n generated by a pair of elements a and b for which $a^{2^{n-2}} = b^2$ and $aba = b$ are isomorphic.

4.40. If G is a group of order 2^n which is generated by elements a and b such that

$$a^{2^{n-2}} = b^2 = (ab)^2,$$

then $G \cong \mathbf{Q}_n$.

4.41. Prove that $\mathbf{Q}_n$ has a unique involution z, and that $Z(\mathbf{Q}_n) = \langle z \rangle$.

4.42. Prove that $\mathbf{Q}_n/Z(\mathbf{Q}_n) \cong D_{2^{n-1}}$.

4.43. Let $G \leq \mathrm{GL}(2, \mathbb{C})$ be generated by

$$A = \begin{bmatrix} 0 & \omega \\ \omega & 0 \end{bmatrix} \quad \text{and} \quad B = \begin{bmatrix} 0 & 1 \\ -1 & 0 \end{bmatrix},$$

where ω is a primitive 2^{n-1}th root of unity for $n \geq 3$. Prove that $G \cong \mathbf{Q}_n$. Conclude that $\mathbf{Q}_n$ exists.

CHAPTER 5

Normal Series

We begin this chapter with a brief history of the study of roots of polynomials. Mathematicians of the Middle Ages, and probably those in Babylonia, knew the ***quadratic formula*** giving the roots of a quadratic polynomial $f(X) = X^2 + bX + c$. Setting $X = x - \frac{1}{2}b$ transforms $f(X)$ into a polynomial $g(x)$ with no x term:

$$g(x) = x^2 + c - \tfrac{1}{4}b^2.$$

Note that a number α is a root of $g(x)$ if and only if $\alpha - \frac{1}{2}b$ is a root of $f(X)$. The roots of $g(x)$ are $\pm\frac{1}{2}\sqrt{b^2 - 4c}$, and so the roots of $f(X)$ are $\frac{1}{2}(-b \pm \sqrt{b^2 - 4c})$.

Here is a derivation of the cubic formula[1] (due to Scipione del Ferro,

[1] Negative coefficients (and negative roots) were not accepted by mathematicians of the early 1500s. There are several types of cubics if one allows only positive coefficients. About 1515, Scipione del Ferro solved some instances of $x^3 + px = q$, but he kept his solution secret. In 1535, Tartaglia (Niccolò Fontana) was challenged by one of Scipione's students, and he rediscovered the solution of $x^3 + px = q$ as well as the solution of $x^3 = px + q$. Eventually Tartaglia told his solutions to Cardano, and Cardano completed the remaining cases (this was no small task, for all of this occurred before the invention of notation for variables, exponents, $+$, $-$, $\times$, $/$, $\sqrt{\ }$, and $=$). A more complete account can be found in J.-P. Tignol, *Galois' Theory of Equations.*

We should not underestimate the importance of Cardano's formula. First of all, the physicist R.P. Feynmann suggested that its arising at the beginning of the Renaissance and Reformation must have contributed to the development of modern science. After all, it is a genuine example of a solution unknown to the Greeks, and one component contributing to the Dark Ages was the belief that contemporary man was less able than his classical Greek and Roman ancestors. Second, it forced mathematicians to take the complex numbers seriously. While complex roots of quadratics were just ignored, the *casus irreducibilis*, arising from one of Cardano's cases, describes real roots of cubics with imaginary numbers (see the next footnote). Complex numbers could no longer be dismissed, and thus the cubic formula had great impact on the development of mathematics.

Tartaglia (Niccolò Fontana), and Cardano). A cubic $f(X) = X^3 + aX^2 + bX + c$ can be transformed, by setting $X = x - \frac{1}{3}a$, into a polynomial $g(x)$ with no x^2 term:

$$g(x) = x^3 + qx + r,$$

and a number α is a root of $g(x)$ if and only if $\alpha - \frac{1}{3}a$ is a root of $f(X)$. If α is a root of $g(x)$, write $\alpha = \beta + \gamma$, where β and γ are to be found. Now

$$\begin{aligned} \alpha^3 = (\beta + \gamma)^3 &= \beta^3 + \gamma^3 + 3(\beta^2\gamma + \beta\gamma^2) \\ &= \beta^3 + \gamma^3 + 3\alpha\beta\gamma, \end{aligned}$$

and so evaluating $g(\alpha)$ gives

$$\beta^3 + \gamma^3 + (3\beta\gamma + q)\alpha + r = 0. \tag{1}$$

Impose the condition that $\beta\gamma = -q/3$ (forcing the middle term of (1) to vanish). Thus,

$$\beta^3 + \gamma^3 = -r.$$

We also know that

$$\beta^3\gamma^3 = -q^3/27,$$

and we proceed to find β^3 and γ^3. Substituting,

$$\beta^3 - q^3/27\beta^3 = -r,$$

and so the quadratic formula yields

$$\beta^3 = \tfrac{1}{2}[-r \pm (r^2 + 4q^3/27)^{1/2}].$$

Similarly,

$$\gamma^3 = \tfrac{1}{2}[-r \mp (r^2 + 4q^3/27)^{1/2}].$$

If $\omega = e^{2\pi i/3}$ is a primitive cube root of unity, there are now six cube roots available: β, $\omega\beta$, $\omega^2\beta$, γ, $\omega\gamma$, $\omega^2\gamma$; these may be paired to give product $-q/3$:

$$-q/3 = \beta\gamma = (\omega\beta)(\omega^2\gamma) = (\omega^2\beta)(\omega\gamma).$$

It follows that the roots of $g(x)$ are $\beta + \gamma$, $\omega\beta + \omega^2\gamma$, and $\omega^2\beta + \omega\gamma$; this is the ***cubic formula***.[2]

The ***quartic formula*** was discovered by Lodovici Ferrari, about 1545; we present the derivation of this formula due to Descartes in 1637. A quartic $f(X) = X^4 + aX^3 + bX^2 + cX + d$ can be transformed, by setting $X = x - \frac{1}{4}a$, into a polynomial $g(x)$ with no x^3 term:

$$g(x) = x^4 + qx^2 + rx + s;$$

moreover, a number α is a root of $g(x)$ if and only if $\alpha - \frac{1}{4}a$ is a root of $f(X)$.

[2] The roots of $f(x) = (x-1)(x-2)(x+3) = x^3 - 7x + 6$ are obviously 1, 2, and -3. However, the cubic formula gives

$$\beta + \gamma = \sqrt[3]{\tfrac{1}{2}(-6 + \sqrt{-400/27})} + \sqrt[3]{\tfrac{1}{2}(-6 - \sqrt{-400/27})}.$$

Factor $g(x)$ into quadratics:

$$x^4 + qx^2 + rx + s = (x^2 + kx + l)(x^2 - kx + m)$$

(the coefficient of x in the second factor must be $-k$ because there is no cubic term in $g(x)$). If k, l, and m can be found, then the roots of $g(x)$ can be found by the quadratic formula. Expanding the right side and equating coefficients of like terms gives

$$l + m - k^2 = q,$$

$$km - kl = r,$$

$$lm = s.$$

Rewrite the first two equations as

$$m + l = q + k^2,$$

$$m - l = r/k.$$

Adding and subtracting these equations gives

$$2l = k^2 + q - r/k,$$

$$2m = k^2 + q + r/k.$$

These two equations show that we are done if k can be found. But $(k^2 + q - r/k)(k^2 + q + r/k) = 4lm = 4s$ gives

$$k^6 + 2qk^4 + (q^2 - 4s)k^2 - r^2 = 0,$$

a cubic in k^2. The cubic formula allows one to solve for k^2, and it is now easy to determine l, m, and the roots of $g(x)$.

Some Galois Theory

Let us discuss the elements of Galois Theory, the cradle of group theory. We are going to assume in this exposition that every field F is a subfield of an algebraically closed field C. What this means in practice is this. If $f(x) \in F[x]$, the ring of all polynomials with coefficients in F, and if $f(x)$ has degree $n \geq 1$, then there are (not necessarily distinct) elements $\alpha_1, \ldots, \alpha_n$ in C (the ***roots*** of $f(x)$) and nonzero $a \in F$ so that

$$f(x) = a(x - \alpha_1)(x - \alpha_2)\ldots(x - \alpha_n)$$

in $C[x]$. The intersection of any family of subfields of a field is itself a subfield; define the ***smallest subfield*** of C containing a given subset X as the intersection of all those subfields of C containing X. For example, if $\alpha \in C$, the smallest subfield of C containing $X = F \cup \{\alpha\}$ is

$$F(\alpha) = \{f(\alpha)/g(\alpha): f(x), g(x) \in F[x], g(\alpha) \neq 0\};$$

$F(\alpha)$ is called the subfield obtained from F by ***adjoining*** α. Similarly, one can define $F(\alpha_1, \ldots, \alpha_n)$, the subfield obtained from F by ***adjoining*** $\alpha_1, \ldots, \alpha_n$. In particular, if $f(x) \in F[x]$ and $f(x) = (x - \alpha_1)(x - \alpha_2)\ldots(x - \alpha_n) \in C[x]$, then $F(\alpha_1, \ldots, \alpha_n)$, the subfield obtained from F by adjoining all the roots of $f(x)$, is called the ***splitting field*** of $f(x)$ ***over*** F. Notice that the splitting field of $f(x)$ does depend on F. For example, if $f(x) = x^2 + 1 \in \mathbb{Q}[x]$, then the splitting field of $f(x)$ over $\mathbb{Q}$ is $\mathbb{Q}(i)$; on the other hand, if we regard $f(x) \in \mathbb{R}[x]$, then its splitting field of $f(x)$ over $\mathbb{R}$ is $\mathbb{C}$.

It is now possible to give the precise definition we need.

Definition. Let $f(x) \in F[x]$ have splitting field E over F. Then $f(x)$ is ***solvable by radicals*** if there is a chain of subfields

$$F = K_0 \subset K_1 \subset \cdots \subset K_t,$$

in which $E \subset K_t$ and each K_{i+1} is obtained from K_i by adjoining a root of an element of K_i; that is, $K_{i+1} = K_i(\beta_{i+1})$, where $\beta_{i+1} \in K_{i+1}$ and some power of β_{i+1} lies in K_i.

When we say that there is a formula for the roots of $f(x)$, we really mean that $f(x)$ is solvable by radicals. Let us illustrate this by considering the quadratic, cubic, and quartic formulas.

If $f(x) = x^2 + bx + c$, set $F = \mathbb{Q}(b, c)$. If $\beta = \sqrt{b^2 - 4c}$, then $\beta^2 \in F$; define $K_1 = F(\beta)$, and note that K_1 is the splitting field of $f(x)$ over F.

If $f(x) = x^3 + qx + r$, set $F = \mathbb{Q}(q, r)$. Define $\beta_1 = \sqrt{r^2 + 4q^3/27}$, and define $K_1 = F(\beta_1)$; define $\beta_2 = \sqrt[3]{-r + \beta_1}$, and define $K_2 = K_1(\beta_2)$. Finally, set $K_3 = K_2(\omega)$, where ω is a cube root of unity. Notice that the cubic formula implies that K_3 contains the splitting field E of $f(x)$. On the other hand, E need not equal K_3; for example, if all the roots of $f(x)$ are real, then $E \subset \mathbb{R}$ but $K_3 \not\subset \mathbb{R}$.

If $f(x) = x^4 + qx^2 + rx + s$, set $F = \mathbb{Q}(q, r, s)$. Using the notation in our discussion of the quartic, there is a cubic polynomial having k^2 as a root. As above, there is a chain of fields $F = K_0 \subset K_1 \subset K_2 \subset K_3$ with $k^2 \in K_3$. Define $K_4 = K_3(k)$, $K_5 = K_4(\sqrt{\gamma})$, where $\gamma = k^2 - 4l$, and $K_6 = K_5(\sqrt{\delta})$, where $\delta = k^2 - 4m$. The discussion of the quartic formula shows that the splitting field of $f(x)$ is contained in K_6.

Conversely, it is plain that if $f(x)$ is solvable by radicals, then every root of $f(x)$ has some expression in the coefficients of $f(x)$ involving the field operations and extractions of roots.

Definition. If E and E' are fields, then a ***homomorphism*** is a function $\sigma: E \to E'$ such that, for all $\alpha, \beta \in E$,

$$\sigma(1) = 1,$$

$$\sigma(\alpha + \beta) = \sigma(\alpha) + \sigma(\beta),$$

$$\sigma(\alpha\beta) = \sigma(\alpha)\sigma(\beta).$$

If σ is a bijection, then σ is an ***isomorphism***; an isomorphism $\sigma: E \to E$ is called an ***automorphism***.

Lemma 5.1. *Let $f(x) \in F[x]$, let E be its splitting field over F, and let $\sigma: E \to E$ be an automorphism **fixing** F (i.e., $\sigma(a) = a$ for all $a \in F$). If $\alpha \in E$ is a root of $f(x)$, then $\sigma(\alpha)$ is also a root of $f(x)$.*

Proof. If $f(x) = \sum a_i x^i$, then $0 = \sigma(f(\alpha)) = \sigma(\sum a_i \alpha^i) = \sum \sigma(a_i)\sigma(\alpha)^i = \sum a_i \sigma(\alpha)^i$, and so $\sigma(\alpha)$ is a root of $f(x)$. ■

Lemma 5.2. *Let F be a subfield of K, let $\{\alpha_1, \ldots, \alpha_n\} \subset K$, and let $E = F(\alpha_1, \ldots, \alpha_n)$. If K' is a field containing F as a subfield, and if $\sigma: E \to K'$ is a homomorphism fixing F with $\sigma(\alpha_i) = \alpha_i$ for all i, then σ is the identity.*

Proof. The proof is by induction on $n \geq 1$. If $n = 1$, then E consists of all $g(\alpha_1)/h(\alpha_1)$, where $g(x), h(x) \in F[x]$ and $h(\alpha_1) \neq 0$; clearly σ fixes each such element. The inductive step is clear once one realizes that $F(\alpha_1, \ldots, \alpha_n) = F^*(\alpha_n)$, where $F^* = F(\alpha_1, \ldots, \alpha_{n-1})$. ■

It is easy to see that if F is a subfield of a field E, then the set of all automorphisms of E which fix F forms a group under composition.

Definition. If F is a subfield of E, then the ***Galois group***, denoted by $\text{Gal}(E/F)$, is the group under composition of all those automorphisms of E which fix F. If $f(x) \in F[x]$ and $E = F(\alpha_1, \ldots, \alpha_n)$ is the splitting field of $f(x)$ over F, then the ***Galois group*** of $f(x)$ is $\text{Gal}(E/F)$.

Theorem 5.3. *Let $f(x) \in F[x]$ and let $X = \{\alpha_1, \ldots, \alpha_n\}$ be the set of its distinct roots (in its splitting field $E = F(\alpha_1, \ldots, \alpha_n)$ over F). Then the function $\varphi: \text{Gal}(E/F) \to S_X \cong S_n$, given by $\varphi(\sigma) = \sigma|X$, is an imbedding; that is, φ is completely determined by its action on X.*

Proof. If $\sigma \in \text{Gal}(E/F)$, then Lemma 5.1 shows that $\sigma(X) \subset X$; $\sigma|X$ is a bijection because σ is an injection and X is finite. It is easy to see that φ is a homomorphism; it is an injection, by Lemma 5.2. ■

Not every permutation of the roots of a polynomial $f(x)$ need arise from some $\sigma \in \text{Gal}(E/F)$. For example, let $f(x) = (x^2 - 2)(x^2 - 3) \in \mathbb{Q}[x]$. Then $E = \mathbb{Q}(\sqrt{2}, \sqrt{3})$ and there is no $\sigma \in \text{Gal}(E/\mathbb{Q})$ with $\sigma(\sqrt{2}) = \sqrt{3}$.

Definition. If F is a subfield of a field E, then E is a vector space over F (if $a \in F$ and $\alpha \in E$, define scalar multiplication to be the given product $a\alpha$ of two elements of E). The ***degree*** of E over F, denoted by $[E:F]$, is the dimension of E.

EXERCISES

5.1. Show that every homomorphism of fields $\sigma: E \to K$ is an injection.

5.2. Let $p(x) \in F[x]$ be an irreducible polynomial of degree n. If α is a root of $p(x)$ (in a splitting field), prove that $\{1, \alpha, \alpha^2, \dots, \alpha^{n-1}\}$ is a basis of $F(\alpha)$ (viewed as a vector space over F). Conclude that $[F(\alpha) : F] = n$. (*Hint.* The rings $F[x]/(p(x))$ and $F(\alpha)$ are isomorphic via $g(x) + (p(x)) \mapsto g(\alpha)$.)

5.3. Let $F \subset E \subset K$ be fields, where $[K : E]$ and $[E : F]$ are finite. Prove that $[K : F] = [K : E][E : F]$. (*Hint.* If $\{\alpha_1, \dots, \alpha_n\}$ is a basis of E over F and if $\{\beta_1, \dots, \beta_m\}$ is a basis of K over E, then the set of mn elements of the form $\alpha_i\beta_j$ is a basis of K over F.)

5.4. Let E be a splitting field over F of some $f(x) \in F[x]$, and let K be a splitting field over E of some $g(x) \in E[x]$. If $\sigma \in \mathrm{Gal}(K/F)$, then $\sigma|E \in \mathrm{Gal}(E/F)$. (*Hint.* Lemmas 5.1 and 5.2.)

5.5. Let $f(x) = x^n - a \in F[x]$, let E be the splitting field of $f(x)$ over F, and let $\alpha \in E$ be an nth root of a. Prove that there are subfields

$$F = K_0 \subset K_1 \subset \cdots \subset K_t = F(\alpha)$$

with $K_{i+1} = K_i(\beta_{i+1})$, $\beta_{i+1}^{p(i)} \in K_i$, and $p(i)$ prime for all i. (*Hint.* If $n = st$, then $\alpha^n = (\alpha^s)^t$.)

Lemma 5.4. *Let $p(x) \in F[x]$ be irreducible, and let α and β be roots of $p(x)$ in a splitting field of $p(x)$ over F. Then there exists an isomorphism $\lambda^*: F(\alpha) \to F(\beta)$ which fixes F and with $\lambda^*(\alpha) = \beta$.*

Proof. By Exercise 5.2, every element of $F(\alpha)$ has a unique expression of the form $a_0 + a_1\alpha + \cdots + a_{n-1}\alpha^{n-1}$. Define λ^* by

$$\lambda^*(a_0 + a_1\alpha + \cdots + a_{n-1}\alpha^{n-1}) = a_0 + a_1\beta + \cdots + a_{n-1}\beta^{n-1}.$$

It is easy to see that λ^* is a field homomorphism; it is an isomorphism because its inverse can be constructed in the same manner. ■

Remark. There is a generalization of this lemma having the same proof. Let $\lambda: F \to F'$ be an isomorphism of fields, let $p(x) = a_0 + a_1x + \cdots + a_nx^n \in F[x]$ be an irreducible polynomial, and let $p'(x) = \lambda(a_0) + \lambda(a_1)x + \cdots + \lambda(a_n)x^n \in F'[x]$. Finally, let α be a root of $p(x)$ and let β be a root of $p'(x)$ (in appropriate splitting fields). Then there is an isomorphism $\lambda^*: F(\alpha) \to F'(\beta)$ with $\lambda^*(\alpha) = \beta$ and with $\lambda^*|F = \lambda$.

Lemma 5.5. *Let $f(x) \in F[x]$, and let E be its splitting field over F. If K is an "intermediate field," that is, $F \subset K \subset E$, and if $\lambda: K \to K$ is an automorphism fixing F, then there is an automorphism $\lambda^*: E \to E$ with $\lambda^*|K = \lambda$.*

Proof. The proof is by induction on $d = [E : F]$. If $d = 1$, then $E = K$, every root $\alpha_1, \dots, \alpha_n$ of $f(x)$ lies in K, and we may take $\lambda^* = \lambda$. If $d > 1$, then $E \neq K$

and there is some root α of $f(x)$ not lying in K (of course, $\alpha \in E$, by definition of splitting field). Now α is a root of some irreducible factor $p(x)$ of $f(x)$; since $\alpha \notin K$, degree $p(x) = k > 1$. By the generalized version of Lemma 5.4, there is $\beta \in E$ and an isomorphism $\lambda_1: K(\alpha) \to K(\beta)$ which extends λ and with $\lambda_1(\alpha) = \beta$. By Exercise 5.3, $[E : K(\alpha)] = d/k < d$. Now E is the splitting field of $f(x)$ over $K(\alpha)$, for it arises from $K(\alpha)$ by adjoining all the roots of $f(x)$. Since all the inductive hypotheses have been verified, we conclude that λ_1, and hence λ, can be extended to an automorphism of E. ■

Remark. As with the previous lemma, Lemma 5.5 has a more general version having the same proof. It says that if $f(x) \in F[x]$, then any two (abstract) splitting fields of $f(x)$ over F are isomorphic.

Theorem 5.6. *Let p be a prime, let F be a field containing a primitive pth root of unity, say, ω, and let $f(x) = x^p - a \in F[x]$.*

(i) *If α is a root of $f(x)$ (in some splitting field), then $f(x)$ is irreducible if and only if $\alpha \notin F$.*
(ii) *The splitting field E of $f(x)$ over F is $F(\alpha)$.*
(iii) *If $f(x)$ is irreducible, then* $\mathrm{Gal}(E/F) \cong \mathbb{Z}_p$.

Proof. (i) If $\alpha \in F$, then $f(x)$ is not irreducible, for it has $x - \alpha$ as a factor. Conversely, assume that $f(x) = g(x)h(x)$, where degree $g(x) = k < p$. Since the roots of $f(x)$ are $\alpha, \omega\alpha, \omega^2\alpha, \ldots, \omega^{p-1}\alpha$, every root of $g(x)$ has the form $\omega^i\alpha$ for some i. If the constant term of $g(x)$ is c, then $c = \pm\omega^r\alpha^k$ for some r (for c is, to sign, the product of the roots). As both c and ω lie in F, it follows that $\alpha^k \in F$. But $(k, p) = 1$, because p is prime, and so $1 = ks + tp$ for some integers s and t. Thus

$$\alpha = \alpha^{ks+tp} = (\alpha^k)^s(\alpha^p)^t \in F.$$

(ii) Immediate from the observation that the roots of $f(x)$ are of the form $\omega^i\alpha$.

(iii) If $\sigma \in \mathrm{Gal}(E/F)$, then $\sigma(\alpha) = \omega^i\alpha$ for some i, by Lemma 5.1. Define $\varphi: \mathrm{Gal}(E/F) \to \mathbb{Z}_p$ by $\varphi(\sigma) = [i]$, the congruence class of i mod p. It is easy to check that φ is a homomorphism; it is an injection, by Lemma 5.2. Since $f(x)$ is irreducible, by hypothesis, Lemma 5.4 shows that $\mathrm{Gal}(E/F) \neq 1$. Therefore φ is a surjection, for $\mathbb{Z}_p$ has no proper subgroups. ■

Theorem 5.7. *Let $f(x) \in F[x]$, let E be the splitting field of $f(x)$ over F, and assume that $f(x)$ has **no repeated roots** in E (i.e., $f(x)$ has no factor of the form $(x - \alpha)^2$ in $E[x]$). Then $f(x)$ is irreducible if and only if* $\mathrm{Gal}(E/F)$ *acts transitively on the set X of all the roots of $f(x)$.*

Remark. It can be shown that if F has characteristic 0 or if F is finite, then every irreducible polynomial in $F[x]$ has no repeated roots.

Proof. Note first that Lemma 5.1 shows that $\mathrm{Gal}(E/F)$ does act on X. If $f(x)$ is irreducible, then Lemma 5.4 shows that $\mathrm{Gal}(E/F)$ acts transitively on X. Conversely, assume that there is a factorization $f(x) = g(x)h(x)$ in $F[x]$. In $E[x]$, $g(x) = \prod (x - \alpha_i)$ and $h(x) = \prod (x - \beta_j)$; since $f(x)$ has no repeated roots, $\alpha_i \neq \beta_j$ for all i, j. But $\mathrm{Gal}(E/F)$ acts transitively on the roots of $f(x)$, so there exists $\sigma \in \mathrm{Gal}(E/F)$ with $\sigma(\alpha_1) = \beta_1$, and this contradicts Lemma 5.1. ■

It is easy to see that if α_1 is a root of $f(x)$, then the stabilizer of α_1 is $\mathrm{Gal}(E/F(\alpha_1)) \leq \mathrm{Gal}(E/F)$, and $\mathrm{Gal}(E/F(\alpha_1))$ is the Galois group of $f(x)/(x - \alpha_1)$ over $F(\alpha_1)$. Thus, $f(x)/(x - \alpha_1)$ is irreducible (over $F(\alpha_1)$) if and only if $\mathrm{Gal}(E/F(\alpha_1))$ acts transitively on the remaining roots. We shall return to this observation in a later chapter (Example 9.3) when we discuss *multiple transitivity*.

Theorem 5.8. *Let $F \subset K \subset E$ be fields, where K and E are splitting fields of polynomials over F. Then* $\mathrm{Gal}(E/K) \lhd \mathrm{Gal}(E/F)$ *and*

$$\mathrm{Gal}(E/F)/\mathrm{Gal}(E/K) \cong \mathrm{Gal}(K/F).$$

Proof. The function Φ: $\mathrm{Gal}(E/F) \to \mathrm{Gal}(K/F)$, given by $\Phi(\sigma) = \sigma|K$, is well defined (by Exercise 5.4, for K is a splitting field), and it is easily seen to be a homomorphism. The kernel of Φ consists of all those automorphisms which fix K; that is, $\ker \Phi = \mathrm{Gal}(E/K)$, and so this subgroup is normal. We claim that Φ is a surjection. If $\lambda \in \mathrm{Gal}(K/F)$, that is, λ is an automorphism of K which fixes F, then λ can be extended to an automorphism λ^* of E (by Lemma 5.5, for E is a splitting field). Therefore, $\lambda^* \in \mathrm{Gal}(E/F)$ and $\Phi(\lambda^*) = \lambda^*|K = \lambda$. The first isomorphism theorem completes the proof. ■

We summarize this discussion in the next theorem.

Theorem 5.9 (Galois, 1831). *Let $f(x) \in F[x]$ be a polynomial of degree n, let F contain all* pth *roots of unity for every prime p dividing $n!$, and let E be the splitting field of $f(x)$ over F. If $f(x)$ is solvable by radicals, then there exist subgroups $G_i \leq G =$* $\mathrm{Gal}(E/F)$ *such that*:

(i) $G = G_0 \geq G_1 \geq \cdots \geq G_t = 1$;
(ii) $G_{i+1} \lhd G_i$ *for all i; and*
(iii) G_i/G_{i+1} *is cyclic of prime order for all i.*

Proof. Since $f(x)$ is solvable by radicals, there are subfields $F = K_0 \subset K_1 \subset \cdots \subset K_t$ with $E \subset K_t$ and with $K_{i+1} = K_i(\beta_{i+1})$, where $\beta_{i+1} \in K_{i+1}$ and some power of β_{i+1} lies in K_i. By Exercise 5.5, we may assume that some prime power of β_{i+1} is in K_{i+1}. If we define $H_i = \mathrm{Gal}(K_t/K_i)$, then (i) is obvious. Since F contains roots of unity, Theorem 5.6 shows that K_{i+1} is a splitting field over K_i; moreover, it can be shown that there is such a tower of fields in which K_t is a splitting field of some polynomial over F. Theorem 5.8 now applies to show, for all i, that $H_{i+1} = \text{Gal}(K_t/K_{i+1}) \lhd \text{Gal}(K_t/K_i) = H_i$ and $H_{i+1}/H_i \cong \text{Gal}(K_{i+1}/K_i)$; this last group is isomorphic to $\mathbb{Z}_p$, by Theorem 5.6. ■

Remarks. 1. We have shown only that $\text{Gal}(K_t/F)$ satisfies the conclusions of the theorem. By Theorem 5.8, $\text{Gal}(K_t/E) \lhd \text{Gal}(K_t/F)$ and $\text{Gal}(K_t/F)/\text{Gal}(K_t/E) \cong \text{Gal}(E/F)$; that is, $\text{Gal}(E/F)$ is a quotient of $\text{Gal}(K_t/F)$. Theorem 5.15 will show that if a group satisfies the conclusions of Theorem 5.9, then so does any quotient group of it.

2. The hypothesis that F contains various roots of unity can be eliminated.

3. If F has characteristic 0, then the converse of this theorem is also true; it, too, was proved by Galois (1831).

Definition. A ***normal series*** of a group G is a sequence of subgroups

$$G = G_0 \geq G_1 \geq \cdots \geq G_n = 1$$

in which $G_{i+1} \lhd G_1$ for all i. The ***factor groups*** of this normal series are the groups G_i/G_{i+1} for $i = 0, 1, \ldots, n-1$; the ***length*** of the normal series is the number of strict inclusions; that is, the length is the number of nontrivial factor groups.

Note that the factor groups are the only quotient groups that can always be formed from a normal series, for we saw in Exercise 2.45 that normality may not be transitive.

Definition.[3] A finite group G is ***solvable*** if it has a normal series whose factor groups are cyclic of prime order.

In this terminology, Theorem 5.9 and its converse say that a polynomial is solvable by radicals if and only if its Galois group is a solvable group. P. Ruffini (1799) and N.H. Abel (1824) proved the nonexistence of a formula (analogous to the quadratic, cubic, and quartic formulas) for finding the roots of an arbitrary quintic, ending nearly three centuries of searching for a generalization of the work of Scipione, Tartaglia, Cardano, and Lodovici (actually, neither the proof of Ruffini nor that of Abel is correct in all details, but Abel's proof was accepted by his contemporaries and Ruffini's was not). In modern language, they proved that the Galois group of the general quintic is S_5 (nowadays, we know that any irreducible quintic having exactly three real roots [e.g., $f(x) = x^5 - 4x + 2$] will serve); since S_5 is not a solvable group, as we shall soon see, $f(x)$ is not solvable by radicals. In 1829, Abel proved that a polynomial whose Galois group is commutative is solvable by

[3] We shall give another, equivalent, definition of solvable groups later in this chapter.

radicals; this is why abelian groups are so called (of course, Galois groups had not yet been invented). Galois's theorem generalizes Abel's theorem, for every abelian group is solvable. Expositions of Galois Theory can be found in the books of Artin (1955), Birkhoff and Mac Lane (1977), Jacobson (1974), Kaplansky (1972), Rotman (1990), and van der Waerden (1948) listed in the Bibliography.

The Jordan–Hölder Theorem

Not only does Galois Theory enrich the study of polynomials and fields, but it also contributes a new idea, namely, normal series, to the study of groups. Let us give a brief review of what we have learned so far. Our first results arose from examining a single subgroup via Lagrange's theorem. The second, deeper, results arose from examining properties of a family of subgroups via the Sylow theorems; this family of subgroups consists of conjugates of a single subgroup, and so all of them have the same order. Normal series will give results by allowing us to examine a family of subgroups of different orders, thus providing an opening wedge for an inductive proof.

Definition. A normal series

$$G = H_0 \geq H_1 \geq \cdots \geq H_m = 1$$

is a ***refinement*** of a normal series

$$G = G_0 \geq G_1 \geq \cdots \geq G_n = 1$$

if $G_0, G_1, \ldots, G_n$ is a subsequence of $H_0, H_1, \ldots, H_m$.

A refinement is thus a normal series containing each of the terms of the original series.

Definition. A ***composition series*** is a normal series

$$G = G_0 \geq G_1 \geq \cdots \geq G_n = 1$$

in which, for all i, either G_{i+1} is a maximal normal subgroup of G_i or $G_{i+1} = G_i$.

Every refinement of a composition series is also a composition series; it can only repeat some of the original terms.

EXERCISES

5.6. A normal series is a composition series if and only if it has maximal length; that is, every refinement of it has the same length.

5.7. A normal series is a composition series if and only if its factor groups are either simple or trivial.

5.8. Every finite group has a composition series.

5.9. (i) An abelian group has a composition series if and only if it is finite.
(ii) Give an example of an infinite group which has a composition series.

5.10. If G is a finite group having a normal series with factor groups $H_0, H_1, \ldots, H_n$, then $|G| = \prod |H_i|$.

Exercise 5.10 shows that some information about G can be gleaned from a normal series. Let us now consider two composition series of $G = \langle x \rangle \cong \mathbb{Z}_{30}$ (normality is automatic because G is abelian):

$$G \geq \langle x^5 \rangle \geq \langle x^{10} \rangle \geq 1,$$

$$G \geq \langle x^2 \rangle \geq \langle x^6 \rangle \geq 1.$$

The factor groups of the first normal series are $G/\langle x^5 \rangle \cong \mathbb{Z}_5$, $\langle x^5 \rangle / \langle x^{10} \rangle \cong \mathbb{Z}_2$, and $\langle x^{10} \rangle / 1 \cong \langle x^{10} \rangle \cong \mathbb{Z}_3$; the factor groups of the second normal series are $G/\langle x^2 \rangle \cong \mathbb{Z}_2$, $\langle x^2 \rangle / \langle x^6 \rangle \cong \mathbb{Z}_3$, and $\langle x^6 \rangle \cong \mathbb{Z}_5$. In this case, both composition series have the same length and the factor groups can be "paired isomorphically" after rearranging them. We give a name to this phenomenon.

Definition. Two normal series of a group G are ***equivalent*** if there is a bijection between their factor groups such that corresponding factor groups are isomorphic.

Of course, equivalent normal series have the same length. The two composition series for $\mathbb{Z}_{30}$ displayed above are equivalent; the amazing fact is that this is true for every (possibly infinite) group that has a composition series! The next technical result, a generalization of the second isomorphism theorem, will be used in proving this.

Lemma 5.10 (Zassenhaus Lemma, 1934). *Let $A \lhd A^*$ and $B \lhd B^*$ be four subgroups of a group G. Then*

$$A(A^* \cap B) \lhd A(A^* \cap B^*),$$

$$B(B^* \cap A) \lhd B(B^* \cap A^*),$$

and there is an isomorphism

$$\frac{A(A^* \cap B^*)}{A(A^* \cap B)} \cong \frac{B(B^* \cap A^*)}{B(B^* \cap A)}.$$

Remark. Notice that the hypothesis and conclusion are unchanged by transposing the symbols A and B.

Proof. Since $A \lhd A^*$, we have $A \lhd A^* \cap B^*$ and so $A \cap B^* = A \cap (A^* \cap B^*) \lhd A^* \cap B^*$; similarly, $A^* \cap B \lhd A^* \cap B^*$. It follows from Lemma 2.25 and Exercise 2.38 that $D = (A^* \cap B)(A \cap B^*)$ is a normal subgroup of $A^* \cap B^*$.

If $x \in B(B^* \cap A^*)$, then $x = bc$ for $b \in B$ and $c \in B^* \cap A^*$. Define $f: B(B^* \cap A) \to (A^* \cap B^*)/D$ by $f(x) = f(bc) = cD$. To see that f is well defined, assume that $x = bc = b'c'$, where $b' \in B$ and $c' \in B^* \cap A$; then $c'c^{-1} = b'^{-1}b \in (B^* \cap A^*) \cap B = B \cap A^* \leq D$. It is routine to check that f is a surjective homomorphism with kernel $B(B^* \cap A)$. The first isomorphism theorem gives $B(B^* \cap A) \lhd B(B^* \cap A^*)$ and

$$\frac{B(B^* \cap A^*)}{B(B^* \cap A)} \cong \frac{B^* \cap A^*}{D}.$$

Transposing the symbols A and B gives $A(A^* \cap B) \lhd A(A^* \cap B^*)$ and an isomorphism of the corresponding quotient group to $(B^* \cap A^*)/D$. It follows that the two quotient groups in the statement are isomorphic. ∎

Theorem 5.11 (Schreier Refinement Theorem, 1928). *Every two normal series of an arbitrary group G have refinements that are equivalent.*

Proof. Let

$$G = G_0 \geq G_1 \geq \cdots \geq G_n = 1$$

and

$$G = H_0 \geq H_1 \geq \cdots \geq H_m = 1$$

be normal series. Insert a "copy" of the second series between each G_i and G_{i+1}. More precisely, define $G_{i,j} = G_{i+1}(G_i \cap H_j)$ for all $0 \leq j \leq m$. Thus

$$G_{i,j} = G_{i+1}(G_i \cap H_j) \geq G_{i+1}(G_i \cap H_{j+1}) = G_{i,j+1}.$$

Notice that $G_{i,0} = G_i$ because $H_0 = G$ and that $G_{i,m} = G_{i+1}$ because $H_m = 1$. Moreover, setting $A = G_{i+1}$, $A^* = G_i$, $B = H_{j+1}$, and $B^* = H_j$ in the Zassenhaus lemma shows that $G_{i,j+1} \lhd G_{i,j}$. It follows that the sequence

$$G_{0,0} \geq G_{0,1} \geq \cdots \geq G_{0,m} \geq G_{1,0} \geq \cdots \geq G_{n-1,0} \geq \cdots \geq G_{n-1,m} = 1$$

is a refinement of the first normal series (with mn terms). Similarly, if $H_{i,j}$ is defined to be $H_{j+1}(H_j \cap G_i)$, then $H_{i,j} \geq H_{i+1,j}$ and

$$H_{0,0} \geq H_{1,0} \geq \cdots \geq H_{n,0} \geq H_{0,1} \geq \cdots \geq H_{0,m-1} \geq \cdots \geq H_{n,m-1} = 1$$

is a refinement of the second normal series (with mn terms). Finally, the function pairing $G_{i,j}/G_{i,j+1}$ with $H_{i,j}/H_{i+1,j}$ is a bijection, and the Zassenhaus lemma (with $A = G_{i+1}$, $A^* = G_i$, $B = H_{j+1}$, and $B^* = H_j$) shows that corresponding factor groups are isomorphic. Therefore, the two refinements are equivalent. ∎

Theorem 5.12 (Jordan–Hölder). *Every two composition series of a group G are equivalent.*

Remark. C. Jordan (1868) proved that the orders of the composition factors of a finite group depend only on G; O. Hölder (1889) proved that the composition factors themselves, to isomorphism, depend only on G.

Proof. Composition series are normal series, so that every two composition series of G have equivalent refinements. But a composition series is a normal series of maximal length; a refinement of it merely repeats several of its terms, and so its new factor groups have order 1. Therefore, two composition series of G are equivalent. ■

Definition. If G has a composition series, then the factor groups of this series are called the ***composition factors*** of G.

One should regard the Jordan–Hölder theorem as a unique factorization theorem.

Corollary 5.13 (Fundamental Theorem of Arithmetic). *The primes and their multiplicities occurring in the factorization of an integer $n \geq 2$ are determined by n.*

Proof. Let $n = p_1 p_2 \dots p_t$, where the p_i are (not necessarily distinct) primes. If $G = \langle x \rangle$ is cyclic of order n, then

$$G = \langle x \rangle \geq \langle x^{p_1} \rangle \geq \langle x^{p_1 p_2} \rangle \geq \cdots \geq \langle x^{p_1 \cdots p_{t-1}} \rangle \geq 1$$

is a normal series. The factor groups have prime orders $p_1, p_2, \dots, p_t$, respectively, so this is a composition series. The Jordan–Hölder theorem now shows that these numbers depend on n alone. ■

Recall that a finite group G is solvable if it has a normal series (necessarily a composition series) with all its factor groups cyclic of prime order. Thus, one sees that a particular group is not solvable by checking whether every composition series has all its factor groups cyclic. With Jordan–Hölder, one need look only at a single composition series of G. For the next result, however, the Jordan–Hölder theorem is not needed, for we have essentially seen (in Exercise 3.21) that S_n has a unique composition series when $n \geq 5$.

Theorem 5.14. *If $n \geq 5$, then S_n is not solvable.*

Proof. Since A_n is a simple group for $n \geq 5$, the normal series $S_n \geq A_n \geq 1$ is a composition series; its composition factors are $\mathbb{Z}_2$ and A_n, and hence S_n is not solvable. ■

EXERCISES

5.11. Let G and H be finite groups. If there are normal series of G and of H having the same set of factor groups, then G and H have the same composition factors.

5.12. (i) Exhibit a composition series for S_4.
(ii) Show that S_n is solvable for $n \leq 4$.

5.13. Assume that $G = H_1 \times \cdots \times H_n = K_1 \times \cdots \times K_m$, where each H_i and K_j is simple. Prove that $m = n$ and there is a permutation π of $\{1, 2, \ldots, n\}$ with $K_{\pi(i)} \cong H_i$ for all i. (*Hint*. Construct composition series for G.)

5.14. Prove that the dihedral groups D_{2n} are solvable.

Solvable Groups

Even though solvable groups arose in the context of Galois Theory, they comprise a class of groups of purely group-theoretic interest as well. Let us now give another definition of solvability which is more convenient to work with and which is easily seen to agree with the previous definition for finite groups.

Definition. A ***solvable series*** of a group G is a normal series all of whose factor groups are abelian. A group G is ***solvable*** if it has a solvable series.

We are now going to manufacture solvable groups. Afterwards, we will give another characterization of solvability which will give new proofs of these results.

Theorem 5.15. *Every subgroup H of a solvable group G is itself solvable.*

Proof. If $G = G_0 \geq G_1 \geq \cdots \geq G_n = 1$ is a solvable series, consider the series $H = H_0 \geq (H \cap G_1) \geq \cdots \geq (H \cap G_n) = 1$. This is a normal series of H, for the second isomorphism theorem gives $H \cap G_{i+1} = (H \cap G_i) \cap G_{i+1} \lhd H \cap G_i$ for all i. Now $(H \cap G_i)/(H \cap G_{i+1}) \cong G_{i+1}(H \cap G_i)/G_{i+1} \leq G_i/G_{i+1}$; as G_i/G_{i+1} is abelian, so is any of its subgroups. Therefore, H has a solvable series. ■

Theorem 5.16. *Every quotient of a solvable group is solvable.*

Proof. It suffices to prove that if G is a solvable group and $f: G \to H$ is a surjection, then H is a solvable group. If

$$G = G_0 \geq G_1 \geq \cdots \geq G_t = 1$$

is a solvable series, then

$$H = f(G_0) \geq f(G_1) \geq \cdots \geq f(G_t) = 1$$

is a normal series for H: if $f(x_{i+1}) \in f(G_{i+1})$ and $f(x_i) \in f(G_i)$, then $f(x_i)f(x_{i+1})f(x_i)^{-1} = f(x_i x_{i+1} x_i^{-1}) \in f(G_i)$, because $G_{i+1} \lhd G_i$, and so $f(G_{i+1}) \lhd f(G_i)$ for every i. The map $\varphi: G_i \to f(G_i)/f(G_{i+1})$, defined by

$x_i \mapsto f(x_i)f(G_{i+1})$, is a surjection, for it is the composite of the surjections $G_i \to f(G_i)$ and the natural map $f(G_i) \to f(G_i)/f(G_{i+1})$. Since $G_{i+1} \leq \ker \varphi$, this map φ induces a surjection $G_i/G_{i+1} \to f(G_i)/f(G_{i+1})$, namely, $x_i G_{i+1} \mapsto f(x_i)f(G_{i+1})$. Now $f(G_i)/f(G_{i+1})$ is a quotient of the abelian group G_i/G_{i+1}, and so it is abelian. Therefore, H has a solvable series, and hence it is a solvable group. ■

In the proof of Theorem 5.9, we showed only that $\mathrm{Gal}(K_t/F)$ is solvable, whereas we wanted to prove that $\mathrm{Gal}(E/F)$ is solvable. But $\mathrm{Gal}(E/F)$ is a quotient of $\mathrm{Gal}(K_t/F)$, and so Theorem 5.16 completes the proof of Theorem 5.9.

Theorem 5.17. *If $H \lhd G$ and if both H and G/H are solvable, then G is solvable.*

Proof. Let

$$G/H \geq K_1^* \geq K_2^* \geq \cdots \geq K_n^* = 1$$

be a solvable series. By the correspondence theorem, we can construct a sequence looking like the beginning of a solvable series from G to H; that is, there are subgroups K_i (with $H \leq K_i$ and $K_i/H \cong K_i^*$) such that

$$G \geq K_1 \geq K_2 \geq \cdots \geq K_n = H,$$

$K_{i+1} \lhd K_i$, and K_i/K_{i+1} ($\cong K_i^*/K_{i+1}^*$) is abelian. Since H is solvable, it has a solvable series; if we splice these two sequences together at H, we obtain a solvable series for G. ■

Corollary 5.18. *If H and K are solvable groups, then $H \times K$ is solvable.*

Proof. If $G = H \times K$, then $H \lhd G$ and $G/H \cong K$. ■

Theorem 5.19. *Every finite p-group G is solvable.*

Proof. The proof is by induction on $|G|$. By Theorem 4.4, $|Z(G)| \neq 1$. Therefore, $G/Z(G)$ is a p-group of order $< |G|$, and hence it is solvable, by induction. Since every abelian group is solvable, $Z(G)$ is solvable. Therefore, G is solvable, by Theorem 5.17. ■

Here is an alternative proof. The composition factors of G must be simple p-groups. But we remarked (after proving that finite p-groups have nontrivial centers) that there are no finite simple p-groups of order greater than p. It follows that G is solvable.

Another approach to solvability is with commutator subgroups; that we are dealing with abelian quotient groups suggests this approach at once.

Definition. The ***higher commutator subgroups*** of G are defined inductively:

$$G^{(0)} = G; \qquad G^{(i+1)} = G^{(i)\prime};$$

that is, $G^{(i+1)}$ is the commutator subgroup of $G^{(i)}$. The series

$$G = G^{(0)} \geq G^{(1)} \geq G^{(2)} \geq \cdots$$

is called the ***derived series*** of G.

To see that the higher commutator subgroups are normal subgroups of G, it is convenient to introduce a new kind of subgroup.

Definition. An ***automorphism*** of a group G is an isomorphism $\varphi: G \to G$. A subgroup H of G is called ***characteristic*** in G, denoted by H char G, if $\varphi(H) = H$ for every automorphism φ of G.

If $\varphi(H) \leq H$ for every automorphism φ, then H char G: since both φ and φ^{-1} are automorphisms of G, one has $\varphi(H) \leq H$ and $\varphi^{-1}(H) \leq H$; the latter gives the reverse inclusion $H = \varphi\varphi^{-1}(H) \leq \varphi(H)$ and so $\varphi(H) = H$.

For each $a \in G$, conjugation by a (i.e., $x \mapsto axa^{-1}$) is an automorphism of G; it follows at once that every characteristic subgroup is a normal subgroup (but there are normal subgroups which are not characteristic; see Exercise 5.28 below.)

Lemma 5.20.

(i) *If H* char *K and K* char *G, then H* char *G.*
(ii) *If H* char *K and $K \lhd G$, then $H \lhd G$.*

Proof. (i) If φ is an automorphism of G, then $\varphi(K) = K$, and so the restriction $\varphi|K: K \to K$ is an automorphism of K; since H char K, it follows that $\varphi(H) = (\varphi|K)(H) = H$.

(ii) Let $a \in G$ and let $\varphi: G \to G$ be conjugation by a. Since $K \lhd G$, $\varphi|K$ is an automorphism of K; since H char K, $(\varphi|K)(H) \leq H$. This says that if $h \in H$, then $aha^{-1} = \varphi(h) \in H$. ■

Theorem 5.21. *For every group G, the higher commutator subgroups are characteristic, hence normal subgroups.*

Proof. The proof is by induction on $i \geq 1$. Recall that the commutator subgroup $G' = G^{(1)}$ is generated by all commutators; that is, by all elements of the form $aba^{-1}b^{-1}$. If φ is an automorphism of G, then $\varphi(aba^{-1}b^{-1}) = \varphi(a)\varphi(b)\varphi(a)^{-1}\varphi(b)^{-1}$ is also a commutator, and so $\varphi(G') \leq G'$. For the inductive step, we have just shown that $G^{(i+1)}$ char $G^{(i)}$; since $G^{(i)}$ char G, by induction, Lemma 5.20(i) shows that $G^{(i+1)}$ is characteristic in G. ■

It follows that the derived series of a group G is a normal series *if* it ends at 1. The next result shows that if G is solvable, then the derived series descends faster than any other solvable series.

Lemma 5.22. *If $G = G_0 \geq G_1 \geq \cdots \geq G_n = 1$ is a solvable series, then $G_i \geq G^{(i)}$ for all i.*

Proof. The proof is by induction on $i \geq 0$. If $i = 0$, then $G_0 = G = G^{(0)}$. For the inductive step, Theorem 2.23 gives $G_{i+1} \geq G_i'$, since G_i/G_{i+1} is abelian. The inductive hypothesis gives $G_i \geq G^{(i)}$, so that $G_i' \geq G^{(i)\prime} = G^{(i+1)}$. Therefore, $G_{i+1} \geq G^{(i+1)}$, as desired. ■

Theorem 5.23. *A group G is solvable if and only if $G^{(n)} = 1$ for some n.*

Proof. Let $G = G_0 \geq G_1 \geq \cdots \geq G_n = 1$ be a solvable series. By the lemma, $G_i \geq G^{(i)}$ for all i. In particular, $1 = G_n \geq G^{(n)}$, and so $G^{(n)} = 1$.

Conversely, if $G^{(n)} = 1$, then the derived series is a normal series; since it has abelian factor groups, it is a solvable series for G. ■

Thus, the derived series of G is a normal series if and only if G is a solvable group.

The following new proofs of Theorems 5.15, 5.16, and 5.17 should be completed by the reader; they are based on the new criterion for solvability just proved.

Theorem 5.15. If $H \leq G$, then $H^{(i)} \leq G^{(i)}$ for all i; hence $G^{(n)} = 1$ implies $H^{(n)} = 1$, so that every subgroup of a solvable group is solvable.

Theorem 5.16. If $f\colon G \to K$ is surjective, then $f(G^{(i)}) = f(G)^{(i)}$ for all i. Therefore, $G^{(n)} = 1$ implies $1 = f(G^{(n)}) = f(G)^{(n)}$, so that every quotient of a solvable group is solvable.

Theorem 5.17. Let $H \lhd G$, let $K = G/H$, and let $f\colon G \to K$ be the natural map. Then $K^{(n)} = 1$ implies $f(G^{(n)}) = 1$, and hence $G^{(n)} \leq H$. If $H^{(m)} = 1$, then $(G^{(n)})^{(m)} \leq H^{(m)} = 1$. Finally, one proves by induction on $n \geq 1$ that $G^{(m+n)} \leq (G^{(n)})^{(m)}$ for all m, and so G is solvable.

Definition. A normal subgroup H of a group G is a ***minimal normal subgroup*** if $H \neq 1$ and there is no normal subgroup K of G with $1 < K < H$.

Minimal normal subgroups always exist in nontrivial finite groups.

Theorem 5.24. *If G is a finite solvable group, then every minimal normal subgroup is elementary abelian.*

Proof. Let V be a minimal normal subgroup of G. Now Lemma 5.20(ii) says that if H char V, then $H \lhd G$; since V is minimal, either $H = 1$ or $H = V$. In particular, V' char V, so that either $V' = 1$ or $V' = V$. Since G is solvable, so is its subgroup V. It follows from Theorem 5.23 that $V' = V$ cannot occur here, so that $V' = 1$ and so V is abelian. Since V is abelian, a Sylow p-subgroup P of V, for any prime p, is characteristic in V; therefore, V is an abelian p-group. But $\{x \in V: x^p = 1\}$ char V, and so V is elementary abelian. ■

Corollary 5.25. *If V is a minimal normal subgroup of a finite solvable group G, then G acts on V as a group of linear transformations.*

Proof. By the theorem, V is an elementary p-group; by Exercise 2.78, V is a vector space over $\mathbb{Z}_p$ and every homomorphism on V is a linear transformation. Define a homomorphism $G \to \mathrm{GL}(V)$ by $a \mapsto \varphi_a$, where $\varphi_a(v) = ava^{-1}$ for all $v \in V$ (normality of V shows that $\varphi_a(v) \in V$). Moreover, each φ_a is an injection, being the restriction of an automorphism (namely, conjugation), and every injection on a finite-dimensional vector space is a surjection; hence each φ_a is nonsingular. ■

What are the groups G whose only characteristic subgroups are G and 1? Such groups are sometimes called *characteristically simple.*

Theorem 5.26. *A finite group G with no characteristic subgroups other than G and 1 is either simple or a direct product of isomorphic simple groups.*

Proof. Choose a minimal normal subgroup H of G whose order is minimal among all nontrivial normal subgroups. Write $H = H_1$, and consider all subgroups of G of the form $H_1 \times H_2 \times \cdots \times H_n$, where $n \geq 1$, $H_i \lhd G$, and $H_i \cong H$. Let M be such a subgroup of largest possible order. We show that $M = G$ by showing that M char G; to see this, it suffices to show that $\varphi(H_i) \leq M$ for every i and every automorphism φ of G. Of course, $\varphi(H_i) \cong H = H_1$. We show that $\varphi(H_i) \lhd G$. If $a \in G$, then $a = \varphi(b)$ for some $b \in G$, and $a\varphi(H_i)a^{-1} = \varphi(b)\varphi(H_i)\varphi(b)^{-1} = \varphi(aH_ia^{-1}) \leq \varphi(H_i)$, because $H_i \lhd G$. If $\varphi(H_i) \nleq M$, then $\varphi(H_i) \cap M \nleq \varphi(H_i)$ and $|\varphi(H_i) \cap M| < |\varphi(H_i)| = |H|$. But $\varphi(H_i) \cap M \lhd G$, and so the minimality of $|H|$ shows that $\varphi(H_i) \cap M = 1$. The subgroup $\langle M, \varphi(H_i)\rangle = M \times \varphi(H_i)$ is a subgroup of the same type as M but of larger order, a contradiction. We conclude that M char G, and so $M = G$. Finally, $H = H_1$ must be simple: if N is a nontrivial normal subgroup of H, then N is a normal subgroup of $M = H_1 \times H_2 \times \cdots \times H_n = G$, and this contradicts the minimal choice of H. ■

Corollary 5.27. *A minimal normal subgroup H of a finite group G is either simple or a direct product of isomorphic simple subgroups.*

Proof. If N char H, then $N \lhd G$, by Lemma 5.20(ii), so that either $N = 1$ or $N = H$ (because H is a minimal normal subgroup). Therefore, H has no proper characteristic subgroups, and Theorem 5.26 gives the result. ■

This last corollary gives another proof of Theorem 5.24, for a finite simple group G is solvable if and only if it is cyclic of prime order.

EXERCISES

5.15. Every refinement of a solvable series is a solvable series.

5.16. A solvable group having a composition series must be finite.

5.17. If G has a composition series and if $H \lhd G$, then G has a composition series one of whose terms is H.

5.18. (i) If S and T are solvable subgroups of G with $S \lhd G$, then ST is solvable.
(ii) Every finite group G has a unique maximal normal solvable subgroup $\mathscr{S}(G)$; moreover, $G/\mathscr{S}(G)$ has no nontrivial normal solvable subgroups.

5.19. (i) If p and q are primes, then every group of order p^2q is solvable.
(ii) If p and q are primes with $p < q$, then every group of order pq^n is solvable. (*Hint.* Use Sylow's theorems.)

5.20. (i) Show that S_4 has no series

$$S_4 = G_0 \geq G_1 \geq \cdots \geq G_n = 1$$

such that all factor groups are cyclic and each G_i is a normal subgroup of G. (A group G with such a series is called ***supersolvable***, and we now see that not every solvable group is supersolvable.)
(ii) Show that every finite p-group is supersolvable.

5.21. If G is a group with $|G| < 60$, then G is solvable. (*Hint.* Use Exercise 4.36.)

5.22. Burnside proved (using Representation Theory) that the number of elements in a conjugacy class of a finite simple group can never be a prime power larger than 1. Use this fact to prove ***Burnside's theorem***: If p and q are primes, then every group of order $p^m q^n$ is solvable.

5.23. Prove that the following two statements are equivalent:
(i) every group of odd order is solvable;
(ii) every finite simple group has even order.
(In 1963, Feit and Thompson proved (i); the original proof is 274 pages long.)

5.24. Let G be a finite group of order > 1. If G is solvable, then G contains a nontrivial normal abelian subgroup; if G is not solvable, then G contains a nontrivial normal subgroup H with $H = H'$.

5.25. For every group G, its center $Z(G)$ is characteristic in G.

5.26. If $H \lhd G$ and $(|H|, [G : H]) = 1$, then H char G.

5.27. If H char G and $H \leq K \leq G$, then K/H char G/H implies K char G.

5.28. Give an example of a group G containing a normal subgroup that is not a characteristic subgroup. (*Hint*. Let G be abelian.)

Definition. A subgroup H of G is ***fully invariant*** if $\varphi(H) \leq H$ for every homomorphism $\varphi: G \to G$.

Of course, every fully invariant subgroup is characteristic and hence normal.

5.29. Prove that the higher commutator subgroups are fully invariant.

5.30. Show that $Z(G)$ may not be fully invariant. (*Hint*: Let $G = \mathbb{Z}_2 \times S_3$.) (See Exercise 5.25.)

Two Theorems of P. Hall

The main results of this section are generalizations of the Sylow theorems that hold for (and, in fact, characterize) finite solvable groups.

Theorem 5.28 (P. Hall, 1928). *If G is a solvable group of order ab, where $(a, b) = 1$, then G contains a subgroup of order a. Moreover, any two subgroups of order a are conjugate.*

Proof. The proof is by induction on $|G|$; as usual, the base step is trivially true.

Case 1. G contains a normal subgroup H of order $a'b'$, where $a'|a$, $b'|b$, and $b' < b$.

Existence. In this case, G/H is a solvable group of order $(a/a')(b/b')$, which is strictly less than ab; by induction, G/H has a subgroup A/H of order a/a'. Now A has order $(a/a')|H| = ab' < ab$; since A is solvable, it has a subgroup of order a, as desired.

Conjugacy. Let A and A' be subgroups of G of order a. Let us compute $k = |AH|$. Since $AH \leq G$, Lagrange's theorem gives $|AH| = \alpha\beta$, where $\alpha|a$ and $\beta|b$. Since $(a, b) = 1$ and $A \leq AH$, we have $a|\alpha\beta$, so that $a = \alpha$; since $H \leq AH$, we have $a'b'|\alpha\beta$, so that $b'|\beta$. But the second isomorphism theorem (actually, the product formula) gives $k|aa'b'$, so that $\beta|b'$. We conclude that $|AH| = k = ab'$. A similar calculation shows that $|A'H| = ab'$ as well. Thus, AH/H and $A'H/H$ are subgroups of G/H of order a/a'. As $|G/H| = (a/a')(b/b')$, these subgroups are conjugate, by induction, say by $xH \in G/H$. It is quickly checked that $xAHx^{-1} = A'H$. Therefore, xAx^{-1} and A' are subgroups of $A'H$ of order a, and so they are conjugate, by induction. This completes Case 1.

If there is some proper normal subgroup of G whose order is not divisible by b, then the theorem has been proved. We may, therefore, assume that $b \mid |H|$ for every proper normal subgroup H. If H is a minimal normal subgroup, however, then Theorem 5.24 says that H is an (elementary) abelian p-group for some prime p. It may thus be assumed that $b = p^m$, so that H is a Sylow p-subgroup of G. Normality of H forces H to be unique (for all Sylow p-subgroups are conjugate). The problem has now been reduced to the following case.

Case 2. $|G| = ap^m$, where $p \nmid a$, G has a normal abelian Sylow p-subgroup H, and H is the unique minimal normal subgroup in G.

Remark. We shall complete the proof here, but we note that this case follows immediately from the Schur–Zassenhaus lemma to be proved in Chapter 7.

Existence. The group G/H is a solvable group of order a. If K/H is a minimal normal subgroup of G/H, then $|K/H| = q^n$ for some prime $q \neq p$, and so $|K| = p^m q^n$; if Q is a Sylow q-subgroup of K, then $K = HQ$. Let $N^* = N_G(Q)$ and let $N = N^* \cap K = N_K(Q)$. We claim that $|N^*| = a$.

The Frattini argument (Theorem 4.18) gives $G = KN^*$. Since

$$G/K = KN^*/K \cong N^*/N^* \cap K = N^*/N,$$

we have $|N^*| = |G||N|/|K|$. But $K = HQ$ and $Q \leq N \leq K$ gives $K = HN$; hence $|K| = |HN| = |H||N|/|H \cap N|$, so that

$$\begin{aligned}|N^*| = |G||N|/|K| &= |G||N||H \cap N|/|H||N| \\ &= (|G|/|H|)|H \cap N| = a|H \cap N|.\end{aligned}$$

Hence $|N^*| = a$ if $H \cap N = 1$. We show that $H \cap N = 1$ in two stages: (i) $H \cap N \leq Z(K)$; and (ii) $Z(K) = 1$.

(i) Let $x \in H \cap N$. Every $k \in K = HQ$ has the form $k = hs$ for $h \in H$ and $s \in Q$. Now x commutes with h, for H is abelian, and so it suffices to show that x commutes with s. But $(xsx^{-1})s^{-1} \in Q$, because x normalizes Q, and $x(sx^{-1}s^{-1}) \in H$, because H is normal; therefore, $xsx^{-1}s^{-1} \in Q \cap H = 1$.

(ii) By Lemma 5.20(ii), $Z(K) \lhd G$. If $Z(K) \neq 1$, then it contains a minimal subgroup which must be a minimal normal subgroup of G. Hence $H \leq Z(K)$, for H is the unique minimal normal subgroup of G. But since $K = HQ$, it follows that Q char K. Thus $Q \lhd G$, by Lemma 5.20(ii), and so $H \leq Q$, a contradiction. Therefore, $Z(K) = 1$, $H \cap N = 1$, and $|N^*| = a$.

Conjugacy. We keep the notation of the existence proof. Recall that $N^* = N_G(Q)$ has order a; let A be another subgroup of G of order a. Since $|AK|$ is divisible by a and by $|K| = p^m q^n$, it follows that $|AK| = ab = |G|$, $AK = G$,

$$G/K = AK/K \cong A/(A \cap K),$$

and $|A \cap K| = q^n$. By the Sylow theorem, $A \cap K$ is conjugate to Q. As conjugate subgroups have conjugate normalizers, by Exercise 3.9(i), $N^* = N_G(Q)$ is conjugate to $N_G(A \cap K)$, and so $a = |N^*| = |N_G(A \cap K)|$. Since $A \cap K \lhd A$, we have $A \leq N_G(A \cap K)$ and so $A = N_G(A \cap K)$ (both have order a). Therefore, A is conjugate to N^*. ■

The following definition is made in tribute to this theorem.

Definition. If G is a finite group, then a ***Hall subgroup*** H of G is a subgroup whose order and index are relatively prime; that is, $(|H|, [G:H]) = 1$.

It is sometimes convenient to display the prime divisors of the order of a group. If π is a set of primes, then a ***π-number*** is an integer n all of whose prime factors lie in π; the complement of π is denoted by π', and so a ***π'-number*** is an integer n none of whose prime factors lie in π.

Definition. If π is a set of primes, then a group G is a ***π-group*** if the order of each of its elements is a π-number; a group is a ***π'-group*** if the order of each of its elements is a π'-number.

Of course if a is a π-number and b is a π'-number, then a and b are relatively prime. If π consists of a single prime p, then π-groups are just p-groups, while p'-groups have no elements of order a power of p. It follows from (Sylow's) Theorem 4.14 that every Sylow p-subgroup in a finite group is a Hall p-subgroup (Hall's theorem says that Hall π-subgroups always exist in finite solvable groups). In contrast to Sylow p-subgroups, however, Hall π-subgroups (with $|\pi| \geq 2$) of a group G need not exist. For example, let $G = A_5$ and $\pi = \{3, 5\}$; since $|A_5| = 60$, a Hall π-subgroup would have index 4 and order 15, and Corollary 3.16 shows that no such subgroup exists.

Definition. If p is a prime, and G is a finite group of order ap^n, where a is a p'-number, then a ***p-complement*** of G is a subgroup of order a.

Hall's theorem implies that a finite solvable group has a p-complement for every prime p. If G is a group of order $p^m q^n$, then G has a p-complement, namely, a Sylow q-subgroup, and a q-complement, namely, a Sylow p-subgroup. In the coming proof of the converse of Theorem 5.28, we shall use ***Burnside's theorem*** (see Exercise 5.22): Every group of order $p^m q^n$ is solvable.

Theorem 5.29 (P. Hall, 1937). *If G is a finite group having a p-complement for every prime p, then G is solvable.*

Proof. We proceed by induction on $|G|$; assume, on the contrary, that there are nonsolvable groups satisfying the hypotheses, and choose one such, say G, of smallest order. If G has a nontrivial normal subgroup N, and if H is any Hall p'-subgroup of G, then checking orders shows that $H \cap N$ is a Hall p'-subgroup of N and HN/N is a Hall p'-subgroup of G/N. Since both N and G/N have order smaller than $|G|$, it follows that both N and G/N are solvable. But Theorem 5.17 now shows that G is solvable, a contradiction.

We may assume, therefore, that G is simple. Let $|G| = p_1^{e_1} \dots p_n^{e_n}$, where the p_i are distinct primes and $e_i > 0$ for all i. For each i, let H_i be a Hall p_i'-subgroup of G, so that $[G : H_i] = p_i^{e_i}$, and thus $|H_i| = \prod_{j \neq i} p_j^{e_j}$. If $D = H_3 \cap \cdots \cap H_n$, then $[G : D] = \prod_{i=3}^{n} p_i^{e_i}$, by Exercise 3.31(ii), and so $|D| = p_1^{e_1} p_2^{e_2}$. Now D is a solvable group, by Burnside's theorem. If N is a minimal normal subgroup of D, then Theorem 5.24 says that N is elementary abelian; for notation, assume that N is a p_1-group. Exercise 3.31(ii) shows that $[G : D \cap H_2] = \prod_{i=2}^{n} p_i^{e_i}$, so that $|D \cap H_2| = p_1^{e_1}$ and $D \cap H_2$ is a Sylow p_1-subgroup of D. By Exercise 4.16, $N \leq D \cap H_2$ and so $N \leq H_2$. But, as above, $|D \cap H_1| = p_2^{e_2}$, and comparison of orders gives $G = H_2(D \cap H_1)$. If $g \in G$, then $g = hd$, where $h \in H_2$ and $d \in D \cap H_1$; if $x \in N$, then $gxg^{-1} = hdxd^{-1}h^{-1} = hyh^{-1}$ (where $y = dxd^{-1} \in N$, because $N \lhd D$) and $hyh^{-1} \in H_2$ (because $N \leq H_2$). Therefore, $N^G \leq H_2$, where N^G is the normal subgroup of G generated by N. Since $H_2 < G$, $N \neq 1$ is a proper normal subgroup of G, and this contradicts the assumption that G is simple. ∎

This proof exhibits two recurring themes in finite group theory. Many theorems having the form "If a group G has property P, then it also has property Q" are proved by induction on $|G|$ in the following style: assume that G is a group of smallest possible order which has property P but not property Q, and then obtain a contradiction. In R. Baer's suggestive phrase, we assume that G is a "least criminal."

The other theme is the reduction of a problem to the special case when a simple group is involved. Many theorems are proved by such a reduction, and this is one reason why the classificaton of the finite simple groups is so important.

Exercises

5.31. If G is a finite (not necessarily solvable) group and H is a normal Hall subgroup, then H char G.

5.32. (i) If G is a finite solvable group, then for every set of primes π, a maximal π-subgroup is a Hall π-subgroup.
(ii) Let $\pi = \{3, 5\}$. Show that both $\mathbb{Z}_3$ and $\mathbb{Z}_5$ are maximal π-subgroups of S_5. Conclude, when G is not solvable, that maximal π-subgroups may not be isomorphic and hence may not be conjugate.
(iii) If $\pi = \{2, 5\}$, then a maximal π-subgroup of S_5 is not a Hall π-subgroup. (*Hint.* Exercise 3.25.)

Definition. If π is a set of primes, define $O_\pi(G)$ to be the subgroup of G generated by all the normal π-subgroups of G.

5.33. Show that $O_\pi(G)$ is a characteristic subgroup of G.

5.34. Show that $O_\pi(G)$ is the intersection of all the maximal π-subgroups of G.

Central Series and Nilpotent Groups

The Sylow theorems show that knowledge of p-groups gives information about arbitrary finite groups. Moreover, p-groups have a rich supply of normal subgroups, and this suggests that normal series might be a powerful tool in their study. It turns out that the same methods giving theorems about p-groups also apply to a larger class, the *nilpotent groups*, which may be regarded as generalized p-groups.

Definition. If $H, K \leq G$, then

$$[H, K] = \langle [h, k] \colon h \in H \text{ and } k \in K \rangle,$$

where $[h, k]$ is the commutator $hkh^{-1}k^{-1}$.

An example was given, in Exercise 2.43, showing that the *set* of all commutators need not be a subgroup; in order that $[H, K]$ be a subgroup, therefore, we must take the subgroup generated by the indicated commutators. It is obvious that $[H, K] = [K, H]$, for $[h, k]^{-1} = [k, h]$. The commutator subgroup G' is equal to $[G, G]$ and, more generally, the higher commutator subgroup $G^{(i+1)}$ is equal to $[G^{(i)}, G^{(i)}]$.

We say that a subgroup K ***normalizes*** H if $K \leq N_G(H)$; it is easy to see that K normalizes H if and only if $[H, K] \leq H$.

Definition. If $H \leq G$, the ***centralizer*** of H in G is

$$C_G(H) = \{x \in G \colon x \text{ commutes with every } h \in H\};$$

that is, $C_G(H) = \{x \in G \colon [x, h] = 1 \text{ for all } h \in H\}$.

We say that a subgroup K ***centralizes*** H if $K \leq C_G(H)$; it is easy to see that K centralizes H if and only if $[H, K] = 1$.

If $x, y \in G$ and $[x, y] \in K$, where $K \lhd G$, then x and y "commute mod K"; that is, $xKyK = yKxK$ in G/K.

Lemma 5.30.

(i) *If $K \lhd G$ and $K \leq H \leq G$, then $[H, G] \leq K$ if and only if $H/K \leq Z(G/K)$.*
(ii) *If $H, K \leq G$ and $f \colon G \to L$ is a homomorphism, then $f([H, K]) = [f(H), f(K)]$.*

Proof. (i) If $h \in H$ and $g \in G$, then $hKgK = gKhK$ if and only if $[h, g]K = K$ if and only if $[h, g] \in K$.

(ii) Both sides are generated by all $f([h, k]) = [f(h), f(k)]$. ∎

Definition. Define characteristic subgroups $\gamma_i(G)$ of G by induction:

$$\gamma_1(G) = G; \qquad \gamma_{i+1}(G) = [\gamma_i(G), G].$$

Notice that $\gamma_2(G) = [\gamma_1(G), G] = [G, G] = G' = G^{(1)}$. It is easy to check that $\gamma_{i+1}(G) \leq \gamma_i(G)$. Moreover, Lemma 5.30(i) shows that $[\gamma_i(G), G] = \gamma_{i+1}(G)$ gives $\gamma_i(G)/\gamma_{i+1}(G) \leq Z(G/\gamma_{i+1}(G))$.

Definition. The ***lower central series*** (or *descending central series*) of G is the series

$$G = \gamma_1(G) \geq \gamma_2(G) \geq \cdots$$

(this need not be a normal series because it may not reach 1).

There is another series of interest.

Definition. The ***higher centers*** $\zeta^i(G)$ are the characteristic subgroups of G defined by induction:

$$\zeta^0(G) = 1; \qquad \zeta^{i+1}(G)/\zeta^i(G) = Z(G/\zeta^i(G));$$

that is, if $\nu_i\colon G \to G/\zeta^i(G)$ is the natural map, then $\zeta^{i+1}(G)$ is the inverse image of the center.

Of course, $\zeta^1(G) = Z(G)$.

Definition. The ***upper central series*** (or *ascending central series*) of G is

$$1 = \zeta^0(G) \leq \zeta^1(G) \leq \zeta^2(G) \leq \cdots.$$

When no confusion can occur, we may abbreviate $\zeta^1(G)$ by ζ^i and $\gamma_i(G)$ by γ_i.

Theorem 5.31. *If G is a group, then there is an integer c with $\zeta^c(G) = G$ if and only if $\gamma_{c+1}(G) = 1$. Moreover, in this case,*

$$\gamma_{i+1}(G) \leq \zeta^{c-i}(G) \qquad \textit{for all } i.$$

Proof. Assume that $\zeta^c = G$, and let us prove that the inclusion holds by induction on i. If $i = 0$, then $\gamma_1 = G = \zeta^c$. If $\gamma_{i+1} \leq \zeta^{c-i}$, then

$$\gamma_{i+2} = [\gamma_{i+1}, G] \leq [\zeta^{c-i}, G] \leq \zeta^{c-i-1},$$

the last inclusion following from Lemma 5.30. We have shown that the inclusion always holds; in particular, if $i = c$, then $\gamma_{c+1} \leq \zeta^0 = 1$.

Assume that $\gamma_{c+1} = 1$, and let us prove that $\gamma_{c+1-j} \leq \zeta^j$ by induction on j (this is the same inclusion as in the statement: set $j = c - i$). If $j = 0$, then $\gamma_{c+1} = 1 = \zeta^0$. If $\gamma_{c+1-j} \leq \zeta^j$, then the third isomorphism theorem gives a surjective homomorphism $G/\gamma_{c+1-j} \to G/\zeta^j$. Now $[\gamma_{c-j}, G] = \gamma_{c+1-j}$, so that Lemma 5.30 gives $\gamma_{c-j}/\gamma_{c+1-j} \leq Z(G/\gamma_{c+1-j})$. By Exercise 3.10 [if $A \leq Z(G)$

and $f: G \to H$ is surjective, then $f(A) \leq Z(H)$], we have

$$\gamma_{c-j}\zeta^j/\zeta^j \leq Z(G/\zeta^j) = \zeta^{j+1}/\zeta^j.$$

Therefore, $\gamma_{c-j} \leq \gamma_{c-j}\zeta^j \leq \zeta^{j+1}$, as desired. We have shown that the inclusion always holds; in particular, if $j = c$, then $G = \gamma_1 \leq \zeta^c$. ■

The following result reflects another relationship between these two series.

Theorem 5.32 (Schur). *If G is a group with $G/Z(G)$ finite, then G' is also finite.*

Proof **(Ornstein).** Let $g_1, \ldots, g_n$ be representatives of the cosets of $Z(G)$ in G; that is, each $x \in G$ has the form $x = g_i z$ for some i and some $z \in Z(G)$. For all $x, y \in G$, $[x, y] = [g_i z, g_j z'] = [g_i, g_j]$. Hence, every commutator has the form $[g_i, g_j]$ for some i, j, so that G' has a finite number ($< n^2$) of generators.

Each element $g' \in G'$ can be written as a word $c_1 \cdots c_t$, where each c_i is a commutator (no exponents are needed, for $[x, y]^{-1} = [y, x]$). It suffices to prove that if a factorization of g' is chosen so that $t = t(g')$ is minimal, then $t(g') < n^3$ for all $g' \in G'$.

We prove first, by induction on $r \geq 1$, that if $a, b \in G$, then $[a, b]^r = (aba^{-1}b^{-1})^r = (ab)^r(a^{-1}b^{-1})^r u$, where u is a product of $r - 1$ commutators. This is obvious when $r = 1$. Note, for the inductive step, that if $x, y \in G$, then $xy = yxx^{-1}y^{-1}xy = yx[x^{-1}, y^{-1}]$; that is, $xy = yxc$ for some commutator c. Thus, if $r > 1$, then

$$\begin{aligned}
(aba^{-1}b^{-1})^{r+1} &= aba^{-1}b^{-1}(aba^{-1}b^{-1})^r \\
&= ab[a^{-1}b^{-1}]\{(ab)^r(a^{-1}b^{-1})^r\}u \\
&= ab\{(ab)^r(a^{-1}b^{-1})^r\}[a^{-1}b^{-1}]cu
\end{aligned}$$

for some commutator c, as desired.

Since $yx = x^{-1}(xy)x$, we have $(yx)^n = x^{-1}(xy)^n x = (xy)^n$, because $[G : Z(G)] = n$ implies $(ab)^n \in Z(G)$. Therefore, $(a^{-1}b^{-1})^n = ((ba)^{-1})^n = ((ba)^n)^{-1} = ((ab)^n)^{-1}$. It follows that

(∗) $\quad [a, b]^n$ is a product of $n - 1$ commutators.

Now $xyx = (xyx^{-1})x^2$, so that two x's can be brought together at the expense of replacing y by a conjugate of y. Take an expression of an element $g' \in G'$ as a product of commutators $c_1 \ldots c_t$, where t is minimal. If $t \geq n^3$, then there is some commutator c occurring m times, where $m > n$ (for there are fewer than n^2 distinct commutators). By our remark above, all such factors can be brought together to c^m at the harmless expense of replacing commutators by conjugates (which are still commutators); that is, the number of commutator factors in the expression is unchanged. By (∗), the length of the minimal expression for g' is shortened, and this is a contradiction. Therefore, $t < n^3$, and so G' is finite. ■

Definition. A group G is ***nilpotent***[4] if there is an integer c such that $\gamma_{c+1}(G) = 1$; the least such c is called the ***class*** of the nilpotent group G.

Theorem 5.31 shows, for nilpotent groups, that the lower and upper central series are normal series of the same length.

A group is nilpotent of class 1 if and only if it is abelian. By Theorem 5.31, a nilpotent group G of class 2 is described by $\gamma_2(G) = G' \leq Z(G) = \zeta^1(G)$. Every nonabelian group of order p^3 is nilpotent of class 2, by Exercise 4.7.

Theorem 5.33. *Every finite p-group is nilpotent.*

Proof. Recall Theorem 4.4. Every finite p-group has a nontrivial center. If, for some i, we have $\zeta^i(G) < G$, then $Z(G/\zeta^i(G)) \neq 1$ and so $\zeta^i(G) < \zeta^{i+1}(G)$. Since G is finite, there must be an integer i with $\zeta^i(G) = G$; that is, G is nilpotent. ■

This theorem is false without the finiteness hypothesis, for there exist infinite p-groups that are not nilpotent (see Exercise 5.45 below); indeed, there is an example of McLain (1954) of an infinite p-group G with $Z(G) = 1$, with $G' = G$ (so that G is not even solvable), and with no characteristic subgroups other than G and 1.

Theorem 5.34.

(i) *Every nilpotent group G is solvable.*
(ii) *If $G \neq 1$ is nilpotent, then $Z(G) \neq 1$.*
(iii) *S_3 is a solvable group that is not nilpotent.* ■

Proof. (i) An easy induction shows that $G^{(i)} \leq \gamma_i(G)$ for all i. It follows that if $\gamma_{c+1}(G) = 1$, then $G^{(c+1)} = 1$; that is, if G is nilpotent (of class $\leq c$), then G is solvable (with derived length $\leq c + 1$).

(ii) Assume that $G \neq 1$ is nilpotent of class c, so that $\gamma_{c+1}(G) = 1$ and $\gamma_c(G) \neq 1$. By Theorem 5.31, $1 \neq \gamma_c(G) \leq \zeta^1(G) = Z(G)$.

(iii) The group $G = S_3$ is solvable and $Z(S_3) = 1$. ■

Theorem 5.35. *Every subgroup H of a nilpotent group G is nilpotent. Moreover, if G is nilpotent of class c, then H is nilpotent of class $\leq c$.*

Proof. It is easily proved by induction that $H \leq G$ implies $\gamma_i(H) \leq \gamma_i(G)$ for all i. Therefore, $\gamma_{c+1}(G) = 1$ forces $\gamma_{c+1}(H) = 1$. ■

[4] There is an analogue of the descending central series for Lie algebras, and *Engel's theorem* says that if the descending central series of a Lie algebra L reaches 0, then L is isomorphic to a Lie algebra whose elements are nilpotent matrices. This is the reason such Lie algebras are called nilpotent, and the term for groups is taken from Lie algebras.

Theorem 5.36. *If G is nilpotent of class c and $H \lhd G$, then G/H is nilpotent of class $\leq c$.*

Proof. If $f: G \to L$ is a surjective homomorphism, then Lemma 5.30 gives $\gamma_i(L) \leq f(\gamma_i(G))$ for all i. Therefore, $\gamma_{c+1}(G) = 1$ forces $\gamma_{c+1}(L) = 1$. The theorem follows by taking f to be the natural map. ■

We have proved the analogues for nilpotent groups of Theorems 5.15 and 5.16; is the analogue of Theorem 5.17 true? If $H \lhd G$ and both H and G/H are nilpotent, then is G nilpotent? The answer is "no": we have already seen that S_3 is not nilpotent, but both $A_3 \cong \mathbb{Z}_3$ and $S_3/A_3 \cong \mathbb{Z}_2$ are abelian, hence nilpotent. A positive result of this type is due to P. Hall. If $H \lhd G$, then we know that $H' \lhd G$; Hall proved that if both H and G/H' are nilpotent, then G is nilpotent (a much simpler positive result is in Exercise 5.38 below). The analogue of Corollary 5.18 is true, however.

Theorem 5.37. *If H and K are nilpotent, then their direct product $H \times K$ is nilpotent.*

Proof. An easy induction shows that $\gamma_i(H \times K) \leq \gamma_i(H) \times \gamma_i(K)$ for all i. If $M = \max\{c, d\}$, where $\gamma_{c+1}(H) = 1 = \gamma_{d+1}(K)$, then $\gamma_{M+1}(H \times K) = 1$ and $H \times K$ is nilpotent. ■

Theorem 5.38. *If G is nilpotent, then it satisfies the* ***normalizer condition***: *if $H < G$, then $H < N_G(H)$.*

Proof. There exists an integer i with $\gamma_{i+1}(G) \leq H$ and $\gamma_i(G) \nleq H$ (this is true for any descending series of subgroups starting at G and ending at 1). Now $[\gamma_i, H] \leq [\gamma_i, G] = \gamma_{i+1} \leq H$, so that γ_i normalizes H; that is, $\gamma_i \leq N_G(H)$. Therefore, H is a proper subgroup of $N_G(H)$. ■

The converse is also true; it is Exercise 5.37 below.

Theorem 5.39. *A finite group G is nilpotent if and only if it is the direct product of its Sylow subgroups.*

Proof. If G is the direct product of its Sylow subgroups, then it is nilpotent, by Theorems 5.32 and 5.36.

For the converse, let P be a Sylow p-subgroup of G for some prime p. By Exercise 4.11, $N_G(P)$ is equal to its own normalizer. On the other hand, if $N_G(P) < G$, then Theorem 5.38 shows that $N_G(P)$ is a proper subgroup of its own normalizer. Therefore, $N_G(P) = G$ and $P \lhd G$. The result now follows from Exercise 4.12. ■

Of course, in any group, every subgroup of prime index is a maximal subgroup. The converse is false in general (S_4 has a maximal subgroup of index 4, as the reader should check), but it is true for nilpotent groups.

Theorem 5.40. *If G is a nilpotent group, then every maximal subgroup H is normal and has prime index.*

Proof. By Theorem 5.38, $H < N_G(H)$; since H is maximal, $N_G(H) = G$, and so $H \lhd G$. Exercise 2.58 now shows that G/H has prime order. ■

Theorem 5.41. *Let G be a nilpotent group.*

(i) *If H is a nontrivial normal subgroup, then $H \cap Z(G) \neq 1$.*
(ii) *If A is a maximal abelian normal subgroup of G, then $A = C_G(A)$.*

Proof. (i) Since $\zeta^0(G) = 1$ and $G = \zeta^c(G)$ for some c, there is an integer i for which $H \cap \zeta^i(G) \neq 1$; let m be the minimal such i. Now $[H \cap \zeta^m(G), G] \leq H \cap [\zeta^m(G), G] \leq H \cap \zeta^{m-1}(G) = 1$, because $H \lhd G$, and this says that $1 \neq H \cap \zeta^m(G) \leq H \cap Z(G)$.

(ii) Since A is abelian, $A \leq C_G(A)$. For the reverse inclusion, assume that $g \in C_G(A)$ and $g \notin A$. It is easy to see, for any subgroup H (of any group G) and for all $g \in G$, that $gC_G(H)g^{-1} = C_G(g^{-1}Hg)$. Since $A \lhd G$, it follows that $gC_G(A)g^{-1} = C_G(A)$ for all $g \in G$, and so $C_G(A) \lhd G$. Therefore, $C_G(A)/A$ is a nontrivial normal subgroup of the nilpotent group G/A; by (i), there is $Ax \in (C_G(A)/A) \cap Z(G/A)$. The correspondence theorem gives $\langle A, x \rangle$ a normal abelian subgroup of G strictly containing A, and this contradicts the maximality of A. ■

Exercises

5.35. If G is nilpotent of class 2 and if $a \in G$, then the function $G \to G$, defined by $x \mapsto [a, x]$, is a homomorphism. Conclude, in this case, that $C_G(a) \lhd G$.

5.36. If G is nilpotent of class c, then $G/Z(G)$ is nilpotent of class $c - 1$.

5.37. Show that the following conditions on a finite group G are equivalent:
(i) G is nilpotent;
(ii) G satisfies the normalizer condition;
(iii) Every maximal subgroup of G is normal.

5.38. If $H \leq Z(G)$ and if G/H is nilpotent, then G is nilpotent.

Definition. A normal series

$$G = G_1 \geq G_2 \geq \cdots \geq G_n = 1$$

with each $G_i \lhd G$ and $G_i/G_{i+1} \leq Z(G/G_{i+1})$ is called a ***central series***.

5.39. (i) If G is nilpotent, then both the upper and lower central series of G are central series.

(ii) Prove that a group G is nilpotent if and only if it has a central series $G = G_1 \geq G_2 \geq \cdots \geq G_n = 1$. Moreover, if G is nilpotent of class c, then $\gamma_{i+1}(G) \leq G_{i+1} \leq \zeta^{c-i}(G)$ for all i.

5.40. If G is a nilpotent group and H is a minimal normal subgroup of G, then $H \leq Z(G)$.

5.41. The dihedral group D_{2n} is nilpotent if and only if n is a power of 2.

5.42. Let G be a finite nilpotent group of order n. If $m|n$, then G has a subgroup of order m.

5.43. (i) If H and K are normal nilpotent subgroups of a finite group G, then HK is a normal nilpotent subgroup.
(ii) Every finite group G has a unique maximal normal nilpotent subgroup $\mathscr{F}(G)$ (which is called the ***Fitting subgroup*** of G).
(iii) Show that $\mathscr{F}(G)$ char G when G is finite.

5.44. (i) Show $\gamma_i(\mathrm{UT}(n, \mathbb{Z}_p))$ consists of all upper triangular matrices with 1's on the main diagonal and 0's on the $i - 1$ superdiagonals just above the main diagonal (*Hint*. If A is unitriangular, consider powers of $A - E$, where E is the identity matrix.)
(ii) The group $\mathrm{UT}(n, \mathbb{Z}_p)$ of all $n \times n$ unitriangular matrices over $\mathbb{Z}_p$ is a p-group that is nilpotent of class $n - 1$.

5.45. For each $n \geq 1$, let G_n be a finite p-group of class n. Define H to be the group of all sequences $(g_1, g_2, \ldots)$, with $g_n \in G_n$ for all n and with $g_n = 1$ for all large n; that is, $g_n \neq 1$ for only a finite number of g_n. Show that H is an infinite p-group which is not nilpotent.

5.46. If $x, y \in G$, denote yxy^{-1} by x^y. If $x, y, z \in G$, prove

$$[x, yz] = [x, y][x, z]^y \qquad \text{and} \qquad [xy, z] = [y, z]^x[x, z].$$

(Recall that $[x, y] = xyx^{-1}y^{-1}$.)

5.47 **(Jacobi identity).** If $x, y, z \in G$, denote $[x, [y, z]]$ by $[x, y, z]$. Prove that $[x, y^{-1}, z]^y[y, z^{-1}, x]^z[z, x^{-1}, y]^x = 1$.

5.48. (i) Let H, K, L be subgroups of G, and let $[H, K, L] = \langle [h, k, l]: h \in H, k \in K, l \in L \rangle$. Show that if $[H, K, L] = 1 = [K, L, H]$, then $[L, H, K] = 1$.
(ii) **(Three subgroups lemma).** If $N \lhd G$ and $[H, K, L][K, L, H] \leq N$, then $[L, H, K] \leq N$.
(iii) If H, K, and L are all normal subgroups of G, then $[L, H, K] \leq [H, K, L][K, L, H]$. (*Hint*. Set $N = [H, K, L][K, L, H]$.)

5.49. If G is a group with $G = G'$, then $G/Z(G)$ is centerless. (*Hint*. Use the three subgroups lemma with $H = \zeta^2(G)$ and $K = L = G$.)

5.50. Prove that $[\gamma_i(G), \gamma_j(G)] \leq \gamma_{i+j}(G)$ for all i, j. (*Hint*. Use the three subgroups lemma.)

5.51. If $H \lhd G$ and $H \cap G' = 1$, then $H \leq Z(G)$ (and so H is abelian).

p-Groups

There are many ***commutator identities*** that are quite useful even though they are quite elementary.

Lemma 5.42. *Let $x, y \in G$ and assume that both x and y commute with $[x, y]$. Then:*

(i) $[x, y]^n = [x^n, y] = [x, y^n]$ *for all* $n \in \mathbb{Z}$; *and*
(ii) $(xy)^n = [y, x]^{n(n-1)/2} x^n y^n$ *for all* $n \geq 0$.

Proof. (i) We first prove (i) for nonnegative n by induction on $n \geq 0$; of course, it is true when $n = 0$. For the inductive step, note that

$$\begin{aligned}[x, y]^n[x, y] &= x[x, y]^n yx^{-1}y^{-1}, \quad \text{by hypothesis}\\ &= x[x^n, y]yx^{-1}y^{-1}, \quad \text{by induction}\\ &= x(x^n yx^{-n}y^{-1})yx^{-1}y^{-1}\\ &= [x^{n+1}, y].\end{aligned}$$

Now $x[x, y] = [x, y]x$, by hypothesis, so that $xyx^{-1}y^{-1} = yx^{-1}y^{-1}x$; that is, $[x, y]^{-1} = [y, x^{-1}]^{-1} = [x^{-1}, y]$. Therefore, if $n \geq 0$, then $[x, y]^{-n} = [x^{-1}, y]^n = [x^{-n}, y]$, as desired.

(ii) The second identity is also proved by induction on $n \geq 0$.

$$\begin{aligned}(xy)^n(xy) &= [y, x]^{n(n-1)/2} x^n y^n xy\\ &= [y, x]^{n(n-1)/2} x^{n+1}[x^{-1}, y^n]y^{n+1}\\ &= [y, x]^{n(n-1)/2} x^{n+1}[y, x]^n y^{n+1}\\ &= [y, x]^{n(n-1)/2}[y, x]^n x^{n+1} y^{n+1}\\ &= [y, x]^{(n+1)n/2} x^{n+1} y^{n+1}. \quad ■\end{aligned}$$

Theorem 5.43. *If G is a p-group having a unique subgroup of order p and more than one cyclic subgroup of index p, then $G \cong \mathbf{Q}$, the quaternions.*

Proof. If A is a subgroup of G of index p, then $A \lhd G$, by Theorem 5.40. Thus, if $x \in G$, then $Ax \in G/A$, a group of order p, and so $x^p \in A$. Let $A = \langle a \rangle$ and $B = \langle b \rangle$ be distinct subgroups of index p, and let $D = A \cap B$; note that $D \lhd G$, for it is the intersection of normal subgroups. Our initial remarks show that the subset

$$G^p = \{x^p : x \in G\}$$

is contained in D. Since A and B are distinct maximal subgroups, it follows that $AB = G$, and so the product formula gives $[G : D] = p^2$. Hence, G/D is abelian and $G' \leq D$, by Theorem 2.23. As $G = AB$, each $x \in G$ is a product of a power of a and a power of b; but every element of D is simultaneously a power of a and a power of b, and so it commutes with each $x \in G$; that is, $D \leq Z(G)$. We have seen that

$$G' \leq D \leq Z(G),$$

so that the hypothesis of Lemma 5.42(i) holds. Hence, for every $x, y \in G$, $[y, x]^p = [y^p, x]$. But $y^p \in D \leq Z(G)$, and so $[y, x]^p = 1$. Now Lemma 5.42(ii) gives $(xy)^p = [y, x]^{p(p-1)/2} x^p y^p$. If p is odd, then $p | p(p-1)/2$, and $(xy)^p = x^p y^p$. By Exercise 2.55, if $G[p] = \{x \in G: x^p = 1\}$ and $G^p = \{x^p: x \in G\}$ (as defined above), then both these subsets are subgroups and $[G : G[p]] = |G^p|$. Thus,

$$|G[p]| = [G : G^p] = [G : D][D : G^p] \geq p^2,$$

and $G[p]$ contains a subgroup E of order p^2; but E must be elementary abelian, so that $G[p]$, hence G, contains more than one subgroup of order p. We conclude that $p = 2$.

When $p = 2$, we have $D = \langle a^2 \rangle = G^2 \leq Z(G)$, $[G : D] = 4$, and since $[y, x]^2 = 1$ for all $x, y \in G$,

$$(xy)^4 = [y, x]^6 x^4 y^4 = x^4 y^4.$$

Hence $|G[2]| = [G : G^4] = [G : D][D : G^4] = 8$, because $D = \langle a^2 \rangle$ and $G^4 = \langle a^4 \rangle$. If $G[2]$ had only one cyclic subgroup of order 4, then it would contain more than one involution (for every element of $G[2]$ has order either 1, 2, or 4); there are thus two cyclic subgroups $\langle u \rangle$ and $\langle v \rangle$ of order 4 in $G[2]$. If $a^4 \neq 1$, we may take $\langle u \rangle \leq \langle a^2 \rangle \leq Z(G)$, and so $\langle u \rangle \langle v \rangle$ is an abelian subgroup of G. But $\langle u \rangle \langle v \rangle$ contains at least two involutions: either $u^2 \neq v^2$ or $u^2 \neq uv^{-1}$; this contradiction shows that $a^4 = 1$. It follows that $|D| = 2$ and $|G| = 8$. By Exercise 4.34, $G \cong \mathbf{Q}$ or $G \cong \mathbb{Z}_8$; but only $\mathbf{Q}$ has more than one subgroup of index 2. ■

We do an exercise in congruences before giving the next theorem.

Theorem 5.44. *Let $U(\mathbb{Z}_{2^m})$ be the multiplicative group*

$$U(\mathbb{Z}_{2^m}) = \{[a] \in \mathbb{Z}_{2^m}: a \text{ is odd}\}.$$

If $m \geq 3$, then

$$U(\mathbb{Z}_{2^m}) = \langle [-1], [5] \rangle \cong \mathbb{Z}_2 \times \mathbb{Z}_{2^{m-2}}.$$

Remark. $U(\mathbb{Z}_{2^m})$ is the group of units in the ring $\mathbb{Z}_{2^m}$.

Proof. By Exercise 2.23, $|U(\mathbb{Z}_{2^m})| = \varphi(2^m) = 2^{m-1}$. Induction and the binomial theorem show that

$$5^{2^{m-3}} = (1 + 4)^{2^{m-3}} \equiv 1 + 2^{m-1} \mod 2^m.$$

Since $U(\mathbb{Z}_{2^m})$ is a 2-group, $[5]$ has order 2^s, for some $s \geq m - 2$ (because $1 + 2^{m-1} \not\equiv 1 \mod 2^m$). Of course, $[-1]$ has order 2. We claim that $\langle [5] \rangle \cap \langle [-1] \rangle = 1$. If not, then $[5^t] = [-1]$ for some t; that is,

$5^t \equiv -1 \bmod 2^m$. Since $m \geq 3$, this congruence implies $5^t \equiv -1 \bmod 4$; but $5 \equiv 1 \bmod 4$ implies $5^t \equiv 1 \bmod 4$, a contradiction. It follows that these two cyclic subgroups generate their direct product, which is a subgroup of order at least $2 \times 2^s \geq 2 \times 2^{m-2} = 2^{m-1} = \varphi(2^m)$. This subgroup is thus all of $U(\mathbb{Z}_{2^m})$. ■

Corollary 5.45. *Let G be a group containing elements x and y such that x has order 2^m (where $m \geq 3$), $y^2 = x^{2^r}$, and $yxy^{-1} = x^t$. Then*

$$t = \pm 1 \quad \text{or} \quad t = \pm 1 + 2^{m-1}.$$

In the latter two cases, G contains at least two involutions.

Proof. Since $y^2 = x^{2^r}$ commutes with x, we have

$$x = y^2 x y^{-2} = y x^t y^{-1} = x^{t^2},$$

so that $t^2 \equiv 1 \bmod 2^m$, and the congruence class $[t]$ is an element of order 2 in $U(\mathbb{Z}_{2^m})$. If $m \geq 3$, the lemma exhibits the only four such elements, and this gives the first statement.

One involution in G is $x^{2^{m-1}}$. Suppose $t = 1 + 2^{m-1}$. For any integer k,

$$(x^k y)^2 = x^k (y x^k y^{-1}) y^2 = x^{k + kt + 2^r} = x^{2s},$$

where $s = k(1 + 2^{m-2}) + 2^{r-1}$. Since $m \geq 3$, $1 + 2^{m-2}$ is odd, and we can solve the congruence

$$s = k(1 + 2^{m-2}) + 2^{r-1} \equiv 0 \mod 2^{m-1}.$$

For this choice of k, we have $(x^k y)^2 = x^{2s} = x^{2^m} = 1$, so that $x^k y$ is a second involution (lest $y \in \langle x \rangle$).

Suppose that $t = -1 + 2^{m-1}$. As above, for any integer k, $(x^k y)^2 = x^{k + kt + 2^r} = x^{k 2^{m-1} + 2^r}$. Rewrite the exponent

$$k 2^{m-1} + 2^r = 2^r (k 2^{m-r-1} - 1),$$

and choose k so that $k 2^{m-r-1} \equiv 1 \bmod 2^{m-r}$; that is, there is an integer l with $k 2^{m-r-1} - 1 = l 2^{m-r}$. For this choice of k, we have

$$(x^k y)^2 = x^{2^r (k 2^{m-r-1} - 1)} = x^{l 2^m} = 1,$$

and so G contains a second involution. ■

Theorem 5.46. *A finite p-group G having a unique subgroup of order p is either cyclic or generalized quaternion.*

Proof. The proof is by induction on n, where $|G| = p^n$; of course, the theorem is true when $n = 0$.

Assume first that p is odd. If $n > 0$, then G has a subgroup H of index p, by Exercise 4.2, and H is cyclic, by induction. There can be no other subgroup of index p, lest G be the quaternions (Theorem 5.43), which is a 2-group. Therefore, H is the unique maximal subgroup of G, and so it contains every proper subgroup of G. But if G is not cyclic, then $\langle x \rangle$ is a proper subgroup of G for every $x \in G$, and so $G \leq H$, which is absurd.

Assume now that G is a 2-group. If G is abelian, then Theorem 2.19 shows that G is cyclic; therefore, we may assume that G is not abelian. Let A be a maximal normal abelian subgroup of G. Since A has a unique involution, A is cyclic, by Theorem 2.19, say, $A = \langle a \rangle$. We claim that A has index 2. Assume, on the contrary, that $|G/A| \geq 4$. If G/A does not have exponent 2, then there is $Ab \in G/A$ with $b^2 \notin A$. Consider $H = \langle a, b^2 \rangle < \langle a, b \rangle \leq G$. If H is abelian, then b^2 centralizes A, contradicting Theorem 5.41(ii). As H is not abelian, it must be generalized quaternion, by induction. We may thus assume that $b^2ab^{-2} = a^{-1}$. Now $\langle a \rangle \lhd G$ gives $bab^{-1} = a^i$ for some i, so that

$$a^{-1} = b^2ab^{-2} = b(bab^{-1})b^{-1} = ba^ib^{-1} = a^{i^2},$$

and $i^2 \equiv -1 \bmod 2^e$, where 2^e is the order of a. Note that $e \geq 2$, for A properly contains $Z(G)$. But there is no such congruence: if $e \geq 3$, then Theorem 5.44 shows that this congruence never holds; if $e = 2$, then -1 is not a square mod 4. It follows that G/A must have exponent 2. Since $|G/A| \geq 4$, G/A contains a copy of V. Therefore, there are elements c and d with $c, d, c^{-1}d \notin A$ and with $\langle a, c \rangle$, $\langle a, d \rangle$, and $\langle a, c^{-1}d \rangle$ proper subgroups of G. Now none of these can be abelian, lest c, d, or $c^{-1}d$ centralize A, so that all three are generalized quaternion. But there are equations $cac^{-1} = a^{-1} = dad^{-1}$, giving $c^{-1}d \in C_G(A)$, a contradiction. We conclude that $A = \langle a \rangle$ must have index 2 in G.

Choose $b \in G$ with $b^2 \in \langle a \rangle$. Replacing a by another generator of A if necessary, we may assume, by Exercise 2.20, that there is some $r \leq n - 2$ with

$$b^2 = a^{2^r}.$$

Now $bab^{-1} = a^t$ for some t, because $\langle a \rangle \lhd G$. Since G has only one involution, Corollary 5.45 gives $t = \pm 1$. But $t = 1$ says that a and b commute, so that G is abelian, hence cyclic. Therefore, we may assume that $t = -1$ and $G = \langle a, b \rangle$, where

$$a^{2^{n-1}} = 1, \qquad bab^{-1} = a^{-1}, \qquad b^2 = a^{2^r}.$$

To complete the proof, we need only show that $r = n - 2$. This follows from Theorem 5.44: since $t = -1$, we have $2^r \equiv -2^r \bmod 2^{n-1}$, so that $2^{r+1} \equiv 0 \bmod 2^{n-1}$, and $r = n - 2$. ■

It is not unusual that the prime 2 behaves differently than odd primes.

Definition. If G is a group, the its ***Frattini subgroup*** $\Phi(G)$ is defined as the intersection of all the maximal subgroups of G.

If G is finite, then G always has maximal subgroups; if G is infinite, it may have no maximal subgroups. For example, let $G = \mathbb{Q}$, the additive group of rationals. Since G is abelian, a maximal subgroup H of G would be normal, and so G/H would be a simple abelian group; hence G/H would be finite and of prime order. But it is easy to see that $\mathbb{Q}$ has no subgroups of finite index (it has no finite homomorphic images).

If an (infinite) group G has no maximal subgroups, one defines $\Phi(G) = G$. It is clear that $\Phi(G)$ char G, and so $\Phi(G) \lhd G$.

Definition. An element $x \in G$ is called a ***nongenerator*** if it can be omitted from any generating set: if $G = \langle x, Y \rangle$, then $G = \langle Y \rangle$.

Theorem 5.47. *For every group G, the Frattini subgroup $\Phi(G)$ is the set of all nongenerators.*

Proof. Let x be a nongenerator of G, and let M be a maximal subgroup of G. If $x \notin M$, then $G = \langle x, M \rangle = M$, a contradiction. Therefore $x \in M$, for all M, and so $x \in \Phi(G)$. Conversely, if $z \in \Phi(G)$, assume that $G = \langle z, Y \rangle$. If $\langle Y \rangle \neq G$, then there exists a maximal subgroup M with $\langle Y \rangle \leq M$. But $z \in M$, and so $G = \langle z, Y \rangle \leq M$, a contradiction. Therefore, z is a nongenerator. ■

Theorem 5.48. *Let G be a finite group.*

(i) **(Frattini, 1885).** *$\Phi(G)$ is nilpotent.*
(ii) *If G is a finite p-group, then $\Phi(G) = G'G^p$, where G^p is the subgroup of G generated by all pth powers.*
(iii) *If G is a finite p-group, then $G/\Phi(G)$ is a vector space over $\mathbb{Z}_p$.*

Proof. (i) Let P be a Sylow p-subgroup of $\Phi(G)$ for some p. Since $\Phi(G) \lhd G$, the Frattini argument (!) gives $G = \Phi(G)N_G(P)$. But $\Phi(G)$ consists of nongenerators, and so $G = N_G(P)$; that is, $P \lhd G$ and hence $P \lhd \Phi(G)$. Therefore, $\Phi(G)$ is the direct product of its Sylow subgroups; by Theorem 5.39, $\Phi(G)$ is nilpotent.

(ii) If M is a maximal subgroup of G, where G is now a p-group, then Theorem 5.40 gives $M \lhd G$ and $[G : M] = p$. Thus, G/M is abelian, so that $G' \leq M$; moreover, G' has exponent p, so that $x^p \in M$ for all $x \in G$. Therefore, $G'G^p \leq \Phi(G)$.

For the reverse inclusion, observe that $G/G'G^p$ is an abelian group of exponent p, hence is elementary abelian, and hence is a vector space over $\mathbb{Z}_p$. Clearly $\Phi(G/G'G^p) = 1$. If $H \lhd G$ and $H \leq \Phi(G)$, then it is easy to check that $\Phi(G)$ is the inverse image (under the natural map) of $\Phi(G/H)$ (for maximal subgroups correspond). It follows that $\Phi(G) = G'G^p$.

(iii) Since $G'G^p = \Phi(G)$, the quotient group $G/\Phi(G)$ is an abelian group of exponent p; that is, it is a vector space over $\mathbb{Z}_p$. ■

Theorem 5.49 (Gaschütz, 1953). *For every (possibly infinite) group G, one has $G' \cap Z(G) \leq \Phi(G)$.*

Proof. Denote $G' \cap Z(G)$ by D. If $D \not\leq \Phi(G)$, there is a maximal subgroup M of G with $D \not\leq M$. Therefore, $G = MD$, so that each $g \in G$ has a factorization

$g = md$ with $m \in M$ and $d \in D$. Since $d \in Z(G)$, $gMg^{-1} = mdMd^{-1}m^{-1} = mMm^{-1} = M$, and so $M \lhd G$. By Exercise 2.58, G/M has prime order, hence is abelian. Therefore, $G' \leq M$. But $D \leq G' \leq M$, contradicting $D \nleq M$. ∎

Definition. A ***minimal generating set*** of a group G is a generating set X such that no proper subset of X is a generating set of G.

There is a competing definition in a finite group: a generating set of smallest cardinality. Notice that these two notions can be distinct. For example, let $G = \langle a \rangle \times \langle b \rangle$, where a has order 2 and b has order 3. Now $\{a, b\}$ is a minimal generating set, for it generates G and no proper subset of it generates. On the other hand, G is cyclic (of order 6) with generator ab, and so $\{ab\}$ is a minimal generating set of smaller cardinality. In a finite p-group, however, there is no such problem.

Theorem 5.50 (Burnside Basis Theorem, 1912). *If G is a finite p-group, then any two minimal generating sets have the same cardinality, namely,* $\dim G/\Phi(G)$. *Moreover, every $x \notin \Phi(G)$ belongs to some minimal generating set of G.*

Proof. If $\{x_1, \ldots, x_n\}$ is a minimal generating set, then the family of cosets $\{\bar{x}_1, \ldots, \bar{x}_n\}$ spans $G/\Phi(G)$ (where $\bar{x}$ denotes the coset $x\Phi(G)$). If this family is dependent, then one of them, say $\bar{x}_1$, lies in $\langle \bar{x}_2, \ldots, \bar{x}_n \rangle$. There is thus $y \in \langle x_2, \ldots, x_n \rangle \leq G$ with $x_1 y^{-1} \in \Phi(G)$. Clearly, $\{x_1 y^{-1}, x_2, \ldots, x_n\}$ generates G, so that $G = \langle x_2, \ldots, x_n \rangle$, by Theorem 5.47, and this contradicts minimality. Therefore, $n = \dim G/\Phi(G)$, and all minimal generating sets have the same cardinality.

If $x \notin \Phi(G)$, then $\bar{x} \neq 0$ in the vector space $G/\Phi(G)$, and so it is part of a basis $\{\bar{x}, \bar{x}_2, \ldots, \bar{x}_n\}$. If x_i represents the coset $\bar{x}_i$, for $i \geq 2$, then $G = \langle \Phi(G), x, x_2, \ldots, x_n \rangle = \langle x, x_2, \ldots, x_n \rangle$. Moreover, $\{x, x_2, \ldots, x_n\}$ is a minimal generating set, for the cosets of a proper subset do not generate $G/\Phi(G)$. ∎

EXERCISES

5.52. Every subgroup of $\mathbf{Q}_n$ is either cyclic or generalized quaternion.

5.53 **(Wielandt).** A finite group G is nilpotent if and only if $G' \leq \Phi(G)$.

5.54. If G is a finite p-group, then G is cyclic if and only if $G/\Phi(G)$ is cyclic.

Definition. A finite p-group G is ***extra-special*** if $Z(G)$ is cyclic and $\Phi(G) = Z(G) = G'$.

5.55. If G is extra-special, then $G/Z(G)$ is an elementary abelian group.

5.56. Every nonabelian group of order p^3 is extra-special.

5.57. (i) If m is a power of 2, what is the class of nilpotency of D_{2n}?
(ii) What is the class of nilpotency of $\mathbf{Q}_n$? (*Hint.* Exercise 4.42.)

CHAPTER 6

Finite Direct Products

The main result of this chapter is a complete description of all finite abelian groups as direct products of cyclic p-groups. By passing from abelian groups to modules over a principal ideal domain, we show that this result gives canonical forms for matrices. The essential uniqueness of the factorization of a finite abelian group as a direct product of cyclic p-groups is then generalized to nonabelian groups that are direct products of "indecomposable" groups.

The Basis Theorem

For the next few sections, we shall deal exclusively with abelian groups; as is customary, we shift from multiplicative to additive notation. Here is a dictionary of some common terms.

ab	$a + b$
1	0
a^{-1}	$-a$
a^n	na
ab^{-1}	$a - b$
HK	$H + K$
aH	$a + H$
direct product	direct sum
$H \times K$	$H \oplus K$
$\prod H_i$	$\sum H_i$

If a nonabelian group $G = H \times K$ is a direct product, then H is called a ***direct***

factor of G; in additive notation, one writes $G = H \oplus K$, and one calls H a (***direct***) ***summand*** of G.

There are two remarks greatly facilitating the study of abelian groups. First, if $a, b \in G$ and $n \in \mathbb{Z}$, then $n(a + b) = na + nb$ (in multiplicative notation, $(ab)^n = a^n b^n$, for a and b commute). Second, if X is a nonempty subset of G, then $\langle X \rangle$ is the set of all linear combinations of elements in X having coefficients in $\mathbb{Z}$ (see Theorem 2.7: in additive notation, words on X become linear combinations).

Definition. If G is an abelian p-group for some prime p, then G is also called a ***p-primary*** group.

When working wholly in the context of abelian groups, one uses the term *p-primary*; otherwise, the usage of *p-group* is preferred.

We have already proved, in Theorem 5.39, that every finite nilpotent group is the direct product of its Sylow subgroups; since every abelian group is nilpotent, the next theorem is an immediate consequence. However, we give another proof here to put the reader in the abelian mode. The following theorem was attributed to Gauss by G.A. Miller (1901).

Theorem 6.1 (Primary Decomposition). *Every finite abelian group G is a direct sum of p-primary groups.*

Proof. Since G is finite, it has exponent n for some n: we have $nx = 0$ for all $x \in G$. For each prime divisor p of n, define

$$G_p = \{x \in G: p^e x = 0 \text{ for some } e\}.$$

Now G_p is a subgroup of G, for if $p^n x = 0$ and $p^m y = 0$, where $m \leq n$, then $p^n(x - y) = 0$ (because G is abelian). We claim that $G = \sum G_p$, and we use the criterion in Exercise 2.75(i).

Let $n = p_1^{e_1} \dots p_t^{e_t}$, where the p_i are distinct primes and $e_i > 0$ for all i. Set $n_i = n/p_i^{e_i}$, and observe that the gcd $(n_1, \dots, n_t) = 1$ (no p_j divides every n_i). By Theorem VI.2 in Appendix VI, there are integers s_i with $\sum s_i n_i = 1$, and so $x = \sum (s_i n_i x)$. But $s_i n_i x \in G_{p_i}$, because $p_i^{e_i} s_i n_i x = s_i n x = 0$. Therefore, G is generated by the family of G_p's.

Assume that $x \in G_p \cap \langle \bigcup_{q \neq p} G_q \rangle$. On the one hand, $p^e x = 0$ for some $e \geq 0$; on the other hand, $x = \sum x_q$, where $q^{e_q} x_q = 0$ for exponents e_q. If $m = \prod q^{e_q}$, then m and p^e are relatively prime, and there are integers r and s with $1 = rm + sp^e$. Therefore, $x = rmx + sp^e x = 0$, and so $G_p \cap \langle \bigcup_{q \neq p} G_q \rangle = 0$. ∎

Definition. The subgroups G_p are called the ***p-primary components*** of G.

Of course, G_p is the Sylow p-subgroup of G, but the usage of *p-primary component* is preferred when the works wholly in the context of abelian groups.

We are going to show that every finite abelian group is a direct sum of cyclic groups; it now suffices to assume that G is p-primary.

Definition. A set $\{x_1, \ldots, x_r\}$ of nonzero elements in an abelian group is ***independent*** if, whenever there are integers $m_1, \ldots, m_r$ with $\sum_{i=1}^r m_i x_i = 0$, then each $m_i x_i = 0$.

When an abelian group G has exponent p, for some prime p, then it is a vector space over $\mathbb{Z}_p$, and the notion of independence just defined coincides with the usual notion of linear independence: $m_i x_i = 0$ implies $p | m_i$, so that the congruence class $[m_i] = 0$ in $\mathbb{Z}_p$. Of course, if G has no elements of finite order (as is the case, for example, when G is a vector space over $\mathbb{Q}$, $\mathbb{R}$, or $\mathbb{C}$), then $m_i x_i = 0$ implies $m_i = 0$, and so the definition of independence coincides with that of linear independence in this case as well.

Lemma 6.2. *If G is an abelian group, then a subset $\{x_1, \ldots, x_r\}$ of nonzero elements of G is independent if and only if $\langle x_1, \ldots, x_r \rangle = \langle x_1 \rangle \oplus \cdots \oplus \langle x_r \rangle$.*

Proof. Assume independence; if $y \in \langle x_i \rangle \cap \langle \{x_j : j \neq i\} \rangle$, then there are integers $m_1, \ldots, m_r$ with $y = -m_i x_i = \sum_{j \neq i} m_j x_j$, and so $\sum_{k=1}^r m_k x_k = 0$. By independence, $m_k x_k = 0$ for all k; in particular, $m_i x_i = 0$ and so $y = -m_i x_i = 0$. Exercise 2.75(i) now shows that $\langle x_1, \ldots, x_r \rangle = \langle x_1 \rangle \oplus \cdots \oplus \langle x_r \rangle$.

For the converse, assume that $\sum m_i x_i = 0$. For each j, we have $-m_j x_j = \sum_{k \neq j} m_k x_k \in \langle x_j \rangle \cap \langle \{x_k : k \neq j\} \rangle = 0$. Therefore, each $m_j x_j = 0$ and $\{x_1, \ldots, x_r\}$ is independent. ■

Here is a solution to a part of Exercise 2.78.

Corollary 6.3. *Every finite abelian group G of prime exponent p is an elementary abelian p-group.*

Proof. As a vector space over $\mathbb{Z}_p$, G has a basis $\{x_1, \ldots, x_r\}$. Therefore, $G = \langle x_1, \ldots, x_r \rangle$, because a basis spans, and $G = \langle x_1 \rangle \oplus \cdots \oplus \langle x_r \rangle$, because a basis is independent. ■

Lemma 6.4. *Let $\{x_1, \ldots, x_r\}$ be an independent subset of a p-primary abelian group G.*

(i) *If $\{z_1, \ldots, z_r\} \subset G$, where $pz_i = x_i$ for all i, then $\{z_1, \ldots, z_r\}$ is independent.*
(ii) *If $k_1, \ldots, k_r$ are integers with $k_i x_i \neq 0$ for all i, then $\{k_1 x_1, \ldots, k_r x_r\}$ is also independent.*

Proof. An exercise for the reader. ■

Definition. If G is an abelian group and $m > 0$ is an integer, then

$$mG = \{mx : x \in G\}.$$

It is easy to see that mG is a subgroup of G; indeed, since G is abelian, the function $\mu_m: G \to G$, defined by $x \mapsto mx$, is a homomorphism (called ***multiplication by m***), and $mG = \text{im } \mu_m$. We denote $\ker \mu_m$ by $G[m]$; that is,

$$G[m] = \{x \in G: mx = 0\}.$$

Theorem 6.5 (Basis Theorem).[1] *Every finite abelian group G is a direct sum of primary cyclic groups.*

Proof. By Theorem 6.1, we may assume that G is p-primary for some prime p. We prove the theorem by induction on n, where $p^nG = 0$. If $n = 1$, then the theorem is Corollary 6.3.

Suppose that $p^{n+1}G = 0$. If $H = pG$, then $p^nH = 0$, so that induction gives $H = \sum_{i=1}^r \langle y_i \rangle$. Since $y_i \in H = pG$, there are $z_i \in G$ with $pz_i = y_i$. By Lemma 6.2, $\{y_1, \ldots, y_r\}$ is independent; by Lemma 6.4 (i), $\{z_1, \ldots, z_r\}$ is independent, and so $L = \langle z_1, \ldots, z_r \rangle$ is a direct sum: $L = \sum_{i=1}^r \langle z_i \rangle$.

Here is the motivation for the next step. Were the theorem true, then $G = \sum C_k$, where each C_k is cyclic, and $H = pG = \sum pC_k$. In considering pG, therefore, we have neglected all C_k of order p, if any, for multiplication by p destroys them. The construction of L has recaptured the C_k of order greater than p, and we must now revive the C_k of order p.

For each i, let k_i be the order of y_i, so that k_iz_i has order p. The linearly independent subset $\{k_1z_1, \ldots, k_rz_r\}$ of the vector space $G[p]$ can be extended to a basis: there are elements $\{x_1, \ldots, x_s\}$ so that $\{k_1z_1, \ldots, k_rz_r, x_1, \ldots, x_s\}$ is a basis of $G[p]$. If $M = \langle x_1, \ldots, x_s \rangle$, then independence gives $M = \sum \langle x_j \rangle$. We now show that M consists of the resurrected summands of order p; that is, $G = L \oplus M$, and this will complete the proof.

(i) $L \cap M = 0$. If $g \in L \cap M$, then $g = \sum b_iz_i = \sum a_jx_j$. Now $pg = 0$, because $g \in M$, and so $\sum pb_iz_i = 0$. By independence, $pb_iz_i = b_iy_i = 0$ for all i. It follows from Exercise 2.13 that $b_i = b_i'k_i$ for some b_i'. Therefore, $0 = \sum b_i'k_iz_i - \sum a_jx_j$, and so independence of $\{k_1z_1, \ldots, k_rz_r, x_1, \ldots, x_s\}$ gives each term 0; hence $g = \sum a_jx_j = 0$.

(ii) $L + M = G$. If $g \in G$, then $pg \in pG = H$, and so $pg = \sum c_iy_i = \sum pc_iz_i$. Hence, $p(g - \sum c_iz_i) = 0$ and $g - \sum c_iz_i \in G[p]$. Therefore, $g - \sum c_iz_i = \sum b_ik_iz_i + \sum a_jx_j$, so that $g = \sum (c_i + b_ik_i)z_i + \sum a_jx_j \in L + M$. ∎

Corollary 6.6. *Every finite abelian group G is a direct sum of cyclic groups: $G = \sum_{i=1}^t \langle x_i \rangle$, where x_i has order m_i, and*

$$m_1 | m_2 | \ldots | m_t.$$

Proof. Let the primary decomposition of G be $G = \sum_{i=1}^r G_{p_i}$. By the basis theorem, we may assume that each G_{p_i} is a direct sum of cyclic groups; let

[1] The basis theorem was proved by E. Schering (1868) and, independently, by L. Kronecker (1870).

C_i be a cyclic summand of G_{p_i} of largest order, say, $p_i^{e_i}$. It follows that $G = K \oplus (C_1 \oplus \cdots \oplus C_r)$, where K is the direct sum of the remaining cyclic summands. But $C_1 \oplus \cdots \oplus C_t$ is cyclic of order $m = \prod p_i^{e_i}$, by Exercise 2.62(ii). Now repeat this construction: let $K = H \oplus D$, where D is cyclic of order n, say. If there is a cyclic summand S_i in D arising from G_{p_i}, that is, if $G_{p_i} \neq C_i$, then S_i has order $p_i^{f_i} \leq p_i^{e_i}$, so that $p_i^{f_i} | p_i^{e_i}$, for all i, and $n | m$. This process ends in a finite number of steps. ■

Definition. If G has a decomposition as a direct sum $G = \sum C_i$, where C_i is cyclic of order m_i and $m_1 | m_2 | \ldots | m_t$, then one says that G has ***invariant factors*** $(m_1, \ldots, m_t)$.

Theorem 6.7. *If p is an odd prime, the multiplicative group*

$$U(\mathbb{Z}_{p^n}) = \{[a] \in \mathbb{Z}_{p^n}: (a, p) = 1\},$$

is cyclic of order $(p-1)p^{n-1}$.

Remark. Theorem 5.44 computes this group for the prime 2.

Proof. If $n = 1$, the result is Theorem 2.18, and so we may assume that $n \geq 2$. Let us denote $U(\mathbb{Z}_{p^n})$ by G. By Exercise 2.23, $|G| = \varphi(p^n) = (p-1)p^{n-1}$.

It is easy to see that $B = \{[b] \in G: b \equiv 1 \bmod p\}$ is a subgroup of G. Every integer b has a unique expression in base p: if $1 \leq b < p^n$, then

$$b = a_0 + a_1 p + \cdots + a_{n-1}p^{n-1}, \qquad \text{where} \quad 0 \leq a_i < p.$$

Since $[b] \in B$ if and only if $a_0 = 1$, it follows that $|B| = p^{n-1}$, and so B is p-primary. By the primary decomposition, there is a subgroup A of G with $|A| = p - 1$ and with $G = A \oplus B$. If we can show that each of A and B is cyclic, then Exercise 2.62(ii) will show that G is cyclic.

Consider $f: G \to U(\mathbb{Z}_p)$ defined by $f([a]) = \operatorname{cls} a$ (where $[a]$ denotes the congruence class of $a \bmod p^n$, and $\operatorname{cls} a$ denotes the congruence class of $a \bmod p$). Clearly, f is a surjection and $\ker f = B$, so that $G/B \cong U(\mathbb{Z}_p) \cong \mathbb{Z}_{p-1}$. On the other hand, $G/B = (A \oplus B)/B \cong A$, and so $A \cong \mathbb{Z}_{p-1}$.

We shall show that B is cyclic by showing that $[1 + p]$ is a generator. Let us prove, by induction on $m \geq 0$, that

$$(1+p)^{p^m} \equiv 1 \mod p^{m+1} \qquad \text{and} \qquad (1+p)^{p^m} \not\equiv 1 \mod p^{m+2}.$$

If $m = 0$, then $1 + p \equiv 1 \bmod p$ and $1 + p \not\equiv 1 \bmod p^2$. For the inductive step, the assumed congruence gives $(1+p)^{p^{m+1}} = ((1+p)^{p^m})^p = (1 + kp^{m+1})^p$, for some integer k; the assumed incongruence gives $p \nmid k$. The binomial theorem gives $(1 + kp^{m+1})^p = 1 + kp^{m+2} + lp^{m+3}$ for some l. Hence, $(1+p)^{p^{m+1}} \equiv 1 \bmod p^{m+2}$ and $(1+p)^{p^{m+1}} \not\equiv 1 \bmod p^{m+3}$. It follows that $(1+p)^{p^{n-2}} \not\equiv 1 \bmod p^n$, and so $[1 + p]$ has order p^{n-1}. ■

Here is another proof of the basis theorem.

Lemma 6.8. *If $G = \langle x_1, \ldots, x_n \rangle$ and if $a_1, \ldots, a_n$ are relatively prime integers, then there is a generating set of G comprised of n elements one of which is $\sum_{i=1}^{n} a_i x_i$.*

Proof. By Lemma VI.4 (in Appendix VI), there is a unimodular $n \times n$ matrix A with integer entries (i.e., $\det A = 1$) whose first row is $a_1, \ldots, a_n$. Define $Y = AX$, where X is the column vector with entries $x_1, \ldots, x_n$. The entries $y_1, \ldots, y_n$ of the column vector Y are linear combinations of the x_i, hence are elements of G; moreover, $y_1 = \sum_{i=1}^{n} a_i x_i$. Now $X = A^{-1}AX = A^{-1}Y$. Since A is unimodular, all the entries of A^{-1} are also integers. It follows that each x_i is a $\mathbb{Z}$-linear combination of the y's, and so $G = \langle y_1, \ldots, y_n \rangle$. ■

Theorem 6.9 (Basis Theorem). *Every finite abelian group G is a direct sum of cyclic groups.*

***Proof* (E. Schenkman).** Choose n smallest such that G can be generated by a set with n elements. Among all generating sets $\{x_1, \ldots, x_n\}$ of size n, choose one containing an element x_1 of smallest order k; that is, no generating set of size n contains an element of order less than k. If $H = \langle x_2, \ldots, x_n \rangle$, then H is a proper subgroup of G (by the minimal choice of n), so that an induction on $|G|$ gives H a direct sum of cyclic groups. We claim that $G = \langle x_1 \rangle \oplus H$. It suffices to show that $\langle x_1 \rangle \cap H = 0$, for $\langle x_1 \rangle + H = \langle x_1, \ldots, x_n \rangle = G$. If $z \in \langle x_1 \rangle \cap H$ and $z \neq 0$, then $z = a_1 x_1 = \sum_{i=2}^{n} a_i x_i$, for $a_1, \ldots, a_n \in \mathbb{Z}$ and $0 < a_1 < k$. If d is the gcd of $a_1, \ldots, a_n$, define $g = -(a_1/d)x_1 + \sum_{i=2}^{n} (a_i/d)x_i$. Now the order of g is smaller than k, for $dg = 0$ and $d \leq a_1 < k$. But $a_1/d, \ldots, a_n/d$ are relatively prime, so that the lemma gives a generating set of G of size n one of whose elements is g; this contradicts the minimality of k. ■

Exercises

6.1. Use the basis theorem to show that if G is a finite abelian group of order n, and if $k | n$, then G contains a subgroup of order k.

6.2. Use the basis theorem to give a new proof of Theorem 2.19.

6.3. A finite abelian group G is generated by its elements of largest order. Show, by considering D_8, that this may not be true of nonabelian groups.

6.4. If G is a finite p-primary abelian group, and if $x \in G$ has largest order, then $\langle x \rangle$ is a direct summand of G.

6.5. If G is an abelian group with invariant factors $(m_1, \ldots, m_t)$, then the order of G is $\prod m_i$ and the minimal exponent of G is m_t.

6.6. If G is a finite p-primary group, then $\Phi(G) = pG$. Conclude, from the Burnside basis theorem, that $d(G)$, the minimal number of generators of G, is $\dim G/pG$.

6.7. If G and H are elementary abelian p-groups, then $d(G \oplus H) = d(G) + d(H)$.

6.8. Let G be a direct sum of b cyclic groups of order p^m. If $n < m$, then $p^nG/p^{n+1}G$ is elementary and $d(p^nG/p^{n+1}G) = b$.

The Fundamental Theorem of Finite Abelian Groups

We have not yet answered a basic question about finite abelian groups: When are two such groups G and H isomorphic? Since both G and H are direct sums of cyclic groups, your first guess is that $G \cong H$ if they have the same number of summands of each kind; since $\mathbb{Z}_6 \cong \mathbb{Z}_2 \oplus \mathbb{Z}_3$, however, one had better try to count *primary* cyclic summands of each kind. But this leads to a serious problem. How can we count summands at all? To do so would require that the number of primary cyclic summands of any given kind is the same for every decomposition of G. That is, we seek an analogue of the fundamental theorem of arithmetic in which the analogue of a prime number is a primary cyclic group.

Lemma 6.10. *If a p-primary abelian group G has a decomposition $G = \sum C_i$ into a direct sum of cyclic groups, then the number of C_i having order $\geq p^{n+1}$ is $d(p^nG/p^{n+1}G)$, the minimal number of generators of $p^nG/p^{n+1}G$.*

Proof. Let B_k be the direct sum of all C_i, if any, of order p^k; say, there are $b_k \geq 0$ such summands in B_k. Thus,

$$G = B_1 \oplus \cdots \oplus B_t.$$

Now $p^nG = p^nB_{n+1} \oplus \cdots \oplus p^nB_t$, because $p^nB_1 = \cdots = p^nB_n = 0$, and $p^{n+1}G = p^{n+1}B_{n+2} \oplus \cdots \oplus p^{n+1}B_t$. Therefore, $p^nG/p^{n+1}G \cong p^nB_{n+1} \oplus (p^nB_{n+2}/p^{n+1}B_{n+2}) \oplus \cdots \oplus (p^nB_t/p^{n+1}B_t)$, and so Exercise 6.7 gives $d(p^nG/p^{n+1}G) = b_{n+1} + b_{n+2} + \cdots + b_t$. ■

Definition. If G is a finite p-primary abelian group and $n \geq 0$, then

$$U_p(n, G) = d(p^nG/p^{n+1}G) - d(p^{n+1}G/p^{n+2}G).$$

The important thing to notice now is that $U_p(n, G)$ is a number depending on G but not upon any particular decomposition of G into a direct sum of cyclic groups.

Theorem 6.11. *If G is a finite p-primary abelian group, then any two decompositions of G into direct sums of cyclic groups have the same number of summands of each kind. More precisely, for every $n \geq 0$, the number of cyclic summands of order p^{n+1} is $U_p(n, G)$.*

Proof. For any decomposition of G into a direct sum of cyclic groups, the lemma shows that there are exactly $U_p(n, G)$ cyclic summands of order p^{n+1}. The result follows, for $U_p(n, G)$ does not depend on the choice of decomposition. ■

Corollary 6.12. *If G and H are finite p-primary abelian groups, then $G \cong H$ if and only if $U_p(n, G) = U_p(n, H)$ for all $n \geq 0$.*

Proof. If $\varphi: G \to H$ is an isomorphism, then $\varphi(p^n G) = p^n H$ for all $n \geq 0$, and so φ induces isomorphisms $p^n G/p^{n+1}G \cong p^n H/p^{n+1}H$ for all n. Therefore, $U_p(n, G) = U_p(n, H)$ for all n.

Conversely, if G and H each have direct sum decompositions into cyclic groups with the same number of summands of each kind, then it is easy to construct an isomorphism $G \to H$. ■

We have only to delete the adjective *p-primary* to complete this discussion.

Definition. The orders of the primary cyclic summands of G, that is, the numbers p^{n+1} with multiplicity $U_p(n, G) > 0$ for all primes p and all $n \geq 0$, are called the ***elementary divisors*** of G.

For example, the elementary divisors of an elementary abelian group of order p^3 are (p, p, p), and the elementary divisors of $\mathbb{Z}_6$ are $(2, 3)$.

Theorem 6.13 (Fundamental Theorem of Finite Abelian Groups).[2] *If G and H are finite abelian groups, then $G \cong H$ if and only if, for all primes p, they have the same elementary divisors.*

Proof. The proof follows from two facts, whose easy proofs are left to the reader:

(1) If $\varphi: G \to H$ is a homomorphism, then $\varphi(G_p) \leq H_p$ for all primes p;
(2) $G \cong H$ if and only if $G_p \cong H_p$ for all primes p. ■

Corollary 6.14. *Let G be a finite abelian group.*

(i) *If G has invariant factors $(m_1, \ldots, m_t)$ and invariant factors $(k_1, \ldots, k_s)$, then $s = t$ and $k_i = m_i$ for all i.*
(ii) *Two finite abelian groups G and H are isomorphic if and only if they have the same invariant factors.*

Proof. (i) The hypothesis gives two direct sum decompositions: $G = \sum_{i=1}^{t} C_i$ and $G = \sum_{j=1}^{s} D_j$, where C_i is cyclic of order m_i, D_j is cyclic of order k_j, $m_1 | m_2 | \ldots | m_t$, and $k_1 | k_2 | \ldots | k_s$. By Exercise 6.5, $m_t = k_s$, for each is the minimal exponent of G. By Exercise 6.10(i) below, the complementary summands $\sum_{i=1}^{t-1} C_i$ and $G = \sum_{j=1}^{s-1} D_j$, are isomorphic, and the proof is completed by induction on $\max\{s, t\}$.

(ii) This follows at once from (i). ■

If one arranges the elementary divisors of a p-primary group in ascending order, then they coincide with the invariant factors of G. However, elementary divisors and invariant factors can differ for groups G which are not

[2] This theorem was proved in 1879 by G. Frobenius and L. Stickelberger.

p-primary. For example, let

$$G = \mathbb{Z}_2 \oplus \mathbb{Z}_2 \oplus \mathbb{Z}_2 \oplus \mathbb{Z}_4 \oplus \mathbb{Z}_3 \oplus \mathbb{Z}_9.$$

The elementary divisors of G are (2, 2, 2, 4; 3, 9), while the invariant factors of G are (2, 2, 6, 36).

EXERCISES

6.9. If G and H are finite abelian groups, then

$$U_p(n, G \oplus H) = U_p(n, G) + U_p(n, H)$$

for all primes p and all $n \geq 0$.

6.10. (i) If A, B, and C are finite abelian groups with $A \oplus C \cong B \oplus C$, then $A \cong B$. (*Hint*. Exercise 6.9.)
(ii) If A and B are finite abelian groups for which $A \oplus A \cong B \oplus B$, then $A \cong B$.

6.11. (i) If p is a prime and $e \geq 1$, then the number of nonisomorphic abelian groups of order p^e is $\mathscr{P}(e)$, the number of partitions of e.
(ii) The number of nonisomorphic abelian groups of order $n = \prod p_i^{e_i}$ is $\prod_i \mathscr{P}(e_i)$, where the p_i are distinct primes and the e_i are positive integers.
(iii) How many abelian groups are there of order $864 = 2^5 3^3$?

6.12. (i) Let $G = \langle a \rangle \times \langle b \rangle$, where both $\langle a \rangle$ and $\langle b \rangle$ are cyclic of order p^2. If $H = \langle pa \rangle \times \langle pb \rangle$, compare $U_p(n, G)$ with $U_p(n, H)$ and $U_p(n, G/H)$.
(ii) Let G and H be finite abelian groups. If, for each k, both G and H have the same number of elements of order k, then $G \cong H$. (Compare Exercise 4.33.)

6.13. If G is a finite abelian group and $H \leq G$, then G contains a subgroup isomorphic to G/H. (Compare Exercise 4.29.)

Remark. The best solution to this exercise uses *character groups*; see Theorem 10.55.

6.14. What are the elementary divisors of $U(\mathbb{Z}_n)$, the multiplicative group of all congruence classes $[a]$ mod n with $(a, n) = 1$?

6.15. Use the Fundamental Theorem of Finite Abelian Groups to prove the Fundamental Theorem of Arithmetic. (*Hint*. If $n = p_1^{e_1} \dots p_t^{e_t}$, then Exercise 2.62(ii) gives $\mathbb{Z}/n\mathbb{Z} \cong \prod_{i=1}^{t} \mathbb{Z}/p_i^{e_i}\mathbb{Z}$.)

Canonical Forms; Existence

We digress from the study of groups to apply the results of the preceding two sections to Linear Algebra; we shall prove the existence and uniqueness of the rational and Jordan canonical forms of a matrix. This material will not be used until Chapter 8, but the reader will be pleased to see that the difficult portion of a first course in Linear Algebra can be done more easily from a more advanced viewpoint (it is assumed that the reader has already learned much of this, and so our pace is not leisurely). This project is one of translation, and so we first introduce a new vocabulary. The reader unfamiliar with the rudiments of principal ideal domains can consult Appendix VI.

Definition. Let R be a ring. An abelian group V is an ***R-module*** if there is a function $s: R \times V \to V$ (called *scalar multiplication* and denoted by $(\alpha, v) \mapsto \alpha v$) such that, for every $\alpha, \beta, 1 \in R$ and $u, v \in V$:

(i) $(\alpha\beta)v = \alpha(\beta v)$;
(ii) $(\alpha + \beta)v = \alpha v + \beta v$;
(iii) $\alpha(u + v) = \alpha u + \alpha v$; and
(iv) $1v = v$.

When R is a field, an R-module is just a vector space. Thus, one may think of an R-module as a vector space over a ring. Here we are concerned with R-modules for R a principal ideal domain (we shall abbreviate "principal ideal domain" to PID). Our favorite PID's are $\mathbb{Z}$ and $k[x]$, the ring of all polynomials in x with coefficients in a field k.

EXAMPLE 6.1. The terms *abelian group* and *$\mathbb{Z}$-module* are synomyms. Every abelian group is a $\mathbb{Z}$-module, for axioms (i) through (iv) always hold for scalars in $\mathbb{Z}$.

EXAMPLE 6.2. Let k be a field and let $R = k[x]$. If V is a vector space over k and $T: V \to V$ is a linear transformation, then V can be made into a $k[x]$-module, denoted by V^T, by defining $f(x)v = f(T)v$ for all $f(x) \in k[x]$ and $v \in V$. In more detail, if $f(x) = \alpha_0 + \alpha_1 x + \alpha_2 x^2 + \cdots + \alpha_n x^n \in k[x]$, define

$$\begin{aligned} f(x)v &= (\alpha_0 + \alpha_1 x + \alpha_2 x^2 + \cdots + \alpha_n x^n)v \\ &= \alpha_0 v + \alpha_1 Tv + \alpha_2 T^2 v + \cdots + \alpha_n T^n v, \end{aligned}$$

where T^i is the composite of T with itself i times. The reader should check that axioms (i) through (iv) do hold.

Just as a principal ideal domain is a generalization of $\mathbb{Z}$, so is an R-module a generalization of an abelian group. Almost any theorem holding for abelian groups has a true analogue for R-modules when R is a PID; moreover, the proofs of the generalizations are usually translations of the proofs for abelian groups.

Definition. If V is an R-module, then a subgroup W of V is a ***submodule*** if it closed under scalar multiplication: if $w \in W$ and $r \in R$, then $rw \in W$. If W is a submodule of V, then the ***quotient module*** V/W is the abelian group V/W equipped with the scalar multiplication $r(v + W) = rv + W$ (the reader should check that this is well defined).

EXERCISES

6.16. A commutative ring R itself is an R-module (if $r, s \in R$, define scalar multiplication rs to be the given product of two elements in R). Show that the submodules of R are its ideals.

6.17. (i) The intersection of any family of submodules of an R-module V is itself a submodule.
(ii) If X is a subset of V, let $\langle X \rangle$ denote the submodule ***generated*** by X; that is, $\langle X \rangle$ is the intersection of all the submodules of V containing X. If $X \neq \varnothing$, show that $\langle X \rangle$ is the set of all ***R-linear combinations*** of elements of X; that is,

$$\langle X \rangle = \{\text{finite sums } \sum r_i x_i \colon r_i \in R \text{ and } x_i \in X\}.$$

In particular, the ***cyclic submodule*** generated by v, denoted by $\langle v \rangle$, is $\{rv \colon r \in R\}$.

6.18. An R-module V is called ***finitely generated*** if there is a finite subset X wth $V = \langle X \rangle$.
(i) If R is a field, prove that an R-module V is finitely generated if and only if it is finite-dimensional.
(ii) Prove that an abelian group ($\mathbb{Z}$-module) G is finite if and only if G is finitely generated and every element in G has finite order.

6.19. Let V^T be the $k[x]$-module of Example 6.2. Prove that W is a submodule if and only if W is a subspace of V for which $T(W) \leq W$ (W is often called a ***T-invariant subspace*** of V).

Definition. If V and W are R-modules, then their ***direct sum*** is the direct sum $V \oplus W$ of abelian groups equipped with the scalar multiplication $r(v, w) = (rv, rw)$.

6.20. If $W_1, \ldots, W_n$ are submodules of an R-module V, then $V \cong W_1 \oplus \cdots \oplus W_n$ if and only if $V = W_1 + \cdots + W_n$ (i.e., every $v \in V$ is a sum $v = \sum w_i$, where $w_i \in W_i$) and $W_i \cap \langle \bigcup_{j \neq i} W_j \rangle = 0$ for all i.

We are almost finished with the vocabulary lesson.

Definition. Let V be an R-module, where R is a PID, and let $v \in V$. The ***order*** of v, denoted by $\operatorname{ord}(v)$, is $\{r \in R \colon rv = 0\}$ (it is easily checked that $\operatorname{ord}(v)$ is an ideal in R). One says that v has ***finite order*** if $\operatorname{ord}(v) \neq 0$, and one says that V is ***p-primary*** (where $p \in R$ is irreducible) if, for all $v \in V$, $\operatorname{ord}(v) = (p^m)$ for some m (where (p) is the principal ideal generated by p). An R-module V is ***finite*** if it is finitely generated and every element of V has finite order.

If V is an abelian group and $v \in V$, then a generator of $\operatorname{ord}(v)$ is the smallest positive integer m for which $mv = 0$; that is, $\operatorname{ord}(v) = (m)$, where m is the order of v in the usual sense. Exercise 6.18(ii) tells us that we have translated "finite abelian group" correctly into the language of R-modules: a "finite $\mathbb{Z}$-module" is an abelian group of finite order.

We remark that Exercise 6.18(ii) is false if one drops the hypothesis that the group G is abelian. The question (posed in 1902) whether a finitely generated group G of exponent e is finite became known as ***Burnside's problem.***[3] Burnside proved that if $G \leq \mathrm{GL}(n, \mathbb{C})$ is finitely generated and has exponent e, then G is finite. There is an "obvious" candidate for a counterexample, and it was actually shown to be one, when e is a large odd number, by Adian and Novikov in 1968 (in 1975, Adian showed that the group is infinite for all odd $e \geq 665$). The proof is very long and intricate; a much simpler "geometric" proof was found by A. Ol'shanskii in 1982. In 1994, S. Ivanov showed that there are infinite finitely generated groups of exponent $e = 2^k m$, where $k \geq 48$ and $m \geq 1$ is any odd number.

Theorem 6.15. *If V is a finite R-module, where R is a PID, then there are $v_1, \ldots, v_s \in V$ with*

$$V = \langle v_1 \rangle \oplus \cdots \oplus \langle v_s \rangle.$$

Moreover, the cyclic summands may be chosen to satisfy either of the following conditions. If $\mathrm{ord}(v_i) = (r_i)$, *then either*:

(i) *each r_i is a power of some irreducible element in R; or*
(ii) $r_1 | r_2 | \ldots | r_s$.

Proof. The proofs of the corresponding group-theoretic theorems translate routinely to proofs for modules. The decomposition of the first type arises from Corollary 6.12 (using the primary decomposition and elementary divisors), and the decomposition of the second type arises from Corollary 6.6 (using invariant factors). ■

Corollary 6.16. *Let $T: V \to V$ be a linear transformation on a finite-dimensional vector space over a field k. Then $V = W_1 \oplus \cdots \oplus W_s$, where each $W_i = \langle v_i \rangle$ is a cyclic T-invariant subspace. Moreover, the vectors $v_1, \ldots, v_s$ can be chosen with* $\mathrm{ord}(v_i) = (f_i(x))$, *so that either*:

(i) *each $f_i(x)$ is a power of an irreducible polynomial in $k[x]$; or*
(ii) $f_1(x) | f_2(x) | \ldots | f_s(x)$.

Proof. Regard V as a $k[x]$-module V^T, as in Example 6.2. Since V is finite-dimensional, it has a basis $\{w_1, \ldots, w_n\}$; each vector $v \in V$ is a k-linear combi-

[3] The ***restricted Burnside Problem*** asks whether there is a function $f(n, d)$ with $|G| \leq f(n, d)$ for every finite group G having minimal exponent n and d generators. P. Hall and G. Higman (1956) proved that it suffices to find such a function for all prime powes n; A.L. Kostrikin (1959) found such a function for all primes p; E.I. Zelmanov (1989) completed the proof by finding such a function for all prime powers; see Vaughan-Lee (1993).

nation $v = \sum \alpha_j w_j$; *a fortiori*, each v is a $k[x]$-linear combination of the w's, and so V^T is a finitely generated $k[x]$-module. Also, every $v \in V$ is annihilated by some nonzero polynomial (this follows from the Cayley–Hamilton theorem or, more simply, from the observation that since $\dim(V) = n$, the $n + 1$ vectors $v, Tv, T^2v, \ldots, T^nv$ must be linearly dependent, and so there is some nonzero polynomial in T that annihilates every vector v).

Since V^T is a finite $k[x]$-module, Theorem 6.15 shows that it is a direct sum of cyclic submodules. By Exercise 6.19, these summands are cyclic T-invariant subspaces. ■

Lemma 6.17. *Let $T: V \to V$ be a linear transformation on a finite-dimensional vector space V.*

(i) *A subspace W is a cyclic T-invariant subspace of V if and only if there is a vector $v \in W$ and an integer $s \geq 1$ so that $\{v, Tv, T^2v, \ldots, T^{s-1}v\}$ is a basis of W.*
(ii) *Moreover, if $T^sv = \sum_{i=0}^{s-1} \alpha_i T^iv$, then* $\mathrm{ord}(v)$ *is generated by $g(x) = x^s - \sum_{i=0}^{s-1} \alpha_i x^i$.*

Proof. (i) Consider the sequence $v, Tv, T^2v, \ldots$ in W; since V is finite-dimensional, there is an integer $s \geq 1$ and a linearly independent subset $\{v, Tv, T^2v, \ldots, T^{s-1}v\}$ which becomes linearly dependent when T^sv is adjoined. Therefore, there are $\alpha_i \in k$ with $T^sv = \sum_{i=0}^{s-1} \alpha_i T^iv$. If $w \in W$, then $w = f(T)v$ for some $f(x) \in k[x]$, and an easy induction on degree f shows that w lies in the subspace spanned by $\{v, Tv, T^2v, \ldots, T^{s-1}v\}$; it follows that the subset is a basis of W.

(ii) It is clear that $g(x) \in \mathrm{ord}(v)$. If $h(x) \in \mathrm{ord}(v)$, that is, if $h(T)v = 0$, then the division algorithm gives $q(x), r(x) \in k[x]$ with $h(x) = q(x)g(x) + r(x)$ and either $r(x) = 0$ or degree $r(x) <$ degree $g(x) = s$. Now $r(x) \in \mathrm{ord}(v)$; hence $r(x) = \sum_{j=0}^{t} \beta_j x^j$, $t \leq s - 1$, and $\sum_{j=0}^{t} \beta_j T^jv = 0$, contradicting the linear independence of $\{v, Tv, T^2v, \ldots, T^{s-1}v\}$. ■

We remind the reader of the correspondence between linear transformations and matrices. Let V be a vector space over a field k with an ordered basis $\{u_1, \ldots, u_r\}$, and let $T: V \to V$ be a linear transformation. For each j, Tu_j is a linear combination of the u_i: there are $\alpha_{ij} \in k$ with

$$Tu_j = \sum \alpha_{ij} u_i.$$

The ***matrix of T relative to the ordered basis*** $\{u_1, \ldots, u_r\}$ is defined to be $A = [\alpha_{ij}]$. Therefore, for each j, the coordinates of Tu_j form the jth *column* of A.

Definition. If $f(x) = x^r + \alpha_{r-1}x^{r-1} + \cdots + \alpha_0 \in k[x]$, where $r \geq 2$, then the

companion matrix of $f(x)$ is the $r \times r$ matrix $C(f)$:

$$C(f) = \begin{bmatrix} 0 & 0 & 0 & \dots & 0 & -\alpha_0 \\ 1 & 0 & 0 & \dots & 0 & -\alpha_1 \\ 0 & 1 & 0 & \dots & 0 & -\alpha_2 \\ 0 & 0 & 1 & \dots & 0 & -\alpha_3 \\ \dots & \dots & \dots & \dots & \dots & \dots \\ 0 & 0 & 0 & \dots & 1 & -\alpha_{m-1} \end{bmatrix};$$

if $f(x) = x - \alpha$, then $C(f)$ is the 1×1 matrix $[\alpha]$.

Lemma 6.18. *Let $T: W \to W$ be a linear transformation on a finite-dimensional vector space W over k, and let $W = W^T = \langle v \rangle$ be a cyclic $k[x]$-module. The matrix of T relative to the basis $\{v, Tv, T^2v, \dots, T^{s-1}v\}$ of W is the $s \times s$ companion matrix $C(g)$, where $g(x)$ is the monic generator of* $\text{ord}(v)$. *Moreover, the characteristic polynomial of $C(g)$ is $g(x)$.*

Proof. If $0 \le i < s-1$, then $T(T^i v) = T^{i+1}v$, and $T(T^{s-1}v) = T^s v = \sum_{i=0}^{s-1} \alpha_i T^i v$, so that the matrix of T relative to the given basis is a companion matrix. The Cayley–Hamilton theorem gives $\chi(T) = 0$, where $\chi(x)$ is the characteristic polynomial of $C(g)$. Therefore $\chi(x) \in \text{ord}(v) = (g(x))$, so that $g(x) | \chi(x)$. Hence $s = \text{degree } \chi(x) \ge \text{degree } g(x) = s$, and so $\chi(x) = g(x)$. ∎

Definition. Let A be an $r \times r$ matrix and let B be an $s \times s$ matrix; their ***direct sum*** is the $(r + s) \times (r + s)$ matrix

$$\begin{bmatrix} A & 0 \\ 0 & B \end{bmatrix}.$$

Note that the direct sum of A and B is similar to the direct sum of B and A.

Theorem 6.19. *Every $n \times n$ matrix A over a field k is similar to a direct sum of companion matrices $C(f_1), \dots, C(f_q)$. Moreover, the $f_i(x)$ may be chosen so that either:*

(i) *the $f_i(x)$ are powers of irreducible polynomials in $k[x]$; or*
(ii) $f_1(x) | f_2(x) | \dots | f_q(x)$.

Proof. Let $V = k^n$, the vector space of all n-tuples of elements in k (viewed as column vectors). The ***standard basis*** of V is $\{e_1, \dots, e_n\}$, where e_i is the n-tuple with ith coordinate 1 and all other coordinates 0. The matrix A defines a linear transformation $T: V \to V$ by $T(v) = Av$, where v is a column vector. Note that A is the matrix of T relative to the standard basis, for Ae_i is the ith column of A. View V as the $k[x]$-module V^T. By Corollary 6.13, V is a direct sum of cyclic T-invariant subspaces: $V = \langle v_1 \rangle \oplus \dots \oplus \langle v_q \rangle$; by Lemma 6.17, there is a new basis of V:

$$\{v_1, Tv_1, T^2v_1, \ldots; v_2, Tv_2, T^2v_2, \ldots; \ldots; v_q, Tv_q, T^2v_q, \ldots\}.$$

The matrix B of T relative to this new basis is a direct sum of companion matrices, by Lemma 6.18, and A is similar to B, for they represent the same linear transformation relative to different ordered bases of V. Finally, Corollary 6.16 shows that the v_i can be chosen so that the polynomials $f_i(x)$ satisfy either (i) or (ii). ■

Definition. A ***rational canonical form*** is a matrix B that is the direct sum of companion matrices $C(f_1), \ldots, C(f_q)$ with $f_1(x) | f_2(x) | \ldots | f_q(x)$. The polynomials $f_1(x), f_2(x), \ldots, f_q(x)$ are called the ***invariant factors*** of B.

Theorem 6.19(ii) thus says that every matrix over a field k is similar to a rational canonical form.

Recall that the ***minimum polynomial*** of a matrix A is the monic polynomial $m(x)$ of smallest degree with $m(A) = 0$. The reader should look again at Exercise 6.5 to realize that the characteristic polynomial of a rational canonical form B is analogous to the order of a finite abelian group and the minimum polynomial is analogous to the minimal exponent.

In Chapter 8, we will study groups whose elements are nonsingular matrices. Since the order of a group element is the same as the order of any of its conjugates, the order of such a matrix is the order of its rational canonical form. It is difficult to compute powers of companion matrices, and so we introduce another canonical form (when the field of entries is large enough) whose powers are easily calculated.

Definition. An $s \times s$ ***Jordan block*** is an $s \times s$ matrix of the form

$$\begin{bmatrix} \alpha & 0 & 0 & \ldots & 0 & 0 \\ 1 & \alpha & 0 & \ldots & 0 & 0 \\ 0 & 1 & \alpha & \ldots & 0 & 0 \\ 0 & 0 & 1 & \ldots & 0 & 0 \\ \multicolumn{6}{c}{\dotfill} \\ 0 & 0 & 0 & \ldots & \alpha & 0 \\ 0 & 0 & 0 & \ldots & 1 & \alpha \end{bmatrix}.$$

A 1×1 Jordan block has the form $[\alpha]$.

Let K denote the $n \times n$ matrix consisting of all 0's except for 1's on the first subdiagonal below the main diagonal; thus, a Jordan block has the form $\alpha E + K$, where E is the $n \times n$ identity matrix. Note that K^2 is all 0's except for 1's on the second subdiagonal below the main diagonal, K^3 is all 0's except for 1's on the third subdiagonal, etc., and $K^n = 0$.

Lemma 6.20. *If $J = \alpha E + K$ is an $n \times n$ Jordan block, then*

$$J^m = \alpha^m E + \sum_{i=1}^{n-1} \binom{m}{i} \alpha^{m-i} K^i$$

(we agree that $\binom{m}{i} = 0$ if $i > m$.)

Proof. The binomial theorem applies because αE and K commute. The sum is from 1 to $n - 1$ because $K^n = 0$. ■

Lemma 6.20 is very useful because the matrices K^i are "disjoint." For example,

$$\begin{bmatrix} \alpha & 0 \\ 1 & \alpha \end{bmatrix}^m = \begin{bmatrix} \alpha^m & 0 \\ m\alpha^{m-1} & \alpha^m \end{bmatrix}$$

and

$$\begin{bmatrix} \alpha & 0 & 0 \\ 1 & \alpha & 0 \\ 0 & 1 & \alpha \end{bmatrix}^m = \begin{bmatrix} \alpha^m & 0 & 0 \\ m\alpha^{m-1} & \alpha^m & 0 \\ \binom{m}{2}\alpha^{m-2} & m\alpha^{m-1} & \alpha^m \end{bmatrix}.$$

Theorem 6.21. *If A is an $n \times n$ matrix over a field k which contains all the eigenvalues of A, then A is similar to a direct sum of Jordan blocks.*

Proof. Theorem 6.19(i) shows that it suffices to prove that a companion matrix $C(f)$, with $f(x)$ a power of an irreducible polynomial, is similar to a Jordan block. The hypothesis on k gives $f(x) = (x - \alpha)^s$ for some $\alpha \in k$. Let W be the subspace with basis $\{v, Tv, T^2v, \ldots, T^{s-1}v\}$, where T is the linear transformation arising from the companion matrix $C(f)$. Consider the subset $\mathscr{B} = \{u_0, u_1, \ldots, u_{s-1}\}$ of W, where $u_0 = v$, $u_1 = (T - \alpha E)v, \ldots, u_{s-1} = (T - \alpha E)^{s-1}v$. It is plain that $\mathscr{B}$ spans W, for $T^i v \in \langle u_0, \ldots, u_i \rangle$ for all i; since $|\mathscr{B}| = s$, it follows that $\mathscr{B}$ is an ordered basis of W.

Let us compute the matrix J of T relative to $\mathscr{B}$. If $j + 1 \leq s$,

$$\begin{aligned} Tu_j &= T(T - \alpha E)^j v \\ &= (T - \alpha E)^j Tv \\ &= (T - \alpha E)^j [\alpha E + (T - \alpha E)]v \\ &= \alpha(T - \alpha E)^j v + (T - \alpha E)^{j+1} v. \end{aligned}$$

If $j + 1 < s$, then $Tu_j = \alpha u_j + u_{j+1}$; if $j + 1 = s$, then $(T - \alpha E)^{j+1} = (T - \alpha E)^s = 0$, by the Cayley–Hamilton theorem (Lemma 6.18 identifies $f(x) = (x - \alpha)^s$ as the characteristic polynomial of $C(f)$, hence of T). Therefore $Tu_{s-1} = \alpha u_{s-1}$. The matrix J is thus a Jordan block; it is similar to $C(f)$ because both represent the same linear transformation relative to different ordered bases of W. ■

Definition. A ***Jordan canonical form*** is a matrix B that is a direct sum of Jordan blocks $J_1, \ldots, J_q$. Each J_i determines a polynomial $g_i(x)$ which is a power of an irreducible polynomial, and the polynomials $g_1(x), \ldots, g_q(x)$ are called the ***elementary divisors*** of B.

Theorem 6.21 thus says that a matrix A is similar to a Jordan canonical form if the ground field k contains all the eigenvalues of A. In particular, if k is algebraically closed, then this is always the case.

Canonical Forms; Uniqueness

Our discussion is still incomplete, for we have not yet considered uniqueness; can a matrix A be similar to several rational canonical forms? Can A be similar to several Jordan canonical forms? We have used the module analogue of the basis theorem; we are now going to use the module analogue of the fundamental theorem.

Definition. If V and W are R-modules, then a function $\varphi\colon V \to W$ is an ***R-homomorphism*** if

$$\varphi(v + v') = \varphi(v) + \varphi(v')$$

and

$$\varphi(\alpha v) = \alpha\varphi(v)$$

for all $v, v' \in V$ and $\alpha \in R$; if φ is a bijection, then it is called an ***R-isomorphism***. Two modules V and W are called ***R-isomorphic***, denoted by $V \cong W$, if there exists an R-isomorphism $\varphi\colon V \to W$.

If R is a field, then an R-homomorphism is an R-linear transformation; if V and W are abelian groups, then every homomorphism is a $\mathbb{Z}$-homomorphism. If $R = k[x]$ and V and W are $k[x]$-modules, then $\varphi\colon V \to W$ is a k-linear transformation such that

$$\varphi(f(x)v) = f(x)\varphi(v)$$

for all $v \in V$ and $f(x) \in k[x]$.

Exercises

6.21. Prove the first isomorphism theorem for modules. (*Hint.* Since modules are abelian groups, the reader need check only that the isomorphism in Theorem 2.24 is an R-homomorphism.)

6.22. Every cyclic R-module $V = \langle v \rangle$ is R-isomorphic to $R/\operatorname{ord}(v)$. Conclude that two cyclic modules are R-isomorphic if and only if they have generators with the same order ideal.

6.23. If R is a PID and $a, b \in R$ are relatively prime, then $R/(ab) \cong R/(a) \oplus R/(b)$.

Theorem 6.22 (Fundamental Theorem). *If R is a PID and V and W are finite R-modules (i.e., they are finitely generated and every element has finite order), then $V \cong W$ if and only if either they have the same invariant factors or the same elementary divisors.*

Proof. Translate Corollary 6.14 into the language of modules. ■

We continue the analysis of $k[x]$-modules in order to apply this theorem to matrices.

Lemma 6.23. *Let V and W be vector spaces over a field k, let $T: V \to V$ and $S: W \to W$ be linear transformations, and let V^T and W^S be the corresponding $k[x]$-modules. A function $\varphi: V \to W$ is a $k[x]$-homomorphism if and only if φ is a linear transformation such that $\varphi(Tv) = S\varphi(v)$ for all $v \in V$.*

Proof. If $\varphi: V^T \to W^S$ is a $k[x]$-homomorphism, then $\varphi(f(x)v) = f(x)\varphi v$ for all $v \in V$ and all $f(x) \in k[x]$. In particular, this equation holds for every constant $f(x)$, so that φ is a k-linear transformation, and for $f(x) = x$, so that $\varphi(xv) = x\varphi v$. But $xv = Tv$ and $x(\varphi v) = S\varphi v$, as desired.

The converse is easy: if equality $\varphi(f(x)v) = f(x)\varphi v$ holds for every $v \in V$ whenever $f(x)$ constant or $f(x) = x$, then it holds for every polynomial. ■

Theorem 6.24. *If A and B are $n \times n$ matrices over a field k, then A and B are similar if and only if the corresponding $k[x]$-modules they determine are $k[x]$-isomorphic.*

Proof. Construct the modules determined by the matrices: let V be the vector space of all column vectors of n-tuples of elements in k; define $T, S: V \to V$ by $Tv = Av$ and $Sv = Bv$. As usual, write V^T to denote V made into a $k[x]$-module by $xv = T(v) = Av$, and write V^S to denote V made into a module by $xv = S(v) = Bv$.

If A and B are similar, then there is a nonsingular matrix P with $PAP^{-1} = B$. Now P defines an invertible linear transformation $\varphi: V \to V$ by $\varphi v = Pv$. We claim that φ is a $k[x]$-isomorphism. By Lemma 6.23, it suffices to show that $\varphi T = S\varphi$, and this follows from the given matrix equation $PT = SP$.

Conversely, assume that there is a $k[x]$-isomorphism $\varphi: V^T \to V^S$. By Lemma 6.23, $\varphi T = S\varphi$; since φ is a bijection, $\varphi T \varphi^{-1} = S$. If P is the matrix of φ relative to the standard basis of V, this gives $PAP^{-1} = B$, and so A and B are similar. ■

Theorem 6.25. *Two $n \times n$ matrices A and B over a field k are similar if and only if they have the same invariant factors. Moreover, a matrix is similar to exactly one rational canonical form.*

Proof. Only necessity needs proof. Since similarity is an equivalence relation, we may assume that both A and B are rational canonical forms. By Theorem 6.24, A and B are similar if and only if V^T and V^S are $k[x]$-isomorphic (where $T(v) = Av$ and $S(v) = Bv$). Recall that the invariant factors are monic polynomials $f_i(x)$ with $f_1(x)|f_2(x)|\dots|f_q(x)$; if $V^T = \sum C_i$, where C_i is a cyclic module with order ideal $(f_i(x))$, then Theorem 6.22 says that the invariant factors are, indeed, invariant; that is, they do not change after an isomorphism is applied to V^T. Therefore, V^T and V^S have the same invariant factors. But the invariant factors of A and of B are just the polynomials determined by the last columns of their companion matrices, and so A and B are equal. ■

Theorem 6.26. *Two $n \times n$ matrices A and B over a field k containing their eigenvalues are similar if and only if they have the same elementary divisors. Moreover, if a matrix is similar to Jordan canonical forms J and J', then J and J' have the same Jordan blocks.*

Proof. The proof is essentially the same as that of Theorem 6.25, with companion matrices replaced by Jordan blocks. ■

Note that the rational canonical form of a matrix A is absolutely unique, whereas the Jordan canonical form is unique only up to a permutation of the Jordan blocks occurring in it.

This discussion of canonical forms has the disadvantage of not showing how to compute the invariant factors of any particular matrix, and a Linear Algebra course should include a discussion of the ***Smith canonical form*** which provides an algorithm for displaying them; see Cohn (1982). Let A be an $n \times n$ matrix over a field k. Using elementary row and column operations over the ring $k[x]$, one can put the matrix $xE - A$ into diagonal form $(d_1(x), \dots, d_n(x))$, where each $d_i(x)$ is a monic polynomial or 0, and $d_1(x)|d_2(x)|\dots|d_n(x)$. The invariant factors of A turn out to be those $d_i(x)$ which are neither constant nor 0.

There are shorter proofs of Theorems 6.25 and 6.26, involving matrix computations; for example, see Albert (1941).

Exercises

6.24. If A is a companion matrix, then its characteristic and minimum polynomials are equal.

6.25. (i) Give an example of two nonisomorphic finite abelian groups having the same order and the same minimal exponent.

(ii) Give an example of two complex $n \times n$ matrices which have the same characteristic polynomials and the same minimum polynomials, yet which are not similar.

6.26. If b and b' are nonzero elements of a field k, then

$$\begin{bmatrix} a & b \\ 0 & c \end{bmatrix} \quad \text{and} \quad \begin{bmatrix} a & b' \\ 0 & c \end{bmatrix}$$

are similar.

6.27. Let A and B be $n \times n$ matrices with entries in a field k. If k is a subfield of a field K, then A and B are similar over k if and only if they are similar over K. (*Hint.* A rational canonical form for A over k is also a rational canonical form for A over K.)

6.28 **(Jordan Decompositions).** Let k be an algebraically closed field.

(i) Every matrix A over k can be written $A = D + N$, where D is ***diagonalizable*** (i.e., similar to a diagonal matrix), N is nilpotent, and $DN = ND$. (It may be shown that D and N are unique.)

(ii) Every nonsingular matrix A over k can be written $A = DU$, where D is diagonalizable, U is ***unipotent*** (i.e., $U - E$ is nilpotent), and $DU = UD$. (It may be shown that D and U are unique.) (*Hint*: Define $U = E + ND^{-1}$.)

The Krull–Schmidt Theorem

If a (not necessarily abelian) group is a direct product of subgroups, each of which cannot be decomposed further, are the factors unique to isomorphism? The Krull–Schmidt theorem, the affirmative answer for a large class of groups (which contains all the finite groups), is the main result of this section. This is another instance in which the name of a theorem does not coincide with its discoverers. The theorem was first stated by J.M.H. Wedderburn in 1909, but his proof had an error. The first correct proof for finite groups was given by R. Remak in 1911, with a simplification by O.J. Schmidt in 1912. The theorem was extended to modules by W. Krull in 1925 and to operator groups by Schmidt in 1928 (we shall discuss operator groups in the next section).

Let us return to the multiplicative notation for groups.

Definition. An ***endomorphism*** of a group G is a homomorphism $\varphi: G \to G$.

There are certain endomorphisms of a group G that arise quite naturally when G is a direct product (see Exercise 2.73).

Definition. If $G = H_1 \times \cdots \times H_m$, then the maps $\pi_i: G \to H_i$, defined by $\pi_i(h_1 \dots h_m) = h_i$, are called ***projections***.

If the inclusion $H_i \hookrightarrow G$ is denoted by λ_i, then the maps $\lambda_i \pi_i$ are endomorphisms of G. Indeed, they are ***idempotent***: $\lambda_i \pi_i \circ \lambda_i \pi_i = \lambda_i \pi_i$.

Definition. An endomorphism φ of a group G is ***normal*** if $\varphi(axa^{-1}) = a\varphi(x)a^{-1}$ for all $a, x \in G$.

It is easy to see that if G is a direct product, then the maps $\lambda_i \pi_i$ are normal endomorphisms of G.

Lemma 6.27.

(i) *If φ and ψ are normal endomorphisms of a group G, then so is their composite $\varphi \circ \psi$.*
(ii) *If φ is a normal endomorphism of G and if $H \lhd G$, then $\varphi(H) \lhd G$.*
(iii) *If φ is a normal automorphism of a group G, then φ^{-1} is also normal.*

Proof. The proofs are routine calculations. ■

Here is a new way to combine endomorphisms of a group; unfortunately, the new function is not always an endomorphism.

Definition. If φ and ψ are endomorphisms of a group G, then $\varphi + \psi: G \to G$ is the function defined by $x \mapsto \varphi(x)\psi(x)$.

If G is abelian, then $\varphi + \psi$ is always an endomorphism. If $G = S_3$, φ is conjugation by $(1\ 2\ 3)$, and ψ is conjugation by $(1\ 3\ 2)$, then $\varphi + \psi$ is not an endomorphism of S_3.

It is easy to see that if φ and ψ are normal endomorphisms and if $\varphi + \psi$ is an endomorphism, then it is normal. The equation $(i \circ q)(x)(j \circ p)(x) = x$ for all $x \in G$ in Exercise 2.73 may now be written $1_G = iq + jp$.

Lemma 6.28. *Let $G = H_1 \times \cdots \times H_m$ have projections $\pi_i: G \to H_i$ and inclusions $\lambda_i: H_i \hookrightarrow G$. Then the sum of any k distinct $\lambda_i \pi_i$ is a normal endomorphism of G. Moreover, the sum of all the $\lambda_i \pi_i$ is the identity function on G.*

Proof. Note that $\lambda_i \pi_i(h_1 \ldots h_m) = h_i$. If $\varphi = \sum_{i=1}^{k} \lambda_i \pi_i$ (we consider the first k maps for notational convenience), then $\varphi(h_1 \ldots h_m) = h_1 \ldots h_k$; that is, $\varphi = \lambda\pi$, where π is the projection of G onto the direct factor $H_1 \times \cdots \times H_k$ and λ is the inclusion of $H_1 \times \cdots \times H_k$ into G. It follows that φ is a normal endomorphism of G and, if $k = m$, that $\varphi = 1_G$. ■

Definition. A group G is ***indecomposable*** if $G \neq 1$ and if $G = H \times K$, then either $H = 1$ or $K = 1$.

We now consider a condition on a group that will ensure that it is a direct product of indecomposable groups (for there do exist groups without this property).

Definition. A group G has **ACC** (***ascending chain condition***) if every increasing chain of normal subgroups *stops*; that is, if

$$K_1 \leq K_2 \leq K_3 \leq \cdots$$

is a chain of normal subgroups of G, then there is an integer t for which $K_t = K_{t+1} = K_{t+2} = \cdots$.

A group G has **DCC** (***descending chain condition***) if every decreasing chain of normal subgroups stops; that is, if

$$H_1 \geq H_2 \geq H_3 \geq \cdots.$$

is a chain of normal subgroups of G, then there is an integer s for which $H_s = H_{s+1} = H_{s+2} = \cdots$.

A group G has ***both chain conditions*** if it has both chain conditions!

Every finite group has both chain conditions. The group $\mathbb{Z}$ has ACC but not DCC; in Chapter 10, we shall meet a group $\mathbb{Z}(p^\infty)$ with DCC but not ACC; the additive group of rationals $\mathbb{Q}$ has neither ACC nor DCC.

Lemma 6.29.

(i) *If $H \lhd G$ and both H and G/H have both chain conditions, then G has both chain conditions. In particular, if H and K have both chain conditions, then so does $H \times K$.*

(ii) *If $G = H \times K$ and G has both chain conditions, then each of H and K has both chain conditions.*

Proof. (i) If $G_1 \geq G_2 \geq \cdots$ is a chain of normal subgroups of G, then $H \cap G_1 \geq H \cap G_2 \geq \cdots$ is a chain of normal subgroups of H and $HG_1/H \geq HG_2/H \geq \cdots$ is a chain of normal subgroups of G/H. By hypothesis, there is an integer t with $H \cap G_t = H \cap G_{t+1} = \cdots$, and there is an integer s with $HG_s/H = HG_{s+1}/H = \cdots$; that is, $HG_s = HG_{s+1} = \cdots$. Let $l = \max\{s, t\}$. By the Dedekind law (Exercise 2.49), for all $i \geq l$,

$$\begin{aligned} G_i = G_i H \cap G_i &= G_{i+1} H \cap G_i \\ &= G_{i+1}(H \cap G_i) = G_{i+1}(H \cap G_{i+1}) \leq G_{i+1}, \end{aligned}$$

and so $G_l = G_{l+1} = \cdots$. A similar argument holds for ascending chains.

(ii) If $G = H \times K$, then every normal subgroup of H is also a normal subgroup of G. Therefore, every (ascending or descending) chain of normal subgroups of H is a chain of normal subgroups of G, hence must stop. ■

Lemma 6.30. *If G has either chain condition, then G is a direct product of a finite number of indecomposable groups.*

Proof. Call a group *good* if it satisfies the conclusion of this lemma; call it *bad* otherwise. An indecomposable group is good and, if both A and B are good groups, then so is $A \times B$. Therefore, a bad group G is a direct product, say, $G = U \times V$, with both U and V proper subgroups, and with U or V bad.

Suppose there is a bad group G. Define $H_0 = G$. By induction, for every n, there are bad subgroups $H_0, H_1, \ldots, H_n$ such that each H_i is a proper bad direct factor of H_{i-1}. There is thus a strictly decreasing chain of normal subgroups of G

$$G = H_0 > H_1 > H_2 > \cdots;$$

if G has DCC, we have reached a contradiction.

Suppose that G has ACC. Since each H_i is a direct factor of H_{i-1}, there are normal subgroups K_i with $H_{i-1} = H_i \times K_i$. There is thus an ascending chain of normal subgroups

$$K_1 < K_1 \times K_2 < K_1 \times K_2 \times K_3 < \cdots,$$

and we reach a contradiction in this case, too. ■

Lemma 6.31. *Let G have both chain conditions. If φ is a normal endomorphism of G, then φ is an injection if and only if it is a surjection. (Thus, either property ensures that φ is an automorphism.)*

Proof. Suppose that φ is an injection and that $g \notin \varphi(G)$. We prove, by induction, that $\varphi^n(g) \notin \varphi^{n+1}(G)$. Otherwise, there is an element $h \in G$ with $\varphi^n(g) = \varphi^{n+1}(h)$, so that $\varphi(\varphi^{n-1}(g)) = \varphi(\varphi^n(h))$. Since φ is an injection, $\varphi^{n-1}(g)) = \varphi^n(h)$, contradicting the inductive hypothesis. There is thus a strictly decreasing chain of subgroups

$$G > \varphi(G) > \varphi^2(G) > \cdots.$$

Now φ normal implies φ^n is normal; by Lemma 6.27, $\varphi^n(G) \lhd G$ for all n, and so the DCC is violated. Therefore, φ is a surjection.

Assume that φ is a surjection. Define $K_n = \ker \varphi^n$; each K_n is a normal subgroup of G because φ^n is a homomorphism (the normality of φ^n is here irrelevant). There is an ascending chain of normal subgroups

$$1 = K_0 \leq K_1 \leq K_2 \leq \cdots.$$

This chain stops because G has ACC; let t be the smallest integer for which $K_t = K_{t+1} = K_{t+2} = \cdots$. We claim that $t = 0$, which will prove the result. If $t \geq 1$, then there is $x \in K_t$ with $x \notin K_{t-1}$; that is, $\varphi^t(x) = 1$ and $\varphi^{t-1}(x) \neq 1$. Since φ is a surjection, there is $g \in G$ with $x = \varphi(g)$. Hence, $1 = \varphi^t(x) = \varphi^{t+1}(g)$, so that $g \in K_{t+1} = K_t$. Therefore, $1 = \varphi^t(g) = \varphi^{t-1}(\varphi(g)) = \varphi^{t-1}(x)$, a contradiction. Thus, φ is an injection. ■

Definition. An endomorphism φ of G is ***nilpotent*** if there is a positive integer k such that $\varphi^k = 0$, where 0 denotes the endomorphism which sends every element of G into 1.

Theorem 6.32 (Fitting's Lemma, 1934). *Let G have both chain conditions and let φ be a normal endomorphism of G. Then $G = K \times H$, where K and H are each invariant under φ (i.e., $\varphi(K) \leq K$ and $\varphi(H) \leq H$), $\varphi|K$ is nilpotent, and $\varphi|H$ is a surjection.*

Proof. Let $K_n = \ker \varphi^n$ and let $H_n = \operatorname{im} \varphi^n$. As in the proof of Lemma 6.31, there are two chains of normal subgroups of G:

$$G \geq H_1 \geq H_2 \geq \cdots \quad \text{and} \quad 1 \leq K_1 \leq K_2 \leq \cdots.$$

Since G has both chain conditions, each of these chains stops: the H_n after t steps, the K_n after s steps. Let l be the larger of t and s, so that $H_l = H_{l+1} = H_{l+2} = \cdots$ and $K_l = K_{l+1} = K_{l+2} = \cdots$. Define $H = H_l$ and $K = K_l$; it is easy to check that both H and K are invariant under φ.

Let $x \in H \cap K$. Since $x \in H$, there is $g \in G$ with $x = \varphi^l(g)$; since $x \in K$, $\varphi^l(x) = 1$. Therefore, $\varphi^{2l}(g) = \varphi^l(x) = 1$, so that $g \in K_{2l} = K_l$. Hence, $x = \varphi^l(g) = 1$, and so $H \cap K = 1$.

If $g \in G$, then $\varphi^l(g) \in H_l = H_{2l}$, so there is $y \in G$ with $\varphi^l(g) = \varphi^{2l}(y)$. Applying φ^l to $g\varphi^l(y^{-1})$ gives 1, so that $g\varphi^l(y^{-1}) \in K_l = K$. Therefore, $g = [g\varphi^l(y^{-1})]\varphi^l(y) \in KH$, and so $G = K \times H$.

Now $\varphi(H) = \varphi(H_l) = \varphi(\varphi^l(G)) = \varphi^{l+1}(G) = H_{l+1} = H_l = H$, so that φ is a surjection. Finally, if $x \in K$, then $\varphi^l(x) \in K \cap H = 1$, and so $\varphi|K$ is nilpotent. ∎

Corollary 6.33. *If G is an indecomposable group having both chain conditions, then every normal endomorphism φ of G is either nilpotent or an automorphism.*

Proof. By Fitting's lemma, $G = K \times H$ with $\varphi|K$ nilpotent and $\varphi|H$ surjective. Since G is indecomposable, either $G = K$ or $G = H$. In the first case, φ is nilpotent. In the second case, φ is surjective and, by Lemma 6.31, φ is an automorphism. ∎

Lemma 6.34. *Let G be an indecomposable group with both chain conditions, and let φ and ψ be normal nilpotent endomorphisms of G. If $\varphi + \psi$ is an endomorphism of G, then it is nilpotent.*

Proof. We have already observed that if $\varphi + \psi$ is an endomorphism, then it is normal, and so Corollary 6.33 says that it is either nilpotent or an automorphism. If $\varphi + \psi$ is an automorphism, then Lemma 6.27 (iii) says that its inverse γ is also normal. For each $x \in G$, $x = (\varphi + \psi)\gamma x = \varphi\gamma(x)\psi\gamma(x)$, so that if we define $\lambda = \varphi\gamma$ and $\mu = \psi\gamma$, then $1_G = \lambda + \mu$. In particular, $x^{-1} = \lambda(x^{-1})\mu(x^{-1})$ and, taking inverses, $x = \mu(x)\lambda(x)$; that is, $\lambda + \mu = \mu + \lambda$. The equation $\lambda(\lambda + \mu) = (\lambda + \mu)\lambda$ (which holds because $\lambda + \mu = 1_G$) implies that $\lambda\mu = \mu\lambda$. If follows that the set of all endomorphisms of G obtained from λ and μ forms an algebraic system[4] in which the binomial theorem holds: for every integer $m > 0$,

$$(\lambda + \mu)^m = \sum_i \binom{m}{i} \varphi^i \psi^{m-i}.$$

[4] This algebraic system (called a *semiring*) is not a commutative ring because additive inverses need not exist.

Nilpotence of φ and ψ implies nilpotence of $\lambda = \varphi\gamma$ and $\mu = \psi\gamma$ (they cannot be automorphisms because they have nontrivial kernels); there are thus positive integers r and s with $\lambda^r = 0$ and $\mu^s = 0$. If $m = r + s - 1$, then either $i \geq r$ or $m - i \geq s$. It follows that each term $\lambda^i\mu^{m-i}$ in the binomial expansion of $(\lambda + \mu)^m$ is 0: if $i \geq r$, then $\lambda^i = 0$; if $m - i \geq s$, then $\mu^{m-i} = 0$. Hence, $(1_G)^m = (\lambda + \mu)^m = 0$, and $1_G = 0$, forcing $G = 1$. This is a contradiction, for every indecomposable group is nontrivial. ■

Corollary 6.35. *Let G be an indecomposable group having both chain conditions. If $\varphi_1, \ldots, \varphi_n$ is a set of normal nilpotent endomorphisms of G such that every sum of distinct φ's is an endomorphism, then $\varphi_1 + \cdots + \varphi_n$ is nilpotent.*

Proof. Induction on n. ■

Theorem 6.36 (Krull–Schmidt). *Let G be a group having both chain conditions. If*

$$G = H_1 \times \cdots \times H_s = K_1 \times \cdots \times K_t$$

are two decompositions of G into indecomposable factors, then $s = t$ and there is a reindexing so that $H_i \cong K_i$ for all i. Moreover, given any r between 1 and s, the reindexing may be chosen so that

$$G = H_1 \times \cdots \times H_r \times K_{r+1} \times \cdots \times K_s.$$

Remark. The last conclusion is stronger than saying that the factors are determined up to isomorphism; one can replace factors of one decomposition by suitable factors from the other.

Proof. We shall give the proof when $r = 1$; the reader may complete the proof for general r by inducton. Given the first decomposition, we must find a reindexing of the K's so that $H_i \cong K_i$ for all i and $G = H_1 \times K_2 \times \cdots \times K_t$. Let $\pi_i\colon G \to H_i$ and $\lambda_i\colon H_i \hookrightarrow G$ be the projections and inclusions from the first decomposition, and let $\sigma_j\colon G \to K_j$ and $\mu_j\colon K_j \hookrightarrow G$ be the projections and inclusions from the second decomposition. The maps $\lambda_i\pi_i$ and $\mu_j\sigma_j$ are normal endomorphisms of G.

By Lemma 6.28, every partial sum $\sum \mu_j\sigma_j$ is a normal endomorphism of G. Hence, every partial sum of

$$1_{H_1} = \pi_1\lambda_1 = \pi_1 \circ 1_G \circ \lambda_1 = \pi_1 \circ \left(\sum \mu_j\sigma_j\right) \circ \lambda_1 = \sum \pi_1\mu_j\sigma_j\lambda_1$$

is a normal endomorphism of H_1. Since $1_{H_1} = \pi_1\lambda_1$ is not nilpotent, Lemma 6.29 and Corollary 6.35 give an index j with $\pi_1\mu_j\sigma_j\lambda_1$ an automorphism. We reindex so that $\pi_1\mu_1\sigma_1\lambda_1$ is an automorphism of H_1; let γ be its inverse.

We claim that $\sigma_1\lambda_1\colon H_1 \to K_1$ is an isomorphism. The definition of γ gives $(\gamma\pi_1\mu_1)(\sigma_1\lambda_1) = 1_{H_1}$. To compute the composite in the reverse order, let

$\theta = \sigma_1 \lambda_1 \gamma \pi_1 \mu_1 : K_1 \to K_1$, and note that $\theta^2 = \theta$:

$$\theta \circ \theta = \sigma_1 \lambda_1 [\gamma \pi_1 \mu_1 \sigma_1 \lambda_1] \gamma \pi_1 \mu_1 = \theta$$

(the term in brackets is 1_{H_1}). Now $1_{H_1} = 1_{H_1} \circ 1_{H_1} = \gamma\pi_1\mu_1\sigma_1\lambda_1\gamma\pi_1\mu_1\sigma_1\lambda_1 = \gamma\pi_1\mu_1\theta\sigma_1\lambda_1$; it follows that $\theta \neq 0$ (lest $1_{H_1} = 0$). Were θ nilpotent, then $\theta^2 = \theta$ would force $\theta = 0$. Therefore, θ is an automorphism of K_1, and so $\theta^2 = \theta$ gives $\theta = 1$ (multiply each side by θ^{-1}). It follows that $\sigma_1\lambda_1 : H_1 \to K_1$ is an isomorphism (with inverse $\gamma\pi_1\mu_1$).

Now σ_1 sends $K_2 \times \cdots \times K_t$ into 1 while $\sigma_1\lambda_1$ restricts to an isomorphism on H_1. Therefore

$$H_1 \cap (K_2 \times \cdots \times K_t) = 1.$$

If we define $G^* = \langle H_1, K_2 \times \cdots \times K_t \rangle \leq G$, then

$$G^* = H_1 \times K_2 \times \cdots \times K_t.$$

If $x \in G$, then $x = k_1 k_2 \dots k_t$, where $k_j \in K_j$. Since $\pi_1\mu_1$ is an isomorphism, the map $\beta: G \to G$, defined by $x \mapsto \pi_1\mu_1(k_1)k_2 \dots k_t$, is an injection with image G^*. By Lemma 6.31, β is a surjection; that is, $G = G^* = H_1 \times K_2 \times \cdots \times K_t$. Finally,

$$K_2 \times \cdots \times K_t \cong G/H_1 \cong H_2 \times \cdots \times H_s,$$

so that the remaining uniqueness assertions follow by induction on $\max\{s, t\}$. ■

Exercises

6.29. Show that the following groups are indecomposable: $\mathbb{Z}$; $\mathbb{Z}_{p^n}$; $\mathbb{Q}$; S_n; D_{2n}; $\mathbf{Q}_n$; simple groups; nonabelian groups of order p^3; A_4; the group T of order 12 (see Theorem 4.24).

6.30. Assuming the Basis Theorem, use the Krull–Schmidt theorem to prove the Fundamental Theorem of Finite Abelian Groups.

6.31. If G has both chain conditions, then there is no proper subgroup H of G with $G \cong H$, and there is no proper normal subgroup K of G with $G/K \cong G$.

6.32. Assume that G has both chain conditions. If there is a group H with $G \times G \cong H \times H$, then $G \cong H$. (*Hint*. Use Lemma 6.29(ii).)

6.33.[5] Let G have both chain conditions. If $G \cong A \times B$ and $G \cong A \times C$, then $B \cong C$.

6.34. Let G be the additive group of $\mathbb{Z}[x]$. Prove that $G \times \mathbb{Z} \cong G \times \mathbb{Z} \times \mathbb{Z}$, but that $\mathbb{Z} \ncong \mathbb{Z} \times \mathbb{Z}$.

Definition. A subgroup $H \leq G$ is ***subnormal*** if there is a normal series

$$G \geq G_1 \geq G_2 \geq \cdots \geq H \geq 1.$$

[5] R. Hirshon (*Amer. Math. Monthly* **76** (1969), pp. 1037–1039) proves that if A is finite and B and C are arbitrary groups, then $A \times B \cong A \times C$ implies $B \cong C$.

6.35. (i) Give an example of a subgroup that is not subnormal.
(ii) Give an example of a subnormal subgroup that is not a normal subgroup.
(iii) If G has a composition series and if H is subnormal in G, then G has a composition series one of whose terms is H.
(iv) A group G has a composition series if and only if it has both chain conditions on subnormal subgroups.

Operator Groups

There is a generalization of modules that allows us to extend some of our earlier theorems in a straightforward way. We present new results having old proofs.

Definition. Let Ω be a set and let G be a group. Then Ω is a set of ***operators*** on G and G is an ***Ω-group*** if there is a function $\Omega \times G \to G$, denoted by $(\omega, g) \mapsto \omega g$, such that

$$\omega(gh) = (\omega g)(\omega h)$$

for all $\omega \in \Omega$ and $g, h \in G$.

Definition. If G and H are Ω-groups, then a function $\varphi: G \to H$ is an ***Ω-map*** if φ is a homomorphism such that $\varphi(\omega g) = \omega\varphi(g)$ for all $\omega \in \Omega$ and $g \in G$.

If G is an Ω-group, then a subgroup $H \leq G$ is an ***admissible subgroup*** if $\omega h \in H$ for all $\omega \in \Omega$ and $h \in H$.

EXAMPLE 6.3. If $\Omega = \varnothing$, then an Ω-group is just a group, every homomorphism is an Ω-map, and every subgroup is admissible.

EXAMPLE 6.4. If G is an abelian group and Ω is a ring, then every Ω-module is an Ω-group, every Ω-homomorphism is an Ω-map, and every submodule is admissible.

EXAMPLE 6.5. If Ω is the set of all conjugations of a group G, then G is an Ω-group whose admissible subgroups are the normal subgroups. An Ω-map of G to itself is a normal endomorphism. An Ω-isomorphism between Ω-groups is called a ***central isomorphism.***

EXAMPLE 6.6. If Ω is the set of all automorphisms of a group G, then G is an Ω-group whose admissible subgroups are the characteristic subgroups.

EXAMPLE 6.7. If Ω is the set of all endomorphisms of a group G, then G is an Ω-group whose admissible subgroups are the fully invariant subgroups.

All the elementary results of Chapter 2 (and their proofs!) carry over for Ω-groups. For example, the intersection of admissible subgroups is admissible and, if H and K are admissible subgroups at least one of which is normal, then HK is admissible. If H is a normal admissible subgroup of an Ω-group G, then the quotient group G/H is an Ω-group (where one defines $\omega(gH) = (\omega g)H$). The kernel of an Ω-map is an admissible normal subgroup, and the first, hence the second and third, isomorphism theorems hold, as does the correspondence theorem. The direct product of Ω-groups H and K becomes an Ω-group if one defines $\omega(h, k) = (\omega h, \omega k)$.

Definition. An Ω-group G is ***Ω-simple*** if has no admissible subgroups other than 1 and G.

An admissible normal subgroup H of an Ω-group G is maximal such if and only if G/H is Ω-simple.

If Ω is a ring, then an Ω-module V with no submodules other than 0 and V is an Ω-simple group. In particular, when Ω is a field, a one-dimensional Ω-module is Ω-simple. If Ω is the set of all conjugations of G, then Ω-simple groups are just simple groups in the usual sense. If Ω is the set of all automorphisms of a finite group G, then G is Ω-simple (or *characteristically simple*) if and only if it is a direct product of isomorphic simple groups (Theorem 5.26).

We call attention to three generalizations of results from groups to Ω-groups (the proofs are routine adaptations of the proofs we have given for groups).

Definition. Let G be an Ω-groups. An ***Ω-series*** for G is a normal series

$$G = G_0 \geq G_1 \geq \cdots \geq G_n = 1$$

with each G_i admissible; an ***Ω-composition series*** is an Ω-series whose factor groups are Ω-simple.

If V is abelian and Ω is a ring, then every normal series $V = V_0 \geq V_1 \geq \cdots \geq V_n = 1$ in which the V_i are submodules is an Ω-series; it is an Ω-composition series if each V_i/V_{i+1} is Ω-simple. In particular, if Ω is a field, then V is a vector space and the factors are one-dimensional spaces.

If Ω is the set of all conjugations of G, then an Ω-series is a normal series $G = G_0 \geq G_1 \geq \cdots \geq G_n = 1$ in which each G_i is a normal subgroup of G (such a normal series is called a ***chief series*** or a *principal series*). If Ω is the set of all automorphisms of G, then an Ω-series is a normal series in which each term is a characteristic subgroup of G. Both the Zassenhaus lemma and the Schreier refinement theorem carry over for Ω-groups; it follows that a generalized Jordan–Hölder theorem is true.

Theorem 6.37 (Jordan–Hölder). *Every two Ω-composition series of an Ω-group are equivalent.*

One can now prove that if V is a finite-dimensional vector space over a field k (that is, V is spanned by finitely many vectors), then the size n of a basis

$\{v_1, \ldots, v_n\}$ depends only on V. For $i \geq 0$, define subspaces V_i of V by $V_i = \langle v_{i+1}, \ldots, v_n \rangle$, the subspace spanned by $\{v_{i+1}, \ldots, v_n\}$. Then $V = V_0 \geq V_1 \geq \cdots \geq V_n = 0$ is an Ω-composition series for V (where $\Omega = k$) because the factor spaces are all isomorphic to k, hence are Ω-simple. Therefore, the dimension of V is well defined, for it is the length of an Ω-composition series of V.

Here are two more applications. Any two chief composition series of a group have centrally isomorphic factor groups (Ω = conjugations), and any two characteristic series in which the factor groups are products of isomorphic simple groups have isomorphic factor groups (Ω = automorphisms).

Theorem 6.38 (Fitting's Lemma). *Let G be an Ω-group having both chain conditions on admissible subgroups, and let φ be an Ω-endomorphism of G. Then $G = H \times K$, where H and K are admissible subgroups, $\varphi|H$ is nilpotent, and $\varphi|K$ is a surjection.*

Here is an application. Let V be a finite-dimensional vector space over a field k and let $T: V \to V$ be a linear transformation. Then $V = U \oplus W$, where U and W are T-invariant, $T|U$ is nilpotent, and $T|W$ is nonsingular. We have taken $\Omega = k$ and observed that $T|W$ surjective implies $T|W$ nonsingular. The matrix interpretation of Fitting's lemma thus says that every $n \times n$ matrix over a field k is similar to the direct sum of a nilpotent matrix and a nonsingular matrix.

An Ω-group G is ***Ω-indecomposable*** if it is not the direct product of nontrivial admissible subgroups.

Theorem 6.39 (Krull–Schmidt). *Let G be an Ω-group having both chain conditions on admissible subgroups. If*

$$G = H_1 \times \cdots \times H_s = K_1 \times \cdots \times K_t$$

are two decompositions of G into Ω-indecomposable factors, then $s = t$ and there is a reindexing so that $H_i \cong K_i$ for all i. Moreover, given any r between 1 and s, the reindexing may be chosen so that $G = H_1 \times \cdots \times H_r \times K_{r+1} \times \cdots \times K_s$.

CHAPTER 7

Extensions and Cohomology

A group G having a normal subgroup K can be "factored" into K and G/K. The study of extensions involves the inverse question: Given $K \lhd G$ and G/K, to what extent can one recapture G?

The Extension Problem

Definition. If K and Q are groups, then an ***extension*** of K by Q is a group G having a normal subgroup $K_1 \cong K$ with $G/K_1 \cong Q$.

As a mnemonic device, K denotes kernel and Q denotes quotient. We think of G as a "product" of K and Q.

EXAMPLE 7.1. Both $\mathbb{Z}_6$ and S_3 are extensions of $\mathbb{Z}_3$ by $\mathbb{Z}_2$. However, $\mathbb{Z}_6$ is an extension of $\mathbb{Z}_2$ by $\mathbb{Z}_3$, but S_3 is not such an extension (for S_3 has no normal subgroup of order 2).

EXAMPLE 7.2. For any groups K and Q, the direct product $K \times Q$ is an extension of K by Q as well as an extension of Q by K.

The extension problem (formulated by O. Hölder) is to find all extensions of a given group K by a given group Q. We can better understand the Jordan–Hölder theorem in light of this problem. Let a group G have a composition series

$$G = K_0 \geq K_1 \geq \cdots \geq K_{n-1} \geq K_n = 1$$

and corresponding factor groups

$$K_0/K_1 = Q_1, \ldots, K_{n-2}/K_{n-1} = Q_{n-1}, K_{n-1}/K_n = Q_n.$$

Since $K_n = 1$, we have $K_{n-1} = Q_n$, but something more interesting happens at the next stage; $K_{n-2}/K_{n-1} = Q_{n-1}$, so that K_{n-2} is an extension of K_{n-1} by Q_{n-1}. If we could solve the extension problem, then we could recapture K_{n-2} from K_{n-1} and Q_{n-1}; that is, from Q_n and Q_{n-1}. Once we have K_{n-2}, we can attack K_{n-3} in a similar manner, for $K_{n-3}/K_{n-2} = Q_{n-2}$. Thus, a solution of the extension problem would allow us to recapture K_{n-3} from Q_n, Q_{n-1}, and Q_{n-2}. Climbing up the composition series to $K_0 = G$, we could recapture G from $Q_n, \ldots, Q_1$. The group G is thus a "product" of the Q_i, and the Jordan–Hölder theorem says that the simple groups Q_i in this "factorization" of G are uniquely determined by G. We could thus survey all finite groups if we knew all finite simple groups and if we could solve the extension problem. In particular, we could survey all finite solvable groups if we could solve the extension problem.

A solution of the extension problem consists of determining from K and Q all the groups G for which $G/K \cong Q$. But what does "determining" a group mean? We gave two answers to this question at the end of Chapter 1 when we considered "knowing" a group. One answer is that a multiplication table for a group G can be constructed; a second answer is that the isomorphism class of G can be characterized. In 1926, O. Schreier determined all extensions in the first sense (see Theorem 7.34). On the other hand, no solution is known in the second sense. For example, given K and Q, Schreier's solution does not allow us to compute the number of nonisomorphic extensions of K by Q (though it does give an upper bound).

EXERCISES

7.1. If K and Q are finite, then every extension G of K by Q has order $|K||Q|$. If G has a normal series with factor groups $Q_n, \ldots, Q_1$, then $|G| = \prod |Q_i|$.

7.2. (i) Show that A_4 is an extension of $\mathbf{V}$ by $\mathbb{Z}_3$.
(ii) Find all the extensions of $\mathbb{Z}_3$ by $\mathbf{V}$.

7.3. If p is prime, every nonabelian group of order p^3 is an extension of $\mathbb{Z}_p$ by $\mathbb{Z}_p \times \mathbb{Z}_p$. (*Hint.* Exercise 4.7.)

7.4. Give an example of an extension of K by Q that does not contain a subgroup isomorphic to Q.

7.5. If $(a, b) = 1$ and K and Q are abelian groups of orders a and b, respectively, then there is only one (to isomorphism) abelian extension of K by Q.

7.6. Which of the following properties, when enjoyed by both K and Q, is also enjoyed by every extension of K by Q? (i) finite; (ii) p-group; (iii) abelian; (iv) cyclic; (v) solvable; (vi) nilpotent; (vii) ACC; (viii) DCC; (ix) ***periodic*** (every element has finite order); (x) ***torsion-free*** (every element other than 1 has infinite order).

Automorphism Groups

The coming construction is essential for the discussion of the extension problem; it is also of intrinsic interest.

Definition. The ***automorphism group*** of a group G, denoted by $\mathrm{Aut}(G)$, is the set of all the automorphisms of G under the operation of composition.

It is easy to check that $\mathrm{Aut}(G)$ is a group; indeed, it is a subgroup of the symmetric group S_G.

Definition. An automorphism φ of G is ***inner*** if it is conjugation by some element of G; otherwise, it is ***outer***. Denote the set of all inner automorphisms of G by $\mathrm{Inn}(G)$.

Theorem 7.1.

(i) **(N/C Lemma).** *If $H \leq G$, then $C_G(H) \lhd N_G(H)$ and $N_G(H)/C_G(H)$ can be imbedded in* $\mathrm{Aut}(H)$.
(ii) $\mathrm{Inn}(G) \lhd \mathrm{Aut}(G)$ *and* $G/Z(G) \cong \mathrm{Inn}(G)$.

Proof. (i) If $a \in G$, let γ_a denote conjugation by a. Define $\varphi: N_G(H) \to \mathrm{Aut}(H)$ by $a \mapsto \gamma_a|H$ (note that $\gamma_a|H \in \mathrm{Aut}(H)$ because $a \in N_G(H)$); φ is easily seen to be a homomorphism. The following statements are equivalent: $a \in \ker \varphi$; $\gamma_a|H$ is the identity on H; $aha^{-1} = h$ for all $h \in H$; $a \in C_G(H)$. By the first isomorphism theorem, $C_G(H) \lhd N_G(H)$ and $N_G(H)/C_G(H) \cong \mathrm{im}\ \varphi \leq \mathrm{Aut}(H)$.

(ii) If $H = G$, then $N_G(G) = G$, $C_G(G) = Z(G)$, and $\mathrm{im}\ \varphi = \mathrm{Inn}(G)$. Therefore, $G/Z(G) \cong \mathrm{Inn}(G)$ is a special case of the isomorphism just established.

To see that $\mathrm{Inn}(G) \lhd \mathrm{Aut}(G)$, take $\gamma_a \in \mathrm{Inn}(G)$ and $\varphi \in \mathrm{Aut}(G)$. Then $\varphi\gamma_a\varphi^{-1} = \gamma_{\varphi a} \in \mathrm{Inn}(G)$, as the reader can check. ■

Definition. The group $\mathrm{Aut}(G)/\mathrm{Inn}(G)$ is called the ***outer automorphism group*** of G.

EXAMPLE 7.3. $\mathrm{Aut}(\mathbf{V}) \cong S_3 \cong \mathrm{Aut}(S_3)$.

The 4-group $\mathbf{V}$ consists of 3 involutions and 1, and so every $\varphi \in \mathrm{Aut}(\mathbf{V})$ permutes the 3 involutions: if $X = \mathbf{V} - \{1\}$, then the map $\varphi \mapsto \varphi|X$ is a homomorphism $\mathrm{Aut}(\mathbf{V}) \to S_X \cong S_3$. The reader can painlessly check that this map is an isomorphism.

The symmetric group S_3 consists of 3 involutions, 2 elements of order 3, and the identity, and every $\varphi \in \mathrm{Aut}(S_3)$ must permute the involutions: if $Y = \{(1\ 2), (1\ 3), (2\ 3)\}$, then the map $\varphi \mapsto \varphi|Y$ is a homomorphism $\mathrm{Aut}(S_3) \to S_Y \cong S_3$; this map is easily seen to be an isomorphism.

We conclude that nonisomorphic groups can have isomorphic automorphism groups.

EXAMPLE 7.4. If G is an elementary abelian group of order p^n, then $\text{Aut}(G) \cong \text{GL}(n, p)$.

This follows from Exercise 2.78: G is a vector space over $\mathbb{Z}_p$ and every automorphism is a nonsingular linear transformation.

EXAMPLE 7.5. $\text{Aut}(\mathbb{Z}) \cong \mathbb{Z}_2$.

Let $G = \langle x \rangle$ be infinite cyclic. If $\varphi \in \text{Aut}(G)$, then $\varphi(x)$ must be a generator of G. Since the only generators of G are x and x^{-1}, there are only two automorphisms of G, and so $\text{Aut}(\mathbb{Z}) \cong \text{Aut}(G) \cong \mathbb{Z}_2$. Thus, an infinite group can have a finite automorphism group.

EXAMPLE 7.6. $\text{Aut}(G) = 1$ if and only if $|G| \leq 2$.

It is clear that $|G| \leq 2$ implies $\text{Aut}(G) = 1$. Conversely, assume that $\text{Aut}(G) = 1$. If $a \in G$, then $\gamma_a = 1$ if and only if $a \in Z(G)$; it follows that G is abelian. The function $a \mapsto a^{-1}$ is now an automorphism of G, so that G has exponent 2; that is, G is a vector space over $\mathbb{Z}_2$. If $|G| > 2$, then $\dim G \geq 2$ and there exists a nonsingular linear transformation $\varphi: G \to G$ other than 1.

Recall that if R is a ring, then $U(R)$ denotes its group of units:

$$U(R) = \{r \in R\text{: there is } s \in R \text{ with } sr = 1 = rs\}.$$

Lemma 7.2. *If G is a cyclic group of order n, then* $\text{Aut}(G) \cong U(\mathbb{Z}_n)$.

Proof. Let $G = \langle a \rangle$. If $\varphi \in \text{Aut}(G)$, then $\varphi(a) = a^k$ for some k; moreover, a^k must be a generator of G, so that $(k, n) = 1$, by Exercise 2.20, and $[k] \in U(\mathbb{Z}_n)$. It is routine to show that $\Theta: \text{Aut}(G) \to U(\mathbb{Z}_n)$, defined by $\Theta(\varphi) = [k]$, is an isomorphism. ■

Theorem 7.3.

(i) $\text{Aut}(\mathbb{Z}_2) = 1$; $\text{Aut}(\mathbb{Z}_4) \cong \mathbb{Z}_2$; *if* $m \geq 3$, *then* $\text{Aut}(\mathbb{Z}_{2^m}) \cong \mathbb{Z}_2 \times \mathbb{Z}_{2^{m-2}}$.
(ii) *If p is an odd prime, then* $\text{Aut}(\mathbb{Z}_{p^m}) \cong \mathbb{Z}_l$, *where* $l = (p-1)p^{m-1}$.
(iii) *If $n = p_1^{e_1} \dots p_t^{e_t}$, where the p_i are distinct primes and the $e_i > 0$, then* $\text{Aut}(\mathbb{Z}_n) \cong \prod_i \text{Aut}(\mathbb{Z}_{q_i})$, *where* $q_i = p_i^{e_i}$.

Proof. (i) $U(\mathbb{Z}_2) = 1$ and $U(\mathbb{Z}_4) = \{[1], [-1]\} \cong \mathbb{Z}_2$. If $m \geq 3$, then the result is Theorem 5.44.

(ii) This is Theorem 6.7.

(iii) If a ring $R = R_1 \times \cdots \times R_t$ is a direct product of rings (addition and multiplication are coordinatewise), then it is easy to see that $U(R)$ is the direct product of groups $U(R_1) \times \cdots \times U(R_t)$; moreover, the primary decomposition of the cyclic group $\mathbb{Z}_n = \mathbb{Z}_{q_1} \times \cdots \times \mathbb{Z}_{q_t}$ is also a decomposition of $\mathbb{Z}_n$ as a direct product of rings. ∎

Theorem 7.1 suggests the following class of groups:

Definition. A group G is ***complete*** if it is centerless and every automorphism of G is inner.

It follows from Theorem 7.1(ii) that $\mathrm{Aut}(G) \cong G$ for every complete group. We are now going to see that almost every symmetric group is complete.

Lemma 7.4. *An automorphism φ of S_n preserves transpositions ($\varphi(\tau)$ is a transposition whenever τ is) if and only if φ is inner.*

Proof. If φ is inner, then it preserves the cycle structure of every permutation, by Theorem 3.5.

We prove, by induction on $t \geq 2$, that there exist conjugations $\gamma_2, \ldots, \gamma_t$ such that $\gamma_t^{-1} \ldots \gamma_2^{-1}\varphi$ fixes $(1\ 2), \ldots, (1\ t)$. If $\pi \in S_n$, we will denote $\varphi(\pi)$ by π^φ in this proof. By hypothesis, $(1\ 2)^\varphi = (i\ j)$ for some i, j; define γ_2 to be conjugation by $(1\ i)(2\ j)$ (if $i = 1$ or $j = 2$, then interpret $(1\ i)$ or $(2\ j)$ as the identity). By Lemma 3.4, the quick way of computing conjugates in S_n, we see that $(1\ 2)^\varphi = (1\ 2)^{\gamma_2}$, and so $\gamma_2^{-1}\varphi$ fixes $(1\ 2)$.

Let $\gamma_2, \ldots, \gamma_t$ be given by the inductive hypothesis, so that $\psi = \gamma_t^{-1} \ldots \gamma_2^{-1}\varphi$ fixes $(1\ 2), \ldots, (1\ t)$. Since ψ preserves transpositions, $(1\ \ t+1)^\psi = (l\ k)$. Now $(1\ 2)$ and $(l\ k)$ cannot be disjoint, lest $[(1\ 2)(1\ \ t+1)]^\psi = (1\ 2)^\psi(1\ \ t+1)^\psi = (1\ 2)(l\ k)$ have order 2, while $(1\ 2)(1\ \ t+1)$ has order 3. Thus, $(1\ \ t+1)^\psi = (1\ k)$ or $(1\ \ t+1)^\psi = (2\ k)$. If $k \leq t$, then $(1\ \ t+1)^\psi \in \langle (1\ 2), \ldots, (1\ t) \rangle$, and hence it is fixed by ψ; this contradicts ψ being injective, for either $(1\ \ t+1)^\psi = (1\ k) = (1\ k)^\psi$ or $(1\ \ t+1)^\psi = (2\ k) = (2\ k)^\psi$. Hence, $k \geq t+1$. Define γ_{t+1} to be conjugation by $(k\ \ t+1)$. Now γ_{t+1} fixes $(1\ 2), \ldots, (1\ t)$ and $(1\ \ t+1)^{\gamma_{t+1}} = (1\ \ t+1)^\psi$, so that $\gamma_{t+1}^{-1} \cdots \gamma_2^{-1}\varphi$ fixes $(1\ 2), \ldots, (1\ \ t+1)$ and the induction is complete. It follows that $\gamma_n^{-1} \cdots \gamma_2^{-1}\varphi$ fixes $(1\ 2), \ldots, (1\ n)$. But these transpositions generate S_n, by Exercise 2.9(i), and so $\gamma_n^{-1} \ldots \gamma_2^{-1}\varphi$ is the identity. Therefore, $\varphi = \gamma_2 \cdots \gamma_n \in \mathrm{Inn}(S_n)$. ∎

Theorem 7.5. *If $n \neq 2$ or $n \neq 6$, then S_n is complete.*

Remark. $S_2 \cong \mathbb{Z}_2$ is not complete because it has a center; we shall see in Theorem 7.9 that S_6 is not complete.

Proof. Let T_k denote the conjugacy class in S_n consisting of all products of k disjoint transpositions. By Exercise 1.16, a permutation in S_n is an involution if and only if it lies in some T_k. It follows that if $\theta \in \mathrm{Aut}(S_n)$, then $\theta(T_1) = T_k$

for some k. We shall show that if $n \neq 6$, then $|T_k| \neq |T_1|$ for $k \neq 1$. Assuming this, then $\theta(T_1) = T_1$, and Lemma 7.4 completes the proof.

Now $|T_1| = n(n-1)/2$. To count T_k, observe first that there are

$$\tfrac{1}{2}n(n-1) \times \tfrac{1}{2}(n-2)(n-3) \times \cdots \times \tfrac{1}{2}(n-2k+2)(n-2k+1)$$

k-tuples of disjoint transpositions. Since disjoint transpositions commute and there are $k!$ orderings obtained from any k-tuple, we have

$$|T_k| = n(n-1)(n-2)\cdots(n-2k+1)/k!2^k.$$

The question whether $|T_1| = |T_k|$ leads to the question whether there is some $k > 1$ such that

$$(*) \qquad (n-2)(n-3)\cdots(n-2k+1) = k!2^{k-1}.$$

Since the right side of $(*)$ is positive, we must have $n \geq 2k$. Therefore, for fixed n,

$$\text{left side} \geq (2k-2)(2k-3)\cdots(2k-2k+1) = (2k-2)!.$$

An easy induction shows that if $k \geq 4$, then $(2k-2)! > k!2^{k-1}$, and so $(*)$ can hold only if $k = 2$ or $k = 3$. When $k = 2$, the right side is 4, and it easy to see that equality never holds; we may assume, therefore, that $k = 3$. Since $n \geq 2k$, we must have $n \geq 6$. If $n > 6$, then the left side of $(*) \geq 5 \times 4 \times 3 \times 2 = 120$, while the right side is 24. We have shown that if $n \neq 6$, then $|T_1| \neq |T_k|$ for all $k > 1$, as desired. ∎

Corollary 7.6. *If θ is an outer automorphism of S_6, and if $\tau \in S_6$ is a transposition, then $\theta(\tau)$ is a product of three disjoint transpositions.*

Proof. If $n = 6$, then we saw in the proof of the theorem that $(*)$ does not hold if $k \neq 3$. (When $k = 3$, both sides of $(*)$ equal 24.) ∎

Corollary 7.7. *If $n \neq 2$ or $n \neq 6$, then* $\mathrm{Aut}(S_n) \cong S_n$.

Proof. If G is complete, then $\mathrm{Aut}(G) \cong G$. ∎

We now show that S_6 is a genuine exception. Recall that a subgroup $K \leq S_X$ is *transitive* if, for every pair $x, y \in X$, there exists $\sigma \in K$ with $\sigma(x) = y$. In Theorem 3.14, we saw that if $H \leq G$, then the family X of all left cosets of H is a G-set (where $\rho_a\colon gH \mapsto agH$ for each $a \in G$); indeed, X is a transitive G-set: given gH and $g'H$, then $\rho_a(gH) = g'H$, where $a = g'g^{-1}$.

Lemma 7.8. *There exists a transitive subgroup $K \leq S_6$ of order* 120 *which contains no transpositions.*

Proof. If σ is a 5-cycle, then $P = \langle \sigma \rangle$ is a Sylow 5-subgroup of S_5. The Sylow theorem says that if r is the number of conjugates of P, then $r \equiv 1 \bmod 5$ and r is a divisor of 120; it follows easily that $r = 6$. The representation of S_5 on

X, the set of all left cosets of $N = N_{S_5}(P)$, is a homomorphism $\rho: S_5 \to S_X \cong S_6$. Now X is a transitive S_5-set, by Exercise 4.11, and so $|\ker \rho| \leq |S_5|/r = |S_5|/6$, by Exercise 3.44(iii). Since the only normal subgroups of S_5 are S_5, A_5, and 1, it follows that $\ker \rho = 1$ and ρ is an injection. Therefore, $\text{im } \rho \cong S_5$ is a transitive subgroup of S_X of order 120.

For notational convenience, let us write $K \leq S_6$ instead of $\text{im } \rho \leq S_X$. Now K contains an element α of order 5 which must be a 5-cycle; say, $\alpha = (1\ 2\ 3\ 4\ 5)$. If K contains a transposition $(i\ j)$, then transitivity of K provides $\beta \in K$ with $\beta(j) = 6$, and so $\beta(i\ j)\beta^{-1} = (\beta i\ \beta j) = (l\ 6)$ for some $l \neq 6$ (of course, $l = \beta i$). Conjugating $(l\ 6) \in K$ by the powers of α shows that K contains (16), $(2\ 6)$, $(3\ 6)$, $(4\ 6)$, and $(5\ 6)$. But these transpositions generate S_6, by Exercise 2.9(i), and this contradicts K ($\cong S_5$) being a proper subgroup of S_6. ■

The "obvious" copy of S_5 in S_6 consists of all the permutations fixing 6; plainly, it is not transitive, and it does contain transpositions.

Theorem 7.9 (Hölder, 1895). *There exists an outer automorphism of* S_6.

Proof. Let K be a transitive subgroup of S_6 of order 120, and let Y be the family of its left cosets: $Y = \{\alpha_1 K, \ldots, \alpha_6 K\}$. If $\theta: S_6 \to S_Y$ is the representation of S_6 on the left cosets of K, then $\ker \theta \leq K$ is a normal subgroup of S_6. But A_6 is the only proper normal subgroup of S_6, so that $\ker \theta = 1$ and θ is an injection. Since S_6 is finite, θ must be a bijection, and so $\theta \in \text{Aut}(S_6)$, for $S_Y \cong S_6$. Were θ inner, then it would preserve the cycle structure of every permutation in S_6. In particular, $\theta_{(1\,2)}$, defined by $\theta_{(1\,2)}: \alpha_i K \mapsto (1\ 2)\alpha_i K$ for all i, is a transposition, and hence θ fixes $\alpha_i K$ for four different i. But if $\theta_{(1\,2)}$ fixes even one left coset, say $\alpha_i K = (1\ 2)\alpha_i K$, then $\alpha_i^{-1}(1\ 2)\alpha_i$ is a transposition in K. This contradiction shows that θ is an outer automorphism. ■

Theorem 7.10. $\text{Aut}(S_6)/\text{Inn}(S_6) \cong \mathbb{Z}_2$, and so $|\text{Aut}(S_6)| = 1440$.

Proof. Let T_1 be the class of all transpositions in S_6, and let T_3 be the class of all products of 3 disjoint transpositions. If θ and ψ are outer automorphisms of S_6, then both interchange T_1 and T_3, by Corollary 7.6, and so $\theta^{-1}\psi(T_1) = T_1$. Therefore, $\theta^{-1}\psi \in \text{Inn}(S_6)$, by Lemma 7.4, and $\text{Aut}(S_6)/\text{Inn}(S_6)$ has order 2. ■

This last theorem shows that there is essentially only one outer automorphism θ of S_6; given an outer automorphism θ, then every other such has the form $\gamma\theta$ for some inner automorphism γ. It follows that S_6 has exactly 720 outer automorphisms, for they comprise the other coset of $\text{Inn}(S_6)$ in $\text{Aut}(S_6)$.

Definition. A ***syntheme***[1] is a product of 3 disjoint transpositions. A ***pentad*** is a family of 5 synthemes, no two of which have a common transposition.

If two synthemes have a common transposition, say, $(a\ b)(c\ d)(e\ f)$ and $(a\ b)(c\ e)(d\ f)$, then they commute. It is easy to see that the converse holds: two commuting synthemes share a transposition.

Lemma 7.11. *S_6 contains exactly 6 pentads. They are:*

$$(12)(34)(56), (13)(25)(46), (14)(26)(35), (15)(24)(36), (16)(23)(45);$$

$$(12)(34)(56), (13)(26)(45), (14)(25)(36), (15)(23)(46), (16)(24)(35);$$

$$(12)(35)(46), (13)(24)(56), (14)(25)(36), (15)(26)(34), (16)(23)(45);$$

$$(12)(35)(46), (13)(26)(45), (14)(23)(56), (15)(24)(36), (16)(25)(34);$$

$$(12)(36)(45), (13)(24)(56), (14)(26)(35), (15)(23)(46), (16)(25)(34);$$

$$(12)(36)(45), (13)(25)(46), (14)(23)(56), (15)(26)(34), (16)(24)(35).$$

Proof. There are exactly 15 synthemes, and each lies in at most two pentads. There are thus at most 6 pentads, for $2 \times 15 = 30 = 6 \times 5$; there are exactly 6 pentads, for they are displayed above. ■

Theorem 7.12. *If $\{\sigma_2, \ldots, \sigma_6\}$ is a pentad in some ordering, then there is a unique outer automorphism θ of S_6 with θ: $(1\ i) \mapsto \sigma_i$ for $i = 2, 3, 4, 5, 6$. Moreover, every outer automorphism of S_6 has this form.*

Proof. Let $X = \{(1\ 2), (1\ 3), (1\ 4), (1\ 5), (1\ 6)\}$. If θ is an outer automorphism of S_6, then Corollary 7.6 shows that each $\theta((1\ i))$ is a syntheme. Since $(1\ i)$ and $(1\ j)$ do not commute for $i \neq j$, it follows that $\theta((1\ i))$ and $\theta((1\ j))$ do not commute; hence, $\theta(X)$ is a pentad. Let us count the number of possible functions from X to pentads arising from outer automorphisms. Given an outer automorphism θ, there are 6 choices of pentad for $\theta(X)$; given such a pentad P, there are $5! = 120$ bijections $X \to P$. Hence, there are at most 720 bijections from X to pentads which can possibly arise as restrictions of outer automorphisms. But there are exactly 720 outer automorphisms, by Theorem 7.10, and no two of them can restrict to the same bijection because X generates S_6. The statements of the theorem follow. ■

Since every element in S_6 is a product of transpositions, the information in the theorem allows one to evaluate $\theta(\beta)$ for every $\beta \in S_6$.

Corollary 7.13. *There is an outer automorphism of S_6 which has order 2.*

[1] A *syntheme* is a partition of a set X into subsets P_i with $|P_i| = |P_j|$ for all i, j (this term is due to J.J. Sylvester).

Proof. Define[2] $\psi \in \operatorname{Aut}(S_6)$ by

$$(1\ 2) \mapsto (1\ 5)(2\ 3)(4\ 6),$$
$$(1\ 3) \mapsto (1\ 4)(2\ 6)(3\ 5),$$
$$(1\ 4) \mapsto (1\ 3)(2\ 4)(5\ 6),$$
$$(1\ 5) \mapsto (1\ 2)(3\ 6)(4\ 5),$$
$$(1\ 6) \mapsto (1\ 6)(2\ 5)(3\ 4).$$

A routine but long calcuation shows that $\psi^2 = 1$. ■

Here is another source of (possibly infinite) complete groups.

Theorem 7.14. *If G is a simple nonabelian group, then* $\operatorname{Aut}(G)$ *is complete.*

Proof. Let $I = \operatorname{Inn}(G) \lhd \operatorname{Aut}(G) = A$. Now $Z(G) = 1$, because G is simple and nonabelian, and so Theorem 7.1 gives $I \cong G$. Now $Z(A) \leq C_A(I) = \{\varphi \in A: \varphi\gamma_g = \gamma_g\varphi$ for all $g \in G\}$. We claim that $C_A(I) = 1$; it will then follow that A is centerless. If $\varphi\gamma_g = \gamma_g\varphi$ for all $g \in G$, then $\gamma_g = \varphi\gamma_g\varphi^{-1} = \gamma_{\varphi(g)}$. Therefore, $\varphi(g)g^{-1} \in Z(G) = 1$ for all $g \in G$, and so $\varphi = 1$.

It remains to show that every $\sigma \in \operatorname{Aut}(A)$ is inner. Now $\sigma(I) \lhd A$, because $I \lhd A$, and so $I \cap \sigma(I) \lhd \sigma(I)$. But $\sigma(I) \cong I \cong G$ is simple, so that either $I \cap \sigma(I) = 1$ or $I \cap \sigma(I) = \sigma(I)$. Since both I and $\sigma(I)$ are normal, $[I, \sigma(I)] \leq I \cap \sigma(I)$. In the first case, we have $[I, \sigma(I)] = 1$; that is, $\sigma(I) \leq C_A(I) = 1$, and this contradicts $\sigma(I) \cong I$. Hence, $I \cap \sigma(I) = \sigma(I)$, and so $\sigma(I) \leq I$. This inclusion holds for every $\sigma \in \operatorname{Aut}(A)$; in particular, $\sigma^{-1}(I) \leq I$, and so $\sigma(I) = I$ for every $\sigma \in \operatorname{Aut}(A)$. If $g \in G$, then $\gamma_g \in I$; there is thus $\alpha(g) \in G$ with

$$\sigma(\gamma_g) = \gamma_{\alpha(g)}.$$

The reader may check easily that the function $\alpha: G \to G$ is a bijection. We now show that α is an automorphism of G; that is, $\alpha \in A$. If $g, h \in G$, then $\sigma(\gamma_g\gamma_h) = \sigma(\gamma_{gh}) = \gamma_{\alpha(gh)}$. On the other hand, $\sigma(\gamma_g\gamma_h) = \sigma(\gamma_g)\sigma(\gamma_h) = \gamma_{\alpha(g)}\gamma_{\alpha(h)} = \gamma_{\alpha(g)\alpha(h)}$; hence $\alpha(gh) = \alpha(g)\alpha(h)$.

We claim that $\sigma = \Gamma_\alpha$, conjugation by α. To this end, define $\tau = \sigma \circ \Gamma_\alpha^{-1}$. Observe, for all $h \in G$, that

$$\begin{aligned}\tau(\gamma_h) &= \sigma\Gamma_\alpha^{-1}(\gamma_h) = \sigma(\alpha^{-1}\gamma_h\alpha)\\ &= \sigma(\gamma_{\alpha^{-1}(h)})\\ &= \gamma_{\alpha\alpha^{-1}(h)} = \gamma_h.\end{aligned}$$

[2] In Lam, T.Y., and Leep, D.B., Combinatorial structure on the automorphism group of S_6, *Expo. Math.* (1993), it is shown that the order of any outer automorphism φ of S_6 is either 2, 4, 8, or 10, and they show how to determine the order of φ when it is given, as in Theorem 7.12, in terms of its values on $(1\ i)$, for $2 \leq i \leq 6$.

Thus, τ fixes everything in I. If $\beta \in A$, then for every $g \in G$,

$$\begin{aligned} \beta\gamma_g\beta^{-1} &= \tau(\beta\gamma_g\beta^{-1}) \quad &&\text{(because } \beta\gamma_g\beta^{-1} \in I \text{ and } \tau \text{ fixes } I) \\ &= \tau(\beta)\gamma_g\tau(\beta)^{-1} \quad &&\text{(because } \tau \text{ fixes } I). \end{aligned}$$

Hence $\tau(\beta)\beta^{-1} \in C_A(I) = 1$, and $\tau(\beta) = \beta$. Therefore, $\tau = 1$, $\sigma = \Gamma_\alpha$, and A is complete. ■

It follows, for every nonabelian simple group G, that $\mathrm{Aut}(G) \cong \mathrm{Aut}(\mathrm{Aut}(G))$. There is a beautiful theorem of Wielandt (1939) with a similar conclusion. We know, by Theorem 7.1, that every centerless group G can be imbedded in $\mathrm{Aut}(G)$. Moreover, $\mathrm{Aut}(G)$ is also centerless, and so it can be imbedded in its automorphism group $\mathrm{Aut}(\mathrm{Aut}(G))$. This process may thus be iterated to give the ***automorphism tower*** of G:

$$G \le \mathrm{Aut}(G) \le \mathrm{Aut}(\mathrm{Aut}(G)) \le \cdots.$$

Wielandt proved, for every finite centerless group G, that this tower is constant from some point on. Since the last term of an automorphism tower is a complete group, it follows that every finite centerless group can be imbedded is a complete group. Of course, there is an easier proof of a much stronger fact: Cayley's theorem imbeds a finite group in some S_n with $n > 6$, and Theorem 7.5 applies to show that S_n is complete.

The automorphism tower of an infinite centerless group need not stop after a finite number of steps, but a transfinite automorphism tower (indexed by ordinals) can be defined (taking unions at limit ordinals). S. Thomas (1985) proved, for every centerless group, that this automorphism tower eventually stops. As in the finite case, the last term of an automorphism tower is complete, and so every centerless group can be imbedded in a complete group. It is shown, in Exercise 11.56, that every group can be imbedded in a centerless group, and so it follows that every group can be imbedded in a complete group.

Theorem 7.15. *If $K \lhd G$ and K is complete, then K is a direct factor of G; that is, there is a normal subgroup Q of G with $G = K \times Q$.*

Proof. Define $Q = C_G(K) = \{g \in G: gk = kg \text{ for all } k \in K\}$. Now $K \cap Q \le Z(K) = 1$, and so $K \cap Q = 1$. To see that $G = KQ$, take $g \in G$. Now $\gamma_g(K) = K$, because $K \lhd G$, and so $\gamma_g|K \in \mathrm{Aut}(K) = \mathrm{Inn}(K)$. There is thus $k \in K$ with $gxg^{-1} = kxk^{-1}$ for all $x \in K$. Hence $k^{-1}g \in \bigcap_{x \in K} C_G(x) = C_G(K) = Q$, and so $g = k(k^{-1}g) \in KQ$. Finally, $Q \lhd G$, for $gQg^{-1} = k(k^{-1}g)Q(k^{-1}g)^{-1}k^{-1} = kQk^{-1}$ (because $k^{-1}g \in Q$), and $kQk^{-1} = Q$ (because every element of Q commutes with k). Therefore, $G = K \times Q$. ■

The converse of Theorem 7.15 is true, and we introduce a construction in order to prove it. Recall that if $a \in K$, then $L_a: K \to K$ denotes left translation by a; that is, $L_a(x) = ax$ for all $a \in K$. As in the Cayley theorem (Theorem 3.12), K is isomorphic to $K^l = \{L_a: a \in K\}$, which is a subgroup of S_K. Similarly, if $R_a: K \to K$ denotes right translation by a, that is, $R_a: x \mapsto xa^{-1}$, then $K^r = \{R_a: a \in K\}$ is also a subgroup of S_K isomorphic to K.

Definition. The ***holomorph*** of a group K, denoted by $\mathrm{Hol}(K)$, is the subgroup of S_K generated by K^l and $\mathrm{Aut}(K)$.

Notice, for all $a \in K$, that $R_a = L_{a^{-1}}\gamma_a$, so that $K^r \leq \mathrm{Hol}(K)$; indeed, it is easy to see that $\mathrm{Hol}(K) = \langle K^r, \mathrm{Aut}(K)\rangle$.

Lemma 7.16. *Let K be a group.*

(i) $K^l \lhd \mathrm{Hol}(K)$, $K^l\,\mathrm{Aut}(K) = \mathrm{Hol}(K)$, *and* $K^l \cap \mathrm{Aut}(K) = 1$.
(ii) $\mathrm{Hol}(K)/K^l \cong \mathrm{Aut}(K)$.
(iii) $C_{\mathrm{Hol}(K)}(K^l) = K^r$.

Proof. (i) It is easy to see that $\varphi L_a \varphi^{-1} = L_{\varphi(a)}$, and that it lies in K^l for every $a \in K$ and $\varphi \in \mathrm{Aut}(K)$; since $\mathrm{Hol}(K) = \langle K^l, \mathrm{Aut}(K)\rangle$, it follows that $K^l \lhd \mathrm{Hol}(K)$ and that $\mathrm{Hol}(K) = K^l\,\mathrm{Aut}(K)$. If $a \in K$, then $L_a(1) = a$; therefore, $L_a \in \mathrm{Aut}(K)$ if and only if $a = 1$; that is, $K^l \cap \mathrm{Aut}(K) = 1$.

(ii) $\mathrm{Hol}(K)/K^l = K^l\,\mathrm{Aut}(K)/K^l \cong \mathrm{Aut}(K)/(K^l \cap \mathrm{Aut}(K)) \cong \mathrm{Aut}(K)$.

(iii) If $a, b, x \in K$, then $L_a R_b(x) = a(xb^{-1})$ and $R_b L_a(x) = (ax)b^{-1}$, so that associativity gives $K^r \leq C_{\mathrm{Hol}(K)}(K^l)$. For the reverse inclusion, assume that $\eta \in \mathrm{Hol}(K)$ satisfies $\eta L_a = L_a \eta$ for all $a \in K$. Now $\eta = L_b \varphi$ for some $b \in K$ and $\varphi \in \mathrm{Aut}(K)$. If $x \in K$, then $\eta L_a(x) = L_b \varphi L_a(x) = b\varphi(a)\varphi(x)$ and $L_a \eta(x) = L_a L_b \varphi(x) = ab\varphi(x)$. Hence, $b\varphi(a) = ab$ for all $a \in K$; that is, $\varphi = \gamma_{b^{-1}}$. It follows that $\eta(x) = L_b\varphi(x) = b(b^{-1}xb) = xb$, and so $\eta = R_{b^{-1}} \in K^r$, as desired. ∎

Here is the converse of Theorem 7.15.

Theorem 7.17. *If a group K is a direct factor whenever it is (isomorphic to) a normal subgroup of a group, then K is complete.*

Proof. We identify K with the subgroup $K^l \leq \mathrm{Hol}(K)$. Since K^l is normal, the hypothesis gives a subgroup B with $\mathrm{Hol}(K) = K^l \times B$. Now $B \leq C_{\mathrm{Hol}}(K^l) = K^r$, because every element of B commutes with each element of K^l. It follows that if $\varphi \in \mathrm{Aut}(K) \leq \mathrm{Hol}(K)$, then $\varphi = L_a R_b$ for some $a, b \in K$. Hence, $\varphi(x) = axb^{-1}$ for all $x \in K$. But now $axyb^{-1} = \varphi(x)\varphi(y) = axb^{-1}ayb^{-1}$, so that $1 = b^{-1}a$; therefore, $\varphi = \gamma_a \in \mathrm{Inn}(K)$.

Since $\mathrm{Hol}(K) = K^l \times B$ and $B \leq K^r \leq \mathrm{Hol}(K)$, Exercise 7.16 below shows that $K^r = B \times (K^r \cap K^l)$. If $\varphi \in K^l \cap K^r$, then $\varphi = L_a = R_b$, for $a, b \in K$. For all $c \in K$, $L_a(c) = R_b(c)$ gives $ac = cb^{-1}$; if $c = 1$, then $a = b^{-1}$, from which it follows that $a \in Z(K)$. Therefore, $K^r \cap K^l = Z(K^r)$ and $K \cong B \times Z(K)$. If $1 \neq$

$\varphi \in \operatorname{Aut}(Z(K))$, then it is easy to see that $\tilde{\varphi}: B \times Z(K) \to B \times Z(K)$, defined by $(b, z) \mapsto (b, \varphi z)$, is an automorphism of K; $\tilde{\varphi}$ must be outer, for conjugation by $(\beta, \zeta) \in B \times Z(K) \cong K$ sends (b, z) into $(\beta, \zeta)(b, z)(\beta^{-1}, \zeta^{-1}) = (\beta b \beta^{-1}, z)$. But K has no outer automorphisms, so that $\operatorname{Aut}(Z(K)) = 1$ and, by Example 7.6, $|Z(K)| \leq 2$. If $Z(K) \cong \mathbb{Z}_2$, then it is isomorphic to a normal subgroup N of $\mathbb{Z}_4$ which is not a direct factor. But K is isomorphic to the normal subgroup $B \times N$ of $B \times \mathbb{Z}_4$ which is not a direct factor, contradicting the hypothesis. Therefore, $Z(K) = 1$ and K is complete. ■

The holomorph allows one to extend commutator notation. Recall that the commutator $[a, x] = axa^{-1}x^{-1} = x^a x^{-1}$. Now let G be a group and let $A = \operatorname{Aut}(G)$. We may regard G and A as subgroups of $\operatorname{Hol}(G)$ (by identifying G with G^l). For $x \in G$ and $\alpha \in A$, define

$$[\alpha, x] = \alpha(x)x^{-1},$$

and define

$$[A, G] = \langle [\alpha, x]: \alpha \in A, x \in G \rangle.$$

The next lemma will be used to give examples of nilpotent groups arising naturally.

Lemma 7.18. *Let G and A be subgroups of a group H, and let $G = G_0 \geq G_1 \geq \cdots$ be a series of normal subgroups of G such that $[A, G_i] \leq G_{i+1}$ for all i. Define $A_1 = A$ and*

$$A_j = \{\alpha \in A: [\alpha, G_i] \leq G_{i+j} \text{ for all } i\}.$$

Then $[A_j, A_l] \leq A_{j+l}$ for all j and l, and $[\gamma_j(A), G_i] \leq G_{i+j}$ for all i and j.

Proof. The definition of A_j gives $[A_j, G_i] \leq G_{i+j}$ for all i. It follows that $[A_j, A_l, G_i] = [A_j, [A_l, G_i]] \leq [A_j, G_{l+i}] \leq G_{j+l+i}$. Similarly, $[A_l, A_j, G_i] \leq G_{j+l+i}$. Now $G_{j+l+i} \lhd \langle G, A \rangle$, because both G and A normalize each G_i. Since $[A_j, A_l, G_i][A_l, A_j, G_i] \leq G_{j+l+i}$, the three subgroups lemma (Exercise 5.48 (ii)) gives $[G_i, [A_j, A_l]] = [[A_j, A_l], G_i] \leq G_{j+l+i}$. Therefore, $[A_j, A_l] \leq A_{j+l+i}$ for all i, by definition of A_{j+l}. In particular, for $i = 0$, we have $[A_j, A_l] \leq A_{j+l}$. It follows that $A = A_1 \geq A_2 \geq \cdots$ is a central series for A, and so Exercise 5.38(ii) gives $\gamma_j(A) \leq A_j$ for all j. Therefore, $[\gamma_j(A), G] \leq [A_j, G] \leq G_{i+j}$, as desired. ■

Definition. Let $G = G_0 \geq G_1 \geq \cdots \geq G_r = 1$ be a series of normal subgroups of a group G. An automorphism $\alpha \in \operatorname{Aut}(G)$ ***stabilizes*** this series if $\alpha(G_i x) = G_i x$ for all i and all $x \in G_{i-1}$. The ***stabilizer*** A of this series is the subgroup

$$A = \{\alpha \in \operatorname{Aut}(G): \alpha \text{ stabilizes the series}\} \leq \operatorname{Aut}(G).$$

Thus, α stabilizes a normal series $G = G_0 \geq G_1 \geq \cdots \geq G_r = 1$ if and only if $\alpha(G_i) \leq G_i$ and the induced map $G_i/G_{i+1} \to G_i/G_{i+1}$, defined by $G_{i+1}x \mapsto G_{i+1}\alpha(x)$, is the identity map for each i.

Theorem 7.19. *The stabilizer A of a series of normal subgroups $G = G_0 \geq G_1 \geq \cdots \geq G_r = 1$ is a nilpotent group of class $\leq r - 1$.*

Proof. Regard both G and A as subgroups of Hol(G). For all i, if $x \in G_i$ and $\alpha \in A$, then $\alpha(x) = g_{i+1}x$ for some $g_{i+1} \in G_{i+1}$, and so $\alpha(x)x^{-1} \in G_{i+1}$. In commutator notation, $[A, G] \leq G_{i+1}$. By Lemma 7.18, $[\gamma_j(A), G_i] \leq G_{i+j}$ for all i and j. In particular, for $i = 0$ and $j = r$, we have $[\gamma_r(A), G] \leq G_r = 1$; that is, for all $x \in G$ and $\alpha \in \gamma_r(A)$, we have $\alpha(x)x^{-1} = 1$. Therefore, $\gamma_r(A) = 1$ and A is nilpotent of class $\leq r - 1$. ■

For example, let $\{v_1, \ldots, v_n\}$ be a basis of a vector space V over a field k, and define $V_{i-1} = \langle v_i, v_{i+1}, \ldots, v_n \rangle$. Hence,

$$V = V_0 > V_1 > \cdots > V_n = 0$$

is a series of normal subgroups of the (additive abelian) group V. If $A \leq \mathrm{GL}(V)$ is the group of automorphisms stabilizing this series, then A is a nilpotent group of class $\leq n - 1$. If each $\alpha \in A \cap \mathrm{GL}(V)$ is regarded as a matrix (relative to the given basis), then it is easy to see that $A \cap \mathrm{GL}(V) = \mathrm{UT}(n, k)$, the group of all unitriangular matrices. Therefore, $\mathrm{UT}(n, k)$ is also nilpotent of class $\leq n - 1$. Compare this with Exercise 5.44.

If $G = G_0 \geq G_1 \geq \cdots \geq G_r = 1$ is any (not necessarily normal) series of a group G (i.e., G_i need not be a normal subgroup of G_{i-1}), then P. Hall (1958) proved that the stabilizer of this series is always nilpotent of class $\leq \frac{1}{2}r(r - 1)$.

Exercises

7.7. If G is a finite nonabelian p-group, then $p^2 \mid |\mathrm{Aut}(G)|$.

7.8. If G is a finite abelian group, then Aut(G) is abelian if and only if G is cyclic.

7.9. (i) If G is a finite abelian group with $|G| > 2$, then Aut(G) has even order.
(ii) If G is not abelian, then Aut(G) is not cyclic. (*Hint.* Show that Inn(G) is not cyclic.)
(iii) There is no finite group G with Aut(G) cyclic of odd order > 1.

7.10. Show that $|\mathrm{GL}(2, p)| = (p^2 - 1)(p^2 - p)$. (*Hint.* How many ordered bases are in a two-dimensional vector space over $\mathbb{Z}_p$?)

7.11. If G is a finite group and Aut(G) acts transitively on $G^\# = G - \{1\}$, then G is an elementary abelian group.

7.12. If H and K are finite groups whose orders are relatively prime, then $\mathrm{Aut}(H \times K) \cong \mathrm{Aut}(H) \times \mathrm{Aut}(K)$. Show that this may fail if $(|H|, |K|) > 1$. (*Hint.* Take $H = \mathbb{Z}_p = K$.)

7.13. Prove that $\mathrm{Aut}(\mathbf{Q}) \cong S_4$. (*Hint.* $\mathrm{Inn}(\mathbf{Q}) \cong \mathbf{V}$ and it equals its own centralizer in Aut($\mathbf{Q}$); use Theorem 7.1 with $G = \mathrm{Aut}(\mathbf{Q})$ and $H = \mathrm{Inn}(\mathbf{Q})$.)

7.14. (i) Show that $\mathrm{Hol}(\mathbb{Z}_2) \cong \mathbb{Z}_2$, $\mathrm{Hol}(\mathbb{Z}_3) \cong S_3$, $\mathrm{Hol}(\mathbb{Z}_4) \cong D_8$, and $\mathrm{Hol}(\mathbb{Z}_6) \cong D_{12}$.
(ii) If P is a Sylow 5-subgroup of S_5, then $\mathrm{Hol}(\mathbb{Z}_5) \cong N_{S_5}(P)$. (*Hint.* See Exercise 4.20.)

7.15. Prove that $\mathrm{Aut}(D_8) \cong D_8$, but that $\mathrm{Aut}(D_{16}) \not\cong D_{16}$.

7.16. Is $\mathrm{Aut}(A_4) \cong S_4$? Is $\mathrm{Aut}(A_6) \cong S_6$?

7.17. If $G = B \times K$ and $B \leq L \leq G$, then $L = B \times (L \cap K)$.

7.18. If $H \lhd G$, prove that

$$\{\varphi \in \mathrm{Aut}(G)\colon \varphi \text{ fixes } H \text{ pointwise and } \varphi(g)H = gH \text{ for all } g \in G\}$$

is an abelian subgroup of $\mathrm{Aut}(G)$.

7.19. (i) Prove that the alternating groups A_n are never complete.
(ii) Show that if G is a complete group with $G \neq G'$, then G is not the commutator subgroup of any group containing it. Conclude that S_n, for $n \neq 2, 6$, is never a commutator subgroup.

7.20. If G is a complete group, then $\mathrm{Hol}(G) = G^l \times G^r$. Conclude, for $n \neq 2$ and $n \neq 6$, that $\mathrm{Hol}(S_n) \cong S_n \times S_n$.

7.21. Prove that every automorphism of a group G is the restriction of an inner automorphism of $\mathrm{Hol}(G)$.

7.22. Let G be a group and let $f \in S_G$. Prove that $f \in \mathrm{Hol}(G)$ if and only if $f(xy^{-1}z) = f(x)f(y)^{-1}f(z)$ for all $x, y, z \in G$.

Semidirect Products

Definition. Let K be a (not necessarily normal) subgroup of a group G. Then a subgroup $Q \leq G$ is a ***complement*** of K in G if $K \cap Q = 1$ and $KQ = G$.

A subgroup K of a group G need not have a complement and, even if it does, a complement need not be unique. In S_3, for example, every subgroup of order 2 serves as a complement to A_3. On the other hand, if they exist, complements are unique to isomorphism, for

$$G/K = KQ/K \cong Q/(K \cap Q) = Q/1 \cong Q.$$

A group G is the direct product of two normal subgroups K and Q if $K \cap Q = 1$ and $KQ = G$.

Definition. A group G is a ***semidirect product*** of K by Q, denoted by $G = K \rtimes Q$, if $K \lhd G$ and K has a complement $Q_1 \cong Q$. One also says that G ***splits*** over K.

We do not assume that a complement Q_1 is a normal subgroup; indeed, if Q_1 is a normal subgroup, then G is the direct product $K \times Q_1$.

In what follows, we denote elements of K by letters a, b, c in the first half of the alphabet, and we denote elements of Q by letters x, y, z at the end of the alphabet.

Before we give examples of semidirect products, let us give several different descriptions of them.

Lemma 7.20. *If K is a normal subgroup of a group G, then the following statements are equivalent:*

(i) *G is a semidirect product of K by G/K (i.e., K has a complement in G);*
(ii) *there is a subgroup $Q \le G$ so that every element $g \in G$ has a unique expression $g = ax$, where $a \in K$ and $x \in Q$;*
(iii) *there exists a homomorphism $s: G/K \to G$ with $vs = 1_{G/K}$, where $v: G \to G/K$ is the natural map; and*
(iv) *there exists a homomorphism $\pi: G \to G$ with $\ker \pi = K$ and $\pi(x) = x$ for all $x \in \operatorname{im} \pi$ (such a map π is called a **retraction** of G and $\operatorname{im} \pi$ is called a **retract** of G).*

Proof. (i) $\Rightarrow$ (ii) Let Q be a complement of K in G. Let $g \in G$. Since $G = KQ$, there exist $a \in K$ and $x \in Q$ with $g = ax$. If $g = by$ is a second such factorization, then $xy^{-1} = a^{-1}b \in K \cap Q = 1$. Hence $b = a$ and $y = x$.

(ii) $\Rightarrow$ (iii) Each $g \in G$ has a unique expression $g = ax$, where $a \in K$ and $x \in Q$. If $Kg \in G/K$, then $Kg = Kax = Kx$; define $s: G/K \to G$ by $s(Kg) = x$. The routine verification that s is a well defined homomorphism with $vs = 1_{G/K}$ is left as an exercise for the reader.

(iii) $\Rightarrow$ (iv) Define $\pi: G \to G$ by $\pi = sv$. If $x = \pi(g)$, then $\pi(x) = \pi(\pi(g)) = svsv(g) = sv(g) = \pi(g) = x$ (because $vs = 1_{G/K}$). If $a \in K$, then $\pi(a) = sv(a) = 1$, for $K = \ker v$. For the reverse inclusion, assume that $1 = \pi(g) = sv(g) = s(Kg)$. Now s is an injection, by set theory, so that $Kg = 1$ and so $g \in K$.

(iv) $\Rightarrow$ (i) Define $Q = \operatorname{im} \pi$. If $g \in Q$, then $\pi(g) = g$; if $g \in K$, then $\pi(g) = 1$; *a fortiori*, if $g \in K \cap Q$, then $g = 1$. If $g \in G$, then $g\pi(g^{-1}) \in K = \ker \pi$, for $\pi(g\pi(g^{-1})) = 1$. Since $\pi(g) \in Q$, we have $g = [g\pi(g^{-1})]\pi(g) \in KQ$. Therefore, Q is a complement of K in G and G is a semidirect product of K by Q. ∎

Example 7.7. S_n is a semidirect product of A_n by $\mathbb{Z}_2$.

Take $Q = \langle (1\ 2) \rangle$ to be a complement of A_n.

Example 7.8. D_{2n} is a semidirect product of $\mathbb{Z}_n$ by $\mathbb{Z}_2$.

If $D_{2n} = \langle a, x \rangle$, where $\langle a \rangle \cong \mathbb{Z}_n$ and $\langle x \rangle \cong \mathbb{Z}_2$, then $\langle a \rangle$ is normal and $\langle x \rangle$ is a complement of $\langle a \rangle$.

Example 7.9. For any group K, $\operatorname{Hol}(K)$ is a semidirect product of K^l by $\operatorname{Aut}(K)$.

This is contained in Lemma 7.16.

EXAMPLE 7.10. Let G be a solvable group of order mn, where $(m, n) = 1$. If G contains a normal subgroup of order m, then G is a semidirect product of K by a subgroup Q of order n.

This follows from P. Hall's theorem (Theorem 5.28).

EXAMPLE 7.11. $\mathrm{Aut}(S_6)$ is a semidirect product of S_6 by $\mathbb{Z}_2$.

This follows from P. Hall's theorem (Theorem 5.28).

EXAMPLE 7.12. If $G = \langle a \rangle$ is cyclic of order 4 and $K = \langle a^2 \rangle$, then G is not a semidirect product of K by G/K.

Since normality is automatic in an abelian group, an abelian group G is a semidirect product if and only if it is a direct product. But G is not a direct product. Indeed, it is easy to see that no primary cyclic group is a semidirect product.

EXAMPLE 7.13. Both S_3 and $\mathbb{Z}_6$ are semidirect products of $\mathbb{Z}_3$ by $\mathbb{Z}_2$.

Example 7.13 is a bit jarring at first, for it says, in contrast to direct product, that a semidirect product of K by Q is not determined to isomorphism by the two subgroups. When we reflect on this, however, we see that a semidirect product should depend on "how" K is normal in G.

Lemma 7.21. *If G is a semidirect product of K by Q, then there is a homomorphism $\theta: Q \to \mathrm{Aut}(K)$, defined by $\theta_x = \gamma_x | K$; that is, for all $x \in Q$ and $a \in K$,*

$$\theta_x(a) = xax^{-1}.$$

Moreover, for all $x, y, 1 \in Q$ and $a \in K$,

$$\theta_1(a) = a \qquad \textit{and} \qquad \theta_x(\theta_y(a)) = \theta_{xy}(a).$$

Proof. Normality of K gives $\gamma_x(K) = K$ for all $x \in Q$. The rest is routine. ■

Remark. It follows that K is a group with operators Q.

The object of our study is to recapture G from K and Q. It is now clear that G also involves a homomorphism $\theta: Q \to \mathrm{Aut}(K)$.

Definition. Let Q and K be groups, and let $\theta: Q \to \mathrm{Aut}(K)$ be a homomorphism. A semidirect product G of K by Q ***realizes*** θ if, for all $x \in Q$ and $a \in K$,

$$\theta_x(a) = xax^{-1}.$$

In this language, Lemma 7.21 says that every semidirect product G of K by Q determines some θ which it realizes. Intuitively, "realizing θ" is a way of describing how K is normal in G. For example, if θ is the trivial map, that is, $\theta_x = 1_K$ for every $x \in G$, then $a = \theta_x(a) = xax^{-1}$ for every $a \in K$, and so $K \leq C_G(Q)$.

Definition. Given groups Q and K and a homomorphism $\theta: Q \to \mathrm{Aut}(K)$, define $G = K \rtimes_\theta Q$ to be the set of all ordered pairs $(a, x) \in K \times Q$ equipped with the operation

$$(a, x)(b, y) = (a\theta_x(b), xy).$$

Theorem 7.22. *Given groups Q and K and a homomorphism $\theta: Q \to \mathrm{Aut}(K)$, then $G = K \rtimes_\theta Q$ is a semidirect product of K by Q that realizes θ.*

Proof. We first prove that G is a group. Multiplication is associative:

$$\begin{aligned} &[(a, x)(b, y)](c, z) && (a, x)[(b, y)(c, z)] \\ &= (a\theta_x(b), xy)(c, z) && = (a, x)(b\theta_y(c), yz) \\ &= (a\theta_x(b)\theta_{xy}(c), xyz), && = (a\theta_x(b\theta_y(c)), xyz). \end{aligned}$$

The formulas in Lemma 7.21 (K is a group with operators Q) show that the final entries in each column are equal.

The identity element of G is $(1, 1)$, for

$$(1, 1)(a, x) = (1\theta_1(a), 1x) = (a, x);$$

the inverse of (a, x) is $((\theta_{x^{-1}}(a))^{-1}, x^{-1})$, for

$$((\theta_{x^{-1}}(a))^{-1}, x^{-1})(a, x) = ((\theta_{x^{-1}}(a))^{-1}\theta_{x^{-1}}(a), x^{-1}x) = (1, 1).$$

We have shown that G is a group.

Define a function $\pi: G \to Q$ by $(a, x) \mapsto x$. Since the only "twist" occurs in the first coordinate, it is routine to check that π is a surjective homomorphism and that $\ker \pi = \{(a, 1): a \in K\}$; of course, $\ker \pi$ is a normal subgroup of G. We identify K with $\ker \pi$ via the isomorphism $a \mapsto (a, 1)$. It is also easy to check that $\{(1, x): x \in Q\}$ is a subgroup of G isomorphic to Q (via $x \mapsto (1, x)$), and we identify Q with this subgroup. Another easy calculation shows that $KQ = G$ and $K \cap Q = 1$, so that G is a semidirect product of K by Q.

Finally, G does realize θ:

$$(1, x)(a, 1)(1, x)^{-1} = (\theta_x(a), x)(1, x^{-1}) = (\theta_x(a), 1). \quad \blacksquare$$

Since $K \rtimes_\theta Q$ realizes θ, that is, $\theta_x(b) = xbx^{-1}$, there can be no confusion if we write $b^x = xbx^{-1}$ instead of $\theta_x(b)$. The operation in $K \rtimes_\theta Q$ will henceforth be written

$$(a, x)(b, y) = (ab^x, xy).$$

Theorem 7.23. *If G is a semidirect product of K by Q, then there exists $\theta: Q \to \mathrm{Aut}(K)$ with $G \cong K \rtimes_\theta Q$.*

Proof. Define $\theta_x(a) = xax^{-1}$ (as in Lemma 7.21). By Lemma 7.20 (ii), each $g \in G$ has a unique expression $g = ax$ with $a \in K$ and $x \in Q$. Since multiplication in G satisfies

$$(ax)(by) = a(xbx^{-1})xy = ab^x xy,$$

it is easy to see that the map $K \rtimes_\theta Q \to G$, defined by $(a, x) \mapsto ax$, is an isomorphism. ■

We now illustrate how this construction can be used.

EXAMPLE 7.14. The group T of order 12 (see Theorem 4.24) is a semidirect product of $\mathbb{Z}_3$ by $\mathbb{Z}_4$.

Let $\mathbb{Z}_3 = \langle a \rangle$, let $\mathbb{Z}_4 = \langle x \rangle$, and define $\theta: \mathbb{Z}_4 \to \mathrm{Aut}(\mathbb{Z}_3) \cong \mathbb{Z}_2$ by sending a into the generator. In more detail,

$$a^x = a^2 \qquad \text{and} \qquad (a^2)^x = a,$$

while x^2 acts on $\langle a \rangle$ as the identity automorphism: $a^{x^2} = a$.

The group $G = \mathbb{Z}_3 \rtimes_\theta \mathbb{Z}_4$ has order 12. If $s = (a^2, x^2)$ and $t = (1, x)$, then the reader may check that

$$s^6 = 1 \qquad \text{and} \qquad t^2 = s^3 = (st)^2,$$

which are the relations in T.

EXAMPLE 7.15. Let p be a prime, let $K = \langle a, b \rangle$ be an elementary abelian group of order p^2, and let $Q = \langle x \rangle$ be a cyclic group of order p. Define $\theta: Q \to \mathrm{Aut}(K) \cong \mathrm{GL}(2, p)$ by

$$x^i \mapsto \begin{bmatrix} 1 & 0 \\ i & 1 \end{bmatrix}.$$

Thus, $a^x = ab$ and $b^x = b$. The commutator $a^x a^{-1}$ is seen to be b. Therefore, $G = K \rtimes_\theta Q$ is a group of order p^3 with $G = \langle a, b, x \rangle$, and these generators satisfy relations

$$a^p = b^p = x^p = 1, \qquad b = [x, a], \qquad \text{and} \qquad [b, a] = 1 = [b, x].$$

If p is odd, then we have the nonabelian group of order p^3 and exponent p; if $p = 2$, then $G \cong D_8$ (as the reader may check). In Example 7.8, we saw that $D_8 \cong \mathbb{Z}_4 \rtimes_\theta \mathbb{Z}_2$; we have just seen here that $D_8 \cong \mathbf{V} \rtimes_\theta \mathbb{Z}_2$. A group may thus have distinct factorizations into a semidirect product.

EXAMPLE 7.16. Let p be an odd prime, let $K = \langle a \rangle$ be cyclic of order p^2, and let $Q = \langle x \rangle$ be cyclic of order p. By Theorem 7.3, $\mathrm{Aut}(K) \cong \mathbb{Z}_{p(p-1)} \cong \mathbb{Z}_{p-1} \times \mathbb{Z}_p$; indeed, by Theorem 6.9, the cyclic summand $\mathbb{Z}_p = \langle \alpha \rangle$, where $\alpha(a) = a^{1+p}$. If one defines $\theta: Q \to \mathrm{Aut}(K)$ by $\theta_x = \alpha$, then the group $G = K \rtimes_\theta Q$ has order p^3, generators x, a, and relations $x^p = 1$, $a^{p^2} = 1$, and $xax^{-1} = a^x = a^{1+p}$. We have constructed the second nonabelian group of order p^3 (see Exercise 4.32).

EXERCISES

7.23. Show that the group $\mathbf{Q}_n$ of generalized quaternions is not a semidirect product.

7.24. If $|G| = mn$, where $(m, n) = 1$, and if $K \leq G$ has order m, then a subgroup $Q \leq G$ is a complement of K if and only if $|Q| = n$.

7.25. If k is a field, then $\mathrm{GL}(n, k)$ is a semidirect product of $\mathrm{SL}(n, k)$ by $k^\times$, where $k^\times = k - \{0\}$.

7.26. If M is the group of all motions of $\mathbb{R}^n$, then M is a semidirect product of $\mathrm{Tr}(n, \mathbb{R})$ by $\mathrm{O}(n, \mathbb{R})$.

7.27. If K and Q are solvable, then $K \rtimes_\theta Q$ is also solvable.

7.28. Show that $K \rtimes_\theta Q$ is the direct product $K \times Q$ if and only if $\theta: Q \to \mathrm{Aut}(K)$ is ***trivial*** (that is, $\theta_x = 1$ for all $x \in Q$).

7.29. If p and q are distinct primes, construct all semidirect products of $\mathbb{Z}_p$ by $\mathbb{Z}_q$, and compare your results to Theorem 4.20. (The condition $q \nmid p - 1$ in that theorem should now be more understandable.)

Wreath Products

Let D and Q be groups, let Ω be a finite Q-set, and let $\{D_\omega: \omega \in \Omega\}$ be a family of isomorphic copies of D indexed by Ω.

Definition. Let D and Q be groups, let Ω be a finite Q-set, and let $K = \prod_{\omega \in \Omega} D_\omega$, where $D_\omega \cong D$ for all $\omega \in \Omega$. Then the ***wreath product*** of D by Q, denoted by $D \wr Q$ (or by D wr Q), is the semidirect product of K by Q, where Q acts on K by $q \cdot (d_\omega) = (d_{q\omega})$ for $q \in Q$ and $(d_\omega) \in \prod_{\omega \in \Omega} D_\omega$. The normal subgroup K of $D \wr Q$ is called the ***base*** of the wreath product.

The notation $D \wr Q$ is deficient, for it does not display the Q-set Ω; perhaps one should write $D \wr_\Omega Q$.

If D is finite, then $|K| = |D|^{|\Omega|}$; if Q is also finite, then $|D \wr Q| = |K \rtimes Q| = |K||Q| = |D|^{|\Omega|}|Q|$.

If Λ is a D-set, then $\Lambda \times \Omega$ can be made into a $(D \wr Q)$-set. Given $d \in D$ and $\omega \in \Omega$, define a permutation d_ω^* of $\Lambda \times \Omega$ as follows: for each $(\lambda, \omega') \in \Lambda \times \Omega$, set

$$d_\omega^*(\lambda, \omega') = \begin{cases} (d\lambda, \omega') & \text{if } \omega' = \omega, \\ (\lambda, \omega') & \text{if } \omega' \neq \omega. \end{cases}$$

It is easy to see that $d_\omega^* d_\omega'^* = (dd')_\omega^*$, and so D_ω^*, defined by

$$D_\omega^* = \{d_\omega^*: d \in D\},$$

is a subgroup of $S_{\Lambda \times \Omega}$; indeed, for each ω, the map $D \to D_\omega^*$, given by $d \mapsto d_\omega^*$, is an isomorphism.

For each $q \in Q$, define a permutation q^* of $\Lambda \times \Omega$ by

$$q^*(\lambda, \omega') = (\lambda, q\omega'),$$

and define

$$Q^* = \{q^*: q \in Q\}.$$

It is easy to see that Q^* is a subgroup of $S_{\Lambda \times \Omega}$ and that the map $Q \to Q^*$, given by $q \mapsto q^*$, is an isomorphism.

Theorem 7.24. *Given groups D and Q, a finite Q-set Ω, and a D-set Λ, then the wreath product $D \wr Q$ is isomorphic to the subgroup*

$$W = \langle Q^*, D_\omega^*: \omega \in \Omega \rangle \leq S_{\Lambda \times \Omega},$$

and hence $\Lambda \times \Omega$ is a $(D \wr Q)$-set.

Proof. We show first that $K^* = \langle \bigcup_{\omega \in \Omega} D_\omega^* \rangle$ is the direct product $\prod_{\omega \in \Omega} D_\omega^*$. It is easy to see that D_ω^* centralizes $D_{\omega'}^*$ for all $\omega' \neq \omega$, and so $D_\omega^* \lhd K^*$ for every ω. Each $d_\omega^* \in D_\omega^*$ fixes all $(\lambda, \omega') \in \Lambda \times \Omega$ with $\omega' \neq \omega$, while each element of $\langle \bigcup_{\omega' \neq \omega} D_{\omega'}^* \rangle$ fixes all (λ, ω) for all $\lambda \in \Lambda$. It follows that if $d_\omega^* \in D_\omega^* \cap \langle \bigcup_{\omega' \neq \omega} D_{\omega'}^* \rangle$, then $d_\omega^* = 1$.

If $q \in Q$ and $\omega \in \Omega$, then a routine computation gives

$$q^* d_\omega^* q^{*-1} = d_{q\omega}^*$$

for each $\omega \in \Omega$. Hence $q^* K^* q^{*-1} \leq K^*$ for each $q \in Q$, so that $K^* \lhd W$ (because $W = \langle K^*, Q^* \rangle$); it follows that $W = K^* Q^*$. To see that W is a semidirect product of K^* by Q^*, it suffices to show that $K^* \cap Q^* = 1$. Now $d_\omega^*(\lambda, \omega') = (d\lambda, \omega')$ or (λ, ω'); in either case, d_ω^* fixes the second coordinate. If $q^* \in Q^*$, then $q^*(\lambda, \omega') = (\lambda, q\omega')$ and q^* fixes the first coordinate. Therefore, any $g \in K^* \cap Q^*$ fixes every (λ, ω') and hence is the identity.

It is now a simple matter to check that the map $D \wr Q \to W$, given by $(d_\omega)q \mapsto (d_\omega^*)q^*$, is an isomorphism. ■

Call the subgroup W of $S_{\Lambda \times \Omega}$ the ***permutation version*** of $D \wr Q$; when we wish to view $D \wr Q$ acting on $\Lambda \times \Omega$, then we will think of it as W.

Theorem 7.25. *Let D and Q be groups, let Ω be a finite Q-set, let Λ be a D-set, and let $W \leq S_{\Lambda \times \Omega}$ be the permutation version of $D \wr Q$.*

(i) *If Ω is a transitive Q-set and Λ is a transitive D-set, then $\Lambda \times \Omega$ is a transitive $(D \wr Q)$-set.*
(ii) *If $\omega \in \Omega$, then its stabilizer Q_ω acts on $\Omega - \{\omega\}$. If $(\lambda, \omega) \in \Lambda \times \Omega$ and $D(\lambda) \leq D$ is the stabilizer of λ, then the stabilizer $W_{(\lambda, \omega)}$ of (λ, ω) is isomorphic to $D(\lambda) \times (D \wr Q_\omega)$, and $[W : W_{(\lambda, \omega)}] = [D : D(\lambda)][Q : Q_\omega]$.*

Proof. (i) Let $(\lambda, \omega), (\lambda', \omega') \in \Lambda \times \Omega$. Since D acts transitively, there is $d \in D$ with $d\lambda = \lambda'$; since Q acts transitively, there is $q \in Q$ with $q\omega = \omega'$. The reader may now check that $q^* d_\omega^*(\lambda, \omega) = (\lambda', \omega')$.

(ii) Each element of W has the form $(d_\omega^*)q^*$, and $(d_\omega^*)q^*(\lambda, \omega) = (\prod_{\omega' \in \Omega} d_{\omega'}^*)(\lambda, q\omega) = d_{q\omega}^*(\lambda, q\omega) = (d_{q\omega}\lambda, q\omega)$. It follows that $(d_\omega^*)q^*$ fixes (λ, ω) if and only if q fixes ω and d_ω fixes λ. Let $D_\omega^*(\lambda) = \{d_\omega^*: d \in D(\lambda)\}$. Now $D_\omega^*(\lambda)$ is disjoint from $\langle \prod_{\omega' \neq \omega} D_{\omega'}^*, Q_\omega^* \rangle$ and centralizes it: if $q^* \in Q_\omega^*$, then $q^* d_\omega^* q^{*-1} = d_{q\omega}^* = d_\omega^*$); hence

$$\begin{aligned} W_{(\lambda,\omega)} &= \left\langle D_\omega^*(\lambda), \prod_{\omega' \neq \omega} D_{\omega'}^*, Q_\omega^* \right\rangle \\ &= D_\omega^*(\lambda) \times \left\langle \prod_{\omega' \neq \omega} D_{\omega'}^*, Q_\omega^* \right\rangle \\ &\cong D(\lambda) \times (D \wr Q_\omega). \end{aligned}$$

It follows that $|W_{(\lambda,\omega)}| = |D(\lambda)||D|^{|\Omega|-1}|Q_\omega|$ and

$$[W : W_{(\lambda,\omega)}] = |D|^{|\Omega|}|Q|/|D(\lambda)||D|^{|\Omega|-1}|Q_\omega| = [D : D(\lambda)][Q : Q_\omega]. \quad \blacksquare$$

Theorem 7.26. *Wreath product is associative: if both Ω and Λ are finite, if T is a group, and if Δ is a T-set, then $T \wr (D \wr Q) \cong (T \wr D) \wr Q$.*

Proof. The permutation versions of both $T \wr (D \wr Q)$ and $(T \wr D) \wr Q$ are subgroups of $S_{\Delta \times \Lambda \times \Omega}$; we claim that they coincide. The group $T \wr (D \wr Q)$ is generated by all $t_{(\lambda,\omega)}^*$ (for $t \in T$ and $(\lambda, \omega) \in \Lambda \times \Omega$) and all f^* (for $f \in D \wr Q$). Note that $t_{(\lambda,\omega)}^*: (\delta', \lambda', \omega') \mapsto (t\delta', \lambda', \omega')$ if $(\lambda', \omega') = (\lambda, \omega)$, and fixes it otherwise; also, $f^*: (\delta', \lambda', \omega') \mapsto (\delta', f(\lambda', \omega'))$. Specializing f^* to d_ω^* and to q^*, we see that $T \wr (D \wr Q)$ is generated by all $t_{(\lambda,\omega)}^*$, d_ω^*, and q^{**}, where $d_\omega^*: (\delta', \lambda', \omega') \mapsto (\delta', d\lambda', \omega')$ if $\omega' = \omega$, and fixes it otherwise, and $q^{**}: (\delta', \lambda', \omega') \mapsto (\delta', \lambda', q\omega')$.

A similar analysis of $(T \wr D) \wr Q$ shows that it is generated by all q^{**}, d_ω^*, and $(t_\lambda)_\omega^*$, where $(t_\lambda)_\omega^*: (\delta', \lambda', \omega') \mapsto (t\delta', \lambda', \omega')$ if $\omega' = \omega$ and $\lambda' = \lambda$, and fixes it otherwise. Since $(t_\lambda)_\omega^* = t_{(\lambda,\omega)}^*$, the two wreath products coincide. $\blacksquare$

The best way to understand wreath products is by considering graphs.

Definition. A ***graph*** Γ is a nonempty set V, called ***vertices***, together with an ***adjacency*** relation on V, denoted by $v \sim u$, that is symmetric ($v \sim u$ implies $u \sim v$ for all $u, v \in V$) and irreflexive ($v \nsim v$ for all $v \in V$).

One can draw pictures of finite graphs; regard the vertices as points and join each adjacent pair of vertices with a line segment or ***edge***. Notice that our graphs are *nondirected*; that is, one can traverse an edge in either direction; moreover, there are no "loops"; every edge has two distinct endpoints. An ***automorphism*** of a graph Γ with vertices V is a bijection $\varphi: V \to V$ such that $u, v \in V$ are adjacent if and only if $\varphi(u)$ and $\varphi(v)$ are adjacent. It is plain that the set of all automorphisms of a graph Γ, denoted by $\mathrm{Aut}(\Gamma)$, is a group under composition.

For example, consider the following graph Γ:

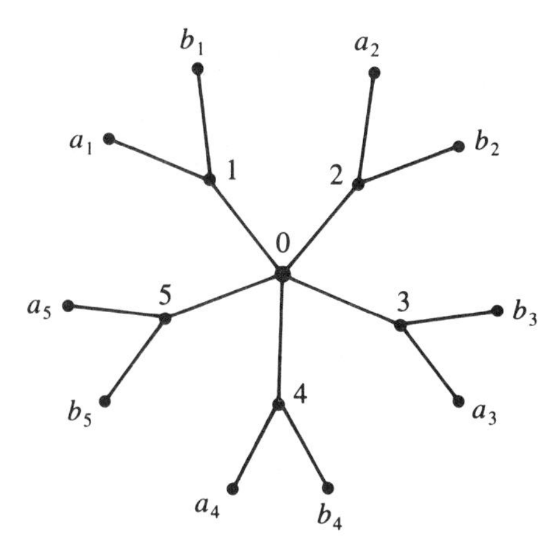

Figure 7.1

If $\varphi \in \mathrm{Aut}(\Gamma)$, then φ fixes vertex 0 (it is the only vertex adjacent to 5 vertices), φ permutes the "inner ring" $\Omega = \{1, 2, 3, 4, 5\}$, and, for each i, either $\varphi(a_i) = a_{\varphi i}$ and $\varphi(b_i) = b_{\varphi i}$ or $\varphi(a_i) = b_{\varphi i}$ and $\varphi(b_i) = a_{\varphi i}$. It is now easy to see that $|\mathrm{Aut}(\Gamma)| = 2^5 \times 5!$. Regard S_5 as acting on Ω and regard S_2 as acting on $\Lambda = \{a, b\}$. Identify the outer ring of all vertices $\{a_i, b_i: i \in \Omega\}$ with $\Lambda \times \Omega$ by writing a_i as (a, i) and b_i as (b, i). If $q \in S_5$, then q permutes the inner ring: $q^*(a_i) = a_{qi}$ and $q^*(b_i) = b_{qi}$; that is, $q^*(a, i) = (a, qi)$ and $q^*(b, i) = (b, qi)$. If $d \in S_2$ and $i \in \Omega$, then $d_i^*(a, i) = (da, i)$, $d_i^*(b, i) = (db, i)$, while d_i^* fixes (a, j) and (b, j) for $j \neq i$. For example, if d interchanges a and b, then $d_i^*(a_i) = b_i$ and $d_i^*(b_i) = a_i$, while d_i^* fixes a_j and b_j for all $j \neq i$. Thus, both q^* and d_i^* correspond to automorphisms of Γ. In Exercise 7.30 below, you will show that $\mathrm{Aut}(\Gamma) \cong S_2 \wr S_5$.

A special case of the wreath product construction has $\Omega = Q$ regarded as a Q-set acting on itself by left multiplication. In this case, we write $W = D \wr_r Q$, and we call W the ***regular wreath product***. Thus, the base is the direct product of $|Q|$ copies of D, indexed by the elements of Q, and $q \in Q$ sends a $|Q|$-tuple $(d_x) \in \prod_{x \in Q} D_x$ into (d_{qx}). Note that $|D \wr_r Q| = |D|^{|Q|}|Q|$. It is easy to see that the formation of regular wreath products is *not* associative when all groups are finite, for $|T \wr_r (D \wr_r Q)| \neq |(T \wr_r D) \wr_r Q|$.

If Ω is an infinite set and $\{D_\omega: \omega \in \Omega\}$ is a family of groups, then there are two direct product constructions. The first, sometimes called the *complete direct product*, consists of all "vectors" (d_ω) in the cartesian product $\prod_{\omega \in \Omega} D_\omega$ with "coordinatewise" multiplication: $(d_\omega)(d'_\omega) = (d_\omega d'_\omega)$. The second, called the *restricted direct product*, is the subgroup of the first consisting of all those (d_ω) with only finitely many coordinates $d_\omega \neq 1$. Both versions coincide when the index set Ω is finite. The wreath product using the complete direct product is called the ***complete wreath product***; the wreath product using the restricted direct product is called the ***restricted wreath product***. We shall see a use for the complete wreath product at the end of the next section. The first example of a (necessarily infinite) centerless p-group was given by D.H. McLain (1954); it is a restricted wreath product of a group of prime order p by $\mathbb{Z}(p^\infty)$ (the latter group is discussed in Chapter 10; it is the multiplicative group of all pth power roots of unity). McLain's example is thus a p-group that is not nilpotent.

What is the order of a Sylow p-subgroup of the symmetric group S_m? If $k \leq m$ are positive integers, define $t = [m/k]$, the greatest integer in m/k. Thus, $k, 2k, \ldots, tk \leq m$, while $(t+1)k > m$, so that t is the number of integers $i \leq m$ which are divisible by k. If p is a prime, what is the largest power μ of p dividing $m!$? Think of $m!$ as factored: $m! = 2 \times 3 \times 4 \times \cdots \times m$. By our initial remark, $[m/p]$ factors of $m!$ are divisible by p, $[m/p^2]$ factors are divisible by p^2, etc. Hence, if $m! = p^\mu m'$, where $(m', p) = 1$, then

$$\mu = [m/p] + [m/p^2] + [m/p^3] + \cdots.$$

For example, if $p = 2$, then $[m/2]$ is the number of even integers $\leq m$, $[m/4]$ is the number of multiples of $4 \leq m$, and so forth. (Notice, for example, that $8 = 2^3$ is counted three times by the formula for μ.) In particular, if $m = p^n$, then the largest power of p dividing $p^n!$ is

$$\mu = \mu(n) = p^{n-1} + p^{n-2} + \cdots + p + 1,$$

and so the order of a Sylow p-subgroup of the symmetric group S_{p^n} is $p^{\mu(n)}$.

Theorem 7.27 (Kaloujnine, 1948). *If p is a prime, then a Sylow p-subgroup of S_{p^n} is an iterated regular wreath product $W_n = \mathbb{Z}_p \wr_r \mathbb{Z}_p \wr_r \cdots \wr_r \mathbb{Z}_p$ of n copies of $\mathbb{Z}_p$, where $W_{n+1} = W_n \wr_r \mathbb{Z}_p$.*

Proof. The proof is by induction on n, the case $n = 1$ holding because a Sylow p-subgroup of S_p has order p. Assume that $n > 1$. Let Λ be a set with p^n elements and let D be a Sylow p-subgroup of S_Λ; thus, Λ is a D-set. Let $\Omega = \{0, 1, \ldots, p-1\}$, and let $Q = \langle q \rangle$ be a cyclic group of order p acting on Ω by $qi = i + 1 \bmod p$. The permutation version of the wreath product $P = D \wr_r \mathbb{Z}_p$ is a subgroup of $S_{\Lambda \times \Omega}$; of course, $|\Lambda \times \Omega| = p^{n+1}$. By induction, D is a wreath product of n copies of $\mathbb{Z}_p$, and so P is a wreath product of $n + 1$ copies of $\mathbb{Z}_p$. To see that P is a Sylow p-subgroup, it suffices to see that its order is $p^{\mu(n+1)}$, where $\mu(n+1) = p^n + p^{n-1} + \cdots + p + 1$. Now $|D| = p^{\mu(n)}$, so that $|P| = |D \wr_r \mathbb{Z}_p| = (p^{\mu(n)})^p p = p^{p\mu(n)+1} = p^{\mu(n+1)}$. ■

Theorem 7.27 may be used to compute the Sylow p-subgroup of S_m for any m (not necessarily a power of p). First write m in base p:

$$m = a_0 + a_1 p + a_2 p^2 + \cdots a_t p^t, \qquad \text{where} \quad 0 \leq a_i \leq p - 1.$$

Partition $X = \{1, 2, \ldots, m\}$ into a_0 singletons, a_1 p-subsets, a_2 p^2-subsets, $\ldots$, and a_t p^t-subsets. On each of these p^i-subsets Y, construct a Sylow p-subgroup of S_Y. Since disjoint permutations commute, the direct product of all these Sylow subgroups is a subgroup of S_X of order p^N, where $N = a_1 + a_2\mu(2) + \cdots + a_t\mu(t)$ (recall that $\mu(i) = p^{i-1} + p^{i-2} + \cdots + p + 1$). But p^N is the highest power of p dividing $m!$, for

$$m = a_0 + a_1 p + a_2 p^2 + \cdots + a_t p^t,$$

and so

$$\begin{aligned}[m/p] + [m/p^2] + [m/p^3] + \cdots &= (a_1 + a_2 p + a_3 p^2 + \cdots + a_t p^{t-1}) \\ &\quad + (a_2 + a_3 p + a_4 p^2 + \cdots + a_t p^{t-2}) \\ &\quad + (a_3 + a_4 p + \cdots + a_t p^{t-3}) + \cdots \\ &= a_1 + a_2(p+1) + a_3(p^2 + p + 1) + \cdots \\ &= a_1 + a_2 \mu(2) + \cdots + a_t \mu(t) = N.\end{aligned}$$

Thus, the direct product has the right order, and so it must be a Sylow p-subgroup of $S_X \cong S_m$.

For example, let us compute a Sylow 2-subgroup of S_6 (this has been done by hand in Exercise 4.15 (ii)). In base 2, we have $6 = 0 \times 1 + 1 \times 2 + 1 \times 4$. A Sylow 2-subgroup of S_2 is $\mathbb{Z}_2$; a Sylow 2-subgroup of S_4 is $\mathbb{Z}_2 \wr \mathbb{Z}_2$. We conclude that a Sylow 2-subgroup P of S_6 is $\mathbb{Z}_2 \times (\mathbb{Z}_2 \wr \mathbb{Z}_2)$. By Exercise 7.31 below, $\mathbb{Z}_2 \wr \mathbb{Z}_2 \cong D_8$, so that $P \cong \mathbb{Z}_2 \times D_8$.

Exercises

7.30. Prove that $\mathrm{Aut}(\Gamma) \cong S_2 \wr S_5$, where Γ is the graph in Figure 7.1. (*Hint.* Every $\varphi \in \mathrm{Aut}(\Gamma)$ is completely determined by its behavior on the outer ring consisting of all vertices of the form a_i or b_i.)

7.31. Prove that $\mathbb{Z}_2 \wr \mathbb{Z}_2 \cong D_8$. (*Hint.* $\mathbb{Z}_2 \wr \mathbb{Z}_2$ has several involutions.)

7.32. If both D and Q are solvable, then $D \wr Q$ is solvable.

7.33. **Definition.** Let D be a (multiplicative) group. A ***monomial matrix*** μ ***over*** D is a permutation matrix P whose nonzero entries have been replaced by elements of D; we say that P is the ***support*** of μ. If Q is a group of $n \times n$ permutation matrices, then

$$M(D, Q) = \{\text{all monomial matrices } \mu \text{ over } D \text{ with support in } Q\}.$$

(i) Prove that $M(D, Q)$ is a group under matrix multiplication.
(ii) Prove that the subgroup $Q \cong M(1, Q) \leq M(D, Q)$.
(iii) Prove that the diagonal $M(D, 1)$ is isomorphic to the direct product $D \times \cdots \times D$ (n times).
(iv) Prove that $M(D, 1) \lhd M(D, Q)$ and that $M(D, Q)$ is a semidirect product of $M(D, 1)$ by $M(1, Q)$.
(v) Prove that $M(D, Q) \cong D \wr Q$.

7.34. (i) Fix a group Q and a finite Q-set Ω. For all groups D and A and all homomorphisms $f: D \to A$, there is a homomorphism $M(f): M(D, Q) \to M(A, Q)$ such that $M(1_D) = 1_{M(D,Q)}$ and, whenever $g: A \to B$, then $M(gf) = M(g)M(f)$. (*Hint.* Just replace every nonzero entry x of a monomial matrix over D by $f(x)$.) (In categorical language, this exercise shows that wreath product is a functor.)

(ii) If D is abelian, show that determinant $d: M(D, Q) \to D$ is a (well defined) homomorphism.

7.35. If $(a, x) \in D \wr Q$ (so that $a \in K = \prod D_\omega$), then

$$(a, x)^n = (aa^x a^{x^2} \dots a^{x^{n-1}}, x^n).$$

7.36. Let $X = B_1 \cup \dots \cup B_m$ be a partition of a set X in which each B_i has k elements. If

$$G = \{g \in S_X\text{: for each } i, \text{ there is } j \text{ with } g(B_i) = B_j\},$$

then $G \cong S_k \wr S_m$.

Factor Sets

Since there are nonsimple groups that are not semidirect products, our survey of extensions is still incomplete. Notice the kind of survey we already have: if we know Q, K, and θ, then we know the semidirect product $K \rtimes_\theta Q$ in the sense that we can write a multiplication table for it (its elements are ordered pairs and we know how to multiply any two of them).

In discussing general extensions G of K by Q, it is convenient to use the additive notation for G and its subgroup K (this is one of the rare instances in which one uses additive notation for a nonabelian group). For example, if $k \in K$ and $g \in G$, we shall write the conjugate of k by g as $g + k - g$.

Definition. If $K \leq G$, then a ***(right) transversal*** of K in G (or a *complete set of right coset representatives*) is a subset T of G consisting of one element from each right coset of K in G.

If T is a right transversal, then G is the disjoint union $G = \bigcup_{t \in T} K + t$. Thus, every element $g \in G$ has a unique factorization $g = k + t$ for $k \in K$ and $t \in T$. There is a similar definition of left transversal; of course, these two notions coincide when K is normal.

If G is a semidirect product and Q is a complement of K, then Q is a transversal of K in G.

Definition. If $\pi: G \to Q$ is surjective, then a ***lifting*** of $x \in Q$ is an element $l(x) \in G$ with $\pi(l(x)) = x$.

If one chooses a lifting $l(x)$ for each $x \in Q$, then the set of all such is a transversal of $\ker \pi$. In this case, the function $l: Q \to G$ is also called a ***right transversal*** (thus, both l and its image $l(Q)$ are called right transversals).

Theorem 7.28. *Let G be an extension of K by Q, and let $l: Q \to G$ be a transversal.*

If K is abelian, then there is a homomorphism $\theta: Q \to$ Aut(K) with

$$\theta_x(a) = l(x) + a - l(x)$$

for every $a \in K$. *Moreover, if* $l_1: Q \to G$ *is another transversal, then* $l(x) + a - l(x) = l_1(x) + a - l_1(x)$ *for all* $a \in K$ *and* $x \in Q$.

Proof. Since $K \lhd G$, the restriction $\gamma_g|K$ is an automorphism of K for all $g \in G$, where γ_g is conjugation by g. The function $\mu: G \to \mathrm{Aut}(K)$, given by $g \mapsto \gamma_g|K$, is easily seen to be a homomorphism; moreover, $K \leq \ker \mu$, for K being abelian implies that each conjugation by $a \in K$ is the identity. Therefore, μ induces a homomorphism $\mu_\#: G/K \to \mathrm{Aut}(K)$, namely, $Kg \mapsto \mu(g)$.

The first isomorphism theorem says more than $Q \cong G/K$; it gives an explicit isomorphism $\lambda: Q \to G/K$: if $l: Q \to G$ is a transversal, then $\lambda(x) = K + l(x)$. If $l_1: Q \to G$ is another transversal, then $l(x) - l_1(x) \in K$, so that $K + l(x) = K + l_1(x)$ for every $x \in Q$. It follows that λ does not depend on the choice of transversal. Let $\theta: Q \to \mathrm{Aut}(K)$ be the composite: $\theta = \mu_\# \lambda$. If $x \in Q$, then $\theta_x = \mu_\# \lambda(x) = \mu_\#(K + l(x)) = \mu(l(x)) \in \mathrm{Aut}(K)$; therefore, if $a \in K$,

$$\theta_x(a) = \mu(l(x))(a) = l(x) + a - l(x)$$

does not depend on the choice of lifting $l(x)$. ■

There is a version of Theorem 7.28 that provides a homomorphism θ when K is not assumed to be abelian (in this case, $\theta: Q \to \mathrm{Aut}(K)/\mathrm{Inn}(K)$), but this more general situation is rather complicated. Thus, we shall assume that K is abelian for the rest of this chapter.

A homomorphism $\theta: Q \to \mathrm{Aut}(K)$ makes K into a Q-set, where the action is given by $xa = \theta_x(a)$. (For semidirect products, we denoted $\theta_x(a)$ by a^x; since we are now writing K additively, however, the notation xa is more appropriate.) The following formulas are valid for all $x, y, 1 \in Q$ and $a, b \in K$:

$$x(a + b) = xa + xb,$$

$$(xy)a = x(ya),$$

$$1a = a.$$

Definition. Call an ordered triple (Q, K, θ) ***data*** if K is an abelian group, Q is a group, and $\theta: Q \to \mathrm{Aut}(K)$ is a homomorphism. We say that a group G ***realizes*** this data if G is an extension of K by Q and, for every transversal $l: Q \to G$,

$$xa = \theta_x(a) = l(x) + a - l(x)$$

for all $x \in Q$ and $a \in K$.

Using these terms, Theorem 7.28 says that when K is abelian, every extension G of K by Q determines a homomorphism $\theta: Q \to \mathrm{Aut}(K)$, and G realizes the data. The intuitive meaning of θ is that it describes how K is a normal subgroup of G. For example, if the abelian group K is a subgroup of the center $Z(G)$, then θ is the trivial homomorphism with $\theta_x = 1$ for all $x \in Q$ (for then $a = xa = l(x) + a - l(x)$, and a commutes with all $l(x)$, hence with all $g = b + l(x)$ for $b \in K$). The extension problem is now posed more precisely as follows: find all the extensions G which realize given data (Q, K, θ). Our aim is to write a multiplication table (rather, an addition table!) for all such G.

Let $\pi: G \to Q$ be a surjective homomorphism with kernel K, and choose a transversal $l: Q \to G$ with $l(1) = 0$. Once this transversal has been chosen, every element $g \in G$ has a unique expression of the form

$$g = a + l(x), \qquad a \in K, \quad x \in Q,$$

(after all, $l(x)$ is a representative of a coset of K in G, and G is the disjoint union of these cosets). There is a formula: for all $x, y \in Q$,

$$(1) \qquad l(x) + l(y) = f(x, y) + l(xy) \qquad \text{for some} \quad f(x, y) \in K,$$

because both $l(x) + l(y)$ and $l(xy)$ represent the same coset of K.

Definition. If $\pi: G \to Q$ is a surjective homomorphism with kernel K, and if $l: Q \to G$ is a transversal with $l(1) = 0$, then the function $f: Q \times Q \to K$, defined by (1) above, is called a ***factor set*** (or *cocycle*). (Of course, the factor set f depends on the transversal l.)

Consider the special case of a semidirect product G. Theorem 7.23 shows that $G \cong K \rtimes_\theta Q$, so that we may assume that G consists of all ordered pairs $(a, x) \in K \times Q$ and that multiplication is given by

$$(a, x)(b, y) = (ab^x, xy).$$

In additive notation, this becomes

$$(a, x) + (b, y) = (a + xb, xy).$$

If $l: Q \to G$ is the transversal defined by $l(x) = (0, x)$, then l is a homomorphism—$l(xy) = l(x) + l(y)$—and so the factor set determined by this l is identically zero. Thus, one may think of a factor set as a "measure" of G's deviation from being a semidirect product, for it describes the obstruction to the transversal l being a homomorphism.

Theorem 7.29. *Let $\pi: G \to Q$ be a surjective homomorphism with kernel K, let $l: Q \to G$ be a transversal with $l(1) = 0$, and let $f: Q \times Q \to K$ be the corresponding factor set. Then*:

(i) *for all $x, y \in Q$,*
$$f(1, y) = 0 = f(x, 1);$$

(ii) *the **cocycle identity** holds for every $x, y, z \in Q$:*
$$f(x, y) + f(xy, z) = xf(y, z) + f(x, yz).$$

Proof. The definition of f gives $l(x) + l(y) = f(x, y) + l(xy)$. In particular, $l(1) + l(y) = f(1, y) + l(y)$; since we are assuming that $l(1) = 0$, we have $f(1, y) = 0$. A similar calculation shows that $f(x, 1) = 0$. The cocycle identity follows from associativity:

$$[l(x) + l(y)] + l(z) = f(x, y) + f(xy, z) + l(xyz);$$

on the other hand,

$$\begin{aligned} l(x) + [l(y) + l(z)] &= xf(y, z) + l(xy) + l(z) \quad \text{by (i)} \\ &= xf(y, z) + f(xy, z) + l(xyz). \end{aligned}$$

The cocycle identity follows. ■

A more interesting result is that the converse of Theorem 7.29 is true when K is abelian.

Theorem 7.30. *Given data (Q, K, θ), a function $f\colon Q \times Q \to K$ is a factor set if and only if it satisfies the cocycle identity*

$$xf(y, z) - f(xy, z) + f(x, yz) - f(x, y) = 0$$

as well as $f(1, y) = 0 = f(x, 1)$ for all $x, y, z \in Q$. More precisely, there is an extension G realizing the data and a transversal $l\colon Q \to G$ such that f is the corresponding factor set.

Proof. Necessity is Theorem 7.29. To prove sufficiency, let G be the set of all ordered pairs $(a, x) \in K \times Q$ equipped with the operation

$$(a, x) + (b, y) = (a + xb + f(x, y), xy)$$

(note that if f is identically 0, then this is the semidirect product $K \rtimes_\theta Q$).

The proof that G is a group is similar to the proof of Theorem 7.22. The cocycle identity is needed to prove associativity; the identity is $(0, 1)$; inverses are given by

$$-(a, x) = (-x^{-1}a - x^{-1}f(x, x^{-1}), x^{-1}).$$

Define $\pi\colon G \to Q$ by $(a, x) \mapsto x$. It is easy to see that π is a surjective homomorphism with kernel $\{(a, 1)\colon k \in K\}$. If we identify K with $\ker \pi$ via $a \mapsto (a, 1)$, then $K \lhd G$ and G is an extension of K by Q.

Does G realize the data? We must show, for every transversal $l\colon Q \to G$, that $xa = l(x) + a - l(x)$ for all $x \in Q$ and $a \in K$. Now we must have $l(x) = (b, x)$ for some $b \in K$. Therefore,

$$\begin{aligned} l(x) + a - l(x) &= (b, x) + (a, 1) - (b, x) \\ &= (b + xa, x) + (-x^{-1}b - x^{-1}f(x, x^{-1}), x^{-1}) \\ &= (b + xa + x[-x^{-1}b - x^{-1}f(x, x^{-1})] + f(x, x^{-1}), 1). \end{aligned}$$

Since K is abelian, the last term simplifies to $(xa, 1)$. As any element of K, we identify xa with $(xa, 1)$, and so G does realize the data.

Finally, define a transversal $l: Q \to K$ by $l(x) = (0, x)$ for all $x \in Q$. The factor set F corresponding to this transversal satisfies $F(x, y) = l(x) + l(y) - l(xy)$. But a straightforward calculation shows that $F(x, y) = (f(x, y), 1)$, and so f is a factor set, as desired. ■

Notation. Denote the extension G constructed in the proof of Theorem 7.30 by G_f; it realizes (Q, K, θ) and it has f as a factor set (arising from the transversal $l(x) = (0, x)$).

Definition. $Z^2(Q, K, \theta)$ is the set of all factor sets $f: Q \times Q \to K$.

Theorem 7.30 shows that $Z^2(Q, K, \theta)$ is an abelian group under pointwise addition: $f + g: (x, y) \mapsto f(x, y) + g(x, y)$. If f and g are factor sets, then so is $f + g$ (for $f + g$ also satisfies the cocycle identity and vanishes on $(1, y)$ and $(x, 1)$). If G_f and G_g are the extensions constructed from them, then G_{f+g} is also an extension; it follows that there is an abelian group structure on the family of all extensions realizing the data (Q, K, θ) whose identity element is the semidirect product (which is G_0)! This group of all extensions is extravagantly large, however, because the same extension occurs many times. After all, take a fixed extension G realizing the data, and choose two different transversals, say, l and l'. Each transversal gives a factor set:

$$l(x) + l(y) = f(x, y) + l(xy),$$

$$l'(x) + l'(y) = f'(x, y) + l'(xy).$$

Now the factor sets f and f' are distinct, but both of them have arisen from the same extension.

Lemma 7.31. *Let G be an extension realizing (Q, K, θ), and let l and l' be transversals with $l(1) = 0 = l'(1)$ giving rise to factor sets f and f', respectively. Then there is a function $h: Q \to K$ with $h(1) = 0$ such that*

$$f'(x, y) - f(x, y) = xh(y) - h(xy) + h(x)$$

for all $x, y \in Q$.

Proof. For each $x \in Q$, both $l(x)$ and $l'(x)$ are representatives of the same coset of K in G; there is thus an element $h(x) \in K$ with

$$l'(x) = h(x) + l(x).$$

Since $l'(1) = 0 = l(1)$, we have $h(1) = 0$. The main formula is derived as follows.

$$\begin{aligned} l'(x) + l'(y) &= [h(x) + l(x)] + [h(y) + l(y)] \\ &= h(x) + xh(y) + l(x) + l(y) \quad (G \text{ realizes the data}) \\ &= h(x) + xh(y) + f(x, y) + l(xy) \\ &= h(x) + xh(y) + f(x, y) - h(xy) + l'(xy). \end{aligned}$$

Therefore, $f'(x, y) = h(x) + xh(y) + f(x, y) - h(xy)$. The desired formula follows because each term lies in the abelian group K. ∎

Definition. Given data (Q, K, θ), a ***coboundary*** is a function $g: Q \times Q \to K$ for which there exists $h: Q \to K$ with $h(1) = 0$ such that

$$g(x, y) = xh(y) - h(xy) + h(x).$$

The set of all coboundaries is denoted by $B^2(Q, K, \theta)$.

It is easy to check that $B^2(Q, K, \theta)$ is a subgroup of $Z^2(Q, K, \theta)$; that is, every coboundary g satisfies the cocycle identity and $g(x, 1) = 0 = g(1, x)$ for all $x \in Q$. Moreover, Lemma 7.31 says that factor sets f and f' arising from different transversals of the same extension satisfy $f' - f \in B^2(Q, K, \theta)$; that is, they lie in the same coset of $B^2(Q, K, \theta)$ in $Z^2(Q, K, \theta)$. We have been led to the following group and equivalence relation.

Definition. Given data (Q, K, θ), then

$$H^2(Q, K, \theta) = Z^2(Q, K, \theta)/B^2(Q, K, \theta);$$

it is called the ***second cohomology group*** of the data.

Definition. Two extensions G and G' realizing data (Q, K, θ) are ***equivalent*** if there are factor sets f of G and f' of G' with $f' - f \in B^2(Q, K, \theta)$; that is, the factor sets determine the same element of $H^2(Q, K, \theta)$.

Here is a characterization of equivalence in terms of the extensions.

A diagram of groups and homomorphisms ***commutes*** if, for each ordered pair of groups G and H in the diagram, all composites of arrows from G *to* H are equal. For example,

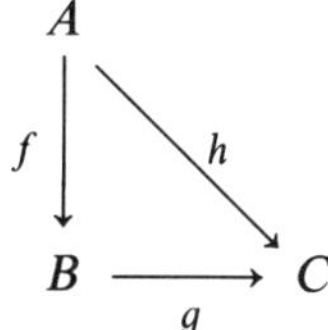

commutes if and only if $h = gf$.

Theorem 7.32. *Two extensions G and G' realizing data (Q, K, θ) are equivalent if and only if there exists an isomorphism γ making the following diagram commute:*

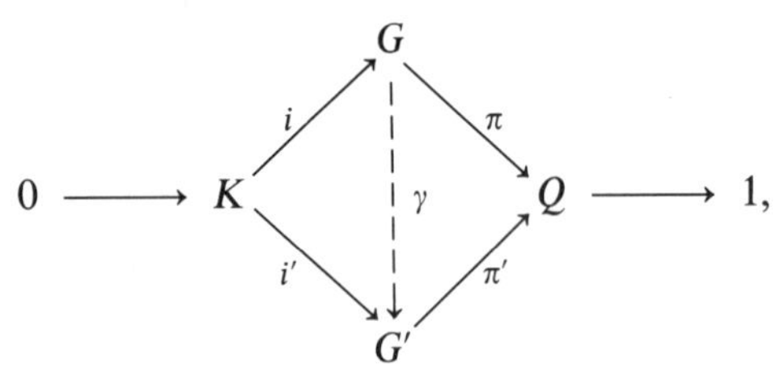

where i and i' are injective, π and π' are surjective, $\operatorname{im} i = \ker \pi$, *and* $\operatorname{im} i' = \ker \pi'$.

Remark. A homomorphism γ making the diagram commute is necessarily an isomorphism.

Proof. Assume that G and G' are equivalent. There are thus factor sets f, $f'\colon Q \times Q \to K$, arising from liftings l, l', respectively, and a function $h\colon Q \to K$ with $h(1) = 0$ such that

$$f'(x, y) - f(x, y) = xh(y) - h(xy) + h(x) \tag{2}$$

for all $x, y \in Q$. Each element of G has a unique expression of the form $a + l(x)$, where $a \in K$ and $x \in Q$, and addition is given by

$$[a + l(x)] + [b + l(y)] = a + xb + f(x, y) + l(xy);$$

there is a similar description of addition in G'. Define $\gamma\colon G \to G'$ by

$$\gamma(a + l(x)) = a + h(x) + l'(x).$$

Since $l(1) = 0$, we have $\gamma(a) = \gamma(a + l(1)) = a + h(1) + l'(1) = a$, for all $a \in K$, because $h(1) = 0$; that is, γ fixes K pointwise. Also, $x = \pi(a + l(x))$, while $\pi'\gamma(a + l(x)) = \pi'(a + h(x) + l'(x)) = \pi'(l'(x)) = x$. We have shown that the diagram commutes. It remains to show that γ is a homomorphism. Now

$$\begin{aligned}\gamma([a + l(x)] + [b + l(y)]) &= \gamma(a + xb + f(x, y) + l(xy)) \\ &= a + xb + f(x, y) + h(xy) + l'(xy),\end{aligned}$$

while

$$\begin{aligned}\gamma(a + l(x)) + \gamma(b + l(y)) &= [a + h(x) + l'(x)] + [b + h(y) + l'(y)] \\ &= a + h(x) + xb + xh(y) + f'(x, y) + l'(xy).\end{aligned}$$

The element $l'(xy)$ is common to both expressions, and (2) shows that the remaining elements of the abelian group K are equal; thus, γ is a homomorphism.

Conversely, assume that there exists an homomorphism γ as in the statement. Commutativity of the diagram gives $\gamma(a) = a$ for all $a \in K$. Moreover, if $x \in Q$, then $x = \pi(l(x)) = \pi'\gamma(l(x))$; that is, $\gamma l: Q \to G'$ is a lifting. Applying γ to the equation $l(x) + l(y) = f(x, y) + l(xy)$ shows that γf is the factor set determined by the lifting γl. But $\gamma f(x, y) = f(x, y)$ for all $x, y \in Q$, because $f(x, y) \in K$. Therefore $\gamma f = f$; that is, f is a factor set of G'. But f' is also a factor set of G' (arising from another lifting), and so Lemma 7.31 gives $f' - f \in B^2$; that is, G and G' are equivalent. ■

Definition. If G is an extension of K by Q, then $\gamma \in \text{Aut}(G)$ ***stabilizes*** the extension if the diagram in Theorem 7.32 (with G' replaced by G) commutes. The group A of all such γ is called the ***stabilizer*** of the extension.

Theorem 7.33. *If K and Q are (not necessarily abelian) groups and G is an extension of K by Q, then the stabilizer A of the extension is abelian.*

Proof. We give two proofs.

(i) If $\gamma \in A$, then the hypothesis that γ makes the diagram commute is precisely the statement that γ stabilizes the series $G > K > 1$. It follows from Theorem 7.19 that the group A of all such γ is nilpotent of class ≤ 1; that is, A is abelian.

(ii) Here is a proof without Theorem 7.19. We first show that if $\varphi \in A$ and $g \in G$, then $g\varphi(g)^{-1} \in K$. Let T be a transversal of K in G, and write $g = at$ for $a \in K$ and $t \in T$. By hypothesis, $\varphi(g) = a't$ for some $a' \in K$, and $g\varphi(g)^{-1} = att^{-1}a'^{-1} = aa'^{-1} \in K$. Next, we show that $g\varphi(g)^{-1} \in Z(K)$. If $b \in K$, then

$$\begin{aligned}
[g\varphi(g)^{-1}, b] &= g\varphi(g)^{-1}b\varphi(g)g^{-1}b^{-1} \\
&= g\varphi(g)^{-1}\varphi(b)\varphi(g)g^{-1}b^{-1} \quad (\varphi(b) = b \text{ since } b \in K) \\
&= g\varphi(g^{-1}bg)g^{-1}b^{-1} \\
&= g(g^{-1}bg)g^{-1}b^{-1} \quad (gbg^{-1} \in K) \\
&= 1.
\end{aligned}$$

The map $\Theta: A \to \prod_{g \in G} Z_g$ (where $Z_g \cong Z(K)$ for all $g \in G$) defined by $\Theta: \varphi \mapsto (g^{-1}\varphi(g))$, has kernel $\{\varphi \in A: g^{-1}\varphi(g) = 1 \text{ for all } g \in G\}$. Thus, $\varphi \in \ker \Theta$ if and only if $\varphi(g) = g$ for all $g \in G$; that is, $\varphi = 1_G$. Therefore, A is imbedded in the abelian group $\prod Z_g$, and hence A is abelian. ■

The next theorem summarizes the results of this section.

Theorem 7.34 (Schreier, 1926). *There is a bijection from $H^2(Q, K, \theta)$ to the set E of all equivalence classes of extensions realizing data (Q, K, θ) taking the identity 0 to the class of the semidirect product.*

Proof. Denote the equivalence class of an extension G realizing the data (Q, K, θ) by $[G]$. Define $\varphi: H^2(Q, K, \theta) \to E$ by $\varphi(f + B^2(Q, K, \theta)) = [G_f]$, where G_f is the extension constructed in Theorem 7.30 from a factor set f.

First, φ is well defined, for if f and g are factor sets with $f + B^2 = g + B^2$, then $f - g \in B^2$, G_f and G_g are equivalent, and $[G_f] = [G_g]$. Conversely, φ is an injection: if $\varphi(f + B^2) = \varphi(g + B^2)$, then $[G_f] = [G_g]$, $f - g \in B^2$, and $f + B^2 = g + B^2$. Now φ is a surjection: if $[G] \in E$ and f is a factor set (arising from some choice of transversal), then $[G] = [G_f]$ and $[G] = \varphi(f + B^2)$. The last part of the theorem follows from Exercise 7.39(ii) below: an extension is a semidirect product if and only if it has a factor set in B^2. ■

By Exercise 1.44, there is a unique group structure on E making φ an isomorphism, namely, $[G_f] + [G_g] = [G_{f+g}]$.

Corollary 7.35. $H^2(Q, K, \theta) = 0$ *if and only if every extension G realizing data* (Q, K, θ) *is a semidirect product.*

Definition. If both Q and K are abelian and $\theta: Q \to \mathrm{Aut}(K)$ is trivial, then $\mathrm{Ext}(Q, K)$ is the set of all equivalence classes of abelian extensions G of K by Q.

Corollary 7.36. *If both Q and K are abelian and* $\theta: Q \to \mathrm{Aut}(K)$ *is trivial, then* $\mathrm{Ext}(Q, K) \leq H^2(Q, K, \theta)$.

Proof. If $f: Q \times Q \to K$ is a factor set, then the corresponding extension G_f is abelian if and only if $f(x, y) = f(y, x)$ for all $x, y \in Q$: since θ is trivial,

$$
\begin{aligned}
(a, x) + (b, y) &= (a + b + f(x, y), xy) \\
&= (a + b + f(y, x), xy) \\
&= (b + a + f(y, x), yx) \\
&= (b, y) + (a, x).
\end{aligned}
$$

It is easy to see that the set $S^2(Q, K, \theta)$ of all such "symmetric" factor sets forms a subgroup of $Z^2(Q, K, \theta)$, and

$$
\begin{aligned}
\mathrm{Ext}(Q, K) &= (Z^2 \cap S^2)/(B^2 \cap S^2) \\
&\cong [(Z^2 \cap S^2) + B^2]/B^2 \leq H^2(Q, K, \theta). \quad ■
\end{aligned}
$$

The groups $H^2(Q, K, \theta)$ and $\mathrm{Ext}(Q, K)$ (when Q is abelian and θ is trivial) are studied and computed in Homological Algebra and, in particular, in Cohomology of Groups.

Here is an interesting use of the cocycle identity. Given a Q-set Ω, the wreath product $D \wr Q$ is the semidirect product $K \rtimes Q$, where $K = \prod_{\omega \in \Omega} D_\omega$ and $q \in Q$ acts on $(d_\omega) \in \prod_{\omega \in \Omega} D_\omega$ by $q(d_\omega) = (d_{q\omega})$. Now a formal description of the elements of the base $K = \prod_{\omega \in \Omega} D_\omega$ is as functions $\sigma: \Omega \to D$; that is, σ is the $|\Omega|$-tuple whose ωth coordinate is $\sigma(\omega)$; the multiplication in K is $(\sigma\tau)(\omega) = \sigma(\omega)\tau(\omega)$: the product of the ωth coordinate of σ with the ωth coordinate of τ. Now the action of Q on K is given by $\tau^q(\omega) = \tau(q\omega)$ for $q \in Q$ and $\omega \in \Omega$. It follows that the ωth coordinate of τ^q is the $q^{-1}\omega$th coordinate of τ. Thus,

$$(\sigma\tau^q)(\omega) = \sigma(\omega)\tau(q^{-1}\omega),$$

and multiplication in the wreath product is

$$(\sigma, q)(\tau, q') = (\sigma\tau^q, qq').$$

Theorem 7.37 (Kaloujnine and Krasner, 1951). *If D and Q are groups with Q finite, then the regular wreath product $D \wr_r Q$ contains an isomorphic copy of every extension of D by Q.*

Remark. The group D may not be abelian.

Proof. If G is an extension of D by Q, then there is a surjective homomorphism $G \to Q$ with kernel D, which we denote by $a \mapsto \bar{a}$. Choose a transversal $l: Q \to G$.

For $a \in G$, define $\sigma_a: Q \to D$ by

$$\sigma_a(x) = l(x)^{-1}al(\overline{a^{-1}}x).$$

If $a, b \in G$, then

$$\begin{aligned} \sigma_a(x)\sigma_b^{\bar{a}}(x) &= \sigma_a(x)\sigma_b(\overline{a^{-1}}x) \\ &= l(x)^{-1}al(\overline{a^{-1}}x)l(\overline{a^{-1}}x)^{-1}bl(\overline{b^{-1}}\,\overline{a^{-1}}x) \\ &= l(x)^{-1}bl((\bar{a}\bar{b})^{-1}x) = \sigma_{ab}(x). \end{aligned}$$

Define $\varphi: G \to D \wr_r Q$ by

$$\varphi(a) = (\sigma_a, \bar{a})$$

for every $a \in G$. We see that φ is a homomorphism, for

$$\begin{aligned} \varphi(a)\varphi(b) &= (\sigma_a, \bar{a})(\sigma_b, \bar{b}) \\ &= (\sigma_a\sigma_b^{\bar{a}}, \bar{a}\bar{b}) \\ &= (\sigma_{ab}, \overline{ab}). \end{aligned}$$

Finally, φ is injective. If $a \in \ker \varphi$, then $\bar{a} = 1$ and $\sigma_a(x) = 1$ for all $x \in Q$. The first equation gives $a \in D$; the second equation gives $\sigma_a(x) = l(x)^{-1}al(\overline{a^{-1}}x) = 1$ for every $x \in Q$. Since $\overline{a^{-1}} = 1$, we have $l(x)^{-1}al(x) = 1$, and so $a = 1$. ■

Remark. Here is a way to view the proof just given. The lifting l determines a factor set $f: Q \times Q \to D$ (which determines the extension G to isomorphism). For each $a \in G$, fixing one variable gives a function $f_a = f(\ ,\bar{a}): Q \to D$, namely, $f_a(x) = l(x)l(\bar{a})l(x\bar{a})^{-1}$. Since $l(\bar{a})$ is almost a, we see that σ_a is just a variation of f_a.

Corollary 7.38. *If $\mathscr{C}$ is a class of finite groups closed under subgroups and* *semidirect products (i.e., if $A \in \mathscr{C}$ and $S \leq A$, then $S \in \mathscr{C}$; if $A, B \in \mathscr{C}$, then $A \rtimes_\theta B \in \mathscr{C}$ for all θ), then $\mathscr{C}$ is closed under extensions.*

Proof. Assume that G is an extension of D by Q, where both $D, Q \in \mathscr{C}$. Since $\mathscr{C}$ is closed under semidirect products, it is closed under finite direct products; hence, $\prod_{\omega \in \Omega} D_\omega \in \mathscr{C}$. Since $\mathscr{C}$ is closed under semidirect products, the wreath product $D \wr_r Q = (\prod_{\omega \in \Omega} D_\omega) \rtimes Q \in \mathscr{C}$; since $\mathscr{C}$ is closed under subgroups, the theorem gives $G \in \mathscr{C}$. ∎

One may remove the finiteness hypothesis from Theorem 7.37 by using the complete wreath product.

EXERCISES

7.37. If H is a subgroup of a group G and if U is a right transversal of H in G, then $U^{-1} = \{u^{-1}: u \in U\}$ is a left transversal of H in G. Conclude, for $H \lhd G$, that both U and U^{-1} are transversals of H in G.

7.38. Prove that an extension G of K by Q is a semidirect product if and only if there is a transversal $l: Q \to G$ that is a homomorphism.

7.39. (i) Prove that any two semidirect products realizing data (Q, K, θ) are equivalent.
(ii) Prove that an extension G realizing data (Q, K, θ) is a semidirect product if and only if it has a factor set $f \in B^2$.

7.40. Let p be an odd prime. Give an example of two extensions G and G' realizing $(K, Q, \theta) = (\mathbb{Z}_p, \mathbb{Z}_p, \theta)$ with $G \cong G'$ but with G and G' not equivalent. (*Hint.* Let $K = \langle a \rangle$ be cyclic of order p, and let $G = \langle g \rangle$ and $G' = \langle h \rangle$ be cyclic of order p^2. Define $i: K \to G$ by $i(a) = pg$, and define $i': K \to G'$ by $i'(a) = 2ph$. Show that there is no isomorphism $\gamma: G \to G'$ making the diagram commute.)

Remark. It is plain that $|H^2(K, Q, \theta)|$ is an upper bound for the number of nonisomorphic extensions G of K by Q realizing θ. This last exercise shows that this bound need not be attained. After all, every extension of $\mathbb{Z}_p$ by $\mathbb{Z}_p$ has order p^2, hence is abelian, so that either $G \cong \mathbb{Z}_{p^2}$ or $G \cong \mathbb{Z}_p \times \mathbb{Z}_p$; but it can be shown that $|H^2(\mathbb{Z}_p, \mathbb{Z}_p, \theta)| = |\mathrm{Ext}(\mathbb{Z}_p, \mathbb{Z}_p)| = p$.

Theorems of Schur–Zassenhaus and Gaschütz

We now apply Corollary 7.35 of Schreier's theorem.

Recall that a Hall subgroup of a finite group (when it exists) is a subgroup whose order and index are relatively prime. Also recall Exercise 7.24: If $|G| = mn$, where $(m, n) = 1$, and if $K \leq G$ has order m, then a subgroup $Q \leq G$ is a complement of K if and only if $|Q| = n$.

Theorem 7.39. *If K is an abelian normal Hall subgroup of a finite group G (i.e., $(|K|, [G:K]) = 1$), then K has a complement.*

Proof. Let $|K| = m$, let $Q = G/K$, and let $|Q| = n$, so that $(m, n) = 1$. It suffices to prove, by Corollary 7.35, that every factor set $f: Q \times Q \to K$ is a coboundary. Define $\sigma: Q \to K$ by

$$\sigma(x) = \sum_{y \in Q} f(x, y);$$

σ is well defined since Q is finite and K is abelian. Sum the cocycle identity

$$xf(y, z) - f(xy, z) + f(x, yz) = f(x, y)$$

over all $z \in Q$ to obtain

$$x\sigma(y) - \sigma(xy) + \sigma(x) = nf(x, y)$$

(as z ranges over all of Q, so does yz). Since $(m, n) = 1$, there are integers s and t with $sm + tn = 1$. Define $h: Q \to K$ by $h(x) = t\sigma(x)$. Then $h(1) = 0$ and

$$xh(y) - h(xy) + h(x) = f(x, y) - msf(x, y).$$

But $sf(x, y) \in K$, so that $msf(x, y) = 0$. Therefore, f is a coboundary. ■

If $K \leq G$ has complements Q and Q', then $Q \cong Q'$ (each is isomorphic to G/K). One can say more if the orders of K and Q are relatively prime.

Theorem 7.40. *If K is an abelian normal Hall subgroup of a finite group G, then any two complements of K are conjugate.*

Proof. Denote $|K|$ by m and $|G/K|$ by n, so that $(m, n) = 1$. Let Q_1 and Q_2 be subgroups of G of order n. As we observed above, each of these subgroups is a complement of K. By Exercise 7.38, there are transversals $l_i: G/K \to G$, for $i = 1, 2$, with $l_i(G/K) = Q_i$ and with each l_i a homomorphism. It follows that the factor sets f_i determined by l_i are identically zero. If $h(x)$ is defined by $l_1(x) = h(x) + l_2(x)$, then

$$0 = f_1(x, y) - f_2(x, y) = xh(y) - h(xy) + h(x).$$

Summing over all $y \in G/K$ gives the equation in K:

$$0 = xa_0 - a_0 + nh(x),$$

where $a_0 = \sum_{y \in G/K} h(y)$. Let $sm + tn = 1$ and define $b_0 = ta_0$. Since K has exponent m,

$$h(x) = h(x) - smh(x) = -tnh(x) = xta_0 - ta_0 = xb_0 - b_0$$

for all $x \in G/K$. We claim that $-b_0 + Q_1 + b_0 = Q_2$. If $l_1(x) \in Q_1$, then $-b_0 + l_1(x) + b_0 = -b_0 + xb_0 + l_1(x) = -h(x) + l_1(x) = l_2(x) - l_1(x) + l_1(x) = l_2(x)$. ■

We now remove the hypothesis that K be abelian.

Theorem 7.41 (Schur–Zassenhaus Lemma, 1937).[3] *A normal Hall subgroup K of a finite group G has a complement (and so G is a semidirect product of K by G/K).*

Proof. Let $|K| = m$ and let $|G| = mn$, where $(m, n) = 1$. We prove, by induction on $m \geq 1$, that G contains a subgroup of order n. The base step is trivially true. If K contains a proper subgroup T which is also normal in G, then $K/T \lhd G/T$ and $(G/T)/(K/T) \cong G/K$ has order n; that is, K/T is a normal Hall subgroup of G/T. If $|K/T| = m'$, then $m' < m$ and $[G/T : K/T]| = n$. The inductive hypothesis gives a subgroup $N/T \leq G/T$ of order n. Now $|N| = n|T|$ and $(n, |T|) = 1$ (for $|T|$ divides m), so that T is a normal Hall subgroup of N (with $|T| < m$ and with index $[N : T] = n$). By induction, N and hence G contains a subgroup of order n.

We may now assume that K is a minimal normal subgroup of G. If p is a prime dividing m and if P is a Sylow p-subgroup of K, then the Frattini argument (Theorem 4.18) gives $G = KN_G(P)$. By the second isomorphism theorem,

$$G/K = KN_G(P)/K \cong N_G(P)/(K \cap N_G(P)) = N_G(P)/N_K(P),$$

so that $|N_K(P)|n = |N_K(P)||G/K| = |N_G(P)|$. If $N_G(P)$ is a proper subgroup of G, then $|N_G(P)| < m$, and induction shows that $N_G(P)$ contains a subgroup of order n. We may assume, therefore, that $N_G(P) = G$; that is, $P \lhd G$.

Since $K \geq P$ and K is a minimal normal subgroup of G, we have $K = P$. Lemma 5.20(ii) now applies: $Z(P)$ char P and $P \lhd G$ imply $Z(P) \lhd G$. Minimality applies again, and $Z(P) = P$ ($Z(P) \neq 1$ because P is a finite p-group). But now $P = K$ is abelian, and the proof follows from Theorem 7.39. ■

It follows that if a finite group G has a normal Sylow p-subgroup P, for some prime p, then P has a complement and G is a semidirect product of P by G/P.

We can prove part of the generalization of Theorem 7.41 for K nonabelian.

Theorem 7.42. *Let K be a normal Hall subgroup of a finite group G. If either K or G/K is solvable, then any two complements of K in G are conjugate.*

Remark. The Feit–Thompson theorem says that every group of odd order is solvable. Since $|K|$ and $|G/K|$ are relatively prime, at least one of them has odd order, and so complements of normal Hall subgroups are always conjugate.

Proof. Let $|K| = m$, let $|G/K| = n$, and let Q_1 and Q_2 be complements of K in G; of course, $Q_1 \cong G/K \cong Q_2$.

Assume first that K is solvable. By Lemma 5.20(ii), $K' \lhd G$; moreover,

[3] Schur (1904) proved this in the special case Q cyclic.

$Q_1K'/K' \cong Q_1/(Q_1 \cap K') \cong Q_1$ (because $Q_1 \cap K' \leq Q_1 \cap K = 1$), so that $|Q_1K'/K'| = n$. Now $K' < K$, because K is solvable. If $K' = 1$, then K is abelian and the result is Theorem 7.40; otherwise, $|G/K'| < |G|$, and induction on $|G|$ shows that the subgroups Q_1K'/K' and Q_2K'/K' are conjugate in G/K'. Thus, there is $\bar{g} \in G/K'$ with $\bar{g}(Q_1K'/K')\bar{g}^{-1} = Q_2K'/K'$; that is, $gQ_1g^{-1} \leq Q_2K'$ (where $K'g = \bar{g}$). But $K' < K$ gives $|Q_1K'| < |G|$, and so the subgroups gQ_1g^{-1} and Q_2 of order n are conjugate in Q_2K', hence are conjugate in G.

Assume now that G/K is solvable. We do an induction on $|G|$ that any two complements of K are conjugate. Let M/K be a minimal normal subgroup of G/K. Since $K \leq M$, the Dedekind law (Exercise 2.49) gives

$$(*) \qquad M = M \cap G = M \cap Q_iK = (M \cap Q_i)K \qquad \text{for} \quad i = 1, 2;$$

note also that $M \cap Q_i \lhd Q_i$. Now solvability of G/K gives M/K a p-group for some prime p, by Theorem 5.24. If $M = G$, then G/K is a p-group (indeed, since M/K is a minimal normal subgroup of G/K, we must have $|M/K| = p$). Therefore, Q_1 and Q_2 ($\cong G/K$) are Sylow p-subgroups of G, and hence are conjugate, by the Sylow theorem.

We may assume, therefore, that $M < G$. Now $M \cap Q_i$ is a complement of K in M, for $i = 1, 2$ (because $M = (M \cap Q_i)K$, by $(*)$, and $(M \cap Q_i) \cap K \leq Q_i \cap K = 1$). By induction, there is $x \in M \leq G$ with $M \cap Q_1 = x(M \cap Q_2)x^{-1} = M \cap xQ_2x^{-1}$. Replacing Q_2 by its conjugate xQ_2x^{-1} if necessary, we may assume that

$$M \cap Q_1 = M \cap Q_2.$$

If $J = M \cap Q_1 = M \cap Q_2$, then $J \lhd Q_i$ for $i = 1, 2$, and so

$$Q_i \leq N_G(J).$$

Two applications of the Dedekind law give

$$N_G(J) = N_G(J) \cap KQ_i = (N_G(J) \cap K)Q_i$$

and

$$J[N_G(J) \cap K] \cap Q_i = J([N_G(J) \cap K] \cap Q_i) = J$$

(because $(N_G(J) \cap K) \cap Q_i \leq K \cap Q_i = 1$). Therefore, Q_1/J and Q_2/J are complements of $J(N_G(J) \cap K)/J$ in $N_G(J)/J$. By induction, there is $\bar{y} \in N_G(J)/J$ with $Q_1/J = \bar{y}(Q_2/J)\bar{y}^{-1}$; it follows that $Q_1 = yQ_2y^{-1}$, where $Jy = \bar{y}$, as desired. ■

The proof of the following theorem is a variation on the proof of Theorem 7.39. Recall Exercise 4.16; a normal p-subgroup K of a finite group G is contained in every Sylow p-subgroup P of G.

Theorem 7.43 (Gaschütz, 1952). *Let K be a normal abelian p-subgroup of a finite group G, and let P be a Sylow p-subgroup of G. Then K has a complement in G if and only if K has a complement in P.*

Proof. Exercise 7.17 shows that if Q is a complement of K in G, then $Q \cap P$ is a complement of K in P.

For the converse, assume that Q is a complement of K in P, so that Q is a transversal of K in P. All groups in this proof will be written additively. If U is a transversal of P in G (which need not be a subgroup!), then $P = \bigcup_{q \in Q} K + q$ and $G = \bigcup_{u \in U} P + u = \bigcup_{q,u} K + q + u$; thus, $Q + U = \{q + u : q \in Q, u \in U\}$ is a transversal of K in G. By Exercise 7.37, $-U - Q = -U + Q = \{-u + q : u \in U, q \in Q\}$ is also a transversal of K in G ($-Q = Q$ because Q is a subgroup). Let us denote $-U$ by T, so that $|T| = [G : P]$ and $T + Q$ is a transversal of K in G.

Define $l: G/K \to G$ by $l(K + t + q) = t + q$. The corresponding factor set $f: G/K \times G/K \to K$ is defined by

$$l(K + t' + q') + l(K + t + q) = f(K + t' + q', K + t + q) + l(K + t' + q' + t + q).$$

In particular, if $t = 0$, then

$$(*) \qquad f(x, K + q) = 0 \qquad \text{for all} \quad x \in G/K, \quad \text{all } q \in Q$$

($l(K + t' + q' + q) = t' + q' + q = l(K + t' + q') + l(K + q)$ because $q' + q \in Q$). Consider the cocycle identity

$$xf(y, z) - f(x + y, z) + f(x, y + z) - f(x, y) = 0$$

for $x, y \in G/K$ and $z = K + q$ with $q \in Q$. Equation $(*)$ shows that the first two terms are zero, and so

$$(**) \qquad f(x, y + z) = f(x, y) \qquad \text{for} \quad x, y \in G/K \quad \text{and} \quad z = K + q.$$

Let $T = \{t_1, \ldots, t_n\}$, where $n = [G : P]$. For fixed $y = K + g \in G/K$ and for any t_i, $K + g + t_i$ lies in G/K; since $T + Q$ is a transversal of K in G, there is $t_{\pi i} \in T$ and $q_i \in Q$ with $K + g + t_i = K + t_{\pi i} + q_i$. We claim that π is a permutation. If $K + g + t_j = K + t_{\pi i} + q_j$, then

$$g + t_i - (t_{\pi i} + q_i) \in K \qquad \text{and} \qquad g + t_j - (t_{\pi i} + q_j) \in K$$

give

$$g + t_i - q_i - t_{\pi i} + t_{\pi i} + q_j - t_j - g \in K.$$

Since $K \lhd G$, we have $t_i - q_i + q_j - t_j \in K$; since Q is a subgroup, $-q_i + q_j = q \in Q$, so that $t_i + q - t_j \in K$ and $K + t_j = K + t_i + q$. It follows that $t_j = t_i + q$, so that $j = i$ (and $q = 0$). Therefore, π is an injection and hence a permutation.

Let $\overline{T} = \{K + t : t \in T\}$ For $x \in G/K$, define

$$\sigma(x) = \sum_{z \in \overline{T}} f(x, z);$$

σ is well defined because G/K is finite and K is abelian. Summing the cocycle identity gives

$$x\sigma(y) - \sigma(x) + \sum_{z \in \overline{T}} f(x, y + z) = [G : P] f(x, y).$$

But $y + z = K + g + K + t_i = K + g + t_i = K + t_{\pi i} + q_i$, so that $f(x, y + z) = f(x, K + t_{\pi i} + q_i) = f(x, K + t_{\pi i})$, by (**); since π is a permutation, $\sum_{z \in T} f(x, y + z) = \sigma(x)$. Therefore,

$$x\sigma(y) - \sigma(x + y) + \sigma(x) = [G : P]f(x, y).$$

This is an equation in $K \leq P$. Since $([G : P], |K|) = 1$, there are integers a and b with $a|K| + b[G : P] = 1$. Define $h: G/K \to K$ by $h(x) = b\sigma(x)$. Then $h(1) = 0$ and

$$xh(y) - h(x + y) + h(x) = f(x, y);$$

that is, f is a coboundary and so G is a semidirect product. ■

EXERCISES

7.41. Use the Schur–Zassenhaus lemma and Exercise 7.29 to reclassify all groups of order pq, where p and q are distinct primes.

7.42. Prove that every group of order p^2q, where $p > q$ are primes, has a normal Sylow p-subgroup, and classify all such groups.

7.43. Using Lemma 4.23, reclassify all groups of order 12.

7.44. Use factor sets to prove the existence of the generalized quaternions $\mathbf{Q}_n$. (*Hint.* Exercise 4.42.)

Transfer and Burnside's Theorem

We have seen several conditions guaranteeing that a group be a semidirect product: P. Hall's theorem (Theorem 5.28) for solvable groups; the Schur–Zassenhaus lemma; Gaschütz's theorem. In each of these theorems, one starts with a normal subgroup K of G and constructs a complement Q ($\cong G/K$). We now aim for a companion theorem, due to Burnside, that begins with a Sylow subgroup Q of G and, in certain cases, constructs a normal complement K.[4] It is natural to seek a homomorphism whose kernel is such a normal subgroup; it is called the *transfer*. The next lemma is in the spirit of a portion of the proof of Gaschütz's theorem given above.

Lemma 7.44. *Let Q be a subgroup of finite index n in G, and let $\{l_1, \ldots, l_n\}$ and $\{h_1, \ldots, h_n\}$ be left transversals of Q in G. For fixed $g \in G$ and each i, there is a unique $\sigma(i)$ $(1 \leq \sigma(i) \leq n)$ and a unique $x_i \in Q$ with*

$$gh_i = l_{\sigma i}x_i.$$

Moreover, σ is a permutation of $\{1, \ldots, n\}$.

[4] Some other such theorems are quoted at the end of this section.

Proof. Since the left cosets of Q partition G, there is a unique left coset l_jQ containing gh_i; the first statement follows by defining $\sigma i = j$. Assume that $\sigma i = \sigma k = j$. Then $gh_i = l_j x_i$ and $gh_k = l_j x_k$; thus $gh_i x_i^{-1} = gh_k x_k^{-1}$, $h_i^{-1}h_k = x_i^{-1}x_k \in Q$, $h_iQ = h_kQ$, and $i = k$. Therefore, σ is an injection of a finite set to itself, hence is a permutation. ■

This lemma will be used in two cases. The first has $l_i = h_i$ for all i; that is, the transversals coincide. In this case,

$$gl_i = l_{\sigma i} x_i,$$

where $\sigma \in S_n$ and $x_i \in Q$. The second case has two transversals, but we set $g = 1$. Now

$$h_j = l_{\alpha j} y_j,$$

where $\alpha \in S_n$ and $y_j \in Q$.

Definition. If Q is a subgroup of finite index n in a group G, then the ***transfer*** is the function $V\colon G \to Q/Q'$ defined by

$$V(g) = \prod_{i=1}^{n} x_i Q',$$

where $\{l_1, \ldots, l_n\}$ is a left transversal of Q in G and $gl_i = l_j x_i$.

Remark. The transfer $V\colon G \to Q/Q'$ is often denoted by $V_{G \to Q}$, the letter V abbreviating the original German term *Verlagerung*.

Theorem 7.45. *If Q is a subgroup of finite index in a group G, then the transfer $V\colon G \to Q/Q'$ is a homomorphism whose definition is independent of the choice of left transversal of Q in G.*

Remark. See Exercise 7.45 below which shows that a transfer defined via right transversals coincides with V.

Proof. Let $\{l_1, \ldots, l_n\}$ and $\{h_1, \ldots, h_n\}$ be left transversals of Q in G. By Lemma 7.44, there are equations for each $g \in G$:

$$\begin{aligned} gl_i &= l_{\sigma i} x_i & \sigma &\in S_n, & x_i &\in Q,\\ gh_i &= h_{\tau i} y_i, & \tau &\in S_n, & y_i &\in Q,\\ h_i &= l_{\alpha i} z_i, & \alpha &\in S_n, & z_i &\in Q. \end{aligned}$$

Now

$$gh_i = gl_{\alpha i} z_i = l_{\sigma\alpha i} x_{\alpha i} z_i.$$

Defining j by $\alpha j = \sigma\alpha i$, we have $h_j = l_{\sigma\alpha i} z_j$, whence

$$gh_i = h_j z_j^{-1} x_{\alpha i} z_i.$$

The uniqueness assertion of Lemma 7.44 and the definition of j give

$$y_i = z_j^{-1} x_{\alpha i} z_i = z_{\alpha^{-1}\sigma\alpha i}^{-1} x_{\alpha i} z_i.$$

Factors may be rearranged in the abelian group Q/Q': thus

$$\prod y_i Q' = \prod z_{\alpha^{-1}\sigma\alpha i}^{-1} x_{\alpha i} z_i Q' = \prod x_{\alpha i} Q',$$

because $\alpha^{-1}\sigma\alpha \in S_n$, and so the inverse of each z_i occurs and cancels z_i. Finally, $\prod x_{\alpha i} Q' = \prod x_i Q'$ since $\alpha \in S_n$. We have shown that V is independent of the choice of transversal.

Let $g, g' \in G$ and let $\{l_1, \ldots, l_n\}$ be a left transversal of Q in G. Thus, $gl_i = l_{\sigma i} x_i$ and $g'l_i = l_{\tau i} y_i$, where $x_i, y_i \in Q$. Then

$$gg'l_i = gl_{\tau i} y_i = l_{\sigma\tau i} x_{\tau i} y_i.$$

Therefore,

$$V(gg') = \prod x_{\tau i} y_i Q' = \left(\prod x_{\tau i} Q'\right)\left(\prod y_i Q'\right)$$

$$= \left(\prod x_i Q'\right)\left(\prod y_i Q'\right) = V(g)V(g'). \quad \blacksquare$$

If a subgroup Q of finite index in a group G has a (not necessarily normal) complement K, then $K = \{a_1, \ldots, a_n\}$ is a left transversal of Q in G. If $b \in K$, then $ba_i \in K$ for all i: $ba_i = a_{\sigma i}$. But the general formula (Lemma 7.44) is $ba_i = a_{\sigma i} x_i$, so that each $x_i = 1$. We conclude that if $b \in K$, then $V(b) = 1$; that is, $K \leq \ker V$. If Q is abelian, then $Q' = 1$ and we may identify Q/Q' with Q; thus, $\operatorname{im} V \leq Q$ in this case. These remarks indicate that $K = \ker V$ is a reasonable candidate for a normal complement of Q.

The following formula for the transfer says that $V(g)$ is a product of conjugates of certain powers of g.

Lemma 7.46. *Let Q be a subgroup of finite index n in G, and let $\{l_1, \ldots, l_n\}$ be a left transversal of Q in G. For each $g \in G$, there exist elements $h_1, \ldots, h_m$ of G and positive integers $n_1, \ldots, n_m$ (all depending on g) such that:*

(i) *each $h_i \in \{l_1, \ldots, l_n\}$;*
(ii) *$h_i^{-1} g^{n_i} h_i \in Q$;*
(iii) *$\sum n_i = n = [G : Q]$; and*
(iv) *$V(g) = \prod (h_i^{-1} g^{n_i} h_i) Q'$.*

Proof. We know that $gl_i = l_{\sigma i} x_i$, where $\sigma \in S_n$ and $x_i \in Q$. Write the complete factorization of σ as a product of disjoint cycles (so there is one 1-cycle for each fixed point): $\sigma = \alpha_1 \ldots \alpha_m$. If $\alpha_i = (j_1, \ldots, j_r)$, then

$$gl_{j_1} = l_{\sigma j_1} x_{j_1} = l_{j_2} x_{j_1}, \quad gl_{j_2} = l_{j_3} x_{j_3}, \ldots, \quad gl_{j_r} = l_{j_1} x_{j_r},$$

and

$$l_{j_1}^{-1} g^r l_{j_1} = x_{j_r} \dots x_{j_1} \in Q.$$

Define $h_i = l_{j_1}$ and $n_i = r$; all the conclusions now follow. ∎

Theorem 7.47. *If Q is an abelian subgroup of finite index n in a group G and if $Q \leq Z(G)$, then $V(g) = g^n$ for all $g \in G$.*

Proof. Since Q is abelian, we may regard the transfer as a homomorphism $V: G \to Q$. The condition $Q \leq Z(G)$ implies that Q is a normal subgroup of G. If $g \in G$ and $h^{-1}g^r h \in Q$, then normality of Q gives $g^r = h(h^{-1}g^r h)h^{-1} \in Q$. But $Q \leq Z(G)$ now gives $h^{-1}g^r h = g^r$. The result now follows from formulas (iii) and (iv) of Theorem 7.45. ∎

Corollary 7.48. *If a group G has a subgroup Q of finite index n with $Q \leq Z(G)$, then $g \mapsto g^n$ is a homomorphism.*

Proof. We have just seen that this function is the transfer. ∎

The reader should try to prove this last corollary without using the transfer; I do not know a simpler proof.

Lemma 7.49. *Let Q be a Sylow p-subgroup of a finite group G (for some prime p). If $g, h \in C_G(Q)$ are conjugate in G, then they are conjugate in $N_G(Q)$.*

Proof. If $h = \gamma^{-1}g\gamma$ for some $\gamma \in G$, then $h \in \gamma^{-1}C_G(Q)\gamma = C_G(\gamma^{-1}Q\gamma)$. Since Q and $\gamma^{-1}Q\gamma$ are contained in $C_G(h)$, both are Sylow subgroups of it. The Sylow theorem gives $c \in C_G(b)$ with $Q = c^{-1}\gamma^{-1}Q\gamma c$. Clearly $\gamma c \in N_G(Q)$ and $c^{-1}\gamma^{-1}g\gamma c = c^{-1}hc = h$. ∎

Theorem 7.50 (Burnside Normal Complement Theorem, 1900). *Let G be a finite group and let Q be an abelian Sylow subgroup contained in the center of its normalizer: $Q \leq Z(N_G(Q))$. Then Q has a normal complement K (indeed, K is even a characteristic subgroup of G).*

Proof. Since Q is abelian, we may regard the transfer V as a homomorphism from G to Q. Let us compute $V(g)$ for each $g \in Q$. By Lemma 7.46, $V(g) = \prod h_i^{-1}g^{n_i}h_i$. For each i, if g^{n_i} and $h_i^{-1}g^{n_i}h_i$ lie in Q, then they are conjugate elements which lie in $C_G(Q)$ (for Q abelian implies $Q \leq C_G(Q)$). By Lemma 7.49, there is $c_i \in N_G(Q)$ with $h_i^{-1}g^{n_i}h_i = c_i^{-1}g^{n_i}c_i$. But $Q \leq Z(N_G(Q))$ implies $c_i^{-1}g^{n_i}c_i = g^{n_i}$. Hence, if $n = [G : Q]$, then $V(g) = g^n$ for all $g \in Q$. If $|Q| = q$, then $(n, q) = 1$ (because Q is a Sylow subgroup), and there are integers α and β with $\alpha n + \beta q = 1$. Therefore, when $g \in Q$, we have $g = g^{\alpha n}g^{\beta q} = (g^\alpha)^n$, so that $V: G \to Q$ is surjective: if $g \in Q$, then $V(g^\alpha) = g^{\alpha n} = g$. The first isomorphism theorem gives $G/K \cong Q$, where $K = \ker V$. It follows that $G = KQ$

and $K \cap Q = 1$ (because $|K| = n$ and so $(|K|, |Q|) = 1$). Therefore, K is a normal complement of Q. Indeed, K char G because it is a normal Hall subgroup. ■

Definition. If a Sylow p-subgroup of a finite group G has a normal p-complement, then G is called ***p-nilpotent***.

Thus, Burnside's theorem says that if $Q \leq Z(N_G(Q))$, then G is p-nilpotent. Here are some consequences of Burnside's theorem.

Theorem 7.51. *Let G be a finite group and let p be the smallest prime divisor of $|G|$. If a Sylow p-subgroup Q of G is cyclic, then G is p-nilpotent.*

Proof. By Theorem 7.1, there is an imbedding $N_G(Q)/C_G(Q) \hookrightarrow \mathrm{Aut}(Q)$. Obviously, $|N/C|$ divides $|G|$. Now Q is cyclic of order p^m, say, and so Theorem 7.3 gives $|\mathrm{Aut}(Q)| = p^{m-1}(p-1)$ (this is true even for $p = 2$). Since Q is a Sylow subgroup of G and $Q \leq C = C_G(Q)$, p does not divide $|N/C|$. Hence, $|N/C|$ divides $p - 1$. But $(p - 1, |G|) = 1$, because p is the smallest prime divisor of $|G|$, and so $|N/C| = 1$. Therefore $N_G(Q) = C_G(Q)$. Since $Q \leq G$ is abelian, $Q \leq Z(C_G(Q))$, and since $N_G(Q) = C_G(Q)$, we have $Q \leq Z(C_G(Q)) = Z(N_G(Q))$. Thus, the hypothesis of Burnside's theorem is satisfied, and so Q has a normal complement. ■

Corollary 7.52. *A nonabelian simple group cannot have a cyclic Sylow 2-subgroup.*

Remark. We have already seen this in Exercise 3.30.

Theorem 7.53 (Hölder, 1895). *If every Sylow subgroup of a finite group G is cyclic, then G is solvable.*

Proof. If p is the smallest prime divisor of $|G|$, then Theorem 7.51 provides a normal complement K to a Sylow p-subgroup of G. and $Q \cong G/K$. By induction on $|G|$, G/K is solvable. Since K is solvable (it is cyclic, hence abelian), Theorem 5.17 shows that G is solvable. ■

Corollary 7.54. *Every group G of squarefree order is solvable.*

Proof. Every Sylow subgroup of G must be cyclic. ■

Corollary 7.55. *Let G be a nonabelian simple group, and let p be the smallest prime divisor of $|G|$. Then either p^3 divides $|G|$ or 12 divides $|G|$.*

Proof. Let Q be a Sylow p-subgroup of G. By Corollary 7.52, Q is not cyclic. Hence, $|Q| \geq p^2$, so that if p^3 does not divide $|Q|$, then Q is elementary abelian of order p^2. Since Q is a two-dimensional vector space over $\mathbb{Z}_p$, Exercise 7.10

gives $|\mathrm{Aut}(Q)| = (p^2 - 1)(p^2 - p) = p(p + 1)(p - 1)^2$. Now $N_G(Q)/C_G(Q) \hookrightarrow \mathrm{Aut}(Q)$, so that $|N/C|$ is a divisor of $|\mathrm{Aut}(Q)|$, and $|N/C| \neq 1$ lest Burnside's theorem apply. Since $Q \leq C$, p does not divide $|N/C|$; since p is the smallest prime divisor of $|G|$, $|N/C|$ must divide $p + 1$. But this is impossible if p is odd, for the smallest prime divisor of $|N/C|$ is $\geq p + 2$. (We have shown that if p is odd, then p^3 must divide $|G|$.) Moreover, if $p = 2$, then $|\mathrm{Aut}(Q)| = 6$ and $|N/C| = 3$, so that $|G|$ is divisible by $2^2 \times 3 = 12$. ■

The simple group A_5 has order 60, and 60 is divisible by 12 but not by 8. There is an infinite class of simple groups, the Suzuki groups, whose orders are not divisible by 3, hence not by 12 (their orders are divisible by 8). The magnificent result of Feit and Thompson says that every simple nonabelian group G has even order, and so $|G|$ is divisible by either 8 or 12.

In Chapter 5, we gave an elementary but ingenious proof of a theorem of Schur (Theorem 5.32). We now give a straightforward proof of this theorem using the transfer; indeed, it was Schur who invented the transfer in order to give the forthcoming proof of Theorem 7.57.

It is not generally true (see Theorem 11.48) that a subgroup of a finitely generated group G is itself finitely generated.

Lemma 7.56. *If G is a finitely generated group and H is a subgroup of G of finite index, then H is finitely generated.*

Proof. Let $G = \langle g_1, \ldots, g_m \rangle$ and let $t_1, \ldots, t_n$ be right coset representatives of H with $t_1 = 1$; thus, $G = \bigcup_{i=1}^n Ht_i$. Enlarging the generating set if necessary, we may assume that if g_i is a generator, then $g_i^{-1} = g_k$ for some k.

For all i, j, there is $h(i, j) \in H$ with $t_i g_j = h(i, j) t_{k(i,j)}$. We claim that H is generated by all the $h(i, j)$. If $a \in H$, then $a = g_{i_1} g_{i_2} \cdots g_{i_s}$; no exponents are needed since the inverse of a generator is also a generator. Now

$$\begin{aligned} a = g_{i_1} g_{i_2} \cdots g_{i_s} &= (t_1 g_{i_1}) g_{i_2} \cdots g_{i_s} \\ &= h(1, i_1) t_{1'} g_{i_2} \cdots g_{i_s} && \text{some } 1' \\ &= h(1, i_1) h(1', i_2) t_{2'} g_{i_3} \cdots g_{i_s} && \text{some } 2' \\ &= h(1, i_1) h(1', i_2) h(2', i_3) t_{3'} g_{i_4} \cdots g_{i_s} && \text{some } 3' \\ &= h(1, i_1) h(1', i_2) \ldots h((s-1)', i_s) t'_s. \end{aligned}$$

Since a and all the h's lie in H, we have $t'_s \in H$; therefore, $t'_s = t_1 = 1$, as desired. ■

Theorem 7.57 (Schur). *If $Z(G)$ has finite index in a group G, then G' is finite.*

Proof. As in the proof of Theorem 5.32, G' has a finite number of generators. Now $G'/(G' \cap Z(G)) \cong G'Z(G)/Z(G) \leq G/Z(G)$, so that $G'/(G' \cap Z(G))$ is finite. By the lemma, $G' \cap Z(G)$ is finitely generated. As $G' \cap Z(G) \leq Z(G)$ is abelian, it is finite if it has finite exponent (this is not true for nonabelian groups). Let $V: G \to Z(G)$ be the transfer. If $a \in G'$, then $V(a) = 1$; if $a \in Z(G)$, then $V(a) = a^n$, where $n = [G : Z(G)]$ (Theorem 7.47). Therefore, $G' \cap Z(G)$ has exponent n, and hence is finite. As G' is an extension of one finite group by another finite group, it, too, is finite. ■

The reader should be aware of two theorems of Grün whose proofs involve the transfer; if G is a finite group and P is a Sylow subgroup, then one may often compute $P \cap G'$ (see Robinson (1982, pp. 283–286)).

We state some other theorems guaranteeing p-nilpotence.

Theorem (Tate, 1964). *Let G be a finite group, and let $P \leq G$ be a Sylow p-subgroup of G. If $N \lhd G$ and $N \cap P \leq \Phi(P)$, then N is p-nilpotent.*

Proof. The original proof is short, using the 5-term exact sequence in cohomology of groups; a longer proof using the transfer is in Huppert (1967), p. 431. ■

Call a subgroup H of a group G ***p-local***, for some prime p, if there is some nontrivial p-subgroup Q of G with $H = N_G(Q)$.

Theorem (Frobenius). *A group G is p-nilpotent if and only if every p-local subgroup H of G is p-nilpotent.*

Proof. See [Aschbacher, p. 203]. ■

If p is a prime and P is a Sylow p-subgroup of a group G, define $E(P) = \{x \in Z(P): x^p = 1\}$. If $p^{d(P)}$ is the largest order of an elementary abelian subgroup of P, then the ***Thompson subgroup*** $J(P)$ is defined as the subgroup of G generated by all the elementary abelian p-subgroups of G of order $p^{d(P)}$.

Theorem (Thompson). *Let p be an odd prime, and let P be a Sylow p-subgroup of a group G. If $C_G(E(P))$ and $N_G(J(P))$ are p-nilpotent, then G is p-nilpotent.*

Proof. See [Aschbacher, p. 203]. ■

EXERCISES

7.45. Let Q be a subgroup of finite index n in a group G, and let $\{y_1, \ldots, y_n\}$ be a *right* transversal of Q in G. For $a \in G$, $y_i a = p_i y_{\tau i}$ for $p_i \in Q$ and $\tau \in S_n$. Prove that $R: G \to Q/Q'$, defined by $R(a) = \prod p_i Q'$, is the transfer; that is, $R(a) = V(a)$ for all $a \in G$. (*Hint.* Exercise 7.37.)

7.46. (i) Let $p_1 < p_2 < \cdots < p_t$ be primes and let $n = p_1 \ldots p_t$. Prove that every group G of order n has a normal Sylow p_t-subgroup. (*Hint*. Exercise 4.38.)

(ii) If, in addition, $(p_i, p_j - 1) = 1$ for all $i < j$, then G must be cyclic. (One can characterize such integers n by $(n, \varphi(n)) = 1$, where φ is the Euler φ-function.)

(iii) Find all integers n such that every group G of order n is abelian.

Remark. We sketch a proof that there exist arbitrarily large sets of primes $\{p_1, \ldots, p_t\}$ which satisfy condition (ii) of Exercise 7.46. Let $p_1 = 3$, and suppose that $p_1 < p_2 < \cdots < p_t$ are primes with $(p_i, p_j - 1) = 1$ for all $i < j$. A theorem of Dirichlet states: If $(a, b) = 1$, then the arithmetic progression $a, a + b, a + 2b, \ldots$ contains infinitely many primes. Since $(2, p_1 \ldots p_t) = 1$, there is a positive integer m such that $p_{t+1} = 2 + mp_1 \ldots p_t$ is prime. The set $\{p_1, \ldots, p_{t+1}\}$ satisfies all the desired conditions.

7.47. (i) Let q and p be primes such that $q \equiv 1 \bmod p^e$. Show that the multiplicative group $\mathbb{Z}_p^\times$ contains a cyclic subgroup isomorphic to $\mathbb{Z}_{p^e}$. (*Hint*: Theorem 2.18.)

(ii) Let $G = C_1 \times \cdots \times C_n$, where C_i is cyclic of order $p_i^{e_i}$ for (not necessarily distinct) primes p_i. Use Dirichlet's theorem (see the remark following Exercise 7.46) to show that there are n distinct primes q_i with $q_i \equiv 1 \bmod p_i^{e_i}$ for $i = 1, \ldots, n$.

(iii) If G is a finite abelian group, then G can be imbedded in $\mathbb{Z}_m^\times$ for some m. (*Hint*. Let $m = \prod q_i$, and use Theorem 7.3 (iii).) (This proof is due to G. McCormick.)

7.48. (i) If $V: G \to Q/Q'$ is the transfer, then $G' \le \ker V$ and V induces a homomorphism $\bar{V}: G/G' \to Q/Q'$, namely, $G'a \mapsto V(a)$.

(ii) Prove that the transfer is transitive: if $P \le Q \le G$ are subgroups of finite index, and if $T: G \to Q/Q'$, $U: G \to P/P'$, and $V: Q \to P/P'$ are transfers, then $\bar{U} = \bar{V}\bar{T}$.

7.49. If Q has index n in G, and if $K \le G$ satisfies $G = KQ$ and $Q \le C_G(K)$, then $V(a) = a^n Q'$ for all $a \in G$. (*Hint*. There is a transversal of Q contained in K; if $a \in K$ and $g \in K$, then $g^{-1}a^r g \in Q \le C_G(K)$, and this implies $a^r \in C_G(K)g^{-1} = C_G(gKg^{-1}) = C_G(K)$.)

7.50. If G is a finite group of order mn, where $(m, n) = 1$, and if $Q \le Z(G)$ is a Hall subgroup of order m, then $K = \ker V$ is a normal complement of Q (where $V: G \to Q/Q'$ is the transfer), and $G = K \times Q$.

7.51. Let G be a torsion-free group having a cyclic subgroup of finite index. Prove that G is cyclic.

7.52. If p and q are primes, prove that every group of order p^2q^2 is solvable.

7.53. If G is a nonabelian group with $|G| \le 100$ and $|G| \ne 60$, then G is solvable. (See Exercise 4.36.) (The next order of a nonabelian simple group is 168.)

Projective Representations and the Schur Multiplier

We have already seen the usefulness of $H^2(Q, K, \theta)$, where K is abelian and $\theta: Q \to \text{Aut}(K)$ is a homomorphism. When θ is ***trivial***, that is, $\theta(x) = 1_K$ for all $x \in Q$, then we drop it from the notation and write $H^2(Q, K)$.

Definition. A ***central extension*** of K by Q is an extension G of K by Q with $K \leq Z(G)$.

It is easy to see that if $G = K \rtimes_\theta Q$ is a central extension, then G is the direct product $K \times Q$.

Lemma 7.58. *Given data (Q, K, θ), then θ is trivial if and only if every extension realizing the data is a central extension.*

Proof. Recall that θ arises from the equation: for all $x \in Q$ and $a \in K$,

$$\theta_x(a) = l(x) + a - l(x),$$

where $l(x)$ is a lifting of x. Assume that θ is trivial. Every $g \in G$ has the form $g = b + l(x)$ for some $b \in K$ and $x \in Q$. If $a \in K$, then a commutes with $l(x)$ for all $x \in Q$; since K is abelian, a commutes with g, and so $a \in Z(G)$. Conversely, if G is a central extension, each $a \in K$ commutes with every $l(x)$ and so $\theta_x(a) = a$ for all $x \in Q$ and $a \in K$. ■

Theorem 7.59. *There is a bijection from the set of all equivalence classes of central extensions realizing data (Q, K, θ), where θ is trivial, to $H^2(Q, K)$.*

Proof. Theorem 7.34 specializes to this result once we take account of the lemma. ■

Definition. If Q is a group, then its ***Schur multiplier*** (or *multiplicator*) is the abelian group

$$M(Q) = H^2(Q, \mathbb{C}^\times),$$

where $\mathbb{C}^\times$ denotes the multiplicative group of nonzero complex numbers.

Since $\mathbb{C}^\times$ is written multiplicatively, it is more convenient to write $M(Q) = H^2(Q, \mathbb{C}^\times)$ multiplicatively as well. Thus, with θ trivial, a function $f: Q \times Q \to \mathbb{C}^\times$ is a factor set if and only if, for all $x, y \in Q$:

$$f(1, y) = 1 = f(x, 1);$$

$$f(x, y)f(xy, z)^{-1}f(x, yz)f(x, y)^{-1} = 1;$$

a function $g: Q \times Q \to \mathbb{C}^\times$ is a coboundary if and only if there is a function

$h: Q \to \mathbb{C}^\times$ with $h(1) = 1$ such that

$$g(x, y) = h(y)h(xy)^{-1}h(x).$$

Two factor sets f and g are equivalent if and only if fg^{-1} is a coboundary.

Definition. If G is a finite group, its ***minimal exponent***, denoted by $\exp(G)$, is the least positive integer e for which $x^e = 1$ for all $x \in G$.

Theorem 7.60. *If Q is a finite group, then $M(Q)$ is a finite abelian group and $\exp(M(Q))$ divides $|Q|$.*

Remark. The first paragraph of the proof is just a repetition, in multiplicative notation, of a portion of the proof given in Theorem 7.39.

Proof. If $f: Q \times Q \to \mathbb{C}^\times$ is a factor set, define $\sigma: Q \to \mathbb{C}^\times$ by

$$\sigma(x) = \prod_{z \in Q} f(x, z).$$

Note that $\sigma(1) = 1$. Now multiply the cocycle identity

$$f(y, z)f(xy, z)^{-1}f(x, yz)f(x, y)^{-1} = 1$$

over all $z \in Q$ to obtain

$$\sigma(y)\sigma(xy)^{-1}\sigma(x) = f(x, y)^n,$$

where $n = |Q|$ (as z ranges over all of Q, so does yz). But this equation says that $(fB^2(Q, \mathbb{C}^\times))^n = 1$; that is, $M(Q)^n = 1$. Thus, the minimal exponent of $M(Q)$ divides n when Q is finite.

For each $x \in Q$, define $h: Q \to \mathbb{C}^\times$ by choosing $h(1) = 1$ and $h(x)$ to be some nth root of $\sigma(x)^{-1}$: $h(x)^n = \sigma(x)^{-1}$. Define $g: Q \times Q \to \mathbb{C}^\times$ by $g(x, y) = f(x, y)h(y)h(xy)^{-1}h(x)$. Clearly, f and g are equivalent, for they differ by a coboundary. On the other hand,

$$\begin{aligned} g(x, y)^n &= f(x, y)^n h(y)^n h(xy)^{-n} h(x)^n \\ &= \sigma(y)\sigma(xy)^{-1}\sigma(x)\sigma(y)^{-1}\sigma(xy)\sigma(x)^{-1} = 1. \end{aligned}$$

Therefore, each element $[f] \in M(Q)$ determines a function $g: Q \times Q \to Z_n$, where Z_n denotes the subgroup of $\mathbb{C}^\times$ consisting of all the nth roots of unity. The result follows, for there are only finitely many such functions g. ∎

Corollary 7.61. *If Q is a finite p-group, then $M(Q)$ is a finite abelian p-group.*

The Schur multiplier arises in Representation Theory; more precisely, in the study of projective representations.

Definition. A ***projective representation*** of a group Q is a homomorphism $\tau: Q \to \mathrm{PGL}(n, \mathbb{C}) = \mathrm{GL}(n, \mathbb{C})/\mathrm{Z}(n, \mathbb{C})$, the group of all nonsingular $n \times n$

complex matrices modulo its center, the normal subgroup $Z(n, \mathbb{C})$ of all nonzero scalar matrices.

Just as $GL(n, \mathbb{C})$ consists of automorphisms of a vector space, we shall see in Chapter 9 that $PGL(n, \mathbb{C})$ consists of automorphisms of a projective space. An important example of a central extension is provided by $GL(n, \mathbb{C})$, which is a central extension of $\mathbb{C}^\times$ by $PGL(n, \mathbb{C})$. Note that $Z(n, \mathbb{C}) \cong \mathbb{C}^\times$.

"Ordinary" representations of Q are homomorphisms $Q \to GL(n, \mathbb{C})$; they have been well studied and contain valuable information about Q. The question whether a given irreducible representation φ of a normal subgroup $H \lhd Q$ can be extended to a representation of Q leads, in a natural way, to a projective representation of Q. If $x \in Q$, one can define a new representation $\varphi^x: H \to GL(n, \mathbb{C})$ by $\varphi^x(h) = \varphi(xhx^{-1})$; this makes sense because $xhx^{-1} \in H$ for all $h \in H$. To make it easier to extend φ, let us assume (as would be the case were φ extendable to G) that the representations φ and φ^x are all similar; that is, there is a nonsingular matrix P_x with $\varphi(xhx^{-1}) = P_x\varphi(h)P_x^{-1}$ for all $h \in H$ (of course, we may choose P_1 to be the identity matrix E). Now choose a transversal T of H in Q, so that each $x \in Q$ has a unique expression of the form $x = th$ for $t \in T$ and $h \in H$. The obvious candidate for an extension $\Phi: Q \to GL(n, \mathbb{C})$ of φ is $\Phi(th) = P_t\varphi(h)$; it is clear that Φ is a well defined function that extends φ. If $h' \in H$ and $x = th \in Q$ (where $t \in T$ and $h \in H$), then

$$\begin{aligned}\varphi(xh'x^{-1}) &= \varphi(thh'h^{-1}t^{-1}) \\ &= P_t\varphi(hh'h^{-1})P_t^{-1} \\ &= P_t\varphi(h)\varphi(h')\varphi(h)^{-1}P_t^{-1} \\ &= \Phi(x)\varphi(h')\Phi(x)^{-1}.\end{aligned}$$

It follows that if $x, y \in Q$, then

$$\varphi(xyh'y^{-1}x^{-1}) = \Phi(xy)\varphi(h')\Phi(xy)^{-1}.$$

On the other hand,

$$\begin{aligned}\varphi(x(yh'y^{-1})x^{-1}) &= \Phi(x)\varphi(yh'y^{-1})\Phi(x)^{-1} \\ &= \Phi(x)\Phi(y)\varphi(h')\Phi(y^{-1})\Phi(x)^{-1}.\end{aligned}$$

Hence, $\Phi(xy)^{-1}\Phi(x)\Phi(y)$ centralizes $\varphi(h')$ for all $h' \in H$. Now Schur's lemma (!) states that if φ is irreducible, then any matrix A centralizing every $\varphi(h')$ must be a scalar matrix. Therefore, $\Phi(xy)^{-1}\Phi(x)\Phi(y)$ is a scalar; that is, there is $f(x, y) \in \mathbb{C}^\times$ with

$$\Phi(x)\Phi(y) = \Phi(xy)f(x, y)E,$$

where E is the $n \times n$ identity matrix. This equation says that Φ defines a homomorphism $Q \to PGL(n, \mathbb{C})$, namely, $x \mapsto \Phi(x)Z(n, \mathbb{C})$.

It will be more convenient for the coming computations to think of projective representations $\tau: Q \to \mathrm{PGL}(n, \mathbb{C})$ not as homomorphisms whose values are cosets mod $\mathrm{Z}(n, \mathbb{C})$ but, as they have just occurred above, as ***matrix-valued functions*** T satisfying the equation

$$T(x)T(y) = f_\tau(x, y)T(xy)$$

for all $x, y \in Q$, where $f_\tau: Q \times Q \to \mathbb{C}^\times$ is a function. Of course, one can always obtain this latter form by arbitrarily choosing (matrix) representatives of the cosets $\tau(x)$ in $\mathrm{PGL}(n, \mathbb{C})$; that is, if $\pi: \mathrm{GL}(n, \mathbb{C}) \to \mathrm{PGL}(n, \mathbb{C})$ is the natural map, choose $T(x)$ with $\pi T(x) = \tau(x)$. We always choose the identity matrix E as the representative of $\tau(1)$; that is, $T(1) = E$.

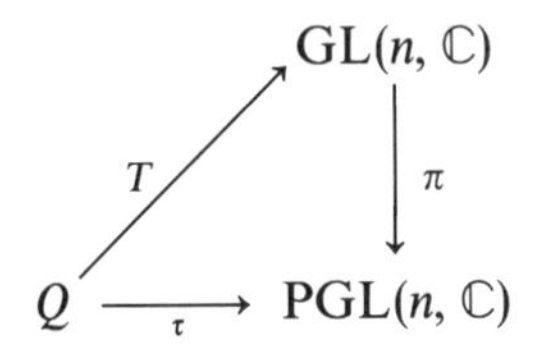

Lemma 7.62. *Let $\tau: Q \to \mathrm{PGL}(n, \mathbb{C})$ be a projective representation. If $T: Q \to \mathrm{GL}(n, \mathbb{C})$ is a matrix-valued function of τ, that is, $T(x)T(y) = f_\tau(x, y)T(xy)$, then f_τ is a factor set. If T' is another matrix-valued function of τ, so that $T'(x)T'(y) = g_\tau(x, y)T'(xy)$, then $f_\tau g_\tau^{-1}$ is a coboundary. Therefore, τ determines a unique element $[f_\tau] \in M(Q)$.*

Remark. This result is the reason factor sets are so called.

Proof. Since $T(1) = E$, we have $f_\tau(1, y) = 1 = f_\tau(x, 1)$, while the cocycle identity follows from associativity: $T(x(yz)) = T((xy)z)$.

Since $\pi T(x) = \tau(x) = \pi T'(x)$ for all $x \in Q$, $T(x)T'(x)^{-1} = h(x)E$, where $h(x) \in \mathbb{C}^\times$. It is easy to see, as in Lemma 7.31, that $f_\tau(x, y)g_\tau(x, y)^{-1} = h(y)h(xy)^{-1}h(x)$; that is, $f_\tau g_\tau^{-1}$ is a coboundary. ∎

Schur proved, for every finite group Q, that there is a "cover" U with $Q = U/M$ such that every projective representation of Q determines an ordinary representation of U; moreover, it turns out that $M \cong M(Q)$ and U is a central extension of $M(Q)$ by Q. The reader can now see why $M(Q)$ is of interest; for example, if $M(Q) = 1$, then $U = Q$, and every projective representation of Q gives a representation of Q itself. We now begin the proof of these assertions.

Definition. Let $\nu: U \to Q$ be a surjective homomorphism with kernel K, and assume that U is a central extension of K by Q. If $\tau: Q \to \mathrm{PGL}(n, \mathbb{C})$ is a projective representation, then τ can be ***lifted*** to U if there exists a homomorphism $\tilde{\tau}$ making the following diagram commute:

$$\begin{array}{ccc} U & \xrightarrow{\nu} & Q \\ {\scriptstyle \tilde{\tau}}\downarrow & & \downarrow{\scriptstyle \tau} \\ \mathrm{GL} & \xrightarrow[\pi]{} & \mathrm{PGL}. \end{array}$$

We say that U has the ***projective lifting property*** if every projective representation of Q can be lifted to U.

By Theorem 7.30, the elements of the central extension U may be viewed as ordered pairs $(a, x) \in K \times Q$, where $K = \ker \nu$. With this understanding, τ can be lifted to U if and only if $\pi\tilde{\tau}(a, x) = \tau\nu(a, x) = \tau(x)$ for all $(a, x) \in U$. As usual, we replace τ by a matrix-valued function T; the equation $\pi\tilde{\tau}((a, x)) = \tau(x)$ in $\mathrm{PGL}(n, \mathbb{C})$ is now replaced by a matrix equaton in $\mathrm{GL}(n, \mathbb{C})$:

$$\tilde{\tau}(a, x) = \mu(a, x)T(x)$$

for some function $\mu: U \to \mathbb{C}^\times$, where $U = K \times Q$.

Definition. If K and A are groups with A abelian, then

$$\mathrm{Hom}(K, A) = \{\text{all homomorphism } K \to A\}.$$

If $\varphi, \psi \in \mathrm{Hom}(K, A)$, define $\varphi\psi: K \to A$ by $\varphi\psi: x \mapsto \varphi(x) + \psi(x)$ for all $x \in K$; note that $\varphi\psi$ is a homomorphism because A is abelian. It is easy to see that $\mathrm{Hom}(K, A)$ is an abelian group under this operation.

In particular, if K is also an abelian group, define its ***character group*** $K^* = \mathrm{Hom}(K, \mathbb{C}^\times)$.

Since $\mathbb{C}^\times$ is a multiplicative group, it is more convenient to write K^* as a multiplicative abelian group: if $a \in K$ and $\varphi, \psi \in K^*$, then $\varphi\psi: a \mapsto \varphi(a)\psi(a)$. We shall prove, in Theorem 10.56, that $K \cong K^*$ for every finite abelian group K.

If U is any central extension of Q with kernel K, define a function $\delta: K^* \to H^2(Q, \mathbb{C}^\times) = M(Q)$ as follows. By Theorem 7.34, U corresponds to an element $[e] \in H^2(Q, K)$, where $e: Q \times Q \to K$ is a factor set, and U consists of all $(a, x) \in K \times Q$, where $(a, x)(b, y) = (abe(x, y), xy)$. If $\varphi \in K^*$, it is routine to check that the composite $\varphi \circ e: Q \times Q \to \mathbb{C}^\times$ is a factor set, and so $[\varphi \circ e] \in H^2(Q, \mathbb{C}^\times)$. Moreover, if e and e' are equivalent factor sets, then $\varphi \circ e$ and $\varphi \circ e'$ are also equivalent: if $e'(x, y) = e(x, y)h(y)h(xy)^{-1}h(x)$, then

$$\varphi \circ e'(x, y) = \varphi \circ e(x, y)\varphi h(y)\varphi h(xy)^{-1}\varphi h(x).$$

Definition. Let U be a central extension of K by Q. If $e: Q \times Q \to K$ is a factor set of U, then the ***transgression*** $\delta = \delta^U$ is the homomorphism $\delta: K^* \to M(Q)$ defined by $\delta(\varphi) = [\varphi \circ e]$.

The preceding discussion shows that the transgression does not depend on the choice of factor set e arising from the central extension U.

Lemma 7.63. *Let U be a central extension of K by Q. Then the transgression $\delta: K^* \to M(Q)$ is surjective if and only if U has the projective lifting property.*

Proof. By Theorem 7.34, the central extension U of K by Q determines $[e] \in H^2(Q, K)$, where $e: Q \times Q \to K$ is a factor set. The strategy is to show that if

$\tau: Q \to \mathrm{PGL}$ corresponds to a matrix-valued function T with $T(x)T(y) = f_\tau(x, y)T(xy)$, then τ can be lifted to U if and only if $[f_\tau] \in \operatorname{im} \delta$; that is, there is $\varphi \in K^*$ with $\delta(\varphi) = [\varphi \circ e] = [f_\tau]$.

If $[f_\tau] \in \operatorname{im} \delta$, then, for all $x, y \in Q$, there is $\varphi \in K^*$ and $h: Q \to \mathbb{C}^\times$ with

$$\varphi \circ e(x, y) = f_\tau(x, y)h(y)h(xy)^{-1}h(x). \tag{3}$$

As in Theorem 7.30, regard the elements of U as ordered pairs $(a, x) \in K \times Q$. Define $\tilde{\tau}: U \to \mathrm{GL}(n, \mathbb{C})$ by

$$\tilde{\tau}(a, x) = \varphi(a)h(x)T(x).$$

We compute:

$$\begin{aligned}\tilde{\tau}((a, x)(b, y)) &= \tilde{\tau}((abe(x, y), xy)) \\ &= \varphi(abe(x, y))h(xy)T(xy) \\ &= \varphi(ab)\varphi \circ e(x, y)h(xy)f_\tau(x, y)T(x)T(y).\end{aligned}$$

On the other hand, since $\mathbb{C}^\times$ is abelian,

$$\begin{aligned}\tilde{\tau}((a, x))\tilde{\tau}((b, y)) &= \varphi(a)h(x)T(x)\varphi(b)h(y)T(y) \\ &= \varphi(ab)h(x)h(y)T(x)T(y).\end{aligned}$$

By (3), $\tilde{\tau}$ is a homomorphism. But $\tau\nu(a, x) = \tau(x)$ and $\pi\tilde{\tau}(a, x) = \pi\varphi(a)h(x)T(x) = \pi T(x) = \tau(x)$, and so τ can be lifted to U.

Conversely, assume that τ can be lifted to U: there is a homomorphism $\tilde{\tau}: U \to \mathrm{GL}(n, \mathbb{C})$ and a function $\mu: K \times Q \to \mathbb{C}^\times$ with

$$\tilde{\tau}(a, x) = \mu(a, x)T(x)$$

for all $a \in K$ and $x \in Q$. Since $\tilde{\tau}$ is a homomorphism, we may evaluate $\tilde{\tau}(a, x)\tilde{\tau}(b, y)$ in two ways: on the one hand, it is

$$\mu(a, x)T(x)\mu(b, y)T(y);$$

on the other hand, it is $\tilde{\tau}(abe(x, y), xy)$, which is

$$\mu(abe(x, y), xy)T(xy).$$

But $T(x)T(y) = f_\tau(x, y)T(xy)$, so that, for all $a, b \in K$ and $x, y \in Q$,

$$\mu(abe(x, y), xy) = f_\tau(x, y)\mu(a, x)\mu(b, y). \tag{4}$$

If $x = y = 1$, then (4) gives

$$\mu(ab, 1) = \mu(a, 1)\mu(b, 1);$$

that is, the function $\varphi: K \to \mathbb{C}^\times$, defined by

$$\varphi(a) = \mu(a, 1),$$

lies in the character group K^*. Define $p: Q \to \mathbb{C}^\times$ by

$$p(x) = \mu(1, x).$$

If $a = b = 1$ in (4), then

$$\mu(e(x, y), xy) = f_\tau(x, y)p(x)p(y).$$

Finally, setting $b = 1$ and $x = 1$ gives $\mu(a, y) = \varphi(a)p(y)$. Hence

$$\varphi \circ e(x, y)p(xy) = f_\tau(x, y)p(x)p(y),$$

and so $\varphi \circ e(x, y)f_\tau(x, y)^{-1}$ is a factor set. Therefore, $[f_\tau] = [\varphi \circ e] = \delta(\varphi)$, as desired. ■

Several properties of character groups, proved in Chapter 10, will be used in the next proof. A (multiplicative) abelian group D is called ***divisible*** if every element $d \in D$ has an nth root in D; that is, for every $n > 0$, there exsts $x \in D$ with $x^n = d$. For example, $\mathbb{C}^\times$ is a divisible group.

Lemma 7.64. *Let U be a central extension of K by Q. Then the transgression $\delta: K^* \to M(Q)$ is injective if and only if $K \leq U'$.*

Proof. We first show that if $\varphi \in K^*$ and $\varphi(K \cap U') = 1$, then $\delta(\varphi) = [\varphi \circ e] = 1$ in $M(Q)$. The second isomorphism theorem gives $K/K \cap U' \cong KU'/U'$. Define $\psi: KU'/U' \to \mathbb{C}^\times$ by $\psi(aU') = \varphi(a)$ for all $a \in K$; ψ is a well defined homomorphism because $K \cap U' \leq \ker \varphi$. But KU'/U' is a subgroup of the abelian group U/U', so that the injective property (Theorem 10.23) of the divisible group $\mathbb{C}^\times$ gives a homomorphism $\Psi: U/U' \to \mathbb{C}^\times$ extending ψ. Now, for all $x, y \in Q$,

$$\begin{aligned}\Psi((1, x)U')\Psi(1, y)U') &= \Psi((1, x)(1, x)U') \\ &= \Psi((e(x, y), xy)U') \\ &= \varphi(e(x, y))\Psi((1, xy)U').\end{aligned}$$

Define $h: Q \to \mathbb{C}^\times$ by $h(x) = \Psi((1, x)U')$. A routine calculation shows that

$$\varphi \circ e(x, y) = h(y)h(xy)^{-1}h(x),$$

so that $\delta(\varphi) = [\varphi \circ e] = 1$ in $M(Q)$. Since δ is injective, $\varphi = 1$. But Theorem 10.58 shows that if $K \cap U' < K$, then there exists $\varphi \in K^*$ with $\varphi \neq 1$ and $K \cap U' \leq \ker \varphi$. We conclude that $K \cap U' = K$; that is, $K \leq U'$.

Conversely, if $\varphi \in \ker \delta$, then $1 = \delta(\varphi) = [\varphi \circ e]$, where e is a factor set of the extension U of K by Q. Thus, $\varphi \circ e$ is a coboundary: there is a function $h: Q \to \mathbb{C}^\times$ with $h(1) = 1$ such that, for all $x, y \in Q$,

$$\varphi \circ e(x, y) = h(y)h(xy)^{-1}h(x).$$

If we define $\Phi: U \to \mathbb{C}^\times$ by $\Phi(a, x) = \varphi(a)h(x)$, then a routine calculation shows that Φ is a homomorphism with $\Phi|K = \varphi$. As $\mathbb{C}^\times$ is abelian, $U' \leq \ker \Phi$, and so $U' \cap K \leq \ker \varphi$. The hypothesis $K \leq U'$ gives $U' \cap K = K$, and so φ is trivial; that is, δ is injective. ■

Definition. If Q is a group, then a ***cover*** (or *representation group*) of Q is a central extension U of K by Q (for some abelian group K) with the projective lifting property and with $K \leq U'$.

The following lemma will be used in proving the existence of covers. It follows from the preceding two lemmas that if a cover U of a finite group Q exists, where U is a central extension of K by Q, then $K \cong M(Q)$.

Lemma 7.65. *If Q is a finite group, then the subgroup $B^2(Q, \mathbb{C}^\times)$ has a finite complement $M \cong M(Q)$ in $Z^2(Q, \mathbb{C}^\times)$.*

Proof. We first show that $B^2(Q, \mathbb{C}^\times)$ is a divisible group. If $f \in B^2(Q, \mathbb{C}^\times)$, then there is $h: Q \to \mathbb{C}^\times$ with $h(1) = 1$ and, for all $x, y \in Q$,

$$f(x, y) = h(y)h(xy)^{-1}h(x).$$

For $n > 0$ and each $x \in Q$, let $k(x)$ be an nth root of $h(x)$ with $k(1) = 1$. Then $g: Q \times Q \to \mathbb{C}^\times$, defined by $g(x, y) = k(y)k(xy)^{-1}k(x)$, is a coboundary with $g^n = f$. By Corollary 10.24, $B^2(Q, \mathbb{C}^\times)$ has a complement M in $Z^2(Q, \mathbb{C}^\times)$. But $M \cong Z^2(Q, \mathbb{C}^\times)/B^2(Q, \mathbb{C}^\times) = M(Q)$, which is finite, by Theorem 7.60. ∎

Theorem 7.66 (Schur, 1904). *Every finite group Q has a cover U which is a central extension of $M(Q)$ by Q.*

Proof. Let M be a complement of $B^2(Q, \mathbb{C}^\times)$ in $Z^2(Q, \mathbb{C}^\times)$, as in the lemma; note that the elements of M are factor sets $Q \times Q \to \mathbb{C}^\times$. Define $K = M^*$. We first define a certain factor set $s: Q \times Q \to M^*$, and then construct the desired central extension from it.

For each $(x, y) \in Q \times Q$, define $s(x, y): M \to \mathbb{C}^\times$ by $f \mapsto f(x, y)$; thus, $s(x, y) \in M^*$, and there is a function $s: Q \times Q \to M^*$ given by $(x, y) \mapsto s(x, y)$. Now s satisfies the cocycle identity: for all $x, y, z \in Q$.

$$s(y, z)s(xy, z)^{-1}s(x, yz)s(x, y)^{-1}: f \mapsto f(y, z)f(xy, z)^{-1}f(x, yz)f(x, y)^{-1},$$

and the last term is 1 because every $f \in M$ is a factor set. Similarly, $s(1, y) = 1 = s(x, 1)$ for all $x, y \in Q$ because, for example, $s(1, y): f \mapsto f(1, y) = 1$. Therefore, $s \in Z^2(Q, M^*)$. Let U be the corresponding central extension of M^* by Q. To see that U is a cover of Q, it suffices, by Lemmas 7.61 and 7.62, to prove that the transgression $\delta: M^{**} \to M(Q)$ is an isomorphism. If $[f] \in M(Q)$, then $f \in Z^2(Q, \mathbb{C}^\times)$; since $Z^2 = B^2 \times M$, f has a unique expression $f = bf'$, where $b \in B^2(Q, \mathbb{C}^\times)$, $f' \in M$, and $[f] = [bf'] = [f']$. Now $\mu(f') \in \mathbb{C}^\times$ makes sense for all $\mu \in M^*$ (since $f' \in M$), and it is easy to check that $\varphi: M^{**} \to \mathbb{C}^\times$, given by $\varphi: \mu \mapsto \mu(f')$, is a homomorphism. By definition, $\delta(\varphi) = [\varphi \circ s]$. The composite $\varphi \circ s: Q \times Q \to M^* \to \mathbb{C}^\times$ sends $(x, y) \mapsto s(x, y) \mapsto \varphi(s(x, y)) = s(x, y)(f') = f'(x, y)$; that is, $\varphi \circ s = f'$ and $\delta(\varphi) = [\varphi \circ s] = [f'] = [f]$. Therefore, δ is surjective. But $|M^{**}| = |M(Q)|$, since $M \cong M(Q)$ is finite, and so δ must be an isomorphism. ∎

It should not be surprising that calculations of the multiplier are best done using techniques of Homological Algebra, Cohomology of Groups, and Representation Theory; for example, see Huppert (1967, §V.25) and Karpilovsky (1987). Indeed, the proofs given above fit into a general scheme (e.g., the transgression homomorphism δ arises in the "five-term exact sequence" as the map $H^1(K, \mathbb{C}^\times) \to H^2(Q, \mathbb{C}^\times)$). When Q is finite, we mention that $M(Q)$ has been calculated in many cases; sometimes it is 1; sometimes it is not.

EXAMPLE 7.17. Covers U of a finite group Q need not be unique.

If $\mathbf{V}$ is the 4-group, consider the central extensions U of $K \cong \mathbb{Z}_2$ by $\mathbf{V}$, where $U \cong D_8$ or $U \cong \mathbf{Q}$, the quaternions. In each case, $K = Z(U) = U'$, so that U is a central extension with $K \leq U'$. It follows from Lemma 7.64 that $M(\mathbf{V}) \neq 1$. We shall see in Chapter 11 that $M(\mathbf{V}) = \mathbb{Z}_2$; it will then follow from Lemma 7.63 that both D_8 and $\mathbf{Q}$ are (nonisomorphic) covers of $\mathbf{V}$.

The following discussion comparing two central extensions of a group Q will be completed in Chapter 11 when we show that covers of perfect groups, in particular, covers of simple groups, are unique. Consider the commutative diagram

$$\begin{array}{ccccccccc} 1 & \longrightarrow & K & \hookrightarrow & U & \xrightarrow{\nu} & Q & \longrightarrow & 1 \\ & & & & \downarrow{\scriptstyle\alpha} & & \downarrow{\scriptstyle 1_Q} & & \\ 1 & \longrightarrow & L & \hookrightarrow & V & \xrightarrow[\mu]{} & Q & \longrightarrow & 1, \end{array}$$

where both rows are central extensions. We claim that $\alpha(K) \leq L$. If $a \in K$, then $1 = \nu(a) = \mu\alpha(a)$, so that $\alpha(a) \in \ker \mu = L$. Denote the restriction of α by $\beta: K \to L$. If $\beta: K \to L$ is any homomorphism, then there is a homomorphism $\beta^*: L^* \to K^*$ given by $\psi \mapsto \psi \circ \beta$, where $\psi: L \to \mathbb{C}^\times$ is in L^*.

The diagram gives two transgressions: denote them by $\delta^U: K^* \to M(Q)$ and by $\delta^V: L^* \to M(Q)$.

Lemma 7.67. *Consider the commutative diagram*

$$\begin{array}{ccccccccc} 1 & \longrightarrow & K & \hookrightarrow & U & \xrightarrow{\nu} & Q & \longrightarrow & 1 \\ & & \downarrow{\scriptstyle\beta} & & \downarrow{\scriptstyle\alpha} & & \downarrow{\scriptstyle 1_Q} & & \\ 1 & \longrightarrow & L & \hookrightarrow & V & \xrightarrow[\mu]{} & Q & \longrightarrow & 1 \end{array}$$

whose rows are central extensions. Then $\delta^U\beta^* = \delta^V$.

Proof. Let $e: Q \times Q \to K$ be a factor set of the top extension; we may assume that U consists of all $(a, x) \in K \times Q$ with multiplication $(a, x)(b, y) = (abe(x, y), xy)$ and with $\nu(a, x) = x$. Similarly, let $f: Q \times Q \to L$ be a factor set of the bottom extension, so that V consists of all $(c, x) \in L \times Q$ with $(c, x)(d, y) = (cdf(x, y), xy)$ and with $\mu(c, x) = x$.

Now $\alpha((1, x)) = (h(x), l(x)) \in V$, and $x = v(1, x) = \mu\alpha((1, x)) = \mu((h(x), l(x)) = l(x)$; hence, $x = l(x)$, and $\alpha(1, x) = (h(x), x)$. Note also that $(a, x) = (a, 1)(1, x)$ for all $a \in K$ and $x \in Q$.

For all $x, y \in Q$,

$$\begin{aligned}\alpha((1, x)(1, y)) &= \alpha((e(x, y), xy))\\ &= \alpha((e(x, y), 1)(1, xy))\\ &= \alpha((e(x, y), 1))\alpha((1, xy))\\ &= (\beta \circ e(x, y), 1)(h(xy), xy),\end{aligned}$$

because $a \in K$ is identified with $(a, 1)$ and $\beta((a, 1)) \in L$. On the other hand,

$$\begin{aligned}\alpha((1, x)(1, y)) &= \alpha((1, x))\alpha((1, y))\\ &= (h(x), x)(h(y), y)\\ &= (h(x)h(y)f(x, y), xy).\end{aligned}$$

Therefore,

$$\beta \circ e(x, y) = f(x, y)h(y)h(xy)^{-1}h(x);$$

that is,

$$[\beta \circ e] = [f] \in M(Q).$$

It follows easily, for any $\psi \in L^*$, that $[\psi \circ \beta \circ e] = [\psi \circ f]$. Hence, $\delta^U\beta^*(\psi) = \delta^U(\psi \circ \beta) = [\psi \circ \beta \circ e] = [\psi \circ f] = \delta^V(\psi)$, as desired. ■

Theorem 7.68 (Alperin–Kuo, 1967). *If Q is a finite group, $e = \exp(M(Q))$, and $e' = \exp(Q)$, then ee' divides $|Q|$.*

***Proof* (Brandis).** We first show that if U is a finite group with subgroup $K \le U' \cap Z(U)$, then $\exp(K)\exp(U/K)$ divides $|U/K|$. Let A be an abelian subgroup with $K \le A \le U$, and let $V: U \to A$ be the transfer. Since A is abelian, $K \le U' \le \ker V$, so that $u \in K$ implies $V(u) = 1$. By Theorem 7.47, if $u \in Z(U)$, then $V(u) = u^n$. Therefore, if $u \in K \le U' \cap Z(U)$, then $x^n = 1$, and so $e = \exp(K)$ divides $n = [U : A] = [U/K : A/K]$. Hence, $e|A/K|$ divides $|U/K|$. In particular, this holds for all cyclic subgroups $A/K \le U/K$ (for $|U/K|$ does not depend on n). But $e' = \exp(U/K)$ divides $|A/K|$ for all cyclic A/K, and so ee' divides $|U/K|$.

If U is a cover of Q, then $M(Q) \cong K \le Z(U)$, for U is a central extension. Now $K \le U'$, by definition of cover, so that $K \le U' \cap Z(U)$. By the first paragraph, $\exp(K)\exp(U/K)$ divides $|U/K|$. But $K \cong M(Q)$ and $U/K \cong Q$, and this gives the result. ■

We remark that Schur proved that if $e = \exp(M(Q))$, then e^2 divides $|Q|$.

Theorem 7.69. *If every Sylow subgroup of a finite group Q is cyclic, then* $M(Q) = 1$.

Proof. For such a group, $\exp(Q) = |Q|$. By Theorem 7.68, $\exp(M(Q)) = 1$, and so $M(Q) = 1$. ■

Corollary 7.70.

(i) *For every* $n \geq 1$, $M(\mathbb{Z}_n) = 1$.
(ii) *If Q has squarefree order, then* $M(Q) = 1$.

There is a more general result which implies Theorem 7.69: if p is a prime divisor of $|Q|$ and P is a Sylow p-subgroup of Q, then the p-primary component of $M(Q)$ can be imbedded in $M(P)$.

We shall return to the Schur multiplier in Chapter 11, for it is also related to presentations of groups.

Derivations

We again consider general data (Q, K, θ) consisting of a group Q, an (additive) abelian group K, and a not necessarily trivial homomorphism $\theta: Q \to \text{Aut}(K)$ giving an action of Q on K.

Definition. Given data (Q, K, θ), a ***derivation*** (or *crossed homomorphism*) is a function $d: Q \to K$ such that

$$d(xy) = xd(y) + d(x).$$

The set $\text{Der}(Q, K, \theta)$ of all derivations is an abelian group under the following operation: if $d, d' \in \text{Der}(Q, K, \theta)$, then $d + d': x \mapsto d(x) + d'(x)$. In particular, if θ is trivial, then $\text{Der}(Q, K) = \text{Hom}(Q, K)$.

EXAMPLE 7.18. Let data (Q, K, θ) be given. For fixed $a \in K$, the function $d_a: Q \to K$, defined by $x \mapsto a - xa$ is easily checked to be a derivation: it is called the ***principal derivation*** determined by a.

EXAMPLE 7.19. Let G be an extension realizing data (Q, K, θ), so that G consists of all $(a, x) \in K \times Q$ with $(a, x) + (b, y) = (a + xb + f(x, y), xy)$ for a factor set f. If $\gamma: G \to G$ is a stabilizing automorphism, then $\gamma: (a, x) \mapsto (a + d(x), x)$ for some $d(x) \in K$, and the reader may check that d is a derivation.

Theorem 7.71. *Let G be an extension realizing data (Q, K, θ). If A is the stabilizer of this extension, then*

$$A \cong \text{Der}(Q, K, \theta).$$

Proof. Regard G as all $(a, x) \in K \times Q$. As in Example 7.19, if $\gamma \in A$, then $\gamma\colon (a, x) \mapsto (a + d(x), x)$, where d is a derivation. Define $\varphi\colon A \to \text{Der}(Q, K, \theta)$ by $\varphi(\gamma) = d$. Now φ is a homomorphism: if $\gamma' \in A$, then $\gamma'\colon (a, x) \mapsto (a + d'(x), x)$, and $\gamma'\gamma\colon (a, x) \mapsto (a + d(x) + d'(x), x)$. We see that φ is an isomorphism by constructing its inverse: if d is a derivation, define $\gamma\colon G \to G$ by $\gamma\colon (a, x) \mapsto (a + d(x), x)$. ■

Lemma 7.72. *Let G be a semidirect product $K \rtimes_\theta Q$, and let $\gamma\colon (a, x) \mapsto (a + d(x), x)$ be a stabilizing automorphism of G. Then γ is an inner automorphism if and only if d is a principal derivation.*

Proof. If d is a principal derivation, then there is $b \in K$ with $d(x) = b - xb$. Hence $\gamma(a, x) = (a + b - xb, x)$. But

$$\begin{aligned}(b, 1) + (a, x) - (b, 1) &= (b + a, x)\\ &= (b + a - xb, x),\end{aligned}$$

so that γ is conjugation by $(b, 1)$.

Conversely, if γ is conjugation by (b, y), then for all $(a, x) \in G$,

$$\begin{aligned}\gamma(a, x) &= (b, y) + (a, x) - (b, y)\\ &= (b + ya, yx) + (-y^{-1}b, y^{-1})\\ &= (b + ya - yxy^{-1}b, yxy^{-1}).\end{aligned}$$

Since γ is stabilizing, we must have $yxy^{-1} = x$, so that

$$\gamma(a, x) = (ya + b - xb, x).$$

Finally, if we choose $x = 1$, then $(a, 1) = \gamma(a, 1)$ (because γ is stabilizing), and so $(a, 1) = (ya, 1)$. Therefore, $ya = a$ for all a, and so $\gamma(a, x) = (a + b - xb, x) = (a + d(x), x)$, where $d(x) = b - xb$ is a principal derivation. ■

It is easy to see that $\text{PDer}(Q, K, \theta)$, the set of all principal derivations, is a subgroup of $\text{Der}(Q, K, \theta)$.

Definition. Given data (Q, K, θ), the ***first cohomology group*** $H^1(Q, K, \theta) = \text{Der}(Q, K, \theta)/\text{PDer}(Q, K, \theta)$.

If θ is trivial, then $H^1(Q, K, \theta) = \text{Hom}(Q, K)$. In particular, if Q is abelian and $K = \mathbb{C}^\times$, then $H^1(Q, K)$ is the character group $\text{Hom}(Q, \mathbb{C}^\times) = Q^*$.

Theorem 7.73. *If (Q, K, θ) are data and $G = H \rtimes_\theta Q$, then*

$$H^1(Q, K, \theta) \leq \text{Aut}(G)/\text{Inn}(G).$$

Proof. Let A be the stabilizer of the extension and let $\varphi\colon A \to \text{Der}(Q, K, \theta)$ be the isomorphism of Theorem 7.71. By Lemma 7.72, $\varphi(A \cap \text{Inn}(G)) =$

PDer(Q, K, θ). Hence

$$\begin{aligned} H^1(Q, K, \theta) &= \mathrm{Der}(Q, K, \theta)/\mathrm{PDer}(Q, K, \theta) \\ &\cong A/(A \cap \mathrm{Inn}(G)) \\ &\cong A\,\mathrm{Inn}(G)/\mathrm{Inn}(G) \leq \mathrm{Aut}(G)/\mathrm{Inn}(G). \quad \blacksquare \end{aligned}$$

Since we will be dealing with two transversals in the next theorem, let us write the elements of a semidirect product $G = K \rtimes_\theta Q$ not as ordered pairs (a, x), but rather as sums $a + l(x)$, where $l: Q \to G$ is a transversal.

Theorem 7.74. *Let $G = K \rtimes_\theta Q$, where K is abelian, and let C and C' be complements of K in G. If $H^1(Q, K, \theta) = 0$, then C and C' are conjugate.*

Remark. By Theorem 7.34, the hypothesis is satisfied if $H^2(Q, K, \theta) = 0$.

Proof. Since C and C' are complements of K, they are isomorphic: $C \cong G/K \cong Q \cong C'$; choose isomorphisms $l: Q \to C$ and $l': Q \to C'$. Both C and C' are also transversals of K in G, and so they determine factor sets f and f' (where, for example, $l(x) + l(y) = f(x, y) + l(xy)$). By Lemma 7.31, there is a function $h: Q \to K$, namely, $h(x) = l'(x) - l(x)$, with

$$f'(x, y) - f(x, y) = xh(y) - h(xy) + h(x).$$

Since l and l' are homomorphisms, it follows that both f and f' are identically zero. Thus, $f'(x, y) - f(x, y)$ is identically zero, $h(xy) = xh(y) + h(x)$, and h is a derivation. The hypothesis $H^1(Q, K, \theta) = 0$ says that h is a principal derivation; that is, there is $b \in K$ with

$$l'(x) - l(x) = h(x) = b - xb.$$

Therefore,

$$l'(x) = b + (-xb + l(x)) = b + l(x) - b;$$

that is, $C' = b + C - b$ is a conjugate of C. $\blacksquare$

Definition. If $d: Q \to K$ is a derivation, then its ***kernel*** is

$$\ker d = \{x \in Q: d(x) = 0\}.$$

For example, if d_a is a principal derivation, then $\ker d_a$ is the set of all $x \in Q$ which fix a.

Lemma 7.75. *Let (Q, K, θ) be data, and let $d: Q \to K$ be a derivation.*

(i) $d(1) = 0$.
(ii) $d(x^{-1}) = -x^{-1}d(x)$.
(iii) $\ker d$ *is a subgroup of* Q.
(iv) $d(x) = d(y)$ *if and only if* $x^{-1}y \in \ker d$.

Proof. (i) $d(1) = d(1 \cdot 1) = 1d(1) + d(1)$.

(ii) $0 = d(1) = d(x^{-1}x) = x^{-1}d(x) + d(x^{-1})$.

(iii) $1 \in \ker d$, by (i). If $x, y \in \ker d$, then $d(x) = 0 = d(y)$. Hence, $d(xy^{-1}) = xd(y^{-1}) + d(x) = -xy^{-1}d(y) + d(x) = 0$. Therefore, $xy^{-1} \in \ker d$ and so $\ker d$ is a subgroup of Q.

(iv) Now $d(x^{-1}y) = x^{-1}d(y) + d(x^{-1}) = x^{-1}d(y) - x^{-1}d(x)$. Hence, if $d(x) = d(y)$, then $d(x^{-1}y) = 0$. Conversely, if $d(x^{-1}y) = 0$, then $x^{-1}d(y) = x^{-1}d(x)$ and $d(y) = d(x)$. ∎

The subgroup $\ker d$ need not be a normal subgroup.

We are going to use derivations to give another proof of the Schur–Zassenhaus lemma. Given data (Q, K, θ), it is now more convenient to write the abelian group K multiplicatively. Assume that an abelian normal subgroup K has finite index n in a group E, and let $l_1, \ldots, l_n$ be a left transversal of K in G; that is, $E = \bigcup_{i=1}^n l_i K$. As in Lemma 7.44, if $e \in E$, then there is a unique $\kappa_i(e) \in K$ with $el_i = l_{\sigma i}\kappa_i(e)$, where σ is a permutation (depending on e). Define $a_i(e) = l_{\sigma i}\kappa_i(e)l_{\sigma i}^{-1}$, so that

$$el_i = a_i(e)l_{\sigma i},$$

and define a function $d: E \to K$ by

$$d(e) = \prod_{i=1}^{n} a_i(e).$$

Lemma 7.76 (Gruenberg). *Let K be an abelian normal subgroup of finite index in a group E, and let $L = \{l_1, \ldots, l_n\}$ be a left transversal of K in E.*

(i) *If $\theta: E \to \mathrm{Aut}(K)$ is defined by $\theta_e(k) = eke^{-1}$, then the function d defined above is a derivation.*
(ii) *If $k \in K$, then $d(k) = k^n$.*
(iii) *If L is a complement of K in E, then $L \leq \ker d$.*
(iv) *If $k \in K$ and $e \in E$, then $d(k^{-1}ek) = d(k)^e d(e) d(k)^{-1}$.*

Proof. (i) If $f \in E$, then $fl_i = a_i(f)l_{\tau i}$, where τ is a permutation. Now

$$\begin{aligned} e(fl_i) &= ea_i(f)l_{\tau i} = ea_i(f)e^{-1}el_{\tau i} \\ &= a_i(f)^e el_{\tau i} = a_i(f)^e a_{\tau i}(e) l_{\sigma\tau i}; \end{aligned}$$

since $(ef)l_i = a_i(ef)l_{\omega i}$, for some permutation ω, so that $a_i(ef) = a_i(f)^e a_{\tau i}(e)$. Since each $a_i(e)$ lies in the abelian group K, for $e \in E$ and $1 \leq i \leq n$, and since τ is a permutation, it follows that

$$\begin{aligned} d(ef) &= \prod_{i=1}^{n} a_i(ef) = \prod_{i=1}^{n} a_i(f)^e a_{\tau i}(e) \\ &= \left(\prod_{i=1}^{n} a_i(f)^e\right)\left(\prod_{i=1}^{n} a_{\tau i}(e)\right) = d(f)^e d(e). \end{aligned}$$

Therefore, d is a derivation.

(ii) If $k \in K$, then $K \lhd E$ implies $l_i^{-1}kl_i \in K$. Thus, $kl_i = l_i(l_i^{-1}kl_i)$ so that $\kappa_i(k) = l_i^{-1}kl_i$ and the permutation σ determined by k is the identity. But $a_i(k) = l_i\kappa_i(k)l_i^{-1} = k$, for all i, and so $d(k) = \prod_{i=1}^n a_i(k) = k^n$.

(iii) Recall that $el_i = l_{\sigma i}\kappa_i(e)$; if $e = l_j$, then $l_jl_i = l_r1$ for some $l_r \in L$, because L is now assumed to be a subgroup. Hence, $\kappa_i(l_j) = 1 = a_i(l_j)$ for all i (because a_i is a conjugate of κ_i), and so $d(l_j) = 1$ for every j.

(iv) First, note that $d(k^{-1}) = d(k)^{-1}$, by Lemma 7.75 (ii), because both k^{-1} and $d(k)^{-1}$ lie in the abelian group K. Now, $d(k^{-1}ek) = d(ek)^{k^{-1}}d(k^{-1}) = d(ek)d(k^{-1})$, because both $d(ek)$ and k^{-1} lie in the abelian group K, and $d(ek)d(k^{-1}) = d(k)^e d(e)d(k^{-1}) = d(k)^e d(e)d(k)^{-1}$. ■

The definition of the derivation d is reminiscent of the transfer; indeed, it is now easy to see that $d|K$ is the transfer.

The following proof is due to K. Gruenberg and B.A.F. Wehrfritz.

Theorem 7.77 (= Theorems 7.40 and 7.41). *Given data (Q, K, θ) with Q a finite group, K a finite abelian group, and $(|K|, |Q|) = 1$, then every extension G of K by Q realizing the data is a semidirect product; moreover, any two complements of K in G are conjugate.*

Proof. Let E be an extension of K by Q realizing the data and let $|Q| = n$. Define $L = \ker d$, where $d: Q \to K$ is Gruenberg's derivation. By Lemma 7.75 (iii), L is a subgroup of Q. We claim that $L \cap K = 1$. If $a \in K$, then $d(a) = a^n$, by Lemma 7.76 (ii); if $a \in L$, then $d(a) = 1$, by definition of kernel; hence, $a^n = 1$ for all $a \in L \cap K$. But $(|K|, n) = 1$, by hypothesis, so that $a = 1$. Let us now see that $E = KL$. If $e \in E$, then $d(e) \in K$. Since $(|K|, n) = 1$, there is $k \in K$ with $d(e) = k^{-n}$, by Exercise 1.31. Hence $d(e) = k^{-n} = f(k^{-1})$, by Lemma 7.76 (ii), and so $ke \in L = \ker d$, by Lemma 7.75 (iv). Therefore, $e = k^{-1}(ke) \in KL$, and E is a semidirect product.

We now show that if Y is another complement of K in E, then Y and L are conjugate. Note that each of Y and L is a transversal of K in E. Let $Y = \{y_1, \ldots, y_n\}$ and $L = \{l_1, \ldots, l_n\}$, and write $y_i = c_il_i$ for $c_i \in K$. The derivation determined by L is $d(e) = \prod_i a_i(e)$; the derivation determined by Y is $\delta(e) = \prod_i \alpha_i(e)$, where $\alpha_i(e) \in K$ and $ey_i = \alpha_i(e)y_{\tau i}$. But

$$ey_i = ec_il_i = (ec_ie^{-1})el_i = c_i^e a_i(e)l_{\sigma i} = c_i^e a_i(e)c_{\sigma i}^{-1}y_{\sigma i};$$

it follows that $\tau = \sigma$, $\alpha_i(e) = c_i^e a_i(e)c_{\sigma i}^{-1}$, and

$$\delta(e) = \prod_i \alpha_i(e) = \prod_i c_i^e a_i(e)c_{\sigma i}^{-1}.$$

If we define $c = \prod c_i$, then $c \in K$ and

$$\delta(e) = c^e d(e)c^{-1}.$$

Since $(|K|, n) = 1$, there is $k \in K$ with $c = k^n$, by Exercise 1.31, and so $c = d(k)$, by Lemma 7.76 (ii). Hence, $d(k^{-1}ek) = d(k)^e df(e)d(k)^{-1} = c^e d(e)c^{-1}$, by Lemma

7.76 (iv). Therefore,

$$\delta(e) = d(k^{-1}ek),$$

so that $L = \ker d = k(\ker \delta)k^{-1} \geq kYk^{-1}$, by Lemma 7.75 (iii). But $|L| = |Y|$, for both are complements of K, and so $L = kYk^{-1}$. ■

Exercises

7.54. If $G = K \rtimes_\theta Q$, then every transversal $l\colon Q \to G$ has the form $l(x) = (d(x), x)$ for some $d(x) \in K$. Show that l is a homomorphism if and only if d is a derivation.

7.55. Give an example of a derivation $d\colon Q \to K$ whose kernel is not a normal subgroup of Q.

7.56. If $\theta\colon Q \to \mathrm{Aut}(K)$ is trivial, then $H^1(Q, K) \cong \mathrm{Hom}(Q, K)$. Conclude that if B is an abelian group, then $H^1(B, \mathbb{C}^\times) \cong B^*$, the character group of B.

7.57. Regard the elements of an extension G realizing data (Q, K, θ) as ordered pairs $(a, x) \in K \times Q$, let $l\colon Q \to G$ be the transversal $l(x) = (0, x)$, and let $f\colon Q \times Q \to K$ be the corresponding factor set. If $d\colon G \to K$ is Gruenberg's derivation, show that $d(a, x) = na + \sum_{y \in Q} f(x, y)$, where $n = |Q|$.

7.58. Let $\mathbb{Z}_3 = \langle a \rangle$, let $\mathbb{Z}_2 = \langle x \rangle$, and define an action θ of $\mathbb{Z}_2$ on $\mathbb{Z}_3$ by $a^x = a^{-1}$. Show that $H^1(\mathbb{Z}_2, \mathbb{Z}_3, \theta) = 0$. (*Hint.* $S_3 \cong \mathbb{Z}_3 \rtimes \mathbb{Z}_2$, and $\mathrm{Aut}(S_3)/\mathrm{Inn}(S_3) = 1$.)

7.59. Let $\varphi\colon G \to G$ be a homomorphism, and let G act on itself by conjugation: if $g, x \in G$, then $g^x = xgx^{-1}$. Show that $d\colon G \to G$, defined by $d(g) = g\varphi(g)^{-1}$, satisfies the equation

$$d(gg') = d(g')^g d(g).$$

(Thus, d would be a derivation if G were abelian. This function d has already arisen in Exercise 1.50 and in the proof of Theorem 7.33.)

Given data (Q, K, θ), there are *cohomology groups* $H^i(Q, K, \theta)$ for all $i \geq 0$ and *homology groups* $H_i(Q, K, \theta)$ for all $i \geq 0$. Cohomology of Groups is the interpretation and computation of these groups. We have already discussed the cohomology groups $H^i(Q, K, \theta)$ for $i = 1, 2$. The group $H^0(Q, K, \theta)$ consists of *fixed points*:

$$H^0(Q, K, \theta) = \{a \in K\colon xa = a \text{ for all } x \in Q\};$$

the group $H^3(Q, K, \theta)$ is involved with *obstructions*: what sorts of data can extensions with nonabelian kernels realize. Now $H_0(Q, K)$ is the maximal Q-trivial quotient of K. If $K = \mathbb{Z}$ and θ is trivial, then the first homology group $H_1(Q, \mathbb{Z}) \cong Q/Q'$ and, when Q is finite, the second homology group $H_2(Q, \mathbb{Z}) \cong M(Q)$. There are also some standard homomorphisms between these groups, one of which, *corestriction*, generalizes the transfer. A fundamental idea is to construct the *group ring* $\mathbb{Z}G$ of a group G and to observe that G acting on an abelian group K corresponds to K being a $\mathbb{Z}G$-module. Indeed, our notation $H^i(Q, K, \theta)$ is usually abbreviated to $H^i(Q, K)$ by assuming at the outset that K is a $\mathbb{Z}G$-module.

CHAPTER 8
Some Simple Linear Groups

The Jordan–Hölder theorem tells us that once we know extensions and simple groups, then we know all finite groups. There are several infinite families of finite simple groups (in addition to the cyclic groups of prime order and the large alternating groups), and our main concern in this chapter is the most "obvious" of these, the *projective unimodular groups*, which arise naturally from the group of all matrices of determinant 1 over a field K. Since these groups are finite only when the field K is finite, let us begin by examining the finite fields.

Finite Fields

Definition. If K is a field, a ***subfield*** of K is a subring k of K which contains the inverse of every nonzero element of k. A ***prime field*** is a field k having no proper subfields.

Theorem 8.1. *Every field K contains a unique prime subfield k, and either $k \cong \mathbb{Q}$ or $k \cong \mathbb{Z}_p$ for some prime p.*

Proof. If k is the intersection of all the subfields of K, then it is easy to check that k is the unique prime subfield of K. Define $\chi\colon \mathbb{Z} \to K$ by $\chi(n) = n1$, where 1 denotes the "one" in K. It is easily checked that χ is a ring homomorphism with im $\chi \subset k$. Since K is a field, im χ is a domain and ker χ must be a prime ideal in $\mathbb{Z}$. Therefore, either ker $\chi = 0$ or ker $\chi = (p)$ for some prime p. In the first case, im $\chi \cong \mathbb{Z}$ and k contains an isomorphic copy F of the fraction field of $\mathbb{Z}$, namely, $\mathbb{Q}$; as k is a prime field, $k = F \cong \mathbb{Q}$. In the second case, k

contains an isomorphic copy E of $\mathbb{Z}/\ker \chi = \mathbb{Z}_p$, which is a field; as k is a prime field, $k = E \cong \mathbb{Z}_p$. ■

Definition. If K is a field with prime field k, then K has ***characteristic 0*** if $k \cong \mathbb{Q}$ and K has ***characteristic p*** if $k \cong \mathbb{Z}_p$.

Observe that if K has characteristic $p \geq 0$, then $pa = 0$ for all $a \in K$.

Corollary 8.2. *If K is a finite field, then $|K| = p^n$ for some prime p and some $n \geq 1$.*

Proof. If k is the prime field of K, then $k \ncong \mathbb{Q}$ because K is finite; therefore, $k \cong \mathbb{Z}_p$ for some prime p. We may view K as a vector space over $\mathbb{Z}_p$ (the "vectors" are the elements of K, the "scalars" are the elements of k, and the "scalar multiplication" $a\alpha$, for $a \in k$ and $\alpha \in K$, is just their product in K); if K has dimension n, then $|K| = p^n$. ■

There exist infinite fields of prime characteristic; for example, the field of all rational functions over $\mathbb{Z}_p$ (i.e., the fraction field of $\mathbb{Z}_p[x]$) is such a field.

The existence and uniqueness of finite fields are proven in Appendix VI: for every prime p and every integer $n \geq 1$, there exists a field with p^n elements (Theorem VI.19); two finite fields are isomorphic if and only if they have the same number of elements (Theorem VI.20). Finite fields are called ***Galois fields*** after their discoverer; we thus denote the field with $q = p^n$ elements by $\mathrm{GF}(q)$ (another common notation for this field is $\mathbb{F}_q$), though we usually denote $\mathrm{GF}(p)$ by $\mathbb{Z}_p$.

Recall that if E is a field, k is a subfield, and $\pi \in E$, then $k(\pi)$, the subfield of E obtained by ***adjoining*** π to k, is the smallest subfield of E containing k and π; that is, $k(\pi)$ is the intersection of all the subfields of E containing k and π.

Definition. A ***primitive element*** of a finite field K is an element $\pi \in K$ with $K = k(\pi)$, where k is the prime field.

Lemma 8.3. *There exists a primitive element π of $\mathrm{GF}(p^n)$; moreover, π may be chosen to be a root of an irreducible polynomial $g(x) \in \mathbb{Z}_p[x]$ of degree n.*

Proof. Let $q = p^n$ *and let* $K = \mathrm{GF}(q)$. By Theorem 2.18(ii), the multiplicative group $K^\times$ is cyclic; clearly, any generator π of $K^\times$ is a primitive element of K (there can be primitive elements of K that are not generators of $K^\times$). By Lagrange's theorem, $\pi^{q-1} = 1$ (for $|K^\times| = q - 1$), and so π is a root of $f(x) = x^{q-1} - 1$. Factoring $f(x)$ into a product of irreducible polynomials in $k[x]$ (where $k \cong \mathbb{Z}_p$ is the prime field) provides an irreducible $g(x) \in k[x]$ having π as a root. If $g(x)$ has degree d, then $k(\pi)$ is a subfield of K with $[k(\pi) : k] = d$ (Theorem VI.21 in Appendix VI); therefore, $|k(\pi)| = p^d$. But $k(\pi) = K$ (because π is a primitive element) and so $d = n$. ■

Theorem 8.4. *If p is a prime, then the group* Aut(GF(p^n)) *of all field automorphisms of* GF(p^n) *is cyclic of order n.*

Proof. Let k be the prime field of $K = \mathrm{GF}(p^n)$. If π is a primitive element of K, as in the lemma, then there is an irreducible polynomial $g(x) \in k[x]$ of degree n having π as a root. Since every $\varphi \in \mathrm{Aut}(K)$ must fix k pointwise (because $\varphi(1) = 1$), Lemma 5.1 shows that $\varphi(\pi)$ is also a root of $g(x)$. As $K = k(\pi)$, Lemma 5.2 shows that φ is completely determined by $\varphi(\pi)$. It follows that $|\mathrm{Aut}(K)| \leq n$, because $g(x)$, having degree n, has at most n roots. The map $\sigma: K \to K$, given by $\sigma(\alpha) = \alpha^p$, is an automorphism of K. If $1 \leq i < n$ and $\sigma^i = 1$, then $\alpha = \alpha^{p^i}$ for every $\alpha \in K$. In particular, $\pi^{p^i - 1} = 1$, contradicting π having order $p^n - 1$ in $K^\times$. Therefore, $\langle \sigma \rangle \leq \mathrm{Aut}(K)$ is cyclic of order n, and so $\mathrm{Aut}(K) = \langle \sigma \rangle$. ■

We remark that Aut(GF(p^n)) is the Galois group Gal(GF(p^n)/$\mathbb{Z}_p$), for every $\varphi \in$ Aut(GF(p^n)) fixes the prime field $\mathbb{Z}_p$ pointwise.

The General Linear Group

Groups of nonsingular matrices are as natural an object of study as groups of permutations: the latter consists of "automorphisms" of a set; the former consists of automorphisms of a vector space.

Definition. If V is an m-dimensional vector space over a field K, then the ***general linear group*** GL(V) is the group of all nonsingular linear transformations on V (with composite as operation).

If one chooses an ordered basis $\{e_1, \ldots, e_m\}$ of V, then each $T \in \mathrm{GL}(V)$ determines a matrix $A = [\alpha_{ij}]$, where $Te_j = \sum_i \alpha_{ij} e_i$ (the jth *column* of A consists of the coordinates of Te_j). The function $T \mapsto A$ is an isomorphism $\mathrm{GL}(V) \to \mathrm{GL}(m, K)$, where $\mathrm{GL}(m, K)$ is the multiplicative group of all $m \times m$ nonsingular matrices over K. When $K = \mathrm{GF}(q)$, we may write $\mathrm{GL}(m, q)$ instead of $\mathrm{GL}(m, K)$.

Theorem 8.5. $|\mathrm{GL}(m, q)| = (q^m - 1)(q^m - q) \ldots (q^m - q^{m-1})$.

Proof. Let V be an m-dimensional vector space over a field K, and let $\{e_1, \ldots, e_m\}$ be an ordered basis of V. If $\mathscr{B}$ denotes the family of all ordered bases of V, then there is a bijection $\mathrm{GL}(V) \to \mathscr{B}$: if T is nonsingular, then $\{Te_1, \ldots, Te_m\}$ is an ordered basis of V; if $\{v_1, \ldots, v_m\}$ is an ordered basis, then there exists a unique nonsingular T with $Te_i = v_i$ for all i.

Let $\{v_1, \ldots, v_m\}$ be an ordered basis of V. Since there are q^m vectors in V, there are $q^m - 1$ candidates for v_1 (the zero vector is not a candidate). Having chosen v_1, the candidates for v_2 are those vectors in V not in $\langle v_1 \rangle$, the subspace spanned by v_1; there are thus $q^m - q$ candidates for v_2. More generally, having chosen an independent set $\{v_1, \ldots, v_i\}$, we may choose v_{i+1} to be any vector not in $\langle v_1, \ldots, v_i \rangle$, and so there are $q^m - q^i$ candidates for v_{i+1}. The result follows. ■

Notation. If V is an m-dimensional vector space over $K = \mathrm{GF}(q)$, if t is a nonnegative integer, and if π is a primitive element of K, then

$$M(t) = \{A \in \mathrm{GL}(V)\colon \det A \text{ is a power of } \pi^t\}.$$

Lemma 8.6. *If $\Omega = |\mathrm{GL}(m, q)|$ and if t is a divisor of $q - 1$, then $M(t)$ is a normal subgroup of $\mathrm{GL}(m, q)$ of order Ω/t. Moreover, if $q - 1 = p_1 \ldots p_r$, where the p_i are (not necessarily distinct) primes, then the following normal series is the beginning of a composition series:*

$$\mathrm{GL}(m, q) = M(1) > M(p_1) > M(p_1 p_2) > \cdots > M(q-1) > 1.$$

Proof. Let $K = \mathrm{GF}(q)$. Use the correspondence theorem in the setting

$$\det\colon \mathrm{GL}(m, q) \to K^\times.$$

If t divides $q - 1 = |K^\times|$, then the cyclic subgroup $\langle \pi^t \rangle$ of $K^\times$ is normal ($K^\times$ is abelian), has order $(q-1)/t$, and has index t. Since $M(t)$ is the subgroup of $\mathrm{GL}(m, q)$ corresponding to $\langle \pi^t \rangle$, it is a normal subgroup of index t hence order Ω/t. Now $|M(p_1 \ldots p_i)/M(p_1 \ldots p_{i+1})| = (\Omega/p_1 \ldots p_i)/(\Omega/p_1 \ldots p_{i+1}) = p_{i+1}$; since the factor groups have prime order, they are simple. ■

Definition. A matrix (or linear transformation) having determinant 1 is called ***unimodular***.

The subgroup $M(q-1)$ consists of all the unimodular matrices, for $\pi^{q-1} = 1$.

Definition. If V is an m-dimensional vector space over a field K, then the ***special linear group*** $\mathrm{SL}(V)$ is the subgroup of $\mathrm{GL}(V)$ consisting of all the unimodular transformations.

Choosing an ordered basis of V gives an isomorphism $\mathrm{SL}(V) \to \mathrm{SL}(m, K)$, the group of all unimodular matrices. If $K = \mathrm{GF}(q)$, we may denote $\mathrm{SL}(m, K)$ by $\mathrm{SL}(m, q)$.

The following elementary matrices are introduced to analyze the structure of $\mathrm{SL}(m, K)$.

Definition. Let λ be a nonzero element of a field K, and let $i \neq j$ be integers between 1 and m. An ***elementary transvection*** $B_{ij}(\lambda)$ is the $m \times m$ matrix differing from the identity matrix E in that it has λ as its ij entry. A ***transvection*** is a matrix B that is similar to some $B_{ij}(\lambda)$; that is, B is a conjugate of some $B_{ij}(\lambda)$ in $\mathrm{GL}(m, K)$.

Every transvection is unimodular. Note that the inverse of an elementary transvection is another such: $B_{ij}(\lambda)^{-1} = B_{ij}(-\lambda)$; it follows that the inverse of any transvection is also a transvection.

If $A \in \mathrm{GL}(m, K)$, then $B_{ij}(\lambda)A$ is the matrix obtained from A by adding λ times its jth row to its ith row.

Lemma 8.7. *Let K be a field. If $A \in \mathrm{GL}(m, K)$ and* $\det A = \mu$, *then $A = UD(\mu)$, where U is a product of elementary transvections and $D =$* $\mathrm{diag}\{1, 1, \ldots, 1, \mu\}$.

Proof. We prove, by induction on $t \leq m - 1$, that A can be transformed, by a sequence of elementary operations which add a multiple of one row to another, into a matrix of the form

$$A_t = \begin{bmatrix} E_t & * \\ 0 & C \end{bmatrix},$$

where E_t is the $t \times t$ identity matrix.

For the base step, note that the first column of A is not zero (A is nonsingular). Adding some row to the second row if necessary, we may assume that $\alpha_{21} \neq 0$. Now add $\alpha_{21}^{-1}(1 - \alpha_{11})$ times row 2 to row 1 get entry 1 in the upper left corner. We may now make the other entries in column 1 equal corner. We may now make the other entries in column 1 equal to zero by adding suitable multiples of row 1 to the other rows, and so A has been transformed into A_1.

For the inductive step, we may assume that A has been transformed into a matrix A_t as displayed above. Note that C is nonsingular (for $\det A_t = \det C$). Assuming that C has at least two rows, we may further assume, as in the base step, that its upper left corner $\gamma_{t+1,t+1} = 1$ (this involves only the rows of C, hence does not disturb the top t rows of A_t). Adding on a suitable multiple of row $t + 1$ to the other rows of A_t yields a matrix A_{t+1}.

We may now assume that A has been transformed into

$$\begin{bmatrix} E_t & * \\ 0 & \mu \end{bmatrix}$$

where $\mu \in K$ and $\mu \neq 0$. Adding suitable multiples of the last row to the other rows cleans out the last column, leaving $D(\mu)$.

In terms of matrix multiplication, we have shown that there is a matrix P, which is a product of elementary transvections, with $PA = D(\mu)$. Therefore, $A = P^{-1}D(\mu)$; this completes the proof because the inverse of an elementary transvection is another such. ■

Theorem 8.8.

(i) $\mathrm{GL}(m, K)$ *is a semidirect product of* $\mathrm{SL}(m, K)$ *by* $K^\times$.
(ii) $\mathrm{SL}(m, K)$ *is generated by elementary transvections.*

Proof. (i) We know that $\mathrm{SL} \lhd \mathrm{GL}$ (because $\mathrm{SL} = \ker \det$), and it is easy to see that $\Delta = \{D(\mu): \mu \in K^\times\}$ ($\cong K^\times$) is a complement of SL.

(ii) By (i), each $A \in \mathrm{GL}$ has a unique factorization $A = UD(\mu)$, where $U \in \mathrm{SL}$, $D(\mu) \in \Delta$, and $\det A = \mu$. Therefore, A is unimodular if and only if $A = U$. The result now follows from Lemma 8.7. ■

Notation. If V is an m-dimensional vector space over a field K, let $\mathrm{Z}(V)$ denote the subgroup of $\mathrm{GL}(V)$ consisting of all scalar transformations, and let $\mathrm{SZ}(V)$ be the subgroup of $\mathrm{Z}(V)$ consisting of all unimodular scalar transformations.

Let $\mathrm{Z}(m, K) \cong \mathrm{Z}(V)$ denote the subgroup of all $m \times m$ scalar matrices αE, and let $\mathrm{SZ}(m, K) \cong \mathrm{SZ}(V)$ denote the subgroup of all αE with $\alpha^m = 1$. If $K = \mathrm{GF}(q)$, we may also denote these subgroups by $\mathrm{Z}(m, q)$ and $\mathrm{SZ}(m, q)$, respectively.

Theorem 8.9.

(i) *The center of* $\mathrm{GL}(V)$ *is* $\mathrm{Z}(V)$.
(ii) *The center of* $\mathrm{SL}(m, K)$ *is* $\mathrm{SZ}(m, K)$.

Proof. (i) If $T \in \mathrm{GL}(V)$ is not a scalar transformation, then there is $v \in V$ with $\{v, Tv\}$ independent; extend this to a basis $\{v, Tv, u_3, \ldots, u_m\}$ of V. It is easy to see that $\{v, v + Tv, u_3, \ldots, u_m\}$ is also a basis of V, so that there is a (nonsingular) linear transformation $S: V \to V$ with $Sv = v$, $S(Tv) = v + Tv$, and $Su_i = u_i$ for all $i \geq 3$. Now T and S do not commute, for $TS(v) = Tv$ while $ST(v) = v + TV$. Therefore, $T \notin \mathrm{Z}(\mathrm{GL}(V))$, and it follows that $\mathrm{Z}(\mathrm{GL}(V)) = \mathrm{Z}(V)$.

(ii) Assume now that $T \in \mathrm{SL}(V)$, that T is not scalar, and that S is the linear transformation constructed in (i). The matrix of S relative to the basis $\{v, Tv, u_3, \ldots, u_m\}$ is the elementary transvection $B_{12}(1)$, so that $\det(S) = 1$ and $S \in \mathrm{SL}(V)$. As in (i), $T \notin \mathrm{Z}(\mathrm{SL}(V))$; that is, if $T \in \mathrm{Z}(\mathrm{SL}(V))$, then $T = \alpha E$ for some $\alpha \in K$. Finally, $\det(\alpha E) = \alpha^m$, and so $\alpha^m = 1$, so that $\mathrm{SZ}(V) = \mathrm{Z}(\mathrm{SL}(V))$. ■

Theorem 8.10. $|\mathrm{SZ}(m, q)| = d$, *where* $d = (m, q - 1)$.

Proof. Let $K = \mathrm{GF}(q)$. We first show, for all $\alpha \in K^\times$, that $\alpha^m = 1$ if and only if $\alpha^d = 1$. Since d divides m, $\alpha^d = 1$ implies $\alpha^m = 1$. Conversely, there are integers a and b with $d = am + b(q-1)$. Thus

$$\alpha^d = \alpha^{am+b(q-1)} = \alpha^{ma}\alpha^{(q-1)b} = \alpha^{ma},$$

because $\alpha^{q-1} = 1$. Hence $\alpha^m = 1$ gives $1 = \alpha^{ma} = \alpha^d$.

It follows that $\mathrm{SZ}(m, q) \cong \{\alpha \in K^\times : \alpha^m = 1\} \cong \{\alpha \in K^\times : \alpha^d = 1\}$. Therefore, if π is a generator of $K^\times$, then $\mathrm{SZ}(m, q) \cong \langle \pi^{n/d} \rangle$ and hence $|\mathrm{SZ}(m, q)| = d$. ∎

Our preceding discussion allows us to lengthen the normal series in Lemma 8.6 as follows:

$$\mathrm{GL}(m, q) > M(p_1) > M(p_1 p_2) > \cdots > \mathrm{SL}(m, q) > \mathrm{SZ}(m, q) > 1.$$

The center $\mathrm{SZ}(m, q)$ is abelian and so its composition factors are no secret (they are cyclic groups of prime order, occurring with multiplicity, for all primes dividing $q - 1$). We now consider the last factor group in this series.

Definition. If V is an m-dimensional vector spaces over a field K, the ***projective unimodular group*** $\mathrm{PSL}(V)$ is the group $\mathrm{SL}(V)/\mathrm{SZ}(V)$.

A choice of ordered basis of V induces an isomorphism $\varphi: \mathrm{SL}(V) \xrightarrow{\sim} \mathrm{SL}(m, K)$ with $\varphi(\mathrm{SZ}(V)) = \mathrm{SZ}(m, K)$, so that $\mathrm{PSL}(V) \cong \mathrm{SL}(m, K)/\mathrm{SZ}(m, K)$. The latter group is denoted by $\mathrm{PSL}(m, K)$. When $K = \mathrm{GF}(q)$, we may denote $\mathrm{PSL}(m, K)$ by $\mathrm{PSL}(m, q)$.

We shall see, in Chapter 9, that these groups are intimately related to projective geometry, whence their name.

Theorem 8.11. *If* $d = (m, q-1)$, *then*

$$|\mathrm{PSL}(m, q)| = (q^m - 1)(q^m - q)\ldots(q^m - q^{m-1})/d.$$

Proof. Immediate from Theorems 8.5 and 8.10. ∎

EXERCISES

8.1. Let $H \lhd \mathrm{SL}(2, K)$, and let $A \in H$. Using the factorization $A = UD(\mu)$ (in the proof of Theorem 8.8), show that if A is similar to $\begin{bmatrix} \alpha & \beta \\ \gamma & \delta \end{bmatrix}$, then there is $\mu \in K^\times$ such that H contains $\begin{bmatrix} \alpha & \mu^{-1}\beta \\ \mu\gamma & \delta \end{bmatrix}$.

8.2. Let $B = B_{ij}(1) \in \mathrm{GL}(m, K) = G$. Prove that $C_G(B)$ consists of all those nonsingular matrices $A = [a_{ij}]$ whose ith column, aside from a_{ii}, and whose jth row, aside from a_{jj}, consist of all 0's.

8.3. Let $\Delta \leq \mathrm{GL}(m, K)$ be the subgroup of all nonsingular diagonal matrices.
(i) Show that Δ is an abelian *self-centralizing* subgroup; that is, if $A \in \mathrm{GL}(m, K)$ commutes with every $D \in \Delta$, then $A \in \Delta$.
(ii) Use part (i) to give another proof that $\mathrm{Z}(\mathrm{GL}(m, K)) = \mathrm{Z}(m, K)$ consists of the scalar matrices.

PSL(2, *K*)

In this section, we concentrate on the case $m = 2$ with the aim of proving that $\mathrm{PSL}(2, q)$ is simple whenever $q > 3$. We are going to see that elementary transvections play the same role here that 3-cycles play in the analysis of the alternating groups.

Definition. A field K is ***perfect*** if either it has characteristic 0 or it has prime characteristic p and every $\lambda \in K$ has a pth root in K.

If K has prime characteristic p, then the map $F: K \to K$, given by $\lambda \mapsto \lambda^p$, is an injective homomorphism. If K is finite, then F must be surjective; that is, every finite field is perfect. Clearly, every algebraically closed field K is perfect. An example of a nonperfect field is $K = \mathbb{Z}_p(x)$, the field of all rational functions with coefficients in $\mathbb{Z}_p$; the indeterminate x does not have a pth root in $\mathbb{Z}_p(x)$.

Lemma 8.12. *Let K be a field which either has characteristic $\neq 2$ or is perfect of characteristic 2. If a normal subgroup H of* $\mathrm{SL}(2, K)$ *contains an elementary transvection $B_{12}(\lambda)$ or $B_{21}(\lambda)$, then $H = \mathrm{SL}(2, K)$.*

Proof. Note first that if $B_{21}(\lambda) \in H$, then $UB_{21}(\mu)U^{-1} = B_{12}(-\mu)$, where

$$U = \begin{bmatrix} 0 & -1 \\ 1 & 0 \end{bmatrix}.$$

By Theorem 8.8(ii), it suffices to prove that H contains every elementary transvection. Conjugate $B_{12}(\lambda)$ by a unimodular matrix:

$$\begin{bmatrix} \alpha & \beta \\ \gamma & \delta \end{bmatrix}\begin{bmatrix} 1 & \lambda \\ 0 & 1 \end{bmatrix}\begin{bmatrix} \delta & -\beta \\ -\gamma & \alpha \end{bmatrix} = \begin{bmatrix} 1 - \lambda\alpha\gamma & \lambda\alpha^2 \\ -\lambda\gamma^2 & 1 + \lambda\alpha\gamma \end{bmatrix}.$$

In particular, if $\gamma = 0$, then $\alpha \neq 0$ and this conjugate is $B_{12}(\lambda\alpha^2)$. Since H is normal in SL, these conjugates lie in H. Define

$$\Gamma = \{0\} \cup \{\mu \in K: B_{12}(\mu) \in H\},$$

It is easy to see that Γ is a subgroup of the additive group K, and so it contains all elements of the form $\lambda(\alpha^2 - \beta^2)$, where $\alpha, \beta \in K$.

We claim that $\Gamma = K$, and this will complete the proof. If K has characteristic $\neq 2$, then each $\mu \in K$ is a difference of squares:

$$\mu = [\tfrac{1}{2}(\mu + 1)]^2 - [\tfrac{1}{2}(\mu - 1)]^2.$$

For each $\mu \in K$, therefore, there are $\alpha, \beta \in K$ with $\lambda^{-1}\mu = \alpha^2 - \beta^2$, so that $\mu = \lambda(\alpha^2 - \beta^2) \in \Gamma$, and $\Gamma = K$. If K has characteristic 2 and is perfect, then every element in K has a square root in K. In particular, there is $\alpha \in K$ with $\lambda^{-1}\mu = \alpha^2$, and Γ contains $\lambda\alpha^2 = \mu$. ■

The next theorem was proved by C. Jordan (1870) for q prime. In 1893, after F. Cole had discovered a simple group G of order 504, E.H. Moore recognized G as PSL(2, 8), and then proved the simplicity of PSL(2, q) for all prime powers $q > 3$.

Theorem 8.13 (Jordan–Moore). *The groups* PSL(2, q) *are simple if and only if* $q > 3$.

Proof. By Theorem 8.11,

$$|\mathrm{PSL}(2, q)| = \begin{cases} (q^2 - 1)(q^2 - q) & \text{if } q = 2^n, \\ (q^2 - 1)(q^2 - q)/2 & \text{if } q = p^n, p \text{ an odd prime.} \end{cases}$$

Therefore, PSL(2, 2) has order 6 and PSL(2, 3) has order 12, and there are no simple groups of these orders.

Assume now that $q \geq 4$. It suffices to prove that a normal subgroup H of SL(2, q) which contains a matrix not in SZ(2, q) must be all of SL(2, q).

Suppose that H contains a matrix

$$A = \begin{bmatrix} \alpha & 0 \\ \beta & \alpha^{-1} \end{bmatrix},$$

where $\alpha \neq \pm 1$; that is, $\alpha^2 \neq 1$. If $B = B_{21}(1)$, then H contains the commutator $BAB^{-1}A^{-1} = B_{21}(1 - \alpha^{-2})$, which is an elementary transvection. Therefore, $H = \mathrm{SL}(2, q)$, by Lemma 8.12.

To complete the proof, we need only display a matrix in H whose top row is $[\alpha\ 0]$, where $\alpha \neq \pm 1$. By hypothesis, there is a matrix M in H, not in SZ(2, q), and M is similar to either a diagonal matrix or a matrix of the form

$$\begin{bmatrix} 0 & -1 \\ 1 & \beta \end{bmatrix},$$

for the only rational canonical forms for a 2×2 matrix are: two 1×1 blocks (i.e., a diagonal matrix) or a 2×2 companion matrix (which has the above form because it is unimodular). In the first case, Exercise 8.1 shows that

$$C = \begin{bmatrix} \alpha & 0 \\ 0 & \beta \end{bmatrix} \in H;$$

since C is unimodular, $\alpha\beta = 1$; since M is not in SZ(m, q), $\alpha \neq \beta$. It follows that $\alpha \neq \pm 1$, and C is the desired matrix. In the second case, Exercise 8.1 shows that H contains

$$D = \begin{bmatrix} 0 & -\mu^{-1} \\ \mu & \beta \end{bmatrix}.$$

If $T = \operatorname{diag}\{\alpha^{-1}, \alpha\}$, where α is to be chosen, then H contains the commutator

$$U = TDT^{-1}D^{-1} = \begin{bmatrix} \alpha^{-2} & 0 \\ \mu\beta(\alpha^2 - 1) & \alpha^2 \end{bmatrix}.$$

We are done if $\alpha^{-2} \neq \pm 1$; that is, if $\alpha^4 \neq 1$. If $q > 5$, then such an α exists, for a field contains at most four roots of $x^4 - 1$. If $q = 4$, then every $\mu \in K$ satisfies the equation $x^4 - x = 0$, so that $\alpha \neq 1$ implies $\alpha^4 \neq 1$.

Only the case $\mathrm{GF}(5) \cong \mathbb{Z}_5$ remains. Consider the factor β occurring in the lower left corner $\lambda = \mu\beta(\alpha^2 - 1)$ of U. If $\beta \neq 0$, choose $\alpha = [2] \in \mathbb{Z}_5$; then $\alpha^2 - 1 \neq 0$ and $U = B_{21}(\lambda)$. Hence H contains the elementary transvection $U^2 = B_{21}(-2\lambda)$ and we are done. If $\beta = 0$, then

$$D = \begin{bmatrix} 0 & -\mu^{-1} \\ \mu & 0 \end{bmatrix} \in H.$$

Therefore, the normal subgroup H contains

$$B_{12}(\nu)DB_{12}(-\nu) = \begin{bmatrix} \mu\nu & -\mu\nu^2 - \mu^{-1} \\ * & * \end{bmatrix}$$

for all $\nu \in \mathbb{Z}_5$. If $\nu = 2\mu^{-1}$, then the top row of this last matrix is $[2 \;\; 0]$, and the theorem is proved. ■

Corollary 8.14. *If K is an infinite field which either has characteristic $\neq 2$ or is perfect of characteristic 2, then* $\mathrm{PSL}(2, K)$ *is a simple group.*

Proof. The finiteness of K in the proof of the theorem was used only to satisfy the hypotheses of Lemma 8.12. ■

Remark. In Theorem 9.48, we will prove that $\mathrm{PSL}(2, K)$ is simple for every infinite field.

Corollary 8.15. $\mathrm{SL}(2, 5)$ *is not solvable.*

Proof. Every quotient of a solvable group is solvable. ■

We have exhibited an infinite family of simple groups. Are any of its members distinct from simple groups we already know? Using Theorem 8.11, we see that both $\mathrm{PSL}(2, 4)$ and $\mathrm{PSL}(2, 5)$ have order 60. By Exercise 4.37, all simple groups of order 60 are isomorphic:

$$\mathrm{PSL}(2, 4) \cong A_5 \cong \mathrm{PSL}(2, 5).$$

If $q = 7$, however, then we do get a new simple group, for $|\mathrm{PSL}(2, 7)| = 168$, which is neither prime nor $\frac{1}{2}n!$. If we take $q = 8$, we see that there is a simple group of order 504; if $q = 11$, we see a simple group of order 660. (It is known that the only other isomorphisms involving A_n's and PSLs, aside from those displayed above, are Exercise 8.12: $\mathrm{PSL}(2, 9) \cong A_6$ (these groups have order 360); Exercise 9.26: $\mathrm{PSL}(2, 7) \cong \mathrm{PSL}(3, 2)$ (these groups have order 168); Theorem 9.73: $\mathrm{PSL}(4, 2) \cong A_8$ (these groups have order 20, 160).)

Exercises

8.4. Show that the Sylow p-subgroups of SL(2, 5) are either cyclic (when p is odd) or quaternion (when $p = 2$). Conclude that $\mathrm{SL}(2, 5) \ncong S_5$.

8.5. What is the Sylow 2-subgroup of SL(2, 3)?

8.6. (i) Show that $\mathrm{PSL}(2, 2) \cong S_3$.
(ii) Show that $\mathrm{SL}(2, 3) \ncong S_4$ but that $\mathrm{PSL}(2, 3) \cong A_4$.

8.7. What are the composition factors of GL(2, 7)?

8.8. Show that if $H \lhd \mathrm{GL}(2, K)$, where K has more than three elements, then either $H \leq Z(\mathrm{GL}(2, K))$ or $\mathrm{SL}(2, K) \leq H$.

8.9. (i) What is the commutator subgroup of GL(2, 2)?
(ii) What is the commutator subgroup of GL(2, 3)?
(iii) If $q > 3$, prove that the commutator subgroup of $\mathrm{GL}(2, q)$ is $\mathrm{SL}(2, q)$.

8.10. Prove, for every field K, that all transvections are conjugate in $\mathrm{GL}(2, K)$.

8.11. Let A be a unimodular matrix. Show that A determines an involution in $\mathrm{PSL}(2, K)$ if and only if A has trace 0, and that A determines an element of order 3 in $\mathrm{PSL}(2, K)$ if and only if A has trace ± 1. (*Hint.* Use canonical forms.)

8.12. Prove that any two simple groups of order 360 are isomorphic, and conclude that $\mathrm{PSL}(2, 9) \cong A_6$. (*Hint.* Show that a Sylow 5-subgroup has six conjugates.)

PSL(m, K)

The simplicity of $\mathrm{PSL}(m, K)$ for all $m \geq 3$ and all fields K will be proved in this section. In 1870, C. Jordan proved this theorem for $K = \mathbb{Z}_p$, and L.E. Dickson extended the result to all finite fields K in 1897, four years after Moore had proved the result for $m = 2$. The proof we present, due to E. Artin, is much more elegant than matrix manipulations (though we prefer matrices when $m = 2$).

An $m \times m$ elementary transvection $B_{ij}(\lambda)$ represents a linear transformation T on an m-dimensional vector space V over K. There is an ordered basis $\{v_1, \ldots, v_m\}$ of V with $Tv_l = v_l$ for all $l \neq i$ and with $Tv_i = v_i + \lambda v_j$. Note that T fixes every vector in the $(m - 1)$-dimensional subspace H spanned by all $v_l \neq v_i$.

Definition. If V is an m-dimensional vector space over a field K, then a ***hyperplane*** H in V is a subspace of dimension $m - 1$.

The linear transformation T arising from an elementary transvection fixes the hyperplane H pointwise. If $w \in V$ and $w \notin H$, then $\langle w \rangle = \{\mu w : \mu \in K\}$ is a transversal of H in V: the vector space V, considered as an additive group, is the disjoint union of the cosets $H + \mu w$. Hence, every vector $v \in V$ has a unique expression of the form

$$v = \mu w + h, \qquad \mu \in K, \quad h \in H.$$

Lemma 8.16. *Let H be a hyperplane in V and let $T \in \mathrm{GL}(V)$ fix H pointwise. If $w \in V$ and $w \notin H$, then*

$$T(w) = \mu w + h_0$$

for some $\mu \in K$ and $h_0 \in H$. Moreover, given any $v \in V$,

$$T(v) = \mu w + h',$$

for some $h' \in H$.

Proof. We observed above that every vector in V has an expression of the form $\lambda w + h$. In particular, $T(w)$ has such an expression. If $v \in V$, then $v = \lambda w + h''$ for some $\lambda \in K$ and $h'' \in H$. Since T fixes H,

$$\begin{aligned} T(v) &= \lambda T(w) + h'' = \lambda(\mu w + h_0) + h'' \\ &= \mu(\lambda w + h'') + [(1 - \mu)h'' + \lambda h_0] \\ &= \mu v + h'. \quad \blacksquare \end{aligned}$$

The scalar $\mu = \mu(T)$ in Lemma 8.16 is thus determined uniquely by any T fixing a hyperplane pointwise.

Definition. Let $T \in \mathrm{GL}(V)$ fix a hyperplane H pointwise, and let $\mu = \mu(T)$. If $\mu \neq 1$, then T is called a ***dilatation***; if $\mu = 1$ and if $T \neq 1_V$, then T is called a ***transvection***.

The next theorem and its corollary show that the transvections just defined are precisely those linear transformations arising from matrix transvections.

Theorem 8.17. *Let $T \in \mathrm{GL}(V)$ fix a hyperplane H pointwise, and let $\mu = \mu(T)$.*

(i) *If T is a dilatation, then T has a matrix $D(\mu) = \mathrm{diag}\{1, \ldots, 1, \mu\}$ (relative to a suitable basis of V).*
(ii) *If T is a transvection, then T has matrix $B_{12}(1)$ (relative to a suitable basis of V). Moreover, T has no eigenvectors outside of H in this case.*

Proof. Every nonzero vector in H is an eigenvector of T (with eigenvalue 1);

are there any others? Choose $w \in V$ with $w \notin H$; since T fixes H pointwise,

$$Tw = \mu w + h, \qquad \text{where} \quad h \in H.$$

If $v \in V$ and $v \notin H$, the lemma gives

$$Tv = \mu v + h',$$

where $h' = (1 - \mu)h'' + \lambda h_0 \in H$. If v is an eigenvector of T, then $Tv = \beta v$ for some $\beta \in K$. But $Tv = \beta v$ if and only if $\beta = \mu$ and $\lambda h = (\mu - 1)h''$: sufficiency is obvious; conversely, if $\beta v = \mu v + h'$, then $(\beta - \mu)v = h' \in \langle v \rangle \cap H = 0$.

(i) If T is a dilatation, then $\mu - 1 \neq 0$ and $h'' = \lambda(\mu - 1)^{-1}h$. It follows that $v = w + (\mu - 1)^{-1}h$ is an eigenvector of T for the eigenvalue μ. If $\{v_1, \ldots, v_{m-1}\}$ is a basis of H, then adjoining v gives a basis of V, and the matrix of T relative to this basis is $D(\mu) = \text{diag}\{1, \ldots, 1, \mu\}$.

(ii) If T is a transvection, then $\mu = 1$. Choose $w \notin H$ so that $Tw = w + h$, where $h \in H$ and $h \neq 0$. If $v \notin H$ is an eigenvector of T, then $\alpha v = Tv = v + h$ for some $\alpha \in K$; hence, $(\alpha - 1)v \in \langle v \rangle \cap H = 0$, so that $\alpha = 1$ and $Tv = v$. It follows that $T = 1_V$, contradicting the proviso in the definition of transvection excluding the identity. Therefore, T has no eigenvectors outside of H. If $\{h, h_3, \ldots, h_m\}$ is a basis of H, then adjoining w as the first vector gives an ordered basis of V, and the matrix of T relative to this basis is $B_{12}(1)$. ■

Corollary 8.18. *All transvections in* GL(m, K) *are conjugate.*

Proof. Since transvections are, by definition, conjugates of elementary transvections, it suffices to prove that any two elementary transvections are conjugate to $B_{21}(1)$. Let V be an m-dimensional vector space over K with basis $\{v_1, \ldots, v_m\}$, and let T be the linear transformation with $Tv_1 = v_1 + v_2$ and $Tv_l = v_l$ for all $l \geq 2$. If $i \neq j$ and $\lambda \neq 0$, define a new ordered basis $\{u_1, \ldots, u_m\}$ of V as follows: put v_1 in position i, put $\lambda^{-1}v_2$ in position j, and fill the remaining $m - 2$ positions with $v_3, \ldots, v_m$ in this order (e.g., if $m = 5$, $i = 2$, and $j = 4$, then $\{u_1, \ldots, u_5\} = \{v_3, v_1, v_4, \lambda^{-1}v_2, v_5\}$). The matrix of T relative to this new ordered basis is easily seen to be $B_{ij}(\lambda)$. Therefore $B_{21}(1)$ and $B_{ij}(\lambda)$ are similar, for they represent the same linear transformation relative to different choices of ordered basis. ■

If $T \in \text{GL}(V)$ is a transvection fixing a hyperplane H and if $w \notin H$, then $Tw = w + h$ for some nonzero $h \in H$. If $v \in V$, then $v = \lambda w + h''$ for some $\lambda \in K$ and $h'' \in H$, and $(*)$ in the proof of Lemma 8.16 gives $Tv = v + \lambda h$ (because $1 - \mu = 0$). The function $\varphi: V \to K$, defined by $\varphi(v) = \varphi(\lambda w + h) = \lambda$ is a K-linear transformation (i.e., it is a ***linear functional***) with kernel H. For each transvection T, there is thus a linear functional φ and a vector $h \in \ker \varphi$ with

$$Tv = v + \varphi(v)h \qquad \text{for all} \quad v \in V.$$

Notation. Given a nonzero linear functional φ on V and a nonzero vector

$h \in \ker \varphi$, define $\{\varphi, h\}: V \to V$ by

$$\{\varphi, h\}: v \mapsto v + \varphi(v)h.$$

It is clear that $\{\varphi, h\}$ is a transvection; moreover, for every transvection T, there exist $\varphi \neq 0$ and $h \neq 0$ with $T = \{\varphi, h\}$.

Lemma 8.19. *Let V be a vector space over K.*

(i) *If φ and ψ are linear functionals on V, and if $h, l \in V$ satisfy $\varphi(h) = \psi(h) = \varphi(l)$, then*

$$\{\varphi, h\} \circ \{\varphi, l\} = \{\varphi, h + l\} \qquad \textit{and} \qquad \{\varphi, h\} \circ \{\psi, h\} = \{\varphi + \psi, h\}.$$

(ii) *For all $\alpha \in K^\times$,*

$$\{\alpha\varphi, h\} = \{\varphi, \alpha h\}.$$

(iii) *$\{\varphi, h\} = \{\psi, l\}$ if and only if there is a scalar $\alpha \in K^\times$ with*

$$\psi = \alpha\varphi \qquad \textit{and} \qquad h = \alpha l.$$

(iv) *If $S \in \mathrm{GL}(V)$, then*

$$S\{\varphi, h\}S^{-1} = \{\varphi S^{-1}, Sh\}.$$

Proof. All are routine. For example, let us prove half of (iii). If $\{\varphi, h\} = \{\psi, l\}$, then $\varphi(v)h = \psi(v)l$ for all $v \in V$. Since $\varphi \neq 0$, there is $v \in V$ with $\varphi(v) \neq 0$, so that $h = \varphi(v)^{-1}\psi(v)l$; if $\alpha = \varphi(v)^{-1}\psi(v)$, then $h = \alpha l$. To see that $\psi(u) = \alpha\varphi(u)$ for all $u \in V$, note that $\varphi(u) = 0$ if and only if $\psi(u) = 0$ (because both $h, l \neq 0$). If $\psi(u)$ and $\varphi(u)$ are nonzero, then $h = \varphi(u)^{-1}\psi(u)l$ implies $\varphi(u)^{-1}\psi(u) = \varphi(v)^{-1}\psi(v) = \alpha$, and so $\psi = \alpha\varphi$. ■

Theorem 8.20. *The commutator subgroup of* GL(V) *is* SL(V) *unless V is a two-dimensional vector space over $\mathbb{Z}_2$.*

Proof. Now det: GL $\to K^\times$ has kernel SL and GL/SL $\cong K^\times$; since $K^\times$ is abelian, (GL)$' \leq$ SL.

For the reverse inclusion, let ν: GL $\to$ GL/(GL)$'$ be the natural map. By Corollary 8.18, all transvections are conjugate in GL, and so $\nu(T) = \nu(T')$ for all transvections T and T'; let d denote their common value. Let $T = \{\varphi, h\}$ be a transvection. If we avoid the exceptional case in the statement, then H contains a nonzero vector l (not necessarily distinct from h) with $h + l \neq 0$. By the lemma, $\{\varphi, h\} \circ \{\varphi, l\} = \{\varphi, h + l\}$ (these are transvections because $l \neq 0$ and $h + l \neq 0$). Applying ν to this equation gives $d^2 = d$ in GL/(GL)$'$, whence $d = 1$. Thus, every transvection $T \in \ker \nu =$ (GL)$'$. But SL is generated by the transvections, by Theorem 8.8(ii), and so SL $\leq$ (GL)$'$. ■

If V is a two-dimensional vector space over $\mathbb{Z}_2$, then GL(V) is a genuine exception to the theorem. In this case,

$$\mathrm{GL}(V) = \mathrm{SL}(V) \cong \mathrm{SL}(2, 2) \cong \mathrm{PSL}(2, 2) \cong S_3,$$

and $(S_3)' = A_3$, a proper subgroup.

We have seen that any two transvections are conjugate in GL. It is easy to see that

$$\begin{bmatrix} 1 & 1 \\ 0 & 1 \end{bmatrix} \quad \text{and} \quad \begin{bmatrix} 1 & -1 \\ 0 & 1 \end{bmatrix}$$

are not conjugate in SL(2, 3); indeed, these transvections are not conjugate in SL(2, K) for any field K in which -1 is not a square. The assumption $m \geq 3$ in the next result is thus essential.

Theorem 8.21. *If $m \geq 3$, then all transvections are conjugate in* SL(V).

Proof. Let $\{\varphi, h\}$ and $\{\psi, l\}$ be transvections, and let $H = \ker \varphi$ and $L = \ker \psi$ be the hyperplanes fixed by each. Choose $v, u \in V$ with $\varphi(v) = 1 = \psi(u)$ (hence $v \notin H$ and $u \notin L$). There are bases $\{h, h_2, \ldots, h_{m-1}\}$ and $\{l, l_2, \ldots, l_{m-1}\}$ of H and L, respectively, and adjoining v and u gives bases $\{v, h, h_2, \ldots, h_{m-1}\}$ and $\{u, l, l_2, \ldots, l_{m-1}\}$ of V. If $S \in \mathrm{GL}(V)$ takes the first of these ordered bases to the second, then

$$(*) \qquad S(v) = u, \qquad S(H) = L, \qquad \text{and} \qquad S(h) = l.$$

Let $\det S = d$; we now show that we can force S to have determinant 1. Since $m \geq 3$, the first basis of V constructed above contains at least one other vector (say, h_{m-1}) besides v and h. Redefine S so that $S(h_{m-1}) = d^{-1}l_{m-1}$. Relative to the basis $\{v, h, h_2, \ldots, h_{m-1}\}$, the matrix of the new transformation differs from the matrix of the original one in that its last column is multiplied by d^{-1}. The new S thus has determinant 1 as well as the other properties $(*)$ of S.

Now $S\{\varphi, h\}S^{-1} = \{\varphi S^{-1}, Sh\} = \{\varphi S^{-1}, l\}$, by Lemma 8.19(iv). Since φS^{-1} and ψ agree on the basis $\{u, l, l_2, \ldots, l_{m-1}\}$ of V, they are equal. Therefore $\{\varphi, h\}$ and $\{\psi, l\}$ are conjugate in SL, as desired. ∎

Notation. If H is a hyperplane in a vector space V, then

$$\mathscr{T}(H) = \{\text{all transvections fixing } H\} \cup \{1_V\}.$$

Lemma 8.22. *Let H be a hyperplane in an m-dimensional vector space V over K.*

(i) *There is a linear functional φ with $H = \ker \varphi$ so that*

$$\mathscr{T}(H) = \{\{\varphi, h\} : h \in H\} \cup \{1_V\}.$$

(ii) *$\mathscr{T}(H)$ is an (abelian) subgroup of* SL(V), *and $\mathscr{T}(H) \cong H$.*

(iii) *The centralizer $C_{\mathrm{SL}}(\mathscr{T}(H)) = \mathrm{SZ}(V)\mathscr{T}(H)$.*

Proof. (i) Observe that linear functionals φ and ψ have the same kernel if and only if there is a nonzero $\alpha \in K$ with $\psi = \alpha\varphi$. Clearly $\psi = \alpha\varphi$ implies $\ker \psi = \ker \varphi$. Conversely, if H is their common kernel, choose $w \in V$ with $w \notin H$. Now $\psi(w) = \alpha\varphi(w)$ for some $\alpha \in K^\times$. If $v \in V$, then $v = \lambda w + h$, for some $\lambda \in K$ and $h \in H$, and $\psi(v) = \lambda\psi(w) = \lambda\alpha\varphi(w) = \alpha\varphi(\lambda w + h) = \alpha\varphi(v)$.

If $\{\varphi, h\}$, $\{\psi, l\} \in \mathscr{T}(H)$, then Lemma 8.19(ii) gives $\{\psi, l\} = \{\alpha\varphi, l\} = \{\varphi, \alpha l\}$. Since $\{\varphi, h\}^{-1} = \{\varphi, -h\}$, Lemma 8.19(i) gives $\{\varphi, h\} \circ \{\psi, l\}^{-1} = \{\varphi, h - \alpha l\} \in \mathscr{T}(H)$. Therefore, $\mathscr{T}(H) \leq \mathrm{SL}(V)$.

(ii) Let φ be a linear functional with $H = \ker \varphi$. By (i), each $T \in \mathscr{T}(H)$ has the form $T = \{\varphi, h\}$ for some $h \in H$, and this form is unique, by Lemma 8.19(iii). It is now easy to see that the function $\mathscr{T}(H) \to H$, given by $\{\varphi, h\} \mapsto h$, is an isomorphism.

(iii) Since $\mathscr{T}(H)$ is abelian, $\mathrm{SZ}(V)\mathscr{T}(H) \leq C_{\mathrm{SL}}(\mathscr{T}(H))$. For the reverse inclusion, assume that $S \in \mathrm{SL}(V)$ commutes with every $\{\varphi, h\}$: for all $h \in H$, $S\{\varphi, h\}S^{-1} = \{\varphi, h\}$. By Lemma 8.19(iv), $S\{\varphi, h\}S^{-1} = \{\varphi S^{-1}, Sh\}$, and so Lemma 8.19(iii) gives $\alpha \in K^\times$ with

$$(**) \qquad \varphi S^{-1} = \alpha\varphi \qquad \text{and} \qquad Sh = \alpha^{-1}h.$$

Hence αS fixes H pointwise, so that αS is either a transvection or a dilatation. If αS is a transvection, then $\alpha S \in \mathscr{T}(H)$, and so $S = \alpha^{-1}(\alpha S) \in \mathrm{SZ}(V)\mathscr{T}(H)$. If αS is a dilatation, then it has an eigenvector w outside of H, and $\alpha Sw = \mu w$, where $1 \neq \mu = \det \alpha S = \alpha^m$ (for $\det S = 1$); hence, $Sw = \alpha^{m-1}w$. But $\varphi S^{-1}w = \varphi(\alpha^{-m+1}w) = \alpha^{-m+1}\varphi(w)$, so that $(**)$ give $\varphi(w) = \alpha^m\varphi(w)$. Since $\varphi(w) \neq 0$ (because $w \notin H$), we reach the contradiction $\alpha^m = 1$. ∎

Theorem 8.23 (Jordan–Dickson). *If $m \geq 3$ and V is an m-dimensional vector space over a field K, then the groups* $\mathrm{PSL}(V)$ *are simple.*

Proof. We show that if N is a normal subgroup of $\mathrm{SL}(V)$ containing some A not in $\mathrm{SZ}(V)$, then $N = \mathrm{SL}(V)$; by Theorem 8.17, it suffices to show that N contains a transvection.

Since $\mathrm{SL}(V)$ is generated by transvections, there exists a transvection T which does not commute with A: the commutator $B = T^{-1}A^{-1}TA \neq 1$. Note that $N \lhd \mathrm{SL}$ gives $B \in N$. Thus

$$B = T^{-1}(A^{-1}TA) = T_1T_2,$$

where each T_i is a transvection. Now $T_i = \{\varphi_i, h_i\}$, where $h_i \in H_i = \ker \varphi_i$ for $i = 1, 2$; that is,

$$T_i(v) = v + \varphi_i(v)h_i \qquad \text{for all} \quad v \in V.$$

Let W be the subspace $\langle h_1, h_2 \rangle \leq V$, so that $\dim W \leq 2$. Since $\dim V \geq 3$, there is a hyperplane L of V containing W. We claim that $B(L) \leq L$. If $l \in L$, then

$$\begin{aligned} B(l) &= T_1T_2(l) = T_2(l) + \varphi_1(T_2(l))h_1 \\ &= l + \varphi_2(l)h_2 + \varphi_1(T_2(l))h_1 \in L + W \leq L. \end{aligned}$$

We now claim that $H_1 \cap H_2 \neq 0$. This is surely true if $H_1 = H_2$. If $H_1 \neq H_2$, then $H_1 + H_2 = V$ (hyperplanes are maximal subspaces) and $\dim(H_1 + H_2) = m$. Since

$$\dim H_1 + \dim H_2 = \dim(H_1 + H_2) + \dim(H_1 \cap H_2),$$

we have $\dim(H_1 \cap H_2) = m - 2 \geq 1$.

If $z \in H_1 \cap H_2$ with $z \neq 0$, then

$$B(z) = T_1 T_2(z) = z.$$

We may assume that B is not a transvection (or we are done); therefore, $B \notin \mathcal{T}(L)$, which is wholly comprised of transvections. If $B = \alpha S$, where $S \in \mathcal{T}(L)$, then z is an eigenvector of S ($z = Bz = \alpha Sz$, and so $Sz = \alpha^{-1}z$). As eigenvectors of transvections lie in the fixed hyperplane, $z \in L$ and so $\alpha = 1$, giving the contradiction $S = B$. Therefore, $B \notin \mathrm{SZ}(V)\mathcal{T}(L) = C_{\mathrm{SL}}(\mathcal{T}(L))$, so there exists $U \in \mathcal{T}(L)$ not commuting with B:

$$C = UBU^{-1}B^{-1} \neq 1;$$

of course, $C = (UBU^{-1})B^{-1} \in N$. If $l \in L$, then

$$C(l) = UBU^{-1}B^{-1}(l) = UB(B^{-1}(l)) = l,$$

because $B^{-1}(l) \in L$ and $U^{-1} \in \mathcal{T}(L)$ fixes L. Therefore, the transformation C fixes the hyperplane L, and so C is either a transvection or a dilatation. But C is not a dilatation because $\det C = 1$. Therefore C is a transvection in N, and the proof is complete. ■

We shall give different proofs of Theorems 8.13 and 8.22 in Chapter 9.

Observe that $|\mathrm{PSL}(3, 4)| = 20{,}160 = \frac{1}{2}8!$, so that PSL(3, 4) and A_8 are simple groups of the same order.

Theorem 8.24 (Schottenfels, 1900). PSL(3, 4) *and* A_8 *are nonisomorphic simple groups of the same order.*

Proof. The permutations $(1\ 2)(3\ 4)$ and $(1\ 2)(3\ 4)(5\ 6)(7\ 8)$ are even (hence lie in A_8), are involutions, and are not conjugate in A_8 (indeed, they are not even conjugate in S_8 for they have different cycle structures). We prove the theorem by showing that all involutions in PSL(3, 4) are conjugate.

A nonscalar matrix $A \in \mathrm{SL}(3, 4)$ corresponds to an involution in PSL(3, 4) if and only if A^2 is scalar, and A^2 is scalar if and only if $(PAP^{-1})^2$ is scalar for every nonsingular matrix P. Thus A can be replaced by anything similar to it, and so we may assume that A is a rational canonical form. If A is a direct sum of 1×1 companion matrices, then $A = \mathrm{diag}\{\alpha, \beta, \gamma\}$. But A^2 scalar implies $\alpha^2 = \beta^2 = \gamma^2$; as GF(4) has characteristic 2, this gives $\alpha = \beta = \gamma$ and A is scalar, a contradiction. If A is a 3×3 companion matrix,

$$A = \begin{bmatrix} \alpha & 0 & 0 \\ 1 & \alpha & 0 \\ 0 & 1 & \alpha \end{bmatrix},$$

then A^2 has 1 as the entry in position (3, 1), and so A^2 is not scalar. We conclude that A is a direct sum of a 1×1 companion matrix and a 2×2 companion matrix:

$$A = \begin{bmatrix} \alpha & 0 & 0 \\ 0 & 0 & \beta \\ 0 & 1 & \gamma \end{bmatrix}.$$

Now $\det A = 1 = \alpha\beta$ (remember that $-1 = 1$ here), so that $\beta = \alpha^{-1}$, and A^2 scalar forces $\gamma = 0$. Thus,

$$A = \begin{bmatrix} \alpha & 0 & 0 \\ 0 & 0 & \alpha^{-1} \\ 0 & 1 & 0 \end{bmatrix}.$$

There are only three such matrices; if π is a primitive element of GF(4), they are

$$A = \begin{bmatrix} 1 & 0 & 0 \\ 0 & 0 & 1 \\ 0 & 1 & 0 \end{bmatrix}; \quad B = \begin{bmatrix} \pi & 0 & 0 \\ 0 & 0 & \pi^2 \\ 0 & 1 & 0 \end{bmatrix}; \quad C = \begin{bmatrix} \pi^2 & 0 & 0 \\ 0 & 0 & \pi \\ 0 & 1 & 0 \end{bmatrix}.$$

Note that $A^2 = E$, $B^2 = \pi^2 E$, and $C^2 = \pi E$. It follows that if $M \in \mathrm{SL}(2, 3)$ and $M^2 = E$ (a stronger condition, of course, than M^2 being scalar), then M is similar to A; that is, $M = PAP^{-1}$ for some $P \in \mathrm{GL}(3, 4)$. In particular, $\pi^2 B$ and πC are involutions, so there are $P, Q \in \mathrm{GL}(3, 4)$ with

$$PAP^{-1} = \pi^2 B \quad \text{and} \quad QAQ^{-1} = \pi C.$$

Since $[\mathrm{GL}(3, 4): \mathrm{SL}(3, 4)] = 3$ (for $\mathrm{GL}/\mathrm{SL} \cong \mathrm{GF}(4)^\times$) and since the matrix $\mathrm{diag}\{\pi, 1, 1\}$ of determinant $\pi \neq 1$ commutes with A, Exercise 3.7 allows us to assume that P and Q lie in SL(3, 4). It follows that A, B, and C become conjugate in PSL(3, 4), as desired. ■

Theorem 8.24 can also be proved by showing that PSL(3, 4) contains no element of order 15, while A_8 does contain such an element, namely, (1 2 3)(4 5 6 7 8).

One can display infinitely many pairs of nonisomorphic simple groups having the same finite order, but the classification of the finite simple groups shows that there do not exist three nonisomorphic simple groups of the same order.

Classical Groups

At the end of the nineteenth century, the investigation of solutions of systems of differential equations led to complex Lie groups which are intimately related to simple Lie algebras of matrices over $\mathbb{C}$. There are analogues of these

Lie groups and Lie algebras which are defined over more general fields, and we now discuss them (not proving all results).

In what follows, all vector spaces are assumed to be finite-dimensional.

Definition. If V is a vector space over a field K, a function $f: V \times V \to K$ is called a ***bilinear form*** if, for each $v \in V$, the functions $f(v, \ )$ and $f(\ , v)$ are linear functionals on V.

A bilinear form f is called ***symmetric*** if $f(v, u) = f(u, v)$ for all $u, v \in V$, and it is called ***alternating*** if $f(v, v) = 0$ for all $v \in V$.

If f is alternating and $u, v \in V$, then $0 = f(u + v, u + v) = f(u, u) + f(u, v) + f(v, u) + f(v, v) = f(u, v) + f(v, u)$, so that $f(v, u) = -f(u, v)$. Conversely, if f is a bilinear form for which $f(v, u) = -f(u, v)$, then $2f(v, v) = 0$ for all $v \in V$. If K has characteristic $\neq 2$, then f is alternating; if K has characteristic 2, then f is symmetric.

There is another interesting type of form, not quite bilinear (Bourbaki calls it "sesquilinear").

Definition. If K is a field having an automorphism σ of order 2 (denoted by $\sigma: \alpha \mapsto \alpha^\sigma$), then a ***hermitian form*** on a vector space V over K is a function $h: V \times V \to K$ such that, for all $u, v \in V$:

(i) $h(u, v) = h(v, u)^\sigma$;
(ii) $h(\alpha u, v) = \alpha h(u, v)$ for all $\alpha \in K$; and
(iii) $h(u + v, w) = h(u, w) + h(v, w)$.

Note that if h is hermitian, then $h(u, \beta v) = h(\beta v, u)^\sigma = (\beta h(v, u))^\sigma = \beta^\sigma h(v, u)^\sigma = \beta^\sigma h(u, v)$. Moreover, h is additive in the second variable, for $h(u, v + w) = h(v + w, u)^\sigma = (h(v, u) + h(w, u))^\sigma = h(v, u)^\sigma + h(w, u)^\sigma = h(u, v) + h(u, w)$.

Complex conjugation $z \mapsto \bar{z}$ is an automorphism of $\mathbb{C}$ of order 2. If V is a complex vector space with basis $\{x_1, \ldots, x_n\}$, if $x = \sum \alpha_j x_j$, and if $y = \sum \beta_j x_j$ (where $\alpha_j, \beta_j \in \mathbb{C}$), then

$$h(x, y) = \sum \alpha_j \overline{\beta}_j$$

is a hermitian form.

If K is a finite field, then it has an automorphism σ of order 2 if and only if $K \cong \mathrm{GF}(q^2)$ for some prime power q, in which case $\alpha^\sigma = \alpha^q$ (this can be shown using Theorem 8.4).

Definition. If $f: V \times V \to K$ is either symmetric, alternating, or hermitian, then we call the ordered pair (V, f) an ***inner product space***.

Definition. Let (V, f) be an inner product space. If $\{v_1, \ldots, v_n\}$ is an ordered basis of V, then the ***inner product matrix*** of f relative to this basis is

$$A = [f(v_i, v_j)].$$

It is clear that f is completely determined by an inner product matrix, for if $u = \sum \alpha_i v_i$ and $w = \sum \beta_i v_i$, then

$$(u, w) = \sum_{i,j} \alpha_i \beta_j f(v_i, v_j).$$

What happens to the inner product matrix after changing basis?

Lemma 8.25. *Let (V, f) be an inner product space, let $\{v_1, \dots, v_n\}$ and $\{u_1, \dots, u_n\}$ be ordered bases of V, and let the corresponding inner product matrices be A and B.*

(i) *If f is bilinear, then A and B are* ***congruent****; that is, there is a nonsingular matrix P with*

$$B = P^t A P.$$

(ii) *If f is hermitian, then A and B are* ***σ-congruent****; that is, there is a nonsingular matrix $P = [p_{ij}]$ with*

$$B = P^t A P^\sigma,$$

where $P^\sigma = [(p_{ij})^\sigma]$.
In either case, B is nonsingular if and only if A is nonsingular.

Proof. (i) Write $u_j = \sum p_{ij} v_i$; the matrix $P = [p_{ij}]$, being a transition matrix between bases, is nonsingular. Now

$$f(u_i, u_j) = f\left(\sum_k p_{ki} v_k, \sum_l p_{lj} v_l\right) = \sum_{k,l} p_{ki} f(v_k, v_l) p_{lj};$$

in matrix terms, this is the desired equation (the transpose is needed because the indices k, i in the first factor must be switched to make the equation correspond to matrix multiplication). The last statement follows from $\det B = \det(P^t A P) = \det(P)^2 \det(A)$.

(ii) $$f(u_i, u_j) = \sum_{k,l} p_{ki} f(v_k, v_l) p_{lj}^\sigma. \quad ∎$$

Definition. An inner product space (V, f) is ***nondegenerate*** (or *nonsingular*) if one (and hence any) of the inner product matrices of f is nonsingular.

Lemma 8.26. *An inner product space (V, f) over a field K is nondegenerate if and only if $f(u, v) = 0$ for all $v \in V$ implies $u = 0$.*

Proof. Let $\{v_1, \dots, v_n\}$ be a basis of V. If an inner product matrix A of f is singular, then there is a nonzero column vector Y with $AY = 0$; that is, if $Y = (\mu_1, \dots, \mu_n)$, where $\mu_i \in K$, then $u = \sum \mu_i v_i$ is a nonzero vector with

$$f(v, u) = X^t A Y = 0$$

for all $v = \sum \lambda_i v_i$ (where $X = (\lambda_1, \dots, \lambda_n)$).

Conversely, if u satisfies $f(u, v) = 0$ for all $v \in V$, then $f(u, v_i) = 0$ for all v_i (where $\{v_1, \ldots, v_n\}$ is a basis of V). If $u = \sum \mu_j v_j$, then $\sum_j \mu_j f(v_j, v_i) = 0$; if $Y = (\mu_1, \ldots, \mu_n)$ is the column vector of u, then $Y \neq 0$ and $AY = 0$. Hence A is singular. ■

Definition. An ***isometry*** of a nondegenerate space (V, f) is a linear transformation $T: V \to V$ such that

$$f(Tu, Tv) = f(u, v)$$

for all $u, v \in V$.

Lemma 8.27. *If (V, f) is a nondegenerate space, then every isometry is nonsingular, and so all the isometries form a subgroup* $\mathrm{Isom}(V, f) \leq \mathrm{GL}(V)$.

Proof. If T is an isometry and $Tu = 0$, then $f(u, v) = f(Tu, Tv) = f(0, Tv) = 0$ for all $v \in V$. Since f is nondegenerate, it follows that $u = 0$ and T is an injection; since V is finite-dimensional, T is nonsingular, ■

Lemma 8.28. *Let (V, f) be a nondegenerate space, let A be the inner product matrix of f relative to an ordered basis $\{v_1, \ldots, v_n\}$ of V, and let T be a linear transformation on V.*

(i) *If f is bilinear, then T is an isometry if and only if its matrix $M = [m_{ij}]$ relative to the ordered basis satisfies*

$$M^t A M = A;$$

in this case, $\det M = \pm 1$.

(ii) *If f is hermitian, then T is an isometry if and only if*

$$M^t A M^\sigma = A,$$

where $M^\sigma = [(m_{ij})^\sigma]$; in this case, $(\det M)(\det M^\sigma) = 1$.

Proof. (i)

$$f(Tv_i, Tv_j) = f\left(\sum_l m_{li} v_l, \sum_k m_{kj} v_k\right)$$

$$= \sum_{l,k} m_{li} f(v_l, v_k) m_{kj} = f(v_i, v_j).$$

After translating into matrix terms, this is the desired equation. It follows that $(\det M)^2(\det A) = \det A$. Moreover, nondegeneracy of (V, f) gives nonsingularity of A, and so $\det M = \pm 1$.

(ii) The obvious modification of the equations above leads to the desired matrix equation and its consequence for determinants. ■

The group $\mathrm{GL}(V)$ acts on $\mathscr{F}(V)$, the set of all functions $V \times V \to K$: if f is a function and $P \in \mathrm{GL}(V)$, define

$$f^P(u, v) = f(P^{-1}u, P^{-1}v);$$

it is easily checked that if $Q \in \mathrm{GL}(V)$, then $f^{PQ} = (f^P)^Q$ (this is the reason for the inverse). Notice that if f is either symmetric, alternating, or hermitian, then so is f^P.

Theorem 8.29. *Let V be a vector space over a field K, and let $\mathscr{F}(V)$ be the $\mathrm{GL}(V)$-set of all functions $V \times V \to K$. When f is either symmetric, alternating, or hermitian, then the stabilizer of f is* $\mathrm{Isom}(V, f)$; *moreover, if g is in the orbit of f, then* $\mathrm{Isom}(V, g)$ *is isomorphic to* $\mathrm{Isom}(V, f)$ (*indeed, they are conjugate subgroups of* $\mathrm{GL}(V)$).

Proof. The stabilizer $\mathrm{GL}(V)_f$ of f is $\{P \in \mathrm{GL}(V): f^P = f\}$, so that $P \in \mathrm{GL}(V)_f$ if and only if $f(u, v) = f(P^{-1}u, P^{-1}v)$ for all $u, v \in V$; that is, $P^{-1} \in \mathrm{Isom}(V, f)$ and hence $P \in \mathrm{Isom}(V, f)$. By Exercise 3.37, $\mathrm{Isom}(V, f^P) = P\,\mathrm{Isom}(V, f)P^{-1}$. ■

Definition. Two functions $f, g \in \mathscr{F}(V)$ are called ***equivalent*** if $g = f^P$ for some $P \in \mathrm{GL}(V)$.

If follows from the theorem that equivalent symmetric, alternating, or hermitian forms determine isomorphic groups of isometries.

Lemma 8.30. *Two bilinear forms $f, g \in \mathscr{F}(V)$ are equivalent if and only if they have inner product matrices A and B which are congruent. Two hermitian forms (relative to the same automorphism σ) are equivalent if and only if their inner product matrices are σ-congruent.*

Proof. By Lemma 8.25(i), we may assume that all inner product matrices are determined by the same basis of V.

Let X and Y be column vectors. If f and g are bilinear and $g = f^P$, then $g(X, Y) = X^t BY$ (see the proof of Lemma 8.26). By definition, $f^P(X, Y) = f(P^{-1}X, P^{-1}Y) = (P^{-1}X)^t A(P^{-1}Y) = X^t[(P^{-1})^t AP^{-1}]Y$. Since this equation holds for all X and Y, it follows that $B = (P^{-1})^t AP^{-1}$; hence, A and B are congruent.

Conversely, if $B = Q^t AQ$ for some nonsingular Q, then $X^t BY = X^t Q^t AQY = (QX)^t A(QY)$, so that $g(X, Y) = f(Q^{-1}X, Q^{-1}Y)$.

This argument, *mutatis mutandis*, works in the hermitian case as well. ■

Let (V, f) be a nondegenerate space. We are going to see that all alternating forms f are equivalent, and so there is, to isomorphism, just one isometry group $\mathrm{Isom}(V, f)$; it is called the ***symplectic group***. If V is an n-dimensional vector space over K, then $\mathrm{Isom}(V, f)$ is denoted by $\mathrm{Sp}(V)$ or by $\mathrm{Sp}(n, K)$; if $K = \mathrm{GF}(q)$, one writes $\mathrm{Sp}(n, q)$.

It is true that all hermitian forms are equivalent, and so there is, to isomorphism, just one isometry group $\mathrm{Isom}(V, f)$ in this case as well; it is called the ***unitary group***. The group $\mathrm{Isom}(V, f)$ is now denoted by $\mathrm{U}(V)$ or by $\mathrm{U}(n, K)$;

when $K = \mathrm{GF}(q^2)$, one writes $U(n, q^2)$ (recall that the only finite fields for which hermitian forms are defined are of the form $\mathrm{GF}(q^2)$).

It is not true that all symmetric forms over a finite field are equivalent, and it turns out that inequivalent forms give nonisomorphic groups. The groups $\mathrm{Isom}(V, f)$ are called ***orthogonal groups***; in odd dimensions over a finite field of odd characteristic, there is only one orthogonal group, but in even dimensions over finite fields of any characteristic, there are two orthogonal groups, denoted by $\mathrm{O}^+(V)$ and by $\mathrm{O}^-(V)$.

Definition. A group is a ***classical group*** if it is either general linear, symplectic, unitary, or orthogonal.

We remark that the term *classical group* is usually not so precisely defined; for most authors, it also encompasses important groups closely related to these as, say, $\mathrm{SL}(V)$ or $\mathrm{PSL}(V)$.

We now discuss symplectic groups; afterwards we will describe corresponding results for the unitary and orthogonal groups.

Lemma 8.31. *If (V, f) is a nondegenerate space, then for every linear functional $g \in V^*$ (the dual space of V), there exists a unique $x \in V$ with $g = f(x,\ )$.*

Proof. We first prove that if $\{v_1, \ldots, v_n\}$ is a basis of V, then $\{f(v_1,\ ), \ldots, f(v_n,\ )\}$ is a basis of V^*. Since $\dim V^* = n$ (by standard linear algebra), it suffices to prove that these n linear functionals are independent. Otherwise, there are scalars λ_i, not all 0, with $\sum \lambda_i f(v_i,\ ) = 0$; that is, $\sum \lambda_i f(v_i, x) = 0$ for all $x \in V$. If $z = \sum \lambda_i v_i$, then $f(z, x) = 0$ for all $x \in V$. Thus, $z = 0$, because f is nondegenerate, and this contradicts the independence of the v_i.

Since $g \in V^*$, there are scalars μ_i with $g = \sum \mu_i f(v_i,\ )$, and $g(v) = f(x, v)$ for all $v \in V$. To prove uniqueness of x, suppose that $f(x, v) = f(y, v)$ for all $v \in V$. Then $f(x - y, v) = 0$ for all $v \in V$, and so nondegeneracy gives $x - y = 0$. ■

Definition. Let (V, f) be an inner product space. If $x, y \in V$, then x and y are ***orthogonal*** if $f(x, y) = 0$. If Y is a nonempty subset of V, then the ***orthogonal complement*** of Y, denoted by $Y^\perp$, is defined by

$$Y^\perp = \{v \in V: f(v, y) = 0 \text{ for all } y \in Y\}.$$

It is easy to see that $Y^\perp$ is always a subspace of V. Using this notation, an inner product space (V, f) is nondegenerate if and only if $V^\perp = 0$.

Let (V, f) be a nondegenerate space. If $W \le V$ is a subspace, then it is possible that the restriction $f|(W \times W)$ is degenerate. For example, let $V = \langle x, y \rangle$ be a two-dimensional space and let f have inner product matrix A relative to this basis:

$$A = \begin{bmatrix} 0 & 1 \\ 1 & 0 \end{bmatrix};$$

thus, f is symmetric and nondegenerate. However, if $W = \langle x \rangle$, the restriction $f|(W \times W)$ is identically zero.

Lemma 8.32. *Let (V, f) be an inner product space, and let W be a subspace of V.*

(i) *If $f|(W \times W)$ is nondegenerate, then*
$$V = W \oplus W^{\perp}.$$

(ii) *If (V, f) is a nondegenerate space and $V = W \oplus W^{\perp}$, then $f|(W^{\perp} \times W^{\perp})$ is nondegenerate.*

Proof. (i) If $x \in W \cap W^{\perp}$, then $f(x, W) = 0$, and so nondegeneracy gives $x = 0$. If $v \in V$, then the restriction $g = f(v, \ )|W$ is a linear functional on W, and so there is $w_0 \in W$ with $g(w) = f(v, w) = f(w_0, w)$ for all $w \in W$. But $v = w_0 + (v - w_0)$, where $w_0 \in W$, and $v - w_0 \in W^{\perp}$.

(ii) If $\{v_1, \ldots, v_r\}$ is a basis of W and $\{v_{r+1}, \ldots, v_n\}$ is a basis of $W^{\perp}$, then the inner product matrix A of f relative to the basis $\{v_1, \ldots, v_n\}$ has the form
$$A = \begin{bmatrix} B & 0 \\ 0 & C \end{bmatrix},$$
so that $\det A = (\det B)(\det C)$. But A nonsingular implies C nonsingular; that is, the restriction of f is nondegenerate. ■

Assume that (V, f) is an inner product space with f alternating. If f is not identically zero, there are vectors x and y with $f(x, y) = \alpha \neq 0$. Replacing x by $\alpha^{-1}x$ if necessary, we may assume that $f(x, y) = 1$. If $\dim V = 2$, then its inner product matrix is thus
$$A = \begin{bmatrix} 0 & 1 \\ -1 & 0 \end{bmatrix}.$$

Definition. A ***hyperbolic plane*** is a two-dimensional nondegenerate space (V, f) with f alternating.

We have just seen that every two-dimensional inner product space (V, f) with f alternating and not identically zero is a hyperbolic plane.

Theorem 8.33. *If (V, f) is a nondegenerate space with f alternating, then V is even-dimensional. Indeed,*
$$V = W_1 \oplus \cdots \oplus W_l,$$
where each W_i is a hyperbolic plane and the summands are ***pairwise orthogonal****; that is, if $i \neq j$, then $f(W_i, W_j) = 0$.*

Proof. We proceed by induction on $\dim V \geq 0$, the base step being trivial. If dim $V > 0$, then our discussion above shows that there exist $x_1, y_1 \in V$ with $f(x_1, y_1) = 1$. If $W = \langle x_1, y_1 \rangle$, then W is a hyperbolic plane and $f|(W \times W)$ is nondegenerate. By Lemma 8.32(i), $V = W \oplus W^\perp$; by Lemma 8.32(ii), $f|(W^\perp \times W^\perp)$ is nondegenerate, and so the inductive hypothesis shows that $W^\perp$ is an orthogonal direct sum of hyperbolic planes. ■

Definition. If (V, f) is a nondegenerate space with f alternating, then a ***symplectic basis*** of V is an ordered basis

$$\{x_1, y_1, x_2, y_2, \dots, x_l, y_l\}$$

with $f(x_i, x_j) = 0 = f(y_i, y_j)$ for all i, j, and $f(x_i, y_j) = \delta_{ij} = -f(y_j, x_i)$, where $\delta_{ij} = 1$ if $i = j$ and $\delta_{ij} = 0$ if $i \neq j$.

Thus, all inner products are 0 except $f(x_i, y_i) = 1 = -f(y_i, x_i)$ for all i.

Theorem 8.34. *Let (V, f) be a nondegenerate space with f alternating.*

(i) *V has a symplectic basis $\{x_1, y_1, x_2, y_2, \dots, x_l, y_l\}$.*
(ii) *The inner product matrix A of f relative to this ordered basis is the matrix J which is the direct sum of 2×2 blocks*

$$\begin{bmatrix} 0 & 1 \\ -1 & 0 \end{bmatrix}.$$

(iii) *If $u = \sum (\alpha_i x_i + \beta_i y_i)$ and $v = \sum (\gamma_i x_i + \delta_i y_i)$, then*

$$f(u, v) = \sum (\alpha_i \delta_i - \beta_i \gamma_i).$$

(iv) *All nondegenerate alternating forms on V are equivalent, and so the symplectic groups* Isom(V, f), *to isomorphism, do not depend on f.*

Proof. (i) If V is the orthogonal direct sum of hyperbolic planes W_i, then the union of the bases of the W_i is a symplectic basis of V.

(ii) and (iii) are now routine calculations.

(iv) If g is a nondegenerate alternating form, then there is a symplectic basis of V relative to g, so that any inner product matrix of g is also congruent to J, and hence to any inner product matrix of f. Therefore, the isometry group Isom(V, f) does not depend on f, by Theorem 8.29. ■

Notice that alternating bilinear forms are thus sums of 2×2 determinants. Note also that if one reorders a symplectic basis so that all the x_i precede all the y_i, then the matrix J is congruent to

$$\begin{bmatrix} 0 & E \\ -E & 0 \end{bmatrix},$$

where E is the $l \times l$ identity matrix.

Definition. If (V, f) is a nondegenerate space, then the ***adjoint*** of T, denoted by T^*, is a linear transformation on V for which

$$f(Tx, y) = f(x, T^*y) \qquad \text{for all} \quad x, y \in V.$$

Lemma 8.35. *Let (V, f) be a nondegenerate space, and let T be a linear transformation on V having an adjoint T^*. Then T is an isometry if and only if $T^*T = 1_V$.*

Proof. If T has an adjoint with $T^*T = 1_V$, then, for all $x, y \in V$, $f(Tx, Ty) = f(x, T^*Ty) = f(x, y)$, so that T is an isometry.

Conversely, for all $x, y \in V$,

$$\begin{aligned} f(x, T^*Ty - y) &= f(x, T^*Ty) - f(x, y) \\ &= f(Tx, Ty) - f(x, y) = 0, \end{aligned}$$

because T is an isometry. Since f is nondegenerate, $T^*Ty = y$ for all $y \in V$, and so $T^*T = 1_V$. ■

One can prove uniqueness of adjoints in general, but we may observe here that uniqueness holds for adjoints of isometries T because $T^* = T^{-1}$. It follows that $TT^* = 1_V$.

How can one recognize a symplectic matrix?

Theorem 8.36. *Let $T \in \mathrm{GL}(V)$ and let $\{x_1, y_1, \ldots, x_l, y_l\}$ be a symplectic basis of V. If the matrix Q of T relative to this basis is decomposed into $l \times l$ blocks*

$$Q = \begin{bmatrix} A & \Gamma \\ B & \Delta \end{bmatrix},$$

then T^* exists and has matrix

$$Q^* = \begin{bmatrix} \Delta^t & -\Gamma^t \\ -B^t & A^t \end{bmatrix};$$

moreover, $Q \in \mathrm{Sp}(2l, K)$ if and only if $Q^*Q = E$.

Proof. Assume that T^* *exists. If* $Tx_i = \sum_\nu (\alpha_{\nu i} x_\nu + \beta_{\nu i} y_\nu)$, then

$$\begin{aligned} f(x_i, T^*x_j) &= f(Tx_i, x_j) \\ &= \sum_\nu \alpha_{\nu i} f(x_\nu, x_j) + \sum_\nu \beta_{\nu i} f(y_\nu, x_j) = -\beta_{ji}. \end{aligned}$$

On the other hand, if $T^*x_j = \sum_r (\lambda_{rj} x_r + \mu_{rj} y_r)$, then

$$f(x_i, T^*x_j) = f\left(x_i, \sum_r (\lambda_{rj} x_r + \mu_{rj} y_r)\right) = \mu_{ij}.$$

It follows that if T^* has matrix

$$\begin{bmatrix} \Lambda & X \\ M & Y \end{bmatrix},$$

then $M = -B^t$. Similar calculations give the other three blocks. For existence of T^*, it is routine to check that the matrix construction in the statement defines a linear transformation that behaves as an adjoint must. The last statement follows from Lemma 8.35. ■

By Lemma 8.28, symplectic matrices have determinant ± 1, but it can be shown, in fact, that every symplectic matrix has determinant 1; that is, $\mathrm{Sp}(V) \leq \mathrm{SL}(V)$.

EXERCISES

8.13. Show that $\mathrm{Sp}(2, K) = \mathrm{SL}(2, K)$ for all fields K.

8.14. Let (V, f) be a nondegenerate space with f alternating.
  (i) Show that $T \in \mathrm{GL}(V)$ is symplectic if and only if T carries symplectic bases to symplectic bases.
  (ii) If V is a vector space over a finite field K, show that $|\mathrm{Sp}(V)|$ is the number of ordered symplectic bases.

The isometry group $\mathrm{Isom}(V, f)$ is called a ***unitary group*** when f is nondegenerate hermitian, and it is denoted by $U(V)$, $U(n, K)$, or $U(n, q^2)$ (when K is finite, a hermitian form requires $|K|$ to be a square).

We now state without proofs the analogues of Theorem 8.34 and 8.35 for unitary groups.

Definition. If (V, f) is an inner product space, then an ***orthonormal basis*** is an ordered basis $\{v_1, \ldots, v_n\}$ with $f(v_i, v_j) = \delta_{ij}$.

Theorem. *Let (V, f) be a nondegenerate space with f hermitian.*

(i) *V has an orthonormal basis $\{x_1, \ldots, x_n\}$.*
(ii) *The inner product matrix of f relative to this basis is the identity matrix.*
(iii) *If $u = \sum \lambda_i x_i$ and $v = \sum \mu_i x_i$, then*

$$f(u, v) = \sum \lambda_i \mu_i^\sigma.$$

(iv) *All nondegenerate alternating forms on V are equivalent, and so unitary groups* $\mathrm{Isom}(V, f)$ *do not depend on f.*
(v) *If $M \in \mathrm{GL}(V)$, then $M^* = (M^t)^\sigma$, and so $M \in U(V)$ if and only if $M(M^t)^\sigma = E$.*
(vi) *A linear transformation $T \in \mathrm{GL}(V)$ is unitary if and only if it takes orthonormal bases into orthonormal bases.*

The isometry groups $\mathrm{Isom}(V, f)$ are called *orthogonal groups* when f is nondegenerate symmetric; we restrict the discussion to finite fields of scalars

K. If $K = \mathrm{GF}(q)$ has odd characteristic and $\dim V = 2l + 1$ is odd, there is a basis of V relative to which f has inner product matrix

$$J = \begin{bmatrix} 1 & 0 & 0 \\ 0 & 0 & E \\ 0 & E & 0 \end{bmatrix},$$

where E is the $l \times l$ identity matrix. All forms are thus equivalent (for all inner product matrices are congruent to J), and so there is, to isomorphism, just one group $\mathrm{Isom}(V, f)$ in this case. It is denoted by $\mathrm{O}(V)$, $\mathrm{O}(2l + 1, K)$, or by $\mathrm{O}(2l + 1, q)$.

If $\dim V = 2l$ is even and $K = \mathrm{GF}(q)$ has odd characteristic, then there are exactly two inequivalent forms: every inner product matrix is congruent to

$$J^\varepsilon = \begin{bmatrix} 0 & E & 0 & 0 \\ E & 0 & 0 & 0 \\ 0 & 0 & 1 & 0 \\ 0 & 0 & 0 & \varepsilon \end{bmatrix},$$

where E is the $(l - 1) \times (l - 1)$ identity matrix and $\varepsilon = \pm 1$. The corresponding isometry groups are not isomorphic; they are denoted by $\mathrm{O}^+(2l, K)$ and by $\mathrm{O}^-(2l, K)$ (or by $\mathrm{O}^+(2l, q)$ and by $\mathrm{O}^-(2l, q)$ when $K = \mathrm{GF}(q)$).

Before describing the orthogonal groups in characteristic 2, let us recall the definition of the real orthogonal group $\mathrm{O}(n, \mathbb{R})$ given in Chapter 3 in terms of distance: if $T \in \mathrm{O}(n, \mathbb{R})$, then $\|Tv\| = \|v\|$ for all $v \in \mathbb{R}^n$. In Theorem 3.29, however, it was shown that T is orthogonal in this sense if and only if $(u, v) = (Tu, Tv)$ for all $u, v \in \mathbb{R}^n$ (where (u, v) is the usual dot product of vectors in $\mathbb{R}^n$). Given any symmetric bilinear form f, define the analogue of $\|v\|^2 = (v, v)$ to be $Q(v) = f(v, v)$. If $u, v \in V$, then

$$f(u + v, u + v) = f(u, u) + 2f(u, v) + f(v, v);$$

that is, $Q(u + v) - Q(u) - Q(v) = 2f(u, v)$.

Definition. A ***quadratic form*** on a vector space V over a field K is a function $Q\colon V \to K$ such that:

(i) $Q(\alpha u) = \alpha^2 Q(u)$ for all $u \in V$ and $\alpha \in K$; and
(ii) $Q(u + v) - Q(u) - Q(v) = g(u, v)$, where g is a bilinear form on V (one calls g a bilinear form ***associated*** to Q).

A quadratic form Q is ***nondegenerate*** if (V, f) is nondegenerate, where f is an associated bilinear form.

If K has characteristic $\neq 2$, one can recapture f from Q: $f(u, v) = \frac{1}{2}(Q(u + v) - Q(u) - Q(v))$, and there is just one bilinear form associated to Q. The proof of Theorem 3.28 shows that a linear transformation T on V lies in $\mathrm{O}(V)$ (i.e., $f(Tu, Tv) = f(u, v)$ for all $u, v \in V$) if and only if $Q(Tu) = Q(u)$ for all $u \in V$. Thus, orthogonal groups in odd characteristic could have been defined in terms of quadratic forms.

Assume now that $K = \mathrm{GF}(q)$ does have characteristic 2. Bilinear forms g associated to a given quadratic form Q are no longer uniquely determined by Q; however, g satisfies two extra conditions: first, $g(u, u) = 0$ (because $u + u = 0$); second, g is symmetric. Since $-\alpha = \alpha$ for all $\alpha \in K$, it follows that g is alternating. Theorem 8.33 shows that if (V, g) is to be nondegenerate, then V must be even-dimensional, say, $\dim V = 2l$. If $K = \mathrm{GF}(q)$ has characteristic 2 and Q is a nondegenerate quadratic form on a $2l$-dimensional vector space V over K, then the orthogonal group $\mathrm{O}(V, Q)$ is defined as

$$\mathrm{O}(V, Q) = \{T \in \mathrm{GL}(V)\colon Q(Tu) = Q(u) \text{ for all } u \in V\}.$$

Given a nondegenerate quadratic form Q, there is always a symplectic basis $\{x_1, y_1, \ldots, x_l, y_l\}$ such that $Q(x_i) = 0 = Q(y_i)$ for all $i < l$; moreover, either $(+)$: $Q(x_l) = 0 = Q(y_l)$ or $(-)$: $Q(y_l) = 1$ and $Q(x_l) = \gamma$, where $t^2 + t + \gamma$ is irreducible in $K[t]$. One proves that there are only two nonisomorphic groups for a given V arising from different quadratic forms Q (one from each possibility just described); they are denoted by $\mathrm{O}^+(2l, K)$ and by $\mathrm{O}^-(2l, K)$; as usual, one may replace K by q in the notation when $K = \mathrm{GF}(q)$.

Each type of classical group gives rise to a family of simple groups; we describe the finite such.

The center of $\mathrm{Sp}(2l, q)$ consists of $\pm E$. Define

$$\mathrm{PSp}(2l, q) = \mathrm{Sp}(2l, q)/\{\pm E\}.$$

These groups are simple unless $(2l, q) = (2, 2), (2, 3)$, or $(4, 2)$. Moreover,

$$|\mathrm{PSp}(2l, q)| = d^{-1}q^{l^2}(q^2 - 1)(q^4 - 1)\ldots(q^{2l} - 1),$$

where $d = (2, q - 1)$.

The center of $\mathrm{U}(n, q^2)$ consists of all scalar transformations λE, where $\lambda\lambda^\sigma = 1$. Let $\mathrm{SU}(n, q^2) \leq \mathrm{U}(n, q^2)$ consist of all unitary transformations of determinant 1, and define

$$\mathrm{PSU}(n, q^2) = \mathrm{SU}(n, q^2)/\text{center}.$$

These groups are simple unless $(n, q^2) = (2, 4), (2, 9)$, or $(3, 4)$. Moreover,

$$|\mathrm{PSU}(n, q^2)| = e^{-1}q^{n(n-1)/2}(q^2 - 1)(q^3 + 1)(q^4 - 1)\ldots(q^n - (-1)^n),$$

where $e = (n, q + 1)$.

Assume that q is a power of an odd prime. Let the commutator subgroup of $\mathrm{O}(n, q)$ be denoted by $\Omega(n, q)$ (it is usually a proper subgroup of $\mathrm{SO}(n, q)$, the subgroup consisting of all orthogonal transformations of determinant 1). The center of $\Omega(n, q)$ consists of diagonal matrices having diagonal entries ± 1, and one defines

$$\mathrm{P}\Omega(n, q) = \Omega(n, q)/\text{center}.$$

When $n \geq 5$ is odd, then these groups are simple, and

$$|\mathrm{P}\Omega(2l + 1, q)| = d^{-1}q^{l^2}(q^2 - 1)(q^4 - 1)\ldots(q^{2l} - 1), \qquad q \text{ odd},$$

where $d = (2, q - 1)$. Notice that $|P\Omega(2l + 1, q)| = |PSp(2l, q^2)|$; these two groups are not isomorphic if $l > 2$ (Theorem 8.24, proved in 1900, exhibited the first pair of nonisomorphic simple groups of the same order; here is an infinite family of such examples).

When $\dim V = 2l$ is even, there are two orthogonal groups. If we denote the commutator subgroup by Ω^ε, where $\varepsilon = \pm 1$, and the quotient Ω^ε/center by $P\Omega^\varepsilon$, then the orders are:

$$|P\Omega^\varepsilon(2l, q)| = g^{-1}q^{l(l-1)}(q^2 - 1)(q^4 - 1)\dots(q^{2l-2} - 1)(q^l - \varepsilon),$$

where $g = (4, q^l - \varepsilon)$.

When q is a power of 2, then the groups $P\Omega^\varepsilon(2l, q)$ are simple for $l \geq 3$, and their orders are given by the same formulas as when q is odd.

For proofs of these results, the reader is referred to Artin (1957), Carter (1972), and Dieudonné (1958).

Every complex Lie group G determines a finite-dimensional Lie algebra $L(G)$ over $\mathbb{C}$, and G simple implies $L(G)$ simple. These simple Lie algebras were classified by E. Cartan and W. Killing about 100 years ago; there are four infinite classes of them and five "sporadic" such. In 1955, Chevalley showed, for every finite field $GF(q)$, how to construct analogues of these simple Lie algebras over $\mathbb{C}$; he also showed how to construct simple finite groups from them. These families of simple groups (four are doubly indexed by the dimension and the field $GF(q)$; five of them, arising from the sporadic Lie algebras, are singly indexed by q) are now called ***Chevalley groups***. There are thus nine such families, which include all those arising from the classical groups (special linear, symplectic orthogonal) except those arising from unitary groups (however, the finite simple groups arising from the unitary groups were known to Dickson in 1900). Simple Lie algebras over $\mathbb{C}$ are classified by certain graphs, called Dynkin diagrams. In 1959, Steinberg showed that automorphisms of these graphs can be used to construct new finite simple groups. There are four infinite classes of these simple groups, called ***Steinberg groups***, two of which are the families $P\Omega^\varepsilon(m, q)$ for $\varepsilon = \pm 1$ arising from the unitary groups. In 1960, Suzuki discovered a new class $Sz(q)$ of simple groups, where q is a power of 2 (these are the only simple groups whose orders are not divisible by 3); in 1961, Ree discovered two more infinite classes whose construction is related to that of the Suzuki groups; these are the ***Suzuki groups*** and the ***Ree groups***. (The interested reader should consult the books of Carter and of Gorenstein for more details.) Collectively, these 16 classes of simple groups are called the groups of ***Lie type***.

The ***classification theorem of finite simple groups*** says that there are exactly 18 infinite classes of them: the cyclic groups of prime order, the alternating groups, and the groups of Lie type; moreover, there are exactly 26 "sporadic" simple groups, 5 of which are the Mathieu groups to be discussed in the next chapter. This theorem is one of the highest achievements of mathematics; it is the culmination of the work of about 100 mathematicians between 1955 and 1985, and it consists of thousands of journal pages.

CHAPTER 9

Permutations and the Mathieu Groups

The Mathieu groups are five remarkable simple groups discovered by E. Mathieu in 1861 and 1873; they belong to no infinite family of simple groups, as do all the simple groups we have so far exhibited, and they are the first examples of what are nowadays called *sporadic* simple groups. This chapter is devoted to proving their existence and displaying some of their interesting properties.

Multiple Transitivity

Even though permutation groups, orbits, and stabilizers were discussed in Chapter 3, we repeat the basic definitions here for the reader's convenience.

Definition. If X is a set and G is a group, then X is a ***G-set*** if there is a function $\alpha: G \times X \to X$ (called an ***action***), denoted by $\alpha: (g, x) \mapsto gx$, such that:

(i) $1x = x$ for all $x \in X$; and
(ii) $g(hx) = (gh)x$ for all $g, h \in G$ and $x \in X$.

One also says that G ***acts*** on X. If $|X| = n$, then n is called the ***degree*** of the G-set X.

We assume throughout this chapter that all groups G and all G-sets X are finite; moreover, all vector spaces considered are assumed to be finite-dimensional.

It is clear that if X is a G-set and $H \leq G$, then X is also an H-set (just restrict the action $G \times X \to X$ to $H \times X$). Let us also recall that G-sets are just another way of viewing permutation representations.

Theorem 3.19. *If X is a G-set with action α, then there is a homomorphism $\tilde{\alpha}: G \to S_X$ given by $\tilde{\alpha}(g): x \mapsto gx = \alpha(g, x)$. Conversely, every homomorphism $\varphi: G \to S_X$ defines an action, namely, $gx = \varphi(g)(x)$.*

Definition. A G-set X with action α is ***faithful*** if $\tilde{\alpha}: G \to S_X$ is injective.

Thus, a G-set X is faithful if and only if $gx = x$ for all $x \in X$ implies $g = 1$.

If X is a G-set with action α, the subgroup $\operatorname{im} \tilde{\alpha} \leq S_X$ is a permutation group; hence, if X is a faithful G-set, then G can be identified with $\operatorname{im} \tilde{\alpha}$, and we may view G itself as a permutation group. Cayley's theorem says that every group G of order n has a faithful representation $\varphi: G \to S_G$; in this case, G itself is a faithful G-set of degree n. Since $|S_n| = n!$ is so much larger than n, however, faithful G-sets of smaller degree are more valuable.

EXAMPLE 9.1. This is a generic example. Assume that a set X has some "structure" and that $\operatorname{Aut}(X)$ is the group of all permutations of X which preserve the structure. Then X is a faithful $\operatorname{Aut}(X)$-set; indeed, X is a faithful G-set for every $G \leq \operatorname{Aut}(X)$.

If X is a G-set and $gx = x$, where $g \in G$ and $x \in X$, then one says that g ***fixes*** x. A G-set X in which each $g \in G$ fixes every $x \in X$ is called a ***trivial*** G-set.

EXAMPLE 9.2. If X is a G-set with action α, let $N = \{g \in G: gx = x \text{ for all } x \in X\}$. Then $N = \ker \tilde{\alpha}$, so that $N \lhd G$, and it is easy to see that X is a faithful (G/N)-set (where $(Ng)x = gx$).

Recall that a G-set X is ***transitive*** if, for every $x, y \in X$, there exists $g \in G$ with $y = gx$. If $x \in X$, we denote its ***G-orbit*** $\{gx: g \in G\}$ either by Gx or by $\mathcal{O}(x)$; thus, X is transitive if $X = Gx$ for some $x \in X$. The first observation when analyzing G-sets is that one may focus on transitive G-sets.

Theorem 9.1. *Every G-set X has a unique decomposition into transitive G-sets. More precisely, X is partitioned into its G-orbits, each of which is a transitive G-set. Conversely, if a G-set X is partitioned into transitive G-sets $\{X_i: i \in I\}$, then the X_i are the G-orbits of X.*

Proof. It is clear that the orbits partition X into transitive G-sets. Conversely, it suffices to prove that each X_i is an orbit. If $x_i \in X_i$, then the orbit $Gx_i \subset X_i$ because X_i is a G-set. For the reverse inclusion, if $y \in X_i$, then transitivity of X_i gives $y = gx_i$ for some $g \in G$. Hence, $y \in Gx_i$, and $X_i \subset Gx_i$. ■

Recall that the ***stabilizer of*** x, denoted by G_x, is defined by

$$G_x = \{g \in G: gx = x\}.$$

Theorem 9.2. *If X is a transitive G-set of degree n, and if $x \in X$, then*

$$|G| = n|G_x|.$$

If X is a faithful G-set, then $|G_x|$ is a divisor of $(n-1)!$.

Proof. By Theorem 3.20, $|Gx| = [G : G_x]$. Since X is transitive, $Gx = X$, and so $n = |G|/|G_x|$. If X is faithful, then $G_x \leq S_{X-\{x\}}$, and the latter group has order $(n-1)!$. ■

Here is a solution to Exercise 3.4. If a finite group G has only two conjugacy classes, then every two nonidentity elements are conjugate; that is, $G^\# = G - \{1\}$ is a transitive G-set (where G acts on $G^\#$ by conjugation). Theorem 9.2 says that $|G| - 1$ is a divisor of $|G|$; this can happen only when $|G| = 2$.

Theorem 9.3. *Let X be a transitive G-set, and let $x, y \in X$.*

(i) *If $tx = y$ for some $t \in G$, then $G_y = G_{tx} = tG_xt^{-1}$.*
(ii) *X has the same number of G_x-orbits as of G_y-orbits.*

Proof. (i) If g fixes x, then $tgt^{-1}y = tgx = tx = y$ and tgt^{-1} fixes y. Therefore, $tG_xt^{-1} \leq G_y$. Similarly, $t^{-1}G_yt \leq G_x$, and this gives the reverse inclusion.

(ii) Since X is transitive, there is $t \in G$ with $tx = y$. Denote the G_x-orbits by $\{G_xz_i : i \in I\}$, where $z_i \in X$. If $w_i = tz_i \in X$, then we shall show that the subsets G_yw_i are G_y-orbits of X. Clearly, these subsets are transitive G_y-sets. Further,

$$G_yw_i = tG_xt^{-1}w_i = tG_xz_i.$$

Since t is a permutation of X, it carries partitions into partitions. The result now follows from Theorem 9.1. ■

Definition. If X is a transitive G-set, then the ***rank*** of X is the number of G_x-orbits of X.

Theorem 9.3 shows that the rank of X does not depend on the choice of $x \in X$. Of course, $\{x\}$ is a G_x-orbit of X, and so rank X is really describing the behavior of G_x on $X - \{x\}$.

Theorem 9.4. *If X is a transitive G-set and $x \in X$, then* rank X *is the number of $(G_x$-$G_x)$-double cosets in G.*

Proof. Define $f: \{G_x\text{-orbits}\} \to \{(G_x\text{-}G_x)\text{-double cosets}\}$ by $f(G_xy) = G_xgG_x$, where $gx = y$. Now f is well defined, for if $hx = y$, then $gx = hx$, $g^{-1}h \in G_x$, $h = 1g(g^{-1}h) \in G_xgG_x$, and $G_xgG_x = G_xhG_x$. Let us show that f is injective. If $f(G_xy) = G_xgG_x = G_xhG_x = f(G_xz)$, where $gx = y$ and $hx = z$, then there are $a, b \in G_x$ with $g = ahb$. Now $y = gx = ahbx = ahx = az \in G_xz$; it follows that $G_x y = G_x z$. Finally, f is surjective: if $g \in G$, then $gx = y$ and $f(G_x y) = G_x g G_x$. ■

See Exercise 9.12 below for another characterization of rank X.

Observe that if X is transitive and $|X| \geq 2$, then rank $X \geq 2$ (otherwise $G = G_x G_x$, and every $g \in G$ fixes x, contradicting transitivity). Let us now consider the minimal case when rank $X = 2$; the G_x-orbits are $\{x\}$ and $X - \{x\}$, and so G_x acts transitively on $X - \{x\}$.

Definition. Let X be a G-set of degree n and let $k \leq n$ be a positive integer. Then X is ***k-transitive*** if, for every pair of k-tuples having distinct entries in X, say, $(x_1, \ldots, x_k)$ and $(y_1, \ldots, y_k)$, there is $g \in G$ with $gx_i = y_i$ for $i = 1, \ldots, k$.

Of course, 1-transitivity is ordinary transitivity. If $k > 1$, then every k-transitive G-set is $(k-1)$-transitive. A k-transitive G-set X is called ***doubly transitive*** (or ***multiply transitive***) if $k \geq 2$, ***triply transitive*** if $k \geq 3, \ldots,$ and so forth.

It is quite common to say that a *group* G is k-transitive if there exists a k-transitive G-set. This usage applies to other adjectives as well.

The easiest example of a multiply transitive G-set is provided by $X = \{1, \ldots, k\}$ and $G = S_k$; it is plain that X is a k-transitive S_k-set.

Lemma 9.5. *Let X be a G-set. If $k \geq 2$, then X is k-transitive if and only if, for each $x \in X$, the G_x-set $X - \{x\}$ is $(k-1)$-transitive.*

Proof. Assume that X is k-transitive, and let $(x_1, \ldots, x_{k-1})$ and $(y_1, \ldots, y_{k-1})$ be $(k-1)$-tuples of distinct elements of $X - \{x\}$. There is thus $g \in G$ with $gx = x$ (so that $g \in G_x$) and $gx_i = y_i$ for all $i \geq 1$.

Conversely, let $(x_1, \ldots, x_k)$ and $(y_1, \ldots, y_k)$ be k-tuples of distinct elements of X. By hypothesis, there is $g \in G_{x_k}$ with $g(x_1, \ldots, x_{k-1}, x_k) = (y_1, \ldots, y_{k-1}, x_k)$, and there is $h \in G_{y_1}$ with $h(y_1, y_2, \ldots, y_{k-1}, x_k) = (y_1, y_2, \ldots, y_{k-1}, y_k)$. Therefore, $hg \in G$ carries $(x_1, \ldots, x_k)$ to $(y_1, \ldots, y_k)$. ■

Theorem 9.6. *Every doubly transitive G-set has* rank 2, *and if $x \in X$ and $g \notin G_x$, then $G = G_x \cup G_x g G_x$.*

Proof. Since G acts k-transitively on X for $k \geq 2$, we have G_x acting $(k-1)$-transitively on $X - \{x\}$; hence G_x acts transitively on $X - \{x\}$. Therefore, as a G_x-set, X has two orbits: $\{x\}$ and $X - \{x\}$; thus rank $X = 2$. By Theorem 9.4, there are exactly two $(G_x\text{-}G_x)$-double cosets in G. ■

Definition. If X is a G-set and $x_1, \ldots, x_t \in X$, then the ***stabilizer*** is the subgroup

$$G_{x_1, \ldots, x_t} = \{g \in G \colon gx_i = x_i \text{ for } i = 1, \ldots, t\}.$$

It is easy to see that $G_{x_1,\ldots,x_t} = \bigcap_i G_{x_i}$. In particular, if $x, y \in X$, then X is also a G_x-set and a G_y-set, and

$$(G_x)_y = G_{x,y} = (G_y)_x = G_x \cap G_y.$$

The proof of Theorem 9.3(i) generalizes: if X is a k-transitive G-set, then stabilizers of k distinct points are conjugate; that is, if $(x_1, \ldots, x_k)$ and $(y_1, \ldots, y_k)$ are k-tuples of distinct points and $tx_i = y_i$ for all i, then $G_{y_1,\ldots,y_k} = tG_{x_1,\ldots,x_k}t^{-1}$.

There is another stabilizer not to be confused with $G_{x_1,\ldots,x_t}$: if $Y = \{x_1, \ldots, x_t\}$, define $G_Y = \{g \in G: g(Y) = Y\}$. Thus, $g \in G_Y$ if and only if g permutes Y, whereas $g \in G_{x_1,\ldots,x_t}$ if and only if g fixes Y pointwise. Hence, $G_{x_1,\ldots,x_t} \leq G_Y$, with strict inclusion almost always.

The next result generalizes Theorem 9.2.

Theorem 9.7. *If X is a k-transitive G-set of degree n, then*

$$|G| = n(n-1)(n-2)\ldots(n-k+1)|G_{x_1,\ldots,x_k}|$$

for every choice of k distinct elements $x_1, \ldots, x_k$. If X is faithful, then $|G_{x_1,\ldots,x_k}|$ is a divisor of $(n-k)!$.

Proof. If $x_1 \in X$, then $|G| = n|G_{x_1}|$, by Theorem 9.2. Since G_{x_1} acts $(k-1)$-transitively on $X - \{x_1\}$, by Lemma 9.5, induction gives

$$|G_{x_1}| = (n-1)\ldots(n-k+1)|G_{x_1,\ldots,x_k}|,$$

and this gives the result. When G acts faithfully, then the result follows from G being imbedded in $S_{X-\{x_1,\ldots,x_k\}}$. ■

Definition. A k-transitive G-set X is ***sharply k-transitive*** if only the identity fixes k distinct elements of X.

Theorem 9.8. *The following conditions are equivalent for a faithful k-transitive G-set X of degree n.*

(i) *X is sharply k-transitive.*
(ii) *If $(x_1, \ldots, x_k)$ and $(y_1, \ldots, y_k)$ are k-tuples of distinct elements in X, then there is a unique $g \in G$ with $gx_i = y_i$ for all i.*
(iii) $|G| = n(n-1)\ldots(n-k+1)$.
(iv) *The stabilizer of any k elements in X is trivial.*

If $k \geq 2$, then these conditions are equivalent to:

(v) *For every $x \in X$, the G_x-set $X - \{x\}$ is sharply $(k-1)$-transitive.*

Proof. All verifications are routine and are left are exercises for the reader. ■

Theorem 9.9. *For every n, the symmetric group S_n acts sharply n-transitively on*

$X = \{1, \ldots, n\}$; *for every* $n \geq 3$, *the alternating group* A_n *acts sharply* $(n-2)$-*transitively on* X.

Proof. The first statement is obvious, for S_n contains every permutation of X. We prove the second statement by induction on $n \geq 3$. If $n = 3$, then $A_3 = \langle(1\ 2\ 3)\rangle$ acts sharply transitively on $X = \{1, 2, 3\}$. If $n > 3$, then $(A_n)_i$, the stabilizer of i, where $1 \leq i \leq n$, is isomorphic to A_{n-1}; by induction, it acts sharply $(n-3)$-transitively on $X - \{i\}$. Theorem 9.8(v) now completes the proof. ■

EXAMPLE 9.3. Let $f(x) \in \mathbb{Q}[x]$ be a polynomial of degree n with distinct roots $X = \{\alpha_1, \ldots, \alpha_n\} \subset \mathbb{C}$; let $E = \mathbb{Q}(\alpha_1, \ldots, \alpha_n)$ be the splitting field of $f(x)$, and let $G = \mathrm{Gal}(E/\mathbb{Q})$ be the Galois group of $f(x)$. By Lemma 5.2, X is a G-set; by Theorem 5.7, X is transitive if and only if $f(x)$ is irreducible over $\mathbb{Q}$. Now $f(x)$ factors in $\mathbb{Q}(\alpha_1)[x]$: $f(x) = (x - \alpha_1)f_1(x)$. Moreover, $G_1 = \mathrm{Gal}(E/\mathbb{Q}(\alpha_1)) \leq \mathrm{Gal}(E/\mathbb{Q}) = G$ is the stabilizer of α_1, and so G_1 acts on $X - \{\alpha_1\}$; indeed, G_1 is also the Galois group of $f_1(x)$. By Theorem 5.7, G_1 acts transitively if and only if $f_1(x)$ is irreducible (over $\mathbb{Q}(\alpha_1)$). Hence, $G = \mathrm{Gal}(E/\mathbb{Q})$ acts doubly transitively on X if and only if both $f(x)$ and $f_1(x)$ are irreducible (over $\mathbb{Q}$ and $\mathbb{Q}(\alpha_1)$, respectively). Of course, this procedure can be iterated. The set $X = \{\alpha_1, \ldots, \alpha_n\}$ of all roots of $f(x)$ is a k-transitive G-set if and only if the polynomials $f(x), f_1(x), \ldots, f_{k-1}(x)$ are all irreducible (over $\mathbb{Q}, \mathbb{Q}(\alpha_1), \ldots$, and $\mathbb{Q}(\alpha_1, \ldots, \alpha_{k-1})$, respectively).

It can be proved that there are no faithful k-transitive groups for $k > 5$ other than the symmetric and alternating groups. It follows that if the Galois group G of a polynomial is 6-transitive, then G is either a symmetric group or an alternating group. Indeed, if G is 4-transitive, then the only additional possibilities for G are four of the Mathieu groups.

Sharp k-transitivity is interesting for small k.

Definition. A sharply 1-transitive G-set X is called ***regular***.

A faithful G-set X is regular if and only if it is transitive and only the identity has a fixed point. It is now easy to see that the left regular representation of a group G in the Cayley theorem makes G itself into a regular G-set.

Our discussion of sharply 2-transitive G-sets begins with a technical definition.

Definition. If X is a G-set, then the ***Frobenius kernel*** N of G is the subset

$$N = \{1\} \cup \{g \in G\colon g \text{ has no fixed points}\}.$$

The Frobenius kernel may not be a subgroup of G, for it may not be closed under multiplication.

Lemma 9.10. *If X is a faithful sharply 2-transitive G-set of degree n, then the Frobenius kernel N of G has exactly n elements.*

Proof. By Theorem 9.8(iii), $|G| = n(n-1)$. For each $x \in X$, the stabilizer G_x has order $n-1$ (because $|G| = n|G_x|$), so that $|G_x^\#| = n-2$, where $G_x^\# = G_x - \{1\}$. If $x \neq y$, then $G_x \cap G_y = G_{x,y} = 1$, by Theorem 9.8(iv). Hence, $\{G_x^\#: x \in X\}$ is a disjoint family, and $|\bigcup_{x \in X} G_x^\#| = n(n-2)$. Since N is the complement of this union, $|N| = n(n-1) - n(n-2) = n$. ■

The basic philosophy is that permutation representations of a group G can yield important information about G. We now illustrate this by classifying all those groups G having a G-set X as in Lemma 9.10 in the special case when $n = |X|$ is odd (the classification when n is even is known, but it is more difficult).

Theorem 9.11. *Let X be a faithful sharply 2-transitive G-set of odd degree n.*

(i) *Each G_x contains a unique involution.*
(ii) *G_x has a center of even order, and a Sylow 2-subgroup of G_x is either cyclic or generalized quaternion.*
(iii) *The Frobenius kernel N of G is a normal subgroup of G.*
(iv) *The degree n is a power of an odd prime p.*
(v) *N is an elementary abelian p-group.*
(vi) *G is a semidirect product of N by G_x.*

Proof. (i) Since $|G| = n(n-1)$ is even, G contains an involution g. Now g, being a permutation of the n points of X, has a cycle decomposition; indeed, Exercise 1.16 shows that g is a product of disjoint transpositions $\tau_1 \ldots \tau_m$. Since $|X|$ is odd, g must fix some $x \in X$; that is, $g \in G_x$; because G acts sharply, g can fix nothing else, so that $m = \frac{1}{2}(n-1)$. If h is another involution in G, then $h \in G_y$ for some $y \in X$ (perhaps $y = x$) and $h = \sigma_1 \ldots \sigma_m$, a product of disjoint transpositions. Note that g and h can have no factors in common: otherwise, we may assume that $\tau_m = \sigma_m = (a\ b)$ (because disjoint transpositions commute); then gh^{-1} fixes a and b, hence is the identity. Therefore, $1 = gh^{-1}$ and $g = h$.

Assume that G_x has t involutions. Sharp 2-transitivity implies that $\{G_x^\#: x \in X\}$ is a disjoint family, and so G contains nt such elements, each of which involves $m = \frac{1}{2}(n-1)$ transpositions. Collectively, there are thus $tm = \frac{1}{2}nt(n-1)$ distinct transpositions occurring as factors of these involutions. But there are only $\frac{1}{2}n(n-1)$ transpositions in $S_X \cong S_n$, and so $t = 1$, as desired.

(ii) Let $g, h \in G_x$, and assume that g is an involution. Then hgh^{-1} is an involution, so that the uniqueness in (i) gives $hgh^{-1} = g$; that is, $g \in Z(G_x)$. The second statement follows from Theorem 5.46.

(iii) If T is the set of all involutions in G, then we claim that $TT \subset N$ (the

Frobenius kernel). Otherwise, there exist $g, h \in T$ with $gh \neq 1$ and $gh \notin N$; that is, gh fixes some $y \in X$. If $z = hy$, then both g and h have the transposition $(y\ z)$ as a factor: $hy = z$ and $hz = hhy = y$ ($h^2 = 1$); $gy = hy = z$ (for $ghy = y$ gives $hy = g^{-1}y = gy$) and $gz = ghy = y$. By (i), $g = h$, giving the contradiction $gh = 1$.

For fixed $g \in T$, the functions $T \to N$, defined by $h \mapsto gh$ and by $h \mapsto hg$, are both injective. By (i), $|T| = n$, for there is exactly one involution in each $G_x^\#$ as x varies over X; by Lemma 9.10, $|N| = n = |T|$. It follows that both injections above must be surjections; that is, $gT = N = Tg$ for all $g \in T$.

The Frobenius kernel always contains 1 and it is closed under inverses and conjugations by elements in G. Here N is closed under multiplication: choose $g \in T$ and observe that

$$NN = (Tg)(gT) = Tg^2T = TT \subset N.$$

(iv) Choose an element $h \in N$ of prime order p; since N is a subgroup of odd order n, p is odd. We claim that $C_G(h) \leq N$. If $f \in G^\#$ commutes with h, then $hfh^{-1} = f$. If $f \in G_x$ for some $x \in X$, then

$$f \in G_x \cap hG_xh^{-1} = G_x \cap G_{hx} = 1,$$

because $hx \neq x$. Therefore, $f \notin \bigcup_{x \in X} G_x$, so that $f \in N$. We conclude that $C_G(h) \leq N$, and so $[G : C_G(h)] = [G : N][N : C_G(h)] \geq [G : N] = n - 1$. But $[G : C_G(h)]$ is the number of conjugates of h, all of which lie in the normal subgroup N of order n. It follows that $N^\#$ consists precisely of all the conjugates of h; that is, N is a p-group (of exponent p), and so $n = |N|$ is a power of p.

(v) If $g \in G$ is an involution, then conjugation by g is an automorphism of N satisfying the conditions in Exercise 1.50(ii). Therefore, N is an abelian group of exponent p, hence it is a vector space over $\mathbb{Z}_p$, and hence it is an elementary abelian p-group.

(vi) We know that $N \lhd G$, $N \cap G_x = 1$, and $NG_x = G$ (by the product rule). (One could also use the Schur–Zassenhaus lemma, for the orders of N and G_x are consecutive integers, hence are relatively prime.) ■

Here is the proper context for this theorem. Let X be a faithful transitive G-set with each $g \in G^\#$ having at most one fixed point. If no such g has a fixed point, then X is a regular G-set; if some g does have a fixed point, then G is called a ***Frobenius group***. (An equivalent description of a Frobenius group is given in Exercise 9.9 below: a group G is a Frobenius group if and only if it contains a proper subgroup $H \neq 1$, called a ***Frobenius complement***, such that $H \cap gHg^{-1} = 1$ for all $g \notin H$.) It is easy to see that if there exists a faithful sharply 2-transitive G-set X of any (not necessarily odd) degree, then G is a Frobenius group: set $H = G_x$, and note that $G_x \cap gG_xg^{-1} = G_x \cap G_{gx} = G_{x,gx} = 1$ for all $g \notin G_x$. In 1901, Frobenius proved that Frobenius kernels of Frobenius groups are (normal) subgroups (in 1959, Thompson proved that Frobenius kernels are always nilpotent), and that every Frobenius group G is a semidirect product of its Frobenius kernel by G_x (which is a Frobenius complement) (see Isaacs (1976), p. 100). Every Sylow subgroup of a Frobenius complement is either cyclic or generalized quaternion; such groups are usually solvable (see Theorem 7.53), but there do exist nonsolvable ones. The group SL(2, 5), which is not solvable, by Corollary 8.15, is a Frobenius complement. Let V be a two-dimensional vector space over GF(29), and let GL(2, 29) act on V by matrix multiplication on the elements of V regarded as column vectors. There is a copy H of SL(2, 5) imbedded in GL(2, 29), and the group $G = V \rtimes H$ is a Frobenius group with Frobenius complement $H \cong$ SL(2, 5) (see Passman (1968), p. 202).

EXERCISES

9.1. If V is a vector space, then GL(V) acts faithfully on V and on $V^\# = V - \{0\}$.

9.2. Let H be a proper subgroup of G.
(i) Show that the representation of G on the cosets of H (Theorem 3.14) makes the set of cosets G/H into a transitive G-set of degree $[G:H]$. Show that G/H need not be faithful.
(ii) Show that the representation of G on the conjugates of H (Theorem 3.17) makes the family of all conjugates of H into a transitive G-set of degree $[G:N_G(H)]$ which need not be faithful.

9.3. Let $n \geq 5$ and $2 < t < n$.
(i) Show that S_n cannot act transitively on a set with t elements. Conclude that every orbit of an S_n-set with more than two elements has at least n elements.
(ii) Show that S_n has no subgroups of index t.

9.4. If X is a G-set and $H \leq G$, then every G-orbit is a disjoint union of H-orbits.

9.5. (i) If G is a finite group with $G^\# = G - \{1\}$ a transitive Aut(G)-set, then G is elementary abelian.
(ii) Show that $\mathbb{Q}^\#$ is a transitive Aut($\mathbb{Q}$)-set.

9.6. If X is a transitive G-set and $N \lhd G$, then X is an N-set and $|N_x| = |N_y|$ for all $x, y \in X$.

9.7. Let G be a group with $|G| < n$. Prove that G is isomorphic to a transitive subgroup of S_n if and only if G contains a subgroup H of index n such that neither H nor any proper subgroup of H is normal in G. (*Hint*. For necessity, use Theorem 3.12; for sufficiency, take H to be the stabilizer of any symbol.)

9.8. Let X be a G-set and let $H \leq G$. If the H-orbits of X are $\{\mathcal{O}_1, \ldots, \mathcal{O}_l\}$, then the orbits of gHg^{-1} are $\{g\mathcal{O}_1, \ldots, g\mathcal{O}_l\}$. Use this result to give a new proof of Theorem 9.3(ii).

9.9. A finite group G acting on a set X is a Frobenius group if and only if there is a subgroup H with $1 < H < G$ and $H \cap gHg^{-1} = 1$ for all $g \notin H$. (*Hint*. Take H to be the stabilizer of a point.)

9.10. If G is abelian, then a faithful transitive G-set is regular. (*Hint*. If $x, y \in X$, then $G_x = G_y$.)

9.11. If X is a sharply k-transitive G-set, then X is not $(k + 1)$-transitive. (In particular, no regular G-set is doubly transitive.)

9.12. If X is a transitive G-set, prove that $|G|$ rank $X = \sum_{g \in G} F(g)^2$, where $F(g)$ is the number of $x \in X$ fixed by g. (*Hint*. Count the set $\{(g, x, y) \in G \times X \times X: gx = x, gy = y\}$ in two different ways.)

Primitive G-Sets

If G is a group of order n, then Cayley's theorem gives a faithful permutation representation of G of degree n; that is, G may be regarded as a subgroup of S_n. The coming discussion can often give representations of smaller degree.

Definition. If X is a G-set, then a ***block*** is a subset B of X such that, for each $g \in G$, either $gB = B$ or $gB \cap B = \varnothing$ (of course, $gB = \{gx: x \in B\}$).

Uninteresting examples of blocks are $\varnothing$, X, and one-point subsets; any other block is called ***nontrivial***.

Consider the following example of a G-set of degree 6, where $G \cong S_3$.

$$G = \{1, (1\ 2\ 3)(a\ b\ c), (1\ 3\ 2)(a\ c\ b), (1\ b)(2\ a)(3\ c), (1\ a)(2\ c)(3\ b), (1\ c)(2\ b)(3\ a)\}$$

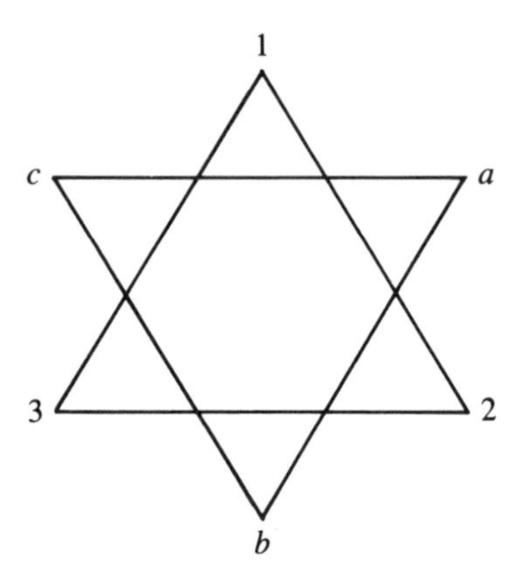

Figure 9.1

It is easy to see that either "triangle" $\{1, 2, 3\}$ or $\{a, b, c\}$ is a block; moreover, $\{1, a\}$, $\{2, b\}$, and $\{3, c\}$ are also blocks.

Definition. A transitive G-set X is ***primitive*** if it contains no nontrivial block; otherwise, it is ***imprimitive***.

Every partition $\{B_1, \ldots, B_m\}$ of a set X determines an equivalence relation $\equiv$ on X whose equivalence classes are the B_i. In particular, if X is a G-set and B is a block, then there is an equivalence relation on X given by $x \equiv y$ if there is some $g_i B$ containing both x and y. This equivalence relation is ***G-invariant***: that is, if $x \equiv y$, then $gx \equiv gy$ for all $g \in G$. Conversely, if $\equiv$ is a G-invariant equivalence relation on a G-set X, then any one of its equivalence classes is a block. Thus, a G-set is primitive if and only if it admits no nontrivial G-invariant equivalence relations.

Theorem 9.12. *Every doubly transitive G-set X is primitive.*

Proof. If X has a nontrivial block B, then there are elements $x, y, z \in X$ with $x, y \in B$ and $z \notin B$. Since X is doubly transitive, there is $g \in G$ with $gx = x$ and $gy = z$. Hence, $x \in B \cap gB$ and $B \neq gB$, a contradiction. ∎

Theorem 9.13. *Let X be a transitive G-set of degree n and let B be a nontrivial block of X.*

(i) *If $g \in G$, then gB is a block.*
(ii) *There are elements $g_1, \ldots, g_m$ of G such that*

$$Y = \{g_1 B, \ldots, g_m B\}$$

is a partition of X.
(iii) *$|B|$ divides n, and Y is a transitive G-set of degree $m = n/|B|$.*

Proof. (i) If $gB \cap hgB \neq \varnothing$ for some $h \in G$, then $B \cap g^{-1}hgB \neq \varnothing$; since B is a block, $B = g^{-1}hgB$ and $gB = hgB$.

(ii) Choose $b \in B$ and $x_1 \notin B$. Since G acts transitively, there is $g_1 \in G$ with $g_1 b = x_1$. Now $B \neq g_1 B$ implies $B \cap g_1 B = \varnothing$, because B is a block. If $X = B \cup g_1 B$, we are done. Otherwise, choose $x_2 \notin B \cup g_1 B$, and choose $g_2 \in G$ with $g_2 b = x_2$. It is easy to see that $g_2 B$ is distinct from B and from $g_1 B$, so that $g_2 B \cap (B \cup g_1 B) = \varnothing$ because B and $g_1 B$ are blocks. The proof is completed by iterating this procedure.

(iii) Since $|B| = |g_i B|$ for all i, $n = |X| = m|B|$ and $|Y| = m = n/|B|$. To see that Y is a transitive G-set, let $g_i B$, $g_j B \in Y$ and choose $x \in B$. Since X is transitive, there is $g \in G$ with $gg_i x = g_j x$; that is, $\varnothing \neq gg_i B \cap g_j B = (gg_i g_i^{-1}) g_j B \cap g_j B$. Since $g_j B$ is a block, it follows that $gg_i B = g_j B$, as desired. ∎

Definition. If B is a block of a transitive G-set X and if $Y = \{g_1 B, \ldots, g_m B\}$ is a partition of X (where $g_1, \ldots, g_m \in G$), then Y is called the ***imprimitive system*** generated by B.

Corollary 9.14. *A transitive G-set of prime degree is primitive.*

Proof. This follows from Theorem 9.13(iii). ∎

Here is a characterization of primitive G-sets.

Theorem 9.15. *Let X be a transitive G-set. Then X is primitive if and only if, for each $x \in X$, the stabilizer G_x is a maximal subgroup.*

Proof. If G_x is not maximal, there is a subgroup H with $G_x < H < G$, and we will show that $Hx = \{gx: g \in H\}$ is a nontrivial block; that is, X is imprimitive. If $g \in G$ and $Hx \cap gHx \neq \varnothing$, then $hx = gh'x$ for $h, h' \in H$. Since $h^{-1}gh'$ fixes x, we have $h^{-1}gh' \in G_x < H$ and so $g \in H$; hence, $gHx = Hx$, and Hx is a block. It remains to show that Hx is nontrivial. Clearly Hx is nonempty. Choose $g \in G$ with $g \notin H$. If $Hx = X$, then for every $y \in X$, there is $h \in H$ with $y = hx$; in particular, $gx = hx$ for some $h \in H$. Therefore $g^{-1}h \in G_x < H$ and $g \in H$, a contradiction. Finally, if Hx is a singleton, then $H \leq G_x$, contradicting $G_x < H$. Therefore, X is imprimitive.

Assume that every G_x is a maximal subgroup, yet there exists a nontrivial block B in X. Define a subgroup H of G:

$$H = \{g \in G: gB = B\}.$$

Choose $x \in B$. If $gx = x$, then $x \in B \cap gB$ and so $gB = B$ (because B is a block); therefore, $G_x \leq H$. Since B is nontrivial, there is $y \in B$ with $y \neq x$. Transitivity provides $g \in G$ with $gx = y$; hence $y \in B \cap gB$ and so $gB = B$. Thus, $g \in H$ while $g \notin G_x$; that is, $G_x < H$. If $H = G$, then $gB = B$ for all $g \in G$, and this contradicts $X \neq B$ being a transitive G-set. Therefore $G_x < H < G$, contradicting the maximality of G_x. ■

EXAMPLE 9.4. If X is a transitive G-set and $1 \neq H \lhd G$, then it need not be true that X is a transitive H-set. For example, if V is a vector space, then $V^\# = V - \{0\}$ is a transitive $\mathrm{GL}(V)$-set. However, if H is the center of $\mathrm{GL}(V)$ (i.e., if H is the subgroup of all the scalar transformations), then $V^\#$ is not a transitive H-set.

Lemma 9.16. *Let X be a G-set, and let $x, y \in X$.*

(i) *If $H \leq G$, then $Hx \cap Hy \neq \varnothing$ implies $Hx = Hy$.*
(ii) *If $H \lhd G$, then the subsets Hx are blocks of X.*

Proof. (i) It is easy to see that $Hy = Hx$ if and only if $y \in Hx$. If $Hy \cap Hx \neq \varnothing$, then there are $h, h' \in H$ with $hy = h'x$. Hence $y = h^{-1}h'x \in Hx$ and $Hy = Hx$.

(ii) Assume $gHx \cap Hx \neq \varnothing$. Since $H \lhd G$, $gHx \cap Hx = Hgx \cap Hx$. There are $h, h' \in H$ with $hgx = h'x$, and so $gx = h^{-1}h'x \in Hx$. Therefore, $gHx = Hx$. ■

Theorem 9.17.

(i) *If X is a faithful primitive G-set of degree $n \geq 2$, if $H \lhd G$ and if $H \neq 1$, then X is a transitive H-set.*
(ii) *n divides $|H|$.*

Proof. (i) The lemma shows that Hx is a block for every $x \in X$. Since X is primitive, either $Hx = \varnothing$ (plainly impossible), $Hx = \{x\}$, or $Hx = X$. If $Hx = \{x\}$ for some $x \in X$, then $H \leq G_x$. But if $g \in G$, then normality of H gives $H = gHg^{-1} \leq gG_xg^{-1} = G_{gx}$. Since X is transitive, $H \leq \bigcap_{y \in X} G_y = 1$, for X is faithful, and this is a contradiction. Therefore $Hx = X$ and X is a transitive H-set.

(ii) This follows from Theorem 9.2. ■

Using this theorem, we see that the $\mathrm{GL}(V)$-set $V^\#$ in Example 9.4 is transitive but not primitive.

Corollary 9.18. *Let X be a faithful primitive G-set of degree n. If G is solvable, then $n = p^m$ for some prime divisor p of $|G|$; if G is nilpotent, then n is a prime divisor of $|G|$.*

Proof. If G is solvable, a minimal normal subgroup H of G is elementary abelian of order p^k, by Theorem 5.24. The theorem now gives n a divisor of p^k, and so n, too, is a power of p. If G is nilpotent, then G has a normal subgroup H of prime order p (e.g., take $H = \langle g \rangle$, where g is an element of order p in $Z(G)$). The theorem gives n a divisor of p; that is, $n = p$. ■

EXERCISES

9.13. Let X be an imprimitive G-set and let B be a maximal nontrivial block of X; that is, B is not a proper subset of a nontrivial block. Show that the imprimitive system Y generated by B is a primitive G-set. Give an example with X faithful and Y not faithful.

9.14. (i) Let X be a transitive G-set, let $x \in X$, and let A be a nonempty subset of X. Show that the intersection of all gA containing x, where $g \in G$, is a block.
(ii) Let X be a primitive G-set and let A be a nonempty proper subset of X. If x and y are distinct elements of X, then there exists $g \in G$ with $x \in gA$ and $y \notin gA$. (*Hint.* The block in part (i) must be $\{x\}$.)

9.15. (i) Prove that if a group G has a faithful primitive G-set, then its Frattini subgroup $\Phi(G) = 1$.
(ii) Prove that a p-group G that is not elementary abelian has no faithful primitive G-set. (*Hint.* Theorem 5.47.)

Simplicity Criteria

We now prepare the way for new proofs showing that the alternating groups and the projective unimodular groups are almost always simple.

Definition. If X is a G-set and $H \lhd G$, then H is a ***regular normal subgroup*** if X is a regular H-set.

If H is a regular normal subgroup, then $|H| = |X|$. Thus, all regular normal subgroups have the same order.

Theorem 9.19. *Let X be a faithful primitive G-set with G_x a simple group. Then either G is simple or every nontrivial normal subgroup H of G is a regular normal subgroup.*

Proof. If $H \lhd G$ and $H \neq 1$, then Theorem 9.17(i) says that X is a transitive H-set. We have $H \cap G_x \lhd G_x$ for every $x \in X$, so that simplicity of G_x gives either $H \cap G_x = 1$ and X is regular or $H \cap G_x = G_x$; that is, $G_x \leq H$ for some $x \in X$. In the latter event, Theorem 9.15 gives G_x a maximal subgroup of G, so that either $G_x = H$ or $H = G$. The first case cannot occur because H acts transitively, so that $H = G$ and G is simple. ∎

It is proved in Burnside (1911), p. 202, Theorem XIII, that if a group G has a faithful doubly transitive G-set X whose degree is not a prime power, then either G is simple or G has a simple normal subgroup. (This result may be false when the degree is a prime power; S_4 is a counterexample.)

Here is the appropriate notion of homomorphism of G-sets.

Definition. If X and Y are G-sets, then a function $f: X \to Y$ is a ***G-map*** if $f(gx) = gf(x)$ for all $x \in X$ and $g \in G$; if f is also a bijection, then f is called a ***G-isomorphism***. Two G-sets X and Y are ***isomorphic***, denoted by $X \cong Y$, if there is a G-isomorphism $f: X \to Y$.

By Theorem 9.1, every G-set X determines a homomorphism $\varphi: G \to S_X$. Usually there is no confusion in saying that X is a G-set and not displaying φ, but because we now wish to compare two G-sets, let us denote X more precisely by (X, φ). The action of $g \in G$ on $x \in X$ is now denoted by $\varphi_g x$ instead of by gx. The definition of G-map $f: (X, \varphi) \to (Y, \psi)$ now reads, for all $g \in G$ and $x \in X$, as

$$f(\varphi_g(x)) = \psi_g(f(x)).$$

EXAMPLE 9.5. Let G be a group, and let $\lambda, \rho: G \to S_G$ be the left and right regular representations of G (recall that $\lambda_g: x \mapsto gx$ and $\rho_g: x \mapsto xg^{-1}$ for all $x, g \in G$). We claim that (G, λ) and (G, ρ) are isomorphic G-sets. Define $f: G \to G$ by $f(x) = x^{-1}$; clearly f is a bijection. Let us see that f is a G-map.

$$f(\lambda_g(x)) = f(gx) = x^{-1}g^{-1} = f(x)g^{-1} = \rho_g(f(x)).$$

EXAMPLE 9.6. *Chinese Remainder Theorem.*

If $S \leq G$ is any (not necessarily normal) subgroup, we denote the family of all left cosets of S in G by G/S; it is a G-set with action $g(xS) = (gx)S$ (as in Theorem 3.12). If X and Y are G-sets, then their cartesian product $X \times Y$ may be regarded as a G-set with ***diagonal action***: $g(x, y) = (gx, gy)$.

If G is a (finite) group and $H, K \leq G$ are such that $HK = G$, then there is a G-isomorphism $f: G/(H \cap K) \xrightarrow{\sim} (G/H) \times (G/K)$, where the latter has diagonal action. Define f by $x(H \cap K) \mapsto (xH, xK)$. It is straightforward to show that f is a well defined injective G-map. Since $HK = G$, the product formula $|HK||H \cap K| = |H||K|$ gives $|G|/|H||K| = 1/|H \cap K|$; multiplying both sides by $|G|$ gives $[G:H][G:K] = [G:H \cap K]$, and so f must be surjective as well. Therefore, f is a G-isomorphism.

Theorem 9.20. *Every transitive G-set X is isomorphic to the G-set G/G_x of all left cosets of G_x on which G acts by left multiplication.*

Proof. Let $X = \{x_1, \ldots, x_n\}$, let $H = G_{x_1}$, and, for each i, choose $g_i \in G$ with $g_i x_1 = x_i$ (which is possible because X is transitive). The routine argument that $f: X \to G/H$, given by $f(x_i) = g_i H$, is a well defined bijection is left to the reader (recall that $n = |\mathcal{O}(x_1)| = [G:H]$). To check that f is a G-map, note that if $g \in G$, then for all i there is j with $gx_i = x_j$, and so

$$f(gx_i) = f(x_j) = g_j H.$$

On the other hand,

$$gf(x_i) = gg_i H.$$

But $gg_i x_1 = gx_i = x_j = g_j x_1$; hence $g_j^{-1} g g_i \in G_{x_1} = H$, and so $g_j H = gg_i H$, as desired. ■

Theorem 9.21.

(i) *If $H, K \leq G$, then the G-sets G/H and G/K (with G acting by left multiplication) are isomorphic if and only if H and K are conjugate in G.*
(ii) *Two transitive G-sets (X, φ) and (Y, ψ) are isomorphic if and only if stabilizers of points in each are conjugate in G.*

Proof. (i) Assume that there is a G-isomorphism $f: G/H \to G/K$. In particular, there is $g \in G$ with $f(H) = gK$. If $h \in H$, then

$$gK = f(H) = f(hH) = hf(H) = hgK.$$

Therefore, $g^{-1}hg \in K$ and $g^{-1}Hg \leq K$. Now $f(g^{-1}H) = g^{-1}f(H) = g^{-1}gK = K$ gives $f^{-1}(K) = g^{-1}H$. The above argument, using f^{-1} instead of f, gives the reverse inclusion $gKg^{-1} \leq H$.

Conversely, if $g^{-1}Hg = K$, define $f: G/H \to G/K$ by $f(aH) = agK$. It is routine to check that f is a well defined G-isomorphism.

(ii) Let H and K be stabilizers of points in (X, φ) and (Y, ψ), respectively. By Theorem 9.20, $(X, \varphi) \cong G/H$ and $(Y, \psi) \cong G/K$. The result now follows from part (i). ■

Corollary 9.22. *If G is solvable, then every maximal subgroup has index a prime power; if G is nilpotent, then every maximal subgroup has prime index.*

Remark. The second statement was proved in Theorem 5.40.

Proof. If $H \leq G$, then the stabilizer of the point $\{H\}$ in the transitive G-set G/H is the subgroup H. If H is a maximal subgroup of G, then G/H is a primitive G-set, by Theorem 9.15, and so $|G/H| = [G:H]$ is a prime power, by Corollary 9.18. A similar argument gives the result when G is nilpotent. ■

Lemma 9.23. *Let X be a transitive G-set and let H be a regular normal subgroup of G. Choose $x \in X$ and let G_x act on $H^\#$ by conjugation. Then the G_x-sets $H^\#$ and $X - \{x\}$ are isomorphic.*

Proof. Define $f: H^\# \to X - \{x\}$ by $f(h) = hx$ (notice that $hx \neq x$ because H is regular). If $f(h) = f(k)$, then $h^{-1}k \in H_x = 1$ (by regularity), and so f is injective. Now $|X| = |H|$ (regularity again), $|H^\#| = |X - \{x\}|$, and so f is surjective. It remains to show that f is a G_x-map. If $g \in G_x$ and $h \in H^\#$, denote the action of g on h by $g * h = ghg^{-1}$. Therefore,

$$f(g * h) = f(ghg^{-1}) = ghg^{-1}x = ghx,$$

because $g^{-1} \in G_x$; on the other hand, $g \cdot f(h) = g(hx)$, and so $f(g * h) = g \cdot f(h)$. ■

Lemma 9.24. *Let $k \geq 2$ and let X be a k-transitive G-set of degree n. If G has a regular normal subgroup H, then $k \leq 4$. Moreover:*

(i) *if $k \geq 2$, then H is an elementary abelian p-group for some prime p and n is a power of p;*
(ii) *if $k \geq 3$, then either $H \cong \mathbb{Z}_3$ and $n = 3$ or H is an elementary abelian 2-group and n is a power of 2; and*
(iii) *if $k \geq 4$, then $H \cong \mathbf{V}$ and $n = 4$.*

Proof. By Lemma 9.5, the G_x-set $X - \{x\}$ is $(k-1)$-transitive for each fixed $x \in X$; by Lemma 9.23, $H^\#$ is a $(k-1)$-transitive G_x-set, where G_x acts by conjugation.

(i) Since $k \geq 2$, $H^\#$ is a transitive G_x-set. The stabilizer G_x acts by conjugation, which is an automorphism, so that all the elements of $H^\#$ have the same (necessarily prime) order p, and H is a group of exponent p. Now $Z(H) \lhd G$, because $Z(H)$ is a nontrivial characteristic subgroup, so that $|X| = |Z(H)| = |H|$, for $Z(H)$ and H are regular normal subgroups of G'. Therefore, $Z(H) = H$, H is elementary abelian, and $|X|$ is a power of p.

(ii) If $h \in H^\#$, then it is easy to see that $\{h, h^{-1}\}$ is a block. If $k \geq 3$, then $H^\#$ is a doubly transitive, hence primitive, G_x-set, so that either $\{h, h^{-1}\} = H^\#$ or $\{h, h^{-1}\} = \{h\}$. In the first case, $|H| = 3$, $H \cong \mathbb{Z}_3$, and $n = 3$. In the second case, h has order 2, and so the prime p in part (i) must be 2.

(iii) If $k \geq 4$, $k - 1 \geq 3$ and $|H^\#| \geq 3$; it follows that both $H \cong \mathbb{Z}_3$ and $H \cong \mathbb{Z}_2$ are excluded. Therefore, H contains a copy of $\mathbf{V}$; say, $\{1, h, k, hk\}$.

Now $(G_x)_h$ acts doubly transitively, hence primitively, on $H^\# - \{h\}$. It is easy to see, however, that $\{k, hk\}$ is now a block, and so $H^\# - \{h\} = \{k, hk\}$. We conclude that $H = \{1, h, k, hk\} \cong \mathbf{V}$ and $n = 4$.

Finally, we cannot have $k \geq 5$ because $n \leq 4$. ■

Of course, the case $k = 4$ does occur ($G = S_4$ and $H = \mathbf{V}$). Compare the case $k = 2$ with Theorem 9.11.

Theorem 9.25. *Let X be a faithful k-transitive G-set, where $k \geq 2$, and assume that G_x is simple for some $x \in X$.*

(i) *If $k \geq 4$, then G is simple.*
(ii) *If $k \geq 3$ and $|X|$ is not a power of 2, then either $G \cong S_3$ or G is simple.*
(iii) *If $k \geq 2$ and $|X|$ is not a prime power, then G is simple.*

Proof. By Theorem 9.19, either G is simple or G has a regular normal subgroup H. In the latter case, Lemma 9.24 gives $k \leq 4$; moreover, if $k = 4$, then $H \cong \mathbf{V}$ and $|X| = 4$. Now the only 4-transitive subgroup of S_4 is S_4 itself, but the stabilizer of a point is the nonsimple group S_3. Therefore, no such H exists, and so G must be simple. The other two cases are also easy consequences of the lemma (note that the stabilizer of a point of an S_3-set is the simple group $S_2 \cong \mathbb{Z}_2$ so that S_3 is a genuine exception in part (ii)). ■

Here is another proof of the simplicity of the large alternating groups.

Theorem 9.26. *A_n is simple for all $n \geq 5$.*

Proof. The proof is by induction on $n \geq 5$. If $n = 5$, then the result is Lemma 3.8. By Theorem 9.9, A_n acts $(n - 2)$-transitively on $X = \{1, 2, \ldots, n\}$; hence, if $n \geq 6$, then A_n acts k-transitively, where $k \geq 4$. The stabilizer $(A_n)_n$ of n is just A_{n-1} (for it consists of all the even permutations of $\{1, \ldots, n - 1\}$), and so it is simple, by induction. Therefore, A_n is simple, by Theorem 9.25(i). ■

Here is another simplicity criterion. It shall be used later to give another proof of the simplicity of the PSLs.

Theorem 9.27 (Iwasawa, 1941). *Let $G = G'$ (such a group is called **perfect**) and let X be a faithful primitive G-set. If there is $x \in X$ and an abelian normal subgroup $K \lhd G_x$ whose conjugates $\{gKg^{-1}: g \in G\}$ generate G, then G is simple.*

Proof. Let $H \neq 1$ be a normal subgroup of G. By Theorem 9.17, H acts transitively on X. By hypothesis, each $g \in G$ has the form $g = \prod g_i k_i g_i^{-1}$, where $g_i \in G$ and $k_i \in K$. Now $G = HG_x$, by Exercise 4.9(i), so that $g_i = h_i s_i$ for each i, where $h_i \in H$ and $s_i \in G_x$. Normality of K in G_x now gives

$$g = \prod h_i s_i k_i s_i^{-1} h_i^{-1} \in HKH \leq HK$$

(because H lies in the subgroup HK), and so $G = HK$. Since K is abelian, $G/H = HK/H \cong K/(H \cap K)$ is abelian, and $H \geq G' = G$. Therefore, G is simple. ■

EXERCISES

9.16. If X is a G-set, let $\mathrm{Aut}(X)$ be the group of all G-isomorphisms of X with itself. Prove that if X is a transitive G-set and $x \in X$, then $\mathrm{Aut}(X) \cong N_G(G_x)/G_x$. (*Hint.* If $\varphi \in \mathrm{Aut}(X)$, there is $g \in G$ with $gx = \varphi(x)$; the desired isomorphism is $\varphi \mapsto g^{-1}G_x$.)

9.17. Let X be a transitive G-set, and let $x, y \in X$. Prove that $G_x = G_y$ if and only if there is $\varphi \in \mathrm{Aut}(X)$ with $\varphi(x) = y$.

Affine Geometry

All vector spaces in this section are assumed to be finite-dimensional.

Theorem 9.28. *If V is an n-dimensional vector space over a field K, then $V^\# = V - \{0\}$ is a transitive $\mathrm{GL}(V)$-set that is regular when $n = 1$. If $n \geq 2$, then $V^\#$ is doubly transitive if and only if $K = \mathbb{Z}_2$.*

Proof. $\mathrm{GL}(V)$ acts transitively on $V^\#$, for every nonzero vector is part of a basis and $\mathrm{GL}(V)$ acts transitively on the set of all ordered bases of V. If $n = 1$, only the identity can fix a nonzero vector, and so $V^\# = K^\times$ is regular.

Assume that $n \geq 2$, and that $\{y, z\}$ is a linearly independent subset. If $K \neq \mathbb{Z}_2$, there exists $\lambda \in K$ with $\lambda \neq 0, 1$; if $x \in V^\#$, then $\{x, \lambda x\}$ is a linearly dependent set, and there is no $g \in G$ with $gx = y$ and $g\lambda x = z$. Therefore, $\mathrm{GL}(V)$ does not act doubly transitively in this case. If $K = \mathbb{Z}_2$, then every pair of distinct nonzero vectors is linearly independent, hence is part of a basis, and double transitivity follows from $\mathrm{GL}(V)$ acting transitively on the set of all ordered bases of V. ■

Definition. If V is a vector space and $y \in V$, then ***translation by y*** is the function $t_y\colon V \to V$ defined by

$$t_y(x) = x + y$$

for all $x \in V$. Let $\mathrm{Tr}(V)$ denote the group of all translations under composition (we may also write $\mathrm{Tr}(n, K)$ or $\mathrm{Tr}(n, q)$).

Definition. If V is a vector space over K, then the ***affine group***, denoted by $\mathrm{Aff}(V)$, is the group (under composition) of all functions $a\colon V \to V$ (called ***affinities***) for which there is $y \in V$ and $g \in \mathrm{GL}(V)$ such that

$$a(x) = gx + y$$

for all $x \in V$. (Of course, a is the composite $t_y g$.) We may also denote $\text{Aff}(V)$ by $\text{Aff}(n, K)$ or by $\text{Aff}(n, q)$.

Since $gt_y g^{-1} = t_{gy}$, we have $\text{Tr}(V) \lhd \text{Aff}(V)$; moreover, the map $t_y \mapsto y$ is an isomorphism of the group $\text{Tr}(V)$ with the additive abelian group V.

Theorem 9.29. *A vector space V of dimension n over a field K is a doubly transitive* $\text{Aff}(V)$*-space; it is sharply 2-transitive when $n = 1$.*

Proof. Since $\text{Tr}(V)$ acts transitively on V (as is easily seen), it suffices to show that if $x \neq 0$ and if $y \neq z$ are vectors, then there exists $a \in \text{Aff}(V)$ with $a(x) = y$ and $a(0) = z$. There exists $g \in \text{GL}(V)$ with $g(x) = y - z$, by Theorem 9.28, and one checks that $a \in \text{Aff}(V)$, defined by $a: v \mapsto g(v) + z$, sends $(x, 0)$ to (y, z). If $n = 1$, then g is unique, hence a is unique, and so the action is sharp. ■

When $n = 1$ and $K = \text{GF}(q)$, then $|\text{Aff}(1, q)| = q(q - 1)$, for $\text{Aff}(1, q)$ acts sharply 2-transitively on a set with q elements. Indeed, $\text{Aff}(1, q)$ is the semidirect product $K \rtimes_\theta K^\times$, where $\theta: K^\times \to K$ is defined by $\theta: x \mapsto$ multiplication by x. If q is a power of an odd prime, then $K = \text{GF}(q)$ is a sharply 2-transitive $\text{Aff}(1, q)$-set of odd degree, and so $\text{Aff}(1, q)$ is described by Theorem 9.11.

Let us now consider linear subsets of a vector space V: lines, planes, etc., not necessarily passing through the origin.

Definition. If S is an m-dimensional subspace of a vector space V, then a coset $W = S + v$, where $v \in V$, is called an ***affine m-subspace*** of V. The ***dimension*** of $S + v$ is defined to be the dimension of the subspace S.

There are special names for certain affine m-subspaces: if $m = 0$, 1, or 2, they are called ***points***, ***affine lines***, and ***affine planes***, respectively; if $\dim V = n$, then affine $(n - 1)$-subspaces of V are called ***affine hyperplanes***.

We are going to focus attention on the affine subspaces of a vector space V—its "geometry (that is, we focus on the lattice of all affine subspaces)." Here is a description of this aspect of V.

Definition. Let V be an n-dimensional vector space over a field K, let A be a set, and, for $0 \leq m \leq n$, let $\mathscr{L}_m(A)$ be a family of subsets of A (called ***affine m-subspaces***). If $\alpha: V \to A$ is a bijection such that a subset W of V is an affine m-subspace ($W = S + v$) if and only if $\alpha(W) \in \mathscr{L}_m(A)$, then $(A, \mathscr{L}_*(A), \alpha)$ is called an ***affine n-space*** over K with ***associated vector space*** V.

Example 9.7. If V is an n-dimensional vector space over a field K, define $\mathscr{L}_m(V)$ to be the family of all affine m-subspaces $S + v$ in V (where S is a sub-vector space) and define $\alpha: V \to V$ to be the identity. Then $(V, \mathscr{L}_*(V), 1_V)$ is an affine space, called the ***standard affine space*** of V over K.

Example 9.8. Let $(V, \mathscr{L}_*(V), 1_V)$ be a standard affine space over K and let

$v \in V$. Then $(V, \mathscr{L}_*(V), t_v)$ is also an affine space over K, where t_v is translation by v.

We have not given the most important examples of affine spaces: certain subsets of *projective spaces* (to be defined in the next section). A projective space X is obtained from a vector space V by adjoining a "hyperplane at infinity"; the original V inside of X is called the "finite" portion of X; that is, it is called *affine*.

Definition. If $(A, \mathscr{L}_*(A), \alpha)$ and $(B, \mathscr{L}_*(B), \beta)$ are affine spaces over K, then a bijection $f: A \to B$ is an ***affine isomorphism*** if, for all m, a subset W of A lies in $\mathscr{L}_m(A)$ if and only if $f(W)$ lies in $\mathscr{L}_m(B)$. One says that $(A, \mathscr{L}_*(A), \alpha)$ and $(B, \mathscr{L}_*(B), \beta)$ are ***isomorphic*** if there is an affine isomorphism between them.

If $(A, \mathscr{L}_*(A), \alpha)$ is an affine space over K, then the set of all affine automorphisms of A is a group under composition, denoted by

$$\mathrm{Aut}(A, \mathscr{L}_*(A), \alpha).$$

If $(A, \mathscr{L}_*(A), \alpha)$ is an affine space over K, then $\alpha: V \to A$ is an affine isomorphism from $(V, \mathscr{L}_*(V), 1_V)$ to $(A, \mathscr{L}_*(A), \alpha)$, and so every affine space is isomorphic to a standard affine space.

Theorem 9.30. *If $(A, \mathscr{L}_*(A), \alpha)$ is an affine space associated to a vector space V, then*

$$\mathrm{Aut}(V, \mathscr{L}_*(V), 1_V) \cong \mathrm{Aut}(A, \mathscr{L}_*(A), \alpha).$$

Moreover, two affine spaces $(A, \mathscr{L}_*(A), \alpha)$ and $(B, \mathscr{L}_*(B), \beta)$ over K are isomorphic if and only if they have associated vector spaces of the same dimension.

Proof. It is easy to see that $f \mapsto \alpha f \alpha^{-1}$ is an isomorphism $\mathrm{Aut}(V, \mathscr{L}_*(V), 1_V) \to \mathrm{Aut}(A, \mathscr{L}_*(A), \alpha)$.

If $(A, \mathscr{L}_*(A), \alpha)$ is an affine space with associated vector space V, then $\dim V$ is the largest m for which $\mathscr{L}_m(A)$ is defined. Therefore, isomorphic affine spaces have associated vector spaces of the same dimension.

Conversely, if $\dim V = \dim U$, then there is a nonsingular linear transformation $g: V \to U$, and $\beta g \alpha^{-1}: A \to B$ is an affine isomorphism, for it is a composite of such. ■

As a consequence of Theorem 9.30, we abbreviate notation and write

$$\mathrm{Aut}(V, \mathscr{L}_*(V), 1_V) = \mathrm{Aut}(V).$$

If V is an n-dimensional vector space over a field K, we may write $\mathrm{Aut}(n, K)$, and if $K = \mathrm{GF}(q)$, we may write $\mathrm{Aut}(n, q)$.

It is clear that $\mathrm{Aff}(V) \leq \mathrm{Aut}(V)$, for affinities preserve affine subspaces. Are there any other affine automorphisms?

EXAMPLE 9.9. If $K = \mathbb{C}$, the map $f: \mathbb{C}^n \to \mathbb{C}^n$ (the vector space of all n-tuples of complex numbers), given by $(z_1, \ldots, z_n) \mapsto (\bar{z}_1, \ldots, \bar{z}_n)$ (where $\bar{z}$ denotes the complex conjugate of z), is easily seen to be an affine automorphism. More generally, if σ is any automorphism of a field K, then $\sigma_*: K^n \to K^n$, defined by

$$\sigma_*(\lambda_1, \ldots, \lambda_n) = (\lambda_1^\sigma, \ldots, \lambda_n^\sigma),$$

is an affine automorphism, where λ^σ denotes the image of λ under σ.

Here is a coordinate-free description of σ_*.

Definition. Let V and V' be vector spaces over K. A function $f: V \to V'$ is a ***semilinear transformation*** if there exists $\sigma \in \mathrm{Aut}(K)$ such that, for all $x, y \in V$ and all $\lambda \in K$,

$$f(x + y) = f(x) + f(y)$$

and

$$f(\lambda x) = \lambda^\sigma f(x).$$

A semilinear transformation f is ***nonsingular*** if it is a bijection.

All linear transformations are semilinear (take $\sigma = 1_K$), as are all the functions σ_* of Example 9.9, and it is easy to see that every nonsingular semilinear transformation $V \to V$ is an affine automorphism of the standard affine space.

We claim that each nonzero semilinear transformation $f: V \to U$ determines a unique automorphism σ of K. Assume that there is an automorphism τ of K for which $f(\lambda v) = \lambda^\sigma f(v) = \lambda^\tau f(v)$. Since f is nonzero, there is a vector $v_0 \in V$ with $f(v_0) \neq 0$. For each $\lambda \in K$, $\lambda^\sigma f(v_0) = \lambda^\tau f(v_0)$; hence, $\lambda^\sigma = \lambda^\tau$ and $\sigma = \tau$. If f and g are semilinear transformations $V \to V$ with field automorphisms σ and τ, respectively, then their composite fg is semilinear with field automorphism $\sigma\tau$; if f is a nonsingular semilinear transformation with field automorphism σ, then f^{-1} is semilinear with field automorphism σ^{-1}.

Definition. All nonsingular semilinear transformations on a vector space V form a group under composition, denoted by $\Gamma\mathrm{L}(V)$. If V is an n-dimensional vector space over K, then we may write $\Gamma\mathrm{L}(n, K)$, and if $K = \mathrm{GF}(q)$, we may write $\Gamma\mathrm{L}(n, q)$.

Of course, $\Gamma\mathrm{L}(V) \leq \mathrm{Aut}(V)$.

The remarkable fact is that we have essentially displayed all possible affine automorphisms. Here are two lemmas needed to prove this.

Lemma 9.31. *Let V be a vector space of dimension ≥ 2 over a field K.*

(i) *Two distinct lines $\ell_1 = Kx_1 + y_1$ and $\ell_2 = Kx_2 + y_2$ are either disjoint or intersect in a unique point.*
(ii) *Two distinct points $x, y \in V$ lie in a unique line, namely, $K(x - y) + y$.*

(iii) *If* $\{x, y\}$ *is linearly independent, then*

$$\{x + y\} = (Kx + y) \cap (Ky + x).$$

(iv) *Two distinct hyperplanes* $H + x$ *and* $J + y$ *are disjoint if and only if* $H = J$. *In particular, if* $\dim V = 2$, *distinct lines* $Kx + y$ *and* $Kz + w$ *are disjoint if and only if* $Kx = Kz$.

Proof. (i) If $z \in \ell_1 \cap \ell_2$, then $\ell_1 \cap \ell_2 = (Kx_1 + y_1) \cap (Kx_2 + y_2) = (Kx_1 \cap Kx_2) + z$, by Exercise 2.26. Now $Kx_1 = Kx_2$ cannot occur lest the lines ℓ_1 and ℓ_2 be distinct cosets of the same subgroup Kx_1 and hence disjoint. Therefore $Kx_1 \cap Kx_2 = \{0\}$ and $\ell_1 \cap \ell_2 = \{z\}$.

(ii) Clearly $K(x - y) + y$ is a line containing x and y; a second such line intersects this one in at least two points, contradicting (i).

(iii) Independence of $\{x, y\}$ implies that the lines $Kx + y$ and $Ky + x$ are distinct. But $x + y \in (Kx + y) \cap (Ky + x)$, and (i) now gives the result.

(iv) If $H = J$, then distinct hyperplanes are distinct cosets of H, and hence are disjoint. If $H \neq J$, then $H + J = V$ (because H and J are maximal subspaces). Hence $y - x = h + j$ for some $h \in H$ and $j \in J$. Therefore, $h + x = -j + y \in (H + x) \cap (J + y)$, and the hyperplanes intersect.

The last statement follows, for a hyperplane in a two-dimensional vector space is a line. ■

Lemma 9.32. *Let* V *and* U *be vector spaces over* K, *let* $f\colon V \to U$ *be an affine isomorphism of standard affine spaces with* $f(0) = 0$, *and let* W *be a two-dimensional subspace of* V *with basis* $\{x, y\}$. *Then*:

(i) $f(Kx) = Kf(x)$;
(ii) $\{f(x), f(y)\}$ *is linearly independent*;
(iii) $f(W) = \langle f(x), f(y)\rangle$, *the subspace spanned by* $f(x)$ *and* $f(y)$; *and*
(iv) $f|W\colon W \to f(W)$ *is an affine isomorphism.*

Proof. (i) Kx is the unique line containing x and 0, and $f(Kx)$ is the unique line containing $f(x)$ and $f(0) = 0$.

(ii) If $\{f(x), f(y)\}$ is dependent, then $f(x)$, $f(y)$, and 0 are collinear; applying the affine isomorphism f^{-1} gives x, y, and 0 collinear, contradicting the independence of $\{x, y\}$.

(iii) For each ordered pair $\lambda, \mu \in K$, not both zero, denote the line containing λx and μy by $\ell(\lambda, \mu)$. Since $W = \bigcup_{\lambda,\mu} \ell(\lambda, \mu)$, it follows that $f(W) = \bigcup_{\lambda,\mu} f(\ell(\lambda, \mu))$. Denote $\langle f(x), f(y)\rangle$ by W'. To see that $f(W) = W'$, choose a line $\ell(\lambda, \mu)$. By (i), there are $\alpha, \beta \in K$ with $f(\lambda x) = \alpha f(x)$ and $f(\mu y) = \beta f(y)$. Now $f(\lambda x), f(\mu y) \in W'$ for all λ and μ, hence $f(\ell(\lambda, \mu)) \subset W'$, and so $f(W) \subset W'$. For the reverse inclusion, consider the affine isomorphism f^{-1} and the subset $\{f(x), f(y)\}$, which is independent, by (ii). As above, we see that $f^{-1}(W') \subset W$, so that $W' \subset f(W)$.

(iv) By (iii), $f(W)$ is a two-dimensional vector space, and so its affine structure is the same as that of W, by Theorem 9.30. ■

Theorem 9.33. *If V and U are isomorphic vector spaces of dimension ≥ 2 over a field K, then every affine isomorphism $f: V \to U$ has the form $f = t_u g$ for some $u \in U$ and some nonsingular semilinear transformation g; that is, for all $x \in V$,*

$$f(x) = g(x) + u.$$

Remarks. 1. One must assume that $\dim V \geq 2$, for every bijection between one-dimensional vector spaces is an affine isomorphism.

2. We will not use the full force of the hypothesis, namely, $f(\mathscr{L}_m(V)) = \mathscr{L}_m(U)$ for all $m \leq n$; we shall assume only that $f(\mathscr{L}_1(V)) = \mathscr{L}_1(U)$; that is, f carries lines to lines.

Proof. Composing f with the translation $x \mapsto x - f(0)$, we may assume that $f(0) = 0$; it suffices to prove that such an affine isomorphism is semilinear. Since f is a bijection, it preserves intersections; in particular, if ℓ_1 and ℓ_2 are lines, then $f(\ell_1 \cap \ell_2) = f(\ell_1) \cap f(\ell_2)$.

If $\{x, y\}$ is independent, then we claim that

$$f(Kx + y) = Kf(x) + f(y).$$

By Lemma 9.32(iii), both $f(Kx + y)$ and $f(Kx)$ are lines contained in $\langle f(x), f(y) \rangle$; indeed, they are disjoint because $Kx + y$ and Kx are disjoint. Since $f(Kx) = Kf(x)$, by Lemma 9.31(i), we may apply Lemma 9.31(iv) to $\langle f(x), f(y) \rangle$ to obtain

$$f(Kx + y) = Kf(x) + z$$

for some z. In particular, there are scalars $\alpha, \beta \in K$ with $f(y) = \alpha f(x) + z$ and $f(\lambda x + y) = \beta f(x) + z$, where $\lambda \in K$ and β depends on λ. Thus

$$\begin{aligned} f(\lambda x + y) &= \beta f(x) + f(y) - \alpha f(x) \\ &= (\beta - \alpha) f(x) + f(y) \in Kf(x) + f(y). \end{aligned}$$

Therefore $f(Kx + y) \subset Kf(x) + f(y)$, and equality holds because both are lines.

We now prove that $f(x + y) = f(x) + f(y)$ when $\{x, y\}$ is independent. By Lemma 9.31(iii), $(Kx + y) \cap (Ky + x) = \{x + y\}$. Since f preserves intersections,

$$\begin{aligned} \{f(x + y)\} &= f(Kx + y) \cap f(Ky + x) \\ &= [Kf(x) + f(y)] \cap [Kf(y) + f(x)] \\ &= \{f(x) + f(y)\}, \end{aligned}$$

the last equality because $\{f(x), f(y)\}$ is independent (Lemma 9.32(ii)). Therefore, $f(x + y) = f(x) + f(y)$ if $\{x, y\}$ is independent. It remains to show that $f(\lambda x + \mu x) = f(\lambda x) + f(\mu x)$, and we do this in two steps (as $f(0) = 0$, we may assume that $\lambda \neq 0$ and $\mu \neq 0$). Since $\dim V \geq 2$, we may choose w so that

$\{w, x\}$ is independent; it follows that $\{x + w, -x\}$ is also independent. Now

$$\begin{aligned} f(w) &= f((x + w) - x) \\ &= f(x + w) + f(-x) \qquad \text{(independence of } \{x + w, -x\}) \\ &= f(x) + f(w) + f(-x) \quad \text{(independence of } \{x, w\}) \end{aligned}$$

It follows that $f(-x) = -f(x)$. Consequently, if $\lambda + \mu = 0$, then $f(\lambda x + \mu x) = 0 = f(\lambda x) + f(\mu x)$. Finally, if $\lambda + \mu \neq 0$, then $\{\lambda x + w, \mu x - w\}$, $\{\lambda x, w\}$, and $\{\mu x, -w\}$ are independent sets (for $\lambda \neq 0$ and $\mu \neq 0$), and so

$$\begin{aligned} f(\lambda x + \mu x) &= f(\lambda x + w) + f(\mu x - w) \\ &= f(\lambda x) + f(w) + f(\mu x) + f(-w) \\ &= f(\lambda x) + f(\mu x). \end{aligned}$$

We have proved that f is additive.

If $x \neq 0$ and $\lambda \in K$, then $f(Kx) = Kf(x)$ implies that there is $\sigma_x(\lambda) \in K$ with $f(\lambda x) = \sigma_x(\lambda) f(x)$. The function $\sigma_x \colon K \to K$ is a bijection (because $f(Kx) = Kf(x)$) with $\sigma_x(1) = 1$. Additivity of f implies

$$\begin{aligned} \sigma_x(\lambda + \mu) f(x) &= f((\lambda + \mu)x) = f(\lambda x + \mu x) \\ &= f(\lambda x) + f(\mu x) = [\sigma_x(\lambda) + \sigma_x(\mu)] f(x); \end{aligned}$$

Since $f(x) \neq 0$, we see that σ_x is additive.

Next, we show that σ_x does not depend on x. Choose w so that $\{x, w\}$ is independent. Now

$$f(\lambda x + \lambda w) = f(\lambda x) + f(\lambda w) = \sigma_x(\lambda) f(x) + \sigma_w(\lambda) f(w).$$

On the other hand,

$$f(\lambda x + \lambda w) = \sigma_{x+w}(\lambda) f(x + w) = \sigma_{x+w}(\lambda)[f(x) + f(w)].$$

Equating coefficients,

$$\sigma_x(\lambda) = \sigma_{x+w}(\lambda) = \sigma_w(\lambda).$$

Lastly, if $\mu \in K^\times$, then $\{\mu x, w\}$ is independent and we see, with μx in place of x, that $\sigma_{\mu x}(\lambda) = \sigma_w(\lambda)$.

It remains to show that $\sigma \colon K \to K$ is multiplicative (the subscript may now be omitted). But

$$f(\lambda \mu x) = \sigma_x(\lambda \mu) f(x)$$

and also

$$f(\lambda \mu x) = \sigma_{\mu x}(\lambda) f(\mu x) = \sigma_{\mu x}(\lambda) \sigma_x(\mu) f(x).$$

Therefore, $\sigma_x(\lambda \mu) = \sigma_{\mu x}(\lambda) \sigma_x(\mu)$; as σ does not depend on the subscript, $\sigma(\lambda \mu) = \sigma(\lambda) \sigma(\mu)$, and so $\sigma \in \mathrm{Aut}(K)$. It follows that f is semilinear. ■

Corollary 9.34. *Let V and W be vector spaces over a field K with $\dim(V) = \dim(W) \geq 2$. If $g \colon V \to W$ is an additive function for which $g(v) = \lambda_v v$ for all $v \in V$, where $\lambda_v \in K$, then all the λ_v are equal and g is a scalar transformation.*

Proof. This is essentially contained in the next to last paragraph of the proof of the theorem. ■

Corollary 9.35. *Let $(A, \mathscr{L}_*(A), \alpha)$ and $(B, \mathscr{L}_*(B), \beta)$ be affine spaces of dimension ≥ 2 with associated vector spaces V and U, respectively. If $f: A \to B$ is an affine isomorphism, then*

$$f = \beta t_z g \alpha^{-1},$$

where $g: V \to U$ is a nonsingular semilinear transformation and $z = \beta^{-1} f\alpha(0)$.

Proof. The function $g': V \to U$, defined by $g' = \beta^{-1} f\alpha$, is an affine isomorphism; if g is defined by $g = t_{-z}g'$, then $g: V \to U$ is an affine isomorphism with $g(0) = 0$. By Theorem 9.33, g is a nonsingular semilinear transformation. Therefore $g = t_{-z}g' = t_{-z}\beta^{-1}f\alpha$, and the result follows. ■

If V is a vector space over a field K, we have written $\mathrm{Aut}(V)$ for the group of all affine automorphisms of V. If $\dim V \geq 2$, then Theorem 9.33 shows that $f \in \mathrm{Aut}(V)$ if and only if $f(x) = g(x) + y$, where g is a nonsingular semilinear transformation and $y \in V$. If $\dim V = 1$, then we have already remarked that every permutation of V is an affine automorphism: the group of all affine automorphisms is the symmetric group on V.

Theorem 9.36. *If V is an n-dimensional vector space over a field K, then $\Gamma\mathrm{L}(V)$ is a semidirect product of $\mathrm{GL}(V)$ by $\mathrm{Aut}(K)$. If $K = \mathrm{GL}(q) = \mathrm{GF}(p^r)$, then*

$$|\Gamma\mathrm{L}(n, q)| = |\Gamma\mathrm{L}(n, p^r)| = r|\mathrm{GL}(n, q)|.$$

Proof. We have already seen that each $g \in \Gamma\mathrm{L}(V)$ determines a unique $\sigma \in \mathrm{Aut}(K)$; this function $\Gamma\mathrm{L}(V) \to \mathrm{Aut}(K)$ is a surjective homomorphism with kernel $\mathrm{GL}(V)$. As in Example 9.9, choose a basis of V and, for each $\sigma \in \mathrm{Aut}(K)$, consider the semilinear transformation σ_*. It is easy to see that $Q = \{\sigma_*: \sigma \in \mathrm{Aut}(K)\} \leq \Gamma\mathrm{L}(V)$, $Q \cong \mathrm{Aut}(K)$, and Q is a complement of $\mathrm{GL}(V)$. Therefore, $\Gamma\mathrm{L}(V) \cong \mathrm{GL}(V) \rtimes \mathrm{Aut}(K)$.

When $K = \mathrm{GF}(q)$, then $|\Gamma\mathrm{L}(V)| = |\mathrm{Aut}(K)||\mathrm{GL}(V)|$. The result now follows from Theorem 8.4. ■

When V is finite, $|\mathrm{GL}(V)|$ is given by Theorem 8.5

Theorem 9.37. *Let V be a vector space of dimension n over a field K.*

(i) $\mathrm{Aut}(V)$, *the group of affine automorphisms of V, is a semidirect product of* $\mathrm{Tr}(V)$ *by* $\Gamma\mathrm{L}(V)$.
(ii) *If $K = \mathrm{GF}(q)$, then*
$$|\mathrm{Aut}(V)| = q^n |\Gamma\mathrm{L}(V)|.$$
(iii) *V is a doubly transitive* $\mathrm{Aut}(V)$*-set.*
(iv) $\mathrm{Aff}(V) = \mathrm{Aut}(V)$ *if and only if* $\mathrm{Aut}(K) = 1$.

Proof. (i) By Theorem 9.33, if $f \in \text{Aut}(V)$, then $f(x) = g(x) + y$ for some $g \in \Gamma L(V)$ and some $y \in V$. Define π: $\text{Aut}(V) \to \Gamma L(V)$ by $\pi(f) = g$. Now $\ker \pi = \text{Tr}(V)$ and π fixes $\Gamma L(V)$ pointwise, so that π is a retraction. Lemma 7.20 now gives $\text{Aut}(V) \cong \text{Tr}(V) \rtimes \Gamma L(V)$.

(ii) This follows from (i) after recalling that $\text{Tr}(V) \cong V$.

(iii) This follows from Theorem 9.29, for $\text{Aff}(V) \leq \text{Aut}(V)$.

(iv) $\text{Aut}(K) = 1$ if and only if every semilinear transformation is linear. ■

There exist fields K for which $\text{Aut}(K) = 1$. For example, the prime fields $\mathbb{Q}$ and $\mathbb{Z}_p$, for all primes p, and the real numbers $\mathbb{R}$.

EXERCISES

9.18. Choose a basis of an n-dimensional vector space V over a field K. To the affinity $a \in \text{Aff}(V)$ with $a(x) = g(x) + y$ (where $g \in \text{GL}(V)$ and $y \in V$), assign the $(n + 1) \times (n + 1)$ matrix

$$\begin{bmatrix} A & b \\ 0 & 1 \end{bmatrix},$$

where A is the matrix of g (relative to the chosen basis) and b is the column vector whose entries are the coordinates of y. Show that this assignment gives an injective homomorphism $\text{Aff}(V) = \text{Aff}(n, K) \to \text{GL}(n + 1, K)$ whose image consists of all matrices whose bottom row is $[0 \cdots 0 \ 1]$.

9.19. Show that $\text{Aff}(V)$ is a semidirect product of $\text{Tr}(V)$ by $\text{GL}(V)$ and that $|\text{Aff}(n, q)| = q^n|\text{GL}(n, q)|$.

9.20. Let V and U be n-dimensional vector spaces over a field K, where $n \geq 2$, and let $(V, \mathscr{L}_*(V), 1_V)$ and $(U, \mathscr{L}_*(U), 1_U)$ be standard affine n-spaces over K. Show that every affine isomorphism f: $V \to U$ has the form $f = t_u g t_v$, where $u \in U$, $v \in V$, and g: $V \to U$ is a nonsingular semilinear transformation.

9.21. Show that the center of $\Gamma L(V)$ consists of all the nonzero scalar transformations.

9.22. Choose a basis of a vector space V over K and, for $\sigma \in \text{Aut}(K)$, define σ_*: $V \to V$ as in Example 9.9. If $g \in \text{GL}(V)$, let $[\lambda_{ij}]$ be the matrix of g relative to the chosen basis. Show that $\sigma_* g \sigma_*^{-1}$ is linear and has matrix $[\sigma(\lambda_{ij})]$. Conclude that $\det(\sigma_* g \sigma_*^{-1}) = \sigma(\det g)$, and hence that $\text{SL}(V) \lhd \Gamma L(V)$.

Projective Geometry

We now turn from affine geometry to projective geometry. The need for projective geometry was felt by artists obliged to understand perspective in order to paint replicas of three-dimensional scenes on two-dimensional canvas. If a viewer's eyes are regarded as a vertex, then the problem of drawing in perspective amounts to analyzing conical projections from this vertex onto planes (see Coxeter (1987), §1.2); this is the etymology of the adjective "projective." To a viewer, two parallel lines appear to meet at the horizon, and projective space was invented to actually make this happen. "Points at infinity" (which constitute a "horizon") are adjoined to ordinary space: for each line in ordinary space, there is a new point serving as the common meeting point of all lines parallel to it. It is more efficient for us, however, to reverse this procedure; we begin with the larger space and then observe that ordinary (affine) space exists inside of it.

Let V be a vector space over a field K. Define an equivalence relation on $V^{\#} = V - \{0\}$ by $x \equiv y$ if there exists $\lambda \in K^{\times}$ with $y = \lambda x$; if $x \in V^{\#}$, denote its equivalence class by $[x]$. If $x \in V^{\#}$, then $[x]$ is the family of all the nonzero points on the line through x and the origin.

Definition. If V is an $(n + 1)$-dimensional vector space over a field K, then $\mathrm{P}(V) = \{[x]: x \in V^{\#}\}$ is called ***projective n-space***; one says that $\mathrm{P}(V)$ has ***projective dimension*** n.

If $V = K^{n+1}$, then we denote $\mathrm{P}(V)$ by $\mathrm{P}^n(K)$; if $K = \mathrm{GF}(q)$, we may denote $\mathrm{P}^n(K)$ by $\mathrm{P}^n(q)$. If $x = (x_0, x_1, \ldots, x_n) \in K^{n+1}$, then we may denote $[x]$ by its ***homogeneous coordinates*** $[x_0, x_1, \ldots, x_n]$. Of course, homogeneous coordinates are not coordinates at all; if $\lambda \neq 0$, then $[\lambda x_0, \lambda x_1, \ldots, \lambda x_n] = [x_0, x_1, \ldots, x_n]$. However, it does make sense to say whether the ith homogeneous coordinate is 0, for $x_i = 0$ implies $\lambda x_i = 0$ for all $\lambda \in K^{\times}$.

Affine spaces have no algebraic operations; one can neither add nor scalar multiply. However, using its associated vector space, one can restore these operation to $(A, \mathscr{L}_*(A), \alpha)$ with the bijection α. Projective spaces $\mathrm{P}(V)$ have also lost the algebraic operations of V, but they cannot be restored; the only vestiges are homogeneous coordinates and projective subspaces, defined below.

Definition. If W is a subset of V, define

$$[W] = \{[x]: x \in W^{\#}\} \subset \mathrm{P}(V).$$

If W is an $(m + 1)$-dimensional sub-vector space of V, then $[W]$ is called a ***projective m-subspace***; one says that $[W]$ has ***projective dimension*** m.

Here, too, there are names for special subspaces: if $m = 0$, 1, or 2, projective m-subspaces are called ***projective points***, ***projective lines***, or ***projective planes***; respectively; projective $(n - 1)$-subspaces of a projective n-space are called ***projective hyperplanes***.

The reason for lowering dimension in passing from V to $\mathrm{P}(V)$ should now be apparent: a line in V (through the origin) becomes a projective point, for all the nonzero points on the line are equivalent. A plane in V (through the origin) becomes a projective line, and so forth.

Theorem 9.38. *Let V be a vector space over a field K of dimension $n \geq 2$.*

(i) *If $x, y \in V^{\#} = V - \{0\}$, then $[x] \neq [y]$ if and only if $\{x, y\}$ is linearly independent.*
(ii) *Every pair of distinct points $[x], [y] \in \mathrm{P}(V)$ lie on a unique projective line.*
(iii) *If H is a projective hyperplane and L is a projective line not contained in H, then $H \cap L$ is a projective point.*

Proof. (i) The following are equivalent: $[x] = [y]$; $y = \lambda x$ for some $\lambda \in K^{\times}$; $\{x, y\}$ is linearly dependent.

(ii) Since $[x] \neq [y]$, the subspace $W = \langle x, y \rangle$ of V is two-dimensional, and so $[W]$ is a projective line containing $[x]$ and $[y]$. This line is unique, for if $[W']$ is another such line, then W' is a two-dimensional subspace of V containing the independent set $\{x, y\}$, and so $W' = W$.

(iii) Write $H = [U]$ and $L = [W]$, where U and W are subspaces of V of dimension n and 2, respectively. Now $W \not\subset U$ because $L \not\subset H$. Therefore, $U + W = V$,

$$\begin{aligned}\dim(U \cap W) &= \dim(U) + \dim(W) - \dim(U + W)\\ &= n + 2 - (n + 1) = 1,\end{aligned}$$

and $[U \cap W]$ has projective dimension 0; that is, $[U \cap W] = [U] \cap [W] = H \cap L$ is a projective point. ■

Definition. Let V and V' be vector spaces. A ***collineation*** (or *projective isomorphism*) is a bijection θ: $\mathrm{P}(V) \to \mathrm{P}(V')$ such that a subset S of $\mathrm{P}(V)$ is a projective m-subspace if and only if $\theta(S)$ is a projective m-subspace of $\mathrm{P}(V')$. Two projective spaces are ***isomorphic*** if there is a collineation between them.

Notation. If g: $V \to V'$ is a nonsingular semilinear transformation, then the function

$$\mathrm{P}(g)\colon \mathrm{P}(V) \to \mathrm{P}(V'),$$

defined by $g([x]) = [g(x)]$, is easily seen to be a collineation. If g is a linear transformation, then $\mathrm{P}(g)$ is called a ***projectivity***.

We are going to see that if $\dim(V) \geq 3$, every collineation on $\mathrm{P}(V)$ has the form $\mathrm{P}(g)$ for some $g \in \Gamma\mathrm{L}(V)$.

Theorem 9.39. *Two projective spaces* $\mathrm{P}(V)$ *and* $\mathrm{P}(V')$ *are isomorphic if and only if* $\dim V = \dim V'$.

Proof. Necessity is obvious: for all m, $\mathrm{P}(V)$ has a projective m-subspace if and only if $\mathrm{P}(V')$ does. Conversely, if $\dim V = \dim V'$, there is a (linear) isomorphism g: $V \to V'$, and hence there is a collineation (even a projectivity) $\mathrm{P}(g)$: $\mathrm{P}(V) \to \mathrm{P}(V')$. ■

In view of this theorem, every projective n-space over K is isomorphic to $\mathrm{P}(K^{n+1}) = \mathrm{P}^n(K)$, where K^{n+1} is the vector space of all $(n+1)$-tuples of elements of K.

Theorem 9.40.

(i) *For every $n \geq 0$ and every prime power q,*

$$|\mathrm{P}^n(q)| = q^n + q^{n-1} + \cdots + q + 1.$$

In particular, every projective line in $\mathrm{P}^n(q)$ has $q + 1$ points.

(ii) *The number of projective lines in $\mathrm{P}^2(q)$ is the same as the number of projective points, namely, $q^2 + q + 1$.*

Proof. (i) If $V = K^{n+1}$, where $K = \mathrm{GF}(q)$, then $|V^{\#}| = q^{n+1} - 1$. Since $V^{\#}$ is partitioned into equivalence classes $[x]$ each of which has $q - 1$ elements, we have

$$|\mathrm{P}^n(q)| = (q^{n+1} - 1)/(q - 1) = q^n + q^{n-1} + \cdots + q + 1.$$

(ii) We claim that there are exactly $q + 1$ lines passing through any point $[x]$. Choose a line ℓ not containing $[x]$; for each of the $q + 1$ points $[y]$ on ℓ, there is a line joining it to $[x]$. If ℓ' is any line containing $[x]$, however, Theorem 9.38(iii) shows that there exists some point $[y] \in \ell' \cap \ell$. Since two points determine a line, ℓ' coincides with the line joining $[x]$ and $[y]$ originally counted.

Choose a line ℓ_0; for each of the $q + 1$ points $[x]$ on ℓ_0, there are exactly q lines (other than ℓ_0) passing through $[x]$. We have displayed $q(q + 1) + 1$ distinct lines (the extra 1 counts the line ℓ_0). Since every line ℓ' meets ℓ_0, we have counted all the lines in $\mathrm{P}^2(q)$. ∎

Here is the important example of an affine space.

Theorem 9.41. *If $[W]$ is a projective hyperplane in a projective n-space $\mathrm{P}(V)$ and if $x \in V - W$, then $A = \mathrm{P}(V) - [W]$ can be given the structure of an affine n-space $(A, \mathscr{L}_*(A), \alpha)$ with associated vector space W, where $\alpha\colon W \to A$ is defined by $\alpha(w) = [w + x]$ for all $w \in W$.*

Proof. If $0 \leq m \leq n$, define

$$\mathscr{L}_m(A) = \{[U] \cap A\colon U \text{ is a subspace of } V \text{ with } \dim(U) = m + 1\}.$$

We show that α is a bijection by exhibiting its inverse. Since $x \notin W$, every $v \in V$ has a unique expression of the form $v = \lambda x + w$ for $w \in W$ and $\lambda \in K$. If $v \notin W$, then $\lambda \neq 0$ and there is an element v' equivalent to v with $v' - x \in W$ (namely, $v' = \lambda^{-1}v$). This element v' is the unique such: if $v'' = \mu v$, $\mu \neq \lambda^{-1}$, and $v'' - x \in W$, then $(v' - x) - (v'' - x) = (\lambda^{-1} - \mu)v$ forces $v \in W$, a contradiction. The function $\beta\colon A \to W$, defined by $\beta([v]) = v' - x$ (where v' is the unique scalar multiple of v for which $v' - x$ is in W), is thus well defined, and

β is easily seen to be α^{-1}. It is now routine to show that $(A, \mathscr{L}_m(A), \alpha)$ is an affine n-space with associated vector space W. ■

We illustrate Theorem 9.41 with $K = \text{GF}(2)$ and $V = K^2$. In this case, the complement of a projective line (which we now call a "line at infinity") is an affine plane consisting of 4 points and 6 lines.

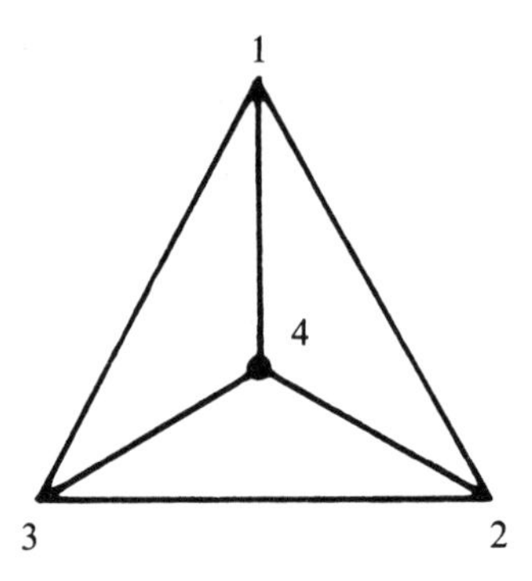

Figure 9.2

There are three pairs of parallel lines in A (e.g., 12 and 34), so that $\text{P}^2(2)$ requires 3 points at infinity to force every pair of lines to meet. Since $|\text{P}^2(2)| = 2^2 + 2 + 1 = 7$, the picture of $\text{P}^2(2)$ is

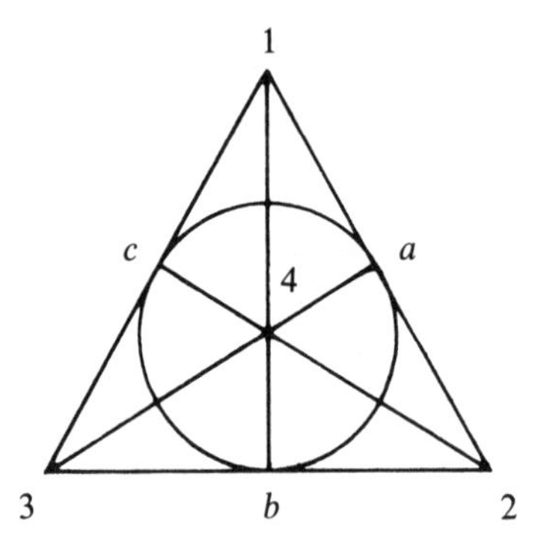

Figure 9.3

There are now 7 lines instead of 6 (the circle $\{a, b, c\}$ is a line; "line" in this pictorial representation of $\text{P}^2(2)$ has nothing to do with euclidean lines drawn on a sheet of paper). We have adjoined one "infinite point" to each of the lines in A so that any pair of extended lines now meet; $\{a, b, c\}$ is the horizon.

Lemma 9.42. *Let V and V' be vector spaces of dimension ≥ 2, let H be a projective hyperplane in $\mathrm{P}(V)$, and let A be the affine space $\mathrm{P}(V) - H$. If $f, h\colon \mathrm{P}(V) \to \mathrm{P}(V')$ are collineations with $f|A = h|A$, then $f = h$.*

Proof. Let ℓ be an affine line in A, and choose two distinct points x and y on ℓ. Since $\mathrm{P}(V)$ is a projective space, there is a unique projective line ℓ^* containing x and y (and ℓ). Since $f(x) = h(x)$ and $f(y) = h(y)$, it follows that $f(\ell^*) = h(\ell^*)$. By Theorem 9.38(iii), $H \cap \ell^*$ is a projective point z. Now $f(z) = f(H \cap \ell^*) = f(H) \cap f(\ell^*) = h(H) \cap h(\ell^*) = h(z)$, so that f and h also agree on all the points of H as well. ■

Theorem 9.43 (Fundamental Theorem of Projective Geometry). *If V and V' are isomorphic vector spaces over a field K of dimension $n + 1 \geq 3$, then every collineation $f\colon \mathrm{P}(V) \to \mathrm{P}(V')$ has the form $f = \mathrm{P}(g)$ for some nonsingular semilinear transformation $g\colon V \to V'$.*

Remarks. 1. One needs $\dim(V) \geq 3$, for then $\mathrm{P}(V)$ has projective dimension ≥ 2. Every bijection between projective lines is a collineation.

2. There is a stronger version of this theorem in which K is assumed only to be a division ring. This version does not affect finite groups, however, for a theorem of Wedderburn (1905) asserts that every finite division ring is a field.

Proof. Let W be an n-dimensional subspace of V, so that $[W]$ is a projective hyperplane in $\mathrm{P}(V)$. As in Theorem 9.41, $A = \mathrm{P}(V) - [W]$ is an affine n-space with affine isomorphism $\alpha\colon W \to A$ given by $\alpha(w) = [w + x]$ for any $x \in V - W$. Now $f([W])$ is a hyperplane in $\mathrm{P}(V')$ (since f is a collineation), so there is an n-dimensional subspace W' of V' such that $A' = \mathrm{P}(V') - f([W]) = [W']$ is an affine n-space with bijection $\beta\colon W' \to A'$ given by $\beta(w') = [w'' + y]$ for $y \in V' - W'$; as any such y will serve, choose y so that $[y] = f([x])$. Since f is a collineation, its restriction $f|A\colon A \to A'$ is an affine isomorphism. By Corollary 9.35, $f|A = \beta t_z g \alpha^{-1}$, where $g\colon W \to W'$ is a nonsingular semilinear transformation and $z = \beta^{-1} f\alpha(0)$. Now

$$z = \beta^{-1} f\alpha(0) = \beta^{-1} f([x]) = \beta^{-1}([y]).$$

If $v' \in V'$, then $\beta^{-1}([v']) = v'' - y$, where v'' is the unique scalar multiple of v' for which $v'' - y \in W'$. If follows that $\beta^{-1}([y]) = y - y = 0$, and so $t_z = t_0 = 1_{W'}$. Therefore,

$$f|A = \beta g \alpha^{-1}.$$

Now $V = \langle x \rangle \oplus W$. If σ is the field automorphism determined by the semilinear transformation g, define $\tilde{g}\colon V \to V'$ by

$$\tilde{g}(\lambda x + w) = \lambda^\sigma y + g(w),$$

where $\lambda \in K$ and $w \in W$. It is routine to check that $\tilde{g}$ is a semilinear transformation with $\tilde{g}(x) = y$ and $\tilde{g}|W = g$; also, $\tilde{g}$ is nonsingular because g is nonsingular and $y \notin W'$. We claim that $f = \mathrm{P}(\tilde{g})$. For all $[v] \in A$,

$$\begin{aligned} f([v]) &= \beta\tilde{g}\alpha^{-1}([v]) \\ &= \beta(\tilde{g}(\lambda v - x)) \quad (\text{where } \lambda v - x \in W) \\ &= \beta(\lambda^\sigma \tilde{g}(v) - \tilde{g}(x)) = \beta(\lambda^\sigma \tilde{g}(v) - y) \\ &= [\lambda^\sigma \tilde{g}(v) - y + y] = [\lambda^\sigma \tilde{g}(v)] = [\tilde{g}(v)] \\ &= \mathrm{P}(\tilde{g})([v]). \end{aligned}$$

Therefore, $f|A = \mathrm{P}(\tilde{g})|A$, and so $f = \mathrm{P}(\tilde{g})$, by Lemma 9.42. ■

There are some interesting groups acting on projective space.

Notation. If V is a vector space over K, denote the group of all nonzero scalar transformations on V by $\mathrm{Z}(V)$. If $\dim(V) = n$, we may denote $\mathrm{Z}(V)$ by $\mathrm{Z}(n, K)$; if $K = \mathrm{GF}(q)$, we may denote $\mathrm{Z}(V)$ by $\mathrm{Z}(n, q)$.

Theorem 9.44. *If* $\dim(V) \geq 3$, *the group* $\Gamma\mathrm{L}(V)/\mathrm{Z}(V)$ *is isomorphic to the group* C *of all collineations of* $\mathrm{P}(V)$ *with itself; if* $\dim(V) = 2$, then $\Gamma\mathrm{L}(V)/\mathrm{Z}(V)$ *is isomorphic to a subgroup of the symmetric group on* $\mathrm{P}(V)$ (*the symmetric group is the full collineation group of* $\mathrm{P}(V)$ *in this case*).

Proof. Define a homomorphism $\pi: \Gamma\mathrm{L}(V) \to C$ by $\pi(g) = \mathrm{P}(g)$. If $g \in \ker \pi$, then $\mathrm{P}(g) = 1$; that is, $[g(v)] = [v]$ for all $v \in V$, so there are scalars λ_v with $g(v) = \lambda_v v$ for all v. By Corollary 9.34, g is scalar, and so π induces an injection $\mathrm{P}\Gamma\mathrm{L}(V) \to C$. If $\dim(V) \geq 3$, then the Fundamental Theorem 9.43 gives π surjective, so that $\Gamma\mathrm{L}(V)/\mathrm{Z}(V) \cong C$. If $\dim(V) = 2$, then C is the symmetric group on $\mathrm{P}(V)$. ■

Definition. If V is a vector space over a field K, then the quotient group $\Gamma\mathrm{L}(V)/\mathrm{Z}(V)$ is denoted by $\mathrm{P}\Gamma\mathrm{L}(V)$ and its subgroup $\mathrm{GL}(V)/\mathrm{Z}(V)$ is denoted by $\mathrm{PGL}(V)$. As usual, one may replace V by (n, K) or by (n, q) when appropriate.

Theorem 9.44 shows that $\mathrm{P}\Gamma\mathrm{L}(V)$ is (isomorphic to) the collineation group of the projective space $\mathrm{P}(V)$ (when $\dim(V) \geq 2$), and it is easy to see that $\mathrm{PGL}(V)$ is the subgroup of all projectivities. The projective unimodular group $\mathrm{PSL}(V) = \mathrm{SL}(V)/\mathrm{SZ}(V)$ can be imbedded in $\mathrm{PGL}(V)$: after all,

$$\begin{aligned} \mathrm{PSL}(V) = \mathrm{SL}(V)/\mathrm{SZ}(V) &= \mathrm{SL}(V)/(\mathrm{Z}(V) \cap \mathrm{SL}(V)) \\ &\cong \mathrm{SL}(V)\mathrm{Z}(V)/\mathrm{Z}(V) \\ &\leq \mathrm{GL}(V)/\mathrm{Z}(V) \\ &= \mathrm{PGL}(V). \end{aligned}$$

EXAMPLE 9.10. $\mathrm{P\Gamma L}(2, 4) \cong S_5$ and $\mathrm{PGL}(2, 4) \cong A_5$.

For all $n \geq 1$, the group $\mathrm{P\Gamma L}(n + 1, q)$ is the group of all collineations of $\mathrm{P}^n(q)$, so that $\mathrm{P\Gamma L}(n + 1, q)$ (and each of its subgroups) acts faithfully on $\mathrm{P}^n(q)$. In particular, $\mathrm{P\Gamma L}(2, 4)$ acts faithfully on $\mathrm{P}^1(4)$. Since $|\mathrm{P}^1(4)| = 5$, $\mathrm{PGL}(2, 4) \leq \mathrm{P\Gamma L}(2, 4) \leq S_5$. By Theorems 9.36 and 8.5, $|\mathrm{P\Gamma L}(2, 4)| = 120$ and $|\mathrm{PGL}(2, 4)| = 60$. The isomorphisms follow.

Theorem 9.45. *For every vector space V, $\mathrm{PSL}(V)$ (and hence the larger groups PGL and $\mathrm{P\Gamma L}$) acts doubly transitively on $\mathrm{P}(V)$.*

Proof. If $([x], [y])$ and $([x'], [y'])$ are ordered pairs of distinct elements of $\mathrm{P}(V)$, then $\{x, y\}$ and $\{x', y'\}$ are linearly independent, by Theorem 9.38. Each of these independent sets can be extended to bases of V, say, $\{x, y, z_3, \ldots, z_n\}$ and $\{x', y', z_3', \ldots, z_n'\}$. There exists $g \in \mathrm{GL}(V)$ with $g(x) = x'$, $g(y) = y'$, and $g(z_i) = z_i'$ for all $i \geq 3$. Hence $\mathrm{P}(g)[x] = [x']$ and $\mathrm{P}(g)[y] = [y']$. If $\det(g) = \lambda$, define $h \in \mathrm{GL}(V)$ by $h(x) = \lambda^{-1}x'$, $h(y) = y'$, and $h(z_i) = z_i'$ for all i. Then $\det(h) = 1$, so that $h \in \mathrm{SL}(V)$, $\mathrm{P}(h)[x] = [\lambda^{-1}x'] = [x']$, and $\mathrm{P}(h)[y] = [y']$. Therefore, $\mathrm{PSL}(V)$ acts doubly transitively on $\mathrm{P}(V)$. ■

We now give a second proof of the simplicity of the PSL's. Recall the discussion preceding Lemma 8.19: for every transvection $T \in \mathrm{SL}(V)$, where V is a vector space over K, there is a linear functional $\varphi: V \to K$ and a nonzero vector $h \in \ker f$ such that $T = \{\varphi, h\}$, where

$$\{\varphi, h\}: v \mapsto v + \varphi(v)h.$$

Theorem 9.46. $\mathrm{PSL}(n, K)$ *is simple if* $(n, K) \neq (2, \mathbb{Z}_2)$ *and* $(n, K) \neq (2, \mathbb{Z}_3)$.

Proof. We use Iwasawa's criterion, Theorem 9.27. Let $G = \mathrm{PSL}(V)$, where V is a vector space over K. By Theorem 9.45, $\mathrm{P}(V)$ is a faithful doubly transitive, hence primitive, G-set.

Choose $h \in V^\#$ and define a subgroup of the stabilizer $G_{[h]}$ by

$$H = \{\mathrm{P}(\{\varphi, h\}): \varphi(h) = 0\} \cup \{1\}.$$

Applying P to the formula $\{\varphi, h\}\{\psi, h\} = \{\varphi + \psi, h\}$ of Lemma 8.19(i), we see that H is abelian. Recall Lemma 8.19(iv): If $S \in \mathrm{GL}(V)$, then $S\{\varphi, h\}S^{-1} = \{\varphi S^{-1}, Sh\}$; this is true, in particular, for $S \in \mathrm{SL}(V)$. Now $\mathrm{P}(S) \in G_{[h]}$ if and only if $Sh = \lambda h$ for some $\lambda \in K$. But $\{\psi, \lambda h\} = \{\lambda\psi, h\}$, by Lemma 8.19(ii), and this shows that $H \lhd G_{[H]}$.

We now show that the conjugates of H generate G, and it suffices, by Lemma 8.8(ii) ($\mathrm{SL}(V)$ is generated by the transvections), to show that every $\mathrm{P}(\{\psi, k\})$ is a conjugate of some $\mathrm{P}(\{\varphi, h\})$. Choose $S \in \mathrm{SL}(V)$ with $Sh = k$. Then, for any $\{\varphi, h\}$,

$$\mathrm{P}(S)\mathrm{P}(\{\varphi, h\})\mathrm{P}(S)^{-1} = \mathrm{P}(\{\varphi S^{-1}, k\}).$$

As φ varies over all linear functionals annihilating h, the linear functional φS^{-1} varies over all those annihilating $k = Sh$. In particular, given ψ, there is φ with $\varphi(h) = 0$ and $\psi = \varphi S^{-1}$.

It remains to prove that G is perfect: $G = G'$. Suppose some transvection T is a commutator: $T = [M, N]$ for some $M, N \in \mathrm{SL}(V)$. If T' is another transvection, then T' and T are conjugate in $\mathrm{GL}(V)$, by Corollary 8.18. Therefore, there is $U \in \mathrm{GL}(V)$ with $T' = T^U = [M, N]^U = [M^U, N^U]$. But both $M^U, N^U \in \mathrm{SL}(V)$, because $\mathrm{SL}(V) \lhd \mathrm{GL}(V)$, and so T', too, is a commutator. It follows that $\mathrm{SL}(V)$ is perfect, and so its quotient $\mathrm{PSL}(V)$ is perfect as well.

Let $n \geq 3$ and let $\{e_1, \ldots, e_n\}$ be a basis of V. Define $T \in \mathrm{GL}(V)$ by $T(e_i) = e_i$ for all $i \neq 3$ and $T(e_3) = e_3 - e_2 - e_1$.

$$T = \begin{bmatrix} 1 & 0 & -1 \\ 0 & 1 & -1 \\ 0 & 0 & 1 \end{bmatrix}.$$

Note that $T = \{\varphi, h\}$, where $h = -e_2 - e_1$ and φ is the linear functional which selects the third coordinate: $\varphi(\sum \lambda_i e_i) = \lambda_3$. Define $M = B_{13}(-1)$ and define N by $Ne_1 = -e_2$, $Ne_2 = e_1$, and $Ne_i = e_i$ for all $i \geq 3$. Both M and N lie in $\mathrm{SL}(V)$, and a routine calculation shows that $[M, N] = MNM^{-1}N^{-1} = T$:

$$\begin{bmatrix} 1 & 0 & -1 \\ 0 & 1 & 0 \\ 0 & 0 & 1 \end{bmatrix} \begin{bmatrix} 0 & 1 & 0 \\ -1 & 0 & 0 \\ 0 & 0 & 1 \end{bmatrix} \begin{bmatrix} 1 & 0 & 1 \\ 0 & 1 & 0 \\ 0 & 0 & 1 \end{bmatrix} \begin{bmatrix} 0 & -1 & 0 \\ 1 & 0 & 0 \\ 0 & 0 & 1 \end{bmatrix} = \begin{bmatrix} 1 & 0 & -1 \\ 0 & 1 & -1 \\ 0 & 0 & 1 \end{bmatrix}.$$

If $n = 2$ and $|K| > 3$, then there exists $\lambda \in K$ with $\lambda^2 \neq 1$. But

$$\begin{bmatrix} \lambda & 0 \\ 0 & \lambda^{-1} \end{bmatrix} \begin{bmatrix} 1 & 1 \\ 0 & 1 \end{bmatrix} \begin{bmatrix} \lambda^{-1} & 0 \\ 0 & \lambda \end{bmatrix} \begin{bmatrix} 1 & -1 \\ 0 & 1 \end{bmatrix} = \begin{bmatrix} 1 & \lambda^2 - 1 \\ 0 & 1 \end{bmatrix},$$

and so the (elementary) transvection $B_{12}(\lambda^2 - 1)$ is a commutator. All the conditions of Iwasawa's criterion have been verified, and we conclude that $\mathrm{PSL}(n, K)$ is simple with two exceptions. ■

Note that the proof of Corollary 8.14 does not work for nonperfect fields of characteristic 2; the proof above has no such limitation.

Here is a chart of the groups that have arisen in this chapter.

$$\begin{array}{ccccc} \mathrm{Tr}(V) & \leq & \mathrm{Aff}(V) & \leq & \mathrm{Aut}(V) \\ & & \cup & & \cup \\ \mathrm{SL}(V) & \leq & \mathrm{GL}(V) & \leq & \Gamma\mathrm{L}(V) \\ \cup & & \cup & & \\ \mathrm{SZ}(V) & \leq & \mathrm{Z}(V). & & \end{array}$$

Here is a summary of the various relations that have arisen among these groups, where V is a vector space over K.

$$\mathrm{SZ}(V) = \mathrm{Z}(V) \cap \mathrm{SL}(V);$$

$$\mathrm{GL}(V) \cong \mathrm{SL}(V) \rtimes K^\times;$$

$$\Gamma\mathrm{L}(V) \cong \mathrm{GL}(V) \rtimes \mathrm{Aut}(K);$$

$$\mathrm{Aut}(V) \cong \Gamma\mathrm{L}(V) \rtimes \mathrm{Tr}(V);$$

$$\mathrm{Aff}(V) \cong \mathrm{GL}(V) \rtimes \mathrm{Tr}(V).$$

EXERCISES

9.23. Show that if $n \geq 2$, then $\mathrm{PSL}(n + 1, K)$ acts faithfully and transitively on the set of all projective lines in $\mathrm{P}^n(K)$. (*Hint.* Two points determine a unique line.)

9.24. Let $f\colon \mathrm{P}(V) \to \mathrm{P}(V)$ be a collineation, where $\dim(V) \geq 3$. If there exists a projective line ℓ in $\mathrm{P}(V)$ for which $f|\ell$ is a projectivity, then f is a projectivity.

9.25. Prove that $\mathrm{PGL}(2, 5) \cong S_5$ and $\mathrm{PSL}(2, 5) \cong A_5$. (*Hint.* $|\mathrm{P}^1(5)| = 6$.)

9.26. Prove that any two simple groups of order 168 are isomorphic, and conclude that $\mathrm{PSL}(2, 7) \cong \mathrm{PSL}(3, 2)$. (*Hint.* If G is a simple group of order 168, then a Sylow 2-subgroup P of G has 7 conjugates and $N_G(P)/P \cong \mathbb{Z}_3$. Construct a projective plane whose points are the conjugates of P and whose lines are those subsets $\{\alpha P\alpha^{-1}, \beta P\beta^{-1}, \gamma P\gamma^{-1}\}$ for which $\{\alpha, \beta, \gamma\}$ is a transversal of P in $N_G(P)$.)

Sharply 3-Transitive Groups

We have seen that the groups $\mathrm{P}\Gamma\mathrm{L}(n, K)$ are interesting for all $n \geq 3$: they are collineation groups of projective $(n - 1)$-space. Let us now see that $\mathrm{P}\Gamma\mathrm{L}(2, K)$ and its subgroup $\mathrm{PGL}(2, K)$ are also interesting.

Definition. If K is a field, let $\hat{K} = K \cup \{\infty\}$, where ∞ is a new symbol. If $\sigma \in \mathrm{Aut}(K)$ and $ad - bc \neq 0$, then a ***semilinear fractional transformation*** is a function $f\colon \hat{K} \to \hat{K}$ given by

$$f(\lambda) = (a\lambda^\sigma + b)/(c\lambda^\sigma + d) \qquad \text{for} \quad \lambda \in \hat{K}.$$

The "extreme" values are defined as follows: $f(\lambda) = \infty$ if $c\lambda^\sigma + d = 0$; $f(\infty) = \infty$ if $c = 0$; $f(\infty) = ac^{-1}$ if $c \neq 0$. If σ is the identity, then f is called a ***linear fractional transformation***.

These functions arise in complex variables; there, $K = \mathbb{C}$ and σ is complex conjugation.

It is easy to see that all the semilinear fractional transformations on $\hat{K}$ form a group under composition.

Notation. $\Gamma\mathrm{LF}(K)$ denotes the group of all semilinear fractional transformations on $\hat{K}$, and $\mathrm{LF}(K)$ denotes the subgroup of all the linear fractional transformations.

Theorem 9.47. *For every field* K, $\mathrm{P\Gamma L}(2, K) \cong \Gamma\mathrm{LF}(K)$ *and* $\mathrm{PGL}(2, K) \cong \mathrm{LF}(K)$.

Proof. Choose a basis of a two-dimensional vector space V over K. Using Theorem 9.36, one sees that each $f \in \Gamma\mathrm{L}(2, K)$ has a unique factorization $f = g\sigma_*$, where $g \in \mathrm{GL}(2, K)$ and $\sigma \in \mathrm{Aut}(K)$. If the matrix of g relative to the chosen basis is

$$\begin{bmatrix} a & b \\ c & d \end{bmatrix},$$

define $\psi\colon \Gamma\mathrm{L}(2, K) \to \Gamma\mathrm{LF}(K)$ by $g\sigma_* \mapsto (a\lambda^\sigma + b)/(c\lambda^\sigma + d)$. It is easy to see that ψ is a surjective homomorphism whose kernel consists of all the nonzero scalar matrices. The second isomorphism is just the restriction of this one. ■

We have seen that $\mathrm{P}^1(K)$ is a $\mathrm{P\Gamma L}(2, K)$-set and that $\hat{K}$ is a $\Gamma\mathrm{LF}(K)$-set. There is an isomorphism $\psi\colon \mathrm{P\Gamma L}(2, K) \to \Gamma\mathrm{LF}(K)$ and there is an obvious bijection $\theta\colon \mathrm{P}^1(K) \to \hat{K}$, namely, $[\lambda, 1] \mapsto \lambda$ if $\lambda \in K$ and $[1, 0] \mapsto \infty$. If $\gamma \in \mathrm{P\Gamma L}(2, K)$ and $\lambda \in K$, when is it reasonable to identify the action of γ on $[\lambda, 1]$ with the action of $\psi(\gamma)$ on $\theta([\lambda, 1])$? More generally, assume that there is a G-set X (with action $\alpha\colon G \to S_X$), an H-set Y (with action $\beta\colon H \to S_Y$), and a bijection $\theta\colon X \to Y$. As in Exercise 1.39, θ gives an isomorphism $\theta_*\colon S_X \to S_Y$ by $\pi \mapsto \theta\pi\theta^{-1}$. Finally, assume that there is an isomorphism $\psi\colon G \to H$. There are now two possible ways to view Y as a G-set: via $\theta_*\alpha$ or via $\beta\psi$.

$$\begin{array}{ccc} G & \xrightarrow{\alpha} & S_X \\ {\scriptstyle\psi}\downarrow & & \downarrow{\scriptstyle\theta_*} \\ H & \xrightarrow[\beta]{} & S_Y \end{array}$$

Definition. With the notation above, the G-set X and the H-set Y are ***isomorphic*** if the diagram commutes; that is, if $\theta_*\alpha = \beta\psi$.

EXAMPLE 9.11. The $\Gamma\mathrm{LF}(K)$-set $\hat{K}$ and the $\mathrm{P\Gamma L}(2, K)$-set $\mathrm{P}^1(K)$ are isomorphic.

Let $\psi\colon \mathrm{P\Gamma L}(2, K) \to \Gamma\mathrm{LF}(K)$ be the isomorphism of Theorem 9.47, and let $\theta\colon \mathrm{P}^1(K) \to \hat{K}$ be the bijection given by $[\lambda, 1] \mapsto \lambda$ and $[0, 1] \mapsto \infty$. Now each $\gamma \in \mathrm{P\Gamma L}(2, K)$ is the coset of some semilinear transformation g; relative to the standard basis of K^2, $g = h\sigma_*$, where σ is the corresponding field automorphism and

$$h = \begin{bmatrix} a & b \\ c & d \end{bmatrix}.$$

The action of γ on $[\lambda, \mu]$ is essentially matrix multiplication: $g[\lambda, \mu] = h\sigma_*[\lambda, \mu] = h[\lambda^\sigma, \mu^\sigma]$ (we are regarding $[\lambda^\sigma, \mu^\sigma]$ as a column vector).

To see that $\theta_*\alpha_\gamma = \beta\psi_\gamma$, it must be shown that $\theta_*\alpha_\gamma$ and $\beta\psi_\gamma$ agree on $K \cup \{\infty\}$. If $\lambda \in K$ and $c\lambda^\sigma + d \neq 0$, then $\theta_*\alpha_\gamma(\lambda) = \theta\alpha_\gamma\theta^{-1}(\lambda) = \theta\alpha_\gamma([\lambda, 1]) = \theta(\mathrm{P}(g)[\lambda, 1]) = \theta(\mathrm{P}(h\sigma_*)[\lambda, 1]) = \theta([a\lambda^\sigma + b, c\lambda^\sigma + d]) = \theta[(a\lambda^\sigma + b)/(c\lambda^\sigma + d), 1] = (a\lambda^\sigma + b)/(c\lambda^\sigma + d) = \beta\psi_\gamma(\lambda)$ (ψ_γ is this semilinear fractional transformation, and $\beta\psi_\gamma(\lambda)$ is its evaluation at λ).

The reader may check that $\theta_*\alpha_\gamma(\lambda) = \beta\psi_\gamma(\lambda)$ when $c\lambda^\sigma + d = 0$ and also that $\theta_*\alpha_\gamma(\infty) = \beta\psi_\gamma(\infty)$.

Theorem 9.48. *For every field K, the projective line $\mathrm{P}^1(K)$ is a faithful sharply 3-transitive $\mathrm{PGL}(2, K)$-set.*

Proof. As in Example 9.11 (with the restriction $\mathrm{PGL}(2, K) \to \mathrm{LF}(K)$ replacing the isomorphism $\mathrm{P\Gamma L}(2, K) \to \Gamma\mathrm{LF}(K)$), it suffices to consider linear fractional transformations acting on $\hat{K}$. The stabilizer of ∞ is $\mathrm{Aff}(1, K) = \{\lambda \mapsto a\lambda + b\}$, the subgroup of $\mathrm{LF}(K)$ consisting of all numerators. By Theorem 9.29, K is a sharply 2-transitive $\mathrm{Aff}(1, K)$-set. It suffices, by Theorem 9.8, to show that $\mathrm{LF}(K)$ acts transitively on $\hat{K}$. But if $\mu \in K$, then $f(\lambda) = \lambda + \mu$ sends 0 to μ and $f(\lambda) = 1/\lambda$ sends 0 to ∞. ■

If $K = \mathrm{GF}(q)$, then Theorem 9.8(iii) gives another proof that $|\mathrm{PGL}(2, q)| = (q + 1)q(q - 1)$.

Let us display a second family of sharply 3-transitive G-sets. The groups occur as subgroups of the semilinear fractional transformations $\Gamma\mathrm{LF}(q)$. If $h = (a\lambda^\sigma + b)/(c\lambda^\sigma + d) \in \Gamma\mathrm{LF}(K)$, then $ad - bc \neq 0$. Multiplying numerator and denominator by $\mu \in K^\times$ does not change h, but it does change the "determinant" to $\mu^2(ad - bc)$. If q is a power of an odd prime p, then the nonzero squares form a subgroup of index 2 in $\mathrm{GF}(q)^\times$, namely, $\langle\pi^2\rangle$, where π is a primitive element of $\mathrm{GF}(q)$ (if q is a power of 2, then every element of $\mathrm{GF}(q)$ is a square). In this odd prime case, it thus makes sense to say that $\det(h)$ is or is not a square.

A second ingredient in the coming construction is an automorphism σ of $\mathrm{GF}(q)$ of order 2; σ exists (uniquely) if and only if $q = p^{2n}$, in which case $\lambda^\sigma = \lambda^{p^n}$.

Definition. Let p be an odd prime, let $q = p^{2n}$, and let $\sigma \in \mathrm{Aut}(\mathrm{GF}(q))$ have order 2. Define $M(q) \leq \Gamma\mathrm{LF}(q) = S \cup T$, where

$$S = \{\lambda \mapsto (a\lambda + b)/(c\lambda + d) \mid ad - bc \text{ is a square}\}$$

and

$$T = \{\lambda \mapsto (a\lambda^\sigma + b)/(c\lambda^\sigma + d) \mid ad - bc \text{ is not a square}\}.$$

It is easy to check that $M(q)$ is a subgroup of $\Gamma\mathrm{LF}(K)$; indeed, S is a subgroup of index 2 in $M(q)$, and T is its other coset. Since it is a subgroup of semilinear fractional transformations, $M(q)$ acts faithfully on $\hat{K}$, where $K = \mathrm{GF}(q)$.

Theorem 9.49. *Let p be an odd prime, let $q = p^{2n}$, and let $K = \mathrm{GF}(q)$. Then $\hat{K}$ is a faithful sharply 3-transitive $M(q)$-set.*

Proof. If $G = M(q)$, then $G = S \cup T$ implies $G_\infty = S_\infty \cup T_\infty$, where

$$S_\infty = \{\lambda \mapsto a\lambda + b \mid a \text{ is a square}\}$$

and

$$T_\infty = \{\lambda \mapsto a\lambda^\sigma + b \mid a \text{ is not a square}\},$$

Let α and β be distinct elements of K. If $\alpha - \beta$ is a square, define $h \in S_\infty$ by $h(\lambda) = (\alpha - \beta)\lambda + \beta$; if $\alpha - \beta$ is not a square, define $h \in T_\infty$ by $h(\lambda) = (\alpha - \beta)\lambda^\sigma + \beta$. In either case, $h(1) = \alpha$ and $h(0) = \beta$, so that G_∞ acts doubly transitively on K. But, in each of S_∞ and T_∞, there are q choices for b and $\frac{1}{2}(q-1)$ choices for a, so that $|G_\infty| = q(q-1)$. By Theorem 9.8(iii), this action is sharp. Finally, G acts transitively on $\hat{K}$, for $\lambda \mapsto -1/\lambda$ lies in G (the negative sign gives determinant 1, and 1 is always a square), and $-1/\lambda$ interchanges 0 and ∞. The result now follows from Theorem 9.8(v). ■

Zassenhaus (1936) proved that the actions of $\mathrm{PGL}(2, q)$ and of $M(q)$ on the projective line (if we identify $\mathrm{GF}(q) \cup \{\infty\}$ with the projective line) are the only faithful sharply 3-transitive G-sets; the first family of groups is defined for all prime powers q; the second is defined for all even powers of odd primes. As each of $\mathrm{PGL}(2, q)$ and $M(q)$ acts sharply 3-transitively on a set with $q + 1$ elements, their common order is $(q+1)q(q-1)$ when $q = p^{2n}$ and p is odd. It is true that $\mathrm{PGL}(2, q) \ncong M(q)$ in this case; here is the smallest instance of this fact.

Notation. The group $M(9)$ is denoted by M_{10} (and it is often called the Mathieu group of degree 10).

In the next section, we shall construct five more Mathieu groups: M_{11}, M_{12}, M_{22}, M_{23}, M_{24} (the subscripts indicate the degree of each group's usual representation as a permutation group). However, the phrase "the Mathieu groups" generally refers to these five groups and not to M_{10}.

Theorem 9.50. $\mathrm{PGL}(2, 9)$ *and M_{10} are nonisomorphic groups of order* 720 *acting sharply 3-transitively on $\hat{K}$, where $K = \mathrm{GF}(9)$.*

Proof. We already know that each of these groups acts sharply 3-transitively on $\hat{K}$, a set with 10 elements, and so each group has order $10 \times 9 \times 8 = 720$.

Let $G = \mathrm{PGL}(2, 9)$. The double stabilizer $G_{0,\infty} = \{h: \lambda \mapsto a\lambda \mid a \neq 0\} \cong$

$GF(9)^{\times} \cong \mathbb{Z}_8$; indeed, a generator of $G_{0,\infty}$ is $g: \lambda \mapsto \pi\lambda$, where π is a primitive element of GF(9). There is an involution $t \in G$, namely, $t: \lambda \mapsto 1/\lambda$, such that $tgt = g^{-1}$. It follows that $\langle t, G_{0,\infty} \rangle \cong D_{16}$ is a Sylow 2-subgroup of G.

Let $H = M_{10}$. The automorphism σ of GF(9) in the definition of M_{10} is $\lambda \mapsto \lambda^3$. The double stabilizer

$$\begin{aligned} H_{0,\infty} &= S_{0,\infty} \cup T_{0,\infty} \\ &= \{h: \lambda \mapsto a^2\lambda \mid a \neq 0\} \cup \{h: \lambda \mapsto a\lambda^3 \mid a \text{ nonsquare}\}. \end{aligned}$$

It is now easy to see that $H_{0,\infty}$ is a nonabelian group of order 8 having a unique involution; hence, $H_{0,\infty} \cong \mathbf{Q}$, the quaternions. Since D_{16} has no quaternion subgroups, it follows that G and H have nonisomorphic Sylow 2-subgroups, and so $G \ncong H$. ■

T.Y. Lam and D.B. Leep found that every subgroup of index 2 in $\mathrm{Aut}(S_6)$, a group of order 1440, is isomorphic to either S_6, M_{10}, or PGL(2, 9), and each of these does occur.

Exercises

The polynomial $p(x) = x^2 + x - 1$ is irreducible in $\mathbb{Z}_3[x]$ (for it is a quadratic having no roots in $\mathbb{Z}_3$). Now GF(9) contains a root π of $p(x)$; indeed, $GF(9) = \mathbb{Z}_3(\pi)$, so that π is a primitive element of GF(9). In the following exercises, π denotes a primitive element of GF(9) for which $\pi^2 + \pi = 1$.

9.27. Prove that a Sylow 3-subgroup of M_{10} is elementary abelian of order 9.

9.28. Prove that $(M_{10})_\infty$ is a group of order 72 having a normal Sylow 3-subgroup. Conclude that $(M_{10})_\infty$ is a semidirect product $(\mathbb{Z}_3 \times \mathbb{Z}_3) \rtimes \mathbf{Q}$. (*Hint.* For every $b \in GF(9)^{\times}$, $\lambda \mapsto \lambda + b$ has order 3; the inverse of $\lambda \mapsto \pi^{2i}\lambda + b$ is $\lambda \mapsto \pi^{6i}\lambda - \pi^{6i}b$, and the inverse of $\lambda \mapsto \pi^{2i+1}\lambda^3 + b$ is $\lambda \mapsto \pi^{2i+5}\lambda - \pi^{2i+5}b^3$.)

9.29. There are exactly 8 elements of $(M_{10})_\infty$ of order 3, and they are conjugate to one another in $(M_{10})_\infty$.

9.30. Prove that the subgroup S of $M(q)$ is isomorphic to PSL(2, q). Conclude that M_{10} is neither simple nor solvable. (*Hint.* If $h(\lambda) = (a\lambda + b)/(c\lambda + d)$ and $ad - bc = \mu^2$, multiply numerator and denominator by μ^{-1}.)

9.31. Show that M_{10} is not a semidirect product of S by $\mathbb{Z}_2$. (*Hint.* T contains no involutions.)

9.32. Regard GF(9) as a two-dimensional vector space over $\mathbb{Z}_3$ with basis $\{1, \pi\}$. Verify the following coordinates for the elements of GF(9):

$$\begin{aligned} 1 &= (1, 0), & \pi^4 &= (-1, 0), \\ \pi &= (0, 1), & \pi^5 &= (0, -1), \\ \pi^2 &= (1, -1), & \pi^6 &= (-1, 1), \\ \pi^3 &= (-1, -1), & \pi^7 &= (1, 1). \end{aligned}$$

(*Hint.* $\pi^4 = -1$.)

9.33. Show that $M_{10} = \langle \sigma_1, \sigma_2, \sigma_3, \sigma_4, \sigma_5 \rangle$, where

$$\sigma_1(\lambda) = -1/\lambda, \qquad \sigma_2(\lambda) = \lambda + 1, \qquad \sigma_3(\lambda) = \lambda + \pi,$$

$$\sigma_4(\lambda) = \pi^2\lambda, \qquad \sigma_5(\lambda) = \pi\lambda^3.$$

9.34. Prove that M_{10} consists of even permutations of $\mathrm{P}^1(9)$. (*Hint.* Write each of the generators σ_i, $1 \leq i \leq 5$, as a product of disjoint cycles.)

9.35. Let σ_6 and σ_7 be the permutations of GF(9) defined by $\sigma_6(\lambda) = \pi^2\lambda + \pi\lambda^3$ and $\sigma_7(\lambda) = \lambda^3$. Regarding GF(9) as a vector space over $\mathbb{Z}_3$, prove that σ_6 and σ_7 are linear transformations.

9.36. Prove that $\mathrm{GL}(2, 3) \cong \langle \sigma_4, \sigma_5, \sigma_6, \sigma_7 \rangle$ (where σ_4 and σ_5 are as in Exercise 9.33, and σ_6 and σ_7 are as in Exercise 9.35). (*Hint.* Using the coordinates in Exercise 9.32, one has

$$\sigma_4 = \begin{bmatrix} 1 & -1 \\ -1 & -1 \end{bmatrix}, \qquad \sigma_5 = \begin{bmatrix} 0 & -1 \\ 1 & 0 \end{bmatrix},$$

$$\sigma_6 = \begin{bmatrix} 1 & 1 \\ 0 & -1 \end{bmatrix}, \qquad \sigma_7 = \begin{bmatrix} 1 & -1 \\ 0 & -1 \end{bmatrix}.)$$

Mathieu Groups

We have already seen some doubly and triply transitive groups. In this section, we construct the five simple Mathieu groups; one is 3-transitive, two are 4-transitive, and two are 5-transitive. In 1873, Jordan proved there are no *sharply* 6-transitive groups (other than the symmetric and alternating groups). One consequence of the classification of all finite simple groups is that no 6-transitive groups exist other than the symmetric and alternating groups; indeed, all multiply transitive groups are now known (see the survey article [P.J. Cameron, Finite permutation groups and finite simple groups, *Bull. London Math. Soc.* **13** (1981), pp. 1–22]).

All G-sets in this section are faithful and, from now on, we shall call such groups G ***permutation groups***; that is, $G \leq S_X$ for some set X. Indeed, we finally succumb to the irresistible urge of applying to groups G those adjectives heretofore reserved for G-sets. For example, we will say "G is a doubly transitive group of degree n" meaning that there is a (faithful) doubly transitive G-set X having n elements.

We know that if X is a k-transitive G-set and if $x \in X$, then $X - \{x\}$ is a $(k-1)$-transitive G_x-set. Is the converse true? Is it possible to begin with a k-transitive G_y-set X and construct a $(k+1)$-transitive G-set $X \cup \{y\}$?

Definition. Let G be a permutation group on X and let $\tilde{X} = X \cup \{\infty\}$, where $\infty \notin X$. A transitive permutation group $\tilde{G}$ on $\tilde{X}$ is a ***transitive extension*** of G if $G \leq \tilde{G}$ and $\tilde{G}_\infty = G$.

Recall Lemma 9.5: If X is a k-transitive G-set, then $\tilde{X}$ is a $(k+1)$-transitive $\tilde{G}$-set (should $\tilde{X}$ exist).

Theorem 9.51. *Let G be a doubly transitive permutation group on a set X. Suppose there is $x \in X$, $\infty \notin X$, $g \in G$, and a permutation h of $\tilde{X} = X \cup \{\infty\}$ such that:*

(i) $g \in G_x$;
(ii) $h(\infty) \in X$;
(iii) $h^2 \in G$ *and* $(gh)^3 \in G$; *and*
(iv) $hG_xh = G_x$.

Then $\tilde{G} = \langle G, h \rangle \leq S_{\tilde{X}}$ is a transitive extension of G.

Proof. Condition (ii) shows that $\tilde{G}$ acts transitively on $\tilde{X}$. It suffices to prove, as Theorem 9.4 predicts, that $\tilde{G} = G \cup GhG$, for then $\tilde{G}_\infty = G$ (because nothing in GhG fixes ∞).

By Corollary 2.4, $G \cup GhG$ is a group if it is closed under multiplication. Now

$$\begin{aligned}(G \cup GhG)(G \cup GhG) &\subset GG \cup GGhG \cup GhGG \cup GhGGhG \\ &\subset G \cup GhG \cup GhGhG,\end{aligned}$$

because $GG = G$. It must be shown that $GhGhG \subset G \cup GhG$, and this will follow if we show that $hGh \subset G \cup GhG$.

Since G acts doubly transitively on X, Theorem 9.4 gives $G = G_x \cup G_xgG_x$ (for $g \notin G_x$). The hypothesis gives $\gamma, \delta \in G$ with $h^2 = \gamma$ and $(gh)^3 = \delta$. It follows that $h\gamma^{-1} = h^{-1} = \gamma^{-1}h$ and $hgh = g^{-1}h^{-1}g^{-1}\delta$. Let us now compute.

$$\begin{aligned}hGh &= h(G_x \cup G_xgG_x)h \\ &= hG_xh \cup hG_xgG_xh \\ &= hG_xh \cup (hG_xh)h^{-1}gh^{-1}(hG_xh) \\ &= G_x \cup G_xh^{-1}gh^{-1}G_x \qquad \text{(condition (iv))} \\ &= G_x \cup G_x(\gamma^{-1}h)g(h\gamma^{-1})G_x \\ &= G_x \cup G_x\gamma^{-1}(g^{-1}h^{-1}g^{-1}\delta)\gamma^{-1}G_x \\ &\subset G \cup Gh^{-1}G \\ &= G \cup G\gamma^{-1}hG \\ &= G \cup GhG. \quad \blacksquare\end{aligned}$$

One can say a bit about the cycle structure of h. If $h(\infty) = a \in X$, then $h^2 \in G = \tilde{G}_\infty$ implies $h(a) = h^2(\infty) = \infty$; hence, $h = (\infty \ a)h'$, where $h' \in G_{a,\infty}$ is disjoint from $(\infty \ a)$. Similarly, one can see that gh has a 3-cycle in its factorization into disjoint cycles.

The reader will better understand the choices in the coming constructions once the relation between the Mathieu groups and Steiner systems is seen.

Theorem 9.52. *There exists a sharply* 4-*transitive group* M_{11} *of degree* 11 *and order* $7920 = 11 \cdot 10 \cdot 9 \cdot 8 = 2^4 \cdot 3^2 \cdot 5 \cdot 11$ *such that the stabilizer of a point is* M_{10}.

Proof. By Theorem 9.49, M_{10} acts sharply 3-transitively on $X = \mathrm{GF}(9) \cup \{\infty\}$. We construct a transitive extension of M_{10} acting on $\tilde{X} = \{X, \omega\}$, where ω is a new symbol. If π is a primitive element of GF(9) with $\pi^2 + \pi = 1$, define

$$x = \infty,$$

$$g = (0\ \ \infty)(\pi\ \ \pi^7)(\pi^2\ \ \pi^6)(\pi^3\ \ \pi^5) = 1/\lambda,$$

and

$$h = (\infty\ \ \omega)(\pi\ \ \pi^2)(\pi^3\ \ \pi^7)(\pi^5\ \ \pi^6) = (\omega\ \ \infty)\sigma_6,$$

where $\sigma_6(\lambda) = \pi^2\lambda + \pi\lambda^3$ (use Exercise 9.32 to verify this).

The element g lies in M_{10}, for $\det(g) = -1 = \pi^4$, which is a square in GF(9). It is clear that $g \notin (M_{10})_\infty$ (for $g(\infty) = 0$), $h(\omega) = \infty \in X$, and $h^2 = 1 \in G$. Moreover, $(gh)^3 = 1$ because $gh = (\omega\ \ 0\ \ \infty)(\pi\ \ \pi^6\ \ \pi^3)(\pi^2\ \ \pi^7\ \ \pi^5)$.

To satisfy the last condition of Theorem 9.51, observe that if $f \in (M_{10})_\infty$, then

$$hfh(\infty) = hf(\omega) = h(\omega) = \infty,$$

so that $h(M_{10})_\infty h = (M_{10})_\infty$ if we can show that $hfh \in M_{10}$. Now $(M_{10})_\infty = S_\infty \cup T_\infty$, so that either $f = \pi^{2i}\lambda + \alpha$ or $f = \pi^{2i+1}\lambda^3 + \alpha$, where $i \geq 0$ and $\alpha \in \mathrm{GF}(9)$. In the first case (computing with the second form of $h = (\omega\ \ \infty)\sigma_6$),

$$hfh(\lambda) = (\pi^{2i+4} + \pi^{6i+4})\lambda + (\pi^{2i+3} + \pi^{6i+7})\lambda^3 + \pi^2\alpha + \pi\alpha^3.$$

The coefficients of λ and λ^3 are $\pi^{2i+4}(1 + \pi^{4i})$ and $\pi^{2i+3}(1 + \pi^{4i+4})$, respectively. When $i = 2j$ is even, the second coefficient is 0 and the first coefficient is π^{4j+4}, which is a square; hence, $hfh \in S_\infty \leq M_{10}$ in this case. When $i = 2j + 1$ is odd, the first coefficient is 0 and the second coefficient is $\pi^{4j}\pi^5$, which is a nonsquare, so that $hfh \in T_\infty \subset M_{10}$. The second case ($f = \pi^{2i+1}\lambda^3 + \alpha$) is similar; the reader may now calculate that

$$hfh(\lambda) = \pi^{2i+6}(1 + \pi^{4i})\lambda + \pi^{2i+1}(1 + \pi^{4i+4})\lambda^3 + \pi^2\alpha + \pi\alpha^3,$$

an expression which can be treated as the similar expression in the first case.

It follows from Theorem 9.8(v) that M_{11}, defined as $\langle M_{10}, h\rangle$, acts sharply 4-transitively on $\tilde{X}$, and so $|M_{11}| = 7920$. ∎

Note, for later use, that both g and h are even permutations, so that Exercise 9.34 gives $M_{11} \leq A_{11}$.

This procedure can be repeated; again, the difficulty is discovering a good permutation to adjoin.

Theorem 9.53. *There exists a sharply 5-transitive group M_{12} of degree 12 and order 95,040 $= 12 \cdot 11 \cdot 10 \cdot 9 \cdot 8 = 2^6 \cdot 3^3 \cdot 5 \cdot 11$ such that the stabilizer of a point is M_{11}.*

Proof. By Theorem 9.52, M_{11} acts sharply 4-transitively on $Y = \{\mathrm{GF}(9), \infty, \omega\}$. We construct a transitive extension of M_{11} acting on $\tilde{Y} = \{Y, \Omega\}$, where Ω is a new symbol. If π is a primitive element of GF(9) with $\pi^2 + \pi = 1$, define

$$x = \omega,$$

$$h = (\infty \ \omega)(\pi \ \pi^2)(\pi^3 \ \pi^7)(\pi^5 \ \pi^6),$$

and

$$k = (\omega \ \Omega)(\pi \ \pi^3)(\pi^2 \ \pi^6)(\pi^5 \ \pi^7) = (\omega \ \Omega)\lambda^3 = (\omega \ \Omega)\sigma_7$$

(note that this is the same h occurring in the construction of M_{11}). Clearly $k(\Omega) = \omega \in Y$ and $h \notin (M_{11})_\omega = M_{10}$. Also, $k^2 = 1$ and $hk = (\omega \ \Omega \ \infty)(\pi \ \pi^7 \ \pi^6)(\pi^2 \ \pi^5 \ \pi^3)$ has order 3. To satisfy the last condition of Theorem 9.51, observe first that if $f \in (M_{11})_\omega = M_{10} = S \cup T$, then kfk also fixes ω. Finally, $kfk \in M_{11}$: if $f(\lambda) = (a\lambda + b)/(c\lambda + d) \in S$, then $kfk(\lambda) = (a^3\lambda + b^3)/(c^3\lambda + d^3)$ has determinant $a^3d^3 - b^3c^3 = (ad - bc)^3$, which is a square because $ad - bc$ is; a similar argument holds when $f \in T$. Thus, $kM_{10}k = M_{10}$.

It follows from Theorem 9.8(v) that M_{12}, defined as $\langle M_{11}, k \rangle$, acts sharply 5-transitively on $\tilde{Y}$, and so $|M_{12}| = 95{,}040$. ■

Note that k is an even permutation, so that $M_{12} \leq A_{12}$.

The theorem of Jordan mentioned at the beginning of this section can now be stated precisely: The only sharply 4-transitive groups are S_4, S_5, A_6, and M_{11}; the only sharply 5-transitive groups are S_5, S_6, A_7, and M_{12}; if $k \geq 6$, then the only sharply k-transitive groups are S_k, S_{k+1}, and A_{k+2}. We remind the reader that Zassenhaus (1936) classified all sharply 3-transitive groups (there are only PGL(2, q) and $M(p^{2n})$ for odd primes p). If p is a prime and $q = p^n$, then Aut(1, q) is a solvable doubly transitive group of degree q. Zassenhaus (1936) proved that every sharply 2-transitive group, with only finitely many exceptions, can be imbedded in Aut(1, q) for some q; Huppert (1957) generalized this by proving that any faithful doubly transitive solvable group can, with only finitely many more exceptions, be imbedded in Aut(1, q) for some q. Thompson completed the classification of sharply 2-transitive groups as certain Frobenius groups. The classification of all finite simple groups can be used to give an explicit enumeration of all faithful doubly transitive groups. The classification of all sharply 1-transitive groups, that is, of all regular groups, is, by Cayley's theorem, the classification of all finite groups.

The "large" Mathieu groups are also constructed as a sequence of transitive extensions, but now beginning with PSL(3, 4) (which acts doubly transitively on $P^2(4)$) instead of with M_{10}. Since $|P^2(4)| = 4^2 + 4 + 1 = 21$, one begins with a permutation group of degree 21. We describe elements of $P^2(4)$ by their homogeneous coordinates.

Lemma 9.54. *Let β be a primitive element of* GF(4). *The functions f_i: $P^2(4) \to P^2(4)$, for $i = 1, 2, 3$, defined by*

$$f_1[\lambda, \mu, \nu] = [\lambda^2 + \mu\nu, \mu^2, \nu^2],$$

$$f_2[\lambda, \mu, \nu] = [\lambda^2, \mu^2, \beta\nu^2],$$

$$f_3[\lambda, \mu, \nu] = [\lambda^2, \mu^2, \nu^2],$$

are involutions which fix $[1, 0, 0]$. *Moreover,*

$$\langle \mathrm{PSL}(3, 4), f_2, f_3 \rangle = \mathrm{P\Gamma L}(3, 4).$$

Proof. The proof is left as an exercise for the reader (with the reminder that all 3-tuples are regarded as column vectors). A hint for the second statement is that $\mathrm{PSL}(3, 4) \lhd \mathrm{P\Gamma L}(3, 4)$, $\mathrm{P\Gamma L}(3, 4)/\mathrm{PSL}(3, 4) \cong S_3$, and, if the unique nontrivial automorphism of GF(4) is σ: $\lambda \mapsto \lambda^2$, then $f_3 = \sigma_*$ and

$$f_2 = \begin{bmatrix} 1 & 0 & 0 \\ 0 & 1 & 0 \\ 0 & 0 & \beta \end{bmatrix} \sigma_*. \quad \blacksquare$$

Theorem 9.55. *There exists a* 3-*transitive group M_{22} of degree* 22 *and order* $443{,}520 = 22 \cdot 21 \cdot 20 \cdot 48 = 2^7 \cdot 3^2 \cdot 5 \cdot 7 \cdot 11$ *such that the stabilizer of a point is* PSL(3, 4).

Proof. We show that $G = \mathrm{PSL}(3, 4)$ acting on $X = P^2(4)$ has a transitive extension. Let

$$x = [1, 0, 0],$$

$$g[\lambda, \mu, \nu] = [\mu, \lambda, \nu],$$

$$h_1 = (\infty \;\; [1, 0, 0])f_1.$$

In matrix form,

$$g = \begin{bmatrix} 0 & 1 & 0 \\ 1 & 0 & 0 \\ 0 & 0 & 1 \end{bmatrix},$$

so that $\det(g) = -1 = 1 \in \mathrm{GF}(4)$ and $g \in \mathrm{PSL}(3, 4)$. It is plain that g does not fix $x = [1, 0, 0]$ and, by the lemma, that $h_1^2 = 1$. The following computation shows that $(gh_1)^3 = 1$. If $[\lambda, \mu, \nu] \neq \infty, [1, 0, 0]$, or $[0, 1, 0]$, then

$$(gh_1)^3[\lambda, \mu, \nu] = [\lambda\nu + \mu^2(\nu^3 + 1), \mu\nu + \lambda^2(\nu^3 + 1), \nu^2].$$

If $\nu \neq 0$, then $\nu^3 = 1$ and $\nu^3 + 1 = 0$, so that the right side is $[\lambda\nu, \mu\nu, \nu^2] =$

$[\lambda, \mu, \nu]$. If $\nu = 0$, then the right side is $[\mu^2, \lambda^2, 0]$; since $\lambda\mu \neq 0$, by our initial choice of $[\lambda, \mu, \nu]$, we have $[\mu^2, \lambda^2, 0] = [(\lambda\mu)\mu^2, (\lambda\mu)\lambda^2, 0] = [\lambda, \mu, 0]$. The reader may show that $(gh_1)^3$ also fixes ∞, $[1, 0, 0]$, and $[0, 1, 0]$, so that $(gh_1)^3 = 1$.

Finally, assume that $k \in G_x \leq \mathrm{PSL}(3, 4)$, so that k is the coset (mod scalar matrices) of

$$k = \begin{bmatrix} 1 & * & * \\ 0 & a & b \\ 0 & c & d \end{bmatrix}$$

(because k fixes $[1, 0, 0]$). Now $\det(k) = 1 = ad - bc$. The reader may now calculate that $h_1 k h_1$, mod scalars, is

$$h_1 k h_1 = \begin{bmatrix} 1 & * & * \\ 0 & a^2 & b^2 \\ 0 & c^2 & d^2 \end{bmatrix}$$

which fixes $[1, 0, 0]$ and whose determinant is $a^2d^2 - b^2c^2 = (ad - bc)^2 = 1$. Thus $h_1 G_x h_1 = G_x$, and Theorem 9.51 shows that $M_{22} = \langle \mathrm{PSL}(3, 4), h_1 \rangle$ acts 3-transitively on $\tilde{X} = \mathrm{P}^2(4) \cup \{\infty\}$ with $(M_{22})_\infty = \mathrm{PSL}(3, 4)$.

By Theorem 9.7, $|M_{22}| = 22 \cdot 21 \cdot 20 \cdot |H|$, where H is the stabilizer in M_{22} of three points. Since $(M_{22})_\infty = \mathrm{PSL}(3, 4)$, we may consider H as the stabilizer in $\mathrm{PSL}(3, 4)$ of two points, say, $[1, 0, 0]$ and $[0, 1, 0]$. If $A \in \mathrm{SL}(3, 4)$ sends $(1, 0, 0)$ to $(\alpha, 0, 0)$ and $(0, 1, 0)$ to $(0, \beta, 0)$, then A has the form

$$A = \begin{bmatrix} \alpha & 0 & \gamma \\ 0 & \beta & \delta \\ 0 & 0 & \eta \end{bmatrix},$$

where $\eta = (\alpha\beta)^{-1}$. There are 3 choices for each of α and β, and 4 choices for each of γ and δ, so that there are 144 such matrices A. Dividing by $\mathrm{SZ}(3, 4)$ (which has order 3), we see that $|H| = 48$. ■

Theorem 9.56. *There exists a 4-transitive group M_{23} of degree 23 and order $10{,}200{,}960 = 23 \cdot 22 \cdot 21 \cdot 20 \cdot 48 = 2^7 \cdot 3^2 \cdot 5 \cdot 7 \cdot 11 \cdot 23$ such that the stabilizer of a point is M_{22}.*

Proof. The proof is similar to that for M_{22}, and so we only provide the necessary ingredients. Adjoin a new symbol ω to $\mathrm{P}^2(4) \cup \{\infty\}$, and let

$$x = \infty,$$

$$g = (\infty \;\; [1, 0, 0])f_1 = \text{the former } h_1,$$

$$h_2 = (\omega \;\; \infty)f_2.$$

The reader may apply Theorem 9.51 to show that $M_{23} = \langle M_{22}, h_2 \rangle$ is a transitive extension of M_{22}. ■

Theorem 9.57. *There exists a 5-transitive group M_{24} of degree 24 and order* $244{,}823{,}040 = 24 \cdot 23 \cdot 22 \cdot 21 \cdot 20 \cdot 48 = 2^{10} \cdot 3^3 \cdot 5 \cdot 7 \cdot 11 \cdot 23$ *such that the stabilizer of a point is M_{23}.*

Proof. Adjoin a new symbol Ω to $P^2(4) \cup \{\infty, \omega\}$, and define

$$x = \omega,$$

$$g = (\omega \ \ \infty)f_2 = \text{the former } h_2,$$

$$h_3 = (\Omega \ \ \omega)f_3.$$

The reader may check that Theorem 9.51 gives $M_{24} = \langle M_{23}, h_3 \rangle$ a transitive extension of M_{23}. ■

Theorem 9.58 (Miller, 1900). *The Mathieu groups M_{22}, M_{23}, and M_{24} are simple groups.*

Proof. Since M_{22} is 3-transitive of degree 22 (which is not a power of 2) and since the stabilizer of a point is the simple group PSL(3, 4), Theorem 9.25(ii) gives simplicity of M_{22}. The group M_{23} is 4-transitive and the stabilizer of a point is the simple group M_{22}, so that Theorem 9.25(i) gives simplicity of M_{23}. Finally, M_{24} is 5-transitive and the stabilizer of a point is the simple group M_{23}, so that Theorem 9.25(i) applies again to give simplicity of M_{24}. ■

Theorem 9.59 (Cole, 1896; Miller, 1899). *The Mathieu groups M_{11} and M_{12} are simple.*

Proof. Theorem 9.25(i) will give simplicity of M_{12} once we prove that M_{11} is simple. The simplicity of M_{11} cannot be proved in this way because the stabilizer of a point is M_{10}, which is not a simple group.

Let H be a nontrivial normal subgroup of M_{11}. By Theorem 9.17, H is transitive of degree 11, so that $|H|$ is divisible by 11. Let P be a Sylow 11-subgroup of H. Since $(11)^2$ does not divide $|M_{11}|$, P is also a Sylow 11-subgroup of M_{11}, and P is cyclic of order 11.

We claim that $P \neq N_H(P)$. Otherwise, P abelian implies $P \leq C_H(P) \leq N_H(P)$ and $N_H(P)/C_H(P) = 1$. Burnside's normal complement theorem (Theorem 7.50) applies: P has a normal complement Q in H. Now $|Q|$ is not divisible by 11, so that $Q \operatorname{char} H$; as $H \lhd M_{11}$, Lemma 5.20(ii) gives $Q \lhd M_{11}$. If $Q \neq 1$, then Theorem 9.17 shows that $|Q|$ is divisible by 11, a contradiction. If $Q = 1$, then $P = H$. In this case, H is abelian, and Exercise 9.10 gives H a regular normal subgroup, contradicting Lemma 9.24.

Let us compute $N_{M_{11}}(P)$. In S_{11}, there are $11!/11 = 10!$ 11-cycles, and hence $9!$ cyclic subgroups of order 11 (each of which consists of 10 11-cycles and the identity). Therefore $[S_{11} : N_{S_{11}}(P)] = 9!$ and $|N_{S_{11}}(P)| = 110$. Now $N_{M_{11}}(P) = N_{S_{11}}(P) \cap M_{11}$. We may assume that $P = \langle \sigma \rangle$, where $\sigma =$

(1 2 ... 10 11); if $\tau = (1\ 11)(2\ 10)(3\ 9)(4\ 8)(5\ 7)$, then τ is an involution with $\tau\sigma\tau = \sigma^{-1}$ and $\tau \in N_{S_{11}}(P)$. But τ is an odd permutation, whereas $M_{11} \leq A_{11}$, so that $|N_{M_{11}}(P)| = 11$ or 55. Now $P \leq N_H(P) \leq N_{M_{11}}(P)$, so that either $P = N_H(P)$ or $N_H(P) = N_{M_{11}}(P)$. The first paragraph eliminated the first possibility, and so $N_H(P) = N_{M_{11}}(P)$ (and their common order is 55). The Frattini argument now gives $M_{11} = HN_{M_{11}}(P) = HN_H(P) = H$ (for $N_H(P) \leq H$), and so M_{11} is simple. ■

Exercises

9.37. Show that the 4-group $\mathbf{V}$ has no transitive extension. (*Hint.* If $h \in S_5$ has order 5, then $\langle \mathbf{V}, h \rangle \geq A_5$.)

9.38. Let $W = \{g \in M_{12}: g \text{ permutes } \{\infty, \omega\ \Omega\}\}$. Show that there is a homomorphism of W onto S_3 with kernel $(M_{12})_{\infty,\omega,\Omega}$. Conclude that $|W| = 6 \times 72$.

9.39. Prove that Aut(2, 3), the group of all affine automorphisms of a two-dimensional vector space over $\mathbb{Z}_3$, is isomorphic to the subgroup W of M_{12} in the previous exercise. (*Hint.* Regard GF(9) as a vector space over $\mathbb{Z}_3$.)

9.40. Show that $\langle \mathrm{PSL}(3, 4), h_2, h_3 \rangle \leq M_{24}$ is isomorphic to $\mathrm{P\Gamma L}(3, 4)$. (*Hint.* Lemma 9.54.)

Steiner Systems

A Steiner system, defined below, is a set together with a family of subsets which can be thought of as generalized lines; it can thus be viewed as a kind of geometry, generalizing the notion of affine space, for example. If X is a set with $|X| = v$, and if $k \leq v$, then a ***k-subset*** of X is a subset $B \subset X$ with $|B| = k$.

Definition. Let $1 < t < k < v$ be integers. A ***Steiner system*** of ***type*** $S(t, k, v)$ is an ordered pair $(X, \mathscr{B})$, where X is a set with v elements, $\mathscr{B}$ is a family of k-subsets of X, called ***blocks***, such that every t elements of X lie in a unique block.

Example 9.12. Let X be an affine plane over the field GF(q), and let $\mathscr{B}$ be the family of all affine lines in X. Then every line has q points and every two points determine a unique line, so that $(X, \mathscr{B})$ is a Steiner system of type $S(2, q, q^2)$.

Example 9.13. Let $X = \mathrm{P}^2(q)$ and let $\mathscr{B}$ be the family of all projective lines in X. Then every line has $q + 1$ points and every two points determine a unique line, so that $(X, \mathscr{B})$ is a Steiner system of type $S(2, q + 1, q^2 + q + 1)$.

Example 9.14. Let X be an m-dimensional vector space over $\mathbb{Z}_2$, where $m \geq 3$, and let $\mathscr{B}$ be the family of all planes (affine 2-subsets of X). Since three distinct points cannot be collinear, it is easy to see that $(X, \mathscr{B})$ is a Steiner system of type $S(3, 4, 2^m)$.

One assumes strict inequalities $1 < t < k < v$ to eliminate uninteresting cases. If $t = 1$, every point lies in a unique block, and so X is just a set partitioned into k-subsets; if $t = k$, then every t-subset is a block; if $k = v$, then there is only one block. In the first case, all "lines" (blocks) are parallel; in the second case, there are too many blocks; in the third case, there are too few blocks.

Given parameters $1 < t < k < v$, it is an open problem whether there exists a Steiner system of type $S(t, k, v)$. For example, one defines a ***projective plane of order n*** to be a Steiner system of type $S(2, n + 1, n^2 + n + 1)$. It is conjectured that n must be a prime power, but it is still unknown whether there exists a projective plane of order 12. (There is a theorem of Bruck and Ryser (1949) saying that if $n \equiv 1$ or $2 \bmod 4$ and n is not a sum of two squares, then there is no projective plane of order n; note that $n = 10$ is the first integer which neither satisfies this hypothesis nor is a prime power. In 1988, C. Lam proved, using massive amounts of computer time, that there is no projective plane of order 10.)

Definition. If $(X, \mathscr{B})$ is a Steiner system and $x \in X$, then

$$\textbf{\textit{star}}(x) = \{B \in \mathscr{B}: x \in \mathscr{B}\}.$$

Theorem 9.60. *Let $(X, \mathscr{B})$ be a Steiner system of type $S(t, k, v)$, where $t \geq 3$. If $x \in X$, define $X' = X - \{x\}$ and $\mathscr{B}' = \{B - \{x\}: B \in \text{star}(x)\}$. Then $(X', \mathscr{B}')$ is a Steiner system of type $S(t - 1, k - 1, v - 1)$ (called the* **contraction** *of $(X, \mathscr{B})$* ***at x****).*

Proof. The routine proof is left to the reader. ∎

A contraction of $(X, \mathscr{B})$ may depend on the point x.

Let Y and Z be finite sets, and let $W \subset Y \times Z$. For each $y \in Y$, define $\#(y, \ ) = |\{z \in Z: (y, z) \in W\}|$ and define $\#(\ , z) = |\{y \in Y: (y, z) \in W\}|$. Clearly,

$$\sum_{y \in Y} \#(y, \ ) = |W| = \sum_{z \in Z} \#(\ , z).$$

We deduce a ***counting principle***: If $\#(y, \ ) = m$ for all $y \in Y$ and if $\#(\ , z) = n$ for all $z \in Z$, then

$$m|Y| = n|Z|.$$

Theorem 9.61. *Let $(X, \mathscr{B})$ be a Steiner system of type $S(t, k, v)$. Then the number of blocks is*

$$|\mathscr{B}| = \frac{v(v - 1)(v - 2)\dots(v - t + 1)}{k(k - 1)(k - 2)\dots(k - t + 1)};$$

if r is the number of blocks containing a point $x \in X$, then r is independent of x

and

$$r = \frac{(v-1)(v-2)\cdots(v-t+1)}{(k-1)(k-2)\cdots(k-t+1)}.$$

Proof. If Y is the family of all t-subsets of X, then $|Y| =$ "v choose t" $= v(v-1)\cdots(v-t+1)/t!$. Define $W \subset Y \times \mathscr{B}$ to consist of all $(\{x_1, \ldots, x_t\}, B)$ with $\{x_1, \ldots, x_t\} \subset B$. Since every t-subset lies in a unique block, $\#(\{x_1, \ldots, x_t\}, \ ) = 1$; since each block B is a k-subset, $\#(\ , B) =$ "k choose t" $= k(k-1)\cdots(k-t+1)/t!$. The counting principle now gives the desired formula for $|\mathscr{B}|$.

The formula for r follows from that for $|\mathscr{B}|$ because r is the number of blocks in the contraction $(X', \mathscr{B}')$ (where $X' = X - \{x\}$), which is a Steiner system of type $S(t-1, k-1, v-1)$. It follows that r does not depend on the choice of x. ■

Remarks. 1. The proof just given holds for all $t \geq 2$ (of course, $(X', \mathscr{B}')$ is not a Steiner system when $t = 2$ since $t - 1 = 1$).

2. The same proof gives a formula for the number of blocks in a Steiner system of type $S(t, k, v)$ containing two points x and y. If $(X', \mathscr{B}')$ is the contraction (with $X' = X - \{x\}$), then the number r' of blocks in $(X', \mathscr{B}')$ containing y is the same as the number of blocks in $(X, \mathscr{B})$ containing x and y. Therefore,

$$r' = \frac{(v-2)(v-3)\cdots(v-t+1)}{(k-2)(k-3)\cdots(k-t+1)}.$$

Similarly, the number $r^{(p)}$ of blocks in $(X, \mathscr{B})$ containing p points, where $1 \leq p \leq t$, is

$$r^{(p)} = \frac{(v-p)(v-p-1)\cdots(v-t+1)}{(k-p)(k-p-1)\cdots(k-t+1)}.$$

3. That the numbers $|\mathscr{B}| = r, r', \ldots r^{(p)}, \ldots, r^{(t)}$ are integers is, of course, a constraint on t, k, v.

Definition. If $(X, \mathscr{B})$ and $(Y, \mathscr{C})$ are Steiner systems, then an ***isomorphism*** is a bijection $f\colon X \to Y$ such that $B \in \mathscr{B}$ if and only if $f(B) \in \mathscr{C}$. If $(X, \mathscr{B}) = (Y, \mathscr{C})$, then f is called an ***automorphism***.

For certain parameters t, k, and v, there is a unique, to isomorphism, Steiner system of type $S(t, k, v)$, but there may exist nonisomorphic Steiner systems of the same type. For example, it is known that there are exactly four projective planes of order 9; that is, there are exactly four Steiner systems of type $S(2, 10, 91)$.

Theorem 9.62. *All the automorphisms of a Steiner system* $(X, \mathscr{B})$ *form a group* $\operatorname{Aut}(X, \mathscr{B}) \leq S_X$.

Proof. The only point needing discussion is whether the inverse of an automorphism h is itself an automorphism. But S_X is finite, and so $h^{-1} = h^m$ for some $m \geq 1$. The result follows, for it is obvious that the composite of automorphisms is an automorphism. ■

Theorem 9.63. *If $(X, \mathscr{B})$ is a Steiner system, then* $\text{Aut}(X, \mathscr{B})$ *acts faithfully on $\mathscr{B}$.*

Proof. If $\varphi \in \text{Aut}(X, \mathscr{B})$ and $\varphi(B) = B$ for all blocks B, then it must be shown that $\varphi = 1_X$.

For $x \in X$, let $r = |\text{star}(x)|$, the number of blocks containing x. Since φ is an automorphism, $\varphi(\text{star}(x)) = \text{star}(\varphi(x))$; since φ fixes every block, $\varphi(\text{star}(x)) = \text{star}(x)$, so that $\text{star}(x) = \text{star}(\varphi(x))$. Thus, $\varphi(x)$ and x lie in exactly the same blocks, and so the number r' of blocks containing $\{\varphi(x), x\}$ is the same as the number r of blocks containing x. If $\varphi(x) \neq x$, however, $r' = r$ gives $k = v$ (using the formulas in Theorem 9.61 and the remark thereafter), contradicting $k < v$. Therefore, $\varphi(x) = x$ for all $x \in X$. ■

Corollary 9.64. *If $(X, \mathscr{B})$ is a Steiner system and $x \in X$, then $\bigcap_{B \in \text{star}(x)} B = \{x\}$.*

Proof. Let $x, y \in X$. If $\text{star}(x) = \text{star}(y)$, then the argument above gives the contradiction $r' = r$. Therefore, if $y \neq x$, there is a block B with $x \in B$ and $y \notin B$, so that $y \notin \bigcap_{B \in \text{star}(x)} B$. ■

We are going to see that multiply transitive groups may determine Steiner systems.

Notation. If X is a G-set and $U \leq G$ is a subgroup, then

$$\mathscr{F}(U) = \{x \in X : gx = x \text{ for all } g \in U\}.$$

Recall that if $U \leq G$ and $g \in G$, then the conjugate gUg^{-1} may be denoted by U^g.

Lemma 9.65. *If X is a G-set and $U \leq G$ is a subgroup, then*

$$\mathscr{F}(U^g) = g\mathscr{F}(U) \qquad \textit{for all} \quad g \in G.$$

Proof. The following statements are equivalent for $x \in X$: $x \in \mathscr{F}(U^g)$; $gug^{-1}(x) = x$ for all $u \in U$; $ug^{-1}(x) = g^{-1}(x)$ for all $u \in U$; $g^{-1}(x) \in \mathscr{F}(U)$; $x \in g\mathscr{F}(U)$. ■

Theorem 9.66. *Let X be a faithful t-transitive G-set, where $t \geq 2$, let H be the stabilizer of t points $x_1, \ldots, x_t$ in X, and let U be a Sylow p-subgroup of H for some prime p.*

(i) *$N_G(U)$ acts t-transitively on $\mathscr{F}(U)$.*

(ii) **(*Carmichael, 1931; Witt, 1938*).** *If $k = |\mathscr{F}(U)| > t$ and U is a nontrivial normal subgroup of H, then $(X, \mathscr{B})$ is a Steiner system of type $S(t, k, v)$, where $|X| = v$ and*

$$\mathscr{B} = \{g\mathscr{F}(U): g \in G\} = \{\mathscr{F}(U^g): g \in G\}.$$

Proof. (i) Note that $\mathscr{F}(U)$ is a $N_G(U)$-set: if $g \in N_G(U)$, then $U = U^g$ and $\mathscr{F}(U) = \mathscr{F}(U^g) = g\mathscr{F}(U)$. Now $\{x_1, \ldots, x_t\} \subset \mathscr{F}(U)$ because $U \leq H$, the stabilizer of $x_1, \ldots, x_t$; hence $k = |\mathscr{F}(U)| \geq t$. If $y_1, \ldots, y_t$ are distinct elements of $\mathscr{F}(U)$, then t-transitivity of G gives $g \in G$ with $gy_i = x_i$ for all i. If $u \in U$, then $gug^{-1}x_i = guy_i = gy_i = x_i$ (because $y_i \in \mathscr{F}(U)$); that is, $U^g \leq H$. By the Sylow theorem, there exists $h \in H$ with $U^g = U^h$. Therefore $h^{-1}g \in N_G(U)$ and $(h^{-1}g)y_i = h^{-1}x_i = x_i$ for all i.

(ii) The hypothesis gives $1 < t < k \leq v$. If $k = v$, then $\mathscr{F}(U) = X$; but $U \neq 1$, contradicting G acting faithfully on X. It is also clear that $k = |\mathscr{F}(U)| = |g\mathscr{F}(U)|$ for all $g \in G$.

If $y_1, \ldots, y_t$ are distinct elements of X, then there is $g \in G$ with $gx_i = y_i$ for all i, and so $\{y_1, \ldots, y_t\} \subset g\mathscr{F}(U)$. It remains to show that $g\mathscr{F}(U)$ is the unique block containing the y_i. If $\{y_1, \ldots, y_t\} \subset h\mathscr{F}(U)$, then there are $z_1, \ldots, z_t \in \mathscr{F}(U)$ with $y_i = hz_i$ for all i. By (i), there is $\sigma \in N_G(U)$ with $z_i = \sigma x_i$ for all i, and so $gx_i = y_i = h\sigma x_i$ for all i. Hence $g^{-1}h\sigma$ fixes all x_i *and* $g^{-1}h\sigma \in H$. Now $H \leq N_G(U)$, because $U \lhd H$, so that $g^{-1}h\sigma \in N_G(U)$ and $g^{-1}h \in N_G(U)$. Therefore, $U^g = U^h$ and $g\mathscr{F}(U) = \mathscr{F}(U^g) = \mathscr{F}(U^h) = h\mathscr{F}(U)$, as desired. ■

Lemma 9.67. *Let $H \leq M_{24}$ be the stabilizer of the five points*

$$\infty, \omega, \Omega, [1, 0, 0], \text{ and } [0, 1, 0].$$

(i) *H is a group of order 48 having a normal elementary abelian Sylow 2-subgroup U of order 16.*

(ii) *$\mathscr{F}(U) = \ell \cup \{\infty, \omega, \Omega\}$, where ℓ is the projective line $v = 0$, and so $|\mathscr{F}(U)| = 8$.*

(iii) *Only the identity of M_{24} fixes more than 8 points.*

Proof. (i) Consider the group $\tilde{H}$ of all matrices over GF(4) of the form

$$A = \lambda \begin{bmatrix} 1 & 0 & \alpha \\ 0 & \gamma & \beta \\ 0 & 0 & \gamma^{-1} \end{bmatrix},$$

where $\lambda, \gamma \neq 0$. There are 3 choices for each of λ and γ, and 4 choices for each of α and β, so that $|\tilde{H}| = 3 \times 48$. Clearly $\tilde{H}/Z(3, 4)$ has order 48, lies in $PSL(3, 4) \leq M_{24}$, and fixes the five listed points, so that $H = \tilde{H}/Z(3, 4)$ (we know that $|H| = 48$ from Theorem 9.57). Define $\tilde{U} \leq \tilde{H}$ to be all those matrices A above for which $\gamma = 1$. Then $U = \tilde{U}/Z(3, 4)$ has order 16 and consists of involutions; that is, U is elementary abelian. But $\tilde{U} \lhd \tilde{H}$, being the kernel of the map $\tilde{H} \to \mathrm{SL}(3, 4)$ given by

$$A \mapsto \begin{bmatrix} \lambda & 0 & 0 \\ 0 & \lambda\gamma & 0 \\ 0 & 0 & \lambda^{-1} \end{bmatrix},$$

so that $U \lhd H$.

(ii) Assume that $[\lambda, \mu, \nu] \in \mathscr{F}(U)$. If $h \in U$, then $\gamma = 1$ and

$$h\begin{bmatrix} \lambda \\ \mu \\ \nu \end{bmatrix} = \begin{bmatrix} 1 & 0 & \alpha \\ 0 & 1 & \beta \\ 0 & 0 & 1 \end{bmatrix}\begin{bmatrix} \lambda \\ \mu \\ \nu \end{bmatrix} = \begin{bmatrix} \lambda + \alpha\nu \\ \mu + \beta\nu \\ \nu \end{bmatrix} = \begin{bmatrix} \xi\lambda \\ \xi\mu \\ \xi\nu \end{bmatrix}$$

for some $\xi \in \mathrm{GF}(4)^\times$. If $\nu = 0$, then all projective points of the form $[\lambda, \mu, 0]$ (which form a projective line ℓ having $4 + 1 = 5$ points) are fixed by h. If $\nu \neq 0$, then these equations have no solution, and so h fixes no other projective points. Therefore, every $h \in U$ fixes ℓ, ∞, ω, Ω, and nothing else, so that $\mathscr{F}(U) = \ell \cup \{\infty, \omega, \Omega\}$ and $|\mathscr{F}(U)| = 8$.

(iii) By 5-transitivity of M_{24}, it suffices to show that $h \in H^\#$ can fix at most 3 projective points in addition to $[1, 0, 0]$ and $[0, 1, 0]$. Consider the equations for $\xi \in \mathrm{GF}(4)^\times$:

$$h\begin{bmatrix} \lambda \\ \mu \\ \nu \end{bmatrix} = \begin{bmatrix} 1 & 0 & \alpha \\ 0 & \gamma & \beta \\ 0 & 0 & \gamma^{-1} \end{bmatrix}\begin{bmatrix} \lambda \\ \mu \\ \nu \end{bmatrix} = \begin{bmatrix} \lambda + \alpha\nu \\ \gamma\mu + \beta\nu \\ \gamma^{-1}\nu \end{bmatrix} = \begin{bmatrix} \xi\lambda \\ \xi\mu \\ \xi\nu \end{bmatrix}.$$

If $\nu = 0$, then we may assume that $\lambda \neq 0$ (for $[0, 1, 0]$ is already on the list of five). Now $\lambda = \lambda + \alpha\nu = \xi\lambda$ and $\mu = \gamma\mu + \beta\nu = \xi\mu$ give $\gamma = 1$; hence $h \in U$ and h fixes exactly 8 elements, as we saw in (ii). If $\nu \neq 0$, then $\nu = \gamma^{-1}\nu = \xi\nu$ implies $\xi = \gamma^{-1}$; we may assume that $\gamma \neq 1$ lest $h \in U$. The equations can now be solved uniquely for λ and μ ($\lambda = (\gamma^{-1} - 1)^{-1}\alpha\nu$ and $\mu = (\gamma^{-1} - \gamma)^{-1}\beta\nu$), so that $h \notin U$ can fix only one projective point other than $[1, 0, 0]$ and $[0, 1, 0]$; that is, such an h can fix at most 6 points. ∎

Theorem 9.68. *Neither M_{12} nor M_{24} has a transitive extension.*

Proof. In order to show that M_{12} has no transitive extension, it suffices to show that there is no sharply 6-transitive group G of degree 13. Now such a group G would have order $13 \cdot 12 \cdot 11 \cdot 10 \cdot 9 \cdot 8$. If $g \in G$ has order 5, then g is a product of two 5-cycles and hence fixes 3 points (g cannot be a 5-cycle lest it fix $8 > 6$ points). Denote these fixed points by $\{a, b, c\}$, and let $H = G_{a,b,c}$. Now $\langle g \rangle$ is a Sylow 5-subgroup of H ($\langle g \rangle$ is even a Sylow 5-subgroup of G), so that Theorem 9.66(i) gives $N = N_G(\langle g \rangle)$ acting 3-transitively on $\mathscr{F}(\langle g \rangle) = \{a, b, c\}$; that is, there is a surjective homomorphism $\varphi: N \to S_3$. We claim that $C = C_G(\langle g \rangle) \not\leq \ker \varphi$. Otherwise, φ induces a surjective map $\varphi_*: N/C \to S_3$. By Theorem 7.1, $N/C \leq \mathrm{Aut}(\langle g \rangle)$, which is abelian, so that N/C and hence S_3 are abelian, a contradiction. Now $C \lhd N$ forces $\varphi(C) \lhd \varphi(N) = S_3$, so that $\varphi(C) = A_3$ (we have just seen that $\varphi(C) \neq 1$) and so 3 divides $|C|$. There is thus an element $h \in C$ of order 3. Since g and h commute, the element gh has order 15. Now gh cannot be a 15-cycle (G has degree 13), and so its cycle structure is either (5, 5, 3), (5, 3, 3), or (5, 3). Hence $(gh)^5$, being either a 3-cycle or a product of 2 disjoint 3-cycles, fixes more than 6 points. This contradiction shows that no such G can exist.

A transitive extension G of M_{24} would have degree 25 and order $25 \cdot 24 \cdot 23 \cdot 22 \cdot 21 \cdot 20 \cdot 48$. If $g \in G$ has order 11, then g is a product of 2 disjoint 11-cycles (it cannot be an 11-cycle lest it fix $14 > 8$ points, contradicting Lemma 9.67(iii)). Arguing as above, there is an element $h \in G$ of order 3 commuting with g, and so gh has order 33. Since G has degree 25, gh is not a 33-cycle, and so its cycle structure is either of the form (11, 11, 3) or one 11-cycle and several 3-cycles. In either case, $(gh)^{11}$ has order 3 and fixes more than 8 points, contradicting Lemma 9.67. ■

Theorem 9.69.

(i) *Let* $X = \mathrm{P}^2(4) \cup \{\infty, \omega, \Omega\}$ *be regarded as an* M_{24}*-set, let* U *be a Sylow 2-subgroup of* H *(the stabilizer of 5 points), and let* $\mathscr{B} = \{g\mathscr{F}(U): g \in M_{24}\}$. *Then* $(X, \mathscr{B})$ *is a Steiner system of type* $S(5, 8, 24)$.

(ii) *If* $g\mathscr{F}(U)$ *contains* $\{\infty, \omega, \Omega\}$, *then its remaining 5 points form a projective line. Conversely, for every projective line* ℓ', *there is* $g \in \mathrm{PSL}(3, 4) \leq M_{24}$ *with* $g\mathscr{F}(U) = \ell' \cup \{\infty, \omega, \Omega\}$.

Proof. (i) Lemma 9.67 verifies that the conditions stated in Theorem 9.66 do hold.

(ii) The remark after Theorem 9.61 gives a formula for the number r'' of blocks containing 3 points; in particular, there are 21 blocks containing $\{\infty, \omega, \Omega\}$. If $\ell \subset \mathscr{F}(U)$ is the projective line $v = 0$, and if $g \in \mathrm{PSL}(3, 4) = (M_{24})_{\infty, \omega, \Omega}$, then $g\mathscr{F}(U) = g(\ell) \cup \{\infty, \omega, \Omega\}$. But PSL(3, 4) acts transitively on the lines of $\mathrm{P}^2(4)$ (Exercise 9.23) and $\mathrm{P}^2(4)$ has exactly 21 lines (Theorem 9.40(ii)). It follows that the 21 blocks containing the 3 infinite points ∞, ω, Ω are as described. ■

The coming results relating Mathieu groups to Steiner systems are due to R.D. Carmichael and E. Witt.

Theorem 9.70. $M_{24} \cong \mathrm{Aut}(X, \mathscr{B})$, *where* $(X, \mathscr{B})$ *is a Steiner system of type* $S(5, 8, 24)$.

Remark. There is only one Steiner system with these parameters.

Proof. Let $(X, \mathscr{B})$ be the Steiner system of Theorem 9.69: $X = \mathrm{P}^2(4) \cup \{\infty, \omega, \Omega\}$ and $\mathscr{B} = \{g\mathscr{F}(U): g \in M_{24}\}$, where $\mathscr{F}(U) = \ell \cup \{\infty, \omega, \Omega\}$ (here ℓ is the projective line $v = 0$).

It is clear that every $g \in M_{24}$ is a permutation of X that carries blocks to blocks, so that $M_{24} \leq \text{Aut}(X, \mathscr{B})$. For the reverse inclusion, let $\varphi \in \text{Aut}(X, \mathscr{B})$. Multiplying φ by an element of M_{24} if necessary, we may assume that φ fixes $\{\infty, \omega, \Omega\}$ and, hence, that $\varphi|\text{P}^2(4)\colon \text{P}^2(4) \to \text{P}^2(4)$. By Theorem 9.69(ii), φ carries projective lines to projective lines, and so φ is a collineation of $\text{P}^2(4)$. But M_{24} contains a copy of $\text{P}\Gamma\text{L}(3, 4)$, the collineation group of $\text{P}^2(4)$, by Exercise 9.40. There is thus $g \in M_{24}$ with $g|\text{P}^2(4) = \varphi|\text{P}^2(4)$, and $\varphi g^{-1} \in \text{Aut}(X, \mathscr{B})$ (because $M_{24} \leq \text{Aut}(X, \mathscr{B})$). Now φg^{-1} can permute only ∞, ω, Ω. Since every block has 8 elements φg^{-1} must fix at least 5 elements; as each block is determined by any 5 of its elements, φg^{-1} must fix every block, and so Theorem 9.63 shows that $\varphi g^{-1} = 1$; that is, $\varphi = g \in M_{24}$, as desired. ■

We interrupt this discussion to prove a result mentioned in Chapter 8.

Theorem 9.71. $\text{PSL}(4, 2) \cong A_8$.

Proof. The Sylow 2-subgroup U in H, the stabilizer of 5 points in M_{24}, is elementary abelian of order 16; thus, U is a 4-dimensional vector space over $\mathbb{Z}_2$. Therefore, $\text{Aut}(U) \cong \text{GL}(4, 2)$ and, by Theorem 8.5, $|\text{Aut}(U)| = (2^4 - 1)(2^4 - 2)(2^4 - 4)(2^4 - 8) = 8!/2$.

Let $N = N_{M_{24}}(U)$. By Theorem 9.66(ii), N acts 5-transitively (and faithfully) on $\mathscr{F}(U)$, a set with 8 elements. Therefore, $|N| = 8 \cdot 7 \cdot 6 \cdot 5 \cdot 4 \cdot s$, where $s \leq 6 = |S_3|$. If we identify the symmetric group on $\mathscr{F}(U)$ with S_8, then $[S_8 : N] = t \leq 6$ (where $t = 6/s$). By Exercise 9.3(ii), S_8 has no subgroups of index t with $2 < t < 8$. Therefore, $t = 1$ or $t = 2$; that is, $N = S_8$ or $N = A_8$. Now there is a homomorphism $\varphi\colon N \to \text{Aut}(U)$ given by $g \mapsto \gamma_g$ = conjugation by g. Since A_8 is simple, the only possibilities for im φ are S_8, A_8, $\mathbb{Z}_2$, or 1. We cannot have im $\varphi \cong S_8$ (since $|\text{Aut}(U)| = 8!/2$); we cannot have $|\text{im } \varphi| \leq 2$ (for $H \leq N$, because $U \lhd H$, and it is easy to find $h \in H$ of odd order and $u \in U$ with $huh^{-1} \neq u$). We conclude that $N = A_8$ and that $\varphi\colon N \to \text{Aut}(U) \cong \text{GL}(4, 2)$ is an isomorphism. ■

Theorem 9.72. $M_{23} \cong \text{Aut}(X', \mathscr{B}')$, *where* $(X', \mathscr{B}')$ *is a Steiner system of type* $S(4, 7, 23)$.

Remark. There is only one Steiner system with these parameters.

Proof. Let $X' = \text{P}^2(4) \cup \{\infty, \omega\}$, let $B' = B'(\ell) = \ell \cup \{\infty, \omega\}$, where ℓ is the projective line $v = 0$, and let $\mathscr{B}' = \{g(B')\colon g \in M_{23}\}$. It is easy to see that $(X', \mathscr{B}')$ is the contraction at Ω of the Steiner system $(X, \mathscr{B})$ in Theorem 9.69, so that it is a Steiner system of type $S(4, 7, 23)$.

It is clear that $M_{23} \leq \text{Aut}(X', \mathscr{B}')$. For the reverse inclusion, let $\varphi \in \text{Aut}(X', \mathscr{B}')$, and regard φ as a permutation of X with $\varphi(\Omega) = \Omega$. Multiplying by an element of M_{23} if necessary, we may assume that φ fixes ∞ and ω.

Since $(X', \mathscr{B}')$ is a contraction of $(X, \mathscr{B})$, a block in $\mathscr{B}'$ containing ∞ and ω has the form $\ell' \cup \{\infty, \omega\}$, where ℓ' is a projective line. As in the proof of Theorem 9.70, $\varphi|\mathrm{P}^2(4)$ preserves lines and hence is a collineation of $\mathrm{P}^2(4)$. Since M_{24} contains a copy of $\mathrm{P\Gamma L}(3, 4)$, there is $g \in M_{24}$ with $g|\mathrm{P}^2(4) = \varphi|\mathrm{P}^2(4)$. Therefore, g and φ can only disagree on the infinite points ∞, ω, and Ω.

If $B \in \operatorname{star}(\Omega)$ (i.e., if B is a block in $\mathscr{B}$ containing Ω), then $\varphi(B)$ and $g(B)$ are blocks; moreover, $|\varphi(B) \cap g(B)| \geq 5$, for blocks have 8 points, while φ and g can disagree on at most 3 points. Since 5 points determine a block, however, $\varphi(B) = g(B)$ for all $B \in \operatorname{star}(\Omega)$. By Corollary 9.64,

$$\begin{aligned}\{\Omega\} = \{\varphi(\Omega)\} &= \varphi\Bigg(\bigcap_{\operatorname{star}(\Omega)} B\Bigg)\\ &= \bigcap_{\operatorname{star}(\Omega)} \varphi(B)\\ &= \bigcap_{\operatorname{star}(\Omega)} g(B) = g\Bigg(\bigcap_{\operatorname{star}(\Omega)} B\Bigg) = \{g(\Omega)\}.\end{aligned}$$

Hence $g(\Omega) = \Omega$ and $g \in (M_{24})_\Omega = M_{23}$. The argument now ends as that in Theorem 9.70: $\varphi g^{-1} \in \operatorname{Aut}(X', \mathscr{B}')$ since $M_{23} \leq \operatorname{Aut}(X', \mathscr{B}')$, φg^{-1} fixes $\mathscr{B}'$, and $\varphi = g \in M_{23}$. ■

Theorem 9.73. *M_{22} is a subgroup of index 2 in* $\operatorname{Aut}(X'', \mathscr{B}'')$, *where $(X'', \mathscr{B}'')$ is a Steiner system of type $S(3, 6, 22)$.*

Remark. There is only one Steiner system with these parameters.

Proof. Let $X'' = X - \{\Omega, \omega\}$, let $b'' = \mathscr{F}(U) - \{\Omega, \omega\}$, and let $\mathscr{B}'' = \{gb'': g \in M_{22}\}$. It is easy to see that $(X'', \mathscr{B}'')$ is doubly contracted from $(X, \mathscr{B})$, so that it is a Steiner system of type $S(3, 6, 22)$.

Clearly $M_{22} \leq \operatorname{Aut}(X'', \mathscr{B}'')$. For the reverse inclusion, let $\varphi \in \operatorname{Aut}(X'', \mathscr{B}'')$ be regarded as a permutation of X which fixes Ω and ω. As in the proof of Theorem 9.72, we may assume that $\varphi(\infty) = \infty$ and that $\varphi|\mathrm{P}^2(4)$ is a collineation. There is thus $g \in M_{24}$ with $g|\mathrm{P}^2(4) = \varphi|\mathrm{P}^2(4)$. Moreover, consideration of $\operatorname{star}(\omega)$, as in the proof of Theorem 9.72, gives $g(\omega) = \omega$. Therefore, φg^{-1} is a permutation of X fixing $\mathrm{P}^2(4) \cup \{\omega\}$. If φg^{-1} fixes Ω, then $\varphi g^{-1} = 1_X$ and $\varphi = g \in (M_{24})_{\Omega,\omega} = M_{22}$. The other possibility is that $\varphi g^{-1} = (\infty\ \Omega)$.

We claim that $[\operatorname{Aut}(X'', \mathscr{B}''): M_{22}] \leq 2$. If $\varphi_1, \varphi_2 \in \operatorname{Aut}(X'', \mathscr{B}'')$ and $\varphi_1, \varphi_2 \notin M_{22}$, then we have just seen that $\varphi_i = (\infty\ \Omega)g_i$ for $i = 1, 2$, where $g_i \in M_{24}$. But $g_1^{-1}g_2 = \varphi_1^{-1}\varphi_2 \in (M_{24})_{\Omega,\omega} = M_{22}$ (since both φ_i fix Ω and ω); there are thus at most two cosets of M_{22} in $\operatorname{Aut}(X'', \mathscr{B}'')$.

Recall the definitions of the elements h_2 and h_3 in M_{24}: $h_2 = (\omega\ \infty)f_2$ and $h_3 = (\Omega\ \omega)f_3$, where f_2, f_3 act on $\mathrm{P}^2(4)$ and fix ∞, ω, and Ω. Note that h_2 fixes Ω and h_3 fixes ∞. Define $g = h_3h_2h_3 = (\Omega\ \infty)f_3f_2f_3$, and define

$\varphi\colon X'' \to X''$ to be the function with $\varphi(\infty) = \infty$ and $\varphi|\mathrm{P}^2(4) = f_3 f_2 f_3$. By Lemma 9.54, $\varphi|\mathrm{P}^2(4)$ is a collineation; since φ fixes ∞, it follows that $\varphi \in \mathrm{Aut}(X'', \mathscr{B}'')$. On the other hand, $\varphi \notin M_{22}$, lest $\varphi g^{-1} = (\Omega\ \ \infty) \in M_{24}$, contradicting Lemma 9.67(iii). We have shown that M_{22} has index 2 in $\mathrm{Aut}(X'', \mathscr{B}'')$. ■

Corollary 9.74. *M_{22} has an outer automorphism of order 2 and* $\mathrm{Aut}(X'', \mathscr{B}'') \cong M_{22} \rtimes \mathbb{Z}_2$.

Proof. The automorphism $\varphi \in \mathrm{Aut}(X'', \mathscr{B}'')$ with $\varphi \notin M_{22}$ constructed at the end of the proof of Theorem 9.73 has order 2, for both f_2 and f_3 are involutions (Lemma 9.54), hence the conjugate $f_3 f_2 f_3$ is also an involution. It follows that $\mathrm{Aut}(X'', \mathscr{B}'')$ is a semidirect product $M_{22} \rtimes \mathbb{Z}_2$. Now φ is an automorphism of M_{22}: if $a \in M_{22}$, then $a^\varphi = \varphi a \varphi^{-1} \in M_{22}$. Were φ an inner automorphism, there would be $b \in M_{22}$ with $\varphi a \varphi^{-1} = bab^{-1}$ for all $a \in M_{22}$; that is, φa^{-1} would centralize M_{22}. But a routine calculation shows that φ does not commute with $h_1 = (\infty\ \ [1, 0, 0])f_1 \in M_{22}$, and so φ is an outer automorphism of M_{22}. ■

The "small" Mathieu groups M_{11} and M_{12} are also intimately related to Steiner systems, but we cannot use Theorem 9.66 because the action is now sharp.

Lemma 9.75. *Regard $X = \mathrm{GF}(9) \cup \{\infty, \omega, \Omega\}$ as an M_{12}-set. There is a subgroup $\Sigma \leq M_{12}$, isomorphic to S_6, having two orbits of size 6, say, Z and Z', and which acts sharply 6-transitively on Z. Moreover,*

$$\Sigma = \{\mu \in M_{12}\colon \mu(Z) = Z\}.$$

Proof. Denote the 5-set $\{\infty, \omega, \Omega, 1, -1\}$ by Y. For each permutation τ of Y, sharp 5-transitivity of M_{12} provides a unique $\tau^* \in M_{12}$ with $\tau^*|Y = \tau$. It is easy to see that the function $S_Y \to M_{12}$, given by $\tau \mapsto \tau^*$, is an injective homomorphism; we denote its image (isomorphic to S_5) by Q.

Let us now compute the Q-orbits of X. One of them, of course, is Y. If τ is the 3-cycle $(\infty\ \ \omega\ \ \Omega)$, then $\tau^* \in Q$ has order 3 and fixes 1 and -1. Now τ^* is a product of three disjoint 3-cycles (fewer than three would fix too many points of X), so that the $\langle\tau^*\rangle$-orbits of the 7-set $X - Y$ have sizes (3, 3, 1). Since the Q-orbits of X (and of $X - Y$) are disjoint unions of $\langle\tau^*\rangle$-orbits (Exercise 9.4), the Q-orbits of $X - Y$ have possible sizes (3, 3, 1), (6, 1), (3, 4), or 7. If Q has one orbit of size 7, then Q acts transitively on $X - Y$; this is impossible, for 7 does not divide $|Q| = 120$. Furthermore, Exercise 9.3(i) says that Q has no orbits of size t, where $2 < t < 5$. We conclude that $X - Y$ has two Q-orbits of sizes 6 and 1, respectively. There is thus a unique point in $X - Y$, namely, the orbit of size 1, that is fixed by every element of Q. If $\sigma \in S_Y$ is the transposition $(1\ \ -1)$, then its correspondent $\sigma^* \in Q$ fixes ∞, ω, Ω and interchanges 1 and -1. But $\zeta\colon \mathrm{GF}(9) \to \mathrm{GF}(9)$, defined by $\zeta\colon \lambda \mapsto -\lambda$, lies in M_{10} (for -1 is a square in GF(9)) and $\zeta | Y = \sigma$, so that $\zeta = \sigma^*$. Since the only other point fixed by ζ is 0, the one-point Q-orbit of $X - Y$ must be $\{0\}$.

Define $Z = Y \cup \{0\} = \{\infty, \omega, \Omega, 1, -1, 0\}$. We saw, in Exercise 9.33, that $M_{10} \leq M_{12}$ contains $\sigma_1\colon \mathrm{P}^1(9) \to \mathrm{P}^1(9)$, where $\sigma_1\colon \lambda \mapsto -1/\lambda$ is $(0\ \infty)(1\ -1)(\pi^3\ \pi)(\pi^5\ \pi^7)$. Let us see that the subgroup $\Sigma = \langle Q, \sigma_1 \rangle \cong S_6$. The set Z is both a Q-set and a $\langle \sigma_1 \rangle$-set, hence it is also a Σ-set. As Σ acts transitively on Z and the stabilizer of 0 is Q (which acts sharply 5-transitively on $Z - \{0\} = Y$), we have Σ acting sharply 6-transitively on the 6-point set Z, and so $\Sigma \cong S_6$. Finally, the 6 points $X - Z$ comprise the other Σ-orbit of X (for we have already seen that $X - Z$ is a Q-orbit).

If $\beta \in Q$, then $\beta(Y) = Y$ and $\beta(0) = 0$, so that $\beta(Z) = Z$. Since $\sigma_1(Z) = Z$, it follows that $\sigma(Z) = Z$ for all $\sigma \in \Sigma$. Conversely, suppose $\mu \in M_{12}$ and $\mu(Z) = Z$. Since Σ acts 6-transitively on Z, there is $\sigma \in \Sigma$ with $\sigma | Z = \mu | Z$. But $\mu\sigma^{-1}$ fixes 6 points, hence is the identity, and $\mu = \sigma \in \Sigma$. ■

Theorem 9.76. *If $X = \mathrm{GF}(9) \cup \{\infty, \omega, \Omega\}$ is regarded as an M_{12}-set and $\mathscr{B} = \{gZ\colon g \in M_{12}\}$, where $Z = \{\infty, \omega, \Omega, 1, -1, 0\}$, then $(X, \mathscr{B})$ is a Steiner system of type $S(5, 6, 12)$.*

Proof. It is clear that every block gZ has 6 points. If $x_1, \ldots, x_5$ are any five distinct points in X, then 5-transitivity of M_{12} provides $g \in M_{12}$ with $\{x_1, \ldots, x_5\} \subset gZ$. It remains to prove uniqueness of a block containing five given points, and it suffices to show that if Z and gZ have five points in common, then $Z = gZ$. Now if $Z = \{z_1, \ldots, z_6\}$, then $gZ = \{gz_1, \ldots, gz_6\}$, where $gz_1, \ldots, gz_5 \in Z$. By Lemma 9.75, there is $\sigma \in \Sigma \leq M_{12}$ with $\sigma z_1 = gz_1, \ldots, \sigma z_5 = gz_5$. Note that $\sigma Z = Z$, for Z is a Σ-orbit. On the other hand, σ and g agree on five points of X, so that sharp 5-transitivity of M_{12} gives $\sigma = g$. Therefore $Z = \sigma Z = gZ$. ■

If GF(9) is regarded as an affine plane over $\mathbb{Z}_3$, then the blocks of the Steiner system constructed above can be examined from a geometric viewpoint.

Lemma 9.77. *Let $(X, \mathscr{B})$ be the Steiner system constructed from M_{12} in Theorem 9.76. A subset B of X containing $T = \{\infty, \omega, \Omega\}$ is a block if and only if $B = T \cup \ell$, where ℓ is a line in GF(9) regarded as an affine plane over $\mathbb{Z}_3$.*

Proof. Note that $Z = T \cup \ell_0$, where $\ell_0 = \{1, -1, 0\}$, and ℓ_0 is the line consisting of the scalar multiples of 1. By Exercises 9.38 and 9.39, M_{12} contains a subgroup $W \cong \mathrm{Aut}(2, 3)$ each of whose elements permutes T. Hence, for every $g \in W$, $gZ = T \cup g\ell_0$, and $g\ell_0$ is an affine line. But one may count exactly 12 affine lines in the affine plane, so that there are 12 blocks of the form $T \cup \ell$. On the other hand, the remark after Theorem 9.61 shows that there exactly 12 blocks containing the 3-point set T. ■

Theorem 9.78. $M_{12} \cong \text{Aut}(X, \mathscr{B})$, *where* $(X, \mathscr{B})$ *is a Steiner system of type* $S(5, 6, 12)$.

Remark. There is only one Steiner system with these parameters.

Proof. Let $(X, \mathscr{B})$ be the Steiner system constructed in Theorem 9.76. Now $M_{12} \leq \text{Aut}(X, \mathscr{B})$ because every $g \in M_{12}$ carries blocks to blocks. For the reverse inclusion, let $\varphi \in \text{Aut}(X, \mathscr{B})$. Composing with an element of M_{12} if necessary, we may assume that φ permutes $T = \{\infty, \omega, \Omega\}$ and φ permutes $\text{GF}(9)$. Regarding $\text{GF}(9)$ as an affine plane over $\mathbb{Z}_3$, we see from Lemma 9.77 that $\varphi|\text{GF}(9)$ is an affine automorphism. By Exercise 9.39, there is $g \in M_{12}$ which permutes T and with $g|\text{GF}(9) = \varphi|\text{GF}(9)$. Now $\varphi g^{-1} \in \text{Aut}(X, \mathscr{B})$, for $M_{12} \leq \text{Aut}(X, \mathscr{B})$, φg^{-1} permutes T, and φg^{-1} fixes the other 9 points of X. We claim that φg^{-1} fixes every block B in $\mathscr{B}$. This is clear if $|B \cap T| = 0, 1$, or 3. In the remaining case, say, $B = \{\infty, \omega, x_1, \ldots, x_4\}$, then $\varphi g^{-1}(B)$ must contain either ∞ or ω as well as the x_i, so that $|B \cap \varphi g^{-1}(B)| \geq 5$. Since 5 points determine a block, $B = \varphi g^{-1}(B)$, as claimed. Theorem 9.63 forces $\varphi g^{-1} = 1$, and so $\varphi = g \in M_{12}$, as desired. ■

Theorem 9.79. $M_{11} \cong \text{Aut}(X', \mathscr{B}')$, *where* $(X', \mathscr{B}')$ *is a Steiner system of type* $S(4, 5, 11)$.

Remark. There is only one Steiner system with these parameters.

Proof. Let $(X', \mathscr{B}')$ be the contraction at Ω of the Steiner system $(X, \mathscr{B})$ of Theorem 9.76. It is clear that $M_{11} \leq \text{Aut}(X', \mathscr{B}')$. For the reverse inclusion, regard $\varphi \in \text{Aut}(X', \mathscr{B}')$ as a permutation of X with $\varphi(\Omega) = \Omega$. Multiplying by an element of M_{11} if necessary, we may assume that φ permutes $\{\infty, \omega\}$. By Lemma 9.77, a block $B' \in \mathscr{B}'$ containing ∞ and ω has the form $B' = \{\infty, \omega\} \cup \ell$, where ℓ is a line in the affine plane over $\mathbb{Z}_3$. As in the proof of Theorem 9.78, $\varphi|\text{GF}(9)$ is an affine isomorphism, so there is $g \in M_{12}$ with $g|\text{GF}(9) = \varphi|\text{GF}(9)$. As in the proof of Theorem 9.72, an examination of $g(\text{star}(\Omega))$ shows that $g(\Omega) = \Omega$, so that $g \in (M_{12})_\Omega = M_{11}$. The argument now finishes as that for Theorem 9.78: $\varphi g^{-1} \in \text{Aut}(X', \mathscr{B}')$; φg^{-1} fixes $\mathscr{B}'$; $\varphi = g \in M_{11}$. ■

The subgroup structures of the Mathieu groups are interesting. There are other simple groups imbedded in them: for example, M_{12} contains copies of A_6, $\text{PSL}(2, 9)$, and $\text{PSL}(2, 11)$, while M_{24} contains copies of M_{12}, A_8, and $\text{PSL}(2, 23)$. The copy Σ of S_6 in M_{12} leads to another proof of the existence of an outer automorphism of S_6.

Theorem 9.80. S_6 *has an outer automorphism of order* 2.

Remark. See Corollary 7.13 for another proof.

Proof. Recall from Lemma 9.75 that if $X = \{\infty, \omega, \Omega\} \cup \mathrm{GF}(9)$ and Σ ($\cong S_6$) is the subgroup of M_{12} in Lemma 9.75, then X has two Σ-orbits, say, $Z = Y \cup \{0\}$ and $Z' = Y' \cup \{0'\}$, each of which has 6 points. If $\sigma \in \Sigma$ has order 5, then σ is a product of two disjoint 5-cycles (only one 5-cycle fixes too many points), hence it fixes, say, 0 and 0′. It follows that if $U = \langle\sigma\rangle$, then each of Z and Z' consists of two U-orbits, one of size 5 and one of size 1. Now $H = (M_{12})_{0,0'} \cong M_{10}$, and U is a Sylow 5-subgroup of H. By Theorem 9.66, $N = N_{M_{12}}(U)$ acts 2-transitively on $\mathscr{F}(U) = \{0, 0'\}$, so there is $\alpha \in N$ of order 2 which interchanges 0 and 0′.

Since α has order 2, $\alpha = \tau_1 \ldots \tau_m$, where the τ_i are disjoint transpositions and $m \leq 6$. But M_{12} is sharply 5-transitive, so that $4 \leq m$; also, $M_{12} \leq A_{12}$, so that $m = 4$ or $m = 6$.

We claim that α interchanges the sets $Z = Y \cup \{0\}$ and $Z' = Y' \cup \{0'\}$. Otherwise, there is $y \in Y$ with $\alpha(y) = z \in Y$. Now $\alpha\sigma\alpha = \sigma^i$ for some i (because α normalizes $\langle\sigma\rangle$). If $\sigma^i(y) = u$ and $\sigma(z) = v$, then $u, v \in Y$ because $Y \cup \{0\}$ is a Σ-orbit. But $u = \sigma^i(y) = \alpha\sigma\alpha(y) = \alpha\sigma(z) = \alpha(v)$, and it is easy to see that y, z, u, and v are all distinct. Therefore, the cycle decomposition of α involves $(0\ 0')$, $(y\ z)$, and $(v\ u)$. There is only one point remaining in Y, say a, and there are two cases: either $\alpha(a) = a$ or $\alpha(a) \in Y'$. If α fixes a, then there is $y' \in Y'$ moved by α, say, $\alpha(y') = z' \in Y'$. Repeat the argument above: there are points $u', v' \in Y'$ with transpositions $(y'\ z')$ and $(v'\ u')$ involved in the cycle decomposition of α. If a' is the remaining point in Y', then the transposition $(a\ a')$ must also occur in the factorization of α because α is not a product of 5 disjoint transpositions. In either case, we have $a \in Y$ and $a' \in Y'$ with $\alpha = (0\ 0')(y\ z)(v\ u)(a\ a')\beta$, where β permutes $Y' - \{a'\}$. But $\alpha\sigma\alpha(a) = \sigma^i(a) \in Z$; on the other hand, if $\sigma(a') = b' \in Y'$, say, then $\alpha\sigma\alpha(a) = \alpha\sigma(a') = \alpha(b')$, so that $\alpha(b') \in Y$. Since a' is the only element of Y' that α moves to Y, $b' = a'$ and $\sigma(a') = b' = a'$; that is, σ fixes a'. This is a contradiction, for σ fixes only 0 and 0′.

It is easy to see that α normalizes Σ. Recall that $\sigma \in \Sigma$ if and only if $\sigma(Z) = Z$ (and hence $\sigma(Z') = Z'$). Now $\alpha\sigma\alpha(Z) = \alpha\sigma(Z') = \alpha(Z') = Z$, so that $\alpha\sigma\alpha \in \Sigma$. Therefore, $\gamma = \gamma_\alpha$ (conjugation by α) is an automorphism of Σ.

Suppose there is $\beta \in \Sigma$ with $\alpha\sigma^*\alpha = \beta\sigma^*\beta^{-1}$ for all $\sigma^* \in \Sigma$; that is, $\beta^{-1}\alpha \in C = C_{M_{12}}(\Sigma)$. If $C = 1$, then $\alpha = \beta \in \Sigma$, and this contradiction would show that γ is an outer automorphism. If $\sigma^* \in \Sigma$, then $\sigma^* = \sigma\sigma'$, where σ permutes Z and fixes Z' and σ' permutes Z' and fixes Z. Schematically,

$$\sigma^* = (z\ x\ \ldots)(z'\ x'\ \ldots);$$

if $\mu \in M_{12}$, then (as any element of S_{12}),

$$\mu\sigma^*\mu^{-1} = (\mu z\ \mu x\ \ldots)(\mu z'\ \mu x'\ \ldots).$$

In particular, if $\mu \in C$ (so that $\mu\sigma^*\mu^{-1} = \sigma^*$), then either $\mu(Z) = Z$ and $\mu(Z') = Z'$ or μ switches Z and Z'. In the first case, $\mu \in \Sigma$, by Lemma 9.75, and $\mu \in C \cap \Sigma = Z(\Sigma) = 1$. In the second case, $\mu\sigma\mu^{-1} = \sigma'$ (and $\mu\sigma'\mu^{-1} = \sigma$), so that σ and σ' have the same cycle structure for all $\sigma^* = \sigma\sigma' \in \Sigma$. But there is $\sigma^* \in \Sigma$ with σ a transposition. If such μ exists, then σ^* would be a product of two disjoint transpositions and hence would fix 8 points, contradicting M_{12} being sharply 5-transitive. ■

There is a similar argument, using an imbedding of M_{12} into M_{24}, which exhibits an outer automorphism of M_{12}. There are several other proofs of the existence of the outer automorphism of S_6; for example, see Conway and Sloane (1993).

The Steiner systems of types $S(5, 6, 12)$ and $S(5, 8, 24)$ arise in algebraic coding theory, being the key ingredients of (ternary and binary) ***Golay codes***. The Steiner system of type $S(5, 8, 24)$ is also used to define the ***Leech lattice***, a configuration in $\mathbb{R}^{24}$ arising in certain sphere-packing problems as well as in the construction of other simple sporadic groups.

CHAPTER 10

Abelian Groups

Commutativity is a strong hypothesis, so strong that all finite abelian groups are completely classified. In this chapter, we focus on finitely generated and, more generally, countable abelian groups.

Basics

A valuable viewpoint in studying an abelian group G is to consider it as an extension of simpler groups. Of course, this reduces the study of G to a study of the simpler groups and an extension problem.

In this chapter, we assume that ***all groups are abelian*** and we again adopt additive notation.

Definition. A sequence of groups and homomorphisms

$$\cdots \to A \xrightarrow{f} B \xrightarrow{g} C \xrightarrow{h} D \to \cdots$$

is an ***exact sequence*** if the image of each map is the kernel of the next map. A ***short exact sequence*** is an exact sequence of the form

$$0 \to A \xrightarrow{f} B \xrightarrow{g} C \to 0.$$

There is no need to label the arrow $0 \to A$, for there is only one such homomorphism, namely, $0 \mapsto 0$; similarly, there is no need to label the only possible homomorphism $C \to 0$: it must be the constant map $x \mapsto 0$. In the short exact sequence above, $0 = \operatorname{im}(0 \to A) = \ker f$ says that f is an injection and $A \cong \operatorname{im} f$; also, $\operatorname{im} g = \ker(C \to 0) = C$ says that g is a surjection. Finally, the first isomorphism theorem gives $B/\operatorname{im} f = B/\ker g \cong \operatorname{im} g = C$.

If $A \leq B$ and f is the inclusion, then $\text{im } f = A$ and $B/A \cong C$. Thus, B is an extension of A by C if and only if there is a short exact sequence $0 \to A \to B \to C \to 0$.

Definition. If G is a group, its ***torsion subgroup*** is

$$tG = \{x \in G: nx = 0 \text{ for some nonzero integer } n\}.$$

Note that tG is a fully invariant subgroup of G.

When G is not abelian, then tG may not be a subgroup. For example, Exercise 2.17 shows that tG is not a subgroup when $G = \mathrm{GL}(2, \mathbb{Q})$.

Definition. A group G is ***torsion*** if $tG = G$; it is ***torsion-free*** if $tG = 0$.

The term torsion is taken from Algebraic Topology; the homology groups of a "twisted" manifold have elements of finite order.

Theorem 10.1. *The quotient group G/tG is torsion-free, and so every group G is an extension of a torsion group by a torsion-free group.*

Proof. If $n(g + tG) = 0$ in G/tG for some $n \neq 0$, then $ng \in tG$, and so there is $m \neq 0$ with $m(ng) = 0$. Since $mn \neq 0$, $g \in tG$, $g + tG = 0$ in G/tG, and G/tG is torsion-free. ■

If an abelian group is a semidirect product, then it is a direct product or, in additive terminology, it is a direct sum. The first question is whether the extension problem above is only virtual or if there exists a group G whose torsion subgroup is not a ***direct summand*** of G (i.e., there is no subgroup $A \leq G$ with $G = tG \oplus A$). Let us first generalize one of the constructions we have already studied.

Definition. Let K be a possibly infinite set and let $\{A_k: k \in K\}$ be a family of groups[1] indexed by K. The ***direct product*** (or *complete direct sum* or *strong direct sum*), denoted by $\prod_{k \in K} A_k$, is the group whose elements are all "vectors" (a_k) in the cartesian product of the A_k and whose operation is

$$(a_k) + (b_k) = (a_k + b_k).$$

The ***direct sum*** (or weak direct sum), denoted by $\sum_{k \in K} A_k$, is the subgroup of $\prod_{k \in K} A_k$ consisting of all those elements (a_k) for which there are only finitely many k with $a_k \neq 0$.

If the index set K is finite, then $\prod_{k \in K} A_k = \sum_{k \in K} A_k$; if the index set K is

[1] These constructions make sense for nonabelian groups as well. They have already arisen, for example, in our remark in Chapter 7 that different wreath products $K \wr Q$ (complete and restricted) arise when Q acts on an infinite set Ω.

infinite and infinitely many $A_k \neq 0$, then the direct sum is a proper subgroup of the direct product.

Definition. If $x \in G$ and n is a nonzero integer, then x is ***divisible by n in G*** if there is $g \in G$ with $ng = x$.

Were the operation in G written multiplicatively, then one would say that x has an nth *root* in G. Exercise 1.31 shows that an element of order m is divisible by every n with $(n, m) = 1$.

Theorem 10.2. *There exists a group G whose torsion subgroup is not a direct summand of G.*

Proof. Let P be the set of all primes, and let $G = \prod_{p \in P} \mathbb{Z}_p$. If q is a prime and $x = (x_p) \in G$ is divisible by q, then there is $y = (y_p)$ with $qy_p = x_p$ for all p; it follows that $x_q = 0$. Therefore, if x is divisible by every prime, then $x = 0$.

We claim that G/tG contains a nonzero element which is divisible by every prime. If this were true, then $G \neq tG \oplus H$ for some subgroup H, because $H \cong G/tG$. If $a_p \in \mathbb{Z}_p$ is a generator, then $a = (a_p)$ has infinite order: if $na = 0$, then $na_p = 0$ for all p, so that p divides n for all p and hence $n = 0$. Therefore $a \notin tG$, and its coset $a + tG$ is a nonzero element of G/tG. If q is a prime, then a_p is divisible by q in $\mathbb{Z}_p$ for all $p \neq q$, by Exercise 1.31; there is thus $y_p \in \mathbb{Z}_p$ with $qy_p = a_p$ for all $p \neq q$. Define $y_q = 0$ and define $y = (y_p)$. Now $a - qy \in tG$ (for its coordinates are all 0 except for a_q in position q). Hence

$$q(y + tG) = qy + tG = a - (a - qy) + tG = a + tG,$$

and so $a + tG$ is divisible by every prime q. ■

We restate Lemma 7.20 for abelian groups.

Lemma 10.3. *If G is an abelian group and $A \leq G$, then the following statements are equivalent.*

(i) *A is a direct summand of G (there is a subgroup $B \leq G$ with $A \cap B = 0$ and $A + B = G$).*
(ii) *There is a subgroup $B \leq G$ so that each $g \in G$ has a unique expression $g = a + b$ with $a \in A$ and $b \in B$.*
(iii) *There exists a homomorphism $s: G/A \to G$ with $vs = 1_{G/A}$, where $v: G \to G/A$ is the natural map.*
(iv) *There exists a retraction $\pi: G \to A$; that is, π is a homomorphism with $\pi(a) = a$ for all $a \in A$.*

The following criterion is a generalization of Exercise 2.75 (which characterizes finite direct sums).

Lemma 10.4. *Let $\{A_k: k \in K\}$ be a family of subgroups of a group G. The following statements are equivalent.*

(i) $G \cong \sum_{k \in K} A_k$.
(ii) *Every $g \in G$ has a unique expression of the form*

$$g = \sum_{k \in K} a_k,$$

where $a_k \in A_k$, the k are distinct, and $a_k \neq 0$ for only finitely many k.
(iii) $G = \langle \bigcup_{k \in K} A_k \rangle$ *and, for each $j \in K$, $A_j \cap \langle \bigcup_{k \neq j} A_k \rangle = 0$.*

Proof. Routine. ■

Theorem 10.5. *If V is a vector space over a field K, then, as an additive group, V is a direct sum of copies of K.*

Proof. Let X be a basis of V. For each $x \in X$, the one-dimensional subspace Kx spanned by x, is, as a group, isomorphic to K. The reader may check, using Lemma 10.4, that $V = \sum_{x \in X} Kx$. ■

There is a notion of independence for abelian groups.

Definition. A finite subset $X = \{x_1, \ldots, x_n\}$ of nonzero elements of a group G is ***independent*** if, for all $m_i \in \mathbb{Z}$, $\sum m_i x_i = 0$ implies $m_i x_i = 0$ for each i. An infinite set X of nonzero elements in G is ***independent*** if every finite subset is independent.

If X is an independent subset of a group G and $\sum m_x x = 0$, then $m_x x = 0$ for each x. If G is torsion-free, then $m_x = 0$ for all x; however, if G has torsion, then one may conclude only that m_x is a multiple of the order of x.

Lemma 10.6. *A set X of nonzero elements of a group G is independent if and only if*

$$\langle X \rangle = \sum_{x \in X} \langle x \rangle.$$

Proof. Assume that X is independent. If $x_0 \in X$ and $y \in \langle x_0 \rangle \cap \langle X - \{x_0\} \rangle$, then $y = mx_0$ and $y = \sum m_i x_i$, where the x_i are distinct elements of X not equal to x_0. Hence

$$-mx_0 + \Sigma m_i x_i = 0,$$

so that independence gives each term 0; in particular, $0 = mx_0 = y$.

The proof of the converse, also routine, is left to the reader. ■

Recall, in the context of abelian groups, that *p-groups* are called *p-primary groups*.

Theorem 10.7 (Primary Decomposition). *Every torsion group G is a direct sum of p-primary groups.*

Proof. For a prime p, define

$$G_p = \{x \in G: p^n x = 0 \text{ for some } n \geq 0\}$$

(G_p is called the ***p-primary component*** of G.) The proof of Theorem 6.1, *mutatis mutandis*, show that $G \cong \sum_p G_p$. ■

Theorem 10.8. *If G and H are torsion groups, then $G \cong H$ if and only if $G_p \cong H_p$ for all primes p.*

Proof. If $\varphi: G \to H$ is a homomorphism, then $\varphi(G_p) \leq H_p$ for all primes p. In particular, if φ is an isomorphism, then $\varphi(G_p) \leq H_p$ and $\varphi^{-1}(H_p) \leq G_p$ for all p. It follows easily that $\varphi | G_p$ is an isomorphism $G_p \to H_p$.

Conversely, assume that there are isomorphisms $\varphi_p: G_p \to H_p$ for all primes p. By Theorem 10.4(ii), each $g \in G$ has a unique expression of the form $g = \sum_p a_p$, where only a finite number of the $a_p \neq 0$. Then $\varphi: G \to H$, defined by $\varphi(\sum a_p) = \sum \varphi_p(a_p)$, is easily seen to be an isomorphism. ■

Because of these last two results, most questions about torsion groups can be reduced to questions about p-primary groups.

Here are two technical results about direct sums and products that will be useful.

Theorem 10.9. *Let G be an abelian group, let $\{A_k: k \in K\}$ be a family of abelian groups, and let $\{i_k: A_k \to G: k \in K\}$ be a family of homomorphisms. Then $G \cong \sum_{k \in K} A_k$ if and only if, given any abelian group H and any family of homomorphisms $\{f_k: A_k \to H: k \in K\}$, then there exists a unique homomorphism $\varphi: G \to H$ making the following diagrams commute ($\varphi i_k = f_k$):*

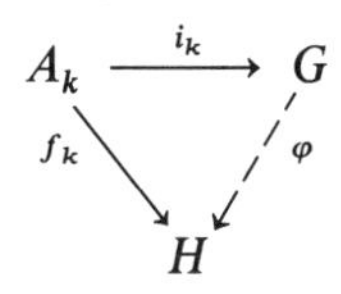

Proof. We show first that $G = \sum A_k$ has the stated property. Define $j_k: A_k \hookrightarrow G$ to be the inclusion. By Lemma 10.4, every $g \in G$ has a unique expression of the form $g = \sum_{k \in K} a_k$, with $a_k \neq 0$ for only finitely many k. It follows that $\psi(g) = \sum f_k(a_k)$ is a well defined function; it is easily checked that ψ is a homomorphism making the kth diagram commute for all k; that is, $\psi j_k = f_k$ for all k.

Assume now that G is any group satisfying the stated property, and choose the diagram with $H = G$ and $f_k = j_k$. By hypothesis, there is a map $\varphi: G \to \sum A_k$ making the diagrams commute.

Finally, we show that $\psi\varphi$ and $\varphi\psi$ are identities. Both $\psi\varphi$ and 1_G complete the diagram

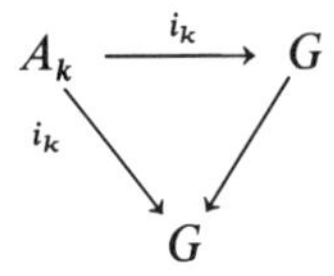

and so the uniqueness hypothesis gives $\psi\varphi = 1_G$. A similar diagram shows that $\varphi\psi$ is the identity on $\sum A_k$. ∎

Theorem 10.10. *Let G be an abelian group, let $\{A_k: k \in K\}$ be a family of abelian groups, and let $\{p_k: G \to A_k: k \in K\}$ be a family of homomorphisms. Then $G \cong \prod_{k \in K} A_k$ if and only if, given any abelian group H and any family of homomorphisms $\{f_k: H \to A_k: k \in K\}$, there exists a unique homomorphism $\varphi: H \to G$ making the following diagrams commute for all k:*

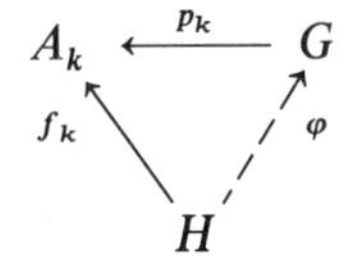

Proof. The argument is similar to the one just given if one defines $p_k: \prod_{l \in K} A_l \to H$ as the projection of a "vector" onto its kth coordinate. ∎

EXERCISES

10.1. Let $\{A_k: k \in K\}$ be a family of torsion groups.
  (i) The direct sum $\sum_{k \in K} A_k$ is torsion.
  (ii) If n is a positive integer and if each A_k has exponent n (i.e., $nA_k = 0$ for all k), then $\prod_{k \in K} A_k$ is torsion.

10.2. If $x \in G$, then any two solutions to the equation $ny = x$ differ by an element z with $nz = 0$. Conclude that y is unique if G is torsion-free.

10.3. If G is a torsion-free group and X is a maximal independent subset, then $G/\langle X \rangle$ is torsion.

10.4. (i) If $G = \sum A_k$, prove that the maps $i_k: A_k \to G$ in Theorem 10.9 are injections.
  (ii) If $G = \prod A_k$, prove that the maps $p_k: G \to A_k$ in Theorem 10.10 are surjections.

Free Abelian Groups

Definition. An abelian group F is ***free abelian*** if it is a direct sum of infinite cyclic groups. More precisely, there is a subset $X \subset F$ of elements of infinite order, called a ***basis*** of F, with $F = \sum_{x \in X} \langle x \rangle$; i.e., $F \cong \sum \mathbb{Z}$.

We allow the possibility $X = \varnothing$, in which case $F = 0$.

It follows at once from Lemma 10.4 that if X is a basis of a free abelian group F, then each $u \in F$ has a unique expression of the form $u = \sum m_x x$, where $m_x \in \mathbb{Z}$ and $m_x = 0$ for "almost all" $x \in X$; that is, $m_x \neq 0$ for only a finite number of x.

Notice that a basis X of a free abelian group is independent, by Lemma 10.6.

Theorem 10.11. *Let F be a free abelian group with basis X, let G be any abelian group, and let $f\colon X \to G$ be any function. Then there is a unique homomorphism $\varphi\colon F \to G$ extending f; that is,*

$$\varphi(x) = f(x) \qquad \text{for all} \quad x \in X.$$

Indeed, if $u = \sum m_x x \in F$, then $\varphi(u) = \sum m_x f(u)$.

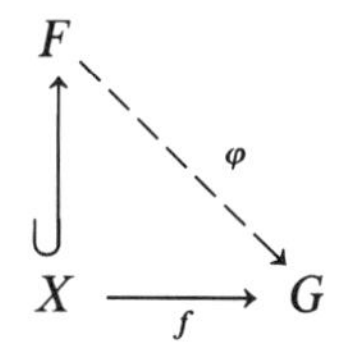

Proof. If $u \in F$, then uniqueness of the expression $u = \sum m_x x$ shows that $\varphi\colon u \mapsto \sum m_x f(u)$ is a well defined function. That φ is a homomorphism extending f is obvious; φ is unique because homomorphisms agreeing on a set of generators must be equal.

Here is a fancy proof. For each $x \in X$, there is a unique homomorphism $\varphi_x\colon \langle x \rangle \to G$ defined by $mx \mapsto mf(x)$. The result now follows from Lemma 10.6 and Theorem 10.10. ∎

Corollary 10.12. *Every (abelian) group G is a quotient of a free abelian group.*

Proof. Let F be the direct sum of $|G|$ copies of $\mathbb{Z}$, and let x_g denote a generator of the gth copy of $\mathbb{Z}$, where $g \in G$. Of course, F is a free abelian group with basis $X = \{x_g\colon g \in G\}$. Define a function $f\colon X \to G$ by $f(x_g) = g$ for all $g \in G$. By Theorem 10.11, there is a homomorphism $\varphi\colon F \to G$ extending f. Now φ is surjective, because f is surjective, and so $G \cong F/\ker \varphi$, as desired. ∎

The construction of a free abelian group in the proof of Corollary 10.12 can be modified: one may identify x_g with g. If X is any set, one may thus construct a free abelian group F having X itself as a basis. This is quite convenient. For example, in Algebraic Topology, one wishes to consider formal $\mathbb{Z}$-linear combinations of continuous maps between topological spaces; this can be done by forming the free abelian group with basis the set of all such functions.

The corollary provides a way of describing abelian groups.

Definition. An abelian group G has ***generators*** X ***and relations*** Δ *if* $G \cong F/R$, where F is the free abelian group with basis X, Δ is a set of $\mathbb{Z}$-linear combinations of elements of X, and R is the subgroup of F generated by Δ. If X can be chosen finite, then G is called ***finitely generated***.

EXAMPLE 10.1. $G = \mathbb{Z}_6$ has generator x and relation $6x$.

EXAMPLE 10.2. $G = \mathbb{Z}_6$ has generators $\{x, y\}$, and relations $\{2x, 3y\}$.

EXAMPLE 10.3. $G = \mathbb{Q}$ has generators $\{x_1, \ldots, x_n, \ldots\}$ and relations $\{x_1 - 2x_2, x_2 - 3x_3, \ldots, x_{n-1} - nx_n, \ldots\}$.

EXAMPLE 10.4. If G is free abelian with basis X, then G has generators X and no relations (recall that 0 is the subgroup generated by the empty set). The etymology of the term *free* should now be apparent.

We have just seen that one can describe a known group by generators and relations. One can also use generators and relations to construct a group with prescribed properties.

Theorem 10.13. *There is an infinite p-primary group* $G = \mathbb{Z}(p^\infty)$ *each of whose proper subgroups is finite (and cyclic).*

Proof. Define a group G having

generators: $X = \{x_0, x_1, \ldots, x_n, \ldots\}$

and

relations: $\{px_0, x_0 - px_1, x_1 - px_2, \ldots, x_{n-1} - px_n, \ldots\}$.

Let F be the free abelian group on X, let $R \leq F$ be generated by the relations, and let $a_n = x_n + R \in F/R = G$. Then $pa_0 = 0$ and $a_{n-1} = pa_n$ for all $n \geq 1$, so that $p^{n+1}a_n = pa_0 = 0$. It follows that G is p-primary, for $p^{t+1}\sum_{n=0}^{t} m_n a_n = 0$, where $m_n \in \mathbb{Z}$. A typical relation (i.e., a typical element of R) has the form:

$$m_0 p x_0 + \sum_{n\geq 1} m_n(x_{n-1} - px_n) = (m_0 p + m_1)x_0 + \sum_{n\geq 1} (m_{n+1} - m_n p)x_n.$$

If $a_0 = 0$, then $x_0 \in R$, and independence of X gives the equations $1 = m_0 p + m_1$ and $m_{n+1} = pm_n$ for all $n \geq 1$. Since $R \leq F$ and F is a direct sum, $m_n = 0$ for large n. But $m_{n+1} = p^n m_1$ for all n, and so $m_1 = 0$. Therefore, $1 = m_0 p$, and this contradicts $p \geq 2$. A similar argument shows that $a_n \neq 0$ for all n. We now show that all a_n are distinct, which will show that G is infinite. If $a_n = a_k$ for $k > n$, then $a_{n-1} = pa_n$ implies $a_k = p^{k-n}a_n$, and this gives $(1 - p^{k-n})a_k = 0$; since G is p-primary, this contradicts $a_k \neq 0$.

Let $H \leq G$. If H contains infinitely many a_n, then it contains all of them, and $H = G$. If H involves only $a_0, \ldots, a_m$, then $H \leq \langle a_0, \ldots, a_m \rangle \leq \langle a_m \rangle$. Thus, H is a subgroup of a finite cyclic group, and hence H is also a finite cyclic group. ■

The group $\mathbb{Z}(p^\infty)$ has other interesting properties (see Exercise 10.5 below), and we shall return to it in a later section.

Theorem 10.14. *Two free abelian groups* $F = \sum_{x \in X} \langle x \rangle$ *and* $G = \sum_{y \in Y} \langle y \rangle$ *are isomorphic if and only if* $|X| = |Y|$.

Proof. Since $|X| = |Y|$, there is a bijection $f\colon X \to Y \subset G$, and f determines a homomorphism $\varphi\colon F \to G$ with $\varphi(x) = f(x)$ for all $x \in X$. Similarly, there is a homomorphism $\psi\colon G \to F$ with $\psi(y) = f^{-1}(y)$ for all $y \in Y$. But $\varphi\psi$ and $\psi\varphi$ are identities because each fixes every element in a basis, and so $\varphi\colon F \to G$ is an isomorphism.

Conversely, if p is a prime, then $V = F/pF$ is a vector space over $\mathbb{Z}_p$. We claim that $\overline{X} = \{x + pF\colon x \in X\}$ is a basis of V. It is clear that $\overline{X}$ spans V. Assume that $\Sigma[m_x](x + pF) = 0$, where $[m_x] \in \mathbb{Z}_p$ and not all $[m_x] = [0]$. If m_x is a representative of $[m_x]$, then $\sum m_x(x + pF) = 0$. In F, this equation becomes $\sum m_x x \in pF$; that is, there are integers n_x with $\sum m_x x = \sum p n_x x$. Independence of a basis gives $m_x = pn_x$ for all x, and so $[m_x] = [0]$ for all x. This contradiction shows that $\overline{X}$ is independent, and hence it is a basis of V. We have shown that dim $F/pF = |\overline{X}| = |X|$. In a similar way, one shows that dim $F/pF = |Y|$, so that $|X| = |Y|$. ∎

Definition. The ***rank*** of a free abelian group is the cardinal of a basis.

Theorem 10.14 says that two free abelian groups are isomorphic if and only if they have the same rank. The reader will not be misled by the analogy: vector space—free abelian group; dimension—rank.

It is clear that if F and G are free abelian, then

$$\operatorname{rank}(F \oplus G) = \operatorname{rank}(F) + \operatorname{rank}(G),$$

for a basis of $F \oplus G$ can be chosen as the union of a basis of F and a basis of G.

Theorem 10.15 (Projective Property). *Let* $\beta\colon B \to C$ *be a surjective homomorphism of groups. If* F *is free abelian and if* $\alpha\colon F \to C$ *is a homomorphism, then there exists a homomorphism* $\gamma\colon F \to B$ *making the diagram below commute (i.e.,* $\beta\gamma = \alpha$*):*

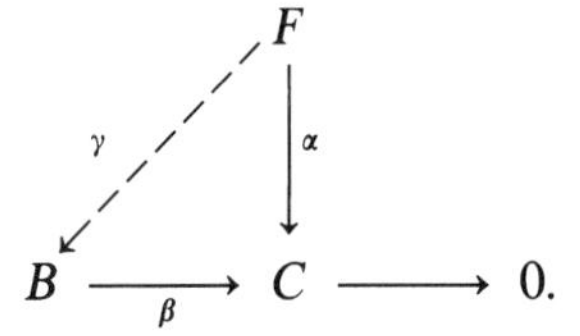

Remark. The converse is also true.

Proof. Let X be a basis of F. For each $x \in X$, surjectivity of α provides $b_x \in B$

with $\beta(b_x) = \alpha(x)$. Define a function $f: X \to B$ by $f(x) = b_x$. By Theorem 10.11 there is a homomorphism $\gamma: F \to B$ with $\gamma(x) = b_x$ for all x. It follows that $\beta\gamma = \alpha$, for they agree on a generating set of F: if $x \in X$, then $\beta\gamma(x) = \beta(b_x) = \alpha(x)$. ■

Corollary 10.16. *If $H \leq G$ and G/H is free abelian, then H is a direct summand of G; that is, $G = H \oplus K$, where $K \leq G$ and $K \cong G/H$.*

Proof. Let $F = G/H$ and let $\beta: G \to F$ be the natural map. Consider the diagram

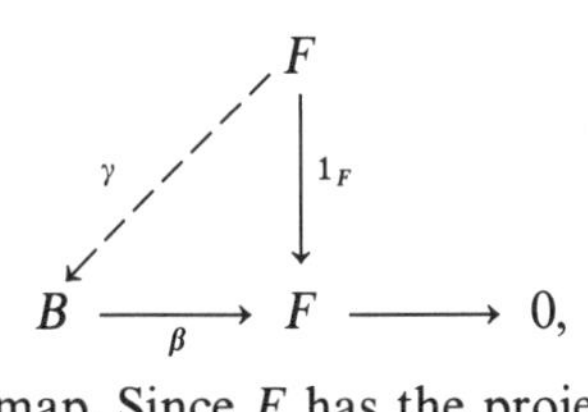

where 1_F is the identity map. Since F has the projective property, there is a homomorphism $\gamma: F \to B$ with $\beta\gamma = 1_F$. Define $K = \text{im } \gamma$. The equivalence of (i) and (iii) in Lemma 10.3 gives $B = \ker \beta \oplus \text{im } \gamma = H \oplus K$. ■

We give two proofs of the next result. The first is a special case of the second, but it contains the essential idea; the second involves infinite methods which, though routine, may obscure the simple idea.

Theorem 10.17. *Every subgroup H of a free abelian group F of finite rank n is itself free abelian; moreover,* $\text{rank}(H) \leq \text{rank}(F)$.

Proof. The proof is by induction on n. If $n = 1$, then $F \cong \mathbb{Z}$. Since every subgroup H of a cyclic group is cyclic, either $H = 0$ or $H \cong \mathbb{Z}$, and so H is free abelian of rank ≤ 1. For the inductive step, let $\{x_1, \ldots, x_{n+1}\}$ be a basis of F. Define $F' = \langle x_1, \ldots, x_n \rangle$ and $H' = H \cap F'$. By induction, H' is free abelian of rank $\leq n$. Now

$$H/H' = H/(H \cap F') \cong (H + F')/F' \leq F/F' \cong \mathbb{Z}.$$

By the base step, either $H/H' = 0$ or $H/H' \cong \mathbb{Z}$. In the first case, $H = H'$ and we are done; in the second case, Corollary 10.16 gives $H = H' \oplus \langle h \rangle$ for some $h \in H$, where $\langle h \rangle \cong \mathbb{Z}$, and so H is free abelian and $\text{rank}(H) = \text{rank}(H' \oplus \mathbb{Z}) = \text{rank}(H') + 1 \leq n + 1$. ■

We now remove the finiteness hypothesis.

Theorem 10.18. *Every subgroup H of a free abelian group F is free abelian, and* $\text{rank}(H) \leq \text{rank}(F)$.

Proof. That every nonempty set can somehow be well-ordered is equivalent to the Axiom of Choice (see Appendix IV). Let $\{x_k: k \in K\}$ be a basis of F, and assume that K is well-ordered.

For each $k \in K$, define $F'_k = \langle x_j: j < k \rangle$ and $F_k = \langle x_j: j \leq k \rangle = F'_k \oplus \langle x_k \rangle$; define $H'_k = H \cap F'_k$ and $H_k = H \cap F_k$. Note that $F = \bigcup F_k$ and $H = \bigcup H_k$. Now $H'_k = H \cap F'_k = H_k \cap F'_k$, and so

$$\begin{aligned} H_k/H'_k &= H_k/(H_k \cap F'_k) \\ &\cong (H_k + F'_k)/F'_k \leq F_k/F'_k \cong \mathbb{Z}. \end{aligned}$$

By Corollary 10.16, either $H_k = H'_k$ or $H_k = H'_k \oplus \langle h_k \rangle$, where $\langle h_k \rangle \cong \mathbb{Z}$. We claim that H is free abelian with basis the set of all h_k; it will then follow that $\text{rank}(H) \leq \text{rank}(F)$, for the set of h_k has cardinal $\leq |K| = \text{rank}(F)$.

Since $F = \bigcup F_k$, each $h \in H$ (as any element of F) lies in some F_k; define $\mu(h)$ to be the smallest index k for which $h \in F_k$ (we are using the fact that K is well-ordered). Let H^* be the subgroup of H generated by all the h_k. Suppose that H^* is a proper subgroup of H. Let j be the smallest index in

$$\{\mu(h): h \in H \text{ and } h \notin H^*\},$$

and choose $h' \in H$, $h' \notin H^*$ with $\mu(h') = j$. Now $\mu(h') = j$ gives $h' \in H \cap F_j$, so that

$$h' = a + mh_j, \qquad a \in H'_j \quad \text{and} \quad m \in \mathbb{Z}.$$

Thus, $a = h' - mh_j \in H$, $a \notin H^*$ (lest $h' \in H^*$), and $\mu(a) < j$, a contradiction. Therefore, $H = H^*$.

By Lemma 10.4(ii), it remains to show that linear combinations of the h_k are unique. It suffices to show that if

$$m_1 h_{k_1} + \cdots + m_n h_{k_n} = 0,$$

where $k_1 < \cdots < k_n$, then each $m_i = 0$. Of course, we may assume that $m_n \neq 0$. But then $m_n h_{k_n} \in \langle h_{k_n} \rangle \cap H'_{k_n} = 0$, a contradiction. It follows that H is free abelian. ■

Exercises

10.5. (i) Prove, for each $n \geq 1$, that $\mathbb{Z}(p^\infty)$ has a unique subgroup of order p^n.
(ii) Prove that the set of all subgroups of $\mathbb{Z}(p^\infty)$ is well-ordered by inclusion.
(iii) Prove that $\mathbb{Z}(p^\infty)$ has the DCC but not the ACC.
(iv) Let $R_p = \{e^{2\pi i k/p^n}: k \in \mathbb{Z}, n \geq 0\} \leq \mathbb{C}$ be the multiplicative group of all pth power roots of unity. Prove that $\mathbb{Z}(p^\infty) \cong R_p$.

10.6. (i) Prove that the group G having generators $\{x_0, x_1, x_2, \ldots\}$ and relations $\{px_0, x_0 - p^n x_n, \text{all } n \geq 1\}$ is an infinite p-primary group with $\bigcap_{n=1}^\infty p^n G \neq 0$.
(ii) Prove that the group G in (i) is not isomorphic to $\mathbb{Z}(p^\infty)$.

10.7. (i) Prove that an abelian group G is finitely generated if and only if it is a quotient of a free abelian group of finite rank.

(ii) Every subgroup H of a finitely generated abelian group G is itself finitely generated. Moreover, if G can be generated by r elements, then H can be generated by r or fewer elements.

10.8. Prove that the multiplicative group of positive rationals is free abelian (of countably infinite rank). (*Hint.* Exercise 1.52(ii).)

10.9. If F is a free abelian group of rank n, then $\mathrm{Aut}(F)$ is isomorphic to the multiplicative group of all $n \times n$ matrices over $\mathbb{Z}$ with determinant $= \pm 1$.

10.10. An abelian group is free abelian if and only if it has the projective property.

10.11. If F is a free abelian group of rank n and H is a subgroup of rank $k < n$, then F/H has an element of infinite order.

10.12. (i) If $A \xrightarrow{f} B \xrightarrow{g} C \xrightarrow{h} D$ is an exact sequence of free abelian groups, prove that $B \cong \operatorname{im} f \oplus \ker h$.

(ii) If $n \geq 1$ and $0 \to F_n \to \cdots \to F_1 \to F_0 \to 0$ is an exact sequence of free abelian groups of finite rank, then $\sum_{i=0}^{n} \operatorname{rank}(F_i) = 0$.

10.13. Prove the converse of Corollary 10.16: If a group G is (isomorphic) to a direct summand whenever it is a homomorphic image, then G is free abelian.

10.14. A torsion-free abelian group G having a free abelian subgroup of finite index is itself free abelian.

Finitely Generated Abelian Groups

We now classify all finitely generated abelian groups.

Theorem 10.19. *Every finitely generated torsion-free abelian group G is free abelian.*

Proof. We prove the theorem by induction on n, where $G = \langle x_1, \ldots, x_n \rangle$. If $n = 1$ and $G \neq 0$, then G is cyclic; $G \cong \mathbb{Z}$ because it is torsion-free.

Define $H = \{g \in G: mg \in \langle x_n \rangle$ for some positive integer $m\}$. Now H is a subgroup of G and G/H is torsion-free: if $x \in G$ and $k(x + H) = 0$, then $kx \in H$, $m(kx) \in \langle x_n \rangle$, and so $x \in H$. Since G/H is a torsion-free group that can be generated by fewer than n elements, it is free abelian, by induction. By Corollary 10.16, $G = F \oplus H$, where $F \cong G/H$, and so it suffices to prove that H is cyclic. Note that H is finitely generated, being a summand (and hence a quotient) of the finitely generated group G.

If $g \in H$ and $g \neq 0$, then $mg = kx_n$ for some nonzero integers m and k. It is routine to check that the function $\varphi: H \to \mathbb{Q}$, given by $g \mapsto k/m$, is a well defined injective homomorphism; that is, H is (isomorphic to) a finitely generated subgroup of $\mathbb{Q}$, say, $H = \langle a_1/b_1, \ldots, a_t/b_t \rangle$. If $b = \prod_{i=1}^{t} b_i$, then the map $\psi: H \to \mathbb{Z}$, given by $h \mapsto bh$, is an injection (because H is torsion-free). Therefore, H is isomorphic to a nonzero subgroup of $\mathbb{Z}$, and hence it is infinite cyclic. ■

Theorem 10.20 (Fundamental Theorem). *Every finitely generated abelian group G is a direct sum of primary and infinite cyclic groups, and the number of summands of each kind depends only on G.*

Proof. Theorem 10.19 shows that G/tG is free abelian, so that Corollary 10.16 gives $G = tG \oplus F$, where $F \cong G/tG$. Now tG is finitely generated, being a summand and hence a quotient of G, and Exercise 6.18(ii) shows that tG is finite. The basis theorem for finite abelian groups says that tG is a direct sum of primary cyclic groups.

The uniqueness of the number of primary cyclic summands is precisely Theorem 6.11; the number of infinite cyclic summands is just $\operatorname{rank}(G/tG)$, and so it, too, depends only on G. ■

The next result will give a second proof of the basis theorem.

Theorem 10.21 (Simultaneous Bases). *Let H be a subgroup of finite index in a free abelian group F of finite rank n. Then there exist bases $\{y_1, \ldots, y_n\}$ of F and $\{h_1, \ldots, h_n\}$ of H such that $h_i \in \langle y_i \rangle$ for all i.*

Proof. If $\{x_1, \ldots, x_n\}$ is an ordered basis of F, then each element $h \in H$ has coordinates. Choose an ordered basis and an element h so that, among all such choices, the first coordinate of h is positive and minimal such. If $h = k_1x_1 + \cdots + k_nx_n$, then we claim that k_1 divides k_i for all $i \geq 2$. The division algorithm gives $k_i = q_ik_1 + r_i$, where $0 \leq r_i < k_1$. Therefore,

$$h = k_1(x_1 + q_2x_2 + \cdots + q_nx_n) + r_2x_2 + \cdots + r_nx_n.$$

Define $y_1 = x_1 + q_2x_2 + \cdots + q_nx_n$, and note that $\{y_1, x_2, \ldots, x_n\}$ is an ordered basis of F. Now $h = k_1y_1 + r_2x_2 + \cdots + r_nx_n$. If $r_i \neq 0$ for some i, then the first coordinate of h relative to the ordered basis $\{x_i, y_1, \ldots, x_n\}$ violates the minimality of our initial choice. Therefore, $r_i = 0$ for all $i \geq 2$ and k_1 divides k_i for all $i \geq 2$.

If $h' = m_1y_1 + m_2x_2 + \cdots + m_nx_n$ is any element of H, we claim that k_1 divides m_1. For if $m_1 = qk_1 + r$, where $0 \leq r < k_1$, then $h' - qh \in H$ has first coordinate $r < k_1$, a contradiction. It follows that the map $\pi\colon H \to H$, given by $h' \mapsto m_1y_1$, is a retraction with image $\langle h \rangle$. By Lemma 10.3, $H = \langle h \rangle \oplus \ker \pi = \langle h \rangle \oplus (H \cap \langle x_2, \ldots, x_n \rangle)$. Since $\langle x_2, \ldots, x_n \rangle$ is free abelian of rank $n - 1$ and $H \cap \langle x_2, \ldots, x_n \rangle$ is a subgroup of finite index, the proof can be completed by induction on n. ■

Corollary 10.22 (Basis Theorem). *Every finite abelian group G is a direct sum of cyclic groups.*

Proof. Write G as F/R, where F is free abelian of finite rank n, say. By Theorem 10.21, there are bases $\{y_1, \ldots, y_n\}$ and $\{h_1, \ldots, h_n\}$ of F and R, respectively, with $h_i = k_i y_i$ for all i. By Theorem 2.30, $G \cong \sum_{i=1}^n \mathbb{Z}_{k_i}$. ■

EXERCISES

10.15. If F is a free abelian group of finite rank n, then a subgroup H of F has finite index if and only if H is free abelian of rank n.

10.16. Let $\{x_1, \ldots, x_n\}$ be a basis of a free abelian group F. If $k_1, \ldots, k_n$ are integers with $\gcd(k_1, \ldots, k_n) = 1$, then there are elements $y_2, \ldots, y_n$ such that $\{k_1 x_1 + \cdots + k_n x_n, y_2, \ldots, y_n\}$ is a basis of F.

10.17. Let F be free abelian of rank n and let H be a subgroup of the same rank. Let $\{x_1, \ldots, x_n\}$ be a basis of F, let $\{y_1, \ldots, y_n\}$ be a basis of H, and let $y_j = \sum m_{ij} x_i$. Prove that
$$[F : H] = |\det[m_{ij}]|.$$
(*Hint*. Show that $|\det[m_{ij}]|$ is independent of the choice of bases of F and of H.)

Divisible and Reduced Groups

A reader of Chapter 1, asked to give examples of infinite abelian groups, probably would have responded with $\mathbb{Z}$, $\mathbb{Q}$, $\mathbb{R}$, and $\mathbb{C}$. We now study a common generalization of the latter three groups.

Definition. A group G is ***divisible*** if each $x \in G$ is divisible by every integer $n \geq 2$; that is, there exists $g_n \in G$ with $ng_n = x$ for all $n \geq 2$.

EXAMPLE 10.5. The following groups are divisible: $\mathbb{Q}$; $\mathbb{R}$; $\mathbb{C}$; the circle group $\mathbf{T}$; $\mathbb{Z}(p^\infty)$; the multiplicative group $F^\times$ of all nonzero elements of an algebraically closed field F (in particular, $\mathbb{C}^\times$).

EXAMPLE 10.6. Every quotient of a divisible group is divisible.

EXAMPLE 10.7. If $\{A_k : k \in K\}$ is a family of groups, Then each of $\sum_{k \in K} A_k$ (and $\prod_{k \in K} A_k$) is divisible if and only if every A_k is divisible.

EXAMPLE 10.8. A torsion-free divisible group G is a vector space over $\mathbb{Q}$.

If $x \in G$ and $n > 0$, then there is a unique $y \in G$ with $ny = x$, by Exercise 10.2. There is thus a function $\mathbb{Q} \times G \to G$, given by $(m/n, x) \mapsto my$ (where $ny = x$), which is a scalar multiplication satisfying the axioms in the definition of vector space.

Theorem 10.23 (Injective Property, Baer, 1940). *Let D be a divisible group and let A be a subgroup of a group B. If $f \colon A \to D$ is a homomorphism, then f*

can be extended to a homomorphism $\varphi: B \to D$; that is, the following diagram commutes:

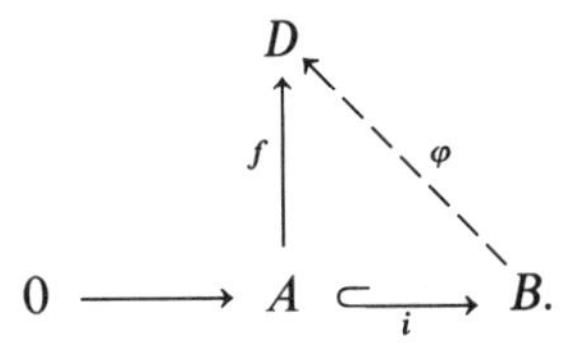

Proof. We use Zorn's lemma. Consider the set $\mathcal{S}$ of all pairs (S, h), where $A \le S \le B$ and $h: S \to D$ is a homomorphism with $h|A = f$. Note that $\mathcal{S} \ne \varnothing$ because $(A, f) \in \mathcal{S}$. Partially order $\mathcal{S}$ by decreeing that $(S, h) \le (S', h')$ if $S \le S'$ and h' extends h; that is, $h'|S = h$. If $\mathcal{C} = \{(S_\alpha, h_\alpha)\}$ is a simply ordered subset of $\mathcal{S}$, define $(\tilde{S}, \tilde{h})$ by $\tilde{S} = \bigcup_\alpha S_\alpha$ and $\tilde{h} = \bigcup_\alpha h_\alpha$ (this makes sense if one realizes that a function *is* a graph; in concrete terms, if $s \in \tilde{S}$, then $s \in S_\alpha$ for some α, and $\tilde{h}(s) = h_\alpha(s)$). The reader may check that $(\tilde{S}, \tilde{h}) \in \mathcal{S}$ and that it is an upper bound of $\mathcal{C}$. By Zorn's lemma, there exists a maximal pair $(M, g) \in \mathcal{S}$. We now show that $M = B$, and this will complete the proof.

Suppose that there is $b \in B$ with $b \notin M$. If $M' = \langle M, b \rangle$, then $M < M'$, and so it suffices to define $h': M' \to D$ extending g to reach a contradiction.

Case 1. $M \cap \langle b \rangle = 0$.

In this case, $M' = M \oplus \langle b \rangle$, and one can define h' as the map $m + kb \mapsto g(m)$.

Case 2. $M \cap \langle b \rangle \ne 0$.

If k is the smallest positive integer for which $kb \in M$, then each $y \in M'$ has a unique expression of the form $y = m + tb$, where $0 \le t < k$. Since D is divisible, there is an element $d \in D$ with $kd = h(kb)$ ($kb \in M$ implies $h(kb)$ is defined). Define $h': M' \to D$ by $m + tb \mapsto g(m) + td$. It is a routine calculation, left for the reader, that h' is a homomorphism extending g. ■

Corollary 10.24. *If a divisible group D is a subgroup of a group G, then D is a direct summand of G.*

Proof. Consider the diagram:

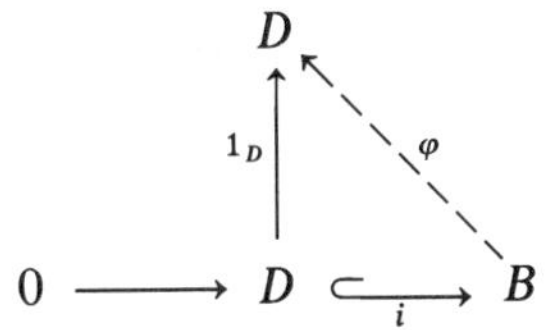

where 1_D is the identity map. By the injective property, there is a homomorphism $\varphi: G \to D$ with $\varphi i = 1_D$ (where i is the inclusion $D \hookrightarrow G$); that is, $\varphi(d) = d$ for all $d \in D$. By Lemma 10.3, D is a direct summand of G. ■

Definition. If G is a group, then dG is the subgroup generated by all the divisible subgroups of G.

Note that dG is a fully invariant subgroup, for every image of a divisible group is divisible.

Lemma 10.25. *For any group G, dG is the unique maximal divisible subgroup of G.*

Proof. It suffices to prove that dG is divisible. Let $x \in dG$ and let $n > 0$. Now $x = d_1 + \cdots + d_t$, where each $d_i \in D_i$, a divisible subgroup of G. Since D_i is divisible, there is $y_i \in D_i$ with $ny_i = d_i$ for all i. Hence $y_1 + \cdots + y_t \in dG$ and $n(y_1 + \cdots + y_t) = x$, as desired. ■

Definition. A group G is ***reduced*** if $dG = 0$.

Of course, G is divisible if and only if $dG = G$.

Theorem 10.26. *For every group G, there is a decomposition*

$$G = dG \oplus R.$$

where R is reduced.

Proof. Since dG is divisible, Corollary 10.24 gives the existence of R. If $D \leq R$ is divisible, then $D \leq R \cap dG = 0$, by Lemma 10.25. ■

The reader should compare the roles of the subgroups tG and dG. Every abelian group G is an extension of the torsion group tG by a torsion-free group (but tG need not be a direct summand); G is an extension of the divisible group dG by a reduced group, and dG is always a direct summand.

Recall that if G is a group, then $G[n] = \{x \in G: nx = 0\}$.

Lemma 10.27. *If G and H are divisible p-primary groups, then $G \cong H$ if and only if $G[p] \cong H[p]$.*

Proof. Necessity follows easily from the fact that $\varphi(G[p]) \leq H[p]$ for every homomorphism $\varphi: G \to H$.

For sufficiency, assume that $\varphi: G[p] \to H[p]$ is an isomorphism; composing with the inclusion $H[p] \hookrightarrow H$, we may assume that $\varphi: G[p] \to H$. The injective property gives the existence of a homomorphism $\Phi: G \to H$ extending φ; we claim that Φ is an isomorphism.

(i) Φ is injective.

We show by induction on $n \geq 1$ that if $x \in G$ has order p^n, then $\Phi(x) = 0$. If $n = 1$, then $x \in G[p]$, so that $\Phi(x) = \varphi(x) = 0$ implies $x = 0$ (because φ is injective). Assume that x has order p^{n+1} and $\Phi(x) = 0$. Now $\Phi(px) = 0$ and px has order p^n, so that $px = 0$, by induction, and this contradicts x having order p^{n+1}.

(ii) Φ is surjective.

We show, by induction on $n \geq 1$, that if $y \in H$ has order p^n, then $y \in \text{im } \Phi$, If $n = 1$, then $y \in H[p] = \text{im } \varphi \leq \text{im } \Phi$. Suppose now that y has order p^{n+1}. Since $p^n y \in H[p]$, there is $x \in G$ with $\Phi(x) = p^n y$; since G is divisible, there is $g \in G$ with $p^n g = x$. Thus, $p^n(y - \Phi(x)) = 0$, so that induction provides $z \in G$ with $\Phi(z) = y - \Phi(g)$. Therefore, $y = \Phi(y + g)$, as desired. ■

Theorem 10.28. *Every divisible group D is a direct sum of copies of $\mathbb{Q}$ and of copies of $\mathbb{Z}(p^\infty)$ for various primes p.*

Proof. It is easy to see that tD is divisible, so that $D = tD \oplus V$, where $V \cong D/tD$. Now V is torsion-free and divisible, so it is a vector space over $\mathbb{Q}$, by Example 10.8; by Theorem 10.5, V is a direct sum of copies of $\mathbb{Q}$.

For every prime p, the p-primary component G of tD is divisible (it is a summand of a divisible group). Let $r = \dim G[p]$ (as a vector space over $\mathbb{Z}_p$), and let H be the direct sum of r copies of $\mathbb{Z}(p^\infty)$. Now H is a p-primary divisible group with $G[p] \cong H[p]$, and so $G \cong H$, by the lemma. ■

Notation. If D is a divisible group, let $\delta_\infty(D) = \dim_{\mathbb{Q}} D/tD$ and let $\delta_p(D) = \dim_{\mathbb{Z}_p} D[p]$.

The proof of the next theorem is left as an exercise.

Theorem 10.29. *If D and D' are divisible groups, then $D \cong D'$ if and only if $\delta_\infty(D) = \delta_\infty(D')$ and, for all primes p, $\delta_p(D) = \delta_p(D')$.*

There is an analogy between theorems about free abelian groups and theorems about divisible groups that may be formalized as follows. Given a commutative diagram containing exact sequences, then its ***dual diagram*** is the diagram obtained from it by reversing all arrows. For example, the dual diagram of $0 \to A \to B$ is $B \to A \to 0$, and this leads one to say that "subgroup" and "quotient group" are dual notions. The notion of short exact sequence is self-dual, Theorems 10.9 and 10.10 show that "direct sum" and "direct product" are dual, and the projective property is dual to the injective property (suggesting that free abelian groups are dual to divisible groups). The next result should be compared to Corollary 10.12.

Theorem 10.30. *Every group G can be imbedded in a divisible group.*

Proof. Write $G = F/R$, where F is free abelian. Now $F = \sum \mathbb{Z}$, so that $F \leq \sum \mathbb{Q}$ (just imbed each copy of $\mathbb{Z}$ into $\mathbb{Q}$). Hence $G = F/R = (\sum \mathbb{Z})/R \leq (\sum \mathbb{Q})/R$, and the last group is divisible, being a quotient of a divisible group. ■

The next result should be compared to Exercise 10.12.

Corollary 10.31. *A group G is divisible if and only if it is a direct summand of any group containing it.*

Proof. Necessity is Corollary 10.24. For sufficiency, Theorem 10.30 says that there is a divisible group D containing G as a subgroup. By hypothesis, G is a direct summand of D, and so G is divisible. ∎

Exercises

10.18. Show that the group G/tG in Theorem 10.2 is divisible.

10.19. If $0 \to A \to B \to C \to 0$ is an exact sequence and if A and C are reduced, then B is reduced.

10.20. (i) If G is the group in Exercise 10.6, then $G/\langle a_0 \rangle$ is a direct sum of cyclic groups.
(ii) Show that G is reduced.

10.21. (i) Prove that $\mathbb{Q}/\mathbb{Z} \cong \sum_p \mathbb{Z}(p^\infty)$. (*Hint.* Use Exercise 10.5.)
(ii) Prove that $(\mathbb{Q}/\mathbb{Z})[n] \cong \mathbb{Z}_n$.

10.22. Prove that a group D is divisible if and only if it has the injective property. (*Hint.* $\langle n \rangle \leq \mathbb{Z}$ for all $n \geq 1$; extend homomorphisms $\langle n \rangle \to D$ to homomorphisms $\mathbb{Z} \to D$.)

10.23. (i) A group G is divisible if and only if $G = pG$ for all primes p.
(ii) A p-primary group is divisible if and only if $G = pG$. (*Hint.* Use Exercise 1.31.)

10.24. If G and H are groups, prove that $G \cong H$ if and only if $dG \cong dH$ and $G/dG \cong H/dH$.

10.25. The following conditions on a group G are equivalent:
(i) G is divisible;
(ii) Every nonzero quotient of G is infinite; and
(iii) G has no maximal subgroups.

10.26. If G and H are divisible groups each of which is isomorphic to a subgroup of the other, then $G \cong H$. Is this true if we drop the adjective "divisible"?

10.27. If G and H are divisible groups for which $G \oplus G \cong H \oplus H$, then $G \cong H$.

10.28. (i) Prove that the following groups are all isomorphic: $\mathbb{R}/\mathbb{Z}$; the circle group $\mathbf{T}$; $\prod_p \mathbb{Z}(p^\infty)$; $\mathbb{R} \oplus (\mathbb{Q}/\mathbb{Z})$; $\mathbb{C}^\times$.
(ii) Prove that $t(\mathbb{C}^\times) \cong \mathbb{Q}/\mathbb{Z}$.

10.29. Prove that every countable abelian group G can be imbedded in $\sum_{i=1}^{\infty} D_i$, where $D_i \cong \mathbb{Q} \oplus (\mathbb{Q}/\mathbb{Z})$ for all i.

10.30. (i) Every torsion-free group G can be imbedded in a vector space over $\mathbb{Q}$. (*Hint.* Imbed G in a divisible group D, consider the natural map $D \to D/tD$, and use Example 10.8.)
(ii) If a maximal independent subset of a torsion-free group G has n elements, then G can be imbedded in a vector space over $\mathbb{Q}$ of dimension n.

10.31. If A is a group and $m \in \mathbb{Z}$, define $m_A: A \to A$ by $a \mapsto ma$.
(i) Show that A is torsion-free if and only if m_A is an injection for all $m \neq 0$.
(ii) Show that A is divisible if and only if m_A is a surjection for every $m \neq 0$.
(iii) Show that A is a vector space over $\mathbb{Q}$ if and only if m_A is an automorphism for every $m \neq 0$.

Torsion Groups

Torsion groups can be quite complicated, but there are two special classes of torsion groups that are quite manageable: divisible groups and direct sums of cyclic groups. We shall prove that every torsion group is an extension of a direct sum of cyclics by a divisible. The proof involves a special kind of subgroup that we now investigate.

Definition. A subgroup $S \geq G$ is a ***pure*** subgroup if, for every integer n,

$$S \cap nG = nS.$$

It is always true that $S \cap nG \geq nS$, and so it is only the reverse inclusion that is significant: if $s \in S \cap nG$, then $s \in nS$; that is, if $s \in S$ and $s = ng$ for some $g \in G$, then there exists $s' \in S$ with $s = ns'$.

EXAMPLE 10.9. Every direct summand is pure.

Let $G = A \oplus B$. If $a \in A$ and $a = ng$, then $g = a' + b'$, for $a' \in A$ and $b' \in B$. Now $nb' = 0$, for $nb' = a - na' \in A \cap B = 0$. Hence $a = na'$ and A is pure.

EXAMPLE 10.10. If $S \leq G$ and G/S is torsion-free, then S is pure.

If $s = ng$, then $g + S \in G/S$ has finite order; since G/S is torsion-free, $g + S = S$, and $g \in S$.

EXAMPLE 10.11. tG is a pure subgroup of G that may not be a direct summand of G.

By Theorem 10.1, G/tG is torsion-free, and so Example 10.10 shows that tG is pure. It follows from Theorem 10.2 that tG need not be a direct summand.

Lemma 10.32. *Let S be a pure subgroup of G, and let v: $G \to G/S$ be the natural map. If $y \in G/S$, then there is $x \in G$ with $v(x) = y$ such that x and y have the same order.*

Proof. Surjectivity of the natural map provides $x \in G$ with $v(x) = y$. If y has infinite order, then so does x. If y has finite order n, then $v(nx) = nv(x) = ny = 0$, so that $nx \in \ker v = S$. Since S is pure, there is $s' \in S$ with $nx = ns'$. If $z = x - s'$, then $nz = 0$ and $v(z) = v(x - s') = y$. But n divides the order of z, by Exercise 2.14, and so z has order n. ■

Lemma 10.33. *Let $T \leq G$ be pure. If $T \leq S \leq G$, then S/T is pure in G/T if and only if S is pure in G.*

Proof. Suppose that S/T is pure in G/T. Assume that $s \in S$ and $s = ng$ for some $g \in G$. In G/T, $\bar{s} = n\bar{g}$ (where bar denotes coset mod T), and so there is $s' \in S$ with $\bar{s} = n\bar{s}'$; that is, there is $t \in T$ with $s = ns' + t$. Thus $t = n(g - s')$, and the purity of T gives $t' \in T$ with $ng - ns' = nt'$. Hence $s = ng = n(s' + t')$. But $s' + t' \in S$, because $T \leq S$, and so S is pure in G.

Conversely, suppose that S is pure in G. If $\bar{s} \in S/T$ and $\bar{s} = n\bar{g}$ in G/T, then $ng = s + t$ for some $t \in T$. Since $T \leq S$, we have $s + t \in S$, and purity gives $s' \in S$ with $s + t = ns'$. Therefore, $\bar{s} = n\bar{s}'$ and S/T is pure. ■

Lemma 10.34. *A p-primary group G that is not divisible contains a pure nonzero cyclic subgroup.*

Proof. Assume first that there is $x \in G[p]$ that is divisible by p^k but not by p^{k+1}, and let $x = p^k y$. We let the reader prove that $\langle y \rangle$ is pure in G (Exercise 1.31 shows that one need check only powers of p).

We may, therefore, assume that every $x \in G[p]$ is divisible by every power of p. In this case, we prove by induction on $k \geq 1$ that if $x \in G$ and $p^k x = 0$, then x is divisible by p. If $k = 1$, then $x \in G[p]$, and the result holds. If $p^{k+1}x = 0$, then $p^k x \in G[p]$, and so there is $z \in G$ with $p^{k+1}z = p^k x$. Hence $p^k(pz - x) = 0$. By induction, there is $w \in G$ with $pw = pz - x$, and $x = p(z - w)$, as desired.

We have shown that $G = pG$, and so Exercise 10.25(ii) gives G divisible, a contradiction. ■

Definition. A subset X of a group G is ***pure-independent*** if it is independent and $\langle X \rangle$ is a pure subgroup of G.

Lemma 10.35. *Let G be a p-primary group. If X is a maximal pure-independent subset of G (i.e., X is contained in no larger such), then $G/\langle X \rangle$ is divisible.*

Proof. If $G/\langle X \rangle$ is not divisible, then Lemma 10.34 shows that it contains a pure nonzero cyclic subgroup $\langle \bar{y} \rangle$; by Lemma 10.32, we may assume that $y \in G$ and $\bar{y} \in G/\langle X \rangle$ have the same order (where $y \mapsto \bar{y}$ under the natural map). We claim that $\{X, y\}$ is pure-independent.

Now $\langle X \rangle \leq \langle X, y \rangle \leq G$, and $\langle X, y \rangle / \langle X \rangle = \langle \bar{y} \rangle$ is pure in $G/\langle X \rangle$; by Lemma 10.32, $\langle X, y \rangle$ is pure in G.

Suppose that $my + \sum m_i x_i = 0$, where $x_i \in X$ and $m, m_i \in \mathbb{Z}$. In $G/\langle X \rangle$, this equation becomes $m\bar{y} = 0$. But y and $\bar{y}$ have the same order, so that $my = 0$. Hence $\sum m_i x_i = 0$, and independence of X gives $m_i x_i = 0$ for all i. Therefore $\{X, y\}$ is independent, and by the preceding paragraph, it is pure-independent, contradicting the maximality of X. ■

Definition. A subgroup B of a torsion group G is a ***basic subgroup*** if:

(i) B is a direct sum of cyclic groups;
(ii) B is a pure subgroup of G; and
(iii) G/B is divisible.

Theorem 10.36 (Kulikov, 1945). *Every torsion group G has a basic subgroup, and so G is an extension of a direct sum of cyclic groups by a divisible group.*

Proof. Let $G = \sum G_p$ be the primary decomposition of G. If G_p has a basic subgroup of B_p, then it is easy to see that $\sum B_p$ is a basic subgroup of G. Thus, we may assume that G is p-primary.

If G is divisible, then $B = 0$ is a basic subgroup. If G is not divisible, then it contains a pure nonzero cyclic subgroup, by Lemma 10.34; that is, G does have pure-independent subsets. Since both purity and independence are preserved by ascending unions (see Exercise 10.28(i)), Zorn's lemma applies to show that there is a maximal pure-independent subset X of G. But Lemmas 10.6 and 10.33 show that $B = \langle X \rangle$ is a basic subgroup. ∎

The following theorem was proved by H. Prüfer in 1923 for G countable; the general case was proved by R. Baer in 1934.

Corollary 10.37 (Prüfer–Baer). *If G is a group of bounded order* (i.e., $nG = 0$ *for some* $n > 0$), *then G is a direct sum of cyclic groups.*

Remark. Were G nonabelian, we would say "G has finite exponent" instead of "G is of bounded order."

Proof. By Theorem 10.28, a bounded divisible group must be 0. Therefore, if B is a basic subgroup of G, then $G/B = 0$ and $B = G$. ∎

Assume that G is a direct sum of p-primary cyclic groups. Let B_n be the direct sum of all those summands of order p^n, if any, so that $G = B_1 \oplus B_2 \oplus \cdots$. When G is finite, we proved (in Chapter 6) that $d_n = \dim p^n G/p^{n+1}G$ is the number of cyclic summands of order $\geq p^{n+1}$, so that the number of cyclic summands of order precisely p^{n+1} is just $d_n - d_{n+1}$. This formula does not generalize to infinite groups because one cannot subtract infinite cardinals.

If G is an infinite direct sum of p-primary cyclic groups, it is still true that d_n is the number of cyclic summands of order $\geq p^{n+1}$. How can we distinguish those elements in $p^n G$ coming from cyclic summands of order p^{n+1} from those cyclic summands of larger order? The elementary observation that a cyclic summand $\langle a \rangle$ has order p^{n+1} if and only if $p^n a$ has order p suggests that we focus on elements of order p; that is, replace $p^n G$ by $p^n G \cap G[p]$.

Definition. If G is a p-primary group and $n \geq 0$, then

$$U\{n, G\} = \dim_{\mathbb{Z}_p} (p^n G \cap G[p])/(p^{n+1} G \cap G[p]).$$

Lemma 10.38. *If G is a direct sum of p-primary cyclic groups, then $U\{n, G\}$ is the number of cyclic summands of order p^{n+1}.*

Proof. Let B_n be the direct sum of all those cyclic summands of order p^n, if any (in the given decomposition of G), so that $G = B_1 \oplus B_2 \oplus \cdots \oplus B_k \oplus \cdots$; let b_k be the number of summands in B_k (of course, b_k may be 0). It is easy to see that

$$G[p] = B_1 \oplus pB_2 \oplus p^2B_3 \oplus \cdots \oplus p^{k-1}B_k \oplus \cdots$$

and

$$p^nG = p^nB_{n+1} \oplus \cdots \oplus p^nB_k \oplus \cdots .$$

Hence, for all $n \geq 0$,

$$p^nG \cap G[p] = p^nB_{n+1} \oplus p^{n+1}B_{n+2} \oplus \cdots,$$

and so

$$(p^nG \cap G[p])/(p^{n+1}G \cap G[p]) \cong p^nB_{n+1}.$$

Therefore, $U\{n, G\} = \dim(p^nB_{n+1}) = b_{n+1}$, as desired. ■

Theorem 10.39. *If G and H are direct sums of p-primary cyclic groups, then $G \cong H$ if and only if $U\{n, G\} = U\{n, H\}$ for all $n \geq 0$.*

Proof. The numbers $U\{n, G\}$ depend only on G and not upon the decomposition. ■

Theorem 10.40. *Any two basic subgroups of a p-primary group G are isomorphic.*

Proof. Let B be a basic subgroup of G. The number b_n of cyclic summands of B of order p^n is equal to the number of such summands of $B/p^{n+1}B$, and so it suffices to show that this latter quotient depends only on G.

We claim, for every $n \geq 1$, that $G = B + p^nG$. If $g \in G$, then divisibility of G/B gives $g + B = p^nx + B$ for some $x \in G$, so there is some $b \in B$ with $g = b + p^nx \in B + p^nG$. It follows that

$$\begin{aligned} G/p^{n+1}G &= (B + p^{n+1}G)/p^{n+1}G \\ &\cong B/(B \cap p^{n+1}G) \\ &= B/p^{n+1}B, \quad \text{by purity.} \end{aligned}$$

Therefore, $B/p^{n+1}B$ is independent of the choice of B. ■

An example is given in Exercise 10.41 below showing that the divisible quotient G/B, where B is a basic subgroup, is *not* determined by G.

Here is a condition forcing a pure subgroup to be a summand.

Corollary 10.41 (Prüfer, 1923). *A pure subgroup S of bounded order is a direct summand.*

Proof. Assume that $S \leq G$ is pure and that $nS = 0$ for some $n > 0$. Let

$\nu: G \to G/(S + nG)$ be the natural map. The group $G/(S + nG)$ is of bounded order (it has exponent n) so that it is a direct sum of cyclic groups, by Corollary 10.33. Write $G/(S + nG) = \sum_i \langle \bar{x}_i \rangle$, where $\bar{x}_i$ has order r_i; for each i, choose $x_i \in G$ with $\nu(x_i) = \bar{x}_i$. Now

$$r_i x_i = s_i + ng_i,$$

where $s_i \in S$ and $g_i \in G$. But r_i divides (the exponent) n, so that

$$s_i = r_i(x_i - (n/r_i)g_i).$$

Since S is pure, there is $t_i \in S$ with $s_i = r_i t_i$. Define

$$y_i = x_i - t_i.$$

Now $\nu(y_i) = \bar{x}_i$ and $r_i y_i = ng_i$. Let K be the subgroup generated by nG and all the y_i; we shall show that $G = S \oplus K$.

(i) $S \cap K = 0$.

Let $s \in S \cap K$; since $s \in K$, $s = \sum m_i y_i + nh$; since $s \in S$, $\nu(s) = 0$. Thus, $0 = \sum m_i \bar{x}_i$, and independence gives $m_i \bar{x}_i = 0$; hence, r_i divides m_i for all i. We have chosen y_i so that $r_i y_i \in nG$, hence $m_i y_i \in nG$. Therefore, $s = \sum m_i y_i + nh \in nG$. Since S is pure, there is $s' \in S$ with $s = ns' \in nS = 0$.

(ii) $S + K = G$.

If $g \in G$, then $\nu(g) = \sum l_i \bar{x}_i$. Since $\nu(\sum l_i y_i) = \sum l_i \bar{x}_i$, we have $g - \sum l_i y_i \in \ker \nu = S + nG$; say, $g - \sum l_i y_i = s + nh$. Thus, $g = s + (nh + \sum l_i y_i) \in S + K$. ∎

Corollary 10.42. *If tG is of bounded order, then tG is a direct summand of G. In particular, tG is a direct summand when it is finite.*

Corollary 10.43. *A torsion group G that is not divisible has a p-primary cyclic direct summand (for some prime p).*

Proof. Since G is not divisible, at least one of its primary components, say, G_p, is not divisible. By Lemma 10.34, G_p has a pure nonzero cyclic summand C, and C must be a summand of G, by the theorem. ∎

Corollary 10.44. *An indecomposable group G is either torsion or torsion-free.*

Proof. Assume that $0 < tG < G$. Now tG is not divisible, lest it be a summand of G, so that Corollary 10.43 shows that G has a (cyclic) summand, contradicting indecomposability. ∎

Here are three lovely results (see Fuchs, Griffith, or Kaplansky for proofs). All countable torsion groups are classified by ***Ulm's Theorem*** (1933): if G is a p-primary group, there is a transfinite version of the numbers $U\{n, G\}$ (the *Ulm invariants*) with n varying over ordinal numbers, and two countable torsion groups are isomorphic if and only if their respective primary components have the same Ulm invariants. (There are uncountable p-primary groups having the same Ulm invariants which are not isomorphic; see Exercise 10.39(ii) below.) A theorem of Prüfer (1923) says that a countable p-primary group G is a direct sum of cyclic groups if and only if $\bigcap_{n=1}^{\infty} p^n G = 0$ (this is false for uncountable groups, as is shown in Exercise 10.39(iii) below). Kulikov (1941) has characterized direct sums of cyclic groups, and one consequence of his criterion is that every subgroup of a direct sum of cyclic groups is another such.

Exercises

10.32. If G is torsion-free, then a subgroup S is pure if and only if G/S is torsion-free.

10.33. (i) Given an example of an intersection of two pure subgroups of a group G not being pure (*Hint*: Take $G = \mathbb{Z}_2 \oplus \mathbb{Z}_8$.)
(ii) Given an example in which the subgroup generated by two pure subgroups is not pure. (*Hint*. Look within a free abelian group of rank 2.)

10.34. (i) If G is torsion-free, then any intersection of pure subgroups is pure, and one can define the ***pure subgroup generated by a subset*** X (as the intersection of all the pure subgroups containing X).
(ii) Let G be torsion-free, and let $x \in G$. Show that the pure subgroup generated by x is:

$$\{g \in G : mg \in \langle x \rangle\}.$$

(We have rediscovered the subgroup H in the proof of Theorem 10.19.)

10.35. A pure subgroup of a divisible group is a direct summand.

10.36. (i) Show that an ascending union of pure subgroups is always pure.
(ii) Show that an ascending union of direct summands need not be a direct summand. (*Hint*. Consider $\prod_{p \in P} \mathbb{Z}_p$.)

10.37. (i) If $G = t(\prod_{n=1}^{\infty} \mathbb{Z}_{p^n})$, then G is an uncountable group with $U\{n, G\} = 1$ for all $n \geq 0$.
(ii) Show that Ulm's theorem does not classify uncountable torsion groups. (*Hint*. $t(\prod_{n=1}^{\infty} \mathbb{Z}_{p^n}) \not\cong \sum_{n=1}^{\infty} \mathbb{Z}_{p^n}$.)
(iii) Prove that $G = t(\prod_{n=1}^{\infty} \mathbb{Z}_{p^n})$ is not a direct sum of cyclic groups.

10.38. Prove that a torsion group is indecomposable if and only if it is isomorphic to a subgroup of $\mathbb{Z}(p^\infty)$ for some prime p.

10.39. Show that $\prod_p \mathbb{Z}_p$ is not a direct sum of (possibly infinitely many) indecomposable groups.

10.40. **(Kaplansky).** In this exercise, G is an infinite group.
(i) If every proper subgroup of G is finite, then $G \cong \mathbb{Z}(p^\infty)$ for some prime p.
(ii) If G is isomorphic to every proper subgroup, then $G \cong \mathbb{Z}$.
(iii) If G is isomorphic to every nonzero quotient, then $G \cong \mathbb{Z}(p^\infty)$.
(iv) If every proper quotient is finite, then $G \cong \mathbb{Z}$.

10.41. Let $G = \sum_{n=0}^{\infty} \langle a_n \rangle$, where a_n has order p^{n+1}, and let $B = \langle pa_1, a_n - pa_{n+1}, n \geq 1 \rangle \leq G$. Show that B is a basic subgroup of G. As G is a basic subgroup of itself, conclude that the quotient G/B, where B is basic in G, is not an invariant of G.

Subgroups of $\mathbb{Q}$

The notion of rank can be generalized from free abelian groups to arbitrary torsion-free groups.

Theorem 10.45. *If G is a torsion-free group, then any two maximal independent sets in G have the same number of elements. If a maximal independent subset of G has r elements, then G is an additive subgroup of an r-dimensional vector space over $\mathbb{Q}$.*

Proof. Let X be a maximal independent subset of G. By Exercise 10.30, G can be imbedded in a vector space W over $\mathbb{Q}$; let V be the subspace of W spanned by G. If X spans V, then it is a basis of V; this will prove both statements in the theorem, for then all maximal independent subsets of G will have dim V elements.

If $v \in V$, then there is a finite sum $v = \sum q_i g_i$, where $q_i \in \mathbb{Q}$ and $g_i \in G$; if b is the product of the denominators of the q_i, then $bv \in G$. By Exercise 10.3, maximality of X gives $G/\langle X \rangle$ torsion There is thus a nonzero integer m with $mbv \in \langle X \rangle$; that is, mbv is a $\mathbb{Z}$-linear combination of elements in X, and so v is a $\mathbb{Q}$-linear combination of elements in X. ∎

Definition. The ***rank*** of a torsion-free group G is the number of elements $\rho(G)$ in a maximal independent subset. Define the ***rank*** $\rho(G)$ of an arbitrary abelian group G to be $\rho(G/tG)$.

Theorem 10.45 shows that the rank is independent of the choice of maximal independent subset; when G is torsion-free, it characterizes $\rho(G)$ as the (minimal) dimension of a vector space V over $\mathbb{Q}$ containing G. Thus, a torsion-free group of finite rank is just a subgroup of a finite-dimensional vector space over $\mathbb{Q}$; in particular, a torsion-free group of rank 1 is just a nonzero subgroup of the additive group of rational numbers $\mathbb{Q}$.

Here are three subgroups of $\mathbb{Q}$.

A: all rationals having squarefree denominators;
B: all ***dyadic rationals***; that is, all rationals of the form $a/2^k$.
C: all rationals whose decimal expansion is finite.

No two of these groups are isomorphic. For example, $A \ncong B$ because B contains a nonzero solution x to the system of equations $2^k y_k = x$, while A does not. One can also describe C as all rationals whose denominators are restricted to be powers of 10, and the same reasoning shows that $A \not\cong C$ and $B \not\cong C$.

Definition. Let G be a torsion-free group and let $x \in G$. If p is a prime, then the ***p-height*** of x, denoted by $h_p(x)$, is the highest power of p dividing x in G: more precisely, if $p^n g_n = x$ is solvable in G for all n, then $h_p(x) = \infty$; if k is the largest integer n for which $p^n g = x$ is solvable in G, then $h_p(x) = k$.

Each nonzero x in a torsion-free group G determines its ***height sequence*** $h(x) = (h_2(x), h_3(x), \ldots, h_p(x), \ldots)$, which is a sequence of nonnegative integers and the symbol ∞. For example, each of the groups $\mathbb{Z}$, $\mathbb{Q}$, A, B, and C contains $x = 1$; its height sequence in each group is:

$$\mathbb{Z}: \quad (0, 0, 0, \ldots);$$

$$\mathbb{Q}: \quad (\infty, \infty, \infty, \ldots);$$

$$A: \quad (1, 1, 1, \ldots);$$

$$B: \quad (\infty, 0, 0, 0, \ldots); \text{ and}$$

$$C: \quad (\infty, 0, \infty, 0, 0, 0, \ldots).$$

Different elements in the same group G of rank 1 may have different height sequences. For example, the height sequence of $x = 168 = 2^3 \cdot 3 \cdot 7$ in each of the groups above is:

$$\mathbb{Z}: \quad (3, 1, 0, 1, 0, 0, \ldots);$$

$$\mathbb{Q}: \quad (\infty, \infty, \infty, \infty, \ldots);$$

$$A: \quad (4, 2, 1, 2, 1, 1, 1, \ldots);$$

$$B: \quad (\infty, 1, 0, 1, 0, 0, 0, \ldots); \text{ and}$$

$$C: \quad (\infty, 1, \infty, 1, 0, 0, 0, \ldots).$$

We have been led to the following definitions.

Definition. A ***characteristic*** is a sequence of nonnegative integers and the symbol ∞. Two characteristics are ***equivalent*** if:

(i) they have ∞ in the same coordinates; and
(ii) they differ in at most a finite number of (other) coordinates.

An equivalence class of characteristics is called a ***type***.

Lemma 10.46. *Let G be a torsion-free group of* rank 1. *If $x, y \in G$ are nonzero, then their height sequences are equivalent.*

Proof. If $y = nx$, where $n = p_1^{e_1} \ldots p_t^{e_t}$, then $h_p(x) = h_p(y)$ for all primes $p \neq$

$p_1, \ldots, p_t$, and $h_{p_i}(y) = e_i + h_{p_i}(x)$ for $i = 1, \ldots, t$ (we agree that $\infty + k = \infty$). Thus, the result is true in this case.

For the general case, note that if G is isomorphic to a subgroup of $\mathbb{Q}$, then there are nonzero integers m and n with $my = nx$. Thus, the height sequences of y, $my = nx$, and x are all equivalent. ■

As a result of the lemma, one may define the ***type*** $\tau(G)$ of a torsion-free group G of rank 1 as the equivalence class of the height sequence of any of its nonzero elements.

Theorem 10.47. *If G and G' are torsion-free groups of rank* 1, *then $G \cong G'$ if and only if $\tau(G) = \tau(G')$.*

Proof. If $x \in G$ and $\varphi: G \to G'$ is a homomorphism, then $h_p(x) \leq h_p(\varphi(x))$ for all primes p; if φ^{-1} exists, then $h_p(\varphi(x)) \leq h_p(\varphi^{-1}(\varphi(x))) = h_p(x)$. Hence, if φ is an isomorphism and $x \in G$, then x and $\varphi(x)$ have the same height sequence, so that $\tau(G) = \tau(G')$.

For the converse, the hypothesis says that nonzero elements $x \in G$ and $x' \in G'$ have equivalent height sequences. Let P be the finite set of primes p for which $h_p(x) < h_p(x')$, and let Q be the finite set of primes for which $h_p(x) > h_p(x')$ (of course, P or Q may be empty). For $p \in P$, define $e_p = h_p(x') - h_p(x)$ (the definition of equivalence says that both $h_p(x)$ and $h_p(x')$ are finite); for $q \in Q$, define $f_q = h_p(x) - h_p(x')$. If $m = \prod_{p \in P} p^{e_p}$ and $n = \prod_{q \in Q} q^{f_q}$, then it is easy to see that mx and nx' have the same height sequence.

Let us now assume that both G and G' are subgroups of $\mathbb{Q}$ containing elements y and y', respectively, having the same height sequence; let $y = a/b$ and $y' = a'/b'$. The subgroup $(b/a)G$ of $\mathbb{Q}$ is isomorphic to G, and the subgroup $(b'/a')G'$ is isomorphic to G'. Replacing G by $(b/a)G$ and G' by $(b'/a')G'$, we may assume that 1 lies in both G and G' and that it has the same height sequence in each group. But it is now an easy exercise that $G = G'$. ■

There is an existence theorem complementing the uniqueness theorem just proved.

Theorem 10.48. *For every type τ, there exists a torsion-free group G of* rank 1 *with $\tau(G) = \tau$.*

Proof. If $(k_2, k_3, \ldots, k_p, \ldots)$ is a characteristic in τ, define G as the subgroup of $\mathbb{Q}$ generated by all rationals of the form $1/p^e$, where $e \leq k_p$ if k_p is finite, and e is any positive integer if $k_p = \infty$. It is easy to see that the height sequence of $1 \in G$ is the given characteristic. ■

Torsion-free groups of rank ≥ 2 are not classified (though there do exist several kinds of description of them). After L. Fuchs (1971) showed that there are indecomposable groups of all ranks r, where r is smaller than the first strongly inaccessible cardinal (should such exist), S. Shelah (1974) showed that indecomposables of every rank r exist.

There are groups of infinite rank that are not direct sums of indecomposable groups. Every group of finite rank is either indecomposable or a direct sum of indecomposables; if all the summands have rank 1, then R. Baer (1937) showed that the summands are unique to isomorphism; otherwise, the summands need not be unique; indeed, not even the ranks of the summands are determined. B. Jónsson (1957) introduced the notion of quasi-isomorphism[2]: two torsion-free groups of finite rank are *quasi-isomorphic* if each is isomorphic to a subgroup of the other having finite index. There is a corresponding notion of indecomposable: a group is *strongly indecomposable* if it is not quasi-isomorphic to a direct sum of two nonzero groups. He proved that every torsion-free group of finite rank is quasi-isomorphic to a direct sum of strongly indecomposable summands and that these summands are unique in the sense of the Krull–Schmidt theorem.

Definition. If G is an abelian group, then its ***endomorphism ring*** $\mathrm{End}(G)$ is the set of all endomorphisms of G with composition as multiplication and pointwise addition (if $\varphi, \psi \in \mathrm{End}(G)$, then $\varphi + \psi: g \mapsto \varphi(g) + \psi(g)$).

There is a remarkable theorem of A.L.S. Corner (1963). Let R be a countable ring whose additive group is torsion-free and reduced; then there exists a countable group G, which is also torsion-free and reduced, with $\mathrm{End}(G) \cong R$. In the proof of the Krull–Schmidt theorem, we saw a close connection between decompositions of a group and endomorphisms, and one can thus use Corner's theorem to produce strange examples of torsion-free groups from pathological rings. For example, there are nonisomorphic countable torsion-free groups, each isomorphic to a direct summand of the other; there is a countable torsion-free group which has no indecomposable direct summands.

Exercises

10.42. If G is torsion-free of rank 1 and $x \in G$ is nonzero, then $G/\langle x \rangle$ is torsion. Describe $G/\langle x \rangle$ in terms of the height sequence of x. (*Hint.* $G/\langle x \rangle \leq \mathbb{Q}/\mathbb{Z}$.)

10.43. If A and B are subgroups of $\mathbb{Q}$, show that there is an exact sequence $0 \to A \cap B \to A \oplus B \to A + B \to 0$.

10.44. If G is a subring of $\mathbb{Q}$, then G is also a torsion-free group of rank 1. Show that the height sequence of 1 consists of 0's and ∞'s.

[2] In his dissertation submitted in 1914 (which was not well known because of World War I), F.W. Levi defined the characteristic of an element in a torsion-free group, classified the subgroups of $\mathbb{Q}$, introduced quasi-isomorphism, and gave the first examples of torsion-free groups having different direct sum decompositions into indecomposables.

10.45. (i) Show that if $\varphi: \mathbb{Q} \to \mathbb{Q}$ is a (group) homomorphism, then there exists $q \in \mathbb{Q}$ with $\varphi(x) = qx$ for all $x \in \mathbb{Q}$.
(ii) If G and G' are subgroups of $\mathbb{Q}$ and if $\varphi: G \to G'$ is a homomorphism, then there is $q \in \mathbb{Q}$ with $\varphi(x) = qx$ for all $x \in G$. (*Hint*. Use the injective property of $\mathbb{Q}$.)

10.46. If G is torsion-free of rank 1 and type τ, prove that $\mathrm{End}(G)$ is a subring of $\mathbb{Q}$ and find its type.

10.47. If R and S are subrings of $\mathbb{Q}$, then $R \cong S$ as rings if and only if $R \cong S$ as abelian groups (by definition, both R and S contain 1). Conclude that there are uncountably many nonisomorphic subrings of $\mathbb{Q}$.

10.48. Give an example of nonisomorphic torsion-free groups of rank 1 having isomorphic endomorphism rings.

10.49. Let A denote the dyadic rationals and let B denote the ***triadic rationals***:

$$B = \{q \in \mathbb{Q}: q = a/3^k, a \in \mathbb{Z} \text{ and } k \geq 0\}.$$

Let G be the subgroup of $\mathbb{Q} \oplus \mathbb{Q}$ generated by

$$\{(a, 0): a \in A\} \cup \{(0, b): b \in B\} \cup \{(\tfrac{1}{5}, \tfrac{1}{5})\}.$$

Prove that G is an indecomposable group of rank 2.

10.50. (i) Use Theorem 10.45 to show that $\rho(A) = \dim(V)$, where V is a vector space over $\mathbb{Q}$ of smallest dimension containing A/tA.
(ii) If $0 \to A \to B \to C \to 0$ is an exact sequence of abelian groups, prove that $\rho(B) = \rho(A) + \rho(C)$.

Character Groups

In Chapter 1, we raised the twin questions of describing groups and of describing homomorphisms, and we now focus on the latter.

Definition. Let $\mathscr{A}$ denote the class of all abelian groups. A function $T: \mathscr{A} \to \mathscr{A}$ is a ***covariant*** (additive) ***functor*** if, for every homomorphism $\varphi: A \to B$, there is a homomorphism $T(\varphi): T(A) \to T(B)$ such that:

(i) $T(1_A) = 1_{T(A)}$;
(ii) if $\theta: B \to C$, then $T(\theta\varphi) = T(\theta)T(\varphi)$; and
(iii) if $\psi: A \to B$, then $T(\varphi + \psi) = T(\varphi) + T(\psi)$.

EXAMPLE 10.12. If G and A are abelian groups, then

$$\mathrm{Hom}(G, A) = \{\text{homomorphisms } \varphi: G \to A\};$$

$\mathrm{Hom}(G, A)$ is an abelian group under pointwise addition: if $\varphi, \psi \in \mathrm{Hom}(G, A)$, then $\varphi + \psi: g \mapsto \varphi(g) + \psi(g)$. (If $A = G$, then $\mathrm{Hom}(G, G)$ is just the additive group of the endomorphism ring $\mathrm{End}(G)$.)

If G is a fixed group, define $T: \mathscr{A} \to \mathscr{A}$ by

$$T(A) = \operatorname{Hom}(G, A);$$

if $\varphi: A \to B$, define $T(\varphi): \operatorname{Hom}(G, A) \to \operatorname{Hom}(G, B)$ by $\alpha \mapsto \varphi\alpha$, where $\alpha \in \operatorname{Hom}(G, A)$. The reader should verify that T is a covariant functor.

EXAMPLE 10.13. The ***identity functor*** $J: \mathscr{A} \to \mathscr{A}$, defined by $J(A) = A$ and $J(\varphi) = \varphi$, is a covariant functor.

EXAMPLE 10.14. The torsion subgroup defines a covariant functor $t: \mathscr{A} \to \mathscr{A}$. Define $t(A)$ to be the torsion subgroup tA, and if $\varphi: A \to B$, then define $t(\varphi) = \varphi | tA$.

In a similar way, the maximal divisible subgroup defines a covariant functor $d: \mathscr{A} \to \mathscr{A}$.

Definition. A function $S: \mathscr{A} \to \mathscr{A}$ is a ***contravariant*** (additive) ***functor*** if, for every homomorphism $\varphi: A \to B$, there is a homomorphism $S(\varphi): S(B) \to S(A)$ such that:

(i) $S(1_A) = 1_{S(A)}$;
(ii) if $\theta: B \to C$, then $S(\theta\varphi) = S(\varphi)S(\theta)$; and
(iii) if $\psi: A \to B$, then $S(\varphi + \psi) = S(\varphi) + S(\psi)$.

Note that contravariant functors reverse the direction of arrows.

EXAMPLE 10.15. If G is a fixed group, define $S: \mathscr{A} \to \mathscr{A}$ by

$$S(A) = \operatorname{Hom}(A, G);$$

if $\varphi: A \to B$, define $S(\varphi): \operatorname{Hom}(B, G) \to \operatorname{Hom}(A, G)$ by $\beta \mapsto \beta\varphi$, where $\beta \in \operatorname{Hom}(B, G)$. The reader can verify that S is a contravariant functor.

To see how functors behave, one must recast definitions in a form recognizable by them. For example, instead of saying that an isomorphism is a homomorphism $\varphi: A \to B$ that is an injection and a surjection, one should say instead that it is a homomorphism for which there exists $\theta: B \to A$ such that

$$\theta\varphi = 1_A \qquad \text{and} \qquad \varphi\theta = 1_B.$$

We can now see that if φ is an isomorphism and T is a functor (contra or co), then $T(\varphi)$ is also an isomorphism: just apply T to the pair of equations above.

Definition. A covariant functor T is ***left exact*** if exactness of $(*)$: $0 \to A \xrightarrow{\alpha} B \xrightarrow{\beta} C$ implies exactness of

$$0 \to T(A) \xrightarrow{T(\alpha)} T(B) \xrightarrow{T(\beta)} T(C);$$

a contravariant functor S is ***left exact*** if exactness of $(**)$: $A \xrightarrow{\alpha} B \xrightarrow{\beta} C \to 0$

implies exactness of

$$0 \to S(C) \xrightarrow{S(\beta)} S(B) \xrightarrow{S(\alpha)} S(A).$$

There are also right exact functors in Homological Algebra.

Theorem 10.49. *If G is a group, then $S = \operatorname{Hom}(\ , G)$ and $T = \operatorname{Hom}(G,\ )$ are left exact functors.*

Proof. We shall prove that S is left exact; the reader may prove that T is left exact.

We must show exactness of

$$0 \to \operatorname{Hom}(C, G) \xrightarrow{S(\beta)} \operatorname{Hom}(B, G) \xrightarrow{S(\alpha)} \operatorname{Hom}(A, G);$$

that is, we must show that $S(\beta)$ is injective, $\operatorname{im} S(\beta) \le \ker S(\alpha)$, and $\ker S(\alpha) \le \operatorname{im} S(\beta)$.

(i) $\ker S(\beta) = 0$.

If $f: C \to G$ and $S(\beta)f = f\beta = 0$, then f annihilates $\operatorname{im} \beta$; as β is surjective, $\operatorname{im} \beta = C$ and $f = 0$.

(ii) $\operatorname{im} S(\beta) \le \ker S(\alpha)$.

If $f: C \to G$, then $S(\alpha)S(\beta)f = f\beta\alpha = 0$ because $\beta\alpha = 0$.

(iii) $\ker S(\alpha) \le \operatorname{im} S(\beta)$.

Suppose that $g: B \to G$ and $S(\alpha) = g\alpha = 0$, so that g annihilates $\operatorname{im} \alpha$. Define $g_\#: C \to G$ by $g_\#(c) = g(b)$, where $\beta(b) = c$ (β is surjective). Now $g_\#$ is well defined: if $\beta(b') = c$, then $b - b' \in \ker \beta = \operatorname{im} \alpha$; that is, $b - b' = \alpha a$, and so $g(b - b') = g\alpha(a) = 0$. But $S(\beta)g_\# = g_\#\beta = g$, for if $b \in B$ and $\beta(b) = c$, then $g_\#\beta(b) = g_\#(c) = g(b)$. ■

Here is the answer to the question when "$\to 0$" occurs at the right end of the functored sequence.

Theorem 10.50. *A group G is free abelian if and only if, for every exact sequence $0 \to A \xrightarrow{\alpha} B \xrightarrow{\beta} C \to 0$, there is an exact sequence*

$$0 \to \operatorname{Hom}(G, A) \xrightarrow{T(\alpha)} \operatorname{Hom}(G, B) \xrightarrow{T(\beta)} \operatorname{Hom}(G, C) \to 0.$$

Proof. Assume that G is free abelian. To prove that $T(\beta)$ is surjective, we must show that if $g \in \operatorname{Hom}(G, C)$, there is $f \in \operatorname{Hom}(G, B)$ with $T(\beta)f = g$; that is, $\beta f = g$. Let us draw a diagram.

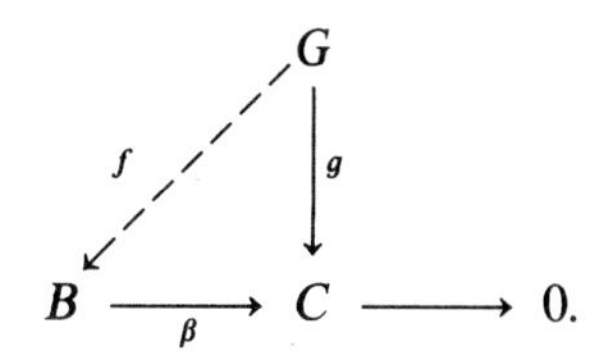

The projective property of G (Theorem 10.15) gives the existence of f (indeed, this is how the projective property was born).

Conversely, a similar argument shows that a group G for which $T(\beta)$ is always surjective must have the projective property. But Exercise 10.10 shows that such a group must be free abelian. ■

Theorem 10.51. *A group G is divisible if and only if, for every exact sequence $0 \to A \xrightarrow{\alpha} B \xrightarrow{\beta} C \to 0$, there is an exact sequence*

$$0 \to \mathrm{Hom}(C, G) \xrightarrow{S(\beta)} \mathrm{Hom}(B, G) \xrightarrow{S(\alpha)} \mathrm{Hom}(A, G) \to 0.$$

Proof. Use the injective property and Exercise 10.22. ■

Theorem 10.52. *Given a group G and a family of groups $\{A_j: j \in J\}$, then*

$$\mathrm{Hom}\left(\sum_{j \in J} A_j, G\right) \cong \prod_{j \in J} \mathrm{Hom}(A_j, G).$$

Proof. For each $j_0 \in J$, let $i_{j_0} \hookrightarrow \sum_{j \in J} A_j$ be the inclusion. Define θ: $\mathrm{Hom}(\sum A_j, G) \to \prod \mathrm{Hom}(A_j, G)$ by $f \mapsto (fi_j)$. Define a map ψ in the reverse direction by $(f_j) \mapsto f$, where f is the unique map $\sum A_j \to G$ for which $fi_j = f_j$ for all j (Theorem 10.10). The reader may check that θ and ψ are inverses. ■

If the index set J is finite, then the functor Hom(, G) sends the finite direct sum $\sum_{j=1}^{n} A_j$ into the direct sum $\sum_{j=1}^{n} A_j$ $\mathrm{Hom}(A_j, G)$; that is,

$$\mathrm{Hom}\left(\sum_{j=1}^{n} A_j, G\right) \cong \sum_{j=1}^{n} \mathrm{Hom}(A_j, G).$$

Theorem 10.53. *Given a group G and a family of groups $\{A_j: j \in J\}$, then*

$$\mathrm{Hom}\left(G, \prod_{j \in J} A_j\right) \cong \prod_{j \in J} \mathrm{Hom}(G, A_j).$$

Proof. For each $j_0 \in J$, let p_{j_0}: $\prod A_j \to A_j$ be the projection onto the j_0th coordinate. An argument similar to that of Theorem 10.52, using Theorem 10.11, shows that the map θ: $\mathrm{Hom}(G, \prod A_j) \to \prod \mathrm{Hom}(G, A_j)$, defined by $f \mapsto (p_j f)$, is an isomorphism. ■

If the index set J is finite, then the functor Hom(G,) sends the (finite) direct sum $\sum_{j=1}^{n} A_j$ into the direction sum $\sum_{j=1}^{n} \mathrm{Hom}(G, A_j)$; that is,

$$\mathrm{Hom}\left(G, \sum_{j=1}^{n} A_j\right) \cong \sum_{j=1}^{n} \mathrm{Hom}(G, A_j).$$

If m is an integer and A is a group, let m_A: $A \to A$ be multiplication by m; that is, m_A: $a \mapsto ma$. Note that $m_A = 1_A + \cdots + 1_A$ (m addends) if $m > 0$.

Theorem 10.54. *If $m \in \mathbb{Z}$ and $m_A: A \to A$ is multiplication by m, then $T(m_A): T(A) \to T(A)$ is also multiplication by m.*

Proof. By Exercise 10.51 below, the result is true when $m = 0$. If $m > 0$, then

$$\begin{aligned} T(m_A) &= T(1_A + \cdots + 1_A) \\ &= T(1_A) + \cdots + T(1_A) \\ &= 1_{T(A)} + \cdots + 1_{T(A)} \\ &= m_{T(A)}. \end{aligned}$$

The reader may easily see that $T(-1_A) = -1_{T(A)}$, so that the result is true for all $m \in \mathbb{Z}$. ■

EXAMPLE 10.16. For every group G, $\operatorname{Hom}(G, \mathbb{Q})$ is torsion-free divisible, and hence it is a vector space over $\mathbb{Q}$.

By Exercise 10.31, $H = \operatorname{Hom}(G, \mathbb{Q})$ is torsion-free divisible if and only if m_H is an automorphism for all $m \neq 0$. Thus, the result follows from Theorem 10.54.

The same argument shows that $\operatorname{Hom}(\mathbb{Q}, G)$ is also a vector space over $\mathbb{Q}$.

EXAMPLE 10.17. $\operatorname{Hom}(\mathbb{Z}, G) \cong G$ for every group G.

It is easy to see that $f \mapsto f(1)$ is an isomorphism.

EXAMPLE 10.18. For every group G,

$$\operatorname{Hom}(\mathbb{Z}_n, G) \cong G[n] = \{g \in G: ng = 0\}.$$

Apply the (contravariant) functor $S = \operatorname{Hom}(\ , G)$ to the exact sequence $0 \to \mathbb{Z} \xrightarrow{n} \mathbb{Z} \to \mathbb{Z}_n \to 0$, where the first map is multiplication by n, to obtain the exact sequence

$$0 \to \operatorname{Hom}(\mathbb{Z}_n, G) \to \operatorname{Hom}(\mathbb{Z}, G) \xrightarrow{S(n)} \operatorname{Hom}(\mathbb{Z}, G).$$

Thus $\operatorname{Hom}(\mathbb{Z}_n, G) \cong \ker S(n)$. But there is a commutative diagram

$$\begin{array}{ccc} \operatorname{Hom}(\mathbb{Z}, G) & \xrightarrow{S(n)} & \operatorname{Hom}(\mathbb{Z}, G) \\ \downarrow & & \downarrow \\ G & \xrightarrow[n]{} & G, \end{array}$$

where the downward arrows are the isomorphisms of Example 10.17. It follows easily that $\ker S(n) \cong \ker n = G[n]$.

Definition. If G is a group, its ***character group*** G^* is

$$G^* = \text{Hom}(G, \mathbb{Q}/\mathbb{Z}).$$

The next lemma shows, when G is finite, that this definition coincides with our earlier definition of character group in Chapter 7.

Lemma 10.55. *If G is finite, then $G^* \cong \text{Hom}(G, \mathbb{C}^\times)$.*

Proof. By Exercise 10.28(ii), the torsion subgroup $t(\mathbb{C}^\times) \cong \mathbb{Q}/\mathbb{Z}$. Therefore, $\mathbb{C}^\times \cong (\mathbb{Q}/\mathbb{Z}) \oplus D$, where D is torsion-free divisible. Hence, $\text{Hom}(G, \mathbb{C}^\times) \cong \text{Hom}(G, \mathbb{Q}/\mathbb{Z}) \oplus \text{Hom}(G, D) \cong \text{Hom}(G, \mathbb{Q}/\mathbb{Z})$, for $\text{Hom}(G, D) = 0$ because G is finite. ∎

Remark. There is another common definition of character group: $\hat{G} = \text{Hom}(G, \mathbf{T})$. Since $\mathbf{T} \cong \mathbb{C}^\times$, by Exercise 10.28(i), we see that $\hat{G} \cong G^*$. The character groups $\hat{G}$ arise in *Pontrjagin duality*, where G is assumed to be a locally compact abelian topological group and $\hat{G}$ is the group of all continuous homomorphisms (when G is a discrete group, then every homomorphism is continuous).

The following properties of character groups were used in our discussion of the Schur multiplier in Chapter 7.

Theorem 10.56. *If G is finite, then $G \cong G^*$.*

Proof. If $G \cong \mathbb{Z}_n$, then Example 10.18 gives $G^* \cong (\mathbb{Q}/\mathbb{Z})[n] \cong \mathbb{Z}_n$ (Exercise 10.21(ii)), as desired. By the basis theorem, $G = \sum C_i$, where C_i is finite cyclic, so that Theorem 10.52 gives $G^* \cong \sum (C_i)^* \cong \sum C_i = G$. ∎

Let us now solve Exercise 6.13.

Theorem 10.57. *If G is a finite abelian group and $S \leq G$, then G contains a subgroup isomorphic to G/S.*

Proof. There is an exact sequence $0 \to S \to G \to G/S \to 0$. Since $\mathbb{Q}/\mathbb{Z}$ is divisible, Theorem 10.51 gives an exact sequence

$$0 \to (G/S)^* \to G^* \to S^* \to 0.$$

Hence $G/S \cong (G/S)^*$ is isomorphic to a subgroup of $G^* \cong G$. ∎

Theorem 10.58. *Let G be an abelian group, let $a \in G$, and let S be a subgroup of G with $a \notin S$. If D is a divisible group with $\mathbb{Q}/\mathbb{Z} \leq D$, then there exists $\varphi\colon G \to D$ with $\varphi(S) = 0$ and $\varphi(a) \neq 0$.*

Remark. The important cases are $D = \mathbb{Q}/\mathbb{Z}$ and $D = \mathbf{T} \cong \mathbb{C}^\times$.

Proof. If $[a] = a + S \in G/S$, then there is a homomorphism $\psi: \langle [a] \rangle \to D$ with $\psi([a]) \neq 0$: if $[a]$ has finite order n, define $\psi([a]) = 1/n + \mathbb{Z} \in \mathbb{Q}/\mathbb{Z} \leq D$; if $[a]$ has infinite order, define $\psi([a]) = 1 + \mathbb{Z}$. By the injective property of D, ψ extends to a homomorphism $\Psi: G/S \to D$. If $\varphi: G \to D$ is defined to be Ψv, where $v: G \to G/S$ is the natural map, then $\varphi(S) = 0$ and $\varphi(a) = \Psi([a]) \neq 0$. ■

If G is finite, then we know that $G \cong G^* \cong (G^*)^*$; let us denote the latter group, the "double dual," by G^{**}. We now exhibit a specific isomorphism.

Definition. Let G be a group. For each $x \in G$, define a homomorphism $E_x: G^* \to \mathbb{Q}/\mathbb{Z}$ by $E_x(\varphi) = \varphi(x)$; thus, $E_x \in G^{**}$. The ***evaluation map*** $E: G \to G^{**}$ is defined by $x \mapsto E_x$.

Theorem 10.59. *For every group G, the evaluation map $E: G \to G^{**}$ is an injection; if G is finite, then E is an isomorphism.*

Proof. If $x \in \ker E$, then $E_x(\varphi) = \varphi(x) = 0$ for all $\varphi \in G^*$; by Theorem 10.58, $x = 0$, and so E is an injection. If G is finite, then $G^{**} \cong G$, by Theorem 10.56, so that $|G^{**}| = |G|$ and E is an isomorphism. ■

Exercises

10.51. Show that if T is an (additive) functor, then $T(0) = 0$, where 0 denotes either the trivial group 0 or the (constant) map which sends every element into 0.

10.52. Prove that $\text{Hom}(A, B) = 0$ in the following cases:
(i) A is torsion and B is torsion-free;
(ii) A is divisible and B is reduced;
(iii) A is p-primary and B is q-primary, where $p \neq q$.

10.53. (i) Show that A is reduced if and only if $\text{Hom}(\mathbb{Q}, A) = 0$.
(ii) Use Theorem 10.50 to show that if $0 \to A \to B \to C \to 0$ is exact and A and C are reduced, then B is reduced.

10.54. If G is finite and $S \leq G$, define

$$S^\perp = \{f \in G^*: f(s) = 0 \text{ for all } s \in S\}.$$

(i) Show that $S^\perp$ is a subgroup of G^* and that $S^\perp \cong (G/S)^*$.
(ii) If G is a finite group of order n and if k is a divisor of n, then G has the same number of subgroups of order n as of index n. (*Hint*: The function $S \mapsto S^\perp$ is a bijection.)

10.55. Here are examples related to Theorems 10.52 and 10.53. Let P denote the set of all primes.
(i) Prove that $\text{Hom}(\prod_{p \in P} \mathbb{Z}_p, \mathbb{Q}) \neq 0$. Conclude that $\text{Hom}(\prod \mathbb{Z}_p, \mathbb{Q})$ is isomorphic to neither $\sum \text{Hom}(\mathbb{Z}_p, \mathbb{Q})$ nor $\prod \text{Hom}(\mathbb{Z}_p, \mathbb{Q})$.

(ii) If $G = \sum_{p \in P} \mathbb{Z}_p$, then $\operatorname{Hom}(G, \sum_{p \in P} \mathbb{Z}_p)$ and $\sum \operatorname{Hom}(G, \mathbb{Z}_p)$ are not isomorphic.

(iii) $\operatorname{Hom}(\mathbb{Z}, \sum_{p \in P} \mathbb{Z}_p)$ and $\prod \operatorname{Hom}(\mathbb{Z}, \mathbb{Z}_p)$ are not isomorphic.

10.56. Let $f: A \to B$ be a homomorphism. Show that $f^*: B^* \to A^*$ is a surjection (where $A^* = \operatorname{Hom}(A, \mathbb{Q}/\mathbb{Z})$ and $f^*: \varphi \mapsto \varphi f$ for every $\varphi \in B^*$) if and only if f is an injection; show that $f^*: B^* \to A^*$ is an injection if and only if f is a surjection. Conclude that f^* is an isomorphism if and only if f is an isomorphism.

10.57. Consider the commutative diagram of not necessarily abelian groups in which the rows are exact sequences (thus, the kernels are assumed to be normal subgroups):

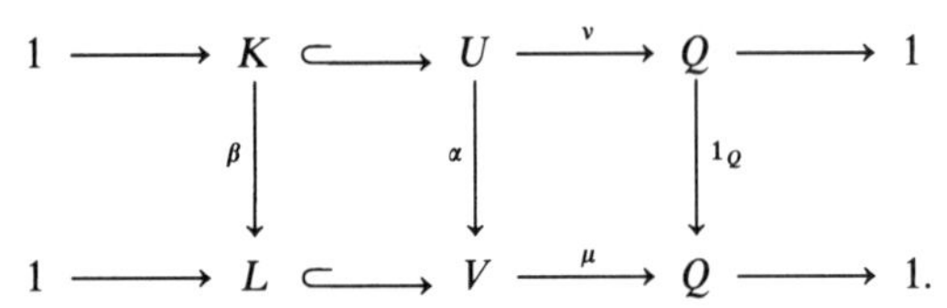

Show that if β is an isomorphism, then α is an isomorphism.

CHAPTER 11

Free Groups and Free Products

Generators and Relations

The notion of generators and relations can be extended from abelian groups to arbitrary groups once we have a nonabelian analogue of free abelian groups. We use the property appearing in Theorem 10.11 as our starting point.

Definition. If X is a subset of a group F, then F is a ***free group*** with ***basis*** X if, for every group G and every function $f: X \to G$, there exists a unique homomorphism $\varphi: F \to G$ extending f

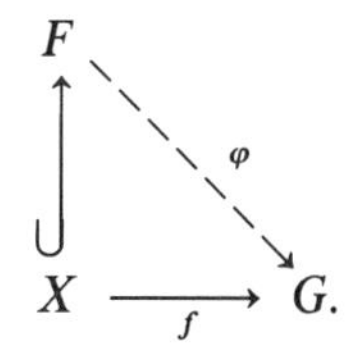

We shall see later that X must generate F.

Observe that a basis in a free group behaves precisely as does a basis $B = \{v_1, \ldots, v_m\}$ of a finite-dimensional vector space V. The theorem of linear algebra showing that matrices correspond to linear transformations rests on the fact that if W is any vector space and $w_1, \ldots, w_m \in W$, then there exists a unique linear transformation $T: V \to W$ with $T(v_i) = w_i$ for all i.

The following construction will be used in proving that free groups exist. Let X be a set and let X^{-1} be a set, disjoint from X, for which there is a bijection $X \to X^{-1}$, which we denote by $x \mapsto x^{-1}$. Let X' be a singleton set disjoint from $X \cup X^{-1}$ whose only element is denoted by 1. If $x \in X$, then x^1 may denote x and x^0 may denote 1.

Definition. A ***word*** on X is a sequence $w = (a_1, a_2, \ldots)$, where $a_i \in X \cup X^{-1} \cup \{1\}$ for all i, such that all $a_i = 1$ from some point on; that is, there is an integer $n \geq 0$ with $a_i = 1$ for all $i > n$. In particular, the constant sequence

$$(1, 1, 1, \ldots)$$

is a word, called the ***empty word***, and it is also denoted by 1.

Since words contain only a finite number of letters before they become constant, we use the more suggestive notation for nonempty words:

$$w = x_1^{\varepsilon_1} x_2^{\varepsilon_2} \ldots x_n^{\varepsilon_n},$$

where $x_i \in X$, $\varepsilon_i = +1, -1$, or 0, and $\varepsilon_n = \pm 1$. Observe that this spelling of a word is unique: two sequences (a_i) and (b_i) are equal if and only if $a_i = b_i$ for all i. The ***length*** of the empty word is defined to be 0; the ***length*** of $w = x_1^{\varepsilon_1} x_2^{\varepsilon_2} \ldots x_n^{\varepsilon_n}$ is defined to be n.

Definition. If $w = x_1^{\varepsilon_1} \ldots x_n^{\varepsilon_n}$ is a word, then its ***inverse*** is the word $w^{-1} = x_n^{-\varepsilon_n} \ldots x_1^{-\varepsilon_1}$.

Definition. A word w on X is ***reduced*** if either w is empty or $w = x_1^{\varepsilon_1} x_2^{\varepsilon_2} \ldots x_n^{\varepsilon_n}$, where all $x_i \in X$, all $\varepsilon_i = \pm 1$, and x and x^{-1} are never adjacent.

The empty word is reduced, and the inverse of a reduced word is reduced.

Definition. A ***subword*** of $w = x_1^{\varepsilon_1} x_2^{\varepsilon_2} \ldots x_n^{\varepsilon_n}$ is either the empty word or a word of the form $v = x_i^{\varepsilon_i} \ldots x_j^{\varepsilon_j}$, where $1 \leq i \leq j \leq n$.

Thus, v is a subword of w if there are (possibly empty) subwords w' and w'' with $w = w'vw''$. A nonempty word w is reduced if and only if it contains no subwords of the form $x^\varepsilon x^{-\varepsilon}$ or x^0.

There is a multiplication of words: if $w = x_1^{\varepsilon_1} x_2^{\varepsilon_2} \ldots x_n^{\varepsilon_n}$ and $u = y_1^{\delta_1} y_2^{\delta_2} \ldots y_m^{\delta_m}$, then $wu = x_1^{\varepsilon_1} x_2^{\varepsilon_2} \ldots x_n^{\varepsilon_n} y_1^{\delta_1} y_2^{\delta_2} \ldots y_m^{\delta_m}$. This multiplication does not define a product on the set of all *reduced* words on X because wu need not be reduced (even when both w and u are). One can define a new multiplication of reduced words w and u as the reduced word obtained from wu after cancellations. More precisely, there is a (possibly empty) subword v of w with $w = w'v$ such that v^{-1} is a subword of u with $u = v^{-1}u''$ and such that $w'u''$ is reduced. Define a product of reduced words, called ***juxtaposition***, by

$$wu = w'u''.$$

Theorem 11.1. *Given a set X, there exists a free group F with basis X.*

Proof. Let F be the set of all the reduced words on X. One can show that F is a group under juxtaposition, but verifying associativity involves tedious case analyses. Instead, we use the ***van der Waerden trick*** (1945).

For each $x \in X$, consider the functions $|x|: F \to F$ and $|x^{-1}|: F \to F$, defined as follows: for $\varepsilon = \pm 1$,

$$|x^\varepsilon|(x_1^{\varepsilon_1} x_2^{\varepsilon_2} \dots x_n^{\varepsilon_n}) = \begin{cases} x^\varepsilon x_1^{\varepsilon_1} x_2^{\varepsilon_2} \dots x_n^{\varepsilon_n} & \text{if } x^\varepsilon \neq x_1^{-\varepsilon_1}, \\ x_2^{\varepsilon_2} \dots x_n^{\varepsilon_n} & \text{if } x^\varepsilon = x_1^{-\varepsilon_1}. \end{cases}$$

Since $|x^\varepsilon| \circ |x^{-\varepsilon}|$ and $|x^{-\varepsilon}| \circ |x^\varepsilon|$ are both equal to the identity $1_F: F \to F$, it follows that $|x^\varepsilon|$ is a permutation of F with inverse $|x^{-\varepsilon}|$. Let S_F be the symmetric group on F, and let $\mathscr{F}$ be the subgroup of S_F generated by $[X] = \{|x|: x \in X\}$. We claim that $\mathscr{F}$ is a free group with basis $[X]$. Note that there is a bijection $\zeta: [X] \to X$, namely, $|x| \mapsto x$.

An arbitrary element $g \in \mathscr{F}$ (other than the identity) has a factorization

$$(*) \qquad g = |x_1^{\varepsilon_1}| \circ |x_2^{\varepsilon_2}| \circ \cdots \circ |x_n^{\varepsilon_n}|,$$

where $\varepsilon_i = \pm 1$ and $|x^\varepsilon|$ and $|x^{-\varepsilon}|$ are never adjacent (or we can cancel). Such a factorization of g is unique, for $g(1) = x_1^{\varepsilon_1} x_2^{\varepsilon_2} \dots x_n^{\varepsilon_n}$, and we have already noted that the spelling of a (reduced) word is unique.

To see that $\mathscr{F}$ is free with basis $[X]$, assume that G is a group and that $f: [X] \to G$ is a function. Since the factorization $(*)$ is unique, the function $\varphi: \mathscr{F} \to G$, given by $\varphi(|x_1^{\varepsilon_1}| \circ |x_2^{\varepsilon_2}| \circ \cdots \circ |x_n^{\varepsilon_n}|) = f(|x_1^{\varepsilon_1}|) f(|x_2^{\varepsilon_2}|) \dots f(|x_n^{\varepsilon_n}|)$, is well defined and extends f. Since $[X]$ generates $\mathscr{F}$, it suffices to show that φ is a homomorphism, for uniqueness of φ would then follow from the fact that two homomorphisms agreeing on a generating set must be equal.

Let w and u be reduced words on $[X]$. It is obvious that $\varphi(w \circ u) = \varphi(w)\varphi(u)$ whenever the word wu (obtained from $w \circ u$ by deleting vertical bars) is reduced. Write $w = w' \circ v$ and $u = v^{-1} \circ u''$, as in the definition of juxtaposition. Now $\varphi(w) = \varphi(w')\varphi(v)$ and $\varphi(u) = \varphi(v^{-1})\varphi(u'') = \varphi(v)^{-1}\varphi(u'')$ (because $w' \circ v$ and $v^{-1} \circ u''$ are reduced). Therefore, $\varphi(w)\varphi(u) = \varphi(w')\varphi(v)\varphi(v)^{-1}\varphi(u'') = \varphi(w')\varphi(u'')$. On the other hand, $\varphi(w \circ u) = \varphi(w' \circ u'') = \varphi(w')\varphi(u'')$ (because $w' \circ u''$ is reduced), and so φ is a homomorphism.

We have shown that $\mathscr{F}$ is a free group with basis $[X]$. Since $\tilde{\zeta}: \mathscr{F} \to F$, defined by $|x_1^{\varepsilon_1}| \circ |x_2^{\varepsilon_2}| \circ \cdots \circ |x_n^{\varepsilon_n}| \mapsto x_1^{\varepsilon_1} x_2^{\varepsilon_2} \dots x_n^{\varepsilon_n}$, is a bijection with $\tilde{\zeta}([X]) = \zeta([X]) = X$, Exercise 1.44 shows that we may regard F as a group isomorphic to $\mathscr{F}$; thus, F is a free group with basis X (moreover, X generates F because $[X]$ generates $\mathscr{F}$). ■

Corollary 11.2. *Every group G is a quotient of a free group.*

Proof. Construct a set $X = \{x_g: g \in G\}$ so that $f: x_g \mapsto g$ is a bijection $X \to G$. If F is free with basis X, then there is a homomorphism $\varphi: F \to G$ extending f, and φ is a surjection because f is. Therefore, $G \cong F/\ker \varphi$. ■

Definition. Let X be a set and let Δ be a family of words on X. A group G has ***generators*** X and ***relations*** Δ if $G \cong F/R$, where F is the free group with basis X and R is the normal subgroup of F generated by Δ. The ordered pair $(X|\Delta)$ is called a ***presentation*** of G.

A relation[1] $r \in \Delta$ is often written as $r = 1$ to convey its significance in the quotient group G being presented.

There are two reasons forcing us to define R as the *normal* subgroup of F generated by Δ: if $r \in \Delta$ and $w \in F$, then $r = 1$ in G implies $wrw^{-1} = 1$ in G; we wish to form a quotient group.

EXAMPLE 11.1. $G = \mathbb{Z}_6$ has generator x and relation $x^6 = 1$.

A free group $F = \langle x \rangle$ on one generator is infinite cyclic, and $\langle x \rangle / \langle x^6 \rangle \cong \mathbb{Z}_6$. A presentation of G is $(x | x^6)$.

EXAMPLE 11.2. Another presentation of $G = \mathbb{Z}_6$ is

$$\mathbb{Z}_6 = (x, y | x^3 = 1, y^2 = 1, xyx^{-1}y^{-1} = 1).$$

When we described a presentation of $\mathbb{Z}_6$ as an abelian group in Example 10.2 (i.e., when we viewed $\mathbb{Z}_6$ as a quotient of a free abelian group), the only relations were x^3 and y^2. Now we must also have the commutator as a relation to force the images of x and y to commute in F/R.

EXAMPLE 11.3. The dihedral group D_{2n} has a presentation

$$D_{2n} = (x, y | x^n = 1, y^2 = 1, yxy = x^{-1}).$$

It is acceptable to write a relation as $yxy = x^{-1}$ instead of $xyxy = 1$. In particular, compare the presentation of D_6 with that of $\mathbb{Z}_6$ in Example 11.2.

We have passed over a point needing more discussion. By definition, D_{2n} is a group *of order* $2n$ having generators S and T satisfying the given relations. If $G = F/R$, where F is the free group with basis $\{x, y\}$ and R is the normal subgroup generated by $\{x^n, y^n, xyxy\}$, does G have order $2n$? We have seen various concrete versions of D_{2n}; for example, Theorem 3.31 displays it as the symmetry group of a regular n-gon. The definition of free group gives a surjective homomorphism $\varphi\colon F \to D_{2n}$ with $\varphi(x) = S$ and $\varphi(y) = T$. Moreover, $R \leq \ker \varphi$, because S and T satisfy the relations, so that the third isomorphism theorem gives a surjection $F/R \to F/\ker \varphi$; that is, there is a surjection[2] $G = F/R \to D_{2n}$. Hence, $|G| \geq 2n$. The reverse inequality also

[1] Many authors use the words "relation" and "relator" interchangeably.

[2] W. von Dyck (1882) invented free groups and used them to give the first precise definition of presentations. The version of the third isomorphism theorem used here is often called ***von Dyck's Theorem***: Let G have a presentation

$$G = (x_1, \ldots, x_n | r_j(x_1, \ldots, x_n), j \in J)$$

so that $G = F/R$, where F is the free group with basis $\{x_1, \ldots, x_n\}$ and R is the normal subgroup generated by the r_j. If H is a group with $H = \langle y_1, \ldots, y_n \rangle$ and if $r_j(y_1, \ldots, y_n) = 1$ for all j, then there is a surjective homomorphism $G \to H$ with $x_i \mapsto y_i$ for all i.

holds, for each element in G has a factorization $x^i y^j R$ with $0 \le i < n$ and $0 \le j < 2$. Thus, $|G| = 2n$, and we are now entitled to write $G \cong D_{2n}$.

A description of a group by generators and relations is flawed in that the order of the presented group is difficult to determine. This is not a minor difficulty, for we shall see in the next chapter that it is even an unsolvable problem (in the logicians' precise sense) to determine, from an arbitrary presentation, the order of the presented group. Indeed, it is an unsolvable problem to determine whether a presentation defines a group of order 1. The reader should also see the next section on coset enumeration.

Let us continue the list of examples.

EXAMPLE 11.4. The group of quaternions has presentations

$$\mathbf{Q} = (a, b \mid a^4 = 1, b^2 = a^2, bab^{-1} = a^{-1})$$

and

$$\mathbf{Q} = (x, y \mid xyx = y, x^2 = y^2).$$

In each case, an argument is needed to show that the presented group has order 8.

EXAMPLE 11.5. Given positive integers l, m, and n, define

$$P(l, m, n) = (s, t \mid s^l = t^m = (st)^n = 1).$$

Example 11.3 shows that $P(n, 2, 2) = D_{2n}$ and, using Exercise 3.52, one can show that $P(2, 3, 3) \cong A_4$, $P(2, 3, 4) \cong S_4$, and $P(2, 3, 5) \cong A_5$. These groups are called *polyhedral groups*, and they are finite only in the cases just listed (see Coxeter–Moser).

EXAMPLE 11.6. The ***braid group*** B_m has the presentation

$$(\sigma_1, \ldots, \sigma_m \mid [\sigma_i, \sigma_j] = 1 \text{ if } j \ne i \pm 1, \sigma_i \sigma_{i+1} \sigma_i = \sigma_{i+1} \sigma_i \sigma_{i+1}).$$

Braid groups were introduced by E. Artin (1925) and are related to knot theory.

EXAMPLE 11.7. A free abelian group G with basis X has presentation

$$G = (X \mid xyx^{-1}y^{-1} = 1 \text{ for all } x, y \in X);$$

a free group F with basis X has presentation

$$F = (X \mid \varnothing).$$

Having proved that free groups exist, let us now consider their uniqueness; that is, when are two free groups isomorphic.

Lemma 11.3. *If F is a free group with basis X, then F/F' is a free abelian group with basis $X_\# = \{xF' : x \in X\}$.*

Proof. Assume that A is an abelian group and that $f: X_\# \to A$ is a function. Define $f_\#: X \to A$ by $x \mapsto f(xF')$. Since F is free with basis X, there is a homomorphism $\varphi: F \to A$ extending $f_\#$. But $F' \leq \ker \varphi$, because A is abelian, so that there is a homomorphism $\tilde{\varphi}: F/F' \to A$, defined by $wF' \mapsto \varphi(w)$, extending f.

We claim that the extension $\tilde{\varphi}$ is unique. Suppose that $\theta: F/F' \to A$ and $\theta(xF') = f(xF')$. If $v: F \to F/F'$ is the natural map, then $\theta v: F \to A$ is a homomorphism with $\theta v(x) = \theta(xF') = f(xF') = \varphi(x)$ for all $x \in X$. Since X is a basis of F, $\theta v = \varphi = \tilde{\varphi} v$; since v is surjective, $\theta = \tilde{\varphi}$. Therefore, F/F' is free abelian with basis $X_\#$. ■

Theorem 11.4. *Let F and G be free groups with bases X and Y, respectively. Then $F \cong G$ if and only if $|X| = |Y|$.*

Proof. If $\varphi: F \to G$ is an isomorphism, then $F/F' \cong G/G'$. By the lemma, F/F' is free abelian with basis $X_\# = \{xF' : x \in X\}$. As $|X_\#| = |X|$, it follows that $|X| = \operatorname{rank}(F/F')$. Similarly, $|Y| = \operatorname{rank}(G/G')$, and so $|X| = |Y|$, by Theorem 10.14.

If $|X| = |Y|$, there is a bijection $f: X \to Y$ which, upon composing with the inclusion $Y \hookrightarrow G$, may be regarded as a function $X \to G$. Since F is free with basis X, there is a unique homomorphism $\varphi: F \to G$ extending f. Similarly, there is a unique homomorphism $\psi: G \to F$ extending $f^{-1}: Y \to X$. The composite $\psi\varphi: F \to F$ is a homomorphism which fixes X pointwise; that is, $\psi\varphi$ extends the inclusion function $\iota: X \hookrightarrow F$. But the identity 1_F also extends ι, and so uniqueness of extension gives $\psi\varphi = 1_F$. Similarly, $\varphi\psi = 1_G$, so that $\varphi: F \to G$ is an isomorphism. ■

Definition. The ***rank*** of a free group F is the number of elements in a basis of F.

Theorem 11.4 says that $\operatorname{rank}(F)$ does not depend on the choice of basis of F.

Corollary 11.5. *If F is free with basis X, then F is generated by X.*

Proof. Choose a set Y with $|Y| = |X|$ and a bijection $f: Y \to X$. The free group G with basis Y constructed in Theorem 11.1 (as the set of all reduced words on Y) is generated by Y. As in the proof of Theorem 11.4, the homomorphism $\psi: G \to F$ extending f is an isomorphism, so that $G = \langle Y \rangle$ implies $F = \langle \psi(Y) \rangle = \langle f(Y) \rangle = \langle X \rangle$. ■

Theorem 11.6 (Projective Property). *Let $\beta: B \to C$ be a surjective homomorphism. If F is free and if $\alpha: F \to C$ is a homomorphism, then there exists a* *homomorphism* $\gamma: F \to B$ *making the diagram below commute* (*i.e.*, $\beta\gamma = \alpha$):

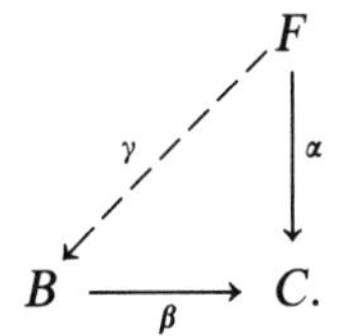

Proof. The proof is identical to that given for free abelian groups in Theorem 10.15. ■

We shall see in Exercise 11.46 below that the converse of Theorem 11.6 is also true: a group G is free if and only if it has the projective property.

Semigroup Interlude

We are now going to construct free semigroups; the formal definition is no surprise.

Definition. If X is a subset of a semigroup Σ, then Σ is a ***free semigroup*** with ***basis*** X if, for every semigroup S and every function $f: X \to S$, there exists a unique homomorphism $\varphi: \Sigma \to S$ extending f.

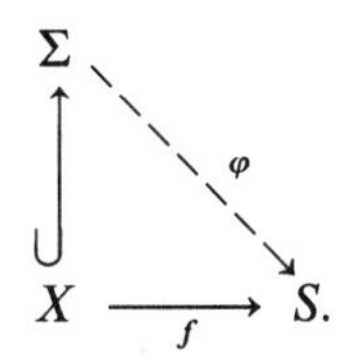

Definition. A word w on X is ***positive*** if either $w = 1$ or $w = x_1^{\varepsilon_1} x_2^{\varepsilon_2} \dots x_n^{\varepsilon_n}$, where all exponents $\varepsilon_i > 0$.

The set Σ of all positive words on X is a free semigroup with basis X (the product of positive words is positive and, with no cancellation possible, it is easy to prove that multiplication is associative). It follows that every semigroup is a homomorphic image of a free semigroup. Before defining presentations of semigroups, however, we first define quotients.

Definition. A ***congruence*** on a semigroup S is an equivalence relation $\equiv$ on S such that

$$a \equiv a' \quad \text{and} \quad b \equiv b' \quad \text{imply} \quad ab \equiv a'b'.$$

If $\equiv$ is a congruence on a semigroup S, then the ***quotient semigroup*** is the set of all equivalence classes, denoted by $S/\equiv$, with the operation

$$[a][b] = [ab],$$

where $[a]$ denotes the equivalence class of $a \in S$ (this operation is well defined because $\equiv$ is a congruence).

There are two general constructions of congruences. The first arises from a homomorphism $\varphi: S \to T$ of semigroups; define $a \equiv b$ if $\varphi(a) = \varphi(b)$. This congruence is called ***ker*** $\boldsymbol{\varphi}$, and it is straightforward to prove the first isomorphism theorem:

$$S/\ker \varphi \cong \operatorname{im} \varphi$$

(if S and T are groups and $K = \{s \in S: \varphi(s) = 1\}$, then $\ker \varphi$ is the equivalence relation on S whose equivalence classes are the cosets of K). Here is a second construction. As any relation on S, a congruence is a subset of $S \times S$. It is easy to see that any intersection of congruences is itself a congruence. Since $S \times S$ is a congruence, one may thus define the congruence *generated* by any subset E of $S \times S$ as the intersection of all the congruences containing E. If Σ is the free semigroup with basis X and if $\{w_i = u_i: i \in I\}$ is a family of equations, where $w_i, u_i \in \Sigma$, then define $\equiv$ to be the congruence generated by $\{(w_i, u_i): i \in I\} \subset \Sigma \times \Sigma$. The quotient semigroup $\Sigma/\equiv$ is said to have the ***presentation***

$$(X \mid w_i = u_i \text{ for all } i \in I).$$

Exercises

11.1. Use presentations to prove the existence of the nonabelian groups of order p^3, where p is prime. (See Exercise 4.32.)

11.2. Prove that a free group of rank ≥ 2 is a centerless torsion-free group.

11.3. Prove that the group $G = (x, y \mid x^m, y^n)$ is infinite when $m, n \geq 2$.

11.4 **(Baer).** Prove that a group E has the injective property if and only if $E = 1$. (*Hint.* ***D.L. Johnson***). Let A be free with basis $\{x, y\}$ and let B be the semidirect product $B = A \rtimes \langle z \rangle$, where z is an involution acting by $zxz = y$ and $zyz = x$.)

11.5. Let X be the disjoint union $X = Y \cup Z$. If F is free with basis X and N is the normal subgroup generated by Y, then F/N is free with basis $\{zN: z \in Z\}$.

11.6. Show that a free group F of rank ≥ 2 has an automorphism φ with $\varphi(\varphi(w)) = w$ for all $w \in F$ and with no fixed points ($\varphi(w) = w$ implies $w = 1$). (Compare Exercise 1.50.)

11.7. If $H \lhd G$ and G/H is free, then G is a semidirect product of H by G/H. (*Hint.* Corollary 10.16 and Lemma 7.20.)

11.8. Let G be a group, let $\{t_i: i \in I\} \subset G$, and let $S = \langle t_i: i \in I \rangle \leq G$. If there is a homomorphism $\varphi: G \to F$ (where F is the free group with basis $X = \{x_i: i \in I\}$) with $\varphi(t_i) = x_i$ for all i, then S is a free group with basis $\{t_i: i \in I\}$.

11.9. The ***binary tetrahedral group*** B is the group having the presentation

$$B = (r, s, t \mid r^2 = s^3 = t^3 = rst).$$

(i) Prove that $rst \in Z(B)$ and that $B/\langle rst \rangle \cong A_4$ (the tetrahedral group).
(ii) Prove that B has order 24.
(iii) Prove that B has no subgroup of order 12.

11.10. The ***dicyclic group*** DC_n is the group having the presentation

$$DC_n = (r, s, t \mid r^2 = s^2 = t^n = rst).$$

(i) If $n = 2^{m-2}$, then $DC_n \cong Q_m$, the generalized quaternion group (see Exercise 4.40).

(ii) Show that DC_n has order $4n$.

11.11. Show that $(\sigma_1 \sigma_2 \dots \sigma_m)^{m+1} \in Z(B_m)$, where B_m is the braid group (see Example 11.6). It is known that $Z(B_m)$ is the infinite cyclic group generated by this element.

11.12. (i) Show that a free semigroup with a basis having at least two elements is not commutative.

(ii) Show that a subsemigroup of a free semigroup need not be free. (*Hint.* Find an appropriate subsemigroup of the multiplicative semigroup of positive integers.)

Coset Enumeration

The method of coset enumeration, distilled by Todd and Coxeter (1936) from earlier particular cases, is a mechanical way to find the order of a given group from a presentation. It does not always work (nor can any such algorithm always work, as we shall see in the next chapter), but it does work whenever the presented group is finite. The method rests on the following elementary lemma.

Lemma 11.7. *Let G be a finite group, X a set of generators of G, $H \leq G$ a subgroup, and $Hw_1, \dots, Hw_n$ some distinct cosets of H. If $\bigcup_{i=1}^{n} Hw_i$ is closed under right multiplication by every $a \in X \cup X^{-1}$, then $G = \bigcup_{i=1}^{n} Hw_i$, $[G : H] = n$, and $|G| = n|H|$.*

Proof. If Y is any nonempty subset of G with $Ya \subset Y$ for all $a \in X \cup X^{-1}$, then $Y = G$ (because X generates G and $w \in Y$ for every word w on X). In particular, $G = \bigcup_{i=1}^{n} Hw_i$, so that every coset of H must appear as Hw_i for some i; that is, $[G : H] = n$. ∎

We illustrate the method in a specific case before describing it in general. Let G be the group having the presentation

$$G = (s, t \mid s^3 = t^2 = 1, tst = s^2).$$

Write each of the relations as a word with all exponents ± 1:

$$sss; \qquad tt; \qquad tsts^{-1}s^{-1}.$$

For each of these relation words, begin making a ***relation table*** by putting a vertical line under each of its letters.

s	s	s

t	t

t	s	t	s^{-1}	s^{-1}

If a word has l letters, there are thus l vertical lines. We regard these lines as being the dividing lines forming $l + 1$ columns, and we now proceed to create rows. In each of the three tables, put 1 at the beginning and at the end of the first row. Draw row 2 (in each table), beginning and ending with 2, and put 2 next to 1 in the first table.

$s \quad s \quad s$

1	2		1
2			2

$t \quad t$

1		1
2		2

$t \quad s \quad t \quad s^{-1} \quad s^{-1}$

1					1
2					2

Build an ***auxiliary table*** containing entries

$$\overset{s}{1 \mid 2} \qquad \text{and} \qquad \overset{s^{-1}}{2 \mid 1}.$$

Now scan each of the tables to see whether there are any empty squares of either of the two forms

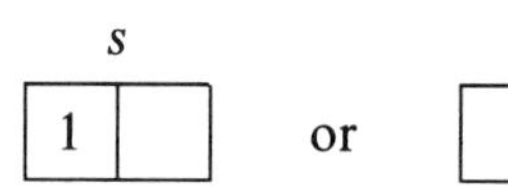

in either case, fill the empty square with 2, obtaining

$s \quad s \quad s$

1	2		1
2			2

$t \quad t$

1		1
2		2

$t \quad s \quad t \quad s^{-1} \quad s^{-1}$

1				2	1
2					2

Having filled all such squares, now draw row 3 (in each table), beginning and ending with 3, and put 3 in the first available square in the first table (next to 2).

$s \quad s \quad s$

1	2	3	1
2			2
3			3

$t \quad t$

1		1
2		2
3		3

$t \quad s \quad t \quad s^{-1} \quad s^{-1}$

1				2	1
2					2
3					3

The auxiliary table receives new entries

$$\overset{s}{2 \mid 3} \qquad \text{and} \qquad \overset{s^{-1}}{3 \mid 2}$$

and, because the first row of table one has been completed, there are bonus entries: the auxiliary table also receives

$$\overset{s}{3 \mid 1} \qquad \text{and} \qquad \overset{s^{-1}}{1 \mid 3}.$$

Now fill more squares using the (enlarged) auxiliary table to obtain

	s	s	s
1	2	3	1
2	3	1	2
3	1	2	3

	t	t
1		1
2		2
3		3

	t	s	t	s^{-1}	s^{-1}
1			3	2	1
2			1	3	2
3			2	1	3

The first table is complete, but we will continue until all the relation tables are complete (if possible). The next step draws row 4 (in all three tables) with 4 in the first row of the second table, yielding auxiliary table entries

$$\begin{array}{c|c} \multicolumn{2}{c}{t} \\ 1 & 4 \end{array} \qquad \text{and} \qquad \begin{array}{c|c} \multicolumn{2}{c}{t^{-1}} \\ 4 & 1 \end{array}$$

as well as bonus entries

$$\begin{array}{c|c} \multicolumn{2}{c}{t} \\ 4 & 1 \end{array} \qquad \text{and} \qquad \begin{array}{c|c} \multicolumn{2}{c}{t^{-1}} \\ 1 & 4. \end{array}$$

Fill in more square using the auxiliary table and obtain

	s	s	s
1	2	3	1
2	3	1	2
3	1	2	3
4			4

	t	t
1	4	1
2		2
3		3
4	1	4

	t	s	t	s^{-1}	s^{-1}
1	4		3	2	1
2		4	1	3	2
3			2	1	3
4	1	2			4

Continue adding rows 5 and 6, filling in squares using all the entries in the auxiliary table.

	s	s	s
1	2	3	1
2	3	1	2
3	1	2	3
4	5		4
5		4	5
6			6

	t	t
1	4	1
2	6	2
3	5	3
4	1	4
5	3	5
6	2	6

	t	s	t	s^{-1}	s^{-1}
1	4	5	3	2	1
2	6	4	1	3	2
3	5	6	2	1	3
4	1	2	6	5	4
5	3	1	4		5
6	2	3	5	4	6

When we try to add row 7, a new feature appears. In row 4 of the first table, the new 7 after 5 gives the auxiliary table entry

$$\begin{array}{c|c} \multicolumn{2}{c}{s} \\ 5 & 7; \end{array}$$

but the auxiliary table already contains

$$5 \overset{s}{\mid} 6.$$

This is an instance of ***coset collapse***; delete row 7 and replace all other occurrences of 7 by the smaller number 6, including the entries in the auxiliary table. Continuing this procedure ultimately leads to the completed tables

	s	s	s
1	2	3	1
2	3	1	2
3	1	2	3
4	5	6	4
5	6	4	5
6	4	5	6

	t	t
1	4	1
2	6	2
3	5	3
4	1	4
5	3	5
6	2	6

	t	s	t	s^{-1}	s^{-1}
1	4	5	3	2	1
2	6	4	1	3	2
3	5	6	2	1	3
4	1	2	6	5	4
5	3	1	4	6	5
6	2	3	1	3	6

The procedure now stops because *all* the relation tables are complete. According to the next theorem, the conclusion is that the presented group G has order 6 (of course, $G \cong S_3$).

Theorem 11.8 (Coset Enumeration). *Let G have a presentation with a finite number of generators and relations. Set up one table for each relation as above, add new integer entries and enlarge the auxiliary table as above whenever possible, and delete any larger numbers involved in coset collapse. If the procedure ends with all relation tables complete and having n rows, then the presented group G has order n.*

Proof. Let 1 denote the identity element of G, and assume that the other integers i in the tables, where $1 < i \leq n$, denote other elements of G. The entry

$$i \overset{a}{\mid} j$$

in any relation table is interpreted as the equation $ia = j$ in G. This explains the twin entries in the auxiliary table: if $ia = j$, then $ja^{-1} = i$. The construction of the relation tables is a naming of elements of G. If there is a blank square to the right of i, with line labeled a between, then j is the element ia; if the blank square is to the left, then j is the element ia^{-1}. Coset collapse occurs when $ia = j$ and $ia = k$, in which case $j = k$.

Let Y be the set of elements in G that have been denoted by some i with $1 \leq i \leq n$. That all the tables are complete says that right multiplication by any $a \in X \cup X^{-1}$ produces only elements of Y. Therefore, Lemma 11.7 applies to Y (with H taken to be the trivial subgroup), and so $|G| = n$. ■

Notice the hypothesis "If the procedure ends"; one does not know in advance whether the procedure will end.

There is a generalization of the algorithm from which the name "coset enumeration" arises. Consider the binary tetrahedral group (of order 24) given in Exercise 11.9:

$$B = (r, s, t \mid r^2 = s^3 = t^3 = rst).$$

First rewrite the presentation to display relations equal to 1:

$$B = (r, s, t \mid r^{-1}st = r^{-2}s^3 = r^{-1}s^{-1}t^2 = 1).$$

One could use Theorem 11.8 to show that B has order 24, but tables with 24 rows are tedious to do. Instead, let us choose a subgroup $H \leq G$ for which generators are known. For example, we might choose $H = \langle s \rangle$ in this example (cyclic subgroups are simplest). The idea is to use a slight variant of Theorem 11.8 to enumerate the cosets of H in G (instead of the elements of G). This is done as follows. In addition to relation tables, draw ***subgroup generator tables***, one for each generator of H. For example, there are two such tables if we choose $H = \langle rst, s \rangle$; there is just one such table if we choose $H = \langle s \rangle$. New tables consist of one row, and they are called complete once all their squares are filled *without drawing any new rows under them.* In our example, there is just one subgroup generator table, and it is already complete.

	s	
1		1

In the general case, the rows of the subgroup generator tables are completed first, giving pairs of entries to the auxiliary table (in our example, the entries in the auxiliary table arising from the subgroup generator table are

$$\begin{array}{c} s \\ 1 \mid 1 \end{array} \quad \text{and} \quad \begin{array}{c} s^{-1} \\ 1 \mid 1 \end{array}).$$

After completing these one-rowed tables, the relation tables are completed as before. The numbers i now denote right cosets of H in G, with 1 denoting H. The entry

$$\begin{array}{c} a \\ i \mid j \end{array}$$

in a table means that if $i = Hw$, then $j = Hwa$. When all the tables are completed, Lemma 11.7 applies to calculate $[G : H]$, and hence $|G|$ is known if $|H|$ is. This version actually does enumerate the cosets of H.

In Exercise 11.13 below, the reader is asked to use coset enumeration to show that the order of the binary tetrahedral group B is 24. One must compute $|H|$; that is, one must compute the order of s (it is 6) and then see that the relation tables are complete with 4 rows.

There are two unexpected consequences of coset enumeration. When $H =$

1, the completed relation tables can be used to construct the regular representation of G. For example, we saw above that the presentation of $G = S_3$,

$$G = (s, t \mid s^3 = t^2 = 1, tst = s^2),$$

has relation tables:

	s	s	s
1	2	3	1
2	3	1	2
3	1	2	3
4	5	6	4
5	6	4	5
6	4	5	6

	t	t
1	4	1
2	6	2
3	5	3
4	1	4
5	3	5
6	2	6

	t	s	t	s^{-1}	s^{-1}
1	4	5	3	2	1
2	6	4	1	3	2
3	5	6	2	1	3
4	1	2	6	5	4
5	3	1	4	6	5
6	2	3	1	3	6

The first column of the first table displays the values of right multiplication by s (as a permutation of $\{1, \ldots, 6\}$), and the first column of the second table does this for t. Right multiplication by s and t are:

$$s \mapsto (1\ 2\ 3)(4\ 5\ 6) \qquad \text{and} \qquad t \mapsto (1\ 4)(2\ 6)(3\ 5),$$

so that the right regular representation has $R_s = (1\ 3\ 2)(4\ 6\ 5)$ (because $R_s\colon i \mapsto is^{-1}$) and $R_t = (1\ 4)(2\ 6)(3\ 5)$. More generally, when one enumerates the cosets of a subgroup H of G, then one obtains the representation of G on the cosets of H (the construction above differs from that of Theorem 3.14 only in giving the representation on the right cosets of H instead of on the left cosets as in that theorem).

The information contained in completed relation tables can also be used to draw a directed graph.

Definition. A ***directed graph*** Γ is a set V, called ***vertices***, together with a subset $E \subset V \times V$; ordered pairs $(u, v) \in E$ are called ***directed edges***. A directed graph yields an ***associated graph*** Γ': both Γ and Γ' have the same vertices, and u and v are called adjacent in Γ' if $u \neq v$ and either (u, v) or (v, u) is a directed edge in Γ.

One can picture a finite directed graph Γ by drawing V as points and drawing an arrow from u to v if $(u, v) \in E$. In contrast to graphs, which have at most one edge between any pair of vertices, a directed graph may have two edges between a pair of vertices, one in each direction (given $u, v \in V$, it may happen that both (u, v) and $(v, u) \in E$). However, even if both (u, v) and (v, u) are directed edges in Γ, there is only edge between them in the associated graph Γ'. (There is a notion of *multigraph*, directed or nondirected, which allows many edges between a given pair of vertices, but we do not need them here.)

Definition. Let G be a group and let X be a set of generators of G. The ***Cayley graph*** $\Gamma = \Gamma(G, X)$ is the directed graph with vertices the elements of G and with a directed edge from g to h if $h = gx$ for some $x \in X$.

If coset enumeration of a presentation $(X|\Delta)$ of a group G yields complete relation tables, then one can record the information in these tables as the Cayley graph $\Gamma(G, X)$. For example, here is the Cayley graph of S_3 obtained from the presentation above.

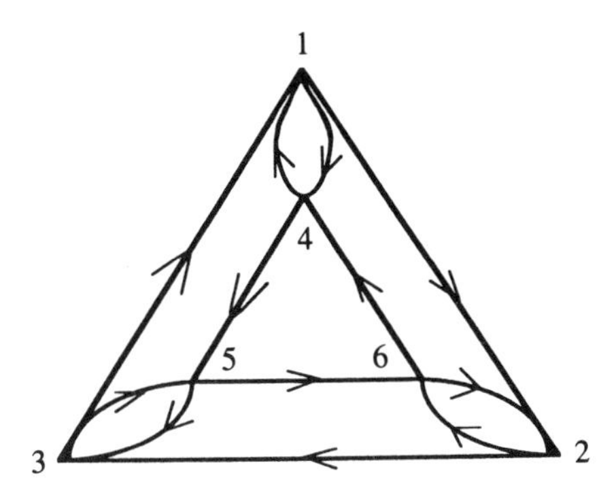

Figure 11.1

The Cayley graph of a group and a generating set is always defined, whether or not coset enumeration can be completed. Notice that the Cayley graph does depend on the choice of generating set. For example, a ***loop*** is an edge of the form (v, v). If we take G itself as a generating set, then $\Gamma(G, G)$ contains the loop $(1, 1)$, while $\Gamma(G, X)$ has no loops if $1 \notin X$. The Cayley graph is the beginning of a rich and fruitful geometric way of viewing presentations (see Burnside (1911), Dicks and Dunwoody (1989), Gersten (1987), Lyndon and Schupp (1977), and Serre (1980)).

Exercises

11.13. (i) In the presentation of the binary tetrahedral group B given above, show that s has order 6 in B.
(ii) Use coset enumeration relative to the subgroup $H = \langle s \rangle$ to compute the order of B.
(iii) Find the representation of B on the (right) cosets of H.

11.14. Describe the group G to isomorphism if G has the presentation

$$(q, r, s, t | rqr^{-1} = q^2, rtr^{-1} = t^2, s^{-1}rs = r^2, tst^{-1} = s^2, rt = tr).$$

11.15. Let $(X|\Delta)$ be a presentation of a group G. Show that the Cayley graph $\Gamma(G, X)$ has no loops if and only if $1 \notin X$.

Definition. The ***degree*** of a vertex v in a graph Γ is the number of vertices adjacent to it; the ***degree*** of a vertex v in a directed graph Γ is its degree in the associated graph Γ'. A graph or directed graph is ***regular*** of degree k if every vertex has the same degree, namely, k.

11.16. If X is a finite generating set of a group G with $1 \notin X$, then the Cayley graph $\Gamma(G, X)$ is regular of degree $2|X|$. (*Hint*. If $g \in G$ and $x \in X$, then (gx^{-1}, g) and (g, gx) are directed edges.)

11.17. Draw the Cayley graph $\Gamma(G, X)$ if G is a free abelian group of rank 2 and X is a basis.

11.18. Draw the Cayley graph $\Gamma(G, X)$ if G is a free group of rank 2 and X is a basis.

Presentations and the Schur Multiplier

The Schur multiplier $M(Q)$ of a group Q is discussed in Chapter 7 (the reader is advised to reread the appropriate section); it is related to presentations of Q because of the following isomorphism.

Hopf's Formula. *If $Q \cong F/R$ is a finite*[3] *group, where F is free, then*

$$M(Q) \cong (R \cap F')/[F, R].$$

Remark. An "aspherical" topological space X has the property that its homology groups are completely determined by its fundamental group $\pi_1(X)$. Hopf (1942) proved that $H_2(X) \cong (R \cap F')/[F, R]$, where F is free and $F/R \cong \pi_1(X)$. Schur (1907) proved that $M(Q) \cong (R \cap F')/[F, R]$ when Q is finite (i.e., Schur proved Hopf's formula in this case!). Comparison of Hopf's formula to Schur's theorem led Eilenberg and Mac Lane to their creation of Cohomology of Groups; the homology group $H_2(X)$ of the aspherical space X is the homology group $H_2(\pi_1(X), \mathbb{Z})$ of the fundamental group $\pi_1(X)$. When $\pi_1(X)$ is finite, $H_2(\pi_1(X), \mathbb{Z})$ is isomorphic to the second cohomology group $H^2(\pi_1(X), \mathbb{C}^\times) = M(\pi_1(X))$.

We will prove Hopf's formula for all finite groups Q, but we first consider a special class of groups.

Definition. A group Q is ***perfect*** if $Q = Q'$.

Every simple group is perfect. The proofs of Theorems 8.13 and 8.23 show that the groups $\mathrm{SL}(n, q)$ are perfect unless $(n, q) = (2, 2)$ or $(2, 3)$.

[3] Let us explain the finiteness hypothesis in Hopf's formula. In Chapter 7, we defined $M(Q)$ as the cohomology group $H^2(Q, \mathbb{C}^\times)$. Nowadays, after defining homology groups of Q, one defines $M(Q)$ as the second homology group $H_2(Q, \mathbb{Z})$. There is always an isomorphism $H_2(Q, \mathbb{Z}) \cong (H^2(Q, \mathbb{C}^\times))^*$, where * denotes character group. When Q is finite, the abelian group H^2 is also finite, and hence it is isomorphic to its own character group, by Theorem 10.54.

The definition of *exact sequence*

$$\cdots \to A \xrightarrow{f} B \xrightarrow{g} C \to \cdots$$

(the image of each homomorphism is equal to the kernel of the next one) makes sense if the groups are nonabelian. Of course, every image, being a kernel of a homomorphism, must be a normal subgroup.

Lemma 11.9. *Let $\nu: U \to Q$ be a central extension of a group $K = \ker \nu$ by a group Q. If Q is perfect, then U' is perfect and $\nu|U'$: $U' \to Q$ is surjective.*

Proof. Since ν is surjective, $\nu(U') = Q'$; as Q is perfect, $\nu(U') = Q$, and $\nu|U'$ is surjective. If $u \in U$, there is thus $u' \in U'$ with $\nu(u') = \nu(u)$; hence, there is $z \in K \leq Z(U)$ with $u = u'z$. To see that U' is perfect, it suffices to show that $U' \leq U''$. But if $[u, v]$ is a generator of U', then there are $u', v' \in U'$ and central elements z_1, z_2 with $[u, v] = [u'z_1, v'z_2] = [u', v'] \in U''$. ∎

Theorem 11.10. *If Q is a perfect finite group and if $Q \cong F/R$, where F is free, then $M(Q) \cong (R \cap F')/[F, R]$. Moreover, $F'/[F, R]$ is a cover of Q.*

Proof. Since $R \lhd F$, we have $[F, R] \leq R$; moreover, $[F, R] \lhd F$. There is thus an exact sequence

$$1 \to R/[F, R] \to F/[F, R] \xrightarrow{\nu} F/R \to 1,$$

which is plainly a central extension. It is easily checked that $(F/[F, R])' = F'/[F, R]$. Since $Q = F/R$ is perfect, Lemma 11.9 gives an exact sequence ξ (with ν' the restriction of ν)

$$\xi: \qquad 1 \to (R \cap F')/[F, R] \to F'/[F, R] \xrightarrow{\nu'} F/R \to 1$$

(for $(R \cap F')/[F, R] = (R/[F, R]) \cap (F'/[F, R]) = \ker \nu'$). Let us denote $(R \cap F')/[F, R]$ by K. As $F'/[F, R]$ is perfect, by Lemma 11.9, we have $K \leq (F'/[F, R])'$. Therefore, the transgression $\delta: K^* \to M(Q)$ is injective, by Lemma 7.64. But Q is finite, by hypothesis, so that $M(Q)$ is also finite, by Theorem 7.60; hence K is finite and $K \cong K^*$, by Theorem 10.56. To see that δ is surjective, it suffices, by Lemma 7.63, to prove that the central extension ξ has the projective lifting property. Consider the diagram with exact rows

$$\begin{array}{ccccccccc} 1 & \longrightarrow & R & \longrightarrow & F & \xrightarrow{\pi} & Q & \longrightarrow & 1 \\ & & & & & & \downarrow{\scriptstyle \tau} & & \\ 1 & \longrightarrow & \mathbb{C}^\times & \longrightarrow & \mathrm{GL} & \longrightarrow & \mathrm{PGL} & \longrightarrow & 1. \end{array}$$

Since F is free, it has the projective property (in the diagram in Theorem 11.6, let $\alpha = \tau\pi$); there exists a homomorphism $\sigma: F \to \mathrm{GL}$ making the diagram commute; moreover, since the bottom extension is central, it is easy to see that $[F, R] \leq \ker \sigma$. There results a commutative diagram with exact rows

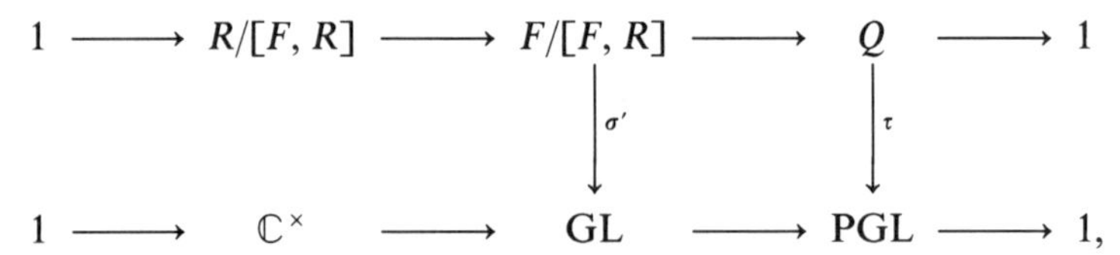

where σ': $u[F, R] \mapsto \sigma(u)$ for all $u \in F$. Since Q is perfect, we may replace the top row by ξ and the downward map σ' by its restriction $\tilde{\tau} = \sigma' | F'/[F, R]$. Thus, τ can be lifted, ξ has the projective lifting property, and the injection δ: $K^* \to M(Q)$ is also a surjection, as desired. Finally, $F'/[F, R]$ is a cover of Q, by Lemmas 7.63 and 7.64. ∎

A central extension U of K by Q has the projective lifting property if, for every homomorphism τ: $Q \to \mathrm{PGL}$, there is a homomorphism $\tilde{\tau}$: $U \to \mathrm{GL}$ making the following diagram commute:

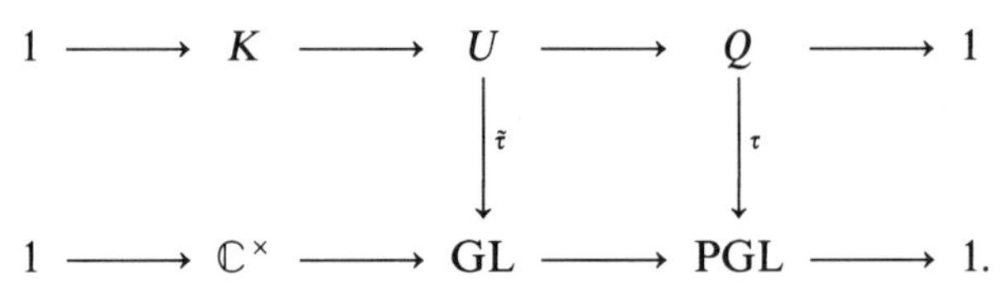

Of course, the bottom row is a central extension. Are there central extensions of Q which have a projective lifting property with respect to other central extensions?

Definition. A central extension U of Q is a ***universal central extension*** if, for every central extension V of Q, there is a unique homomorphism α making the following diagram commute:

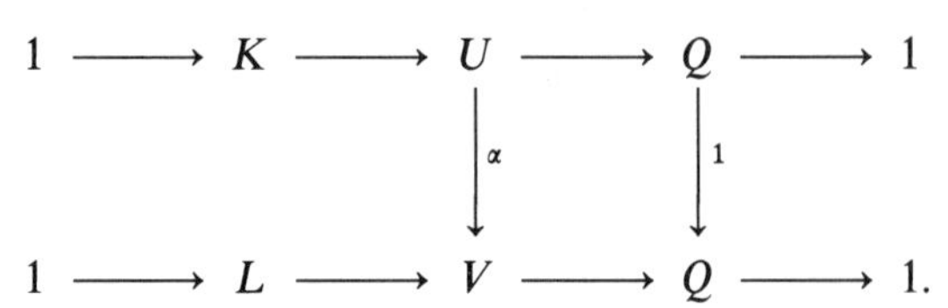

The uniqueness of the homomorphism α will be used to show that if Q has a universal central extension, then U is unique to isomorphism.

Observe that commutativity of the diagram implies that $\alpha(K) \leq L$; that is, insertion of the map $\beta = \alpha | K$: $K \to L$ yields an augmented commutative diagram

$$\begin{array}{ccccccccc} 1 & \longrightarrow & K & \longrightarrow & U & \longrightarrow & Q & \longrightarrow & 1 \\ & & \downarrow{\scriptstyle \beta} & & \downarrow{\scriptstyle \alpha} & & \downarrow{\scriptstyle 1} & & \\ 1 & \longrightarrow & L & \longrightarrow & V & \longrightarrow & Q & \longrightarrow & 1. \end{array}$$

Theorem 11.11. *If Q is a finite perfect group, then its cover $U = F'/[F, R]$ is a universal central extension of Q.*

Proof. Consider the portion of the proof of Theorem 11.10 showing the existence of a map $F'/[F, R] \to \mathrm{GL}$ making the diagram commute; this portion holds if one replaces the central extension $1 \to \mathbb{C}^\times \to \mathrm{GL} \to \mathrm{PGL} \to 1$ by any central extension and the map τ by any map. In particular, one may replace the central extension by $1 \to L \to V \to Q \to 1$ and the map τ by 1_Q, where V is a central extension of L by Q. Thus, it only remains to prove the uniqueness of such a map.

Let $\nu\colon U \to Q$ and $\mu\colon V \to Q$ be the given surjections. Suppose that α, $\beta\colon U \to V$ are homomorphisms with $\mu\alpha = \nu = \mu\beta$. If $u \in U$, then $\mu\alpha(u) = \mu\beta(u)$, so that $\alpha(u)\beta(u)^{-1} \in \ker \mu \leq Z(V)$; there is thus $z \in Z(V)$ with $\alpha(u) = \beta(u)z$. Similarly, if $u' \in U'$, there is $z' \in Z(V)$ with $\alpha(u') = \beta(u')z'$. Therefore, $\alpha([u, u']) = [\alpha(u), \alpha(u')] = [\beta(u)z, \beta(u')z'] = [\beta(u), \beta(u')] = \beta([u, u'])$. Since U is perfect, by Lemma 11.9, it is generated by all commutators. Therefore, $\alpha = \beta$, as desired. ■

The converse of Theorem 11.11 is true: a finite group Q has a universal central extension if and only if Q is perfect (see Milnor (1971), §5). More generally, a possibly infinite group Q has a universal central extension if and only if Q/Q' is free abelian (see Gruenberg (1970), p. 214).

Corollary 11.12. *Every finite perfect group Q has a unique cover U which is itself a finite perfect group.*

Proof. By Theorem 11.11, the cover $U = F'/[F, R]$ is a universal central extension of Q; if a central extension V of L by Q is a cover, then there is a commutative diagram

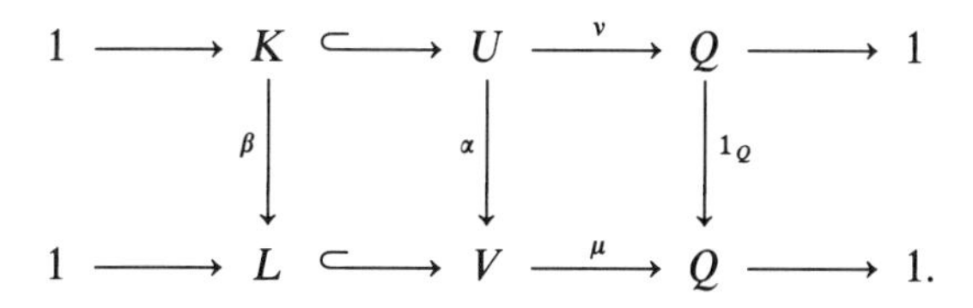

By Lemma 7.67, the transgressions are related by $\delta^U \beta^* = \delta^V$. As both U and V are covers, however, both transgressions are isomorphisms (Lemmas 7.63 and 7.64), and so β^* is an isomorphism. By Exercise 10.55, β is an isomorphism, and by Exercise 10.56, α is an isomorphism. Finally, Lemma 11.9 shows that U is perfect. ■

We are now going to prove that Hopf's formula holds for every (not necessarily perfect) finite group Q.

Lemma 11.13. *If Q is a finite group and*

$$1 \to K \to E \to Q \to 1$$

is a central extension, then $K \cap E'$ is finite.

Proof. Since $K \leq Z(E)$, we have $[E : Z(E)] \leq [E : K] = |Q| < \infty$. Thus $Z(E)$ has finite index in E, and so Schur's Theorem (Theorem 7.57) gives E' finite. Therefore $K \cap E' \leq E'$ is also finite. ■

Lemma 11.14. *If $Q = F/R$ is a finite group, where F is free, then there is a central extension*

$$1 \to R/[F, R] \to F/[F, R] \to Q \to 1.$$

Moreover, if F is finitely generated, then $R/[F, R]$ is also finitely generated.

Proof. We have already noted (in Theorem 11.10) that the sequence is a central extension. Now F finitely generated implies that $F/[F, R]$ is finitely generated. As Q is finite, $R/[F, R] \leq F/[F, R]$ is a subgroup of finite index, and so Lemma 7.56 shows that $R/[F, R]$ is finitely generated. ■

Lemma 11.15. *Let $Q = F/R$ be a finite group, where F is a finitely generated free group. The torsion subgroup of $R/[F, R]$ is $(R \cap F')/[F, R]$, and there is a subgroup S with $[F, R] \leq S \leq R$, with $S \lhd F$, and with*

$$R/[F, R] = (R \cap F')/[F, R] \oplus S/[F, R].$$

Proof. Let $T = (R \cap F')/[F, R]$. Since $T = (R/[F, R]) \cap (F'/[F, R])$, Lemma 11.13 shows that T is finite, and so $T \leq t(R/[F, R])$. For the reverse inclusion, note that $(R/[F, R])/T \cong (R/[F, R])/((R \cap F')/[F, R]) \cong R/(R \cap F') \cong F'R/F' \leq F/F'$, which is free abelian. By Corollary 10.16, there is a subgroup S with $[F, R] \leq S \leq R$ such that $R/[F, R] = T \oplus S/[F, R]$. Therefore, $t(R/[F, R]) = (R \cap F')/[F, R]$. Finally, $R/[F, R] \leq Z(F/[F, R])$, so that all its subgroups are normal in $F/[F, R]$; in particular, $S/[F, R] \lhd F/[F, R]$ and $S \lhd F$. ■

Lemma 11.16. *Let Q be a finite group and let $Q = F/R$, where F is a finitely generated free group.*

(i) *There is a central extension*

$$1 \to K \to E \to Q \to 1$$

with $K \leq E'$ and with $K \cong (R \cap F')/[F, R]$.

(ii) $|(R \cap F')/[F, R]| \leq |M(Q)|$.

Proof. (i) Choose S, as in Lemma 11.15, with $S \lhd F$, $[F, R] \leq S \leq R$, and $R/[F, R] = (R \cap F')/[F, R] \oplus S/[F, R]$. Consider the exact sequence

$$1 \to R/S \to F/S \to Q \to 1.$$

Since $R/[F, R]$ is central in $F/[F, R]$, it follows that $R/S \cong (R/[F, R])/(S/[F, R])$ is central in $F/S \cong (F/[F, R])/(S/[F, R])$; moreover, the definition of S gives $(R/[F, R])/(S/[F, R]) \cong (R \cap F')/[F, R]$. Finally, $R/S = (R \cap F')S/S \leq F'S/S = (F/S)'$.

(ii) By Lemma 11.13, R/S is finite. By Lemma 7.64, the transgression $\delta: (R/S)^* \to M(Q)$ is an injection, so that $|(R \cap F')/[F, R]| = |R/S| = |(R/S)^*| \leq |M(Q)|$. ■

Theorem 11.17 (Schur, 1907). *If Q is a finite group with $Q = F/R$, where F is a finitely generated free group, then*

$$M(Q) \cong (R \cap F')/[F, R].$$

Proof. Let $1 \to L \to U \xrightarrow{\pi} Q \to 1$ be a central extension with $L \leq U'$, let $\{y_1, \ldots, y_n\}$ be a basis of F, and let $\nu: F \to Q$ be a surjective homomorphism. For all i, choose $u_i \in U$ with $\pi(u_i) = \nu(y_i)$. Since F is free with basis $\{y_1, \ldots, y_n\}$, there is a homomorphism $\sigma: F \to U$ with $\pi\sigma = \nu$. Now $L = L \cap U' \leq Z(U) \cap U' \leq \Phi(U)$, by Theorem 5.49. Therefore,

$$U = \langle u_1, \ldots, u_n, L \rangle \leq \langle u_1, \ldots, u_n, \Phi(U) \rangle = \langle u_1, \ldots, u_n \rangle,$$

by Theorem 5.47, so that σ is surjective.

If $a \in L$, then $a = \sigma(w)$ for some $w \in F$ (for σ is surjective), and so $1 = \pi(a) = \pi\sigma(w) = \nu(a)$. Hence, $w \in \ker \nu = R$, and so $a = \sigma(w) \in \sigma(R)$; that is, $L \leq \sigma(R)$. For the reverse inclusion, if $r \in R$, then $\pi\sigma(r) = \nu(r) = 1$, so that $\sigma(r) \in \ker \pi = L$. Thus, $L = \sigma(R)$. Note that $\sigma([F, R]) = [\sigma(F), \sigma(R)] = [U, L] = 1$, because L is central, so that σ induces a homomorphism $\bar{\sigma}: F/[F, R] \to U$. But $\bar{\sigma}((R \cap F')/[F, R])) = \sigma(R) \cap \sigma(F') = L \cap U' = L$, and so $|(R \cap F')/[F, R]| \geq |L|$.

By Theorem 7.66, there is a central extension

$$1 \to L \to U \to Q \to 1$$

with $L \cong M(Q)$ and $L \leq U'$. Therefore, Lemma 11.16(ii) gives

$$|M(Q)| = |L| \leq |(R \cap F')/[F, R]| \leq |M(Q)|.$$

Returning to Lemma 11.16 with $K \cong (R \cap F')/[F, R]$, the injection $\delta: K^* \to M(Q)$ must be surjective, and so $M(Q) \cong (R \cap F')/[F, R]$. ■

Corollary 11.18. *For every finite group Q, if $Q = F/R$, where F is a finitely generated free group, then $(R \cap F')/[F, R]$ is independent of the finite presentation F/R of Q.*

Proof. We have $(R \cap F')/[F, R] \cong M(Q)$, and the Schur multiplier $M(Q)$ is defined independently of a presentation. ■

Definition. If A is a finitely generated abelian group, let $d(A)$ denote the number of elements in a smallest generating set of A; that is, A can be generated by some set of $d(A)$ elements, but it cannot be generated by any set of size $d(A) - 1$.

If $\rho(A)$ denotes the rank of a finitely generated abelian group A, that is, the rank of the free abelian group A/tA, then it is easy to see that

$$\rho(A) \leq d(A);$$

with equality if and only if A is free abelian; moreover, $\rho(A) = 0$ if and only if A is finite. The reader may prove that if A and B are finitely generated abelian groups, then

$$\rho(A \oplus B) = \rho(A) + \rho(B);$$

indeed, Exercise 10.50(ii) shows that if there is an exact sequence $0 \to A \to E \to B \to 0$ of abelian groups, then $\rho(E) = \rho(A) + \rho(B)$. On the other hand, this is not generally true if ρ is replaced by d. For example, $\mathbb{Z}_6 \cong \mathbb{Z}_2 \oplus \mathbb{Z}_3$, and $1 = d(\mathbb{Z}_6) = d(\mathbb{Z}_2 \oplus \mathbb{Z}_3) \neq d(\mathbb{Z}_2) + d(\mathbb{Z}_3) = 2$. However, if F is a finitely generated free abelian group, then

$$d(A \oplus F) = d(A) + d(F).$$

Lemma 11.19. *Assume that Q has a finite presentation*

$$Q = (x_1, \ldots, x_n | y_1, \ldots, y_r).$$

If F is the free group with basis $\{x_1, \ldots, x_n\}$ and R is the normal subgroup generated by $\{y_1, \ldots, y_r\}$, then $R/[F, R]$ is a finitely generated abelian group and $d(R/[F, R]) \leq r$.

Proof. We have $R' = [R, R] \leq [F, R] \leq R$, since $R \lhd F$, so that $R/[F, R]$ is abelian. The proof is completed by showing that it is generated by the cosets of the y's. Now R is generated by $\{fy_if^{-1}: f \in F, i = 1, \ldots, r\}$ (R is the normal subgroup of F generated by $\{y_1, \ldots, y_r\}$). But $fy_if^{-1}y_i^{-1} \in [F, R]$, so that $fy_if^{-1}[F, R] = y_i[F, R]$, as desired. ■

Theorem 11.20. *If Q has a finite presentation*

$$Q = (x_1, \ldots, x_n | y_1, \ldots, y_r),$$

then $M(Q)$ is finitely generated, $d(M(Q)) \leq r$, and

$$n - r \leq \rho(Q/Q') - d(M(Q)).$$

Proof. Let F be free with basis $\{x_1, \ldots, x_n\}$, and let R be the normal subgroup generated by $\{y_1, \ldots, y_r\}$. By Lemma 11.19, there is an exact sequence of finitely generated abelian groups

$$(1) \qquad 0 \to (R \cap F')/[F, R] \to R/[F, R] \to R/(R \cap F') \to 0.$$

Therefore, $d(M(Q)) = d((R \cap F')/[F, R]) \leq r$ (by Exercise 10.7, an easy consequence of Theorem 10.17). Now

$$R/(R \cap F') \cong RF'/F' \le F/F'.$$

By Lemma 11.3, F/F' is free abelian of rank n; by Theorem 10.17, its subgroup $R/(R \cap F')$ is also free abelian, and so Corollary 10.16 shows that the exact sequence (1) splits. Thus,

$$R/[F, R] \cong M(Q) \oplus RF'/F';$$

since RF'/F' is free abelian, $d(M(Q)) + d(RF'/F') = d(R/[F, R]) \le r$, and so

$$d(RF'/F') \le r - d(M(Q)).$$

Since RF'/F' is free abelian, $d(RF'/F') = \rho(RF'/F')$, and so

$$\rho(RF'/F') \le r - d(M(Q)). \tag{2}$$

Now $Q' = (F/R)' = RF'/R$, so that $Q/Q' = (F/R)/(RF'/R) = F/RF'$ and

$$\rho(F/RF') = \rho(Q/Q'). \tag{3}$$

There is another exact sequence

$$0 \to RF'/F' \to F/F' \to F/RF' \to 0,$$

so that $n = \rho(F/F') - \rho(F/RF') = \rho(RF'/F')$. Combining this with (2) and (3) gives

$$n - \rho(Q/Q') = \rho(RF'/F') \le r - d(M(Q)),$$

which is the desired inequality. ∎

Corollary 11.21. *If Q is a finite group having a presentation with n generators and r relations, then*

$$d(M(Q)) \le r - n.$$

Proof. Since Q is finite, Q/Q' is finite and $\rho(Q/Q') = 0$. ∎

Since $d(M(Q)) \ge 0$, it follows that $r \ge n$ for every finite presentation of a finite group Q; that is, there are always more relations than generators. We give a name to the extreme case.

Definition. A group is ***balanced*** if it has a finite presentation with the same number of generators as relations.

Corollary 11.22. *If Q is a finite balanced group, then $M(Q) = 1$.*

The converse of this corollary is false.

Corollary 11.23. *If $\mathbf{V}$ is the 4-group, then $M(\mathbf{V}) \cong \mathbb{Z}_2$.*

Proof. In Example 7.17, we saw that $M(\mathbf{V}) \neq 1$, and Theorem 7.68 shows that $\exp(M(\mathbf{V})) = 2$. There is a presentation $\mathbf{V} = (a, b | a^2 = 1, b^2 = 1, [a, b] = 1)$. By Corollary 11.21, $d(M(\mathbf{V})) \leq 3 - 2 = 1$, and so $M(\mathbf{V})$ is cyclic. ∎

We have now completed Example 7.17: in contrast to perfect groups, the 4-group $\mathbf{V}$ does not have a unique cover.

It is a theorem of J.A. Green (1956) that if p is a prime and Q is a group of order p^n, then $|M(Q)| \leq p^{n(n-1)/2}$. One can also show that this bound is best possible: equality holds if Q is an elementary abelian group of order p^n; of course, a special case of this is $Q = \mathbf{V}$.

We have proved two theorems helping us to compute the Schur multiplier of a finite group Q: the theorem of Alperin–Kuo (Theorem 7.68) giving a bound on $\exp(M(Q))$; Corollary 11.21 giving a bound on $d(M(Q))$.

EXERCISES

11.19. Prove that $M(\mathbf{Q}_n) = 1$, where $\mathbf{Q}_n$ is the group of generalized quaternions.

11.20. Prove that $M(D_{2n}) = 1$ if n is odd and has order ≤ 2 if n is even. (It is known that $M(D_{2n}) \cong \mathbb{Z}_2$ if n is even.)

11.21. Prove that $|M(A_5)| \leq 2$. (It is known that $M(A_5) \cong \mathbb{Z}_2$.)

11.22. (i) As in Example 11.5, show that A_4 has a presentation

$$A_4 = (s, t | s^2 = 1, t^3 = 1, (st)^3 = 1).$$

(ii) Show that the binary tetrahedral group B is a cover of A_4.
(iii) Prove that $M(A_4) \cong \mathbb{Z}_2$.

11.23. Show that $M(S_4)$ is cyclic of order ≤ 2. (*Hint*. Example 11.5.) (It is known that $M(S_n) \cong \mathbb{Z}_2$ for all $n \geq 4$.)

11.24. Show that SL(2, 4) is the cover of PSL(2, 4).

Fundamental Groups of Complexes

The theory of *covering spaces* in Algebraic Topology contains an analogue of Galois Theory: there is a bijection from the family of all covering spaces $\tilde{X}$ mapping onto a topological space X and the family of all subgroups of the fundamental group $\pi_1(X)$. This theory was used by Baer and Levi to prove the Nielsen–Schreier theorem: Every subgroup of a free group is itself free. We mimic the topological theorems here in a purely algebraic setting.

Definition. A ***complex*** K (or *abstract simplicial complex*) is a family of nonempty finite subsets, called ***simplexes***, of a set $V = \operatorname{Vert}(K)$, called ***vertices***, such that:

(i) if $v \in V$, then $\{v\}$ is a simplex;
(ii) if s is a simplex, then so is every nonempty subset of s.

A simplex $s = \{v_0, v_1, \ldots, v_q\}$ with $q + 1$ vertices is called a ***q-simplex***; one says that s has ***dimension*** q, and one writes $\dim(s) = q$. If n is the largest dimension of a simplex in K, then K is called an ***n-complex*** and one writes $\dim(K) = n$ (if there is no simplex of largest dimension, then $\dim(K) = \infty$).

A 0-complex is a set of points, and a 1-complex is a graph: define $u, v \in V$ to be adjacent if and only if $\{u, v\}$ is a 1-simplex. It turns out that 2-complexes are sufficiently complicated to serve all of our needs (see Exercise 11.21 below).

Even though no topology enters into the forthcoming discussion, the reader should know the geometric background behind the definition of complex in that setting. A 0-simplex is a point; regard a 1-simples $\{u, v\}$ as an edge with endpoints u and v; regard a 2-simplex $\{u, v, w\}$ as a (two-dimensional) triangle with vertices u, v, and w; regard a 3-simplex as a (solid) tetrahedron; and so forth. A complex may now be regarded as a space built by gluing simplexes together in a nice way.

Figure 11.2

A complex L is a ***subcomplex*** of a complex K if $\text{Vert}(L) \subset \text{Vert}(K)$ and if every simplex in L is also a simplex in K (we recognize the empty set $\varnothing$ as being a subcomplex). A subcomplex L of K is ***full*** if every simplex in K having all its vertices in L already lies in L. Thus, a full subcomplex L is determined by its vertices $\text{Vert}(L)$.

For example, if s is a simplex, then the subcomplex $|s|$, consisting of s and all its nonempty subsets, is full. For each $q \geq 0$, the ***q-skeleton***, defined by

$$K^q = \{\text{simplexes } s \in K\text{: } \dim(s) \leq q\},$$

is a subcomplex. Thus, $\text{Vert}(K) = K^0 \subset K^1 = K^0 \cup \{\text{all 1-simplexes}\} \subset K^2 \subset K^3 \subset \cdots$. If $\dim(K) = n$ and $q < n$, then K^q is *not* a full subcomplex.

Definition. An ***edge*** in a complex K is an ordered pair $\varepsilon = (u, v)$ of (not necessarily distinct) vertices lying in a simplex. If u and v are vertices in a complex

K, then a ***path*** α of ***length*** n from u to v is a sequence of n edges

$$\alpha = (u, v_1)(v_1, v_2)\dots(v_{n-2}, v_{n-1})(v_{n-1}, v).$$

Call u the ***origin*** of α and denote it by $o(\alpha)$; call v the ***end*** of α and denote it by $e(\alpha)$. A ***closed path at v*** is a path α for which $o(\alpha) = e(\alpha)$.

Definition. A complex K is ***connected*** if there is a path between any pair of its vertices.

Definition. If $\{L_i: i \in I\}$ is a family of subcomplexes of a complex K, then the ***union*** $\bigcup L_i$ is the subcomplex consisting of all those simplexes s lying in at least one L_i, and the ***intersection*** $\bigcap L_i$ is the subcomplex consisting of all those simplexes s lying in every L_i. Two subcomplexes L and L' are ***disjoint*** if $L \cap L' = \varnothing$.

It is easy to see that $\mathrm{Vert}(\bigcup L_i) = \bigcup \mathrm{Vert}(L_i)$ and $\mathrm{Vert}(\bigcap L_i) = \bigcap \mathrm{Vert}(L_i)$; in particular, L and L' are disjoint if and only if $\mathrm{Vert}(L) \cap \mathrm{Vert}(L') = \varnothing$.

Theorem 11.24. *Every complex K is the disjoint union of connected subcomplexes $K = \bigcup K_i$, and the K_i are uniquely determined by K. Moreover, each K_i is a full maximal connected subcomplex.*

Proof. The relation on $V = \mathrm{Vert}(K)$ defined by $u \equiv v$ if there is a path in K from u to v is easily seen to be an equivalence relation; let $\{V_i: i \in I\}$ be its family of equivalence classes, and let K_i be the full subcomplex of K having vertex set V_i. Clearly, K is the union $\bigcup K_i$. If a simplex s in K has a vertex u in K_i, then there is a path from u to each vertex of s, and so $s \subset V_i$; hence, $s \in K_i$ because K_i is full. Now K is the disjoint union $\bigcup K_i$, for if $s \in K_i \cap K_j$, where $i \neq j$, then $s \subset V_i \cap V_j = \varnothing$, a contradiction. To see that K_i is connected, assume that there is an edge (u, v) in K, where $u \in V_i$. Then $s = \{u, v\}$ is a simplex, and so the argument above shows that $s \subset V_i$ and $v \in V_i$. If $u, w \in V_i$, then $u \equiv w$, and so there is a path in K from u to w. An induction on the length of the path shows that the path lies in K_i, and so K_i is connected.

To prove uniqueness, let $K = \bigcup L_j$ be a disjoint union, where each L_j is a connected subcomplex. It is easy to see that each L_j is a full subcomplex; it follows, for each simplex in K, that there is a unique L_j containing all its vertices. In particular, there is no simplex $\{u, v\}$ with $u \in \mathrm{Vert}(L_j)$ and $v \notin \mathrm{Vert}(L_j)$; this shows that each L_j is a maximal connected subcomplex, for there are no paths leading outside of it.

Choose some L_j. If $s \in L_j$, then there is a unique K_i with $s \in K_i$. If $t \in L_j$ is another simplex, then $t \in K_l$ for some l. However, the presence of a path between a vertex of s and a vertex of t shows, as above, that $l = i$. Therefore, $t \in K_i$ and L_j is contained in K_i. Maximality of L_j gives $L_j = K_i$. ■

Definition. The connected subcomplexes K_i occurring in the disjoint union $K = \bigcup K_i$ are called the ***components*** of K.

We are now going to define a multiplication of paths reminiscent of juxtaposition of words in a free group.

Definition. If $\alpha = \varepsilon_1 \ldots \varepsilon_n$ and $\beta = \eta_1 \ldots \eta_m$ are paths in a complex K, where the ε_i and η_j are edges, and if $e(\alpha) = o(\beta)$, then their ***product*** is the path

$$\alpha\beta = \varepsilon_1 \ldots \varepsilon_n \eta_1 \ldots \eta_m.$$

The path $\alpha\beta$ is a path from $o(\alpha)$ to $e(\beta)$. This multiplication is associative when defined, but every other group axiom fails.

Definition. There are two types of ***elementary moves*** on a path α in a complex K. The first replaces a pair of adjacent edges $(u, v)(v, w)$ in α by (u, w) if $\{u, v, w\}$ is a simplex in K; the second is the inverse operation replacing (u, w) by $(u, v)(v, w)$ in this case. Paths α and β in K are ***homotopic***, denoted by $\alpha \simeq \beta$, if one can be obtained from the other by a finite number of elementary moves.

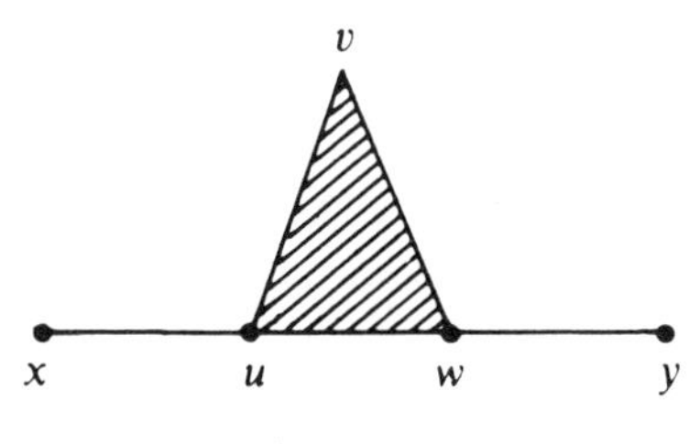

Figure 11.3

For example, let K be the 2-complex drawn above, let $\alpha = (x, u)(u, w)(w, y)$, and let $\beta = (x, u)(u, v)(v, w)(w, y)$. If K contains the simplex $\{u, v, w\}$, then $\alpha \simeq \beta$; if K does not contain this simplex, then $\alpha \not\simeq \beta$.

It is easy to check that homotopy defines an equivalence relation on the family of all paths in K.

Definition. If α is a path in a complex K, then the equivalence class of α, denoted by $[\alpha]$, is called a ***path class***.

If $\alpha \simeq \beta$, then $o(\alpha) = o(\beta)$ and $e(\alpha) = e(\beta)$ (for only "interior" vertices are changed by the elementary moves in a homotopy). Hence, one may define the ***origin*** and ***end*** of a path class $[\alpha]$, denoted by $o[\alpha]$ and $e[\alpha]$, respectively. Homotopy is compatible with the multiplication of paths: if $\alpha \simeq \beta$, $\alpha' \simeq \beta'$, and $e(\alpha) = o(\beta)$, then the reader may check that $\alpha\beta \simeq \alpha'\beta'$; that is, if $e(\alpha) = o(\beta)$, then multiplication of path classes, given by

$$[\alpha][\beta] = [\alpha\beta],$$

is well defined.

If K is a complex and $v \in \text{Vert}(K)$, then the ***trivial path at v*** is (v, v). If $\varepsilon = (u, v)$, define $\varepsilon^{-1} = (v, u)$ and, if $\alpha = \varepsilon_1 \ldots \varepsilon_n$ is a path, define its ***inverse path*** $\alpha^{-1} = \varepsilon_n^{-1} \ldots \varepsilon_1^{-1}$.

Lemma 11.25. *The set of all path classes of a complex K has the following properties*:

(i) *if* $o[\alpha] = u$ *and* $e[\alpha] = v$, *then*

$$[(u, u)][\alpha] = [\alpha] = [\alpha][(v, v)],$$

$$[\alpha][\alpha^{-1}] = [(u, u)],$$

and

$$[\alpha^{-1}][\alpha] = [(v, v)].$$

(ii) *if* α, β, *and* γ *are paths and one of* $([\alpha][\beta])[\gamma]$ or $[\alpha]([\beta][\gamma])$ *is defined, then so is the other and they are equal.*

Proof. Straightforward. ■

The set of all path classes in K with its (not always defined) multiplication is called a ***groupoid***. We extract groups from a groupoid in the obvious way.

Definition. A ***basepoint*** of a complex K is some chosen vertex v. The ***fundamental group*** of a complex with basepoint v is

$$\pi(K, v) = \{[\alpha] \colon \alpha \text{ is a closed path at } v\}.$$

Theorem 11.26. *For every vertex v in a complex K*, $\pi(K, v)$ *is a group with identity the path class of the trivial path at v.*

Proof. This follows at once from the lemma, for multiplication is now always defined. ■

Remark. There is a topological space $|K|$ which is the "geometric realization" of a complex K, and $\pi(K, v)$ is isomorphic to the fundamental group of $|K|$ defined in Algebraic Topology (see Rotman (1988), Theorem 7.36).

The next result shows that the fundamental group of a connected complex does not depend on the choice of basepoint.

Theorem 11.27.

(i) *If (K, v) is a complex with basepoint, and if L is the component of K containing v, then*

$$\pi(K, v) = \pi(L, v).$$

(ii) *If K is a connected complex with basepoints v and v', then*

$$\pi(K, v) \cong \pi(K, v').$$

Proof. (i) Since every (closed) path with origin v has all its vertices in $\mathrm{Vert}(L)$, the underlying sets of the two groups are equal. As the multiplications on each coincide as well, the groups themselves are equal.

(ii) Since K is connected, there is a path γ in K from v to v'. Define $f: \pi(K, v) \to \pi(K, v')$ by $[\alpha] \mapsto [\gamma^{-1}][\alpha][\gamma] = [\gamma^{-1}\alpha\gamma]$. Note that the latter multiplication takes place in the groupoid of all path classes in K; the product, however, lies in $\pi(K, v')$. It is a simple matter, using Lemma 11.25, to check that f is an isomorphism with inverse $[\beta] \mapsto [\gamma][\beta][\gamma^{-1}]$. ■

Definition. If K and L are complexes, then a ***simplicial map*** $\varphi: K \to L$ is a function $\varphi: \mathrm{Vert}(K) \to \mathrm{Vert}(L)$ such that $\{\varphi v_0, \varphi v_1, \ldots, \varphi v_q\}$ is a simplex in L whenever $\{v_0, v_1, \ldots, v_q\}$ is a simplex in K. A simplicial map φ is an ***isomorphism*** if it is a bijection whose inverse is also a simplicial map.

The identity on $\mathrm{Vert}(K)$ is a simplicial map. It is easy to see that the composite of simplicial maps, when defined, is a simplicial map. If $\varphi: K \to L$ is a simplicial map and $\{v_0, v_1, \ldots, v_q\}$ is a simplex, then there may be repeated vertices in the simplex $\{\varphi v_0, \varphi v_1, \ldots, \varphi v_q\}$.

Let $\varphi: K \to L$ be a simplicial map. If $\varepsilon = (u, v)$ is an edge in K, then $\varphi\varepsilon = (\varphi u, \varphi v)$ is an edge in L (because $\{\varphi u, \varphi v\}$ is a simplex in L). If $\alpha = \varepsilon_1 \ldots \varepsilon_n$ is a path, then define

$$\varphi\alpha = \varphi\varepsilon_1 \ldots \varphi\varepsilon_n,$$

which is a path in L. If $\alpha \simeq \beta$ are paths in K, then $\varphi\alpha \simeq \varphi\beta$ in L, for if $\{u, v, w\}$ is a simplex in K, then $\{\varphi u, \varphi v, \varphi w\}$ is a simplex in L.

Theorem 11.28. *If $\varphi: K \to L$ is a simplicial map, then $\varphi_\#: \pi(K, v) \to \pi(L, \varphi v)$, defined by $[\alpha] \mapsto [\varphi\alpha]$, is a homomorphism. Moreover, π is a (covariant) functor: $(1_K)_\#$ is the identity, and if $\psi: L \to M$ is a simplicial map, then $(\psi\varphi)_\# = \psi_\#\varphi_\#: \pi(K, v) \to \pi(M, \psi\varphi v)$.*

Proof. Routine. ■

Definition. A path $\alpha = \varepsilon_1 \ldots \varepsilon_n$ is ***reduced*** if either α is trivial or no $\varepsilon_i = (u, v)$ is adjacent to its inverse (v, u) and no ε_i is a trivial path. A ***circuit*** is a reduced closed path.

Let us show that every path α in a complex is homotopic to either a reduced path or a trivial path. If α contains a subpath $(u, v)(v, u)$, then $\alpha \simeq \alpha'$, where α' is obtained from α by replacing $(u, v)(v, u)$ by the trivial path (u, u). If α' is not trivial and α' contains a trivial path (u, u), then $\alpha' \simeq \alpha''$, where α'' is obtained from α' by deleting (u, u). These steps can be iterated. Since each path obtained is shorter than its predecessor, the process eventually ends, and the last path is either reduced or trivial. In particular, every closed path is homotopic to either a circuit or a trivial path.

Definition. A ***tree*** is a connected complex of dimension ≤ 1 having no circuits (the only zero-dimensional tree has a single vertex).

Let us show that if u and v are distinct vertices in a tree T, then there is a unique reduced path from u to v. Connectivity provides a path α from u to v, which we may assume is reduced. If $\beta \neq \alpha$ is another reduced path from u to v, then α and β contain a (possibly empty) subpath γ such that $\alpha = \alpha'\gamma$, $\beta = \beta'\gamma$, and the last edge of α' is distinct from the last edge of β'. It follows that $\alpha'\beta'^{-1}$ is reduced, and hence it is a circuit in T. This contradiction shows that $\alpha = \beta$.

Definition. A complex K is ***simply connected***[4] if it is connected and $\pi(K, v) = 1$ for some $v \in \text{Vert}(K)$.

By Theorem 11.27(ii), this definition does not depend on the choice of basepoint v in K.

Every tree T is simply connected: we have just noted that every closed path is homotopic to either a circuit or a trivial path, and there are no circuits in a tree.

Theorem 11.29. *Let L be a simply connected subcomplex of a complex K. If α is a closed path in K at v all of whose edges lie in L, then $[\alpha] = 1$ in $\pi(K, v)$. This is true, in particular, when L is a tree.*

Proof. The inclusion φ: $\text{Vert}(L) \hookrightarrow \text{Vert}(K)$ is a simplicial map $L \to K$, and it induces a homomorphism $\varphi_{\#}: \pi(L, v) \to \pi(K, v)$. The hypothesis gives $\varphi_{\#}([\alpha]) = [\varphi\alpha] = [\alpha]$, so that $[\alpha] \in \text{im } \varphi_{\#}$. But L simply connected gives $\pi(L, v) = 1$, hence $[\alpha] = 1$. ■

If L is a subcomplex of a complex K, then the homomorphism $\varphi_{\#}: \pi(L, v) \to \pi(K, v)$ induced by the inclusion φ: $\text{Vert}(L) \hookrightarrow \text{Vert}(K)$ need not be injective. For example, it is easy to see that a 2-simplex K is simply connected, but we shall soon see that its perimeter is not.

[4] Some authors do not insist that simply connected complexes be connected. For them, a complex is simply connected if all its components are simply connected in our sense.

Definition. A subcomplex T of a complex K is a ***maximal tree*** if T is a tree which is contained in no larger tree in K.

Lemma 11.30. *If K is a connected complex, then a tree T in K is a maximal tree if and only if* $\text{Vert}(T) = \text{Vert}(K)$.

Proof. Suppose that T is a maximal tree and there exists a vertex $v \notin \text{Vert}(T)$. Choose a vertex v_0 in T; since K is connected. There is a path $\varepsilon_1 \ldots \varepsilon_n$ in K from v_0 to $v = v_n$; let $\varepsilon_i = (v_{i-1}, v_i)$. Since v_0 is in T and v is not in T, there must be an i with v_{i-1} in T and v_i not in T. Consider the subcomplex T' obtained from T by adjoining the vertex v_i and the simplex $\{v_{i-1}, v_i\}$. Clearly T' is connected, and any possible circuit in T' must involve the new vertex v_i. There are only two nontrivial edges in T' involving v_i, namely, $\varepsilon = (v_{i-1}, v_i)$ and ε^{-1}, and so any closed path involving v_i as an "interior" vertex is not reduced, while any circuit at v_i would yield a circuit in T at v_{i-1}. Thus T' is a tree properly containing T, contradicting the maximality of T.

The proof of the converse, similar to that just given, is left to the reader. ■

Every complex K has a maximal tree (this is obvious when K is finite, and a routine Zorn's lemma argument shows it to be true in general). Usually, a complex has many maximal trees.

Definition. Let K be a complex and let $\mathscr{P} = \{X_i : i \in I\}$ be a partition of $\text{Vert}(K)$. The ***quotient complex*** $K/\mathscr{P}$ has vertices the subsets X_i, and $\{X_{i_0}, \ldots, X_{i_q}\}$ is a simplex if there are vertices $v_{i_j} \in X_{i_j}$ such that $\{v_{i_0}, \ldots, v_{i_q}\}$ is a simplex in K.

Of course, one can construct a quotient complex of K modulo an equivalence relation on $\text{Vert}(K)$, for the equivalence classes partition $\text{Vert}(K)$.

Exercises

11.25. Prove that a complex K is connected if and only if its 1-skeleton K^1 is connected.

11.26. If s is a simplex, then the complex $|s|$ (consisting of s and all its nonempty subsets) is simply connected.

11.27. Prove that the inclusion $K^2 \hookrightarrow K$ induces an isomorphism $\pi(K^2, v) \cong \pi(K, v)$. Conclude that every fundamental group arises as the fundamental group of a 2-complex.

11.28. Let I_n be the 1-complex having vertices $\{t_0, \ldots, t_n\}$ and simplexes $\{t_0, t_1\}$, $\{t_1, t_2\}, \ldots, \{t_{n-1}, t_n\}$. Prove that a path in a complex K of length n is a simplicial map $I_n \to K$.

11.29. Let T be a finite tree. If $v(T)$ is the number of vertices in T and $e(T)$ is the number of edges in T, then
$$v(T) - e(T) = 1.$$

11.30. Prove that a 1-complex K is simply connected if and only if it is a tree.

11.31. (i) Let K be a complex and let T and S be trees in K. If $T \cap S \neq \varnothing$, then $T \cap S$ is a tree if and only if $T \cap S$ is connected.
(ii) If $\{T_i: i \in I\}$ is a family of trees in a complex K with $T_i \cap T_j$ a tree for all i and j, then $\bigcup T_i$ is a tree.

11.32. Let K be a 1-complex with basepoint w, let T be a tree in K, and let (u, v) be an edge not in T. If $\alpha = \alpha'(u, v)\alpha''$ and $\beta = \beta'(u, v)\beta''$ are closed paths in K at w with α', α'', β', and β'' paths in T, then $\alpha \simeq \beta$.

11.33. Let G be a free group of rank 2 with basis X. Show that the graph associated to the Cayley graph $\Gamma(G, X)$ is a tree.

11.34. If $\varphi: K \to L$ is a simplicial map, then im φ is a subcomplex of L; moreover, if K is connected, then im φ is connected.

11.35. If K is a complex and $\mathscr{P} = \{X_i: i \in I\}$ is a partition of its vertices, then the ***natural map*** $\nu: K \to K/\mathscr{P}$, which sends each vertex into the unique X_i containing it, is a simplicial map.

11.36. Let K be a connected complex, and let L be a subcomplex that is a disjoint union of trees. Show that there is a maximal tree of K containing L. (*Hint*. Use Exercise 11.35.)

Tietze's Theorem

Tietze's theorem gives a presentation for the fundamental group of a connected complex.

Definition. If T is a maximal tree in a connected complex K, then $\mathscr{T}(K, T)$ is the group having the presentation:

generators: all edges (u, v) in K;
relations: Type (a): $(u, v) = 1$ if (u, v) is an edge in T;
Type (b): $(u, v)(v, x) = (u, x)$ if $\{u, v, x\}$ is a simplex in K.

Theorem 11.31 (H. Tietze, 1908). *If K is a connected complex and T is a maximal tree in K, then*
$$\pi(K, w) \cong \mathscr{T}(K, T).$$

Remark. Since K is connected, different choices of basepoint w for K yield isomorphic groups.

Proof. Let F be the free group with basis X = all edges (u, v) in K and let R be the normal subgroup of relations, so that $\mathscr{T}(K, T) = F/R$.

Since T is a maximal tree in K, there is a unique reduced path λ_v in T from w to each $v \in \text{Vert}(T) - \{w\} = \text{Vert}(K) - \{w\}$; define $\lambda_w = (w, w)$. Define a function $f: X \to \pi(K, w)$ by

$$(u, v) \mapsto [\lambda_u(u, v)\lambda_v^{-1}]$$

(which is the path class of a closed path at w), and let $\varphi: F \to \pi(K, w)$ be the unique homomorphism it defines. We claim that $R \leq \ker \varphi$.

Type (a): If (u, v) is in T, then the path $\lambda_u(u, v)\lambda_v^{-1}$ lies in T, and hence $[\lambda_u(u, v)\lambda_v^{-1}] = 1$ in $\pi(K, w)$, by Theorem 11.29.

Type (b): If $\{u, v, x\}$ is a simplex in K, then

$$\begin{aligned}[\lambda_u(u, v)\lambda_v^{-1}][\lambda_v(v, x)\lambda_x^{-1}] &= [\lambda_u(u, v)\lambda_v^{-1}\lambda_v(v, x)\lambda_x^{-1}] \\ &= [\lambda_u(u, v)(v, x)\lambda_x^{-1}] \\ &= [\lambda_u(u, x)\lambda_x^{-1}].\end{aligned}$$

Therefore, φ induces a homomorphism $\Phi: \mathscr{T}(K, T) \to \pi(K, w)$ with

$$\Phi: (u, v)R \mapsto [\lambda_u(u, v)\lambda_v^{-1}].$$

We prove that Φ is an isomorphism by constructing its inverse. If $\varepsilon_1 \ldots \varepsilon_n$ is a closed path in K at w, define

$$\theta(\varepsilon_1 \ldots \varepsilon_n) = \varepsilon_1 \ldots \varepsilon_n R \in \mathscr{T}(K, T).$$

Observe that if α and β are homotopic closed paths, then the relations in $\mathscr{T}(K, T)$ of Type (b) show that $\theta(\alpha) = \theta(\beta)$: for example, if $\alpha = \gamma(u, v)(v, x)\delta$ and $\beta = \gamma(u, x)\delta$, where $\{u, v, x\}$ is a simplex in K, then

$$\beta^{-1}\alpha = \delta^{-1}(u, x)^{-1}(u, v)(v, x)\delta \in R.$$

There is thus a homomorphism $\Theta: \pi(K, w) \to \mathscr{T}(K, T)$ given by

$$\Theta: [\varepsilon_1 \ldots \varepsilon_n] \mapsto \theta(\varepsilon_1 \ldots \varepsilon_n) = \varepsilon_1 \ldots \varepsilon_n R.$$

Let us compute composites. If $[\varepsilon_1 \ldots \varepsilon_n] \in \pi(K, w)$, then

$$\begin{aligned}\Phi(\varepsilon_1 \ldots \varepsilon_n R) &= \Phi(\varepsilon_1 R) \ldots \Phi(\varepsilon_n R) \\ &= [\varphi(\varepsilon_1) \ldots \varphi(\varepsilon_n)] \\ &= [\lambda_w \varepsilon_1 \ldots \varepsilon_n \lambda_w^{-1}] \\ &= [\varepsilon_1 \ldots \varepsilon_n],\end{aligned}$$

because λ_w is a trivial path. Therefore, $\Phi\Theta$ is the identity. We now compute the other composite. If (u, v) is an edge in K, then

$$\begin{aligned}\Theta\Phi((u, v)R) &= \Theta(\varphi(u, v)) \\ &= \Theta([\lambda_u(u, v)\lambda_v^{-1}]) \\ &= \lambda_u(u, v)\lambda_v^{-1} R.\end{aligned}$$

But λ_u and λ_v^{-1} lie in R, since their edges do, so that

$$\lambda_u(u, v)\lambda_v^{-1}R = \lambda_u(u, v)R = (u, v)R,$$

the last equation arising from the normality of R (Exercise 2.30(ii)). The composite $\Theta\Phi$ thus fixes a generating set of $\mathscr{T}(K, T)$, hence it is the identity. Therefore, $\pi(K, w) \cong \mathscr{T}(K, T)$. ■

Corollary 11.32. *If K is a connected 1-complex, then $\pi(K, w)$ is a free group. Moreover, if T is a maximal tree in K, then*

$$\text{rank } \pi(K, w) = |\{1\text{-simplexes in } K \text{ not in } T\}|;$$

indeed, F has a basis consisting of all $[\lambda_u(u, v)\lambda_v^{-1}]$, where one edge (u, v) is chosen from each 1-simplex $\{u, v\} \notin T$.

Proof. By the theorem, it suffices to examine $\mathscr{T}(K, T)$. The relations of Type (a) show that $\mathscr{T}(K, T)$ is generated by those edges (u, v) in K which are not in T. A smaller generating set is obtained by discarding, for each 1-simplex $\{u, v\}$, one of the two edges (u, v) or (v, u), for $(v, u)R = (u, v)^{-1}R$ in $\mathscr{T}(K, T)$.

If $\{u, v, w\}$ is a simplex in K, then at least two of the vertices are equal, for $\dim K = 1$. Thus, the relations of Type (b) have the form:

$$(u, u)(u, v) = (u, v),$$

$$(u, v)(v, v) = (v, v),$$

$$(u, v)(v, u) = (u, u).$$

All of these are trivial: since $(v, v) = 1$ and $(u, v) = (v, u)^{-1}$, the subgroup R of relations is 1. Therefore, $\mathscr{T}(K, T)$ is free with basis as described. ■

Here is a simple example of a complex whose fundamental group is free.

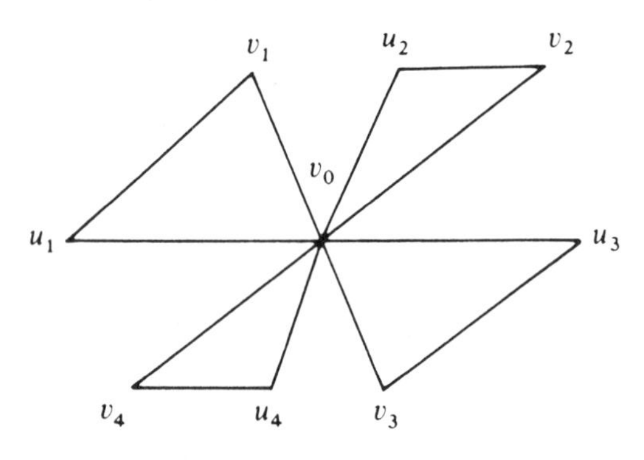

Figure 11.4

Definition. If I is a set, then a ***bouquet of*** $|I|$ ***circles*** is the 1-complex B_I with distinct vertices $\{u_i, v_i: i \in I\} \cup \{w\}$ and 1-simplexes $\{w, u_i\}$, $\{w, v_i\}$, and $\{v_i, u_i\}$ for all $i \in I$.

Corollary 11.33. *If I is a set and B_I is a bouquet of I circles, then $\pi(B_I, w)$ is a free group of rank $|I|$.*

Proof. It is easy to see that B_I is a connected 1-complex and that all the 1-simplexes containing w form a maximal tree. Therefore, $\pi(B_I, w)$ is free with basis

$$\{[(w, u_i)(u_i, v_i)(v_i, w)]: i \in I\}. \quad \blacksquare$$

In Exercise 11.14, we gave a presentation of the trivial group $G = 1$. Using Tietze's theorem, we see that fundamental groups of simply connected complexes produce many such examples.

Covering Complexes

The last section associated a group to a complex; in this section, we associate a complex to a group.

Definition. Let $p: K \to K'$ be a simplicial map. If L' is a subcomplex of K', then its ***inverse image*** is

$$p^{-1}(L') = \{\text{simplexes } s \in K: p(s) \in L'\}.$$

It is easy to check that $p^{-1}(L')$ is a subcomplex of K which is full if L' is full. In particular, if s is a simplex in K', then the subcomplex $|s|$, consisting of s and all its nonempty subsets, is full, and so $p^{-1}(|s|)$ is a full subcomplex of K.

In what follows, we will write s instead of $|s|$.

Definition. Let K be a complex. A connected complex $\tilde{K}$ is a ***covering complex*** of K if there is a simplicial map $p: \tilde{K} \to K$ such that, for every simplex s in K, the inverse image $p^{-1}(s)$ is a disjoint union of simplexes,

$$p^{-1}(s) = \bigcup_{i \in I} \tilde{s}_i,$$

with $p|\tilde{s}_i: \tilde{s}_i \to s$ an isomorphism for each $i \in I$. The map p is called the ***projection*** and the simplexes $\tilde{s}_i$ in $\tilde{K}$ are called the ***sheets*** over s.

If K has a covering space, then K must be connected, for a projection $p: \tilde{K} \to K$ is a surjection, and Exercise 11.28 shows that the image of a connected complex is connected.

Notice that every simplex $\tilde{s}$ in $\tilde{K}$ is isomorphic to a simplex in K; it follows that $\dim \tilde{K} = \dim K$.

The picture to keep in mind is

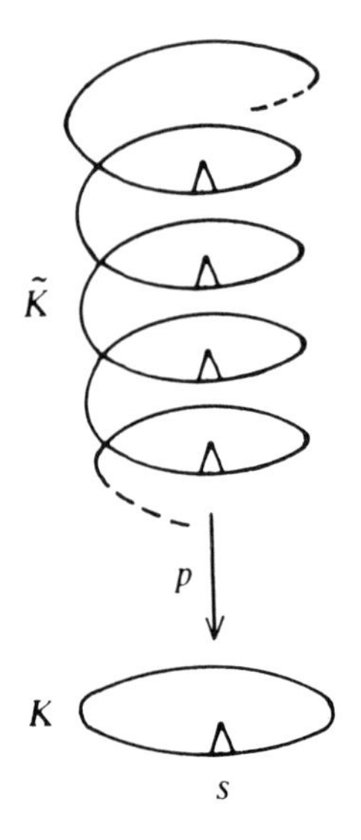

Figure 11.5

EXAMPLE 11.8. Let K be the "circle" having vertices $\{v_0, v_1, v_2\}$ and 1-simplexes $\{v_0, v_1\}$, $\{v_1, v_2\}$, and $\{v_0, v_2\}$, and let $\tilde{K}$ be the "line" having vertices $\{t_i: i \in \mathbb{Z}\}$ and 1-simplexes $\{t_i, t_{i+1}\}$ for all $i \in \mathbb{Z}$. Define $p: \tilde{K} \to K$ by $p(t_i) = v_j$, where $j \equiv i \bmod 3$. The reader may check that $p: \tilde{K} \to K$ is a covering complex.

Theorem 11.34. *Let $p: \tilde{K} \to K$ be a covering complex, and let $\tilde{w}$ be a basepoint in $\tilde{K}$ with $p(\tilde{w}) = w$. Given a path α in K with origin w, there exists a unique path $\tilde{\alpha}$ in $\tilde{K}$ with origin $\tilde{w}$ and with $p\tilde{\alpha} = \alpha$.*

Remark. One calls $\tilde{\alpha}$ the ***lifting*** of α because of the picture

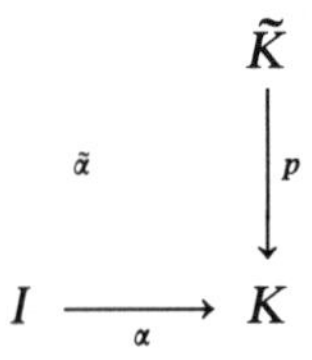

Proof. We prove the lemma by induction on $n = $ length α. If $n = 1$, then $\alpha = (w, v)$, where $s = \{w, v\}$ is a simplex; we may assume that $v \neq w$, so that s is a 1-simplex. If $\tilde{s}$ is the sheet over s containing $\tilde{w}$, then $\tilde{s} = \{\tilde{w}, \tilde{v}\}$ is a

1-simplex, $(\tilde{w}, \tilde{v})$ is an edge, and $p(\tilde{w}, \tilde{v}) = (w, v)$. To prove uniqueness, suppose that $(\tilde{w}, \tilde{u})$ is also a lifting (so that $\{\tilde{w}, \tilde{u}\}$ is a 1-simplex). Then $\{\tilde{w}, \tilde{v}\}$ and $\{\tilde{w}, \tilde{u}\}$ are sheets over $\{w, v\}$ that are not disjoint, contradicting the definition of covering complex. Therefore, $\tilde{u} = \tilde{v}$.

If $n > 1$, then $\alpha = (w, v)\beta$, where β is a path of length $n - 1$ beginning at v. By the base step, there is a unique $(\tilde{w}, \tilde{v})$ lifting (w, v); by induction, there is a unique lifting $\tilde{\beta}$ of β beginning at $\tilde{v}$. Thus, $(\tilde{w}, \tilde{v})\tilde{\beta}$ is the unique lifting of α beginning at $\tilde{w}$. ∎

Lemma 11.35. *Let $p: \tilde{K} \to K$ be a covering complex, and let $\tilde{w}$ be a basepoint in $\tilde{K}$ with $p(\tilde{w}) = w$. If α and β are homotopic paths in K with origin w, then their liftings $\tilde{\alpha}$ and $\tilde{\beta}$ having origin $\tilde{w}$ are homotopic and $e(\tilde{\alpha}) = e(\tilde{\beta})$.*

Proof. It suffices to prove that if $(\tilde{u}, \tilde{v})(\tilde{v}, \tilde{x})$ is a lifting of $(u, v)(v, x)$ and if $s = \{u, v, x\}$ is a simplex in K, then $\{\tilde{u}, \tilde{v}, \tilde{x}\}$ is a simplex in $\tilde{K}$. Let $\tilde{s} = \{\tilde{v}, \tilde{u}', \tilde{x}'\}$ be the sheet over s containing $\tilde{v}$ and let $\tilde{t} = \{\tilde{u}, \tilde{v}'', \tilde{x}''\}$ be the sheet over s containing $\tilde{u}$, where $p\tilde{v} = p\tilde{v}'' = v$ and $p\tilde{u} = p\tilde{u}' = u$. Now $(\tilde{u}, \tilde{v})$ and $(\tilde{u}, \tilde{v}'')$ are liftings of (u, v) beginning at $\tilde{u}$, so that uniqueness of lifting gives $\tilde{v} = \tilde{v}''$. The sheets $\tilde{s}$ and $\tilde{t}$ over s are not disjoint, and so $\tilde{s} = \tilde{t}$; that is, $\tilde{v} = \tilde{v}''$, $\tilde{u}' = \tilde{u}$, and $\tilde{x}' = \tilde{x}''$. A similar argument comparing $\tilde{s}$ with the sheet over s containing $\tilde{x}$ shows that $\tilde{x} = \tilde{x}'$. ∎

Theorem 11.36. *Let $p: \tilde{K} \to K$ be a covering complex, and let $\tilde{w}$ be a basepoint in $\tilde{K}$ with $p(\tilde{w}) = w$. Then $p_{\#}: \pi(\tilde{K}, \tilde{w}) \to \pi(K, w)$ is an injection.*

Proof. Assume that $[A], [B] \in \pi(\tilde{K}, \tilde{w})$ and that $p_{\#}[A] = p_{\#}[B]$; that is, $[pA] = [pB]$. If $\alpha = pA$ and $\beta = pB$, then $A = \tilde{\alpha}$ and $B = \tilde{\beta}$. The hypothesis gives $\alpha \simeq \beta$, and so Lemma 11.35 gives $A = \tilde{\alpha} \simeq \tilde{\beta} = B$; that is, $[A] = [B]$. ∎

What happens to the subgroup $p_{\#}\pi(\tilde{K}, \tilde{w})$ as the basepoint is changed?

Theorem 11.37. *Let $p: \tilde{K} \to K$ be a covering complex, and let $\tilde{w}$ be a basepoint in $\tilde{K}$ with $p(\tilde{w}) = w$. If $p(\tilde{u}) = w$, then $p_{\#}\pi(\tilde{K}, \tilde{w})$ and $p_{\#}\pi(\tilde{K}, \tilde{u})$ are conjugate subgroups of $\pi(K, w)$. Conversely, if $H \leq \pi(K, w)$ is conjugate to $p_{\#}\pi(\tilde{K}, \tilde{w})$, then $H = p_{\#}\pi(\tilde{K}, \tilde{u})$ for some $\tilde{u}$ with $p(\tilde{u}) = w$.*

Proof. Since $\tilde{K}$ is connected, there is a path B from $\tilde{w}$ to $\tilde{u}$. Then $\beta = pB$ is a closed path at w, $[\beta] \in \pi(K, w)$, and Theorem 11.27(ii) gives

$$[B^{-1}]\pi(\tilde{K}, \tilde{w})[B] = \pi(\tilde{K}, \tilde{u});$$

hence

$$[\beta^{-1}]p_{\#}\pi(\tilde{K}, \tilde{w})[\beta] = p_{\#}\pi(\tilde{K}, \tilde{u}).$$

Conversely, assume that $H = [\alpha]p_{\#}\pi(\tilde{K}, \tilde{w})[\alpha]^{-1}$. If $\tilde{\beta}$ is the lifting of α^{-1}

with origin $\tilde{w}$, and if $e(\tilde{\beta}) = \tilde{u}$, then $p(\tilde{u}) = w$. By Theorem 11.27(ii),

$$[\beta^{-1}]\pi(\tilde{K}, \tilde{w})[\beta] = \pi(\tilde{K}, \tilde{u}),$$

so that $p_{\#}\pi(\tilde{K}, \tilde{u}) = p_{\#}([\tilde{\beta}^{-1}]\pi(\tilde{K}, \tilde{w})[\tilde{\beta}]) = [\alpha]p_{\#}\pi(\tilde{K}, \tilde{w})[\alpha]^{-1} = H$, as desired. ■

Definition. If $p: \tilde{K} \to K$ is a simplicial map and if w is a vertex in K, then $p^{-1}(w)$ is called the ***fiber*** over w.

In the next theorem, we observe that the fundamental group $G = \pi(K, w)$ acts on the fiber $p^{-1}(w)$; more precisely, $p^{-1}(w)$ is a right G-set. Recall that one can always convert a right G-set X into a left action by defining gx to be xg^{-1}.

Theorem 11.38. *Let $p: \tilde{K} \to K$ be a covering complex and let w be a vertex in K. Then the fiber $p^{-1}(w)$ is a transitive (right) $\pi(K, w)$-set and the stabilizer of a point $\tilde{w}$ is $p_{\#}\pi(\tilde{K}, \tilde{w})$.*

Proof. If $\tilde{x} \in p^{-1}(w)$ and $[\alpha] \in \pi(K, w)$, define $\tilde{x}[\alpha] = e[\tilde{\alpha}]$, where $\tilde{\alpha}$ is the lifting of α having origin $\tilde{x}$; since homotopic paths have the same end, Lemma 11.35 shows that this definition does not depend on the choice of path in $[\alpha]$.

We now verify the axioms of a G-set. The identity element of $\pi(K, w)$ is $[(w, w)]$; the lifting of (w, w) with origin $\tilde{x}$ is obviously $(\tilde{x}, \tilde{x})$ whose end is $\tilde{x}$. Let $[\alpha], [\beta] \in \pi(K, w)$, let $\tilde{\alpha}$ be the lifting of α having origin $\tilde{x}$, and let $\tilde{y} = e(\tilde{\alpha})$. If $\tilde{\beta}$ is the lifting of β with origin $\tilde{y}$, then $\tilde{\alpha}\tilde{\beta}$ is a lifting of $\alpha\beta$ with origin $\tilde{x}$. By uniqueness of lifting, $\tilde{\alpha}\tilde{\beta}$ is the lifting of $\alpha\beta$ having origin $\tilde{x}$, and so $\tilde{x}[\alpha\beta] = e[\tilde{\alpha}\tilde{\beta}] = e[\tilde{\beta}]$. On the other hand, $\tilde{x}[\alpha][\beta] = (e[\tilde{\alpha}])[\beta] = \tilde{y}[\tilde{\beta}] = e[\tilde{\beta}]$.

Now $\pi(K, w)$ acts transitively: if $\tilde{x}, \tilde{y} \in p^{-1}(w)$, then connectivity of $\tilde{K}$ shows there is a path A in $\tilde{K}$ from $\tilde{x}$ to $\tilde{y}$. If $\alpha = pA$, then $[\alpha] \in \pi(K, w)$ and $\tilde{x}[\alpha] = e[A] = \tilde{y}$.

Finally, the stabilizer of a point $\tilde{x} \in p^{-1}(w)$ consists of all $[\alpha] \in \pi(K, w)$ for which $e[\tilde{\alpha}] = \tilde{x}$. But $e[\tilde{\alpha}] = \tilde{x}$ if and only if $[\tilde{\alpha}] \in \pi(\tilde{K}, \tilde{w})$ if and only if $[\alpha] \in p_{\#}\pi(\tilde{K}, \tilde{w})$. ■

Corollary 11.39. *Let $p: \tilde{K} \to K$ be a covering complex.*

(i) *If w is a basepoint in K and $\tilde{w} \in p^{-1}(w)$, then*

$$[\pi(K, w): p_{\#}\pi(\tilde{K}, \tilde{w})] = |p^{-1}(w)|.$$

(ii) *If w and u are basepoints in K, then $|p^{-1}(w)| = |p^{-1}(u)|$.*

Proof. (i) The number of elements in a transitive G-set is the index of the stabilizer of a point.

(ii) If $\tilde{u} \in p^{-1}(u)$, then there is a path B in $\tilde{K}$ from $\tilde{w}$ to $\tilde{u}$; let $\beta = pB$. Define homomorphisms $\Phi: \pi(\tilde{K}, \tilde{w}) \to \pi(\tilde{K}, \tilde{u})$ and $\varphi: \pi(K, w) \to \pi(K, u)$ by $[A] \mapsto [B^{-1}AB]$ and $[\alpha] \mapsto [\beta^{-1}\alpha\beta]$, respectively. It is easy to check that the following diagram commutes:

$$\begin{array}{ccc} \pi(\tilde{K}, \tilde{w}) & \xrightarrow{\Phi} & \pi(\tilde{K}, \tilde{u}) \\ \downarrow{\scriptstyle p_\#} & & \downarrow{\scriptstyle p_\#} \\ \pi(K, w) & \xrightarrow[\varphi]{} & \pi(K, u). \end{array}$$

Since Φ and φ are isomorphisms, it follows that the index of im $p_\#$ on the left is equal to the index of im $p_\#$ on the right. ∎

We are now going to construct some covering complexes.

Definition. Let K be a complex with basepoint w and let π be a subgroup of $\pi(K, w)$. Define a relation on the set of all paths in K having origin w by

$$\alpha \equiv_\pi \beta \qquad \text{if} \quad e(\alpha) = e(\beta) \quad \text{and} \quad [\alpha\beta^{-1}] \in \pi.$$

Notation. It is easy to see that $\equiv_\pi$ is an equivalence relation. Denote the equivalence class of a path α by $\mathscr{C}\alpha$, and denote the family of all such classes by K_π.

We now make K_π into a complex. Let s be a simplex in K, and let α be a path in K with $o(\alpha) = w$ and $e(\alpha) \in s$. A ***continuation*** of α in s is a path $\beta = \alpha\alpha'$, where α' is a path wholly in s. Define a ***simplex*** in K_π to be

$$[s, \mathscr{C}\alpha] = \{\mathscr{C}\beta \colon \beta \text{ is a continuation of } \alpha \text{ in } s\},$$

where s is a simplex in K and α is a path from w to a vertex in s. Thus, $\mathscr{C}\beta \in [s, \mathscr{C}\alpha]$ if and only if there is a path α' wholly in s with $\beta \equiv_\pi \alpha\alpha'$.

Lemma 11.40. *Let K be a connected complex, let w be a basepoint in K, and let $\pi \le \pi(K, w)$. Then K_π is a complex and the function $p\colon K_\pi \to K$, defined by $\mathscr{C}\alpha \mapsto e(\alpha)$, is a simplicial map.*

Proof. Straightforward. ∎

Define $\tilde{w} = \mathscr{C}(w, w) \in K_\pi$, and choose $\tilde{w}$ as a basepoint.

Lemma 11.41. *If K is a connected complex, then every path α in K with origin w can be lifted to a path A in K_π from $\tilde{w}$ to $\mathscr{C}\alpha$.*

Proof. Let $\alpha = (w, v_1)(v_1, v_2)\dots(v_{n-1}, v)$, and define "partial paths" $\alpha_i = (w, v_1)(v_1, v_2)\dots(v_{i-1}, v_i)$ for $i \ge 1$. If s_i is the simplex $\{v_i, v_{i+1}\}$, then both $\mathscr{C}\alpha_i$ and $\mathscr{C}\alpha_{i+1}$ lie in $[s_i, \mathscr{C}\alpha_i]$, so that $(\mathscr{C}\alpha_i, \mathscr{C}\alpha_{i+1})$ is an edge in K_π. Therefore,

$$A = (\tilde{w}, \mathscr{C}\alpha_1)(\mathscr{C}\alpha_1, \mathscr{C}\alpha_2)\dots(\mathscr{C}\alpha_{n-1}, \mathscr{C}\alpha)$$

is a lifting of α having origin $\tilde{w}$. ∎

Corollary 11.42. *If K is a connected complex, then K_π is a connected complex.*

Proof. There is a path in K_π from $\tilde{w}$ to every $\mathscr{C}\alpha$. ■

Theorem 11.43. *If K is a connected complex with basepoint w, and if $\pi \leq \pi(K, w)$, then $p\colon K_\pi \to K$ is a covering complex with $p_\#(K_\pi, \tilde{w}) = \pi$.*

Proof. If s is a simplex in K, then $p' = p|[s, \mathscr{C}\alpha]$ is an isomorphism $[s, \mathscr{C}\alpha] \to s$ if and only if it is a bijection. To see that p' is an injection, suppose that $\mathscr{C}\beta, \mathscr{C}\gamma \in [s, \mathscr{C}\alpha]$ and $e(\beta) = p(\mathscr{C}\beta) = p(\mathscr{C}\gamma) = e(\gamma)$. Thus, $\beta = \alpha\beta'$ and $\gamma = \alpha\gamma'$, where β' and γ' are paths wholly in s, and

$$\beta\gamma^{-1} = \alpha\beta'\gamma'^{-1}\alpha^{-1} \simeq \alpha\alpha^{-1}$$

($[\alpha][\beta'\gamma'^{-1}][\alpha^{-1}] \in [\alpha]\pi(s, e(\alpha))[\alpha^{-1}] = 1$, by Theorem 11.29). Thus, $[\beta\gamma^{-1}] = 1 \in \pi$, $\beta \equiv_\pi \gamma$, $\mathscr{C}\beta = \mathscr{C}\gamma$, and p' is an injection. To see that p' is a surjection, let v be a vertex in s and let α' be a path in s from $e(\alpha)$ to v. Define $\beta = \alpha\alpha'$, and note that $p(\mathscr{C}\beta) = e(\beta) = v$.

If s is a simplex in K, then it is easy to see that

$$p^{-1}(s) = \bigcup_{\mathscr{C}\alpha \in K_\pi} [s, \mathscr{C}\alpha]$$

as sets; but since s is a full subcomplex of K, its inverse image is a full subcomplex of K_π, and so it is completely determined by its vertices. To prove that $p\colon K_\pi \to K$ is a covering complex, it remains to prove that the sheets are pairwise disjoint. If $\mathscr{C}\gamma \in [s, \mathscr{C}\alpha] \cap [s, \mathscr{C}\beta]$, then $\gamma \equiv_\pi \alpha\alpha'$ and $\gamma \equiv_\pi \beta\beta'$, where α' and β' lie wholly in s. It follows that $e(\alpha') = e(\beta')$ and, as in the preceding paragraph, $[\alpha\alpha'\beta'^{-1}\beta^{-1}] = [\alpha\beta^{-1}]$. But $[\alpha\alpha'\beta'^{-1}\beta^{-1}] = [\gamma\gamma^{-1}] = 1 \in \pi$, so that $\alpha \equiv_\pi \beta$, $\mathscr{C}\alpha = \mathscr{C}\beta$, and $[s, \mathscr{C}\alpha] = [s, \mathscr{C}\beta]$.

Finally, we show that $p_\#\pi(K_\pi, \tilde{w}) = \pi$. If α is a closed path in K at w, then $p\colon K_\pi \to K$ being a covering complex implies, by Theorem 11.34, that there is a unique lifting $\tilde{\alpha}$ of α having origin $\tilde{w}$. But we constructed such a lifting A in Lemma 11.41, so that $\tilde{\alpha} = A$ and $e(\tilde{\alpha}) = e(A) = \mathscr{C}\alpha$. The following statements are equivalent: $[\alpha] \in p_\#\pi(K_\pi, \tilde{w})$; $[\alpha] = [pA]$, where $[A] \in \pi(K_\pi, \tilde{w})$; $e(A) = o(A) = \tilde{w}$; $\mathscr{C}\alpha = \tilde{w}$; $[\alpha(w, w)^{-1}] \in \pi$; $[\alpha] \in \pi$. This completes the proof. ■

Remark. Here is a sketch of the analogy with Galois Theory. If $p\colon \tilde{K} \to K$ is a covering complex, then a simplicial map $\varphi\colon \tilde{K} \to \tilde{K}$ is called a ***covering map*** (or *deck transformation*) if $p\varphi = p$. The set $\text{Cov}(\tilde{K}/K)$ of all covering maps is a group under composition.

Recall that if k is a subfield of a field K, then the Galois group $\text{Gal}(K/k)$ is defined as the set of all automorphisms $\sigma\colon K \to K$ which fix k pointwise. To say it another way, if $i\colon k \hookrightarrow K$ is the inclusion, then $\text{Gal}(K/k)$ consists of all those automorphisms σ of K for which $\sigma i = i$. In the analogy, therefore, all arrows are reversed.

Every connected complex K has a covering complex $p\colon U \to K$ with

U simply connected (Exercise 11.34 below), and it can be shown that $\text{Cov}(U/K) \cong \pi(K, w)$ (note that $\text{Cov}(U/K)$ is defined without choosing a basepoint). It is true that U is a ***universal covering complex*** in the sense that whenever $q: \tilde{K} \to K$ is a covering complex, then there is a projection $r: U \to \tilde{K}$ which is a covering complex of $\tilde{K}$. (One may thus regard U as the analogue of the algebraic closure $\bar{k}$ of a field k, for every algebraic extension of k can be imbedded in $\bar{k}$.) Moreover, the function $\tilde{K} \mapsto \text{Cov}(U/\tilde{K})$ is a bijection from the family of all covering complexes of K to the family of all subgroups of the fundamental group $\pi(K, w)$. For proofs of these results, the reader is referred to my expository paper, *Rocky Mountain Journal of Mathematics*, Covering complexes with applications to algebra, **3** (1973), 641–674.

EXERCISES

11.37. Let $p: \tilde{K} \to K$ be a covering complex. If L is a connected subcomplex of K and if $\tilde{L}$ is a component of $p^{-1}(L)$, then $p|\tilde{L}: \tilde{L} \to L$ is a covering complex.

11.38. (i) if $p: \tilde{K} \to K$ is a covering complex and T is a tree in K, then $p^{-1}(T)$ is a disjoint union of trees. (*Hint*: Show that every component of $p^{-1}(T)$ is a tree.)
(ii) Show that there is a maximal tree of $\tilde{K}$ containing $p^{-1}(T)$. (*Hint*: Use Exercise 11.30.)

11.39. Let $p: \tilde{K} \to K$ be a covering complex, and suppose that there are j points in every fiber $p^{-1}(v)$, where v is a vertex in K. Show that there are exactly j sheets over each simplex s in K.

11.40. (i) Every connected complex K has a universal covering space $p: \tilde{K} \to K$; that is, $\tilde{K}$ is simply connected. (*Hint*. Let π be the trivial subgroup of $\pi(K, w)$.)
(ii) If K is a connected 1-complex, then its universal covering complex is a tree. (This last result may well have been the impetus for the "Bass–Serre theory" of groups acting on trees (see Serre, 1980).

The Nielsen–Schreier Theorem

In 1921, J. Nielsen proved that every finitely generated subgroup H of a free group F is itself free (he also gave an algorithm deciding whether or not a word α in F lies in H). In 1927, O. Schreier eliminated the hypothesis that H be finitely generated. There are, today, several different proofs of this theorem, some "algebraic" and some "geometric." The first geometric proof was given by R. Baer and F.W. Levi in 1936, and this is the proof we present. There is another elegant geometric proof, due to J.-P. Serre (1970), using the notion of groups acting on trees.

Theorem 11.44 (Nielsen–Schreier). *Every subgroup H of a free group F is itself free.*

Proof. If F has rank $|I|$, and if K is a bouquet of $|I|$ circles, then Corollary 11.33 allows us to identify F with $\pi(K, w)$. As in Theorem 11.43, there is a covering complex $p: K_H \to K$ with $p_{\#}\pi(K_H, \tilde{w}) = H$. Now $p_{\#}$ is an injection, by Theorem 11.36, so that $H \cong \pi(K_H, \tilde{w})$. But $\dim(K) = 1$ implies $\dim(K_H) = 1$, and so $\pi(K_H, \tilde{w})$ is free, by Corollary 11.32. ■

Theorem 11.45. *If F is a free group of finite rank n and H is a subgroup of finite index j, then* $\mathrm{rank}(H) = jn - j + 1$.

Proof. If K is a finite connected graph, denote the number of its vertices by $v(K)$ and the number of its 1-simplexes by $e(T)$. If T is a maximal tree in K, the $v(T) = v(K)$, by Lemma 11.30. We saw in Exercise 11.23 that $v(T) - e(T) = 1$. By Lemma 11.30, the number of 1-simplexes in $K - T$ is

$$e(K - T) = e(K) - e(T) = e(K) - v(T) + 1 = e(K) - v(K) + 1.$$

If, now, B_n is a bouquet of n circles, then $v(B_n) = 2n + 1$ and $e(B_n) = 3n$. After identifying F with $\pi(B_n, w)$, let $p: K_H \to B_n$ be the covering complex corresponding to H. By Corollary 11.39, $j = [F : H] = |p^{-1}(w)|$. Therefore, $v(K_H) = jv(B_n)$ and, by Exercise 11.33, $e(K_H) = je(B_n)$. We compute the number of 1-simplexes in K_H outside a maximal tree T:

$$\begin{aligned} e(K_H - T) &= e(K_H) - v(K_H) + 1 \\ &= je(B_n) - jv(B_n) + 1 \\ &= 3jn - j(2n + 1) + 1 \\ &= jn - j + 1. \end{aligned}$$

Corollary 11.32 completes the proof. ■

Remark. If K is an n-complex and b_i denotes the number of its i-simplexes, then its ***Euler–Poincaré characteristic*** $\chi(K)$ is $\sum_{i=0}^{n}(-1)^i b_i$. Thus, $\mathrm{rank}(H) = 1 - \chi(K_H)$.

We have shown that a subgroup H of a free group F is free; can we find a basis of H?

Recall that a right transversal of a subgroup H in a group F consists of one element chosen from each right coset Ha; denote the chosen element by $l(Ha)$. If F is free with basis X and if $x \in X$, then both $l(Hax)$ and $l(Ha)x$ lie in Hax, and so the element $h_{a,x}$ defined by

$$h_{a,x} = l(Ha)xl(Hax)^{-1}$$

lies in H (these elements should be indexed by (Ha, x), but we have abbreviated this to (a, x).) We are going to see that if a transversal of H in F is chosen nicely, then the set of all nontrivial $h_{a,x}$ is a basis of H.

Here is a notion that arose in Schreier's proof of the subgroup theorem.

Definition. Let F be a free group with basis X, and let H be a subgroup of F. A ***Schreier transversal*** of H in F is a right transversal S with the property that whenever $x_1^{\varepsilon_1}x_2^{\varepsilon_2}\ldots x_n^{\varepsilon_n}$ lies in S (where $x_i \in X$ and $\varepsilon_i = \pm 1$), then every initial segment $x_1^{\varepsilon_1}x_2^{\varepsilon_2}\ldots x_k^{\varepsilon_k}$ (for $k \leq n$) also lies in S.

We will soon prove the existence of Schreier transversals.

Lemma 11.46. *Let $p\colon \tilde{K} \to K$ be a covering complex and let T be a maximal tree in K. If w is a basepoint in K, then each component of $p^{-1}(T)$ meets the fiber $p^{-1}(w)$.*

Proof. If C were a component of $p^{-1}(T)$ disjoint from the fiber, then $w \notin p(C)$. There would then exist an edge (u, v) with $u \in p(C)$ and $v \notin p(C)$; of course, $v \in \mathrm{Vert}(T) = \mathrm{Vert}(K)$. By Theorem 11.34, this edge may be lifted to an edge $(\tilde{u}, \tilde{v})$, where $\tilde{u} \in C$ and $\tilde{v} \notin C$. Since $\tilde{v} \in p^{-1}(T)$, this contradicts C being a maximal connected subcomplex of $p^{-1}(T)$. ■

Theorem 11.47. *Let F be a free group with basis X, and let H be a subgroup of F. There exists a Schreier transversal $S = \{l(Ha)\colon a \in F\}$; moreover, a basis for H consists of all those $h_{a,x} = l(Ha)xl(Hax)^{-1}$ that are distinct from 1, where $x \in X$.*

Proof.[5] Identify F with $\pi(K, w)$, where K is a bouquet of circles, and let T be a maximal tree in K. Let $p\colon K_H \to K$ be the covering complex corresponding to H and choose some $\tilde{w} \in p^{-1}(w)$ as the basepoint in K_H.

Let $\tilde{T}$ be a maximal tree in K_H containing $p^{-1}(T)$ (whose existence is guaranteed by Exercise 11.32(ii)). For each vertex V in K_H, there is a unique reduced path λ_V in $\tilde{T}$ from $\tilde{w}$ to V. In particular, there are such paths for every $V \in p^{-1}(w)$, and $p_\#[\lambda_V] \in \pi(K, w)$. We claim that the family

$$S = \{1\} \cup \{\text{all } p_\#[\lambda_V]\colon V \in p^{-1}(T) \text{ and } \lambda_V \text{ not in } \tilde{T}\}$$

is a right transversal of H in $\pi(K, w)$. Given a coset $H[\alpha]$, let $\tilde{\alpha}$ be the lifting of α having origin $\tilde{w}$. If $V = e(\tilde{\alpha})$, then $V \in p^{-1}(w)$ and $[\tilde{\alpha}\lambda_V^{-1}] \in \pi(K_H, \tilde{w})$.

[5] Here is an algebraic proof of the existence of Schreier transversals. Define the *length* $\lambda(Hg)$ of a coset Hg to be the minimum of the lengths of its elements (relative to a basis X of F). We prove, by induction on $n \geq 0$, that every coset Hg with $\lambda(Hg) \leq n$ has a representative of the form $x_1^{\varepsilon_1}x_2^{\varepsilon_2}\ldots x_n^{\varepsilon_n}$ each of whose initial segments is also a representative (of a coset of shorter length). Begin by choosing 1 to be the representative of H. For the inductive step, let $\lambda(Hg) = n + 1$, and let $ux^\varepsilon \in Hg$ be an element of minimal length, where $\varepsilon = \pm 1$. The coset Hu has length n (or we could find a shorter element of Hg), so that it has a representative v of the desired type, by induction. Now $Hg = (Hu)x^\varepsilon = (Hv)x^\varepsilon$, and so vx^ε is a representative of Hg of the desired type.

An algebraic proof that the nontrivial $h_{a,x}$ form a basis of H, however, does not follow as quickly from the existence of a Schreier transversal as it does in the geometric proof above.

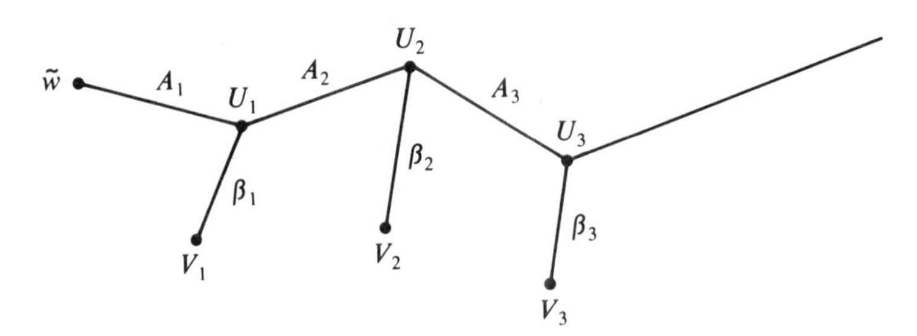

Figure 11.6

Applying $p_\#$ gives $[\alpha]p_\#[\lambda_V^{-1}] \in p_\#\pi(K_H, \tilde{w}) = H$, and so $H[\alpha] = Hp_\#[\lambda_V]$, as desired.

We now show that S is a Schreier transversal. Subdivide λ_V into subpaths: $\lambda_V = A_1 A_2 \dots A_n$, where each A_i contains exactly one edge not in $p^{-1}(T)$ (if there is no such subdivision, then λ_V lies wholly in $p^{-1}(T) \subset \tilde{T}$ and $p_\#[\lambda_V] = 1$). If $U_i = e(\tilde{\alpha}_i)$, then U_i lies in some component C_i of $p^{-1}(T)$. By Lemma 11.46, there is a vertex $V_i \in C_i \cap p^{-1}(w)$ and, since C_i is a tree, there is a unique reduced path β_i in C_i from U_i to V_i. Consider the new path $D_1 D_2 \dots D_n$, where $D_1 = A_1\beta_1$, and $D_i = \beta_{i-1}^{-1} A_i \beta_i$ for $i \geq 2$; of course, $\lambda_V \simeq D_1 D_2 \dots D_n$. For each i, the initial segment $D_1 D_2 \dots D_i$ is a path in the tree $\tilde{T}$ from $\tilde{w}$ to V_i. Since λ_{V_i} is also such a path, these paths are homotopic: $[\lambda_{V_i}] = [D_1 D_2 \dots D_i]$, and so $p_\#[D_1 D_2 \dots D_i]$ lies in S for all i. But Exercise 11.26 shows that each pD_i determines a generator (or its inverse) of $\pi(K, w)$. Therefore, S is a Schreier transversal.

By Corollary 11.32, a basis for $\pi(K_H, \tilde{w})$ is

$$\{[\lambda_U(U, V)\lambda_V^{-1}]: (U, V) \notin \tilde{T}\};$$

since $p_\#$ is an injection, a basis of H is the family of all

$$p_\#[\lambda_U(U, V)\lambda_V^{-1}] = p_\#[D_1 D_2 \dots D_n]p_\#[D_{n+1}]p_\#[D_1 D_2 \dots D_{n+1}]^{-1},$$

where (U, V) is an edge corresponding to a 1-simplex $\{U, V\}$ not in $\tilde{T}$, and $D_{n+1} \simeq \beta(U, V)\beta'$ for paths β and β' having all edges in $\tilde{T}$. ■

Theorem 11.48. *If F is a free group of rank 2, then its commutator subgroup F' is free of infinite rank.*

Proof. If $\{x, y\}$ is a basis of F, then Lemma 11.3 shows that F/F' is free abelian with basis $\{xF', yF'\}$. Therefore, every right coset of $F'a$ has a unique representative of the form $x^m y^n$, where $m, n \in \mathbb{Z}$. The transversal $l(F'a) = l(F'x^m y^n) = x^m y^n$ is a Schreier transversal (e.g., if m and n are positive, write $x^m y^n$ as $x \dots xy \dots y$). If $n > 0$, then $l(F'y^n) = y^n$ while $l(F'y^n x) \neq y^n x$. Therefore $l(F'y^n)xl(F'y^n x)^{-1} \neq 1$ for all $n > 0$, and so there are infinitely many $h_{a,x} \neq 1$. ■

We have seen, in a finitely generated free group F, that a subgroup of finite index is also finitely generated but, in contrast to abelian groups, we now see that there exist subgroups of finitely generated groups that are not finitely generated.

Exercises

11.41. Let G be a noncyclic finite group with $G \cong F/R$, where F is free of finite rank. Prove that $\text{rank}(R) > \text{rank}(F)$.

11.42. Let G have a presentation with n generators and r relations, where $n > r$. Prove that G has an element of infinite order. Conclude that $n \leq r$ when G is finite. (*Hint*. Map a free group on n generators onto a free abelian group on n generators, and use Exercise 10.11.) Equality $n = r$ can occur; for example, the group $\mathbf{Q}$ of quaternions is a finite balanced group.

11.43. Prove that a free group of rank > 1 is not solvable.

11.44. Exhibit infinitely many bases of a free group of rank 2.

11.45. If F is free on $\{x, y\}$, then $\{x, y^{-1}xy, \ldots, y^{-n}xy^n, \ldots\}$ is a basis of the subgroup it generates.

11.46. Prove that a group is free if and only if it has the projective property.

11.47. Use Theorem 11.45 to give another proof of Lemma 7.56: if G is a finitely generated group and H is a subgroup of finite index, then H is finitely generated.

11.48. Show that a finitely generated group G has only finitely many subgroups of any given (finite) index m. (*Hint*. There are only finitely many homomorphisms $\varphi: G \to S_m$ (for there are only finitely many places to send the generators of G). If $H \leq G$ has index m, then the representation of G on the cosets of H is such a homomorphism φ, and $\ker \varphi \leq H$. Apply the correspondence theorem to the finite group $G/\ker \varphi$.)

11.49 **(M. Hall).** If G is a finitely generated group and $H \leq G$ has finite index, then there is $K \leq H$ with $[G : K]$ finite and with K char G. (*Hint*. If $\varphi \in \text{Aut}(G)$, then $[G : \varphi(H)] = [G : H]$. By Exercise 11.48, there are only finitely many subgroups of the form $\varphi(H)$; let K be their intersection.)

11.50. If F is free and $R \lhd F$, then F/R' is torsion-free, where $R' = [R, R]$. (*Hint* (***Rosset***). First reduce to the case F/R cyclic of prime order p. Let $x \in F$ satisfy $x^p \in R'$; if $x \in R$, its coset has finite order in R/R'; if $x \notin R$, then $x \notin F'$ (since $F' \leq R$), and $x^p \notin F'$, hence $x^p \notin R'$.)

11.51. If F is a free group of rank ≥ 2, then its commutator subgroup F' is free of infinite rank.

11.52. Let F be free on $\{x, y\}$, and define $\varphi: F \to S_3$ by $\varphi(x) = (1\ 2)$ and $\varphi(y) = (1\ 2\ 3)$. Exhibit a basis for $\ker \varphi$.

11.53. If F is free on $\{a, b, c, d\}$, prove that $[a, b][c, d]$ is not a commutator.

Free Products

We now generalize the notion of free group to that of free product. As with free groups, free products will be defined with a diagram; that is, they will be defined as solutions to a certain "universal mapping problem." Once existence and uniqueness are settled, then we shall give concrete descriptions of free products in terms of their elements and in terms of presentations.

Definition. Let $\{A_i: i \in I\}$ be a family of groups. A ***free product*** of the A_i is a group P and a family of homomorphisms $j_i: A_i \to P$ such that, for every group G and every family of homomorphisms $f_i: A_i \to G$, there exists a unique homomorphism $\varphi: P \to G$ with $\varphi j_i = f_i$ for all i.

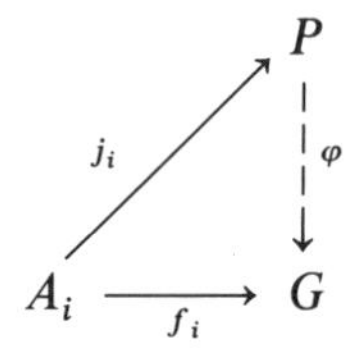

The reader should compare this definition with Theorem 10.9, the analogous property of direct sums of abelian groups.

Lemma 11.49. *If P is a free product of $\{A_i: i \in I\}$, then the homomorphisms j_i are injections.*

Proof. For fixed $i \in I$, consider the diagram in which $G = A_i$, f_i is the identity and, for $k \neq i$, the maps $f_k: A_k \to A_i$ are trivial.

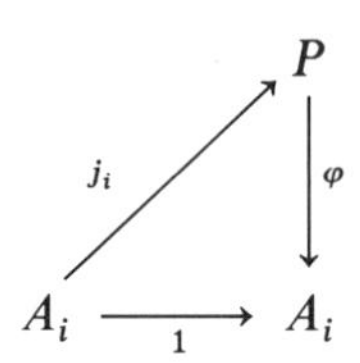

Then $\varphi j_i = 1_{A_i}$, and so j_i is an injection. ∎

In light of this lemma, the maps $j_i: A_i \to P$ are called the ***imbeddings***.

EXAMPLE 11.9. A free group F is a free product of infinite cyclic groups.

If X is a basis of F, then $\langle x \rangle$ is infinite cyclic for each $x \in X$; define $j_x: \langle x \rangle \hookrightarrow F$ to be the inclusion. If G is a group, then a function $f: X \to G$ determines a family of homomorphisms $f_x: \langle x \rangle \to G$, namely, $x^n \mapsto f(x)^n$. Also, the unique homomorphism $\varphi: F \to G$ which extends the function f clearly extends each of the homomorphisms f_x; that is, $\varphi j_x = f_x$ for all $x \in X$.

Here is the uniqueness theorem.

Theorem 11.50. *Let $\{A_i: i \in I\}$ be a family of groups. If P and Q are each a free product of the A_i, then $P \cong Q$.*

Proof. Let $j_i: A_i \to P$ and $k_i: A_i \to Q$ be the imbeddings. Since P is a free product of the A_i, there is a homomorphism $\varphi: P \to Q$ with $\varphi j_i = k_i$ for all i. Similarly, there is a map $\psi: Q \to P$ with $\psi k_i = j_i$ for all i.

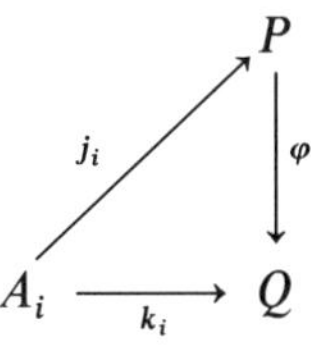

Consider the new diagram

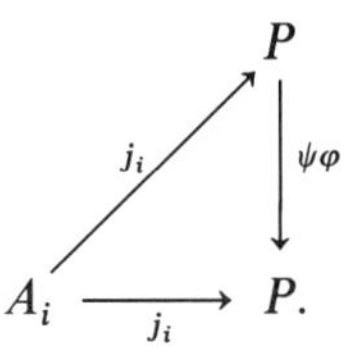

Both $\psi\varphi$ and 1_P are maps making this diagram commute. By hypothesis, there can be only one such map, and so $\psi\varphi = 1_P$. Similarly, $\varphi\psi = 1_Q$, and so $\varphi: P \to Q$ is an isomorphism. ■

Because of Theorem 11.50, we may speak of *the* free product P of $\{A_i: i \in I\}$; it is denoted by

$$P = \mathop{\ast}_{i \in I} A_i;$$

if there are only finitely many A_i's, one usually denotes the free product by

$$A_1 * \cdots * A_n.$$

Theorem 11.51. *Given a family $\{A_i: i \in I\}$ of groups, a free product exists.*

Proof. The proof is so similar to the proof of the existence of a free group (Theorem 11.1) that we present only its highlights.

Assume that the sets $A_i^\# = A_i - \{1\}$ are pairwise disjoint, call $(\bigcup_i A_i^\#) \cup \{1\}$ the *alphabet*, call its elements *letters*, and form *words* w on these letters; that is, $w = a_1 \ldots a_n$, where each a_i lies in some $A_i^\# \cup \{1\}$. A word w is ***reduced*** if either $w = 1$ or $w = a_1 \ldots a_n$, where each letter $a_j \in A_{i_j}^\#$ and adjacent letters lie in distinct $A_i^\#$. Let the elements of the free product be all the reduced words, and let the multiplication be "juxtaposition." In more detail, assume that $a_1 \ldots a_n$ and $b_1 \ldots b_m$ are reduced. If a_n and b_1 lie in distinct $A_i^\#$, then $a_1 \ldots a_n b_1 \ldots b_m$ is reduced, and it is the product. If a_n and b_1 lie in the same $A_i^\#$ and $a_n b_1 \neq 1$ (i.e., $a_n b_1 \in A_i^\#$), then $a_1 \ldots (a_n b_1) \ldots b_m$ is reduced and it is the product. If a_n and b_1 lie in the same $A_i^\#$ and $a_n b_1 = 1$, then cancel and repeat this process for $a_1 \ldots a_{n-1}$ and $b_1 \ldots b_{m-1}$; ultimately, one arrives at a reduced word, and it is the product. It is easy to see that 1 is the identity and that the inverse of a reduced word is reduced; an obvious analogue of the van der Waerden trick can be used to avoid a case analysis in the verification of associativity. ■

If P is the free product of two groups A and B, and if $f: A \to G$ and $g: B \to G$ are homomorphisms, then the homomorphism $\varphi: P \to G$ in the definition of free product is given by

$$\varphi(a_1 b_1 \ldots a_n b_n) = f(a_1)g(b_1) \ldots f(a_n)g(b_n).$$

Uniqueness of the spelling of reduced words shows that φ is a well defined function, and it is not difficult to show (as in the proof of Theorem 11.1) that φ is a homomorphism.

Theorem 11.52 (Normal Form). *If $g \in \ast_{i \in I} A_i$ and $g \neq 1$, then g has a unique factorization*

$$g = a_1 \ldots a_n,$$

where adjacent factors lie in distinct $A_i^\#$.

Proof. The free product constructed in Theorem 11.51 has as its elements all reduced words. ■

Theorem 11.53. *Let $\{A_i: i \in I\}$ be a family of groups, and let a presentation of A_i be $(X_i | \Delta_i)$, where the sets $\{X_i: i \in I\}$ are pairwise disjoint. Then a presentation of $\ast_{i \in I} A_i$ is $(\bigcup X_i | \bigcup \Delta_i)$.*

Proof. Exercise 11.54 below shows that if F_i is the free group with basis X_i, then $F = \ast_{i \in I} F_i$ is the free group with basis $\bigcup_{i \in I} X_i$. Let $\{j_i: A_i \hookrightarrow \ast_{i \in I} A_i\}$ be the imbeddings. If R_i is the normal subgroup of F_i generated by the relations Δ_i, and if $v_i: F_i \to A_i$ is a surjection with $\ker v_i = R_i$, then the map $\varphi: F \to \ast_{i \in I} A_i$ extending all $F_i \to A_i \hookrightarrow \ast_{i \in I} A_i$ has kernel the normal subgroup generated by $\bigcup_{i \in I} \Delta_i$. ■

One can give a second proof of the existence of free products by showing that the group presented by $(\bigcup X_i | \bigcup \Delta_i)$ satisfies the conditions in the definition.

EXERCISES

11.54. Assume that the sets $\{X_i: i \in I\}$ are pairwise disjoint. If F_i is the free group with basis X_i, then $\ast_{i \in I} F_i$ is the free group with basis $\bigcup_{i \in I} X_i$.

11.55. If $a \in A$ and $b \in B$ are nontrivial elements in $A * B$, then $aba^{-1}b^{-1}$ has infinite order. Conclude that $A * B$ is an infinite centerless group.

11.56. Prove that every group E can be imbedded in a centerless group, and use this result to prove that E can be imbedded in $\mathrm{Aut}(H)$ for some H.

11.57 (= 11.4). Prove that a group E has the injective property if and only if $E = 1$. (*Hint* (***Humphreys***). Show that E is not normal in the semidirect product $H \rtimes E$, where H is as in Exercise 11.56.)

11.58. The operation of free product is commutative and associative: for all groups A, B, and C.

$$A * B \cong B * A \qquad \text{and} \qquad A * (B * C) \cong (A * B) * C.$$

11.59. If N is the normal subgroup of $A * B$ generated by A, then $(A * B)/N \cong B$ (compare Exercise 11.5).

11.60. Show that there is a unique homomorphism of $A_1 * \cdots * A_n$ onto $A_1 \times \cdots \times A_n$ which acts as the identity on each A_i.

11.61. Let $A_1, \ldots, A_n$, $B_1, \ldots, B_m$ be indecomposable groups having both chain conditions. If $A_1 * \cdots * A_n \cong B_1 * \cdots * B_m$, then $n = m$ and there is a permutation σ of $\{1, 2, \ldots, n\}$ such that $B_{\sigma(i)} \cong A_i$ for all i.

11.62. If G' is the commutator subgroup of $G = *_{i \in I} A_i$, then $G/G' \cong \sum_{i \in I} (A_i/A_i')$ (compare Lemma 11.3).

Definition. The **infinite dihedral** ***group*** D_∞ is the group with presentation

$$(s, t \mid t^2 = 1, tst^{-1} = s^{-1}).$$

11.63. (i) Prove that $D_\infty \cong \mathbb{Z}_2 * \mathbb{Z}_2$.
(ii) Prove that the Schur multiplier $M(D_\infty) = 1$. (*Hint*. Theorem 11.20.)

Definition. The ***modular group*** is $\mathrm{PSL}(2, \mathbb{Z}) = \mathrm{SL}(2, \mathbb{Z})/\{\pm I\}$.

11.64. (i) The group G with presentation $(a, b \mid a^2 = b^3 = 1)$ is isomorphic to $\mathbb{Z}_2 * \mathbb{Z}_3$.
(ii) Prove that $\mathrm{PSL}(2, \mathbb{Z}) \cong \mathbb{Z}_2 * \mathbb{Z}_3$. (*Hint*. Exercise 2.17(ii).)
(iii) Show that the Schur multiplier $M(\mathrm{PSL}(2, \mathbb{Z})) = 1$. (*Hint*. Theorem 11.20.)

11.65 **(Baer–Levi).** Prove that no group G can be decomposed as a free product and as a direct product (i.e., there are not nontrivial groups with $A * B = G = C \times D$). (*Hint* (***P.M. Neumann***). If $G = A * B$ and $a \in A$ and $b \in B$ are nontrivial, then $C_G(ab) \cong \mathbb{Z}$; if $G = C \times D$, choose $ab = cd$, for $c \in C^\#$ and $d \in D^\#$, to show that $C_G(ab)$ is a direct product.)

The Kurosh Theorem

Kurosh proved that every subgroup of a free product is itself a free product; we will now prove this theorem using covering complexes. To appreciate this geometric proof, the reader should first look at the combinatorial proof in Kurosh, vol. 2, pp. 17–26.

Theorem 11.54. *Let K be a connected complex having connected subcomplexes*

K_i, $i \in I$, such that $K = \bigcup_{i \in I} K_i$. If there is a tree T in K with $T = K_i \cap K_j$ for all $i \neq j$, then

$$\pi(K, w) \cong \mathop{*}_{i \in I} \pi(K_i, w_i)$$

for vertices w in K and w_i in K_i.

Proof. For each i, choose a maximal tree T_i in K_i containing T. By Exercise 11.31(ii), $T^* = \bigcup_{i \in I} T_i$ is a tree in K; it is a maximal tree because it contains every vertex of K.

By Tietze's theorem (Theorem 11.36), $\pi(K_i, w_i)$ has a presentation $(E_i | \Delta_i)$, where E_i is the set of all edges in K_i and Δ_i consists of relations of Type (a): $(u, v) = 1$ if $(u, v) \in T_i$; Type (b): $(u, v)(v, x)(u, x)^{-1} = 1$ if $\{u, v, x\}$ is a simplex in K_i. There is a similar presentation for $\pi(K, w)$, namely, $(E | \Delta)$, where E is the set of all the edges in K, and Δ consists of all edges (u, v) in T^* together with all $(u, v)(v, x)(u, x)^{-1}$ with $\{u, v, x\}$ a simplex in K. Theorem 11.53 says that a presentation for $*_{i \in I} \pi(K_i, w_i)$ is $(\bigcup E_i | \bigcup \Delta_i)$. It follows that $\pi(K, w) \cong *_{i \in I} \pi(K_i, w_i)$, for $E = \bigcup_{i \in I} E_i$, $T^* = \bigcup_{i \in I} T_i$, and $\{u, v, x\}$ lies in a simplex of K if and only if it lies in some K_i. ■

In the next section we will prove that every group G is isomorphic to the fundamental group of some complex. Assuming this result, we now prove the Kurosh theorem.

Theorem 11.55 (Kurosh, 1934). *If $H \leq *_{i \in I} A_i$, then $H = F * (*_{\lambda \in \Lambda} H_\lambda)$ for some (possibly empty) index set Λ, where F is a free group and each H_λ is a conjugate of a subgroup of some A_i.*

Proof. Choose connected complexes K_i with $\pi(K_i, w_i) \cong A_i$. Define a new complex K by adjoining a new vertex w to the disjoint union $\bigcup_{i \in I} \text{Vert}(K_i)$ and new 1-simplexes $\{w, w_i\}$ for all $i \in I$. If T is the tree consisting of these new simplexes, then Theorem 11.54 gives

$$\pi(K, w) \cong \mathop{*}_{i \in I} \pi(K_i \cup T, w_i).$$

But Tietze's theorem (Theorem 11.36) gives $\pi(K_i \cup T, w_i) \cong \pi(K_i, w_i) \cong A_i$ for each i, so that

$$\pi(K, w) \cong \mathop{*}_{i \in I} A_i.$$

Identity $\pi(K, w)$ with $*_{i \in I} A_i$, let $p\colon K_H \to K$ be the covering complex corresponding to the subgroup $H \leq *_{i \in I} A_i$, and choose $\tilde{w} \in p^{-1}(w)$ with $p_\# \pi(K_H, \tilde{w}) = H$. For each i, $p^{-1}(K_i)$ is the disjoint union of its components $\tilde{K}_{ij}$; choose a maximal tree $\tilde{T}_{ij}$ in $\tilde{K}_{ij}$. Define $\tilde{L}$ to be the 1-complex:

$$\tilde{L} = \bigcup \tilde{T}_{ij} \cup p^{-1}(T).$$

Finally, let $\tilde{T}$ be a maximal tree in $\tilde{L}$ containing $p^{-1}(T)$ (which exists, by

Exercise 11.36). Observe that $\tilde{T} \cap \tilde{K}_{ij} = \tilde{T}_{ij}$, lest we violate the maximality of $\tilde{T}_{ij}$ in $\tilde{K}_{ij}$.

Consider, for all i and j, the subcomplexes $\tilde{L}$ and $\tilde{K}_{ij} \cup \tilde{T}$ of K_H. Clearly K_H is the union of these, while the intersection of any pair of them is the tree $\tilde{T}$. By Theorem 11.54,

$$\pi(K_H, \tilde{w}) = \pi(\tilde{L}, \tilde{w}) * \left(\underset{i \in I}{\ast} \pi(\tilde{K}_{ij} \cup \tilde{T}, \tilde{w}_{ij}) \right),$$

where $\tilde{w}_{ij} \in p^{-1}(w) \cap \tilde{K}_{ij}$. Now $\pi(\tilde{L}, \tilde{w})$ is free, because $\dim(\tilde{L}) = 1$. Since $\tilde{T}$ is a maximal tree in $\tilde{K}_{ij} \cup \tilde{T}$, Tietze's theorem gives $\pi(\tilde{K}_{ij} \cup \tilde{T}, \tilde{w}_{ij}) \cong \pi(\tilde{K}_{ij}, \tilde{w}_{ij})$ for each i and j. By Exercise 11.37, $p|\tilde{K}_{ij}: \tilde{K}_{ij} \to K_i$ is a covering complex, and Theorem 11.36 shows that $\pi(\tilde{K}_{ij}, \tilde{w}_{ij})$ is isomorphic (via $(p|\tilde{K}_{ij})_\#$) to a subgroup of $\pi(K_i, w)$. Finally, Theorem 11.37 shows that this subgroup is equal to a conjugate of a subgroup of $\pi(K_i, w)$, as desired. ■

Corollary 11.56. *If $G = \ast_{i \in I} A_i$, where each A_i is torsion, then every torsion-free subgroup of G is a free group.*

Proof. Every nontrivial subgroup of any A_i contains elements of finite order. ■

Corollary 11.57. *If $G = \ast_{i \in I} A_i$, then every finite subgroup is conjugate to a subgroup of some A_i. In particular, every element of finite order in G is conjugate to an element of finite order in some A_i.*

Proof. Every nontrivial free product contains elements of infinite order. ■

The last corollary shows how dramatically the Sylow theorems can fail for infinite groups. If A and B are any two finite p-groups, then each is a Sylow p-subgroup of $A * B$; thus, Sylow subgroups need not be isomorphic, let alone conjugate.

We only state the following important result of Grushko (1940) (see Massey for a proof using fundamental groups). If G is a finitely generated group, define $\mu(G)$ to be the minimal number of generators of G. ***Grushko's theorem*** states that if A and B are finitely generated groups, then

$$\mu(A * B) = \mu(A) + \mu(B).$$

EXERCISES

11.66. Show that if an element $T \in \mathrm{SL}(2, \mathbb{Z})$ has finite order n, then $n = 1, 2, 3, 4$, or 6 (see Exercise 11.64(ii)). Moreover, T is conjugate to either $\pm I$, $\pm A$, or $\pm B$, where

$$A = \begin{bmatrix} 0 & -1 \\ 1 & 0 \end{bmatrix} \quad \text{and} \quad B = \begin{bmatrix} 0 & 1 \\ -1 & -1 \end{bmatrix}.$$

11.67. Prove that the modular group has a free subgroup of index 6. (*Hint.* The kernel of $\mathbb{Z}_2 * \mathbb{Z}_3 \to \mathbb{Z}_2 \times \mathbb{Z}_3$ is torsion-free.)

11.68. Show that the commutator subgroup of the modular group is free.

11.69. Prove that the modular group contains a free subgroup of infinite rank.

11.70. (i) If $f: A \to G$ and $g: B \to H$ are homomorphisms, prove that there is a unique homomorphism $\varphi: A * B \to G * H$ with $\varphi|A = f$ and $\varphi|B = g$. Denote φ by $f * g$.
(ii) Given a group A, show that there is a functor T with $T(G) = A * G$ and, if $g: G \to H$, then $T(g) = 1_A * g: A * G \to A * H$.

The van Kampen Theorem

The van Kampen theorem answers the following natural question: If a complex K is the union of subcomplexes L_1 and L_2, can one compute $\pi(K, w) = \pi(L_1 \cup L_2, w)$ in terms of $\pi(L_1, w)$ and $\pi(L_2, w)$? The key idea is to realize that π is a functor, and since functors recognize only complexes and simplicial maps (but not vertices or simplexes), the notion of union should be described in terms of diagrams.

If S is a set that is a union of two subsets, say, $S = A \cup B$, then there is a commutative diagram (with all arrows inclusions)

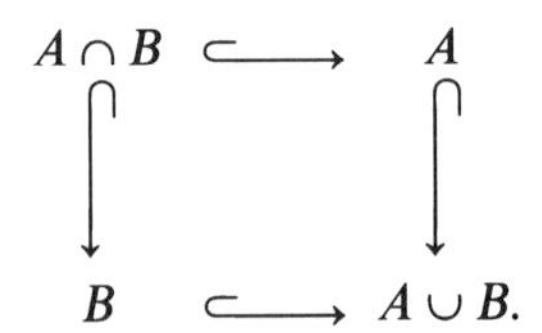

Moreover, $A \cup B$ is the "best" southeast corner of such a diagram in the following sense: given any set X in a commutative diagram

$$\begin{array}{ccc} A \cap B & \hookrightarrow & A \\ \downarrow & & \downarrow f \\ B & \xrightarrow[g]{} & X, \end{array}$$

where f and g are (not necessarily injective) functions agreeing on $A \cap B$, then there is a (unique) function $\varphi: A \cup B \to X$ with $\varphi|A = f$ and $\varphi|B = g$. We have been led to the following definition.

Definition. Let A, B, and C be groups, and let $i: A \to B$ and $j: A \to C$ be (not necessarily injective) homomorphisms. A ***solution*** (G, f, g) of the data is a

commutative diagram

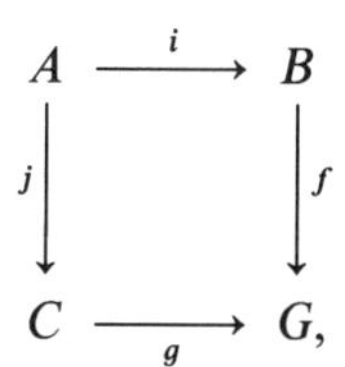

where G is a group and f and g are homomorphisms. A ***pushout*** of this data is a solution (P, j', i')

$$\begin{array}{ccc} A & \xrightarrow{i} & B \\ {\scriptstyle j}\downarrow & & \downarrow{\scriptstyle j'} \\ C & \xrightarrow[i']{} & P \end{array}$$

such that, for every solution (G, f, g), there exists a unique $\varphi: P \to G$ with

$$f = \varphi j' \qquad \text{and} \qquad g = \varphi i'.$$

If a pushout (P, j', i') of the data exists, then P is unique to isomorphism, for if (Q, j'', i'') is another pushout, then the homomorphisms $\varphi: P \to Q$ and $\psi: Q \to P$ provided by the definition are easily seen to be inverses.

Even though a pushout is a triple (P, j', i'), one usually calls the group P a pushout of the data.

Theorem 11.58.

(i) *A pushout (P, j', i') exists for the data $i: A \to B$ and $j: A \to C$.*

(ii) *The pushout P is isomorphic to $(B * C)/N$, where N is the normal subgroup of $B * C$ generated by $\{i(a)j(a^{-1}): a \in A\}$. Indeed, if A has a presentation $(X|\Delta)$ and B has a presentation $(Y|\Gamma)$, then P has a presentation*

$$P = (X \cup Y | \Delta \cup \Gamma \cup \{i(a)j(a^{-1}): a \in A\}).$$

Proof. It is easy to see that (P, j', i') is a solution, where $P = (B * C)/N, j'(b) = bN$, and $i'(c) = cN$.

Suppose that (G, f, g) is a solution of the data. The definition of free product gives a unique homomorphism $\psi: B * C \to G$ with $\psi|B = f$ and $\psi|C = g$: if $b \in B$ and $c \in C$, then $\psi(bc) = f(b)g(c)$. For all $a \in A$,

$$\psi(i(a)j(a^{-1})) = fi(a)gj(a^{-1}) = 1,$$

because $fi = gj$, and so $N \leq \ker \psi$. Therefore, ψ induces a homomorphism $\varphi: P = (B * C)/N \to G$ with $\varphi j'(b) = \varphi(bN) = \psi(b) = f(b)$ and $\varphi i'(c) = \varphi(cN) = \psi(c) = g(c)$. The map φ is unique because P is generated by (the cosets of) $B \cup C$. It is plain that P has the desired presentation. ■

Corollary 11.59.

(i) *If* $i: A \to B$, $j: A \to C$, *and* $C = 1$, *then the pushout* P *of this data is* $P = B/N$, *where* N *is the normal subgroup of* B *generated by* $\operatorname{im} i$.
(ii) *If* $A = 1$, *then* $P \cong B * C$.

If A is an infinite cyclic group with generator x, then $A * A$ is a free group of rank 2 with basis $\{x, y\}$. Obviously, it is necessary to write y for the second generator to avoid confusing it with x. More generally, if groups A_i have presentations $(X_i|\Delta_i)$ for $i = 1, 2$, then we assume that the sets X_i are disjoint when we say that a presentation for $A_1 * A_2$ is $(X_1 \cup X_2|\Delta_1 \cup \Delta_2)$. If the sets X_1 and X_2 are not disjoint, then new notation must be introduced to make them disjoint.

Theorem 11.60 (van Kampen).[6] *Let* K *be a connected complex having connected subcomplexes* L_1 *and* L_2 *with* $K = L_1 \cup L_2$. *If* $L_1 \cap L_2$ *is connected and* $w \in \operatorname{Vert}(L_1 \cap L_2)$, *then* $\pi(K, w)$ *is the pushout of the data*

$$\begin{array}{ccc} \pi(L_1 \cap L_2, w) & \xrightarrow{j_{1\#}} & \pi(L_1, w), \\ {\scriptstyle j_{2\#}} \big\downarrow & & \\ \pi(L_2, w) & & \end{array}$$

where $j_i: L_1 \cap L_2 \hookrightarrow L_i$ *is the inclusion for* $i = 1, 2$.

Moreover, if a presentation of $\pi(L_i, w)$ *is* $(E_i|\Delta_i' \cup \Delta_i'')$ *as in Tietze's theorem* (E_i *is the set of edges in* L_i, Δ_i' *are the relations of Type* (a), *and* Δ_i'' *are the relations of Type* (b)), *then a presentation for* $\pi(K, w)$ *is*

$$(j_1 E_1 \cup j_2 E_2 | j_1 \Delta_1' \cup j_1 \Delta_1'' \cup j_2 \Delta_2' \cup j_2 \Delta_2'' \cup \{(j_1 e)(j_2 e)^{-1}: e \in E_0\}),$$

where E_0 *is the set ofedges in* $L_1 \cap L_2$.

Remark. The hypothesis that K is connected is redundant.

Proof. Choose a maximal tree T in $L_1 \cap L_2$ and, for $i = 1$ and 2, choose a maximal tree T_i in L_i containing T. By Exercise 11.31(ii), $T_1 \cup T_2$ is a tree; it is a maximal tree in K because it contains every vertex. Tietze's theorem gives the presentation $\pi(K, w) = (E|\Delta' \cup \Delta'')$, where E is the set of all edges (u, v) in K, $\Delta' = E \cap (T_1 \cup T_2)$, and

$$\Delta'' = \{(u, v)(v, x)(u, x)^{-1}: \{u, v, x\} \text{ is a simplex in } K\}.$$

There are similar presentations for both $\pi(L_i, w)$, as in the statement of the theorem. Make the sets E_1 and E_2 disjoint by affixing the symbols j_1 and j_2.

[6] This theorem was first proved by Siefert and found later, independently, by van Kampen. This is another instance when the name of a theorem does not coincide with its discoverer.

By Theorem 11.58(ii), a presentation for the pushout is

$$(j_1 E_1 \cup j_2 E_2 | j_1 \Delta_1' \cup j_1 \Delta_1'' \cup j_2 \Delta_2' \cup j_2 \Delta_2'' \cup \{(j_1 e)(j_2 e)^{-1} : e \in E_0\})$$

The generators may be rewritten as

$$j_1 E_0 \cup j_1(E_1 - E_0) \cup j_2 E_0 \cup j_2(E_2 - E_0).$$

The relations include $j_1 E_0 = j_2 E_0$, and so one of these is superfluous. Next, $\Delta_i' = E_i \cap T_i = (E_i \cap T) \cup (E_i \cap (T_i - T))$, and this gives a decomposition of $j_1 \Delta_1' \cup j_2 \Delta_2'$ into four subsets, one of which is superfluous. Further, $\Delta'' = \Delta_1'' \cup \Delta_2''$, for if $(u, v)(v, x)(u, x)^{-1} \in \Delta$, then $s = \{u, v, x\} \in K = L_1 \cup L_2$, and hence $s \in L_i$ for $i = 1$ or $i = 2$. Now transform the presentation as follows:

(i) isolate those generators and relations involving $L_1 \cap L_2$;
(ii) delete superfluous generators and relations involving $L_1 \cap L_2$ (e.g., delete all such having symbol j_2);
(iii) erase the (now unnecessary) symbols j_1 and j_2.

It is now apparent that both $\pi(K, w)$ and the pushout have the same presentation, hence they are isomorphic. ■

Corollary 11.61. *If K is a connected complex having connected subcomplexes L_1 and L_2 such that $L_1 \cup L_2 = K$ and $L_1 \cap L_2$ is simply connected, then*

$$\pi(K, w) \cong \pi(L_1, w) * \pi(L_2, w).$$

Proof. Immediate from the van Kampen theorem and Corollary 11.59(ii). ■

Corollary 11.62. *Let K be a connected complex having connected subcomplexes L_1 and L_2 such that $L_1 \cup L_2 = K$ and $L_1 \cap L_2$ is connected. If $w \in \operatorname{Vert}(L_1 \cap L_2)$ and if L_2 is simply connected, then*

$$\pi(K, w) \cong \pi(L_1, w)/N,$$

where N is the normal subgroup generated by the image of $\pi(L_1 \cap L_2, w)$. Moreover, in the notation of the van Kampen theorem, there is a presentation

$$\pi(K, w) = (E_1 | \Delta_1' \cup \Delta_1'' \cup j_1 E_0).$$

Proof. Since $\pi(L_2, w) = 1$, the result is immediate from the van Kampen theorem and Corollary 11.59(i). ■

We now exploit Corollary 11.62. Let K be a connected 2-complex with basepoint w, and let

$$\alpha = e_1 \ldots e_n = (w, v_1)(v_1, v_2) \ldots (v_{n-1}, w)$$

be a closed path in K at w.

Define a ***triangulated polygon*** $D(\alpha)$ to be the 2-complex with vertices

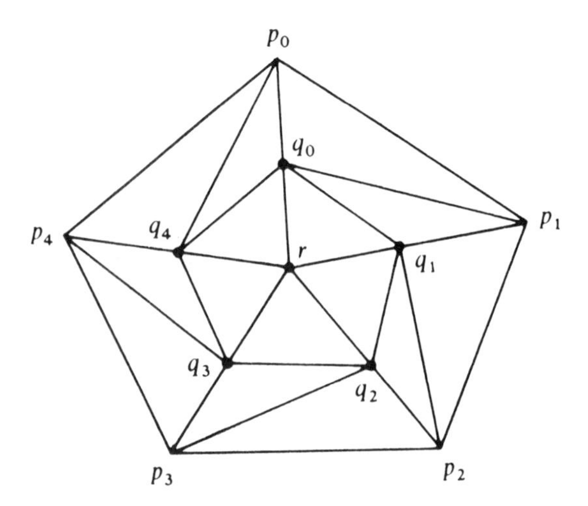

Figure 11.7

$(p_0, \ldots, p_{n-1}, q_0, \ldots, q_{n-1}, r\}$ and 2-simplexes $\{r, q_i, q_{i+1}\}$, $\{q_i, q_{i+1}, p_{i+1}\}$, and $\{q_i, p_i, p_{i+1}\}$, where $0 \le i \le n-1$ and subscripts are read mod n. Let $\partial D(\alpha)$ denote the boundary of $D(\alpha)$; that is, $\partial D(\alpha)$ is the full subcomplex of $D(\alpha)$ having vertices $\{p_0, \ldots, p_{n-1}\}$. Define the ***attaching map*** $\varphi_\alpha \colon \partial D(\alpha) \to K$ by $\varphi_\alpha(p_i) = v_i$ for all $0 \le i \le n-1$ (where we define $v_0 = w$). Clearly, φ_α carries the boundary path $(p_0, p_1)(p_1, p_2)\ldots(p_{n-1}, p_0)$ onto the path α.

Let K be a complex with basepoint w, let α be a closed path in K at w, and let $\varphi_\alpha \colon D(\alpha) \to K$ be the attaching map. Let $\mathscr{P}$ be the partition of the disjoint union $\text{Vert}(K) \cup \text{Vert}(D(\alpha))$ arising from the equivalence relation which identifies each p_i with $\varphi_\alpha(p_i)$. Then the quotient complex $K_\alpha = (K \cup D(\alpha))/\mathscr{P}$ is called the complex obtained from K by ***attaching a 2-cell along*** $\boldsymbol{\alpha}$. Notice that all the p_i in $D(\alpha)$ have been identified with vertices of K, but the "interior" vertices $r, q_0, q_1, \ldots$ are untouched. Thus, if $\hat{Q}$ is the full subcomplex of K_α with vertices $\{r, q_0, q_1, \ldots\}$, then $\hat{Q}$ is simply connected. On the other hand, if we delete the vertex r, and if Q is the full subcomplex with vertices $\{q_0, q_1, \ldots\}$, then $\pi(Q, q_0) \cong \mathbb{Z}$.

Theorem 11.63. *Let α be a closed path at w in a complex K, and let K_α be obtained from K by attaching a 2-cell along α. Then*

$$\pi(K_\alpha, w) \cong \pi(K, w)/N,$$

where N is the normal subgroup generated by $[\alpha]$.

Proof. Define L_1 to be the full subcomplex of K_α with vertices $\text{Vert}(K) \cup \{q_0, \ldots, q_{n-1}\}$, and define L_2 to be the full subcomplex with vertices $\{r, w, q_0, \ldots, q_{n-1}\}$. Note that $L_1 \cup L_2 = K_\alpha$ and that $L_1 \cap L_2$ is the 1-complex with vertices $\{w, q_0, \ldots, q_{n-1}\}$. Since there is just one reduced circuit, it follows that $\pi(L_1 \cap L_2, w) \cong \mathbb{Z}$. Now L_2, being isomorphic to the full subcomplex of $D(\alpha)$ with vertices $\{r, q_0, \ldots, q_{n-1}\}$, is simply connected. The inclusion $j\colon K \hookrightarrow L_1$ induces an isomorphism $\pi(K, w) \to \pi(L_1, w)$. Define a function $\psi\colon \text{Vert}(L_1) \to \text{Vert}(K)$ by $\psi(v) = v$ for all $v \in \text{Vert}(K)$ and $\psi(q_i) = \varphi_\alpha(p_i)$ for all i. It is easy to check that ψ is a simplicial map and that $j\psi\colon L_1 \to L_1$ is homotopic to the identity. Since π is a functor, the induced map ψ_* is the inverse of j_*. The proof is completed by Corollary 11.62, for the image of the infinite cyclic group $\pi(L_1 \cap L_2, w)$ is generated by $[\alpha]$. ∎

The next construction is needed to attach a family of 2-cells.

Definition. Let $\{K_i\colon i \in I\}$ be a family of complexes, and let w_i be a basepoint in K_i. The ***wedge*** $\bigvee_{i \in I} K_i$ is the disjoint union of the K_i in which all the basepoints w_i are identified to a common point b.

For example, a bouquet of circles is a wedge of 1-complexes. Theorem 11.54 shows that $\pi(\bigvee_{i \in I} K_i, b) \cong *_{i \in I}\, \pi(K_i, w_i)$.

The next theorem was used in the proof of the Kurosh theorem.

Theorem 11.64. *Given a group G, there exists a connected 2-complex K with $G \cong \pi(K, w)$.*

Proof. Let $(X \mid \Delta)$ be a presentation of G and let B be a bouquet of $|X|$ circles having vertices $\{w, u^x, v^x\colon x \in X\}$. If each $x \in X$ is identified with the path $(w, u^x)(u^x, v^x)(v^x, w)$, then each relation in Δ may be regarded as a closed path in B at w. For example, xyx^{-1} is viewed as the path

$$(w, u^x)(u^x, v^x)(v^x, w)(w, u^y)(u^y, v^y)(v^y, w)(w, v^x)(v^x, u^x)(u^x, w).$$

For each path $\alpha \in \Delta$, let the triangulated polygon $D(\alpha)$ have vertices $\{r^\alpha, p_0^\alpha, p_1^\alpha, \ldots, q_0^\alpha, q_1^\alpha, \ldots\}$, and let $\varphi_\alpha\colon \partial D(\alpha) \to B$ be the attaching map. Let $D = \bigvee_{\alpha \in \Delta} D(\alpha)$ (in which all the vertices p_0^α are identified to a common basepoint, denoted by p_0). If $\partial(\bigvee_{\alpha \in \Delta} D(\alpha))$ is defined to be the full subcomplex with vertices all the p_i^α, then there is a simplicial map $\varphi\colon \partial(\bigvee_{\alpha \in \Delta} D(\alpha)) \to B$ with $\varphi \mid \partial D(\alpha) = \varphi_\alpha$ for all $\alpha \in \Delta$. Define K as the quotient complex of the disjoint union $B \cup D$ in which p_0 is identified with w and, for all $\alpha \in \Delta$ and all $i > 0$, each vertex p_i^α is identified with $\varphi_\alpha(p_i^\alpha)$. We have "draped" 2-simplexes on a bouquet of circles. Thus,

$$\text{Vert}(K) = \text{Vert}(B) \cup \left(\bigcup_{\alpha \in \Delta} \{r^\alpha, q_0^\alpha, q_1^\alpha, \ldots\} \right).$$

Let T be the tree in K with vertices $\{w, u^x\colon x \in X\}$, let Q^α be the full subcomplex with vertices $\{q_0^\alpha, q_1^\alpha, \ldots\}$, and let $\hat{Q}^\alpha$ be the full subcomplex with vertices $\{r^\alpha, q_0^\alpha, q_1^\alpha, \ldots\}$. Define L_1 to be the full subcomplex with vertices $\text{Vert}(B) \cup (\bigcup_{\alpha \in \Delta} \text{Vert}(Q^\alpha))$, and define L_2 to be full subcomplex with vertices $\text{Vert}(T) \cup (\bigcup \text{Vert}(\hat{Q}^\alpha))$. Each $\hat{Q}^\alpha$ is simply connected, so that (with suitable basepoints $\#$) $\pi(L_2, w) \cong \pi(\bigvee_\alpha \hat{Q}^\alpha, \#)$, by Theorem 11.34, and $\pi(\bigvee \hat{Q}^\alpha, \#) \cong *_\alpha \pi(\hat{Q}^\alpha, \#) = 1$, by Theorem 11.54; thus, $\pi(L_2, w) = 1$. We now show that the inclusion $j: B \hookrightarrow L_1$ induces an isomorphism $\pi(B, v_0) \to \pi(L_1, v_0)$. Define a function ψ: Vert$(L_1) \to$ Vert(B) by $\psi(v) = v$ for all $v \in$ Vert(B) and $\psi(q_i^\alpha) = \varphi_\alpha(p_i^\alpha)$ for all α and i. It is easy to see that ψ is a simplicial map and that $j\psi: L_1 \to L_1$ is homotopic to the identity. Since π is a functor, the induced map $\psi_\#$ is the inverse of $j_\#$. Now $L_1 \cup L_2 = K$ and $L_1 \cap L_2 = T \cup (\bigvee_\alpha Q^\alpha)$, so that $\pi(L_1 \cap L_2, w) \cong \pi(\bigvee_\alpha Q^\alpha, \#) \cong *_\alpha \pi(Q^\alpha, \#)$ is free of rank $|\Delta|$. By Corollary 11.62, $\pi(K, w) \cong \pi(L_1, w)/N$, where N is the image of $\pi(L_1 \cap L_2, w)$. Therefore, $\pi(K, w) \cong G$, for they have the same presentation. ∎

Definition. A group G is ***finitely presented*** (or *finitely related*) if it has a presentation with a finite number of generators and a finite number of relations.

There are uncountably many nonisomorphic finitely generated groups, as we shall see in Theorem 11.73. Since there are only countably many finitely presented groups, there exist finitely generated groups that are not finitely presented. We shall give an explicit example of such a group at the end of this chapter.

Corollary 11.65. *A group G is finitely presented if and only if there is a finite connected 2-complex K (i.e.,* Vert(K) *is finite) with $G \cong \pi(K, w)$.*

Proof. By Tietze's theorem, $\pi(K, w)$ is finitely presented. Conversely, if G is finitely presented, then the complex K constructed in the theorem is finite. ∎

There is a construction dual to that of pushout. A ***solution*** of the diagram

$$(*) \qquad \begin{array}{ccc} & & A \\ & & \downarrow{\scriptstyle \alpha} \\ B & \xrightarrow[\beta]{} & G \end{array}$$

is a triple (D, α', β') making the following diagram commute:

$$\begin{array}{ccc} D & \xrightarrow{\beta'} & A \\ {\scriptstyle \alpha'}\downarrow & & \downarrow{\scriptstyle \alpha} \\ B & \xrightarrow[\beta]{} & G. \end{array}$$

Definition. A ***pullback*** of diagram $(*)$ is a solution (D, α', β') that is "best" in the following sense: if (X, α'', β'') is any other solution, then there is a unique homomorphism $\theta: X \to D$ with $\alpha'' = \alpha'\theta$ and $\beta'' = \beta'\theta$.

It is easy to see that if (D, α', β') is a pullback, then D is unique to isomorphism. One often abuses notation and calls the group D the pullback of diagram (*). Some properties of pullbacks are given in Exercises 11.75 and 11.76 below.

Pullbacks have already arisen in the discussion of the projective lifting property in Chapter 7: given a projective representation $\tau: Q \to \mathrm{PGL}(n, \mathbb{C})$, then the pullback U of τ and the natural map $\pi: \mathrm{GL}(n, \mathbb{C}) \to \mathrm{PGL}(n, \mathbb{C})$ is a central extension which allows τ to be lifted:

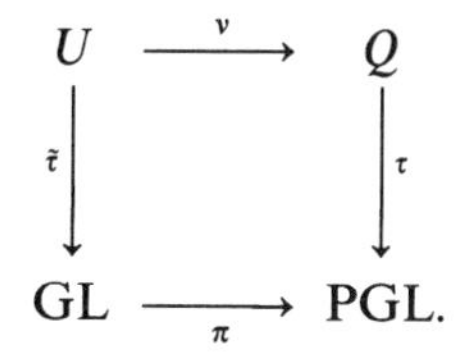

Exercises

11.71. Prove that every finitely generated free group is finitely presented.

11.72. Prove that every finite group is finitely presented.

11.73. Prove that every finitely generated abelian group is finitely presented.

11.74. Prove that a group having a presentation with a finite number of relations is the free product of a finitely presented group and a free group.

11.75. Let A and B be subgroups of a group G, and let α and β be the respective inclusions. Show that $A \cap B$ is the pullback.

11.76. Prove that the pullback of diagram (*) always exists. (*Hint.* Define $D = \{(a, b) | \beta(b) = \alpha(a)\} \leq A \times B$, define $\alpha': (a, b) \mapsto b$, and define $\beta': (a, b) \mapsto a$.)

Amalgams

Amalgams arise from the special case of the van Kampen theorem in which the maps $j_{i\#}: \pi(L_1 \cap L_2, w) \to \pi(L_i, w)$ induced from the inclusions $j_i: L_1 \cap L_2 \hookrightarrow L_i$ (for $i = 1, 2$) are injections. That an inclusion $L \hookrightarrow K$ need not induce an injection between fundamental groups can be seen when K is simply connected and L is not. The advantage of this extra hypothesis, as we shall see, is that a normal form is then available for the elements of the pushout.

Definition. Let A_1 and A_2 be groups having isomorophic subgroups B_1 and B_2, respectively; let $\theta: B_1 \to B_2$ be an isomorphism. The ***amalgam*** of A_1 and

A_2 over θ (often called the *free product with amalgamated subgroup*) is the pushout of the diagram

$$\begin{array}{ccc} B_1 & \xrightarrow{i} & A_1, \\ {\scriptstyle j\theta}\downarrow & & \\ A_2 & & \end{array}$$

where i and j are inclusions.

Recall that pushouts are the diagrammatic version of unions.

We proved in Theorem 11.58 that the amalgam exists: it is $(A_1 * A_2)/N$, where N is the normal subgroup of $A_1 * A_2$ generated by $\{b\theta(b^{-1}): b \in B_1\}$; moreover, any two pushouts of the same data are isomorphic. We denote the amalgam by

$$A_1 *_\theta A_2$$

(a less precise notation replaces the subscript θ by the subscript B_1). An amalgam, as any pushout, is an ordered triple $(A_1 *_\theta A_2, \lambda_1, \lambda_2)$; the maps $\lambda_i: A_i \to A_1 *_\theta A_2$ are given, as in Theorem 11.58, by $a_i \mapsto a_i N$.

It is clear that each $b \in B_1$ is identified in $A_1 *_\theta A_2$ with $\theta(b)$; it is not clear whether other identifications are consequences of the amalgamation. For example, is it obvious whether the maps λ_i are injections? Is it obvious whether $A_1 *_\theta A_2 \neq 1$? These questions can be answered once we give a more concrete description of an amalgam in terms of its elements.

For each $i = 1, 2$, choose a left transversal of B_i in A_i subject only to the condition that the representative of the coset B_i is 1. For $a \in A_i$, denote the chosen representative of aB_i by $l(a)$, so that $a = l(a)b$ for some uniquely determined $b \in B_i$ (depending on a).

Definition. A ***normal form*** is an element of $A_1 * A_2$ of the form

$$l(a_1)l(a_2)\dots l(a_n)b,$$

where $b \in B_1$, $n \geq 0$, the elements $l(a_j)$ lie in the chosen transversals of B_{i_j} in A_{i_j}, and adjacent $l(a_j)$ lie in distinct A_i.

In the special case that B_1 (and hence B_2) is trivial, the amalgam is the free product and every reduced word is a normal form.

Theorem 11.66 (Normal Form). *Let A_1 and A_2 be groups, let B_i be a subgroup of A_i for $i = 1, 2$, and let $\theta: B_1 \to B_2$ be an isomorphism. Then for each element $wN \in A_1 *_\theta A_2$, there is a unique normal form $F(w)$ with $wN = F(w)N$.*

Proof. By Theorem 11.58, $A_1 *_\theta A_2 = (A_1 * A_2)/N$, where N is the normal subgroup of $A_1 * A_2$ generated by $\{b\theta(b^{-1}): b \in B_1\}$. Each coset of N has a representative $w = x_1 y_1 \dots x_n y_n$ in the free product, where $x_i \in A_1$, $y_i \in A_2$, and only x_1 or y_n is allowed to be 1. We now give an algorithm assigning a normal form $F(w)$ to each coset wN in the amalgam such that $F(w)N = wN$.

Let $w = x_1 y_1$. If $y_1 = 1$, then $x_1 = l(x_1)b = F(w)$, where $b \in B_1$, and we are done. If $x_1 = 1$, then $w = y_1 = l(y_1)b$ for $b \in B_2$ and so $l(y_1)\theta^{-1}(b) = F(w)$ is a normal form in wN. If $x_1 \neq 1$ and $y_1 \neq 1$, then

$$x_1 y_1 = l(x_1) b y_1 = l(x_1)\theta(b) y_1 \qquad \text{in} \quad A_1 *_\theta A_2;$$

but $z = \theta(b) y_1 \in A_2$, so that $z = l(z) b_2$ for some $b_2 \in B_2$. Therefore,

$$x_1 y_1 = l(x_1) l(z) b_2 = l(x_1) l(z) \theta^{-1}(b_2) \qquad \text{in} \quad A_1 *_\theta A_2,$$

and the last element is a normal form $F(w)$ in wN. This procedure can be iterated, ending with a normal form. (Observe that the penultimate step produces a last factor b lying in either B_1 or B_2; if $b \in B_1$, one has a normal form; if $b \in B_2$, then it must be replaced by $\theta^{-1}(b) \in B_1$.)

The van der Waerden trick will prove uniqueness of this normal form by constructing a homomorphism Φ with domain $A_1 *_\theta A_2$ which takes different values on different normal forms: if $F(w) \neq F(w')$, then $\Phi(F(w)N) \neq \Phi(F(w')N)$. Let M be the set of all normal forms; by Theorem 11.52, different normal forms have different spellings. If $a \in A_i$, define a function $|a|: M \to M$ by

$$|a|(l(a_1)l(a_2)\dots l(a_n)b) = F(al(a_1)l(a_2)\dots l(a_n)b)$$

(if a and $l(a_1)$ lie in distinct A_i, then $l(a_1)l(a_2)\dots l(a_n)b$ and also $al(a_1)l(a_2)\dots l(a_n)b$ have the form $x_1 y_1 \dots x_n y_n$, and the algorithm F can be applied; if a and $l(a_1)$ lie in the same A_i, then the algorithm applies to $[al(a_1)]l(a_2)\dots l(a_n)b)$. Clearly $|1|$ is the identity function on M, and consideration of several cases (depending on where initial factors live) shows that if $a, a' \in A_1 \cup A_2$, then

$$|a| \circ |a'| = |aa'|.$$

Therefore, $|a^{-1}| = |a|^{-1}$ and each $|a|$ is a permutation of M. If S_M is the group of all permutations of M, then $a \mapsto |a|$ is a homomorphism $A_i \to S_M$ for $i = 1, 2$. In particular, if $b \in B_1 \leq A_1$, $|b|: M \to M$ is defined and $|b| = |\theta(b)|$.

The defining property of free product allows us to assemble these two homomorphisms into a single homomorphism

$$\varphi: A_1 * A_2 \to S_M$$

with $\varphi(l(a_1)l(a_2)\dots l(a_n)b) = |l(a_1)| \circ |l(a_2)| \circ \cdots \circ |l(a_n)| \circ |b|$. For all $b \in B_1$, $|b| = |\theta(b)|$ gives $b\theta(b^{-1}) \in \ker \varphi$, and so φ induces a homomorphism $\Phi: A_1 *_\theta A_2 = (A_1 * A_2)/N \to S_M$ by

$$\begin{aligned} \Phi(l(a_1)l(a_2)\dots l(a_n)bN) &= \varphi(l(a_1)l(a_2)\dots l(a_n)b) \\ &= |l(a_1)| \circ |l(a_2)| \circ \cdots \circ |l(a_n)| \circ |b|. \end{aligned}$$

Now $|l(a_1)| \circ |l(a_2)| \circ \cdots \circ |l(a_n)| \circ |b|(1) = l(a_1)l(a_2)\dots l(a_n)b$; that is, $\Phi(wN): 1 \mapsto F(w)$. Thus, if $G(w)$ were another normal form in wN, then

$\Phi(wN)$: $1 \mapsto G(w)$, and so $G(w) = F(w)$. Therefore, for each element wN of the amalgam, there is a unique normal form $F(w)$ with $wN = F(w)N$. ∎

Theorem 11.67. *Let A_1 and A_2 be groups, let B_1 and B_2 be isomorphic subgroups of A_1 and A_2, respectively, and let θ: $B_1 \to B_2$ be an isomorphism.*

(i) *The (pushout) homomorphisms λ_i: $A_i \to A_1 *_\theta A_2$ are injections for $i = 1, 2$.*
(ii) *If $A'_i = \lambda_i(A_i)$, then $\langle A'_1, A'_2 \rangle = A_1 *_\theta A_2$ and $A'_1 \cap A'_2 = \lambda_1(B_1) = \lambda_2(B_2)$.*

Proof. (i) If $a_i \in A_i$ is not 1, then $F(a_i) \neq 1$ and $\Phi(a_i N) \neq 1$; but $\Phi(a_i N) = \varphi\lambda_i(a_i) \neq 1$ implies $\lambda_i(a_i) \neq 1$, and so λ_i is an injection.

(ii) It follows from $A_1 *_\theta A_2 = (A_1 * A_2)/N$ that $\langle A'_1, A'_2 \rangle = A_1 *_\theta A_2$. If $u \in A'_1 \cap A'_2$, then there are $a_i \in A_i$ with $a_1 N = u = a_2 N$. Now $F(a_1) = l(a_1)b$ and $F(a_2) = l(a_2)b'$, so that the uniqueness of the normal form gives $l(a_1) = l(a_2)$ and $b = b'$. But $l(a_1) = l(a_2)$ can occur only when both are 1, lest we have equality of the distinct normal forms $l(a_1)l(a_2)$ and $l(a_2)l(a_1)$. Hence $bN = a_1 N = u = a_2 N = b'N$, and $u \in \lambda_1(B_1) = \lambda_2(B_2)$. For the reverse inclusion, it is easy to see that if $b \in B_1$, then $bN \in A'_1 \cap A'_2$. ∎

In view of the last theorem, it is customary to regard the elements of $A_1 *_\theta A_2$ as normal forms and the maps λ_i as inclusions. The statement of Theorem 11.67 now simplifies to $\langle A_1, A_2 \rangle = A_1 *_\theta A_2$ and $A_1 \cap A_2 = B_1 = B_2$. This is the point of view taken in the next corollary.

Corollary 11.68. *Let $E \cong A_1 *_\theta A_2$ have amalgamated subgroup B. If $y_1, \ldots, y_r \in E$ with $y_j \in A_{i_j}$, where $i_j \neq i_{j+1}$, and if $y_j \notin B$ for all j, then $y_1 \ldots y_r \neq 1$.*

Proof. Immediate from the normal form theorem. ∎

Theorem 11.69 (Torsion Theorem). *An element in $A_1 *_\theta A_2$ has finite order if and only if it is conjugate to an element of finite order in A_1 or in A_2.*

Proof. This follows easily from the normal form theorem. ∎

We are now going to apply amalgams to prove some imbedding theorems. The following two theorems are due to G. Higman, B.H. Neumann, and Hanna Neumann.

Theorem 11.70 (Higman, Neumann, and Neumann, 1949). *Let G be a group and let φ: $A \to B$ be an isomorphism between subgroups A and B of G. Then there exists a group H containing G and an element $t \in H$ with*

$$\varphi(a) = t^{-1}at \qquad \text{for all} \quad a \in A.$$

Proof. Let $\langle u \rangle$ and $\langle v \rangle$ be (disjoint) infinite cyclic groups, and define groups

$$K_1 = G * \langle u \rangle \qquad \text{and} \qquad K_2 = G * \langle v \rangle.$$

If $L_1 = \langle G, uAu^{-1}\rangle \leq K_1$, then

$$L_1 \cong G * u^{-1}Au,$$

for there can be no equation in K_1, *a fortiori* in L_1, of the form

$$g_1u^{-1}a_1ug_2u^{-1}a_2u\ldots g_nu^{-1}a_nu = 1.$$

Similarly, if $L_2 = \langle G, vBv^{-1}\rangle \leq K_2$, then

$$L_2 \cong G * v^{-1}Bv.$$

By Exercise 11.70, there is an isomorphism $\theta: L_1 \to L_2$ with $\theta|G$ the identity and $\theta(u^{-1}au) = v^{-1}\varphi(a)v$.

Define $H = K_1 *_\theta K_2$. By Theorem 11.67, H contains a subgroup isomorphic to L_1 (and $G \leq L_1$). For each $a \in A$, $u^{-1}au = v^{-1}\varphi(a)v$, so that if $t \in H$ is defined by $t = uv^{-1}$, then

$$t^{-1}at = \varphi(a) \qquad \text{for all} \quad a \in A. \quad ■$$

If G is a countable group, then G is a homomorphic image of a free group F of countable rank: $G \cong F/R$. By Theorem 11.48, F can be imbedded in a free group F^* of rank 2. Were R a normal subgroup of F^* (it is not!), then $G = F/R$ would be imbedded in a group F^*/R having two generators. This proof is fictitious, but the theorem is true.

Theorem 11.71 (Higman, Neumann, and Neumann, 1949). *Every countable group G can be imbedded in a group H having two generators.*

Remark. This follows from Exercise 3.28 when G is finite.

Proof. Let $g_0 = 1$ and let $g_0, g_1, \ldots, g_n, \ldots$ be a list of all the elements of G. Let $H = G * F$, where F is free with basis $\{x, y\}$. Consider the subgroups of H:

$$A = \langle y, x^{-1}yx, \ldots, x^{-n}yx^n, \ldots\rangle$$

and

$$B = \langle x, g_1y^{-1}xy, \ldots, g_ny^{-n}xy^n, \ldots\rangle.$$

Now A is free with basis the displayed generators, by Exercise 11.45, and the map $\varphi: A \to B$ given by

$$\varphi: x^{-n}yx^n \mapsto g_ny^{-n}xy^n \qquad \text{for all} \quad n \geq 0$$

is easily seen to be an isomorphism. Theorem 11.70 gives a group H^Ω containing H and an element $t \in H^\Omega$ such that

$$\varphi(a) = t^{-1}at \qquad \text{for all} \quad a \in A.$$

We claim that $\langle y, t\rangle \leq H^\Omega$ contains G, and this will complete the proof. Now $x = \varphi(y) = t^{-1}yt \in \langle y, t\rangle$. Moreover, $t^{-1}x^{-n}yx^nt = \varphi(x^{-n}yx^n) = g_ny^{-n}xy^n$, and this shows that $g_n \in \langle x, y, t\rangle = \langle y, t\rangle$ for all $n \geq 1$. ■

Corollary 11.72. *If G is a countable group, then there is a 2-generator group E containing it such that, for all $n \geq 1$, E contains an element of order n if and only if G contains an element of order n.*

Proof. Observe, in the proof of Theorem 11.71, that the 2-generator group containing G is obtained in two steps. First, we formed the amalgam $H = G *_\varphi F$, where F is free. By the torsion theorem (Theorem 11.69), the only integers n which are orders of elements of finite order in H are those arising from G. The second step uses Theorem 11.70, where the ultimate group is a subgroup of an amalgam $K_1 *_\theta K_2$, where each $K_i \cong H * \mathbb{Z}$, and the torsion theorem applies again. ■

Theorem 11.73 (B.H. Neumann, 1953). *There are uncountably many nonisomorphic finitely generated groups.*

***Proof* (Schupp).** If S is a set of primes, define $G(S) = \sum_{p \in S} \mathbb{Z}_p$; as in Corollary 11.72, let $H(S)$ be a 2-generator group containing $G(S)$ which has an element of prime order p if and only if $p \in S$. It follows that if T is a subset of the primes, then $T \neq S$ implies $H(T) \ncong H(S)$, for there is a prime p in T that is not in S (or *vice versa*); thus $H(T)$ has an element of order p and $H(S)$ does not. As there are uncountably many subsets of the primes, there are thus uncountably many nonisomorphic 2-generator groups. ■

Exercises

11.77 **(Schupp).** Prove that Corollary 11.68 implies the normal form theorem (Theorem 11.66).

11.78. For every torsion-free group G, there exists a (necessarily simple) group H containing G which has exactly two conjugacy classes. (Compare Exercise 3.4.)

11.79. Prove that there exists a 2-generator group G which contains an isomorphic copy of every countable abelian group. (*Hint.* Exercise 10.31.)

11.80. Prove that there is a 2-generator group containing an isomorphic copy of every finite group.

11.81. Prove that a finitely presented group can be imbedded in a finitely presented group having two generators.

11.82. Consider the diagram

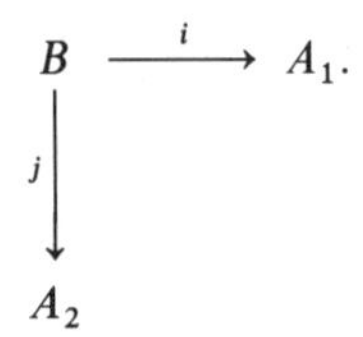

If A_1 and A_2 are finitely presented and if B is finitely generated, then the pushout is finitely presented.

HNN Extensions

There is another construction, closely related to amalgams, that we will use in the next chapter to prove that there exists a finitely presented group having unsolvable word problem.

We adopt the following notation. If a group G has a presentation $(X|\Delta)$, then

$$G^{\Omega} = (G\colon Y|\Delta')$$

denotes the group with presentation $(X \cup Y|\Delta \cup \Delta')$, where it is understood that X and Y are disjoint. In particular, if $Y = \varnothing$, then we are merely adjoining additional relations to G, so that G^{Ω} is a quotient of G. For example, in the notation of Theorem 11.58, the pushout P has the presentation $(B * C|i(a)j(a^{-1}), a \in A)$.

Definition. Let G be a group with isomorphic subgroups A and B, and let $\varphi\colon A \to B$ be an isomorphism. Then the group having the presentation

$$(G; p|p^{-1}ap = \varphi(a) \text{ for all } a \in A)$$

is called an ***HNN extension*** of G; it is denoted by $G\,\Omega_{\varphi}\,A$ or, less precisely, by $G\,\Omega\,A$. The group G is called the ***base*** and the generator p is called the ***stable letter*** of $G\,\Omega_{\varphi}\,A$.

The next theorem shows that HNN extensions appear in Theorem 11.70 of Higman, Neumann, and Neumann.

Theorem 11.74. *The subgroup $\langle G, t\rangle \leq K_1 *_{\theta} K_2$ in Theorem 11.70 is an HNN extension with base G and stable letter t.*

Proof. Let us recall the notation of Theorem 11.70. Begin with a group G and two subgroups A and B isomorphic via an isomorphism $\varphi\colon A \to B$. Let $K_1 = G * \langle u\rangle$, $K_2 = G * \langle v\rangle$, $L_1 = \langle G, u^{-1}Au\rangle \leq G * \langle u\rangle$, $L_2 = \langle G, v^{-1}Bv\rangle \leq G * \langle v\rangle$, and $\theta\colon L_1 \to L_2$ the isomorphism which carries each $g \in G$ into itself and which sends $u^{-1}au$ into $v^{-1}\varphi(a)v$. If a presentation of G is $(X|\Delta)$, then Theorem 11.58(ii) says that a presentation of the amalgam $K_1 *_{\theta} K_2$ is

$$(X, u, v|\Delta, u^{-1}au = v^{-1}\varphi(a)v \text{ for all } a \in A).$$

As $t = uv^{-1}$, we see that a presentation for $\langle G, t\rangle$ is

$$(X, t|\Delta, t^{-1}at = \varphi(a) \text{ for all } a \in A);$$

that is, $\langle G, t\rangle$ is an HNN extension with base G and stable letter t. ■

Note the resemblance between amalgams and HNN extensions. Both begin with a pair of isomorphic subgroups; the amalgam is a group in which the two subgroups are made equal; the HNN extension is a group in which the two subgroups are made conjugate. This observation is important in further study (see Dicks and Dunwoody, Lyndon and Schupp, Serre (1980), and Stallings).

Here is a geometric context in which HNN extensions arise. Consider a connected topological space X with homeomorphic disjoint subspaces A and B, and let $\varphi: A \to B$ be a homeomorphism. We are going to "add a handle" to X. Define a new space X^{Ω} as the quotient space of the disjoint union $X \cup (A \times \mathbf{I})$ (where $\mathbf{I}$ is the closed unit interval) by identifying each $a \in A$ with $(a, 0)$ and each $\varphi(a)$ with $(a, 1)$. The picture is as shown in Figure 11.8.

This construction can be carried out for complexes: the role of the closed unit interval $\mathbf{I}$ is played by the 1-simplex (also denoted by $\mathbf{I}$) having two vertices 0 and 1. If we denote the vertices in $\mathrm{Vert}(A \times \mathbf{I})$ by $a \times 0$ and $a \times 1$, where $a \in \mathrm{Vert}(A)$, then $A \times \mathbf{I}$ is made into a complex by "triangulating" it: if $\dim(A) = n$, then $\dim(A \times \mathbf{I}) = n + 1$ and its $(n + 1)$-simplexes are defined to be all subsets of the form

$$\{a_0 \times 0, \ldots, a_i \times 0, a_i \times 1, \ldots, a_n \times 1\},$$

where $\{a_0, \ldots, a_n\}$ is an n-simplex in A and $0 \le i \le n$. Notice that if $\{a, b\}$ is a 1-simplex in A, then $A \times \mathbf{I}$ has edges $(a \times 0, a \times 1)$, $(b \times 0, b \times 1)$,

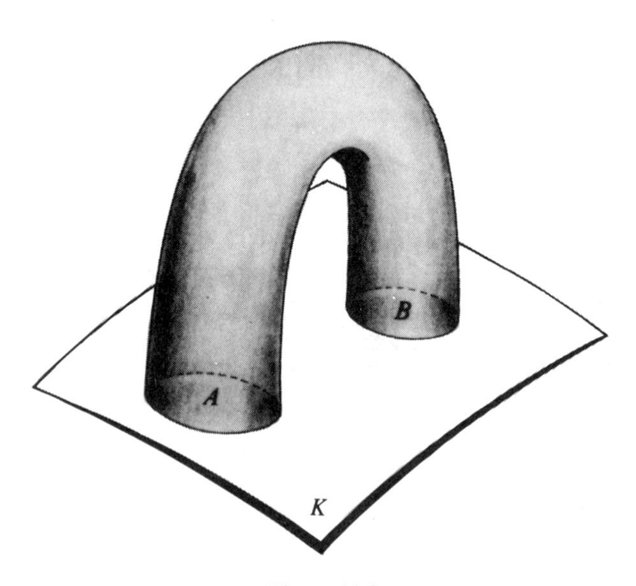

Figure 11.8

$(a \times 0, b \times 0)$, $(a \times 1, b \times 1)$, and $(a \times 0, b \times 1)$. It follows easily that $A \times \mathbf{I}$ is connected if A is connected.

For example, the cartesian product $\mathbf{I} \times \mathbf{I}$ has 2-simplexes pictured below.

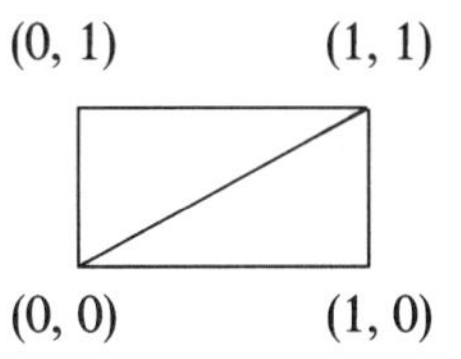

Remark. If $\alpha = (w, a_1)\dots(a_n, w)$ is a closed path in A, then $\alpha \times 0$ and $\alpha \times 1$ are closed paths in $A \times \mathbf{I}$, where $\alpha \times i = (w \times i, a_1 \times i)\dots(a_n \times i, w \times i)$ for $i \in \{0, 1\}$. Moreover, if β is the edge $(w \times 1, w \times 0)$, then

$$\alpha \times 0 \simeq \beta^{-1}(\alpha \times 1)\beta.$$

It is now easy to show that the injection $j\colon A \to A \times \mathbf{I}$, given by $a \mapsto a \times 0$, induces an isomorphism

$$j_{\#}\colon \pi(A, w) \cong \pi(A \times \mathbf{I}, w \times 0).$$

Definition. Let A and B be disjoint subcomplexes of a connected complex K, and let $\varphi\colon A \to B$ be an isomorphism. The complex obtained by ***adding a handle*** to K is

$$K^{\Omega} = (K \cup (A \times \mathbf{I}))/\mathscr{P},$$

where $\mathscr{P}$ identifies $a \in \operatorname{Vert}(A)$ with $a \times 0$ and $\varphi(a)$ with $a \times 1$.

Theorem 11.75. *Let K be a connected complex with disjoint isomorphic subcomplexes A and B; let $w \in \operatorname{Vert}(A)$ be a basepoint, and let $\varphi\colon A \to B$ be an isomorphism. If K^{Ω} is obtained from K by adding a handle according to this data, then $\pi(K^{\Omega}, w)$ is an HNN extension with base $\pi(K, w)$.*

Remark. It is this result that suggests the notation $G\,\Omega\,A$ for HNN extensions.

Proof. Since K is connected, there is a path γ in K from w to $\varphi(w)$; since A, hence $A \times \mathbf{I}$, is connected, there is a path β in the handle from $\varphi(w)$ to w. Define H as the union of γ and the handle (regard a path as the 1-complex consisting of its edges and their vertices); note that $K \cup H = K^{\Omega}$ and $K \cap H = A \cup B \cup \gamma$.

The van Kampen theorem (which applies because $K \cap H$ is connected)

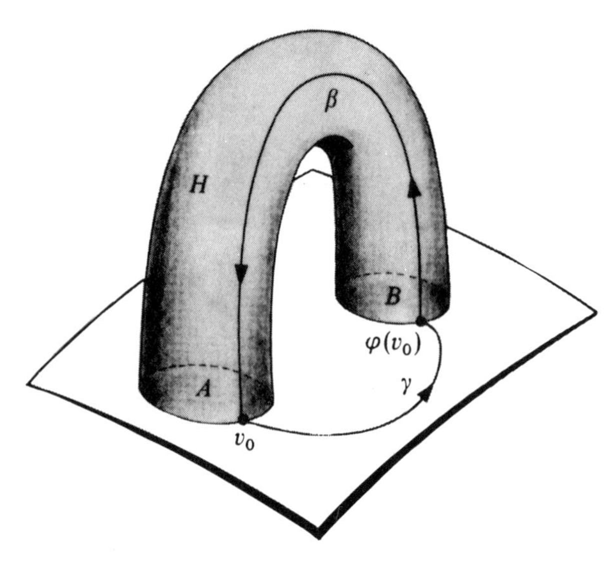

Figure 11.9

shows that $\pi(K^{\Omega}, w)$ is the pushout of the diagram

$$\begin{array}{ccc} \pi(A \cup B \cup \gamma, w) & \longrightarrow & \pi(H, w). \\ \downarrow & & \\ \pi(K, w) & & \end{array}$$

Since $A \cap (B \cup \gamma) = \gamma$, which is simply connected, Corollary 11.61 gives $\pi(A \cup B \cup \gamma, w) \cong \pi(A, w) * \pi(B \cup \gamma, w)$. Now $H = \gamma\beta \cup (H - \gamma)$, so that $\pi(H, w) \cong \pi(H - \gamma, w) * \pi(\gamma\beta, w)$, by Corollary 11.61. As A and B are disjoint, $H - \gamma \cong A \times \mathbf{I}$, where A is identified with $A \times 0$ and B is identified with $A \times 1$. Thus, each closed path α at w in A is identified with the closed path $\alpha \times 0$ at $w \times 0$. By the remark above,

$$\alpha \times 0 \simeq \beta^{-1}(\alpha \times 1)\beta \simeq \beta^{-1}\gamma^{-1}(\gamma(\alpha \times 1)\gamma^{-1})\gamma\beta,$$

where $\gamma(\alpha \times 1)\gamma^{-1}$ is a closed path at $w \times 0$ in $B \cup \gamma$. But $\alpha \times 1 = \varphi_{\#}\alpha$ (because $a = \varphi(a)$ in K^{Ω}), so that $\alpha = \alpha \times 0 \simeq (\gamma\beta)^{-1}\gamma\varphi_{\#}\alpha\gamma^{-1}(\gamma\beta)$. Since $\gamma\beta$ is a loop,

$$\pi(H, w) \cong \pi(H - \gamma, w) * \pi(\gamma\beta, w) \cong \pi(A \times \mathbf{I}, w) * \langle t \rangle,$$

where $t = [\gamma\beta]$ is a generator of the infinite cyclic $\pi(\gamma\beta, w)$. Finally, the injection $j\colon a \mapsto a \times 0$ induces a isomorphism

$$j_{\#}\colon \pi(A, w) \xrightarrow{\sim} \pi(A \times \mathbf{I}, w \times 0).$$

Thus, $\pi(H, w) \cong \pi(A, w) * \langle t \rangle$, and $[\alpha] = t^{-1}\varphi_{\#}[\alpha]t$, where $\varphi_{\#}\colon \pi(A, w) \to \pi(B \cup \gamma, w)$ is induced by the isomorphism φ. Now use the presentation given in van Kampen's theorem. ∎

Notice that the definition of HNN extension involves two isomorphic subgroups A and B of a group G. In contrast to the geometric situation in Theorem 11.75, there is no hypothesis that the subgroups be disjoint in any sense; indeed, $A = B$ is allowed.

We are going to generalize the definition of HNN extension so that it involves a set of stable letters.

Definition. Let a group E have a presentation $(X|\Delta)$, and let $\{p_i: i \in I\}$ be a nonempty set disjoint from X. Assume that there is an index set J and, for each $i \in I$, there is a family $\{a_{ij}, b_{ij}: j \in J\}$ of pairs of words on X. Then the group E^Ω has a presentation with ***base*** E and ***stable letters*** $\{p_i: i \in I\}$ if

$$E^\Omega = (E; p_i, i \in I | p_i^{-1} a_{ij} p_i = b_{ij} \text{ for all } i, j).$$

We allow a_{ij} and b_{ij} to be 1 so that the number of "honest" relations involving p_i (i.e., both a_{ij} and b_{ij} distinct from 1) may be distinct from the number of honest relations involving p_k for some $k \neq i$.

Lemma 11.76. *If E^Ω has a presentation with base E and stable letters $\{p_i: i \in I\}$, then $\langle p_i: i \in I\rangle \leq E^\Omega$ is a free group with basis $\{p_i: i \in I\}$.*

Proof. Let $\{z_i: i \in I\}$ be a set and let F be the free group with basis $\{z_i: i \in I\}$. If E has the presentation $E = (X|\Delta)$, then E^Ω has the presentation

$$E^\Omega = (X, p_i, i \in I | \Delta, p_i^{-1} a_{ij} p_i = b_{ij} \text{ for all } i, j).$$

Define a homomorphism $\varphi: E^\Omega \to F$ by $\varphi(x) = 1$ for all $x \in X$ and $\varphi(p_i) = z_i$ for all i; note that φ is well defined because it sends all the relations into 1. The lemma now follows from Exercise 11.8. ■

Notation. Let E^Ω have a presentation with base E and stable letters $\{p_i: i \in I\}$. For each i, define two subgroups of E:

$$A_i = \langle a_{ij}: j \in J\rangle \qquad \text{and} \qquad B_i = \langle b_{ij}: j \in J\rangle;$$

now define

$$A = \langle \bigcup A_i\rangle \qquad \text{and} \qquad B = \langle \bigcup B_i\rangle.$$

Definition. A group E^Ω having a presentation with base E and stable letters $\{p_i: i \in I\}$ is an ***HNN extension*** if, for each i, there is an isomorphism $\varphi_i: A_i \to B_i$ with $\varphi_i(a_{ij}) = b_{ij}$ for all j.

If there is only one stable letter p, then we can recapture the original definition of HNN extension by setting $\{a_j: j \in J\} = A$ (there is now no need to have the first index i) and $b_j = \varphi(a_j)$ for all j.

EXAMPLE 11.10. If E is a group and F is a free group with basis $\{p_i: i \in I\}$ then $E * F$ is an HNN extension of E with stable letters $\{p_i: i \in I\}$.

This is the "trivial" example in which all a_{ij} and b_{ij} are 1; of course, $E^\Omega = E \times \langle t \rangle \cong E \times \mathbb{Z}$ in this case.

EXAMPLE 11.11. If $E = (X \mid \Delta)$ and $E^\Omega = (E; t \mid t^{-1}\omega_i t = \omega_i, i \in I)$, where the ω_i are words on X, the E^Ω is an HNN extension with base E and stable letter t.

In this case, $A = B = \langle \omega_i, i \in I \rangle$, and the isomorphism $\varphi: A \to B$ is the identity.

EXAMPLE 11.12. Let E be the free group with basis $\{w, x\}$, and let E^Ω have the presentation

$$E^\Omega = (w, x, y, z \mid y^{-1}xy = w, y^{-1}w^{-1}xwy = xw^{-1}, z^{-1}wxz = w).$$

Now E^Ω has a presentation with base E and stable letters $\{y, z\}$. The various subgroups are: $A_y = \langle x, w^{-1}xw \rangle$; $B_y = \langle w, xw^{-1} \rangle$; $A_z = \langle wx \rangle$; $B_z = \langle w \rangle$. There is an isomorphism $\varphi_y: A_y \to B_y$ with $x \mapsto w$ and $w^{-1}xw \mapsto xw^{-1}$, because both groups are free with bases the displayed generating sets; there is also an isomorphism $\varphi_z: A_z \to B_z$ with $wx \mapsto w$, because both $A_z = \langle wx \rangle$ and $B_z = \langle w \rangle$ are infinite cyclic. It follows that E^Ω is an HNN extension with base E and stable letters $\{y, z\}$.

There are two natural questions about an HNN extension E^Ω of E. Is E imbedded in E^Ω? Is there a normal form for the elements of E^Ω?

The next lemma shows that an HNN extension with several stable letters can be viewed as the end result of a sequence of HNN extensions, each involving a single stable letter.

Lemma 11.77. *If E^Ω is an HNN extension of E with stable letters $\{p_1, \ldots, p_n\}$, then there is a group $E^\flat$ which is an HNN extension of E with stable letters $\{p_1, \ldots, p_{n-1}\}$ such that E^Ω is an HNN extension of $E^\flat$ with stable letter p_n.*

Proof. Define

$$E^\flat = (E; p_1, \ldots, p_{n-1} \mid p_i^{-1} a_{ij} p_i = b_{ij}, 1 \le i \le n-1, j \in J). \quad \blacksquare$$

Theorem 11.78. *If E^Ω is an HNN extension with base E and stable letters $\{p_1, \ldots, p_n\}$, then E can be imbedded in E^Ω. In particular, if A and B are isomorphic subgroups of a group E and if $\varphi: A \to B$ is an isomorphism, then $E \le E\,\Omega_\varphi\,A$.*

Proof. We prove the theorem by induction on the number n of stable letters. If $n = 1$, the result follows from Theorem 11.74. For the inductive step, the lemma gives E^Ω an HNN extension with base $E^\flat$ and stable letter p_n. The inductive hypothesis gives $E \le E^\flat$, and the step $n = 1$ gives $E^\flat \le E^\Omega$. $\blacksquare$

Corollary 11.79. *If K is a connected complex with basepoint w and if K^Ω is obtained from K by adding a handle, then*

$$\pi(K, w) \leq \pi(K^\Omega, w).$$

Here is a sharper version of Theorem 11.71.

Corollary 11.80. *Every countable group G can be imbedded in a 2-generator group $H = \langle t, x \rangle$ in which both t and x have infinite order.*

Proof. Let F be a free group with basis $\{x, y\}$. Enumerate the elements of G: $g_0 = 1, g_1, \ldots, g_n, \ldots$, and define

$$H = (G * F; t \mid t^{-1}x^{-n}yx^nt = g_ny^{-n}xy^n, n \geq 0).$$

The group H thus has a presentation with base $G * F$ and stable letter t. In the proof of Theorem 11.71, we saw that

$$A = \langle y, x^{-1}yx, \ldots, x^{-n}yx^n, \ldots \rangle$$

and

$$B = \langle x, g_1y^{-1}xy, \ldots, g_ny^{-n}xy^n, \ldots \rangle$$

are each free with bases the displayed generators; there is thus an isomorphism $\varphi: A \to B$ with $\varphi(x) = y$ and $\varphi(x^{-n}yx^n) = g_ny^{-n}xy^n$ for all $n \geq 1$. Therefore, H is an HNN extension of $G * F$, so that $G \leq G * F \leq H$, and $G \leq \langle t, x \rangle \leq H$, as in the proof of Theorem 11.71. Finally, $x \in F \leq H$ has infinite order, for F is free, while t has infinite order, by Lemma 11.76. ∎

Notation. Let F be a free group with basis X, let N be a normal subgroup of F, and let $G = F/N$. If ω and ω' are (not necessarily reduced) words on X, then we write

$$\omega = \omega' \quad \text{in } G$$

if $\omega N = \omega' N$ in F/N. We write

$$\omega \equiv \omega'$$

if ω and ω' have exactly the same spelling.

For example, if $X = \{a, b, c\}$, $\omega = ac$ and $\omega' = abb^{-1}c$, then $\omega = \omega'$ in G (even in F), but $\omega \not\equiv \omega'$.

In this new notation, if ω is a word on $X = \{x_1, \ldots, x_n\}$, then a word β is a ***subword*** of ω if there are possibly empty words α and γ with $\omega \equiv \alpha\beta\gamma$. A word ω ***involves*** x_i if either x_i or x_i^{-1} is a subword of ω.

Definition. Let E^Ω be an HNN extension with base $E = (X \mid \Delta)$ and stable letters $\{p_i: i \in I\}$. A ***pinch*** is a word of the form

$$p_i^e g p_i^{-e},$$

where g is a word on X such that the element $p_i^e g p_i^{-e}$ lies in $A_i \le A$ if $e = -1$ or the element $p_i^e g p_i^{-e}$ lies in $B_i \le B$ if $e = +1$.

Theorem 11.81 (Britton's Lemma, 1963). *Let E^Ω be an HNN extension with base $E = (X | \Delta)$ and stable letters $\{p_i : i \in I\}$. If ω is a word with $\omega = 1$ in E^Ω and which involves at least one stable letter, then ω has a pinch as a subword.*

***Proof* (Schupp).** Assume first that there is only one stable letter $p_1 = p$, so that the presentation of E^Ω is

$$E^\Omega = (X, p | \Delta, p^{-1} a_j p = b_j, j \in J).$$

Let A and B be isomorphic subgroups of E, and let $\varphi: A \to B$ be an isomorphism, so that the presentation may be rewritten

$$E^\Omega = (X, p | \Delta, p^{-1} a p = \varphi(a), a \in A).$$

We may assume that

$$\omega \equiv g_0 p^{e_1} g_1 \dots p^{e_n} g_n,$$

where $n \ge 1$, each $e_i = \pm 1$, and the g_i are (possibly empty) words on X.

As in Theorem 11.67, we view the amalgam $K_1 *_\theta K_2$ (where $K_1 = E * \langle u \rangle$, $K_2 = E * \langle v \rangle$) as having amalgamated subgroup $H = K_1 \cap K_2 = \langle E, u^{-1} A u \rangle = \langle E, v^{-1} B v \rangle$ satisfying $u^{-1} a u = v^{-1} \varphi(a) v$ for all $a \in A$. By Theorem 11.75, E^Ω can be identified with $\langle E, p \rangle \le K_1 *_\theta K_2$, where $p = uv^{-1}$. We are going to use the normal form theorem for elements in amalgams, as given in Corollary 11.68.

Return to the word ω. Since $p = uv^{-1}$,

$$\omega \equiv g_0 (uv^{-1})^{e_1} g_1 (uv^{-1})^{e_2} g_2 (uv^{-1})^{e_3} g_3 \dots (uv^{-1})^{e_n} g_n,$$

where $g_i \in E$. Reassociate ω according to the following instructions: assume that $e_1, \dots, e_i$ have the same sign, but that e_{i+1} has opposite sign. If $e_1 > 0$, reassociate

$$g_0 (uv^{-1})^{e_1} g_1 (uv^{-1})^{e_2} \dots (uv^{-1})^{e_i} g_i (uv^{-1})^{e_{i+1}}$$

as

$$(g_0 u) v^{-1} (g_1 u) v^{-1} \dots (g_{i-1} u)(v^{-1} g_i v) u^{-1};$$

if $e_1 < 0$, interchange the symbols u and v in the last expression; continue this rewriting process if $e_{i+1}, \dots, e_k$ have the same sign and e_{k+1} has opposite sign, until all of ω has been reassociated according to this scheme. Note that there are conjugates $v^{-1} g_i v$ or $u^{-1} g_i u$ wherever exponents change sign, and that adjacent factors (in the new association) lie in distinct K_i. Since $\omega = 1$ in the amalgam, by hypothesis, Corollary 11.68 says that at least one of the factors must lie in the amalgamated subgroup H. Now the factors u^ε, v^ε, $g_i u$, $g_i v$, where $\varepsilon = \pm 1$, do not lie in $H = \langle E, u^{-1} A u \rangle = \langle E, v^{-1} B v \rangle$, because the sum of the exponents of u or of v in any such word is not 0. Therefore, one of the conjugates $v^{-1} g_i v$ or $u^{-1} g_i u$ lies in H. A conjugate $v^{-1} g_i v$ arises from an element of $H = \langle E, v^{-1}Bv\rangle$ only when g_i lies in B (in the proof of Theorem 11.70, we saw that $\langle E, v^{-1}Bv\rangle \cong E * v^{-1}Bv$); of course, $v^{-1}g_iv$ arises in ω from the subword $uv^{-1}g_ivu^{-1} \equiv pg_ip^{-1}$; that is, from the pinch $p^e g_i p^{-e}$ with $e = +1$. A conjugate $u^{-1}g_iu$ arises from an element of $H = \langle E, u^{-1}Au\rangle$ only when g_i lies in A; of course, $u^{-1}g_iu$ arises from $vu^{-1}g_iuv^{-1} \equiv p^{-1}g_ip$; that is, from the pinch $p^e g_i p^{-e}$ with $e = -1$.

The general case follows by induction. As in the proof of Lemma 11.77, define

$$E^{\flat} = (E; p_1, \ldots, p_{n-1} | p_i^{-1} a_{ij} p_i = b_{ij}, 1 \le i \le n-1, j \in J),$$

and note that E^{Ω} is an HNN extension with base $E^{\flat}$ and stable letter p_n. If $\omega = 1$ does not involve p_n, then it contains a pinch $p_i^e g p_i^{-e}$ for some $i < n$ and induction completes the proof. If $i = n$, then the base step of the induction gives a pinch $p_n^e g p_n^{-e}$ as a subword of ω. If g is a word on X, we are done. If g involves some $p_1, \ldots, p_{n-1}$, then there is a word g' on X, namely, a word on b_{nj}, $j \in J$, with $g = g'$ in $E^{\flat}$. Thus, $gg'^{-1} = 1$ in $E^{\flat}$, and induction provides a pinch π which is a subword of gg'^{-1}. Since g'^{-1} contains no stable letters, π must be a subword of g, as desired. ■

Definition. Let E^{Ω} be an HNN extension with base $E = (X | \Delta)$ and stable letters $\{p_1, \ldots, p_t\}$. A word ω on $X \cup \{p_1, \ldots, p_t\}$ is ***p_i-reduced*** for some i if it contains no pinch $p_i^e w p_i^{-e}$ as a subword.

Note, in particular, that p_i-reduced words contains no subwords of the form $p_i^e p_i^{-e}$.

Corollary 11.82. *Let E^{Ω} be an HNN extension with base $E = (X | \Delta)$ and stable letters $\{p_1, \ldots, p_t\}$. Assume that*

$$\alpha \equiv \gamma_0 p_i^{e_1} \gamma_1 \cdots p_i^{e_m} \gamma_m \qquad \text{and} \qquad \beta \equiv \delta_0 p_i^{f_1} \delta_1 \cdots p_i^{f_n} \delta_n$$

are p_i-reduced words for some i, where each e_j, $f_k = \pm 1$ and none of the (possibly empty) words γ_j or δ_k involve p_i.

If $\alpha = \beta$ in E^{Ω}, then $m = n$, $(e_1, \ldots, e_m) = (f_1, \ldots, f_n)$, and the word $p_i^{e_m} \gamma_m \delta_n^{-1} p_i^{-f_n}$ is a pinch.

Proof. Since $\alpha\beta^{-1} = 1$ in E^{Ω}, Britton's lemma says that the word $\alpha\beta^{-1}$ contains a pinch as a subword. As each of α and β (hence β^{-1}) are p_i-reduced, the pinch must occur at the interface; that is, the subword $p_i^{e_m} \gamma_m \delta_n^{-1} p_i^{-f_n}$ is a pinch. It follows that the exponents e_m and $-f_n$ have opposite sign, and so $e_m = f_n$.

The proof is completed by induction on $\max\{m, n\}$. In the pinch $p_i^{e_m} \gamma_m \delta_n^{-1} p_i^{-f_n}$, Britton's lemma says that

$$\gamma_m \delta_n^{-1} = \begin{cases} a_{ij_1}^{h_1} \cdots a_{ij_t}^{h_t} & \text{if } e_m = -1, \\ b_{ij_1}^{h_1} \cdots b_{ij_t}^{h_t} & \text{if } e_m = +1, \end{cases}$$

where $a_{ij} \in A_i$, $b_{ij} \in B_i$, and $h_v = \pm 1$. If $e_m = -1$,

$$\begin{aligned} p_i^{e_m} \gamma_m \delta_n^{-1} p_i^{-f_n} &= p_i^{-1} a_{ij_1}^{h_1} \dots a_{ij_t}^{h_t} p_i \\ &= (p_i^{-1} a_{ij_1}^{h_1} p_i)(p_i^{-1} a_{ij_2}^{h_2} p_i) \dots (p_i^{-1} a_{ij_t}^{h_t} p_i) \\ &= b_{ij_1}^{h_1} \dots b_{ij_t}^{h_t}. \end{aligned}$$

We have eliminated one p_i from α and one from β, and so the remainder of the proof follows from the inductive hypothesis. The proof in the other case $e_m = +1$ is similar. ■

The normal form theorem for HNN extensions has its most succinct statement in the special case when there is only one stable letter; the statement and proof of the generalization for arbitrary HNN extensions is left to the reader.

Theorem 11.83 (Normal Form). *Let E^Ω be an HNN extension with base $E = (X \mid \Delta)$ and stable letter p. Then each word ω on $\{X, p\}$ is equal in E^Ω to a p-reduced word*

$$\omega_0 p^{e_1} \omega_1 \dots p^{e_n} \omega_n;$$

moreover, the length n and the sequence $(e_1, \dots, e_n)$ of exponents are uniquely determined by ω.

Proof. If ω contains a pinch π as a subword, then the relations in E^Ω allow one to replace π by a subword involving two fewer occurrences of the stable letter p. The uniqueness of the length and exponent sequence follow at once from Corollary 11.82. ■

It follows from Theorem 11.73 that there exist finitely generated groups that are not finitely presented, for there are only countably many finitely presented groups. Here is an explicit example of such a group.

Lemma 11.84. *If a group G has a presentation*

$$(x_1, \dots, x_m \mid \rho_1, \rho_2, \dots, \rho_n, \dots)$$

as well as a finite presentation

$$(y_1, \dots, y_k \mid \sigma_1, \sigma_2, \dots, \sigma_t),$$

then all but a finite number of the relations $\rho_n = 1$ in the first presentation are superfluous.

Proof. Let G_1 be the group defined by the first presentation, let G_2 be the group defined by the second presentation, and let $\varphi\colon G_1 \to G_2$ be an isomorphism with inverse ψ. Since $\psi(\sigma_i) = 1$ in G_1 for each i, it is a word on (finitely many) conjugates of ρ_n's; as there are only finitely many σ_i, only finitely many

ρ_n suffice to prove all $\psi(\sigma_i) = 1$. For notational convenience, let us denote these by $\rho_1, \rho_2, \ldots, \rho_N$.

Since $\varphi(\rho_n) = 1$ in G_2 for each n,

$$\varphi(\rho_n) = \omega_n(\sigma_1, \sigma_2, \ldots, \sigma_t),$$

where ω_n is a word on conjugates of $\sigma_1, \sigma_2, \ldots, \sigma_t$. Therefore,

$$\rho_n = \psi\varphi(\rho_n) = \psi(\omega_n(\sigma_1, \ldots, \sigma_t)) = \omega_n(\psi(\sigma_1), \ldots, \psi(\sigma_t)).$$

But this equation says that every relation ρ_n lies in the normal subgroup generated by $\rho_1, \rho_2, \ldots, \rho_N$, which is what is to be proved. ∎

The following explicit example was found by W.W. Boone.

Theorem 11.85. *Let F be the free group with basis $\{a, b\}$, and let its commutator subgroup F' be free with basis $\{\omega_1, \ldots, \omega_n, \ldots\}$. Then the group G having the presentation*

$$G = (a, b, p \mid p^{-1}\omega_n p = \omega_n, n \geq 1)$$

is a finitely generated group that is not finitely presented.

Proof. Recall first that Theorem 11.48 shows that F' is a free group of infinite rank. Were G finitely presented, then the lemma would say all but a finite number of the relations could be deleted; that is, there is some N with

$$G = (a, b, p \mid p^{-1}\omega_1 p = \omega_1, \ldots, p^{-1}\omega_{N-1} p = \omega_{N-1}).$$

The displayed presentation exhibits G as an HNN extension with base $F = \langle a, b\rangle$ and stable letter p. Since $p^{-1}\omega_N p = \omega_N$ in G. Britton's lemma provides a word β on $\langle \omega_1, \ldots, \omega_{N-1}\rangle$ with $\omega_N = \beta$ in $F = \langle a, b\rangle$, and this contradicts the fact that $\langle \omega_1, \ldots, \omega_N\rangle$ is a free group with basis $\{\omega_1, \ldots, \omega_N\}$. ∎

Exercises

11.83. Let $a \in G$ have infinite order. Show that there is a group H containing G in which $\langle a\rangle$ and $\langle a^2\rangle$ are conjugate. Conclude that a conjugate of a subgroup S may be a proper subgroup of S.

11.84 **(Schupp).** Use the normal form theorem for HNN extensions to prove the normal form theorem for amalgams.

CHAPTER 12

The Word Problem

Introduction

Novikov, Boone, and Britton proved, independently, that there is a finitely presented group $\mathscr{B}$ for which no computer can ever exist that can decide whether an arbitrary word on the generators of $\mathscr{B}$ is 1. We shall prove this remarkable result in this chapter.

Informally, if $\mathscr{L}$ is a list of questions, then a ***decision process*** (or *algorithm*) for $\mathscr{L}$ is a uniform set of directions which, when applied to any of the questions in $\mathscr{L}$, gives the correct answer "yes" or "no" after a finite number of steps, never at any stage of the process leaving the user in doubt as to what to do next.

Suppose now that G is a finitely generated group with the presentation

$$G = (x_1, \ldots, x_n | r_j = 1, j \geq 1);$$

every (not necessarily reduced) word ω on $X = \{x_1, \ldots, x_n\}$ determines an element of G (namely, ωR, where F is the free group with basis X and R is the normal subgroup of F generated by $\{r_j, j \geq 1\}$). We say that G has a ***solvable word problem*** if there exists a decision process for the set $\mathscr{L}$ of all questions of the form: If ω is a word on X, is $\omega = 1$ in G? (It appears that solvability of the word problem depends on the presentation. However, it can be shown that if G is finitely generated and if its word problem is solvable for one presentation, then it is solvable for every presentation with a finite number of generators.)

Arrange all the words on $\{x_1, \ldots, x_n\}$ in a list as follows: Recall that the *length* of a (not necessarily reduced) word $\omega = x_1^{e_1} \ldots x_m^{e_m}$, where $e_i = \pm 1$, is m. For example, the empty word 1 has length 0, but the word xx^{-1} has length 2. Now list all the words on X as follows: first the empty word, then the words of length 1 in the order $x_1, x_1^{-1}, \ldots, x_n, x_n^{-1}$, then the words of length 2 in "lexicographic" order (as in a dictionary): $x_1x_1 < x_1x_1^{-1} < x_1x_2 < \cdots < x_1^{-1}x_1 < x_1^{-1}x_1^{-1} < \cdots < x_n^{-1}x_n^{-1}$, then the words of length 3 in lexicographic order, and so forth. Use this ordering of words: $\omega_0, \omega_1, \omega_2, \ldots$ to define the list $\mathscr{L}$ whose kth question asks whether $\omega_k = 1$ in G.

We illustrate by sketching a proof that a free group

$$G = (x_1, \ldots, x_n | \varnothing)$$

has a solvable word problem. Here is a decision process.

1. If length$(\omega_k) = 0$ or 1, proceed to Step 3. If length$(\omega_k) \geq 2$, underline the first adjacent pair of letters, if any, of the form $x_i x_i^{-1}$ or $x_i^{-1} x_i$; if there is no such pair, underline the final two letters; proceed to Step 2.
2. If the underlined pair of letters has the form $x_i x_i^{-1}$ or $x_i^{-1} x_i$, erase it and proceed to Step 1; otherwise, proceed to Step 3.
3. If the word is empty, write $\omega_k = 1$ and stop; if the word is not empty, write $\omega_k \neq 1$ and stop.

The reader should agree, even without a formal definition, that the set of directions above is a decision process showing that the free group G has a solvable word problem.

The proof of the Novikov–Boone–Britton theorem can be split in half. The initial portion is really Mathematical Logic, and it is a theorem, proved independently by Markov and Post, that there exists a finitely presented semigroup S having an unsolvable word problem. The more difficult portion of the proof consists of constructing a finitely presented group $\mathscr{B}$ and showing that if $\mathscr{B}$ had a solvable word problem, then S would have a solvable word problem. Nowhere in the reduction of the group problem to the semigroup problem is a technical definition of a solvable word problem used, so that the reader knowing only our informal discussion above can follow this part of the proof. Nevertheless, we do include a precise definition below. There are several good reasons for doing so: the word problem can be properly stated; a proof of the Markov–Post theorem can be given (and so the generators and relations of the Markov–Post semigroup can be understood); a beautiful theorem of G. Higman (characterizing the finitely generated subgroups of finitely presented groups) can be given. Here are two interesting consequences: Theorem 12.30 (Boone–Higman): there is a purely algebraic characterization of groups having a solvable word problem; Theorem 12.32 (Adian–Rabin): given almost any interesting property P, there is no decision process which can decide, given an arbitrary finite presentation, whether or not the presented group enjoys P.

Exercises

12.1. Sketch a proof that every finite group has a solvable word problem.

12.2. Sketch a proof that every finitely generated abelian group has a solvable word problem. (*Hint*. Use the fundamental theorem of finitely generated abelian groups.)

12.3. Sketch proofs that if each of G and H have a solvable word problem, then the same is true of their free product $G * H$ and their direct product $G \times H$.

12.4. Sketch a proof that if $G = (x_1, \ldots, x_n | r_j = 1, j \geq 1)$ has a solvable word problem and if H is a finitely generated subgroup of G, then H has a solvable word problem. (*Hint*. If $H = \langle h_1, \ldots, h_m \rangle$, write each h_i as a word in the x.)

Turing Machines

Call a subset E of a (countable) set Ω "enumerable" if there is a computer that can recognize every element of E and no others. Of course, the nature of such a well-behaved subset E should not depend on any accidental physical constraints affecting a real computer; for example, it should not depend on the number of memory cells being less than the total number of atoms in the universe. We thus define an idealized computer, called a *Turing machine*, after its inventor A. Turing (1912–1954), which abstracts the essential features of a real computer and which enumerates only those subsets E that, intuitively, "ought" to be enumerable.

Informally, a Turing machine can be pictured as a box with a tape running through it. The tape consists of a series of squares, which is as long to the left and to the right as desired. The box is capable of printing a finite number of symbols, say, $s_0, s_1, \ldots, s_M$, and of being in any one of a finite number of states, say, $q_0, q_1, \ldots, q_N$. At any fixed moment, the box is in some state q_i as it "scans" a particular square of the tape that bears a single symbol s_j (we agree that s_0 means blank). The next move of the machine is determined by q_i and s_j and its initial structure: it goes into some state q_l after obeying one of the following instructions:

1. Replace the symbol s_j by the symbol s_k and scan the same square.
2. Move one square to the right and scan this square.
3. Move one square to the left and scan this square.

The machine is now ready for its next move. The machine is started in the first place by being given a tape, which may have some nonblank symbols printed on it, one to a square, and by being set to scan some one square while in "starting state" q_1. The machine may eventually stop (we agree that q_0 means "stop"; that is, the machine stops when it enters state q_0) or it may continue working indefinitely.

Here are the formal definitions; after each definition, we shall give an informal interpretation. Choose, once and for all, two infinite lists of letters:

$$s_0, s_1, s_2, \ldots \quad \text{and} \quad q_0, q_1, q_2, \ldots.$$

Definition. A ***quadruple*** is a 4-tuple of one of the following three types:

$$q_i s_j s_k q_l,$$

$$q_i s_j R q_l,$$

$$q_i s_j L q_l.$$

A ***Turing machine*** T is a finite set of quadruples no two of which have the same first two letters. The ***alphabet*** of T is the set $\{s_0, s_1, \ldots, s_M\}$ of all s-letters occurring in its quadruples.

The three types of quadruples correspond to the three types of moves in the informal description given above. For example, $q_i s_j R q_l$ may be interpreted as being the instruction: "When in state q_i and scanning symbol s_j, move right one square and enter state q_l." The "initial structure" of the Turing machine is the set of all such instructions.

Recall that a word is ***positive*** if it is empty or if it has only positive exponents. If an alphabet A is a disjoint union $S \cup T$, where $S = \{s_i : i \in I\}$, then an ***s-word*** is a word on S.

Definition. An ***instantaneous description*** α is a positive word of the form $\alpha = \sigma q_i \tau$, where σ and τ are s-words and τ is not empty.

For example, the instantaneous description $\alpha = s_2 s_0 q_1 s_5 s_2 s_2$ is to be interpreted: the symbols on the tape are $s_2 s_0 s_5 s_2 s_2$, *with blanks everywhere else*, and the machine is in state q_1 scanning s_5.

Definition. Let T be a Turing machine. An ordered pair (α, β) of instantaneous descriptions is a ***basic move*** of T, denoted by

$$\alpha \to \beta,$$

if there are (possibly empty) positive s-words σ and σ' such that one of the following conditions hold:

(i) $\alpha = \sigma q_i s_j \sigma'$ and $\beta = \sigma q_l s_k \sigma'$, where $q_i s_j s_k q_l \in T$;

(ii) $\alpha = \sigma q_i s_j s_k \sigma'$ and $\beta = \sigma s_j q_l s_k \sigma'$, where $q_i s_j R q_l \in T$;

(iii) $\alpha = \sigma q_i s_j$ and $\beta = \sigma s_j q_l s_0$, where $q_i s_j R q_l \in T$;

(iv) $\alpha = \sigma s_k q_i s_j \sigma'$ and $\beta = \sigma q_l s_l s_j \sigma'$, where $q_i s_j L q_l \in T$; and
(v) $\alpha = q_i s_j \sigma'$ and $\beta = q_l s_0 s_k \sigma'$, where $q_i s_j L q_l \in T$.

If α describes the tape at a given time, the state q_i of T, and the symbol s_j being scanned, then β describes the tape, the next state of T, and the symbol being scanned after the machine's next move. The proviso in the definition of a Turing machine that no two quadruples have the same first two symbols means that there is never ambiguity about a machine's next move: if $\alpha \to \beta$ and $\alpha \to \gamma$, then $\beta = \gamma$.

Some further explanation is needed to interpret basic moves of types (iii) and (v). Tapes are finite, but when the machine comes to an end of the tape, the tape is lengthened by adjoining a blank square. Since s_0 means blank, these two rules thus correspond to the case when T is scanning either the last symbol on the tape or the first symbol.

Definition. An instantaneous description α is ***terminal*** if there is no instantaneous description β with $\alpha \to \beta$. If ω is a positive word on the alphabet of T, then T ***computes*** ω if there is a finite sequence of instantaneous descriptions $\alpha_1 = q_1\omega, \alpha_2, \ldots, \alpha_t$, where $\alpha_i \to \alpha_{i+1}$, for all $i \leq t - 1$, and α_t is terminal.

Informally, ω is printed on the tape and T is in starting state q_1 while scanning the first square. The running of T is a possibly infinite sequence of instantaneous descriptions $q_1\omega \to \alpha_2 \to \alpha_3 \to \cdots$. This sequence stops if T computes ω; otherwise, T runs forever.

Definition. Let Ω be the set of all positive words on symbols $S = \{s_1, \ldots, s_M\}$. If T is a Turing machine whose alphabet contains S, define

$$e(T) = \{\omega \in \Omega: T \text{ computes } \omega\},$$

and say that T ***enumerates*** $e(T)$. A subset E of Ω is ***r.e.*** (***recursively enumerable***) if there is some Turing machine T that enumerates E.

The notion of an r.e. subset of Ω can be specialized to subsets of the natural numbers $\mathbb{N} = \{n \in \mathbb{Z}: n \geq 0\}$ by identifying each $n \in \mathbb{N}$ with the positive word s_1^{n+1}. Thus, a subset E of $\mathbb{N}$ is an ***r.e. subset of*** $\mathbb{N}$ if there is a Turing machine T with s_1 in its alphabet such that $E = \{n \in \mathbb{N}: T \text{ computes } s_1^{n+1}\}$.

Every Turing machine T defines an r.e. subset $E = e(T) \subset \Omega$, the set of all positive words on its alphabet. How can we tell whether $\omega \in \Omega$ lies in E? Feed $q_1\omega$ into T and wait; that is, perform the basic moves $q_1\omega \to \alpha_2 \to \alpha_3 \to \cdots$. If $\omega \in E$, then T computes ω and so T will eventually stop. However, for a given ω, there is no way of knowing, *a priori*, whether T will stop. Certainly this is unsatisfactory for an impatient person, but, more important, it suggests a new idea.

Definition. Let Ω be the set of all positive words on $\{s_0, s_1, \ldots, s_M\}$. A subset E of Ω is ***recursive*** if both E and its complement $\Omega - E$ are r.e. subsets.

If E is recursive, there is never an "infinite wait" to decide whether or not a positive word ω lies in E. If T is a Turing machine with $e(T) = E$ and if T' is a Turing machine with $e(T') = \Omega - E$, then, for each $\omega \in \Omega$, either T or T' computes ω. Thus, it can be decided in a finite length of time whether or not a given word ω lies in E: just feed ω into each machine and let T and T' run simultaneously.

Recall the informal discussion in the introduction. If $\mathscr{L}$ is a list of questions, then a decision process for $\mathscr{L}$ is a uniform set of directions which, when applied to any of the questions in $\mathscr{L}$, gives the correct answer "yes" or "no" after a finite number of steps, never at any stage of the process leaving the user in doubt as to what to do next. It is no loss of generality to assume that the list $\mathscr{L}$ has been encoded as positive words on an alphabet, that E consists of all words for which the answer is "yes," and that its complement consists of all words for which the answer is "no." We propose that recursive sets are precisely those subsets admitting a decision process. Of course, this proposition (called ***Church's thesis***) can never be proved, for it is a question of translating an intuitive notion into precise terms. There have been other attempts to formalize the notion of decision process (e.g., using a Turing-like machine that can read a two-dimensional tape; or, avoiding Turing machines altogether and beginning with a notion of computable function). So far, every alternative definition of "decision process" which recognizes all recursive sets has been proved to recognize only these sets.

Theorem 12.1. *There exists an r.e. subset of the natural numbers $\mathbb{N}$ that is not recursive.*

Proof. There are only countably many Turing machines, for a Turing machine is a finite set of quadruples based on the countable set of letters $\{R, L, s_0, s_1, \ldots; q_0, q_1, \ldots\}$. Assign natural numbers to these letters in the following way:

$$R \mapsto 0; \quad L \mapsto 1; \quad q_0 \mapsto 2; \quad q_1 \mapsto 4; \quad q_2 \mapsto 6; \quad \cdots$$
$$s_0 \mapsto 3; \quad s_1 \mapsto 5; \quad s_2 \mapsto 7; \quad \cdots.$$

If T is a Turing machine having m quadruples, juxtapose them in some order to form a word $w(T)$ of length $4m$; note that $T \neq T'$ implies $w(T) \neq w(T')$. Define the ***Gödel number***

$$G(T) = \prod_{i=1}^{4m} p_i^{e_i},$$

where p_i is the ith prime and e_i is the natural number assigned above to the ith letter in $w(T)$. The Fundamental Theorem of Arithmetic implies that distinct Turing machines have distinct Gödel numbers. All Turing machines can now be enumerated: $T_0, T_1, \ldots, T_n, \ldots$: let T precede T' if $G(T) < G(T')$.

Define

$$E = \{n \in \mathbb{N}: T_n \text{ computes } s_1^{n+1}\}$$

(thus, $n \in E$ if and only if the nth Turing machine computes n).

We claim that E is an r.e. set. Consider the following figure reminiscent of the proof that the set of all rational numbers is countable:

$$
\begin{array}{ccccccccc}
T_0 & & T_1 & & T_2 & & T_3 & & T_4 \\
q_1 s_1 & & q_1 s_1^2 & \rightarrow & q_1 s_1^3 & & q_1 s_1^4 & \rightarrow & q_1 s_1^5 \\
\downarrow & \nearrow & & \swarrow & & \nearrow & & \swarrow & \\
\alpha_{12} & & \alpha_{22} & & \alpha_{32} & & \alpha_{42} & & \\
& \swarrow & & \nearrow & & \swarrow & & & \\
\alpha_{13} & & \alpha_{23} & & \alpha_{33} & & & & \\
\downarrow & \nearrow & & \swarrow & & & & & \\
\alpha_{14} & & \alpha_{24} & & & & & &
\end{array}
$$

The nth column consists of the sequence of basic moves of the nth Turing machine T_n beginning with $q_1 s_1^{n+1}$. It is intuitively clear that there is an enumeration of the natural numbers n lying in E: follow the arrows in the figure, and put n in E as soon as one reachers a terminal instantaneous description α_{ni} in column n. A Turing machine T^* can be constructed to carry out these instructions (by Exercise 12.11 below, such a T^* exists having ***stopping state*** q_0; that is, terminal instantaneous descriptions, and only these, involve q_0.) Thus, E is an r.e. subset of $\mathbb{N}$.

The argument showing that E is not recursive is a variation of Cantor's diagonal argument proving that the set of reals is uncountable. It suffices to prove that the complement

$$E' = \{n \in \mathbb{N}: n \notin E\} = \{n \in \mathbb{N}: T_n \text{ does not compute } s_1^{n+1}\}$$

is not an r.e. subset of $\mathbb{N}$. Suppose there were a Turing machine T' enumerating E'; since all Turing machines have been listed, $T' = T_m$ for some $m \in \mathbb{N}$. If $m \in E' = e(T') = e(T_m)$, then T_m computes s_1^{m+1}, and so $m \in E$, a contradiction. If $m \notin E'$, then $m \in E$ and so T_m computes s_1^{m+1} (definition of E); hence $m \in e(T_m) = e(T') = E'$, a contradiction. Therefore, E' is not an r.e. set and E is not recursive. ■

Exercises

12.5. Prove that there are subsets of $\mathbb{N}$ that are not r.e. (*Hint*. There are only countably many Turing machines.)

12.6. Prove that the set of all even natural numbers is r.e.

12.7. Give an example of a Turing machine T, having s_1 in its alphabet, which does not compute s_1.

12.8. Let Ω be the set of all positive words on $\{s_0, s_1, \ldots, s_M\}$. If E_1 and E_2 are r.e. subsets of Ω, then both $E_1 \cup E_2$ and $E_1 \cap E_2$ are also r.e. subsets.

12.9. Let Ω be the set of all positive words on $\{s_0, s_1, \ldots, s_M\}$. If E_1 and E_2 are recursive subsets of Ω, then both $E_1 \cup E_2$ and $E_1 \cap E_2$ are also recursive subsets. Conclude that all recursive subsets of Ω form a Boolean algebra.

12.10. If E_1 and E_2 are recursive subsets of $\mathbb{N}$, then $E_1 \times E_2$ is a recursive subset of $\mathbb{N} \times \mathbb{N}$. (*Hint*. First imbed $\mathbb{N} \times \mathbb{N}$ into $\mathbb{N}$ by "encoding" the ordered pair (m, n) as $2^m 3^n$.)

12.11. If T is a Turing machine enumerating a set E, then there is a Turing machine T^* having the same alphabet and with stopping state q_0 that also enumerates E.

The Markov–Post Theorem

We now link these ideas to algebra.

If Γ is a semigroup with generators $X = \{x_1, \ldots, x_n\}$ and if Ω is the set of all positive words on X, then the semigroup Γ has a ***solvable word problem*** if there is a decision process to determine, for an arbitrary pair of words ω, $\omega' \in \Omega$, whether $\omega = \omega'$ in Γ. This (informal) definition gives a precise definition of unsolvability.

Definition. Let Γ be a semigroup with generators $X = \{x_1, \ldots, x_n\}$, and let Ω be the set of all positive words on X. The semigroup Γ has an ***unsolvable word problem*** if there is a word $\omega_0 \in \Omega$ such that $\{\omega \in \Omega: \omega = \omega_0 \text{ in } \Gamma\}$ is not recursive.

If F is the free group with basis $X = \{x_1, \ldots, x_n\}$, then we shall view the set Ω of all (not necessarily positive) words on X as the set of all positive words on the alphabet

$$\{x_1, x_1^{-1}, \ldots, x_n, x_n^{-1}\}.$$

Definition. Let G be a group with presentation $(x_1, \ldots, x_n | \Delta)$, and let Ω be the set of all words on $x_1, \ldots, x_n$ (viewed as the set of positive words on $\{x_1, x_1^{-1}, \ldots, x_n, x_n^{-1}\}$). Then G has a ***solvable word problem*** if $\{\omega \in \Omega: \omega = 1 \text{ in } G\}$ is recursive.

The distinction between r.e. sets and recursive sets persists in group theory.

Theorem 12.2. *Let G be a finitely presented group with presentation*

$$G = (x_1, \ldots, x_n | r_1, \ldots, r_m).$$

If Ω is the set of all words on $x_1, \ldots, x_n$, then $E = \{\omega \in \Omega: \omega = 1 \text{ in } G\}$ is r.e.

Proof. List the words $\omega_0, \omega_1, \ldots$ in Ω as we did in the Introduction: first the empty word, then the words of length 1 in order $x_1, x_1^{-1}, \ldots, x_n, x_n^{-1}$, then the words of length 2 in lexicographic order, then the words of length 3 in lexicographic order, and so forth. Similarly, list all the words on $\{r_1, \ldots, r_m\}$: ρ_0, $\rho_1, \ldots$. As in the proof of Theorem 12.1, following the arrows in the figure below enumerates E.

$$\begin{array}{ccccccc}
\omega_0\rho_0\omega_0^{-1} & & \omega_0\rho_1\omega_0^{-1} & \rightarrow & \omega_0\rho_2\omega_0^{-1} & & \omega_0\rho_3\omega_0^{-1} \rightarrow \\
\downarrow & \nearrow & & \swarrow & & \nearrow & \\
\omega_1\rho_0\omega_1^{-1} & & \omega_1\rho_1\omega_1^{-1} & & \omega_1\rho_2\omega_1^{-1} & & \\
& \swarrow & & \nearrow & & & \\
\omega_2\rho_0\omega_2^{-1} & & \omega_2\rho_1\omega_2^{-1} & & & & \\
\downarrow & \nearrow & & & & & \\
\omega_3\rho_0\omega_3^{-1} & & \blacksquare & & & &
\end{array}$$

It follows that a finitely presented group G has solvable word problem if and only if $\{\omega \in \Omega: \omega \neq 1 \text{ in } G\}$ is r.e.

Recall the following notation introduced in Chapter 11. If ω and ω' are (not necessarily reduced) words on an alphabet X, then we write

$$\omega \equiv \omega'$$

if ω and ω' have exactly the same spelling.

Suppose that a semigroup Γ has a presentation

$$\Gamma = (X \mid \alpha_j = \beta_j, j \in J).$$

If ω and ω' are positive words on X, then it is easy to see that $\omega = \omega'$ in Γ if and only if there is a finite sequence

$$\omega \equiv \omega_1 \rightarrow \omega_2 \rightarrow \cdots \rightarrow \omega_t \equiv \omega',$$

where $\omega_i \rightarrow \omega_{i+1}$ is an ***elementary operation***; that is, either $\omega_i \equiv \sigma\alpha_j\tau$ and $\omega_{i+1} \equiv \sigma\beta_j\tau$ for some j, where σ and τ are positive words on X or $\omega_{i+1} \equiv \sigma\beta_j\tau$ and $\omega_i \equiv \sigma\alpha_j\tau$.

Let us now associate a semigroup to a Turing machine T having stopping state q_0. For notational convenience, assume that the s-letters and q-letters involved in the quadruples of T are $s_0, s_1, \ldots, s_M$, and $q_0, q_1, \ldots, q_N$. Let q and h be new letters.

Definition. If T is a Turing machine having stopping state q_0, then its ***associated semigroup*** $\Gamma(T)$ has the presentation:

$$\Gamma(T) = (q, h, s_0, s_1, \ldots, s_M, q_0, q_1, \ldots, q_N \mid R(T)),$$

where the relations $R(T)$ are

$$q_i s_j = q_l s_k \qquad \text{if} \quad q_i s_j s_k q_l \in T,$$

for all $\beta = 0, 1, \ldots, M$:

$$q_i s_j s_\beta = s_j q_l s_\beta \qquad \text{if} \quad q_i s_j R q_l \in T,$$

$$q_i s_j h = s_j q_l s_0 h \qquad \text{if} \quad q_i s_j R q_l \in T;$$

$$s_\beta q_i s_j = q_l s_\beta s_j \qquad \text{if} \quad q_i s_j L q_l \in T,$$

$$h q_i s_j = h q_l s_0 s_j \qquad \text{if} \quad q_i s_j L q_l \in T;$$

$$q_0 s_\beta = q_0,$$

$$s_\beta q_0 h = q_0 h,$$

$$h q_0 h = q.$$

The first five types of relations are just the obvious ones suggested by the basic moves of T; the new letter h enables one to distinguish basic move (ii) (in the definition of a Turing machine) from basic move (iii) and to distinguish basic move (iv) from basic move (v). One may thus interpret h as marking the ends of the tape, so that the following words are of interest.

Definition. A word is ***h*-special** if it has the form $h\alpha h$, where α is an instantaneous description.

Since T has stopping state q_0, each $h\alpha h$ (with α terminal) has the form $h\sigma q_0 \tau h$, where σ and τ are s-words and τ is not empty. Therefore, the last three relations allow us to write $h\alpha h = q$ in $\Gamma(T)$ whenever α is terminal.

Lemma 12.3. *Let T be a Turing machine with stopping state q_0 and associated semigroup*

$$\Gamma(T) = (q, h, s_0, s_1, \ldots, s_M, q_0, q_1, \ldots, q_N \mid R(T)).$$

(i) *Let ω and ω' be words on $\{s_0, s_1, \ldots, s_M, q_0, q_1, \ldots, q_N\}$ with $\omega \not\equiv q$ and $\omega' \not\equiv q$. If $\omega \to \omega'$ is an elementary operation, then ω is h-special if and only if ω' is h-special.*

(ii) *If $\omega = h\alpha h$ is h-special, $\omega' \not\equiv q$, and $\omega \to \omega'$ is an elementary operation of one of the first five types, then $\omega' \equiv h\beta h$, where either $\alpha \to \beta$ or $\beta \to \alpha$ is a basic move of T.*

Proof. (i) This is true because the only relation that creates or destroys h is $h q_0 h = q$.

(ii) By the first part, we know that ω' is h-special, say, $\omega' \equiv h\beta h$. Now an elementary move in a semigroup is a substitution using an equation in a defining relation; such a relation in $\Gamma(T)$ of one of the first five types corresponds to a quadruple of T, and a quadruple corresponds to a basic move. Thus, either $\alpha \to \beta$ or $\beta \to \alpha$. ■

Lemma 12.4. *Let T be a Turing machine with stopping state q_0, let Ω be the set* *of all positive words on the alphabet of T, and let $E = e(T)$. If $\omega \in \Omega$, then*

$$\omega \in E \quad \textit{if and only if} \quad hq_1\omega h = q \quad \textit{in } \Gamma(T).$$

Proof. If $\omega \in E$, then there are instantaneous descriptions $\alpha_1 = q_1\omega, \alpha_2, \ldots, \alpha_t$, where $\alpha_i \to \alpha_{i+1}$, and α_t involves q_0. Using the elementary operations in $\Gamma(T)$ of the first five types, one sees that $hq_1\omega h = h\alpha_t h$ in $\Gamma(T)$; using the last three relations, one sees that $h\alpha_t h = q$ in $\Gamma(T)$.

The proof of sufficiency is of a different nature than the proof of necessity just given, for equality in $\Gamma(T)$ is, of course, a symmetric relation, whereas $\alpha \to \beta$ a basic move does not imply that $\beta \to \alpha$ is a basic move.

If $hq_1\omega h = q$ in $\Gamma(T)$, then there are words $\omega_1, \ldots, \omega_t$ on $\{h, s_0, s_1, \ldots, s_M, q_0, q_1, \ldots, q_N\}$ and elementary operations

$$hq_1\omega h \equiv \omega_1 \to \omega_2 \to \cdots \to \omega_t \equiv hq_0 h \to q.$$

By Lemma 12.3(i), each ω_i is h-special: $\omega_i \equiv h\alpha_i h$ for some instantaneous description α_i. By Lemma 12.3(ii), either $\alpha_i \to \alpha_{i+1}$ or $\alpha_{i+1} \to \alpha_i$. We prove, by induction on $t \geq 2$, that all the arrows go to the right; that is, for all $i \leq t - 1$, $\alpha_i \to \alpha_{i+1}$. It will then follow that $q_1\omega \to \alpha_2 \to \cdots \to \alpha_t$ is a sequence of basic moves with α_t terminal (for α_t involves q_0, the stopping state); hence T computes ω and $\omega \in E$. It is always true that $\alpha_{t-1} \to \alpha_t$, for α_t is terminal and hence $\alpha_{t-1} \leftarrow \alpha_t$ cannot occur. In particular, this shows that the induction begins when $t = 2$. Suppose that $t > 2$ and some arrow goes to the left. Since the last arrow $\alpha_{t-1} \to \alpha_t$ points right, moving backward until one reaches an arrow pointing left gives an index i with

$$\alpha_{i-1} \leftarrow \alpha_i \to \alpha_{i+1}.$$

But there is never ambiguity about the next move of a Turing machine, so that $\alpha_{i-1} \equiv \alpha_{i+1}$ and $\omega_{i-1} \equiv h\alpha_{i-1}h \equiv h\alpha_{i+1}h \equiv \omega_{i+1}$. We may thus eliminate ω_i and ω_{i+1}, thereby reducing m, and the proof is completed by induction. ∎

Theorem 12.5 (Markov–Post, 1947).

(i) *There is a finitely presented semigroup*

$$\gamma = (q, h, s_0, s_1, \ldots, s_M, q_0, q_1, \ldots, q_N | R)$$

with an unsolvable word problem.

(ii) *There is no decision process which determines, for an arbitrary h-special word $h\alpha h$, whether $h\alpha h = q$ in γ.*

Proof. (i) If T is a Turing machine with stopping state q_0 and with alphabet $A = \{s_0, s_1, \ldots, s_M\}$, then let Ω be all the positive words on A and let $E = e(T) \subset \Omega$. Define $\overline{\Omega}$ to be all the positive words on $A \cup \{q, h, q_0, q_1, \ldots, q_N\}$, where $q_0, q_1, \ldots, q_N$ are the q-letters occurring in the quadruples of T, and define

$$\bar{E} = \{\bar{\omega} \in \bar{\Omega}: \bar{\omega} = q \text{ in } \Gamma(T)\}.$$

Define $\varphi: \Omega \to \bar{\Omega}$ by $\omega \mapsto hq_1\omega h$, and identify Ω with its image $\Omega_1 \subset \bar{\Omega}$; the subset E of Ω is now identified with

$$E_1 = \{hq_1\omega h: \omega \in E\}.$$

It is plain that E_1 is a recursive subset of Ω_1 if and only if E is a recursive subset of Ω. In this notation, Lemma 12.4 reads:

$$E_1 = \bar{E} \cap \Omega_1.$$

Now assume that T is the Turing machine T^* (with stopping state q_0) of Theorem 12.1, so that E, hence E_1, is r.e. but not recursive. Were $\bar{E}$ recursive, then Exercise 12.9 would give E_1, hence E, recursive, and this is a contradiction. Therefore, $\gamma = \Gamma(T^*)$ has an unsolvable word problem.

(ii) Define

$$\bar{S} = \{h\text{-special words } h\alpha h: h\alpha h = q \text{ in } \Gamma(T^*)\}.$$

Were $\bar{S}$ a recursive subset of $\bar{\Omega}$, then $\bar{S} \cap \Omega_1$ would be a recursive subset of Ω_1, by Exercise 12.9. But $\bar{S} \cap \Omega_1 = E_1$. ■

For later use, we rewrite the generators and relations of the Markov–Post semigroup $\gamma(T^*)$.

Corollary 12.6.

(i) *There is a finitely presented semigroup*

$$\Gamma = (q, q_0, \ldots, q_N, s_0, \ldots, s_M \mid F_i q_{i_1} G_i = H_i q_{i_2} K_i, i \in I),$$

with an unsolvable word problem, where F_i, G_i, H_i, K_i are (possibly empty) positive s-words and $q_{i_1}, q_{i_2} \in \{q, q_0, \ldots, q_N\}$.

(ii) *There is no decision process which determines, for arbitrary q_{i_j} and positive s-words X and Y, whether $Xq_{i_j}Y = q$ in Γ.*

Proof. (i) Regard the generator h of the semigroup $\gamma = \Gamma(T^*)$ as the last s-letter and re-index these s-letters so that $h = s_M$. The rewritten relations in $R(T^*)$ now have the described form.

(ii) Let Ω_2 be the set of all positive words on the rewritten generators of Γ, let

$$\bar{\Lambda} = \{Xq_{i_j}Y: X, Y \text{ are positive words on rewritten } s\text{-letters and } Xq_{i_j}Y = q \text{ in } \Gamma\},$$

and let

$$\bar{S}_2 = \{s_M \alpha s_M: \alpha \equiv \sigma q \tau, \text{ where } \sigma \text{ and } \tau \text{ are positive words on } s_0, \ldots, s_{M-1} \text{ and } s_M \alpha s_M = q \text{ in } \Gamma\}$$

(remember that h has been rewritten as s_M). Of course, $\bar{S}_2$ is just the subset $\bar{S}$ of the theorem rewritten in the new notation. Now $\bar{\Lambda} \cap \Omega_2 = \bar{S}_2$; since $\bar{S}_2$ is not recursive, Exercise 12.9 shows that $\bar{\Lambda}$ is not recursive. ∎

The Novikov–Boone–Britton Theorem: Sufficiency of Boone's Lemma

The word problem for groups was first considered by M. Dehn (1910) and by A. Thue (1914). The solution was given by P.S. Novikov (1955) and, independently, by W.W. Boone (1954–1957) and by J.L. Britton (1958). In 1959, Boone exhibited a much simpler finitely presented group than any of those previously given, and he proved it has an unsolvable word problem. In contrast to the "combinatorial" proofs of Novikov and Boone, Britton's proof relies on properties of HNN extensions (which led him to discover Britton's lemma). In 1963, Britton gave a much simpler and shorter proof for Boone's group; we present his proof here, incorporating later improvements of Boone, D.J. Collins, and C.F. Miller, III. We assure the reader that all the Mathematical Logic required in the proof has already appeared; we need only Corollary 12.6, a paraphrase of the Markov–Post theorem, that exhibits a particular finitely presented semigroup Γ with an unsolvable word problem.

Remember that the proof is going to reduce equality of words in a group to equality of words in a semigroup. It is thus essential to keep track of exponents, for while arbitrary words make sense in a group, only positive words make sense in a semigroup.

Notation. If $X \equiv s_{\beta_1}^{e_1} \dots s_{\beta_m}^{e_m}$ is a (not necessarily positive) s-word, then $X^\# \equiv s_{\beta_1}^{-e_1} \dots s_{\beta_m}^{-e_m}$. Note that if X and Y are s-words, then $(X^\#)^\# \equiv X$ and $(XY)^\# \equiv X^\# Y^\#$.

Recall, for every Turing machine T, that there is a semigroup $\Gamma = \Gamma(T)$ with the presentation

$$\Gamma = (q, q_0, \dots, q_N, s_0, \dots, s_M | F_i q_{i_1} G_i = H_i q_{i_2} K_i, i \in I),$$

where F_i, G_i, H_i, K_i are (possibly empty) positive s-words and $q_{i_1}, q_{i_2} \in \{q, q_0, \dots, q_N\}$.

For every Turing machine T, we now define a group $\mathscr{B} = \mathscr{B}(T)$ that will be shown to have an unsolvable word problem if T is chosen to be the Turing machine T^* in the Markov–Post theorem. The group $\mathscr{B}(T)$ has the presentation:

generators: $q, q_0, \ldots, q_N, s_0, \ldots, s_M, r_i, i \in I, x, t, k$;

relations: for all $i \in I$ and all $\beta = 0, \ldots, M$,

$$\left.\begin{array}{ll} \left.\begin{array}{ll} x s_\beta = s_\beta x^2, & \Delta_1] \\ r_i s_\beta = s_\beta x r_i x, & \\ r_1^{-1} F_i^{\#} q_{i_1} G_i r_i = H_i^{\#} q_{i_2} K_i, & \Delta_2 \end{array}\right] \\ t r_i = r_i t, \\ t x = x t, \quad \Delta_3 \end{array}\right]$$

$$k r_i = r_i k,$$
$$k x = x k,$$
$$k(q^{-1} t q) = (q^{-1} t q) k.$$

The subsets $\Delta_1 \subset \Delta_2 \subset \Delta_3$ of the relations are labeled for future reference.

If X and Y are s-words, define

$$(X q_j Y)^* \equiv X^{\#} q_j Y,$$

where $q_j \in \{q, q_0, \ldots, q_N\}$.

Definition. A word Σ is ***special*** if $\Sigma \equiv X^{\#} q_j Y$, where X and Y are positive s-words and $q_j \in \{q, q_0, \ldots, q_N\}$.

If Σ is special, then $\Sigma \equiv X^{\#} q_j Y$, where X and Y are positive s-words, and so $\Sigma^* \equiv (X^{\#} q_j Y)^* \equiv X q_j Y$ is a positive word; therefore, Σ^* determines an element of the semigroup Γ.

The reduction to the Markov–Post theorem is accomplished by the following lemma:

Lemma 12.7 (Boone). *Let T be a Turing machine with stopping state q_0 and associated semigroup $\Gamma = \Gamma(T)$ (rewritten as in Corollary 12.6). If Σ is a special word, then*

$$k(\Sigma^{-1} t \Sigma) = (\Sigma^{-1} t \Sigma) k \quad \text{in} \quad \mathscr{B} = \mathscr{B}(T)$$

if and only if $\Sigma^* = q$ in $\Gamma(T)$.

Theorem 12.8 (Novikov–Boone–Britton). *There exists a finitely presented group $\mathscr{B}$ with an unsolvable word problem.*

Proof. Choose T to be the Turing machine T^* of the Markov–Post theorem. If there were a decision process to determine, for an arbitrary special word Σ, whether $k \Sigma^{-1} t \Sigma k^{-1} \Sigma^{-1} t^{-1} \Sigma = 1$ in $\mathscr{B}(T^*)$, then this same decision process determines whether $\Sigma^* = q$ in $\Gamma(T^*)$. But Corollary 12.6(ii) asserts that no such decision process for $\Gamma(T^*)$ exists. ∎

Corollary 12.9. *Let T be a Turing machine with stopping state q_0 enumerating a subset E of Ω (the set of all positive words on the alphabet of T). If $\omega \in \Omega$, then $\omega \in E$ if and only if $k(h^{-1}q_1\omega h) = (h^{-1}q_1\omega h)k$ in $\mathscr{B}(T)$.*

Proof. By Lemma 12.4, $\omega \in E$ if and only if $hq_1\omega h = q$ in $\Gamma(T)$. But, in $\mathscr{B}(T)$, $(hq_1\omega h)^* = h^{-1}q_1\omega h$ (which is a special word), and Boone's lemma shows that $(h^{-1}q_1\omega h)^\# = hq_1\omega h = q$ in $\Gamma(T)$ if and only if $k(h^{-1}q_1\omega h) = (h^{-1}q_1\omega h)k$ in $\mathscr{B}(T)$. ■

The proof below is valid for any Turing machine T with stopping state q_0. We abbreviate $\mathscr{B}(T)$ to $\mathscr{B}$ and $\Gamma(T)$ to Γ.

The proof of Boone's lemma in one direction is straightforward.

Lemma 12.10.

(i) *If V is a positive s-word, then*

$$r_i V = VR \quad \text{in } \mathscr{B} \qquad \text{and} \qquad r_i^{-1} V = VR' \quad \text{in } \mathscr{B},$$

where R and R' are words on $\{r_i, x\}$ with R positive.

(ii) *If U is a positive s-word, then*

$$U^\# r_1^{-1} = LU^\# \quad \text{in } \mathscr{B} \qquad \text{and} \qquad U^\# r_i = L'U^\# \quad \text{in } \mathscr{B},$$

where L and L' are words on $\{r_i, x\}$.

Proof. We prove that $r_i V = VR$ in $\mathscr{B}$ by induction on $m \geq 0$, where $V \equiv s_{\beta_1} \dots s_{\beta_m}$. This is certainly true when $m = 0$. If $m > 0$, write $V \equiv V's_{\beta_m}$; by induction, $r_i V \equiv r_i V's_{\beta_m} = V'R's_{\beta_m}$, where R' is a positive word on $\{r_i, x\}$. Using the relations $xs_\beta = s_\beta x^2$ and $r_i s_\beta = s_\beta x r_i x$, we see that there is a positive word R on $\{r_i, x\}$ with $R = R's_{\beta_m}$ in $\mathscr{B}$.

The proofs of the other three equations are similar. ■

Proof of Sufficiency in Boone's Lemma. If Σ is a special word with $\Sigma^* \equiv Xq_jY = q$ in Γ, then there is a sequence of elementary operations

$$\Sigma^* \equiv \omega_1 \to \omega_2 \to \cdots \to \omega_n \equiv q \quad \text{in } \Gamma,$$

where, for each v, one of the words ω_v and ω_{v+1} has the form $UF_iq_{i_1}G_iV$ with U and V positive s-words, and the other has the form $UH_iq_{i_2}K_iV$. By the lemma, there are equations in $\mathscr{B}$:

$$\begin{aligned} U^\#(H_i^\# q_{i_2} K_i)V &= U^\#(r_i^{-1} F_i^\# q_{i_1} G_i r_i)V \\ &= L'U^\#(F_i^\# q_{i_1} G_i)VR', \end{aligned}$$

where L' and R' are words on $\{r_i, x\}$. In a similar manner, one sees that there are words L'' and R'' on $\{r_i, x\}$ with

$$U^\#(F_i^\# q_{i_1} G_i)V = U^\#(r_i H_i^\# q_{i_2} K_i r_i^{-1})V = L''U^\#(H_i^\# q_{i_2} K_i)VR''.$$

Since $\omega_\nu = \omega_{\nu+1}$ in Γ implies $\omega_\nu^* = \omega_{\nu+1}^*$ in $\mathscr{B}$, by the relations labeled Δ_2, it follows, for each ν, that

$$\omega_\nu^* = L_\nu \omega_{\nu+1}^* R_\nu \quad \text{in } \mathscr{B}$$

for words L_ν and R_ν on some r_{i_ν} and x. The words $L \equiv L_1 \ldots L_{n-1}$ and $R \equiv R_{n-1} \ldots R_1$ are thus words on $\{x, r_i, i \in I\}$, and

$$\omega_1^* = L\omega_n^* R \quad \text{in } \mathscr{B}.$$

But $\omega_1^* \equiv (\Sigma^*)^* \equiv \Sigma$ and $\omega_n^* \equiv q^* \equiv q$, so that

$$\Sigma = LqR \quad \text{in } \mathscr{B}.$$

Since the generators t and k commute with x and all the r_i, they commute with L and R. Therefore,

$$\begin{aligned} k\Sigma^{-1}t\Sigma k^{-1}\Sigma^{-1}t^{-1}\Sigma &= kR^{-1}q^{-1}L^{-1}tLqRk^{-1}R^{-1}q^{-1}L^{-1}t^{-1}LqR \\ &= kR^{-1}q^{-1}tqk^{-1}q^{-1}t^{-1}qR \\ &= R^{-1}(kq^{-1}tqk^{-1}q^{-1}t^{-1}q)R \\ &= 1, \end{aligned}$$

because the last word is a conjugate of a relation. ■

Observe that the last relation of the group $\mathscr{B}$ appears only in the last step of the proof.

Cancellation Diagrams

We interupt the proof of Boone's lemma (and the Novikov–Boone–Britton theorem) to discuss a geometric method of studying presentations of groups, essentially due to R. Lyndon, that uses diagrams in the plane. Since we are only going to use diagrams in a descriptive way (and not as steps in a proof), we may write informally. For a more serious account, we refer the reader to Lyndon and Schupp (1977, Chap. V) with the caveat that our terminology does not always coincide with theirs.

When we speak of a ***polygon*** in the plane, we mean the usual geometric figure including its interior; of course, its ***boundary*** (or perimeter) consists of finitely many edges and vertices. A ***directed polygon*** is a polygon each of whose (boundary) edges is given a direction, indicated by an arrow. Finally, given a presentation $(X|\Delta)$ of a group, a ***labeled directed polygon*** is a directed polygon each of whose (directed) edges is labeled by a generator in X.

Given a presentation $(X|\Delta)$ of a group, we are going to construct a labeled directed polygon for (almost) every word

$$\omega \equiv x_1^{e_1} \ldots x_n^{e_n},$$

where $x_1, \ldots, x_n$ are (not necessarily distinct) generators and each $e_i = \pm 1$. For technical reasons mentioned below, ω is restricted a bit.

Definition. Let F be a free group with basis X. A word $\omega \equiv x_1^{e_1} \dots x_n^{e_n}$ on X with each $e_i = \pm 1$ is called ***freely reduced*** if it contains no subwords of the form xx^{-1} or $x^{-1}x$ with $x \in X$.

A ***cyclic permutation*** of $\omega \equiv x_1^{e_1} \dots x_n^{e_n}$ is a word of the form $x_i^{e_i} \dots x_n^{e_n} x_1^{e_1} \dots x_{i-1}^{e_{i-1}}$ (by Exercise 3.8, a cyclic permutation of ω is a conjugate of it). A word ω is ***cyclically reduced*** if every cyclic permutation of it is freely reduced.

If $\omega \equiv x_1^{e_1} \dots x_n^{e_n}$ is cyclically reduced, construct a labeled directed polygon as follows: draw an n-gon in the plane; choose an edge and label it x_1; label successive edges $x_2, x_3, \dots, x_n$ as one proceeds counterclockwise around the boundary; direct the ith edge with an arrow according to the sign of e_i (we agree that the positive direction is counterclockwise). For example, if k and x commute, then the labeled directed polygon is the square in Figure 12.1.

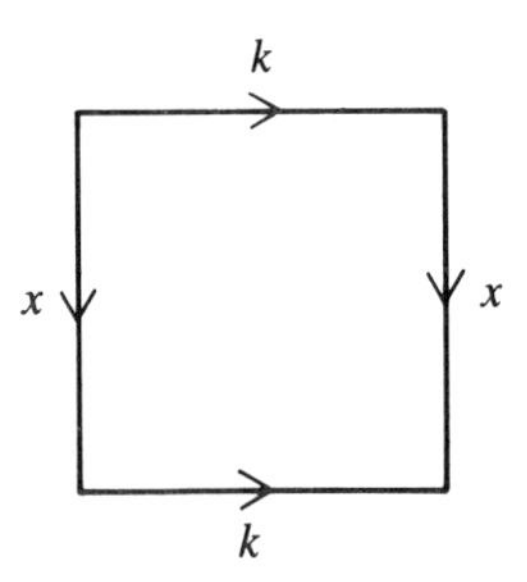

Figure 12.1

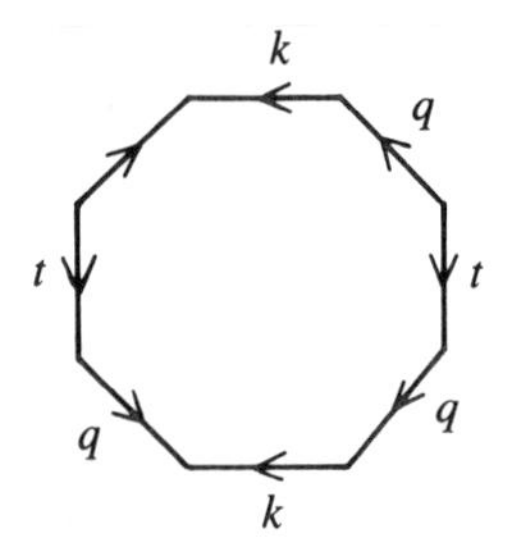

Figure 12.2

As a second example, consider the last relation in Boone's group $\mathscr{B}$: $\omega \equiv kq^{-1}tqk^{-1}q^{-1}t^{-1}q$. The labeled directed polygon for ω is the octagon whose first edge is the top k-edge in Figure 12.2. If ω is not cyclically reduced, this construction gives a polygon having two adjacent edges with the same label and which point in opposite directions, and such polygons complicate proofs. However, there is no loss in generality in assuming that every relation in a presentation is cyclically reduced, for every word has some cyclically reduced conjugate, and one may harmlessly replace a relation by any of its conjugates. Every cyclically reduced relation thus yields a labeled directed polygon called its ***relator polygon***.

We can now draw a picture of a presentation $(X|\Delta)$ of a group G (with cyclically reduced relations Δ) by listing the generators X and by displaying a relator polygon of each relation in Δ. These polygons are easier to grasp (especially when viewing several of them simultaneously) if distinct generators are given distinct colors. The presentation of the group $\mathscr{B}$ in Boone's lemma is pictured in Plate 1 (inside front cover). There are six types of generators: q; s; r; x; t; k, and each has been given a different color.

There is a presentation of a group called $\mathscr{B}_6$ which is pictured in Plate 3. This group will occur in our proof of the Higman imbedding theorem.

Another example is provided by an HNN extension: a relation involving a stable letter p has the form $ap^e bp^{-e}c$, where $e = \pm 1$. If the corresponding relator polygon is drawn so that the p-edges are parallel, then they point in the same direction.

Let D be a labeled directed polygon. Starting at some edge on the boundary of D, we obtain a word ω as we read the edge labels (and the edge directions) while making a complete (counterclockwise) tour of D's boundary. Such a word ω is called a ***boundary word*** of D. (Another choice of starting edge gives another boundary word of D, but it is just a cyclic permutation, hence a conjugate, of ω. A clockwise tour of D's boundary gives a conjugate of ω^{-1}.)

Definition. A ***diagram*** is a labeled directed polygon whose interior may be subdivided into finitely many labeled directed polygons, called ***regions***; we insist that any pair of edges which intersect do so in a vertex.

We quote the fundamental theorem in this context; a proof can be found in Lyndon and Schupp.

Fundamental Theorem of Combinatorial Group Theory. *Let G have a finite presentation $(X|\Delta)$, where Δ satisfies the following conditions:*

(i) *each $\delta \in \Delta$ is cyclically reduced;*
(ii) *if $\delta \in \Delta$, then $\delta^{-1} \in \Delta$;*
(iii) *if $\delta \in \Delta$, then every cyclic permutation of δ lies in Δ.*

If ω is a cyclically reduced word on X, then $\omega = 1$ in G if and only if there is a diagram having a boundary word ω and whose regions are relator polygons of relations in Δ.

An immediate consequence of this theorem is a conjugacy criterion. Assume that ω and ω' are cyclically reduced words on X, and consider the annulus with outer boundary word ω' and inner boundary word ω, as in Figure 12.3.

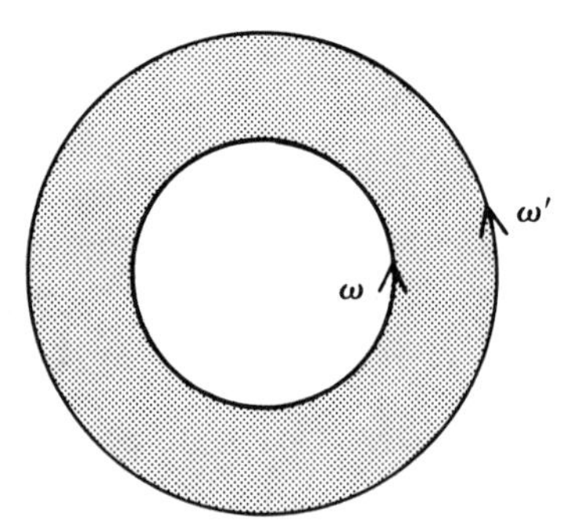

Figure 12.3

Corollary. *The elements ω and ω' are conjugate in G if and only if the interior of the annulus can be subdivided into relator polygons.*

The proof consists in finding a path β from ω' to ω and cutting along β to form a diagram as in Figure 12.4. A boundary word of the new diagram is $\omega'\beta\omega^{-1}\beta^{-1}$, and the fundamental theorem says that this word is 1 in G. Conversely, if $\omega'\beta\omega^{-1}\beta^{-1} = 1$ in G, one may form an annulus by identifying the edges labeled β; that is, start with the diagram on the above right and glue the β's together to obtain the annulus on the left.

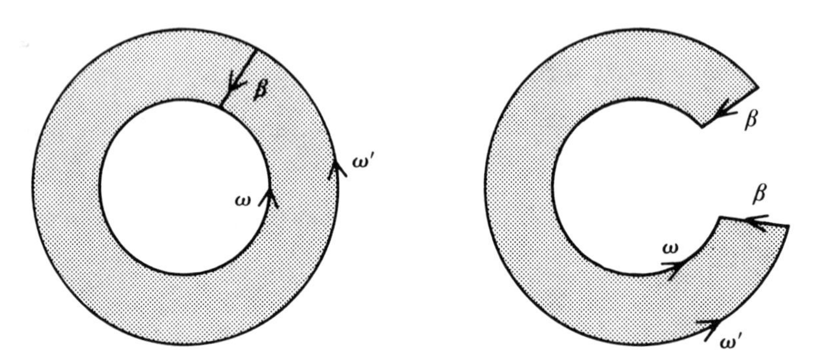

Figure 12.4

An example will reveal how these diagrams can illustrate the various steps taken in rewriting a word using the relations of a given presentation. The proof of sufficiency of Boone's lemma requires one to prove, for a special word Σ, that

$$w(\Sigma) \equiv k\Sigma^{-1}t\Sigma k^{-1}\Sigma^{-1}t^{-1}\Sigma = 1 \quad \text{in } \mathscr{B}.$$

The hypothesis provides a sequence of elementary operations

$$\Sigma^* \equiv \omega_1 \to \omega_2 \to \cdots \to \omega_n \equiv q \quad \text{in } \Gamma.$$

The proof begins by showing that each ω_v^* has the form $U_v^\# q_{i_v} V_v$, where $v \le n - 1$ and U_v and V_v are positive s-words; moreover, there are words L_v and R_v on $\{x, r_i, i \in I\}$ such that, for all v,

$$\omega_v^* = L_v \omega_{v+1}^* R_v \quad \text{in } \mathscr{B}.$$

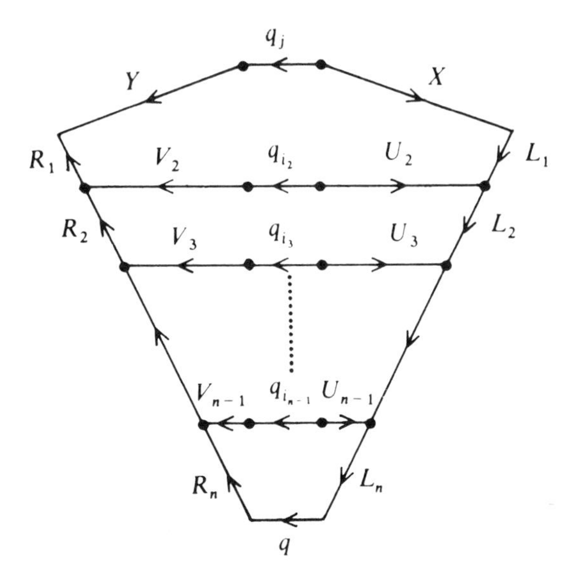

Figure 12.5

Figure 12.5 pictures all of these equations; we have not drawn the subdivision of each interior polygon into relator polygons, and we have taken the liberty of labeling segments comprised of many s-edges by a single label Y, X, V_v, or U_v.

The reader should now look at Plate 2; it is a diagram having $w(\Sigma)$ as a boundary word. In the center is the octagon corresponding to the octagonal relation $w(q) \equiv kq^{-1}tqk^{-1}q^{-1}t^{-1}q$, and there are four (almost identical) quadrants as drawn above, involving either Σ or Σ^{-1} on the outer boundary and q or q^{-1} on the octagon (actually, adjacent quadrants are mirror images). The commutativity of k with x and each r_i allows one to insert sequences of squares connecting k-edges on the outer boundary to k-edges on the octagon; similarly, the commutativity of t with x and each r_i inserts sequences connecting t-edges on the outer boundary with t-edges on the octagon. Since the quadrants have already been subdivided into relator polygons, the four quadrants together with the four border sequences, form a diagram. Therefore, $w(\Sigma) = 1$ in $\mathscr{B}$, as asserted by the fundamental theorem.

Define $\mathscr{B}^{\Delta}$ to be the group having the same presentation as $\mathscr{B}$ except that the octagonal relation is missing. Now regard Plate 2 as an annulus having the octagonal relation as the inner boundary word. This annulus has just been subdivided into relator polygons, and so the corollary of the fundamental theorem says that $w(\Sigma)$ is conjugate to $w(q)$ in $\mathscr{B}^{\Delta}$. This last result is a reflection of the fact that the octagonal relation enters the given proof of the sufficiency of Boone's lemma at the last step.

The Novikov–Boone–Britton Theorem: Necessity of Boone's Lemma

We now turn to the proof of the more difficult half of Boone's lemma. Geometrically, the problem is to subdivide the labeled directed polygon with boundary word $w(\Sigma)$ into a diagram whose regions are relator polygons of $\mathscr{B}$. The conjugacy of $w(\Sigma)$ and the octagonal relation $w(q)$ in the group $\mathscr{B}^{\Delta}$ (mentioned above) suggests a strategy to prove the necessity of Boone's lemma: subdivide the annulus with outer boundary $w(\Sigma)$ and inner boundary $w(q)$ using the relations of $\mathscr{B}^{\Delta}$ (thereby allowing us to avoid further use of the octagonal relation $w(q)$), trying to make the annulus look like Plate 2. We shall give formal algebraic proofs, but, after the proof of each lemma, we shall give informal geometric descriptions. (It was the idea of E. Rips to describe this proof geometrically, and he constructed the diagrams for the Novikov–Boone–Britton theorem as well as for the coming proof of the Higman imbedding theorem. He has kindly allowed me to use his description here.)

Define groups $\mathscr{B}_0$, $\mathscr{B}_1$, $\mathscr{B}_2$, and $\mathscr{B}_3$ as follows:

$$\mathscr{B}_0 = (x \mid \varnothing), \text{ the infinite cyclic group with generator } x;$$

$$\mathscr{B}_1 = (\mathscr{B}_0; s_0, \ldots, s_M \mid \Delta_1)$$

(recall that we labeled certain subsets of the relations of $\mathscr{B}$ as $\Delta_1 \subset \Delta_2 \subset \Delta_3$ when we defined $\mathscr{B}$; recall also that this notation means that we are adjoining the displayed generators and relations to the given presentation of $\mathscr{B}_0$);

$$\mathscr{B}_2 = (\mathscr{B}_1 * Q; r_i, i \in I \mid \Delta_2),$$

where Q is free with basis $\{q, q_0, \ldots, q_N\}$;

$$\mathscr{B}_3 = (\mathscr{B}_2; t \mid \Delta_3).$$

Lemma 12.11. *In the chain*

$$\mathscr{B}_0 \leq \mathscr{B}_1 \leq \mathscr{B}_1 * Q \leq \mathscr{B}_2 \leq \mathscr{B}_3 \leq \mathscr{B},$$

each group is an HNN extension of its predecessor; moreover, $\mathscr{B}_1 * Q$ is an HNN extension of $\mathscr{B}_0$. In more detail:

(i) *$\mathscr{B}_1$ is an HNN extension with base $\mathscr{B}_0$ and stable letters $\{s_0, \ldots, s_M\}$;*
(ii′) *$\mathscr{B}_1 * Q$ is an HNN extension with base $\mathscr{B}_0$ and stable letters $\{s_0, \ldots, s_M\} \cup \{q, q_0, \ldots, q_N\}$;*
(ii) *$\mathscr{B}_1 * Q$ is an HNN extension with base $\mathscr{B}_1$ and stable letters $\{q, q_0, \ldots, q_N\}$;*
(iii) *$\mathscr{B}_2$ is an HNN extension with base $\mathscr{B}_1 * Q$ and stable letters $\{r_i: i \in I\}$;*
(iv) *$\mathscr{B}_3$ is an HNN extension with base $\mathscr{B}_2$ and stable letter t; and*
(v) *$\mathscr{B}$ is an HNN extension with base $\mathscr{B}_3$ and stable letter k.*

Proof. (i) The presentation

$$\mathscr{B}_1 = (x, s_0, \ldots, s_M | s_\beta^{-1} x s_\beta = x^2, \text{all } \beta)$$

shows that $\mathscr{B}_1$ has base $\langle x \rangle = \mathscr{B}_0$ and stable letters $\{s_0, \ldots, s_M\}$. Since x has infinite order, $A_\beta = \langle x \rangle \cong \langle x^2 \rangle = B_\beta$, and so φ_B: $A_\beta \to B_\beta$, defined by $x \mapsto x^2$, is an isomorphism for all β. Therefore, $\mathscr{B}_1$ is an HNN extension.

(ii′) The presentation of $\mathscr{B}_1 * Q$,

$$(x, s_0, \ldots, s_M, q, q_0, \ldots, q_N | s_\beta^{-1} x s_\beta = x^2, q^{-1} x q = x, q_i^{-1} x q_i = x),$$

shows that $\mathscr{B}_1 * Q$ has base $\mathscr{B}_0$ and stable letters $\{s_0, \ldots, s_M\} \cup \{q, q_1, \ldots, q_N\}$. Since x has infinite order, $A_\beta = \langle x \rangle \cong \langle x^2 \rangle = B_\beta$, and so the maps φ_β are isomorphisms, as above; also, the maps φ_{q_j} are identity maps, where $A_{q_j} = \langle x \rangle = B_{q_j}$. Thus, $\mathscr{B}_1 * Q$ is an HNN extension with base $\mathscr{B}_0$ and stable letters $\{s_0, \ldots, s_M\} \cup \{q, q_0, \ldots, q_N\}$.

(ii) Since Q is free with basis $\{q, q_0, \ldots, q_N\}$, Example 11.10 now shows that $\mathscr{B}_1 * Q$ is an HNN extension with base $\mathscr{B}_1$ and stable letters $\{q, q_0, \ldots, q_N\}$.

(iii) The presentation

$$\mathscr{B}_2 = (\mathscr{B}_1 * Q; r_i, i \in I | r_i^{-1}(F_i^{\#} q_{i_1} G_i) r_i = H_i^{\#} q_{i_2} K_i, r_i^{-1}(s_\beta x) r_i = s_\beta x^{-1})$$

shows that $\mathscr{B}_2$ has base $\mathscr{B}_1 * Q$ and stable letters $\{r_i, i \in I\}$. Now, for each i, the subgroup A_i is $\langle F_i^{\#} q_{i_1} G_i, s_\beta x, \text{all } \beta \rangle$ and the subgroup B_i is $\langle H_i^{\#} q_{i_2} K_i, s_\beta x^{-1}, \text{all } \beta \rangle$. We claim that both A_i and B_i are free groups with bases the displayed generating sets. First, use Exercise 11.8 to see that $\langle s_\beta x, \text{all } \beta \rangle$ is free with basis $\{s_\beta x, \text{all } \beta\}$: map $\langle s_\beta x, \text{all } \beta \rangle$ onto the free group with basis $\{s_0, \ldots, s_M\}$ by setting $x = 1$; then observe that $A_i = \langle F_i^{\#} q_{i_1} G_i, s_\beta x, \text{all } \beta \rangle \cong \langle F_i^{\#} q_{i_1} G_i \rangle * \langle s_\beta x, \text{all } \beta \rangle \leq \mathscr{B}_1 * Q$ (because $F_i^{\#} q_{i_1} G_i$ involves a q-letter and elements of the free group $\langle s_\beta x, \text{all } \beta \rangle$ do not). A similar argument applies to B_i, and so there is an isomorphism φ_i: $A_i \to B_i$ with $\varphi_i(F_i^{\#} q_{i_1} G_i) = H_i^{\#} q_{i_2} K_i$ and $\varphi_i(s_\beta x) = s_\beta x^{-1}$ for all β. Thus, $\mathscr{B}_2$ is an HNN extension with base $\mathscr{B}_1 * Q$.

(iv) Note that $\mathscr{B}_3$ has base $\mathscr{B}_2$ and stable letter t:

$$\mathscr{B}_3 = (\mathscr{B}_2; t | t^{-1} r_i t = r_i, t^{-1} x t = x);$$

Since t commutes with the displayed relations, $\mathscr{B}_3$ is an HNN extension of $\mathscr{B}_2$, as in Example 11.11.

(v) Note that $\mathscr{B}$ has base $\mathscr{B}_3$ and stable letter k:

$$\mathscr{B} = (\mathscr{B}_3; k \mid k^{-1}r_ik = r_i, i \in I, k^{-1}xk = x, k^{-1}(q^{-1}tq)k = q^{-1}tq).$$

As in Example 11.11, $\mathscr{B}$ is an HNN extension of $\mathscr{B}_3$. ■

Corollary 12.12.

(i) *The subgroup $\langle s_1 x, \ldots, s_M x\rangle \leq \mathscr{B}_1$ is a free group with basis the displayed letters.*
(ii) *There is an automorphism ψ of $\mathscr{B}_1$ with $\psi(x) = x^{-1}$ and $\psi(s_\beta) = s_\beta$ for all β.*

Proof. (i) This was proved in part (iii) of the above lemma.

(ii) The function on the generators sending $x \mapsto x^{-1}$ and $s_\beta \mapsto s_\beta$ for all β preserves all the relations. ■

The reader should view Lemma 12.11 as preparation for the remainder of the proof; it will allow us to analyze words using Britton's lemma, Theorem 11.81.

Lemma 12.13. *Let Σ be a fixed special word satisfying the hypothesis of Bonne's lemma:*

$$w(\Sigma) \equiv k\Sigma^{-1}t\Sigma k^{-1}\Sigma^{-1}t^{-1}\Sigma = 1 \quad \textit{in } \mathscr{B}.$$

Then there are freely reduced words L_1 and L_2 on $\{x, r_i, i \in I\}$ such that

$$L_1 \Sigma L_2 = q \quad \textit{in } \mathscr{B}_2.$$

Proof. Since $\mathscr{B}$ is an HNN extension with base $\mathscr{B}_3$ and stable letter k, Britton's lemma applies to the word $k\Sigma^{-1}t\Sigma k^{-1}\Sigma^{-1}t^{-1}\Sigma$; it says that $k\Sigma^{-1}t\Sigma k^{-1}$ is a pinch and that $\Sigma^{-1}t\Sigma = C$ in $\mathscr{B}_3$, where C is a word on $\{x, q^{-1}tq, r_i, i \in I\}$. (Since the stable letter k commutes with $\{x, q^{-1}tq, r_i, i \in I\}$, we are in the simple case of Example 11.11 when the subgroups A and B are equal and the isomorphism $\varphi: A \to B$ is the identity.) Therefore, there exist words ω of the form $\Sigma^{-1}t\Sigma C^{-1} = 1$ in $\mathscr{B}_3$; in detail,

$$\omega \equiv \Sigma^{-1}t\Sigma R_0(q^{-1}t^{e_1}q)R_1(q^{-1}t^{e_2}q)R_2 \ldots (q^{-1}t^{e_n}q)R_n = 1 \quad \text{in } \mathscr{B}_3,$$

where the R_j are (possibly empty) freely reduced words on $\{x, r_i, i \in I\}$ and $e_j = \pm 1$. We assume ω is such a word chosen with n minimal.

Since $\mathscr{B}_3$ is an HNN extension with base $\mathscr{B}_2$ and stable letter t, Britton's lemma applies again, showing that ω contains a pinch t^eDt^{-e}, and there is a word R on $\{x, r_i, i \in I\}$ with $D = R$ in $\mathscr{B}_2$.

If the pinch involves the first occurrence of the letter t in ω, then $t^eDt^{-e} \equiv t\Sigma R_0 q^{-1}t^{e_1}$. Hence $e = +1$, $e_1 = -1$, $t\Sigma R_0 q^{-1}t^{e_1} = tRt^{-1}$, and

$$\Sigma R_0 q^{-1} = R \quad \text{in } \mathscr{B}_2;$$

equivalently,

$$R^{-1}\Sigma R_0 = q \quad \text{in } \mathscr{B}_2,$$

which is of the desired form.

If the initial t^e in the pinch is t^{e_j}, where $j \geq 1$, then $t^e D t^{-e} \equiv t^{e_j} q R_j q^{-1} t^{e_{j+1}}$ with $qR_jq^{-1} = R$ in $\mathscr{B}_2$ for some word R on $\{x, r_i, i \in I\}$. Since $\mathscr{B}_2 \leq \mathscr{B}_3$, by Theorem 11.78, we may view this as an equation in $\mathscr{B}_3$:

$$t^{e_j} q R_j q^{-1} t^{e_{j+1}} = t^e q R_j q^{-1} t^{-e} = t^e R t^{-e} \quad \text{in } \mathscr{B}_3.$$

But the stable letter t in $\mathscr{B}_3$ commutes with x and all r_i, so there is an equation

$$qR_jq^{-1} = R \quad \text{in } \mathscr{B}_3.$$

Hence, in $\mathscr{B}_3$,

$$\begin{aligned}(q^{-1} t^{e_j} q) R_j (q^{-1} t^{e_{j+1}} q) &= q^{-1} t^e R t^{-e} q \\ &= q^{-1} R q \quad (\text{for } t \text{ commutes with } x, r_i) \\ &= q^{-1}(qR_jq^{-1})q \\ &= R_j.\end{aligned}$$

There is thus a factorization of ω in $\mathscr{B}_3$ having smaller length, contradicting the choice of n being minimal. Therefore, this case cannot occur. ■

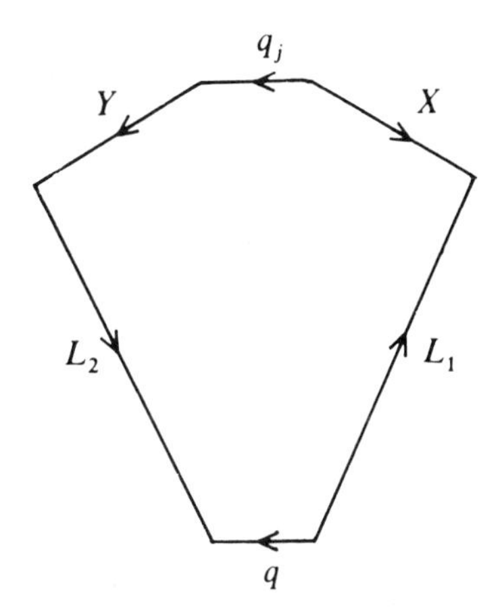

Figure 12.6

Geometrically, we have shown that the labeled directed annulus with outer boundary word $w(\Sigma)$ and inner boundary word the octagon $w(q)$ contains a "quadrant" involving Σ on the outer boundary, q on the inner boundary, and internal paths L_1 and L_2 which are words on $\{x, r_i, i \in I\}$. Of course, there are two such quadrants as well as two "mirror images" of these quadrants which involve Σ^{-1} on the outer boundary and q^{-1} on the inner boundary. Moreover, the regions subdividing these quadrants are relator polygons corresponding to the relations Δ_3; that is, they do not involve k-edges or t-edges.

Finally, there is no problem inserting the "border sequences" connecting k-edges (and t-edges) on the outer boundary with k-edges (and t-edges) on the inner boundary, for the internal paths of the quadrants involve only x and r_i's, all of which commute with k and with t.

Recall that $\Sigma \equiv X^{\#} q_j Y$, where X and Y are positive s-words and $q_j \in \{q, q_0, \ldots, q_N\}$. We have just shown that

$$L_1 X^{\#} q_j Y L_2 = q \quad \text{in } \mathscr{B}_2$$

for some freely reduced words L_1 and L_2 on $\{x, r_i, i \in I\}$. Rewrite this last equation as

$$L_1 X^{\#} q_j = q L_2^{-1} Y^{-1} \quad \text{in } \mathscr{B}_2.$$

Lemma 12.14. *Each of the words $L_1 X^{\#} q_j$ and $q L_2^{-1} Y^{-1}$ is r_i-reduced for every $i \in I$.*

Proof. Suppose, on the contrary, that $L_1 X^{\#} q_j$ contains a pinch $r_k^e C r_k^{-e}$ as a subword. Since $X^{\#}$ is an s-word, this pinch is a subword of L_1, a word on $\{x, r_i, i \in I\}$. Since L_1 is freely reduced, $C \equiv x^m$ for some $m \neq 0$. Since $\mathscr{B}_2$ is an HNN extension with base $\mathscr{B}_1 * Q$ and stable letters $\{r_i, i \in I\}$, Britton's lemma says that there is a word V in $\mathscr{B}_1 * Q$, where $Q = \langle q, q_1, \ldots, q_N \rangle$, such that

$$V \equiv \omega_0 (F_i^{\#} q_{i_1} G_i)^{e_1} \omega_1 \ldots (F_i^{\#} q_{i_1} G_i)^{e_n} \omega_n,$$

$e_j = \pm 1$, ω_j is a word on $\{s_1 x, \ldots, s_M x\}$ for all j, V is reduced as a word in the free product, and

$$x^m = V \quad \text{in } \mathscr{B}_1 * Q.$$

Since $x^m \in \mathscr{B}_1$, one of the free factors of $\mathscr{B}_1 * Q$, we may assume that V does not involve any q-letters; in particular, V does not involve $F_i^{\#} q_{i_1} G_i$. Therefore,

$$x^m = \omega_0 \equiv (s_{\beta_1} x)^{f_1} \ldots (s_{\beta_p} x)^{f_p} \quad \text{in } \mathscr{B}_1,$$

where each $f_\nu = \pm 1$. Since $\mathscr{B}_1$ is an HNN extension with base $\mathscr{B}_0 = \langle x \rangle$ and stable letters $(s_0, \ldots, s_M\}$, another application of Britton's lemma says that the word $x^{-m}\omega_0$, which is 1 in $\mathscr{B}_1$, contains a pinch of the form $s_\beta^f x^\varepsilon s_\beta^{-f}$, where $\varepsilon = \pm 1$. Now inspection of the spelling of ω_0 shows that it contains no such subword; we conclude that $\omega_0 = 1$, hence $x^m = 1$. But x has infinite order (since $\mathscr{B}_0 \leq \mathscr{B}_1$), and this contradicts $m \neq 0$. We conclude that L_1, and hence $L_1 X^{\#} q_j$, is r_i-reduced.

A similar proof shows that $q L_2^{-1} Y^{-1}$ is also r_i-reduced. ∎

We know that the boundary word of each of the four quadrants is 1, so that each quadrant is subdivided into relator polygons. The two words in the lemma are sub-boundary words that do not flank either of the two q-edges; that is, neither of the q-edges is surrounded by other (boundary) edges on both sides. As we are working within $\mathscr{B}^\Delta$, the octagonal relator polygon is not inside a quadrant. The only other relator involving a q-letter is the eight-

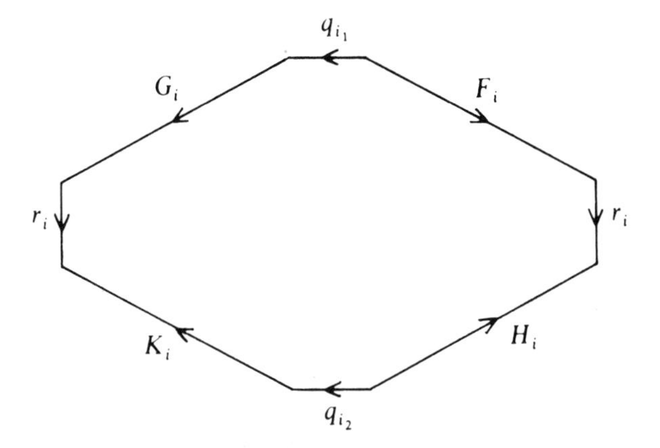

Figure 12.7

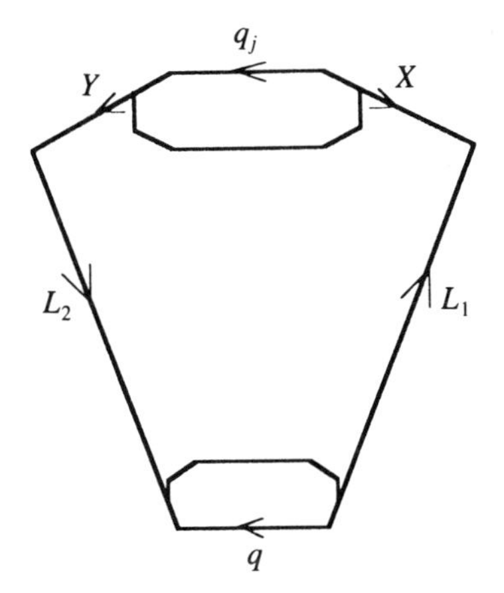

Figure 12.8

sided "petal" in Δ_2 (Figure 12.7). There must be such a petal involving the q-letter on the quadrant's boundary. The lemma shows that the petal's boundary must contain edges in Y and edges in $X^\#$ (Figure 12.8).

The following lemma completes the proof of Boone's lemma and, with it, the Novikov–Boone–Britton theorem. In view of a further application of it in the next section, however, we prove slightly more than we need now.

Lemma 12.15. *Let L_1 and L_2 be words on $\{x, r_i, i \in I\}$ that are r_i-reduced for all $i \in I$. If X and Y are freely reduced words on $\{s_0, \ldots, s_M\}$ and if*

$$L_1 X^\# q_j Y L_2 = q \quad \text{in } \mathscr{B}_2,$$

then both X and Y are positive and

$$Xq_jY \equiv (X^{\#}q_jY)^* = q \quad \text{in } \Gamma.$$

Remark. In our case, both X and Y are freely reduced, for X and Y are positive (because $\Sigma \equiv X^{\#}q_jY$ is special), and positive words are necessarily freely reduced.

Proof. The previous lemma shows that $L_1X^{\#}q_j = qL_2^{-1}Y^{-1}$ in $\mathscr{B}_2$ and that both words are r_i-reduced. By Corollary 11.82, the number $\rho \geq 0$ of r-letters in L_1 is the same as the number of r-letters in L_2 (because no r-letters occur outside of L_1 or L_2); the proof is by induction on ρ.

If $\rho = 0$, then the equation $L_1X^{\#}q_jYL_2 = q$ is

$$x^mX^{\#}q_jYx^n = q \quad \text{in } \mathscr{B}_2.$$

This equation involves no r-letters, and so we may regard it as an equation in $\mathscr{B}_1 * Q \leq \mathscr{B}_3$, where $Q = \langle q, q_0, \ldots, q_N\rangle$. But the normal form theorem for free products (Theorem 11.52) gives $q_j = q$ and $x^mX^{\#} = 1 = Yx^n$ in $\mathscr{B}_1$. Since $\mathscr{B}_1$ is an HNN extension with base $\mathscr{B}_0 = \langle x\rangle$ and stable letters $\{s_0, \ldots, s_M\}$, it follows from Britton's lemma that $m = 0 = n$ and that both X and Y are empty. Thus, X and Y are positive and $Xq_jY \equiv q_j \equiv q$ in Γ.

Assume now that $\rho > 0$. By Lemma 12.14, the words $L_1X^{\#}q_j$ and $qL_2^{-1}Y^{-1}$ are r_i-reduced for all i. Since $\mathscr{B}_2$ is an HNN extension with base $\mathscr{B}_1 * Q$ and stable letters $\{r_i, i \in I\}$, Britton's lemma gives subwords L_3 of L_1 and L_4 of L_2 such that

$$\text{(1)} \qquad L_1X^{\#}q_jYL_2 \equiv L_3(r_i^e x^mX^{\#}q_jYx^nr_i^{-e})L_4 = q \quad \text{in } \mathscr{B}_2,$$

where the word in parentheses is a pinch; moreover, either $e = -1$ and

$$x^mX^{\#}q_jYx^n \in A_i = \langle F_i^{\#}q_{i_1}G_i, s_\beta x, \text{all } \beta\rangle,$$

or $e = +1$ and

$$x^mX^{\#}q_jYx^n \in B_i = \langle H_i^{\#}q_{i_2}K_i, s_\beta x^{-1}, \text{all } \beta\rangle.$$

In the first case,

$$q_j = q_{i_1},$$

for the membership holds in the free product $\mathscr{B}_1 * Q$; in the second case, $q_j = q_{i_2}$. We consider only the case

$$e = -1,$$

leaving the similar case $e = +1$ to the reader. There is a word

$$\omega \equiv x^mX^{\#}q_jYx^nu_0(F_i^{\#}q_jG_i)^{\alpha_1}u_1 \ldots (F_i^{\#}q_jG_i)^{\alpha_t}u_t = 1 \quad \text{in } \mathscr{B}_1 * Q,$$

where $\alpha_j = \pm 1$ and the u_j are possibly empty words on $\{s_\beta x, \text{all } \beta\}$. Of all such words, we assume that ω has been chosen with t minimal. We may further assume that each u_j is a reduced word on $\{s_\beta x, \text{ all } \beta\}$, for Corollary 12.12(i) says that this set freely generates its subgroup. Since $\omega = 1$ in $\mathscr{B}_1 * Q$, the normal form theorem for free products (Theorem 11.52) shows that each "syllable" of ω between consecutive q_j's is equal to 1 in $\mathscr{B}_1$. However, if one views $\mathscr{B}_1 * Q$ as an HNN extension with base $\mathscr{B}_1$ and stable letters $\{q, q_0, \ldots, q_N\}$ (as in Example 11.10, in which case the subgroups A and B are 1), then Britton's lemma says that ω contains a pinch $q_j^\varepsilon C q_j^{-\varepsilon}$ as a subword with $C = 1$ in $\mathscr{B}_1$ (of course, this case of Britton's lemma is very easy to see directly).

If a pinch involves the first occurrence of q_j, then $-\varepsilon = \alpha_1 = -1$ and

$$Yx^n u_0 G_i^{-1} = 1 \quad \text{in } \mathscr{B}_1.$$

We claim that a pinch cannot occur at any other place in ω. Otherwise, there is an index ν with a pinch occurring as a subword of $(F_i^\# q_j G_i)^{\alpha_\nu} u_\nu (F_i^\# q_j G_i)^{\alpha_{\nu+1}}$. If $\alpha_\nu = +1$, then $\alpha_{\nu+1} = -1$, the pinch is $q_j G_i u_\nu G_i^{-1} q_j^{-1}$, and $G_i u_\nu G_i^{-1} = 1$ in $\mathscr{B}_1$; if $\alpha_\nu = -1$, then $\alpha_{\nu+1} = +1$, the pinch is $q_j^{-1} F_i^{\#-1} u_\nu F_i^\# q_j$, and $F_i^{\#-1} u_\nu F_i^\# = 1$ in $\mathscr{B}_1$. In either case, we have $u_\nu = 1$ in $\mathscr{B}_1$. But u_ν is a reduced word on the basis $\{s_\beta x\text{: all } \beta\}$, and so $u_\nu \equiv 1$, contradicting the minimality of t. We conclude that $t = 1$, $\alpha_1 = -1$, and

$$\omega \equiv x^m X^\# q_j Y x^n u_0 G_i^{-1} q_j^{-1} F_i^{\#-1} u_1 = 1 \quad \text{in } \mathscr{B}_1 * Q.$$

We have already seen that

$$Yx^n u_0 G_i^{-1} = 1 \quad \text{in } \mathscr{B}_1,$$

and so it follows from ω being in the free product that

$$F_i^{\#-1} u_1 x^m X^\# = 1 \quad \text{in } \mathscr{B}_1.$$

We rewrite these last two equations, by conjugating, into more convenient form:

$$(2) \qquad \begin{cases} x^n u_0 G_i^{-1} Y = 1 & \text{in } \mathscr{B}_1 \\ X^\# F_i^{\#-1} u_1 x^m = 1 & \text{in } \mathscr{B}_1. \end{cases}$$

Recall that G_i is a positive s-word. Let us show, after canceling all subwords of the form $s_\beta s_\beta^{-1}$ or $s_\beta^{-1} s_\beta$ (if any), that the first surviving letter of $G_i^{-1} Y$ is positive; that is, there is enough cancellation so that the whole of G_i is eaten by Y). Otherwise, after cancellation, $G_i^{-1} Y$ begins with s_β^{-1} for some β. Since $\mathscr{B}_1$ is an HNN extension with base $\langle x \rangle$ and stable letters $\{s_0, \ldots, s_M\}$, then $x^n u_0 G_i^{-1} Y = 1$ in $\mathscr{B}_1$ implies, by Britton's lemma, that its post-cancellation version contains a pinch $s_\lambda^f D s_\lambda^{-f} \equiv s_\lambda^f x^h s_\lambda^{-f}$, where $0 \le \lambda \le M$. Now u_0 is a reduced word on $\{s_0 x, \ldots, s_M x\}$, say,

$$u_0 = (s_{\beta_1} x)^{g_1} \ldots (s_{\beta_p} x)^{g_p},$$

where $g_\nu = \pm 1$. The pinch is not a subword of $x^n u_0$. It follows that the last letter s_λ^{-f} of the pinch $s_\lambda^f x^h s_\lambda^{-f}$ is the first surviving letter s_β^{-1} of $G_i^{-1} Y$. Thus, $\lambda = \beta = \beta_p$, $f = +1 = g_p$, and $s_\beta x^h s_\beta^{-1} \equiv s_\beta x s_\beta^{-1}$; that is, $h = 1$. But $x =$

$s_\beta x^2 s_\beta^{-1}$ in $\mathscr{B}_1$, giving $x \in \langle x^2 \rangle$, a contradiction. The first surviving letter of $G_i^{-1} Y$ is thus positive, and so there is a subword Y_1 of Y beginning with a positive s-letter for which $Y \equiv G_i Y_1$.

In a similar manner, one sees, after canceling all subwords of the form $s_\beta s_\beta^{-1}$ or $s_\beta^{-1} s_\beta$ (if any), that the first surviving letter of $X^\# F_i^{\#-1}$ is negative; that is, there is enough cancellation so that the whole of $F_i^{\#-1}$ is eaten by $X^\#$). The proof is just as above, inverting the original equation $X^\# F_i^{\#-1} u_1 x^m = 1$ in $\mathscr{B}_1$. There is thus a subword X_1 of X with $X_1^\#$ ending in a negative letter and such that $X \equiv X_1 F_i$.

We have proved, in $\mathscr{B}_1$, that $1 = x^n u_0 G_i^{-1} Y = x^n u_0 G_i^{-1} G_i Y_1$, and so

$$u_0^{-1} = Y_1 x^n \quad \text{in } \mathscr{B}_1.$$

Define

$$v_0^{-1} = r_i^{-1} u_0^{-1} r_i.$$

Since u_0 is a word on $s_\beta x$'s and $r_i^{-1} s_\beta x r_i = s_\beta x^{-1}$ for all β, the element v_0^{-1} is a word on $\{s_0 x^{-1}, \ldots, s_M x^{-1}\}$. But we may also regard u_0^{-1} and v_0^{-1} as elements of $\mathscr{B}_1 = \langle x, s_0, \ldots, s_M \rangle$. By Corollary 12.12(ii), there is an automorphism ψ of $\mathscr{B}_1$ with $\psi(x) = x^{-1}$ and $\psi(s_\beta) = s_\beta$ for all β. Hence, $v_0^{-1} = \psi(u_0^{-1}) = \psi(Y_1 x^n) = Y_1 x^{-n}$; that is,

$$v_0^{-1} = Y_1 x^{-n} \quad \text{in } \mathscr{B}_1. \tag{3}$$

If one defines $v_1^{-1} = r_i^{-1} u_1^{-1} r_i$, then a similar argument gives

$$v_1^{-1} = x^{-m} X_1^\#, \tag{4}$$

where X_1 is the subword of X defined above.

Let us return to the induction (remember that we are still in the case $e = -1$ of the beginning equation (1)):

$$L_1 X^\# q_j Y L_2 \equiv L_3 (r_i^e x^m X^\# q_j Y x^n r_i^{-e}) L_4 = q \quad \text{in } \mathscr{B}_2.$$

There are equations in $\mathscr{B}_2$,

$$\begin{aligned} q = L_1 X^\# q_j Y L_2 &\equiv L_3 r_i^{-1} (x^m X^\#) q_j (Y x^n) r_i L_4 \\ &= L_3 r_i^{-1} (u_1^{-1} F_i^\#) q_j (G_i u_0^{-1}) r_i L_4 \qquad \text{Eq. (2)} \\ &= L_3 v_1^{-1} r_i^{-1} (F_i^\# q_j G_i) r_i v_0^{-1} L_4 \\ &= L_3 v_1^{-1} (r_i^{-1} (F_i^\# q_j G_i r_i) v_0^{-1} L_4 \\ &= (L_3 v_1^{-1}) H_i^\# q_{i_2} K_i (v_0^{-1} L_4) \\ &= (L_3 x^{-m} X_1^\#) H_i^\# q_{i_2} K_i (Y_1 x^{-n} L_4) \quad \text{Eqs. (3), (4).} \end{aligned}$$

Therefore,

$$L_3 x^{-m} (X_1^\# H_i^\# q_{i_2} K_i Y_1) x^{-n} L_4 = q \quad \text{in } \mathscr{B}_2.$$

Now $L_3 x^{-m}$ and $x^{-n} L_4$ are words on $\{x, r_i, i \in I\}$ having at most $\rho - 1$ occurrences of various r-letters. In order to apply the inductive hypothesis, we must see that $X_1^{\#}H_i^{\#}$ and K_iY_1 are freely reduced; that is, they contain no "forbidden" subwords of the form $s_\beta s_\beta^{-1}$ or $s_\beta^{-1}s_\beta$. Now K_i is a positive word on s-letters, so that it contains no forbidden subwords; further, $Y_1 \equiv G_i^{-1}Y$ is just a subword of Y (since the whole of G_i is eaten by Y), hence has no forbidden subwords, by hypothesis. Therefore, a forbidden subword can occur in K_iY_1 only at the interface. But this is impossible, for we have seen that Y_1 begins with a positive letter, namely, "the first surviving letter" above. A similar argument shows that $X_1^{\#}H_i^{\#}$ is freely reduced.

By induction, both X_1H_i and K_iY_1 are positive. Hence, their subwords X_1 and Y_1 are also positive, and hence $X \equiv X_1F_i$ and $Y \equiv G_iY_1$ are positive. The inductive hypothesis also gives

$$(X_1^{\#}H_i^{\#}q_{i_2}K_iY_1)^* = q \quad \text{in } \Gamma.$$

Since $(X_1^{\#}H_i^{\#})^{\#} = X_1H_i$, we have

$$X_1H_iq_{i_2}K_iY_1 = q \quad \text{in } \Gamma \tag{5}$$

(it is only now that we see why the "sharp" operation $^{\#}$ was introduced; had we used inversion instead, we would now have $H_iX_1q_{i_2}K_iY_1 = q$ in Γ, and we could not finish the proof). Thus,

$$\begin{aligned} Xq_jY &\equiv X_1F_iq_jG_iY_1 \\ &\equiv X_1F_iq_{i_1}G_iY_1 = X_1H_iq_{i_2}K_iY_1 \quad \text{in } \Gamma. \end{aligned}$$

Combining this with (5) gives

$$Xq_jY = q \quad \text{in } \Gamma,$$

as desired. The case $e = +1$ at the beginning of the inductive step is entirely similar, and the proof of Boone's lemma and the Novikov–Boone–Britton theorem is complete. ∎

Here is some geometric interpretation of the long proof of this last lemma. At the end of the previous lemma, we had shown that a quadrant involving $\Sigma \equiv X^{\#}q_{i_1}Y$ on the outer boundary and a q on the inner boundary must have a "petal" relator polygon next to q_{i_1}. Now there is another q-letter on this petal which is now in the interior of the quadrant. As petals are the only relator polygons involving q-letters (for we are working in $\mathscr{B}^{\Delta}$), there must be a sequence of such petals (involving various q-letters) from the outer boundary of the quadrant to the q on the inner boundary (Figure 12.9).

Do any other q-edges occur on interior regions of the quadrant? The only other possibility is a flower whose eight-sided petals arise from a petal relator regions (Figure 12.10). We have not drawn the relator polygons that subdivide the eye of the flower, but we may assume that the eye contains no relator regions having q-edges (otherwise the eye contains a smaller such flower and we examine it). The boundary word of the flower's eye involves r_i's and s_β's, and this word is 1 in $\mathscr{B}_1$. By Britton's lemma, this word contains a pinch of

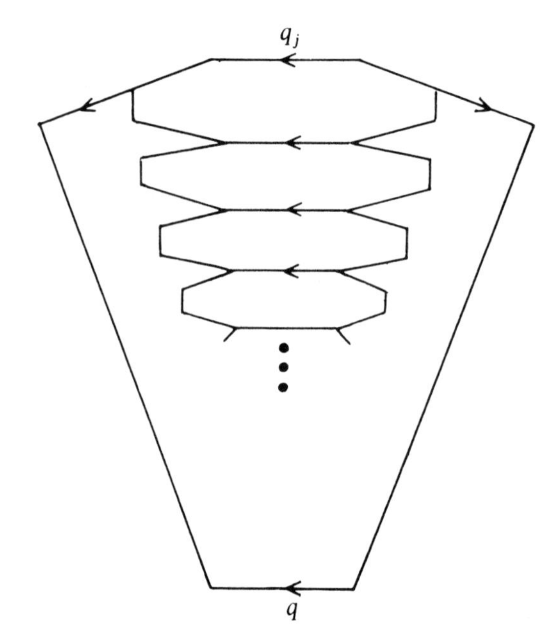

Figure 12.9

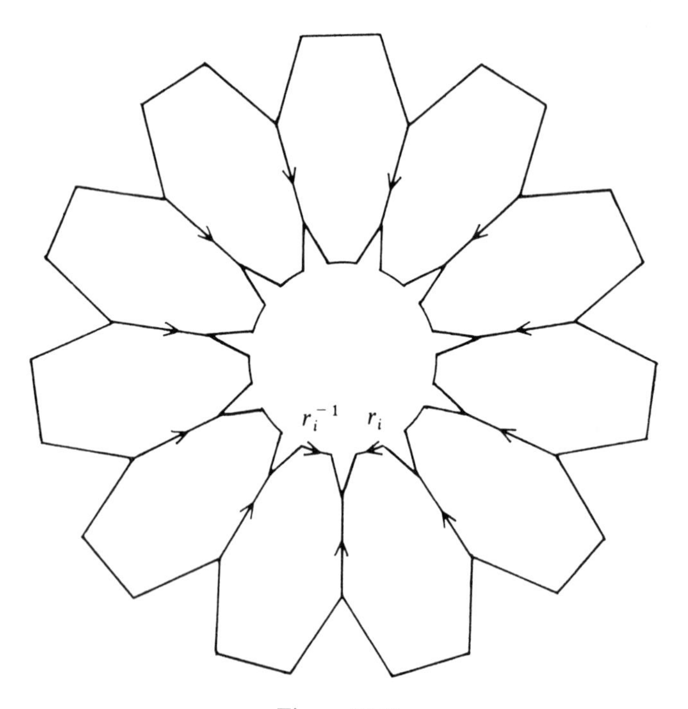

Figure 12.10

the form $r_i^e C r_i^{-e}$. There are thus two adjacent petals whose r-edges point in opposite directions, and this contradicts the orientation of these petals (note how the geometry of the plane enters).

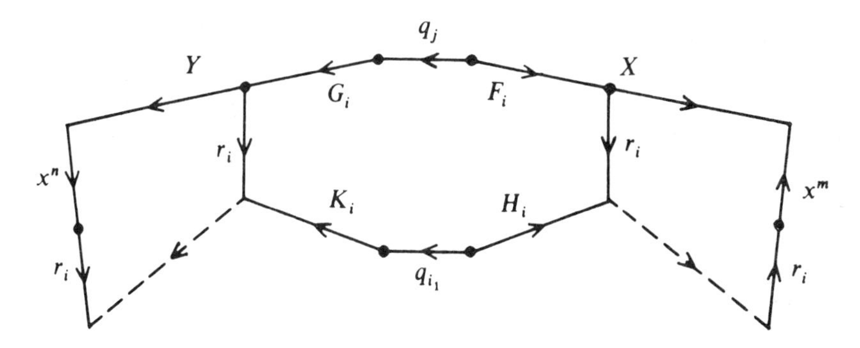

Figure 12.11

Now focus on the top portion of the quadrant. The remainder of the proof shows that the dashed paths comprised of s-edges can be drawn (actually, the proof shows that the rightward path is $X_1^\#$ (followed by x^{-m}, which is incorporated into L_3)and the leftward path is Y_1 (followed by x^n, which is incorporated into L_4). Induction says that one can repeat this construction, so that the petals move down to the bottom q; thus the whole quadrant can be subdivided into relator regions.

Aside from the group-theoretic proof just given (which is a simplification of Britton's original proof), there are several other proofs of the unsolvability of the word problem for groups: the original combinatorial proofs of Novikov and of Boone; a proof of G. Higman's, which is a corollary of his imbedding theorem. The proof of Higman's imbedding theorem that we shall give in the next section uses our development so far, whereas Higman's original proof does not depend on the Novikov–Boone–Britton theorem.

We must mention an important result here (see Lyndon and Schupp (1977) for a proof). W. Magnus (1930) proved the ***Freiheitsatz***. If G is a finitely generated group having only one defining relation r, say, $G = (x_1, \ldots, x_n | r)$, then any subset of $\{x_1, \ldots, x_n\}$ not containing all the x_i involved in r freely generates its subgroup. As a consequence, he showed (1932) that G has a solvable word problem.

There are other group-theoretic questions yielding unsolvable problems; let us consider another such question now.

Definition. A finitely generated group $G = (X | \Delta)$ has a ***solvable conjugacy problem*** if there is a decision process to determine whether an arbitrary pair of words ω and ω' on X are conjugate elements of G.

When G is finitely presented, it can be shown that its having a solvable conjugacy problem does not depend on the choice of finite presentation. A group with a solvable conjugacy problem must have solvable word problem, for one can decide whether an arbitrary word ω is (a conjugate of) 1; the converse is false. We now indicate how this result fits into our account.

Corollary 12.16. *The group $\mathscr{B}^{\Delta}$ has solvable word problem and unsolvable conjugacy problem.*

Proof. Recall that $\mathscr{B}^{\Delta}$ is Boone's group $\mathscr{B}$ without the octagonal relation $w(q)$. A.A. Fridman (1960) and Boone, independently, proved that $\mathscr{B}^{\Delta}$ has solvable word problem (we will not present this argument).

The following three statements are equivalent for any special word Σ:

(i) $w(\Sigma) = 1$ in $\mathscr{B}$;
(ii) $\Sigma^* = 1$ in Γ;
(iii) $w(\Sigma)$ is conjugate to $w(q)$ in $\mathscr{B}^{\Delta}$.

The necessity of Boone's lemma is (i) $\Rightarrow$ (ii); in geometric terms, we have already seen that the labeled directed annulus with outer boundary word $w(\Sigma)$ and inner boundary word $w(q)$ can be subdivided into relator polygons corresponding to relations other than $w(q)$; that is, using relations of $\mathscr{B}^{\Delta}$. This proves (ii) $\Rightarrow$ (iii). Finally, (iii) $\Rightarrow$ (i) is obviously true, because $w(q) = 1$ in $\mathscr{B}$. The equivalence of (i) and (iii) shows that $\mathscr{B}^{\Delta}$ has a solvable conjugacy problem if and only if $\mathscr{B}$ has a solvable word problem. By the Novikov–Boone–Britton theorem, $\mathscr{B}^{\Delta}$ has an unsolvable conjugacy problem. ∎

The Higman Imbedding Theorem

When can a finitely generated group be imbedded in a finitely presented group? The answer to this purely group-theoretic question reveals a harmonic interplay of Group Theory and Mathematical Logic. The proof we present here is due to S. Aanderaa (1970).

The following technical lemma is just a version of the "trick" which allows an arbitrary word on an alphabet to be viewed as a positive word on a larger alphabet.

Lemma 12.17. *Every group G has a presentation*

$$G = (Y \mid \Psi)$$

in which every relation is a positive word on Y. If G is finitely generated (or finitely presented), there is such a presentation in which Y (or both Y and Ψ) is finite.

Proof. If $G = (X \mid \Delta)$ is a presentation, define a new set X' disjoint from X and in bijective correspondence with it via $x \mapsto x'$,

$$X' = \{x' | x \in X\},$$

and define a new presentation of G:

$$G = (X \cup X' | \Delta', xx', x \in X),$$

where Δ' consists of all the words in Δ rewritten by replacing every occurrence of every x^{-1} by x'. ■

Definition. A group R is ***recursively presented*** if it has a presentation

$$R = (u_1, \ldots, u_m | \omega = 1, \omega \in E),$$

where each ω is a positive word on $u_1, \ldots, u_m$, and E is an r.e. set.

The lemma shows that the positivity assumption, convenient for notation, is no real restriction on R.

Exercises

12.12. If a group R is recursively presented, then it has a presentation whose relations form a recursive set of positive words. (*Hint*. If the given presentation is

$$R = (u_1, \ldots, u_m | \omega_k = 1, k \geq 0),$$

where $\{\omega_k = 1, k \geq 0\}$ is an r.e. set of positive words, define a new presentation

$$(u_1, \ldots, u_m, y | y = 1, y^k \omega_k = 1, k \geq 0).)$$

12.13. Every finitely generated subgroup of a finitely presented group is recursively presented. (*Hint*. Consider all words that can be obtained from 1 by a finite number of elementary operations.)

12.14. Every recursively related group can be imbedded in a two-generator recursively presented group. (*Hint*. Corollary 11.80.)

Theorem 12.18 (G. Higman, 1961). *Every recursively presented group R can be imbedded in a finitely presented group.*

With Exercise 12.13, Higman's theorem characterizes finitely generated subgroups of finitely presented groups.

Assume that R has a presentation

$$R = (u_1, \ldots, u_m | \omega = 1, \omega \in E),$$

where E is an r.e. set of positive u-words. There is thus a Turing machine T (with alphabet $\{s_0, \ldots, s_M\}$ containing $\{u_1, \ldots, u_m\}$) enumerating E; moreover, by Exercise 12.11, we may assume that T has stopping state q_0. We are going to use the group $\mathscr{B}(T)$, constructed in Boone's lemma, arising from the semigroup $\Gamma(T)$. Now the original Markov–Post semigroup $\gamma(T)$ was rewritten as $\Gamma(T)$ for the convenience of the proof of the unsolvability of the word problem. For Higman's theorem, we shall rewrite $\gamma(T)$ another way. Of course, this will engender changes in the generators and relations of $\mathscr{B}(T)$, and so we review the construction. Beginning with a Turing machine T with stopping state q_0, we constructed $\gamma(T)$ with generators $q, h, q_0, \ldots, q_N, s_0, \ldots, s_M$, and certain relations. The semigroup $\Gamma(T)$ renamed h as the last s-letter; thus, $\Gamma(T)$ has generators q's and s's and relations those of γ rewritten accordingly. Returning to the original notation (with h no longer an s-letter) gives a group $\mathscr{B}(T)$ with generators:

$$q, h, q_0, \ldots, q_N, s_0, \ldots, s_M, r_i, i \in I, x, k, t$$

and relations those of the original $\mathscr{B}(T)$ but with the relations Δ_2 rewritten accordingly:

Δ_2': for all $\beta = 0, \ldots, M$ and $i \in I$,

$$s_\beta^{-1} x s_\beta = x^2, \qquad h^{-1} x h = x^2,$$

$$r_i^{-1} s_\beta x r_i = x s_\beta^{-1}, \qquad r_i^{-1} h x r_i = x h^{-1},$$

$$r_i^{-1} F_i^{\#} q_{i_1} G_i r_i = H_i^{\#} q_{i_2} K_i.$$

By Corollary 12.9, a positive s-word ω lies in E if and only if $w(h^{-1} q_1 \omega h) = 1$ in $\mathscr{B}(T)$; that is, $\omega \in E$ if and only if

(6) $$k(h^{-1}\omega^{-1} q_1^{-1} h t h^{-1} q_1 \omega h) = (h^{-1}\omega^{-1} q_1^{-1} h t h^{-1} q_1 \omega h)k \quad \text{in } \mathscr{B}(T).$$

Let us introduce new notation to simplify this last equation. First, define $\mathscr{B}_2(T)$ as the group with the presentation

$$\mathscr{B}_2(T) = (q, h, q_0, \ldots, q_N, s_0, \ldots, s_M, r_i, i \in I \mid \Delta_2').$$

Now introduce new symbols:

$$\tau = q_1^{-1} h t h^{-1} q_1 \qquad \text{and} \qquad \kappa = h k h^{-1}.$$

Define a new group $\mathscr{B}_3'(T)$ by the presentation

$$\mathscr{B}_3'(T) = (\mathscr{B}_2(T); \tau \mid \tau^{-1}(q_1^{-1} h r_i h^{-1} q_1)\tau = q_1^{-1} h r_i h^{-1} q_1,$$
$$\tau^{-1}(q_1^{-1} h x h^{-1} q_1)\tau = q_1^{-1} h x h^{-1} q_1).$$

Note that $\mathscr{B}_3'(T)$ is just another presentation of the group

$$\mathscr{B}_3 = (\mathscr{B}_2(T); t \mid t r_i = r_i t, i \in I, tx = xt),$$

as can be quickly seen by replacing τ by its definition. Similarly, we define

$$\mathscr{B}'(T) = (\mathscr{B}_3'(T); \kappa \mid \kappa^{-1}(h r_i h^{-1})\kappa = h r_i h^{-1}, \kappa^{-1}(h x h^{-1})\kappa = h x h^{-1},$$
$$\kappa^{-1}(h q^{-1} h^{-1} q_1 t_0 q_1^{-1} h q h^{-1})\kappa = h q^{-1} h^{-1} q_1 t_0 q_1^{-1} h q h^{-1}).$$

Replacing κ by its definition shows that $\mathscr{B}'(T)$ is another presentation of

$$\mathscr{B}(T) = (\mathscr{B}_3'(T); k \mid k r_i = r_i k, i \in I, kx = xk, k(q^{-1} t q) = (q^{-1} t q)k).$$

Lemma 12.19.

(i) *$\mathscr{B}'_3(T)$ is an HNN extension with base $\mathscr{B}_2(T)$ and stable letter τ.*
(ii) *$\mathscr{B}'(T) \cong \mathscr{B}(T)$, and $\mathscr{B}'(T)$ is an HNN* extension with base *$\mathscr{B}'_3(T)$ and stable letter κ.*

Proof. As the proof of Lemma 12.11. ■

The next lemma shows how $\mathscr{B}'(T)$ simplifies (6).

Lemma 12.20. *If ω is a positive word on $s_0, \ldots, s_M$, then $\omega \in E$ if and only if $\kappa(\omega^{-1}\tau\omega) = (\omega^{-1}\tau\omega)\kappa$ in $\mathscr{B}'(T)$.*

Proof. The equation $\omega(h^{-1}q_1\omega h) = 1$ in $\mathscr{B}(T)$ has this simpler form in $\mathscr{B}'(T)$ once t and k are replaced by τ and κ, respectively. ■

Form the free product $\mathscr{B}'(T) * R$. Recall that R is generated by $\{u_1, \ldots, u_m\}$. At the outset, the Turing machine T enumerating the relations of R was chosen so that its alphabet $\{s_0, \ldots, s_M\}$ contains $\{u_1, \ldots, u_m\}$. Of course, the generating sets of the free factors of $\mathscr{B}'(T) * R$ must be disjoint. Let us, therefore, introduce new letters $\{a_1, \ldots, a_m\} \subset \{s_0, \ldots, s_M\} \subset \mathscr{B}'(T)$ for the replica of $\{u_1, \ldots, u_m\} \subset R$. Henceforth, we will regard the r.e. set E as comprised of certain positive words on $\{a_1, \ldots, a_m\} \subset \{s_0, \ldots, s_M\}$. Our rewriting is completed.

Now define new groups $\mathscr{B}_4$, $\mathscr{B}_5$, and $\mathscr{B}_6$ as follows (these also depend on T, but we abbreviate notation):

$$\mathscr{B}_4 = (\mathscr{B}'(T) * R;\, b_1, \ldots, b_m \mid b_i^{-1}u_jb_i = u_j,\, b_i^{-1}a_jb_i = a_j,$$
$$b_i^{-1}\kappa b_i = \kappa u_i^{-1}, \text{ all } i, j = 1, \ldots, m);$$
$$\mathscr{B}_5 = (\mathscr{B}_4;\, d \mid d^{-1}\kappa d = \kappa,\, d^{-1}a_ib_id = a_i,\, i = 1, \ldots, m);$$
$$\mathscr{B}_6 = (\mathscr{B}_5 \colon \sigma \mid \sigma^{-1}\tau\sigma = \tau d,\, \sigma^{-1}\kappa\sigma = \kappa,\, \sigma^{-1}a_i\sigma = a_i,\, i = 1, \ldots, m).$$

Aanderaa's proof of Higman's theorem is in two steps. The first step shows that each of these groups is an HNN extension of its predecessor:

$$R \leq \mathscr{B}'(T) * R \leq \mathscr{B}_4 \leq \mathscr{B}_5 \leq \mathscr{B}_6;$$

by Theorem 11.78, each group is imbedded in its successor, and so R is a subgroup of $\mathscr{B}_6$. The second step shows that $\mathscr{B}_6$ is finitely presented. After the proof is completed, we shall see that the diagram in Plate 4 partially explains how the generators and relations of the groups $\mathscr{B}_4$, $\mathscr{B}_5$, and $\mathscr{B}_6$ arise.

Lemma 12.21. *The subgroups $\langle a_1, \ldots, a_m, \kappa\rangle$ and $\langle a_1, \ldots, a_m, \tau\rangle$ of $\mathscr{B}'(T) * R$ are free groups with respective bases the displayed generating sets.*

Proof. Recall our analysis of $\mathscr{B}(T)$ in Lemma 12.11: $\mathscr{B}_1$ is an HNN extension with base $\mathscr{B}_0$ and stable letters $\{s_0, \ldots, s_M\}$; in our present notation, $\mathscr{B}_1$ has stable letters $\{h, s_0, \ldots, s_M\}$. It follows from Lemma 11.76 that $\langle h, s_0, \ldots, s_M\rangle$ is a free group with basis $\{h, s_0, \ldots, s_M\}$. Since $\mathscr{B}_1 \leq \mathscr{B}'(T)$, this last statement holds in $\mathscr{B}'(T) * R$. But $\{a_1, \ldots, a_m\} \subset \{s_0, \ldots, s_M\}$, so that $\langle a_1, \ldots, a_m\rangle$ is free with basis $\{a_1, \ldots, a_m\}$.

We now show that $\{a_1, \ldots, a_m, \kappa\}$ freely generates its subgroup (a similar argument that $\{a_1, \ldots, a_m, \tau\}$ freely generates its subgroup is left to the reader). Otherwise, there is a word

$$\omega \equiv c_0 \kappa^{e_1} c_1 \kappa^{e_2} \ldots c_{n-1} \kappa^{e_n} = 1 \quad \text{in } \mathscr{B}'(T) * R,$$

where $e_v = \pm 1$ and c_v are (possibly empty) freely reduced words on $\{a_1, \ldots, a_m\}$; we may further assume, of all such words ω, that n is chosen minimal. Since ω involves no u-letters, we have $\omega = 1$ in $\mathscr{B}'(T)$. As $\mathscr{B}'(T)$ is an HNN extension with base $\mathscr{B}'_4(T)$ and stable letter κ, Britton's lemma says that either ω does not involve κ or ω contains a pinch $\kappa^e c_v \kappa^{-e}$, where c_v is a word on $\{hr_ih^{-1}, i \in I, hxh^{-1}, hq^{-1}h^{-1}q_1\tau q_1^{-1}hqh^{-1}\}$. But the relations in $\mathscr{B}'(T)$ show that κ commutes with c_v, so that $\kappa^e c_v \kappa^{-e} = c_v$ in $\mathscr{B}'(T) \leq \mathscr{B}'(T) * R$, and this contradicts the minimality of n. It follows that ω does not involve κ; that is, ω is a reduced word on $\{a_1, \ldots, a_m\}$. But we have already seen that $\langle a_1, \ldots, a_m\rangle$ is free with basis $\{a_1, \ldots, a_m\}$, so that $\omega \equiv 1$. ■

Lemma 12.22. *$\mathscr{B}_4$ in an HNN extension with base $\mathscr{B}'(T) * R$ and stable letters $\{b_1, \ldots, b_m\}$.*

Proof. It suffices to show there are isomorphisms $\varphi_i: A_i \to B_i$, where

$$A_i = \langle u_1, \ldots, u_m, a_1, \ldots, a_m, \kappa\rangle,$$
$$B_i = \langle u_1, \ldots, u_m, a_1, \ldots, a_m, ku_i^{-1}\rangle,$$

and $\varphi_i(u_j) = u_j$, $\varphi_i(a_j) = a_j$, and $\varphi_i(\kappa) = \kappa u_i^{-1}$. Note that $A_i = B_i$.

It is easy to see, in $\mathscr{B}'(T) * R$, that

$$A_i = \langle a_1, \ldots, a_m, \kappa\rangle * \langle u_1, \ldots, u_m\rangle$$
$$= \langle a_1, \ldots, a_m, \kappa\rangle * R.$$

By Lemma 12.21, $\langle a_1, \ldots, a_m, \kappa\rangle$ is freely generated by $\{a_1, \ldots, a_m, \kappa\}$, so that φ_i is a well defined homomorphism. Similarly, the map $\psi_i: A_i \to A_i$, given by $\psi_i(u_j) = u_j$, $\psi_i(a_j) = a_j$, and $\psi_i(\kappa) = \kappa u_i$, is a well defined homomorphism. But ψ_i is the inverse of φ_i, so that φ_i is an isomorphism and $\mathscr{B}_4$ is an HNN extension. ■

Lemma 12.23. *$\mathscr{B}_5$ is an HNN extension with base $\mathscr{B}_4$ and stable letter d.*

Proof. It suffices to show that there is an isomorphism

$$\varphi: A = \langle \kappa, a_1 b_1, \ldots, a_m b_m\rangle \to B = \langle \kappa, a_1, \ldots, a_m\rangle$$

with $\varphi(\kappa) = \kappa$ and $\varphi(a_i b_i) = a_i$ for all i.

Since $\kappa^{-1}b_i\kappa = b_iu_i$ in $\mathscr{B}_4$, the function $\theta\colon \mathscr{B}_4 \to \mathscr{B}'(T) * R$, defined by sending each b_i to 1, each u_i to 1, and all other generators to themselves, is a well defined homomorphism (it preserves all the relations: $b_i = 1$ implies $1 = \kappa^{-1}b_i\kappa = b_iu_i = u_i$). The map θ takes each of $\langle \kappa, a_1b_1, \ldots, a_mb_m\rangle$ and $\langle \kappa, a_1, \ldots, a_m\rangle$ onto the subgroup $\langle \kappa, a_1, \ldots, a_m\rangle \leq \mathscr{B}'(T) * R$ which, by the preceding lemma, is free on the displayed generators. By Exercise 11.8, each of the two subgroups A and B of $\mathscr{B}_4$ is free on the displayed generators, and so the map $\varphi\colon A \to B$ given above is a well defined isomorphism. ■

The next lemma will be needed in verifying that $\mathscr{B}_6$ is an HNN extension of $\mathscr{B}_5$.

Lemma 12.24. *The subgroup $A = \langle \kappa, a_1, \ldots, a_m, \tau\rangle \leq \mathscr{B}'(T)$ has the presentation*

$$A = (\kappa, a_1, \ldots, a_m, \tau \mid \kappa^{-1}\omega^{-1}\tau\omega\kappa = \omega^{-1}\tau\omega, \omega \in E).$$

Remark. Recall our change in notation: although E was originally given as a set of positive words on $\{u_1, \ldots, u_m\}$, it is now comprised of positive words on $\{a_1, \ldots, a_m\}$.

Proof. By Lemma 12.20, the relations $\kappa^{-1}\omega^{-1}\tau\omega\kappa = \omega^{-1}\tau\omega$, for all $\omega \in E$, do hold in $\mathscr{B}'(T)$, and hence they hold in the subgroup $A \leq \mathscr{B}'(T)$. To see that no other relations are needed, we shall show that if ζ is a freely reduced word on $\{\kappa, a_1, \ldots, a_m, \tau\}$ with $\zeta = 1$ in A, then ζ can be transformed into 1 via elementary operations using only these relations.

Step 1. ζ contains no subword of the form $\tau^\varepsilon\omega\kappa^\eta$, where $\varepsilon = \pm 1$, $\eta = \pm 1$, and $\omega \in E$.

It is easy to see that the given relations imply

$$\tau^\varepsilon\omega\kappa^\eta = \omega\kappa^\eta\omega^{-1}\tau^\varepsilon\omega.$$

If ζ contains a subword $\tau^\varepsilon\omega\kappa^\eta$, then

$$\zeta \equiv \zeta_1\tau^\varepsilon\omega\kappa^\eta\zeta_2 \to \zeta_1\omega\kappa^\eta\omega^{-1}\tau^\varepsilon\omega\zeta_2$$

is an elementary operation. Cancel all subwords (if any) of the form yy^{-1} or $y^{-1}y$, where $y \equiv \tau$, κ, or some a_j. With each such operation, the total number of occurrences of τ^ε which precede some κ^η goes down. Therefore, we may assume that ζ is freely reduced and contains no subwords of the form $\tau^\varepsilon\omega\kappa^\eta$.

Step 2. ζ involves both κ and τ.

If ζ does not involve κ, then it is a word on $\{a_1, \ldots, a_m, \tau\}$. But this set freely generates its subgroup, by Lemma 12.21, and so ζ being freely reduced and $\zeta = 1$ imply $\zeta \equiv 1$. A similar argument shows that ζ involves τ as well.

Since $\mathscr{B}'(T)$ is an HNN extension with base $\mathscr{B}_3'(T)$ and stable letter κ, Britton's lemma says that ζ contains a pinch $\kappa^e V\kappa^{-e}$, where $e = \pm 1$, and there is a word D on $\{hr_ih^{-1}, i \in I, hxh^{-1}, hq^{-1}h^{-1}q_1\tau q_1^{-1}hqh^{-1}\}$ with

$$V = D \quad \text{in } \mathscr{B}_3'(T).$$

Choose D so that the number of occurrences of τ in it is minimal.

Step 3. D is τ-reduced.

Now $\mathscr{B}_3'(T)$ is an HNN extension with base $\mathscr{B}_2(T)$ and stable letter τ. Let us write

$$\delta = hq^{-1}h^{-1}q_1,$$

so that

$$hq^{-1}h^{-1}q_1\tau q_1^{-1}hqh^{-1} \equiv \delta\tau\delta^{-1}.$$

If D, which is now a word on $\{hr_ih^{-1}, i \in I, hxh^{-1}, \delta\tau\delta^{-1}\}$, is not τ-reduced, then it contains a pinch. Since an occurrence of τ can only arise from an occurrence of $\delta\tau\delta^{-1}$, it follows that

$$D \equiv D_1\delta\tau^f\delta^{-1}D_2\delta\tau^{-f}\delta^{-1}D_3,$$

where D_2 does not involve the stable letter τ (just check the cases $f = 1$ and $f = -1$ separately); moreover, there is a word W on $\{q_1^{-1}hr_ih^{-1}q_1, i \in I, q_1^{-1}hxh^{-1}q_1\}$ with

$$\delta^{-1}D_2\delta = W \quad \text{in } \mathscr{B}_2(T)$$

(the subgroups A and B in the HNN extension are here equal, and so we need not pay attention to the sign of f). From the presentation of $\mathscr{B}_2(T)$, we see that τ and W commute. Therefore,

$$D = D_1\delta\tau^f W\tau^{-f}\delta^{-1}D_3 = D_1\delta W\delta^{-1}D_3 \quad \text{in } \mathscr{B}_2(T),$$

contradicting our choice of D having the minimal number of occurrences of τ. It follows that D is τ-reduced.

Step 4. V is τ-reduced.

Otherwise, V contains a pinch $\tau^g C\tau^{-g}$, where

$$C = W \quad \text{in } \mathscr{B}_2(T)$$

and W, a word on $\{q_1^{-1}hr_ih^{-1}q_1, i \in I, q_1^{-1}hxh^{-1}q_1\}$ (as above), commutes with τ in $\mathscr{B}_2(T)$. Now V does not involve κ, so its subword C involves neither κ nor τ. Since ζ, hence its subword V, is a word on $\{\kappa, a_1, \ldots, a_m, \tau\}$, it follows that C is a word on $\{a_1, \ldots, a_m\}$. But $\langle \tau, a_1, \ldots, a_m \rangle \leq \mathscr{B}'(T) * R$ is a free group with basis the displayed generators, by Lemma 12.21, and so C commutes with τ if and only if $C \equiv 1$. Therefore, the pinch $\tau^g C\tau^{-g} \equiv \tau^g\tau^{-g}$, contradicting ζ being freely reduced.

Step 5. Both V and D involve τ.

Since $V = D$ in $\mathscr{B}_3'(T)$ and both are τ-reduced, Corollary 11.82 applies to show that both of them involve the same number of occurrences of the stable letter τ. Assume now that neither V nor D involves τ. Then V is a word on $\{a_1, \ldots, a_m\}$ and D is a word on $\{hr_ih^{-1}, i \in I, hxh^{-1}\}$; we may assume that D has been chosen so that the total number of occurrences of r-letters is minimal; moreover, we may assume that all adjacent factors equal to hxh^{-1} are collected as hx^mh^{-1}. It follows that the equation $V = D$ holds in the subgroup $\mathscr{B}_2(T)$, which is an HNN extension with base $\mathscr{B}_1 * \langle q, q_0, \ldots, q_N \rangle$ and stable letters $\{r_i, i \in I\}$.

Now V is r_i-reduced for all i because V, being a word on $\{a_1, \ldots, a_m\}$, does not even involve any r-letters. We claim that D is also r_i-reduced for all i. Otherwise, D (a word on $\{hr_ih^{-1}, i \in I, hxh^{-1}\}$) contains a pinch: there is thus an index i with

$$D \equiv \Delta_1 hr_i^l h^{-1} \Delta_2 hr_i^{-l} h^{-1} \Delta_3,$$

where $l = \pm 1$ and Δ_2 involves no r-letters (just check the cases $l = 1$ and $l = -1$ separately); hence, $\Delta_2 \equiv hx^mh^{-1}$ (since it is freely reduced), and

$$D \equiv \Delta_1 hr_i^l x^m r_i^{-l} h^{-1} \Delta_3.$$

The pinch in D is thus $r_i^l x^m r_i^{-l}$, and Britton's lemma concludes, depending on the sign of l, that x^m is equal in $\mathscr{B}_1 * \langle q, q_0, \ldots, q_N \rangle$ either to a word on $\{F_i^{\#} q_{i_1} G_i, s_0x, \ldots, s_Mx\}$ or to a word on $\{H_i^{\#} q_{i_2} K_i, s_0x^{-1}, \ldots, s_Mx^{-1}\}$. As D does not involve q-letters, x^m is equal in $\mathscr{B}_1$ to a word on either $\{s_0x, \ldots, s_Mx\}$ or $\{s_0x^{-1}, \ldots, s_Mx^{-1}\}$, and we have already seen, in the proof of Lemma 12.14, that this forces $m = 0$. Therefore, the pinch is $r_i^l r_i^{-l}$, contradicting our choice of D having the minimal number of r-letters. We conclude that both V and D are r_i-reduced for all i. By Corollary 11.82, D involves no r-letters (because V involves none), and D is a word on hxh^{-1}; that is, $D = hx^nh^{-1}$ in $\mathscr{B}_1$. In this step, V is assumed to be a word on $\{a_1, \ldots, a_m\} \subset \{s_0, \ldots, s_M\}$, so that the equation $V = D$ holds in $\mathscr{B}_1$ and

$$Vhx^{-n}h^{-1} = 1 \quad \text{in } \mathscr{B}_1.$$

Recall that $\mathscr{B}_1$ is an HNN extension with base $\langle x \rangle$ and stable letters $\{h, s_0, \ldots, s_M\}$. If V involves a_j for some j, then Britton's lemma gives a pinch $a_j^\nu U a_j^{-\nu}$, where $\nu = \pm 1$ and U is a power of x (for a_j is a stable letter). Now $h \neq a_j$, for $h \notin \{s_0, \ldots, s_M\}$, so that this pinch must be a subword of V. But V does not involve x, and so $U \equiv 1$; therefore $a_j^\nu a_j^{-\nu}$ is a subword of V, contradicting ζ and its subword V being freely reduced. It follows that V involves no a_j; as V is now assumed to be a word on a-letters, we have $V \equiv 1$. Recall that V arose in the pinch $\kappa^e V \kappa^{-e}$, a subword of ζ, and this, too, contradicts ζ being freely reduced.

Step 6. V contains a subword $\tau^\varepsilon V_p$, where V_p is a positive a-word lying in the r.e. set E.

Since D, a word on $\{hr_ih^{-1}, i \in I, hxh^{-1}, hq^{-1}h^{-1}q_1\tau q_1^{-1}hqh^{-1}\}$, involves τ, it must involve $hq^{-1}h^{-1}q_1\tau q_1^{-1}hqh^{-1}$. Write

$$D \equiv \Lambda'(hq^{-1}h^{-1}q_1\tau^\alpha q_1^{-1}hqh^{-1})\Lambda,$$

where $\alpha = \pm 1$ and the word in parentheses is the final occurrence of the long word involving τ in D; thus, Λ is a word on $\{hr_ih^{-1}, i \in I, hxh^{-1}\}$.

Since V involves τ, we may write

$$(7) \qquad V \equiv V_0\tau^{\varepsilon_1}V_1 \ldots V_{p-1}\tau^{\varepsilon_p}V_p,$$

where $\varepsilon_j = \pm 1$ and each V_j is a freely reduced word on a-letters. By Steps 2 and 3, both D and V are τ-reduced, so that Corollary 11.82 says that

$$\tau^{\varepsilon_p}V_p\Lambda^{-1}hq^{-1}h^{-1}q_1\tau^{-\alpha}$$

is a pinch. Thus, there is a word Z on $\{q_1^{-1}hr_ih^{-1}q_1, q_1^{-1}hxh^{-1}q_1\}$ with

$$V_p\Lambda^{-1}hq^{-1}h^{-1}q_1 = Z \quad \text{in } \mathscr{B}_2(T);$$

of course, we may choose

$$Z \equiv q_1^{-1}hL_1h^{-1}q_1,$$

where L_1 is a word on $\{x, r_i, i \in I\}$; similarly, since Λ is a word on $\{hr_ih^{-1}, i \in I, hxh^{-1}\}$, we may write

$$\Lambda^{-1} = hL_2h^{-1} \quad \text{in } \mathscr{B}_2(T),$$

where L_2 is a word on $\{x, r_i, i \in I\}$. Substituting, we see that

$$V_phL_2h^{-1}hq^{-1}h^{-1}q_1 = q_1^{-1}hL_1h^{-1}q_1 \quad \text{in } \mathscr{B}_2(T),$$

and we rewrite this equation as

$$L_1^{-1}h^{-1}q_1V_phL_2 = q \quad \text{in } \mathscr{B}_2(T).$$

Note that V_ph is freely reduced, for V_p is a freely reduced word on a-letters, and h is not an s-letter, hence not an a-letter. By Lemma 12.15 (with $X \equiv h$ and $Y \equiv V_ph$), we have Y a positive word, so that its subword V_p is a positive word on a-letters; moreover,

$$hq_1V_ph = q \quad \text{in } \gamma(T).$$

By Lemma 12.4, $V_p \in E$. Returning to (7), the birthplace of V_p, we see that $\tau^{\varepsilon_p}V_p$ is a subword of V. Indeed, $V \equiv V'\tau^{\varepsilon_p}V_p$, where V' is the initial segment of V.

Step 7. Conclusion.

Recall that V arose inside the pinch $\kappa^eV\kappa^{-e}$, which is a subword of ζ. From the previous step, we see that $\kappa^eV'\tau^{\varepsilon_p}V_p\kappa^{-e}$ is a subword of ζ. In particular, ζ contains a subword of the form $\tau^\varepsilon\omega\kappa^\eta$, where $\varepsilon = \pm 1$, $\eta = \pm 1$, and $\omega \in E$. But we showed, in Step 1, that ζ contains no such subword. This completes the proof. ■

Lemma 12.25.

(i) *$\mathscr{B}_6$ is an HNN extension with base $\mathscr{B}_5$ and stable letter σ.*
(ii) *R is imbedded in $\mathscr{B}_6$.*

Proof. (i) It suffices to show that there is an isomorphism

$$\varphi\colon A = \langle \kappa, \tau, a_1, \ldots, a_m \rangle \to B = \langle \kappa, \tau d, a_1, \ldots, a_m \rangle$$

with $\varphi(\kappa) = \kappa$, $\varphi(\tau) = \tau d$, and $\varphi(a_j) = a_j$ for all j. Since

$$\mathscr{B}'(T) \le \mathscr{B}_4 \le \mathscr{B}_5,$$

the subgroup A is precisely the subgroup whose presentation was determined in the previous lemma:

$$A = (\kappa, a_1, \ldots, a_m, \tau \mid \kappa^{-1}\omega^{-1}\tau\omega\kappa = \omega^{-1}\tau\omega, \omega \in E).$$

To see that φ is a well defined homomorphism, we must show that it preserves all the relations; that is, if $\omega \in E$, then

$$\kappa^{-1}\omega^{-1}\tau d\omega\kappa = \omega^{-1}\tau d\omega \quad \text{in } B.$$

We shall show that this last equation does hold in $\mathscr{B}_5$, and hence it holds in $B \le \mathscr{B}_5$.

Let us introduce notation. If ω is a word on $\{a_1, \ldots, a_m\}$, write ω_b to denote the word obtained from ω by replacing each a_j by b_j, and let ω_u denote the word obtained from ω by replacing each a_j by u_j. If $\omega \in E$, then $\omega_u = 1$, for ω_u is one of the original defining relations of R. For $\omega \in E$, each of the following equations holds in $\mathscr{B}_5$:

$$\begin{aligned}\kappa^{-1}\omega^{-1}\tau d\omega\kappa &= \kappa^{-1}\omega^{-1}\tau(d\omega d^{-1})d\kappa\\ &= \kappa^{-1}\omega^{-1}\tau\omega\omega_b d\kappa\end{aligned}$$

(for $da_id^{-1} = a_ib_i$ in $\mathscr{B}_5$ and a_i and b_j commute in $\mathscr{B}_4 \le \mathscr{B}_5$). Since κ and τ commute in $\mathscr{B}_5$, we have

$$\begin{aligned}\kappa^{-1}\omega^{-1}\tau\omega\omega_b d\kappa &= \kappa^{-1}\omega^{-1}\tau\omega\omega_b\kappa d\\ &= \kappa^{-1}\omega^{-1}\tau\omega\kappa(\kappa^{-1}\omega_b\kappa)d\\ &= \kappa^{-1}\omega^{-1}\tau\omega\kappa\omega_b\omega_u d\end{aligned}$$

(because b_i and u_j commute and $\kappa^{-1}b_i\kappa = b_iu_i$)

$$= \kappa^{-1}\omega^{-1}\tau\omega\kappa\omega_b d$$

(because $\omega_u = 1$). We have shown that

$$\begin{aligned}\kappa^{-1}\omega^{-1}\tau\omega\omega_b d\kappa &= (\kappa^{-1}\omega^{-1}\tau\omega\kappa)\omega_b d\\ &= \omega^{-1}\tau\omega\omega_b d.\end{aligned}$$

On the other hand,

$$\begin{aligned}\omega^{-1}\tau d\omega &= \omega^{-1}(d\omega d^{-1})d\\ &= \omega^{-1}\tau\omega\omega_b d,\end{aligned}$$

as we saw above. Therefore,

$$\kappa^{-1}\omega^{-1}\tau d\omega\kappa = \omega^{-1}\tau d\omega \quad \text{in } \mathscr{B}_5$$

and $\varphi: A \to B$ is a well defined homomorphism.

To see that φ is an isomorphism, we construct a homomorphism $\psi: \mathscr{B}_5 \to \mathscr{B}_5$ whose restriction $\psi|B$ is the inverse of φ. Define ψ by setting $\psi|\mathscr{B}'(T)$ to be the identity map and

$$\psi(d) = \psi(b_i) = \psi(u_i) = 1.$$

Inspection of the various presentations shows that ψ is a well defined homomorphism. Since $\psi(\kappa) = \kappa$, $\psi(a_i) = a_i$, and $\psi(\tau d) = \tau$, we see that $\psi|B$ is the inverse of φ.

(ii) This follows from several applications of Theorem 11.78. ■

The following lemma completes the proof of the Higman imbedding theorem:

Lemma 12.26. *$\mathscr{B}_6$ is finitely presented.*

Proof. The original presentation of R is

$$R = (u_1, \ldots, u_m | \omega = 1, \omega \in E),$$

where E is an r.e. set of positive words on $\{u_1, \ldots, u_m\}$. Recall the notation introduced in the proof of Lemma 12.25: if ω is a word on $\{a_1, \ldots, a_m\}$, then ω_u and ω_b are obtained from ω by replacing each a_i by u_i or b_i, respectively. With this notation, the presentation of R can be rewritten:

$$R = (u_1, \ldots, u_m | \omega_u = 1, \omega \in E).$$

Now $\mathscr{B}'(T) * R$ is a finitely generated group having a finite number of relations occurring in the presentation of $\mathscr{B}'(T)$ together with the (possibly infinitely many) relations above for R. Each step of the construction of $\mathscr{B}_6$ from $\mathscr{B}'(T) * R$ contributes only finitely many new generators and relations. Thus, $\mathscr{B}_6$ is finitely generated, and it is finitely presented if we can show that every relation of the form $\omega_u = 1$, for $\omega \in E$, is a consequence of the remaining relations in $\mathscr{B}_6$.

By Lemma 12.20, $\kappa^{-1}\omega^{-1}\tau\omega\kappa = \omega^{-1}\tau\omega$ for all $\omega \in E$. Hence

$$\sigma^{-1}\kappa^{-1}\omega^{-1}\tau\omega\kappa\sigma = \sigma^{-1}\omega^{-1}\tau\omega\sigma.$$

Since σ commutes with κ and with all a_i, this gives

$$\kappa^{-1}\omega^{-1}\sigma^{-1}\tau\sigma\omega\kappa = \omega^{-1}\sigma^{-1}\tau\sigma\omega.$$

As $\sigma^{-1}\tau\sigma = \tau d$, this gives

$$\kappa^{-1}\omega^{-1}\tau d\omega\kappa = \omega^{-1}\tau d\omega.$$

Inserting $\omega\kappa\kappa^{-1}\omega^{-1}$ and $\omega\omega^{-1}$ gives

$$(\kappa^{-1}\omega^{-1}\tau\omega\kappa)\kappa^{-1}\omega^{-1}d\omega\kappa = (\omega^{-1}\tau\omega)\omega^{-1}d\omega.$$

The terms in parentheses are equal (Lemma 12.20 again), so that canceling gives

$$(8) \qquad \kappa^{-1}\omega^{-1}d\omega\kappa = \omega^{-1}d\omega.$$

Now the relations $da_id^{-1} = a_ib_i$ and $a_ib_j = b_ja_i$, all i, j, give

$$d\omega d^{-1} = \omega\omega_b,$$

hence

$$\omega = d^{-1}\omega\omega_b d,$$

for every word ω on $\{a_1, \ldots, a_m\}$, and so

$$(9) \qquad \omega^{-1}d\omega = \omega^{-1}dd^{-1}\omega\omega_b d = \omega_b d.$$

Substituting (9) into (8) gives

$$\omega_b = \kappa^{-1}(\omega^{-1}d\omega)\kappa d^{-1} = \kappa^{-1}\omega_b d\kappa d^{-1}.$$

Since κ and d commute,

$$(10) \qquad \kappa^{-1}\omega_b\kappa = \omega_b.$$

On the other hand, the relations $\kappa^{-1}b_i\kappa = b_iu_i$ and $b_iu_j = u_jb_i$, all i, j, give

$$\kappa^{-1}\omega_b\kappa = \omega_b\omega_u.$$

This last equation coupled with (10) gives

$$\omega_b\omega_u = \omega_b,$$

and so $\omega_u = 1$, as desired. ∎

Let us review the proof of Higman's theorem to try to understand Aanderaa's construction. Certainly, some of the relations of $\mathscr{B}_6$ are present to guarantee a chain of HNN extensions, for this gives an imbedding of R into $\mathscr{B}_6$. The proof of the last lemma, showing that $\mathscr{B}_6$ is finitely presented, amounts to proving, for $\omega \in E$, that $\omega_u = 1$ follows from the other relations; that is, one can subdivide the labeled directed polygon with boundary word ω_u into relator polygons corresponding to the other relations in $\mathscr{B}_6$. E. Rips has drawn a diagram (Plate 4) that helps explain the construction of $\mathscr{B}_6$.

Before we examine Plate 4, let us discuss diagrams in the plane from a different viewpoint. Regard the plane as lying on the surface of a sphere, and assume that the north pole, denoted by ∞, lies outside a given diagram. Otherwise said, we may regard a given planar diagram D having n regions to actually have $n + 1$ regions, the new "unbounded" region (containing ∞) being the outside of D. We now propose redrawing a diagram so that the unbounded region is drawn as an interior region. For example, assume that

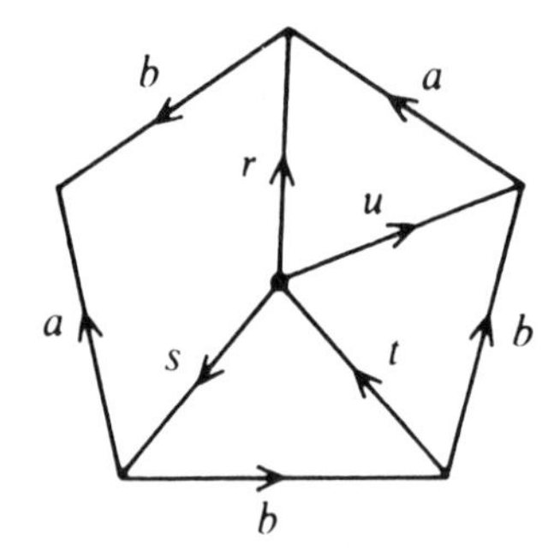

Figure 12.12

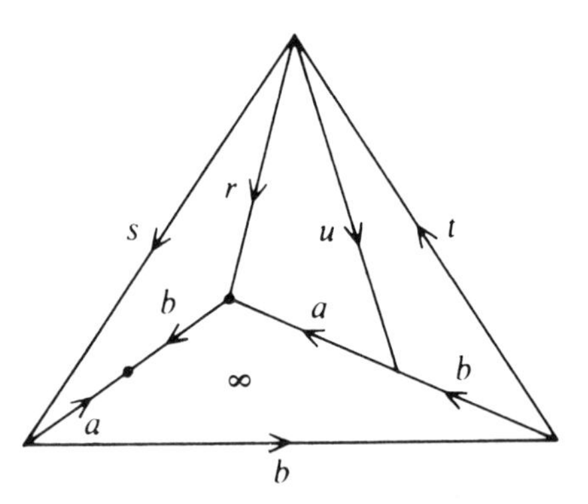

Figure 12.13

Figure 12.12 shows that $\omega \equiv aba^{-1}b^2 = 1$ in some group. Figure 12.13 is a redrawn version of Figure 12.12 with ∞ marking the old unbounded region.

To redraw, first number all the vertices, then connect them as they are connected in the original diagram. Note that all the (bounded) regions are relator regions corresponding to the inverses of the original relations, with the exception of that containing ∞. The boundary word of the region with ∞ is *sbt*, as in the original diagram. In general, every (not necessarily bounded) region in the redrawn diagram is a relator polygon save the new region containing ∞ whose boundary word is ω. Such a diagram will show that $\omega = 1$ if every region (aside from that containing ∞) is a relator polygon and the boundary word of the diagram is 1 in the group.

Let us return to $\mathscr{B}_6$. For a word $\omega \in E$, draw a diagram, new version, showing that $\omega_u = 1$ in $\mathscr{B}_6$ (using only the other relations of $\mathscr{B}_6$). By Lemma 12.20, $\omega \in E$ gives

$$\kappa^{-1}\omega^{-1}\tau\omega\kappa^{-1}\omega^{-1}\tau^{-1}\omega = 1.$$

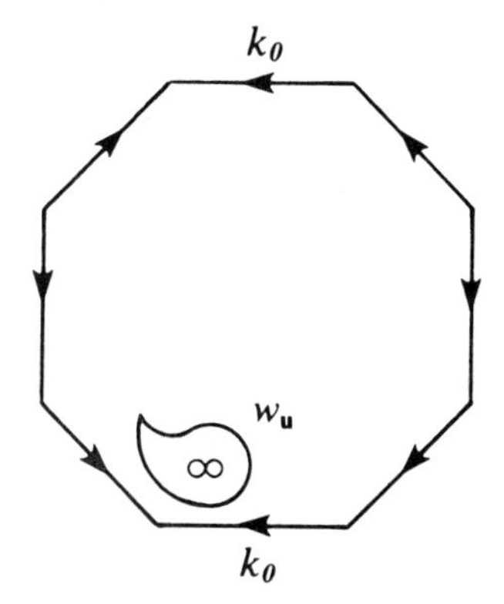

Figure 12.14

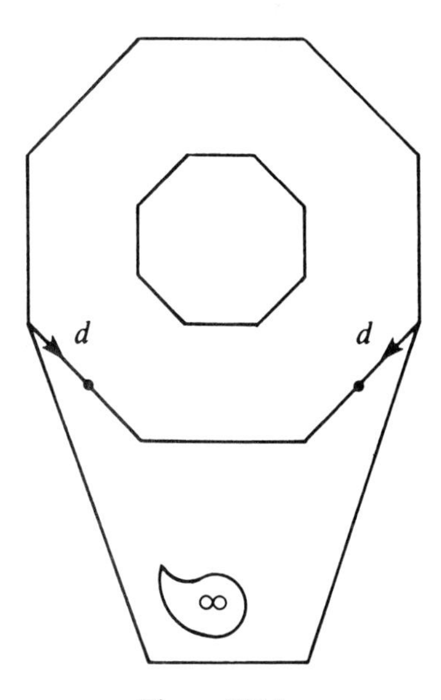

Figure 12.15

We begin, therefore, with a labeled directed octagon for this word as well as with a "balloon" region (containing ∞) inside having boundary word ω_u. To subdivide, draw a second octagon inside it, and yet a third octagon perturbed by two d-edges. Now complete this picture, adding σ-edges and the subdivision of the bottom, to obtain Plate 4, the diagram showing that $\omega_u = 1$ follows from the other relations.

Let us indicate, briefly, how the Novikov–Boone–Britton theorem follows from the Higman theorem. It is not difficult to construct a recursively presented group G having an unsolvable word problem. For example, let G be a variant of the group in Theorem 11.85: if F is a free group with basis $\{a, b\}$, let

$$G = (a, b, p \mid p^{-1}\omega_n p = \omega_n \text{ for all } n \in E),$$

where the commutator subgroup F' is free with basis $\{\omega_0, \omega_1, \ldots, \omega_n, \ldots\}$ and E is an r.e. set in $\mathbb{Z}$ that is not recursive. By Higman's theorem, there is a finitely presented group G^* containing G. If G^* had a solvable word problem, then so would all its finitely generated subgroups, by Exercise 12.4, and this contradicts the choice of G.

Some Applications

Higman's theorem characterizes those finitely generated groups G that can be imbedded in finitely presented groups. Of course, any (perhaps nonfinitely generated) group G that can be so imbedded must be countable. In Theorem 11.71, we saw that every countable group G can be imbedded in a two-generator group G^{II}.

Lemma 12.27. *If G is a countable group for which G^{II} is recursively presented, then G can be imbedded in a finitely presented group.*

Proof. Higman's theorem shows that G^{II} can be imbedded in a finitely presented group. ■

At this point, we omit some details which essentially require accurate bookkeeping in order to give an explicit presentation of G^{II} from a given presentation of G. We assert that there is a presentation of the abelian group

$$G = \sum_{i=1}^{\infty} D_i,$$

where $D_i \cong \mathbb{Q} \oplus (\mathbb{Q}/\mathbb{Z})$ for all i, such that G^{II} is recursively presented.

Theorem 12.28. *There exists a finitely presented group containing an isomorphic copy of every countable abelian group as a subgroup.*

Proof. By Exercise 10.29, every countable abelian group can be imbedded in $G = \sum_{i=1}^{\infty} D_i$, where $D_i \cong \mathbb{Q} \oplus (\mathbb{Q}/\mathbb{Z})$ for all i. Lemma 12.27, with the assertion that G^{II} is recursively presented, gives the result. ■

There are only countably many finitely presented groups, and their free product is a countable group H having a presentation for which H^{II} is recursively presented.

Theorem 12.29. *There exists a **universal finitely presented group** $\mathscr{U}$; that is, $\mathscr{U}$ is a finitely presented group and $\mathscr{U}$ contains an isomorphic copy of every finitely presented group as a subgroup.*

Proof. The result follows from Lemma 12.27 and our assertion about the group H^{II}. ■

Groups with a solvable word problem admit an algebraic characterization. In the course of proving this, we shall encounter groups which are not finitely generated, yet over whose presentations we still have some control. Let G be a group with presentation

$$G = (x_i, i \geq 0 | \Delta)$$

in which each $\delta \in \Delta$ is a (not necessarily positive) word on $\{x_i, i \geq 0\}$, let Ω be the set of all words on $\{x_i, i \geq 0\}$, and let $R = \{\omega \in \Omega: \omega = 1 \text{ in } G\}$. Encode Ω in $\mathbb{N}$ using Gödel numbers: associate to the word $\omega \equiv x_{i_1}^{e_1} \dots x_{i_n}^{e_n}$ the positive integer $g(\omega) = \prod_{k=1}^{n} p_{2k}^{i_k} p_{2k+1}^{1+e_k}$, where $p_0 < p_1 < \cdots$ is the sequence of primes (note that $1 + e_k \geq 0$). The ***Gödel image*** of this presentation is

$$g(R) = \{g(\omega): \omega \in R\}.$$

Definition. A presentation $(x_i, i \geq 0 | \Delta)$ is **r.e.** if its Gödel image $g(R)$ is an r.e. subset of $\mathbb{N}$; this presentation has a ***solvable word problem*** if $g(R)$ is recursive.

Definition. A group G is **r.e.** or has a ***solvable word problem*** if it has some presentation which is either r.e. or has a solvable word problem.

We remarked at the beginning of this chapter that a finitely generated group G having a solvable word problem relative to one presentation with a finite number of generators has a solvable word problem relative to any other such presentation. The analogue of this statement is no longer true when we allow nonfinitely generated groups. For example, let G be a free group of infinite rank with basis $\{x_i, i \geq 0\}$. Now $g(R)$ is recursive (this is not instantly obvious, for R is an infinite set of nonreduced words; list its elements lexicographically and according to length), and so this presentation, and hence G, has a solvable word problem. On the other hand, if E is an r.e. subset of $\mathbb{N}$ that is not recursive, then

$$(x_i, i \geq 0 | x_i = 1 \text{ if and only if } i \in E)$$

is another presentation of G, but $g(R)$ is not recursive; this second presentation has an unsolvable word problem.

We wish to avoid some technicalities of Mathematical Logic (this is not the appropriate book for them), and so we shall shamelessly declare that certain groups arising in the next proof are either r.e. or have a solvable word problem; of course, the serious reader cannot be so cavalier.

A.V. Kuznetsov (1958) proved that every recursively presented simple group has a solvable word problem.

Theorem 12.30 (Boone–Higman, 1974). *A finitely generated group G has a solvable word problem if and only if G can be imbedded in a simple subgroup of some finitely presented group.*

Sketch of Proof. Assume that

$$G = \langle g_1, \ldots, g_n \rangle \leq S \leq H,$$

where S is a simple group and H is finitely presented; say,

$$H = (h_1, \ldots, h_m | \rho_1, \ldots, \rho_q).$$

Let Ω' denote the set of all words on $\{h_1, \ldots, h_m\}$, and let $R' = \{\omega' \in \Omega' : \omega' = 1\}$; let Ω denote the set of all words on $\{g_1, \ldots, g_n\}$, and let $R = \{\omega \in \Omega : \omega = 1\}$. Theorem 12.2 shows that R' is r.e. If one writes each g_i as a word in the h_j, then one sees that $R = \Omega \cap R'$; since the intersection of r.e. sets is r.e. (Exercise 12.8), it follows that R is r.e. We must show that its complement $\{\omega \in \Omega : \omega \neq 1\}$ is also r.e. Choose $s \in S$ with $s \neq 1$. For each $\omega \in \Omega$, define $N(\omega)$ to be the normal subgroup of H generated by $\{\omega, \rho_1, \ldots, \rho_q\}$. Since S is a simple group, the following statements are equivalent for $\omega \in \Omega$: $\omega \neq 1$ in G; $N(\omega) \cap S \neq 1$; $S \leq N(\omega)$; $s = 1$ in $H/N(\omega)$. As $H/N(\omega)$ is finitely presented, Theorem 12.2 shows that the set of all words in Ω which are equal to 1 in $H/N(\omega)$ is an r.e. set. A decision process determining whether $\omega = 1$ in G thus consists in checking whether $s = 1$ in $H/N(\omega)$.

To prove the converse, assume that $G = \langle g_1, \ldots, g_m \rangle$ has a solvable word problem. By Exercise 12.10,

$$\{(u, v) \in \Omega \times \Omega : u \neq 1 \text{ and } v \neq 1\}$$

is a recursive set; enumerate this set $(u_0, v_0), (u_1, v_1), \ldots$ (each word u or v has many subscripts in this enumeration). Define $G_0 = G$, and define

$$G_1 = (G; x_1, t_i, i \geq 0 | t_i^{-1} u_i x_1^{-1} u_i x_1 t_i = v_i x_1^{-1} u_i x_1, i \geq 0).$$

It is plain that G_1 has base $G * \langle x_1 \rangle$ and stable letters $\{t_i, i \geq 0\}$; it is an HNN extension because, for each i, both $A_i = \langle u_i x_1^{-1} u_i x_1 \rangle$ and $B_i = \langle v_i x_1^{-1} u_i x_1 \rangle$ are infinite cyclic. Thus, $G \leq G_1$. One can show that this presentation of G_1 has a solvable word problem (note that G_1 is no longer finitely generated). We now iterate this construction. For each k, there is an HNN extension G_k with base $G_{k-1} * \langle x_k \rangle$, and we define $S = \bigcup_{k \geq 1} G_k$; clearly $G \leq S$. To see that S is simple, choose $u, v \in S$ with $u \neq 1$ and $v \neq 1$. There is an integer k with $u, v \in G_{k-1}$. By construction, there is a stable letter p in G_k with

$$p^{-1} u x_k^{-1} u x_k p = v x_k^{-1} u x_k.$$

Therefore,

$$v = (p^{-1} u p)(p^{-1} x_k^{-1} u x_k p)(x_k^{-1} u^{-1} x_k)$$

lies in the normal subgroup generated by u. Since u and v are arbitrary nontrivial elements of S, it follows that S is simple.

It can be shown that S is recursively presented. By Theorem 11.71, there is a two-generator group S^{II} containing S; moreover, S^{II} is recursively presented. The Higman imbedding theorem shows that S^{II}, hence S, and hence G, can be imbedded in a finitely presented group H. ■

It is an open question whether a group with a solvable word problem can be imbedded in a finitely presented simple group (the simple group S in the proof is unlikely to be finitely generated, let alone finitely presented).

Our final result explains why it is often difficult to extract information about groups from presentations of them. Before giving the next lemma, let us explain a phrase occurring in its statement. We will be dealing with a set of words Ω on a given alphabet and, for each $\omega \in \Omega$, we shall construct a presentation $\mathscr{P}(\omega)$ involving the word ω. This family of presentations is ***uniform in*** ω if, for each $\omega' \in \Omega$, the presentation $\mathscr{P}(\omega')$ is obtained from $\mathscr{P}(\omega)$ by substituting ω' for each occurrence of ω. A presentation $(X|\Delta)$ is called ***finite*** if both X and Δ are finite sets; of course, a group is finitely presented if and only if it has such a presentation.

Lemma 12.31 (Rabin, 1958). *Let $G = (\Sigma|\Delta)$ be a finite presentation of a group and let Ω be the set of all words on Σ. There are finite presentations $\{\mathscr{P}(\omega)$: $\omega \in \Omega\}$, uniform in ω, such that if $R(\omega)$ is the group presented by $\mathscr{P}(\omega)$, then*

(i) if $\omega \neq 1$ in G, then $G \leq R(\omega)$; and
(ii) *if $\omega = 1$ in G, then $\mathscr{P}(\omega)$ presents the trivial group* 1.

Proof **(C.F. Miller, III).** Let $\langle x \rangle$ be an infinite cyclic group; by Corollary 11.80, $G * \langle x \rangle$ can be imbedded in a two-generator group $A = \langle a_1, a_2 \rangle$ in which both generators have infinite order. Moreover, one can argue, as in Exercise 11.80, that A can be chosen to be finitely presented: there is a finite set Δ of words on $\{a_1, a_2\}$ with

$$A = (a_1, a_2|\Delta).$$

Define

$$B = (A; b_1, b_2|b_1^{-1}a_1b_1 = a_1^2, b_2^{-1}a_2b_2 = a_2^2).$$

It is easy to see that B is an HNN extension with base A and stable letters $\{b_1, b_2\}$, so that $G \leq A \leq B$. Define

$$C = (B; c|c^{-1}b_1c = b_1^2, c^{-1}b_2c = b_2^2).$$

Clearly C has base B and stable letter c; C is an HNN extension because b_1 and b_2, being stable letters in B, have infinite order. Thus, $G \leq A \leq B \leq C$.

If $\omega \in \Omega$ and $\omega \neq 1$ in G, then the commutator

$$[\omega, x] = \omega x \omega^{-1} x^{-1}$$

has infinite order in A (because $G * \langle x \rangle \leq A$).

We claim that $\langle c, [\omega, x]\rangle \leq C$ is a free group with basis $\{c, [\omega, x]\}$. Suppose that V is a nontrivial freely reduced word on $\{c, [\omega, x]\}$ with $V = 1$ in C. If V does not involve the stable letter c, then $V \equiv [\omega, x]^n$ for some $n \neq 0$, and this contradicts $[\omega, x]$ having infinite order. If V does involve c, then Britton's lemma shows that V contains a pinch $c^e W c^{-e}$ as a subword, where $e = \pm 1$ and $W \in \langle b_1, b_2\rangle$. But V involves neither b_1 nor b_2, so that $W \equiv 1$ and $c^e W c^{-e} \equiv c^e c^{-e}$, contradicting V being freely reduced. Therefore, V must be trivial.

We now construct a second tower of HNN extensions. Begin with an infinite cyclic group $\langle r\rangle$, define

$$S = (r, s \mid s^{-1} r s = r^2),$$

and define

$$T = (S; t \mid t^{-1} s t = s^2).$$

Since both r and s have infinite order, S is an HNN extension with base $\langle r\rangle$ and stable letter s, and T is an HNN extension with base S and stable letter t. Britton's lemma can be used, as above, to show that $\{r, t\}$ freely generates its subgroup in T.

Since both $\langle r, t\rangle \leq T$ and $\langle c, [\omega, x]\rangle \leq C$ are free groups of rank 2, there is an isomorphism φ between them with $\varphi(r) = c$ and $\varphi(t) = [\omega, x]$. Form the amalgam $R(\omega) = T *_\varphi C$ with presentation

$$\mathscr{P}(\omega) = (T * C \mid r = c, t = [\omega, x]).$$

We conclude from Theorem 11.67(i) that if $\omega \neq 1$ in G, then $G \leq C \leq R(\omega)$.

If $\omega = 1$ in G, the presentation $\mathscr{P}(\omega)$ is still defined (though it need not be an amalgam). The presentations $\mathscr{P}(\omega)$ are uniform in ω:

$$\begin{aligned}\mathscr{P}(\omega) = (a_1, a_2, b_1, b_2, c, r, s, t \mid \Delta, b_i^{-1} a_i b_i = a_i^2, c^{-1} b_i c = b_i^2,\\ i = 1, 2, s^{-1} r s = r^2, t^{-1} s t = s^2, r = c, t = [\omega, x]).\end{aligned}$$

We claim that $\mathscr{P}(\omega)$ is a presentation of the trivial group if $\omega = 1$ in G. Watch the dominoes fall: $\omega = 1 \Rightarrow [\omega, x] = 1 \Rightarrow t = 1 \Rightarrow s = 1 \Rightarrow r = 1 \Rightarrow c = 1 \Rightarrow b_1 = 1 = b_2 \Rightarrow a_1 = 1 = a_2$. ■

Definition. A property $\mathscr{M}$ of finitely presented groups is called a ***Markov property*** if:

(i) every group isomorphic to a group with property $\mathscr{M}$ also has property $\mathscr{M}$;
(ii) there exists a finitely presented group G_1 with property $\mathscr{M}$; and
(iii) there exists a finitely presented group G_2 which cannot be imbedded in a finitely presented group having property $\mathscr{M}$.

Here are some examples of Markov properties: order 1; finite; finite exponent; p-group; abelian; solvable; nilpotent; torsion; torsion-free; free; having a solvable word problem. Being simple is also a Markov property, for the Boone–Higman theorem shows that finitely presented simple groups must have a solvable word problem (and hence so do all their finitely presented subgroups). Having a solvable conjugacy problem is also a Markov property: a finitely presented group G_2 with an unsolvable word problem cannot be imbedded in a finitely presented group H having a solvable conjugacy problem, for H and all its finitely presented subgroups have a solvable word problem. It is fair to say that most interesting group-theoretic properties are Markov properties.

The following result was proved for semigroups by Markov (1950).

Theorem 12.32 (Adian–Rabin, 1958). *If $\mathscr{M}$ is a Markov property, then there does not exist a decision process which will determine, for an arbitrary finite presentation, whether the group presented has property $\mathscr{M}$.*

Proof. Let G_1 and G_2 be finitely presented groups as in the definition of Markov property, and let $\mathscr{B}$ be a finitely presented group with an unsolvable word problem. Define $G = G_2 * \mathscr{B}$, construct groups $R(\omega)$ as in Rabin's lemma, and define (finitely presented) groups $\mathscr{Q}(\omega) = G_1 * R(\omega)$.

Restrict attention to words ω on the generators of $\mathscr{B}$. If such a word $\omega \neq 1$ in $\mathscr{B}$, then $G_2 \leq G \leq R(\omega) \leq \mathscr{Q}(\omega)$. But the defining property of G_2 implies that $\mathscr{Q}(\omega)$ does not have property $\mathscr{M}$. If, on the other hand, $\omega = 1$ in $\mathscr{B}$, then $R(\omega) = 1$ and $\mathscr{Q}(\omega) \cong G_1$ which does have property $\mathscr{M}$. Therefore, any decision process determining whether $\mathscr{Q}(\omega)$ has property $\mathscr{M}$ can also determine whether $\omega = 1$ in $\mathscr{B}$; that is, any such decision process would solve the word problem in $\mathscr{B}$. ■

Corollary 12.33. *There is no decision process to determine, for an arbitrary finite presentation, whether the presented group has any of the following properties: order* 1; *finite*; *finite exponent*; *p-group*; *abelian*; *solvable*; *nilpotent*; *simple*; *torsion*; *torsion-free*; *free*; *solvable word problem*; *solvable conjugacy problem.*

Proof. Each of the listed properties is Markov. ■

Corollary 12.34. *There is no decision process to determine, for an arbitrary pair of finite presentations, whether the two presented groups are isomorphic.*

Proof. Enumerate the presentations $\mathscr{P}_1, \mathscr{P}_2, \ldots$ and the groups $G_1, G_2, \ldots$ they present. If there were a decision process to determine whether $G_i \cong G_j$ for all i and j, then, in particular, there would be a decision process to determine whether $\mathscr{P}_n$ presents the trivial group. ■

While a property of finitely presented groups being Markov is sufficient for the nonexistence of a decision process as in the Adian–Rabin theorem, it is not necessary. For example, the property of being infinite is not a Markov property. However, a decision process that could determine whether the group given by an arbitrary finite presentation is infinite would obviously determine whether the group is finite, contradicting Corollary 12.33. Indeed, this example generalizes to show that the Adian–Rabin theorem also holds for the "complement" of a Markov property.

Does every finitely presented group have some Markov property?

Theorem 12.35. *A finitely presented group H satisfies no Markov property if and only if it is a universal finitely presented group* (*i.e., H contains an isomorphic copy of every finitely presented group as a subgroup*).

Proof. Recall that the existence of universal finitely presented groups was proved in Theorem 12.29.

Let H be a universal finitely presented group, and assume that H has some Markov property $\mathcal{M}$. There is some finitely presented group G_2 that cannot be imbedded in a finitely presented group with property $\mathcal{M}$. But G_2 can be imbedded in H, and this is a contradiction. The converse follows from the observation that "not universal" is a Markov property. ■

Epilogue

Any reader wanting to study Group Theory more deeply must first learn ***Representation Theory***, the analysis of homomorphisms $\varphi: G \to \mathrm{GL}(n, K)$, where K is an algebraically closed field. There are two initial approaches to this subject, and both approaches must eventually be mastered. One approach, historically the first, is *character theory*. If $\varphi: G \to \mathrm{GL}(n, \mathbb{C})$ is a homomorphism, then $\varphi(g)$ is an $n \times n$ complex matrix for each $g \in G$; its ***character*** $\chi(\varphi): G \to \mathbb{C}$ is defined to be the trace function $g \mapsto \mathrm{tr}(\varphi(g))$ (the values of $\chi(\varphi)$ are actually algebraic integers). Of course, if g and g' are conjugate in G, then $\mathrm{tr}(\varphi(g)) = \mathrm{tr}(\varphi(g'))$, so that $\chi(\varphi)$ is really a *class function*; that is, $\chi(\varphi)$ can be regarded as a complex-valued function on the family of conjugacy classes of G. Each character can be uniquely written as a linear combination of *irreducible* characters, and the number c of such irreducible characters is equal to the number of conjugacy classes of G. The $c \times c$ matrix containing the values of all the irreducible characters is called the ***character table*** of G. It contains much important information about G, and sufficient machinery has been developed to allow explicit calculation, in many cases, of its entries. There are wonderful applications that arise quite early: ***Burnside's $p^\alpha q^\beta$-theorem***: Every group of order $p^\alpha q^\beta$, where p and q are primes, is solvable; a ***theorem of Frobenius***: If H is a subgroup of a finite group G such that $H \cap xHx^{-1} = 1$ for all $x \notin H$, then $N = \{1\} \cup (G - \bigcup_{x \in G} xHx^{-1})$ is a (normal) subgroup of G (in a Frobenius group G, this shows that the Frobenius kernel is actually a subgroup). The further one goes into Group Theory, the more Representation Theory arises, and many of the best theorems involve some use of representations.

The theory still works when $\mathbb{C}$ is replaced by any algebraically closed field K whose characteristic does not divide $|G|$; this is the so-called *ordinary representation theory*. When the characteristic p of K divides $|G|$, the study, called *modular representation theory*, becomes more intricate, but it, too, is an essential tool.

Let us now discuss the second approach to representations. If K is a field, then the ***group algebra*** KG of a finite group G over K is the vector space over K having the elements of G as a basis and which is equipped with the multiplication (called *convolution*) determined by the given (group) multiplication of its basis elements. If $\varphi: G \to \mathrm{GL}(n, K)$ is a homomorphism and if V is an n-dimensional vector space over K, then one may view V as a KG-module (and conversely). When $K = \mathbb{C}$, one sees, for example, that $\chi(\varphi)$ is irreducible if and only if V is an indecomposable module. This point of view is quite valuable; for example, it allows ideas and techniques of Homological Algebra to be used.

There are many excellent books on Representation Theory. For example: Alperin, Benson, Curtis and Reiner, Dornhoff, Feit (1967 and 1982), Isaacs, James and Liebeck, Puttaswamaiah and Dixon, and Serre (1977).

A Personal Note. If Representation Theory is so important, why have I not included it in this book? It is not because the beginnings of the subject require greater sophistication on the part of the reader.

Let me explain with an analogy. I have long felt that many entering university students who have seen some Calculus in high school are at a disadvantage. There are, to be sure, good Calculus courses taught in high schools, and those students who have done well in such courses are well prepared. But, too often, high school Calculus courses are inadequate, so that, upon completion, even good students (with good teachers) are poorly prepared. As a consequence, many students must start learning the subject anew when they enter the university. Their time has been wasted and their enthusiasm has been dampened.

I feel that one chapter on Representation Theory is necessarily inadequate; it is like a bad high school Calculus course that leaves one unprepared. After a longish excursion into Ring Theory (including the theorems of Wedderburn and Maschke), one learns the first properties of characters and how to compute them, and one proves the theorems of Burnside and Frobenius mentioned above. However, a group theorist must have a more thorough course in order to feel comfortable with both characters and modules. Most likely, a student having read only one chapter in a text like this one would still have to begin the subject anew, and this would be a waste of valuable time.

Here are some suggestions of other topics in Group Theory that the reader may wish to pursue. For general group theory, see Huppert, Huppert and Blackburn (1981 and 1982), Robinson (1982), and Suzuki (1982 and 1986).

Simple Groups. All finite simple groups were classified by the 1980s, and there is an explicit description of them all. This is the most profound and sophisticated part of Group Theory, using every known technique. Introductions to this study are Artin (1957), Aschbacher (1994), Borel, Carter (1972 and 1985), Conway et al., Dieudonné, and Gorenstein (1982 and 1983). For some applications of the classification theorem, see P.J. Cameron, Finite permutation groups and finite simple groups, *Bull. London Math. Soc.* **13** (1981), pp. 1–22.

Solvable Groups. See Doerk and Hawkes, Huppert and Blackburn (1981), and Robinson (1972).

p-Groups. We recommend P. Hall's notes "Nilpotent Groups" in his *Collected Works*, Dixon and du Sautoy and Mann and Segal, Huppert, Huppert and Blackburn (1981), Khukhro, and Vaughan-Lee.

Cohomology of Groups. For a general account of Homological Algebra, the reader may look at Cartan and Eilenberg, Mac Lane, and Rotman (1979). For Cohomology of Groups, which is Homological Algebra specialized to a group-theoretic context, see Benson, Brown, Evens, Karpilovsky, and Weiss.

Combinatorial Group Theory. This is the study of presentations of groups. Suggested books are Coxeter and Moser, Johnson, Lyndon and Schupp, Magnus and Karrass and Solitar, and Zieschang and Vogt and Coldewey. There is another aspect involving groups acting on trees; we suggest Dicks and Dunwoody, and Serre (1980). The Cayley graph of a finitely generated group can be made into a metric space, and the hyperbolic groups introduced by Gromov can be found in Gersten.

See Higman for further development of his imbedding theorem, Miller for group-theoretic decision problems, and Epstein et al. for a treatment of automatic groups.

Abelian Groups. We suggest Fuchs (1970 and 1973), Griffith, and Kaplansky.

Finitely Generated Groups. We suggest Kegel and Wehrfritz, Kurosh, Robinson (1972), and Wehrfritz.

History. We suggest Chandler and Magnus, and Wussing.

There are several computer systems devoted to group theory: for example, MAGMA (nee CAYLEY) and GAP.

Certainly, there are other valuable books, as well as other valuable areas of Group Theory (e.g., crystallographic groups, Möbius groups, knot groups, varieties of groups) that I have not even mentioned. I apologize to their authors and their practitioners.

Primary Sources. One must always look at the masters. The following books contain extensive bibliographies of journal articles: Carter, Coxeter and Moser, Curtis and Reiner, Fuchs, Gorenstein (1982), Huppert, Huppert and Blackburn, Lyndon and Schupp, Magnus and Karass and Solitar, Robinson (1982), Scott, and Suzuki. Both Baumslag, and Gorenstein (1974) contain reviews of the all the articles on Group Theory written between 1940 and 1970.

APPENDIX I

Some Major Algebraic Systems

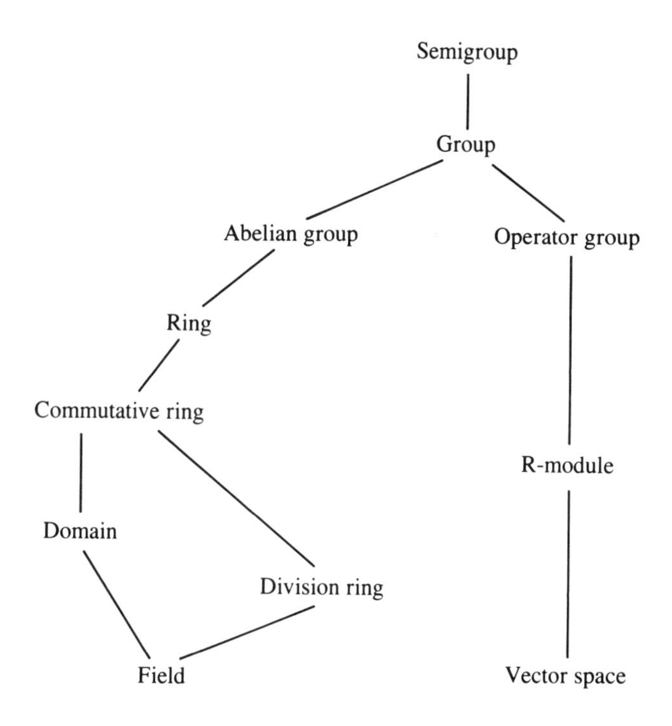

A ***ring*** (always containing $1 \neq 0$) is a set with two operations, addition and multiplication. It is an abelian group under addition, a semigroup with 1 under multiplication, and the two operations are linked by the distributive laws.

A ***commutative ring*** is a ring in which multiplication is commutative.

A ***domain*** (or *integral domain*) is a commutative ring in which $ab = 0$ implies $a = 0$ or $b = 0$; equivalently, the ***cancellation law*** holds: if $ab = ac$ and $a \neq 0$, then $b = c$.

A ***division ring*** (or *skew field*) is a (not necessarily commutative) ring in which every nonzero element has a multiplicative inverse: if $a \neq 0$, then there is $b \in K$ with $ab = 1 = ba$. The set of nonzero elements $K^\times = K - \{0\}$ is thus a multiplicative group.

A ***field*** K is a commutative division ring.

It is a theorem of Wedderburn (1905) that every finite division ring is a field.

APPENDIX II

Equivalence Relations and Equivalence Classes

A ***relation*** on a set X is a subset $\equiv$ of $X \times X$. One usually writes $x \equiv y$ instead of $(x, y) \in \equiv$; for example, the relation $<$ on the set of real numbers $\mathbb{R}$ consists of all points in the plane $\mathbb{R} \times \mathbb{R}$ lying above the line with equation $y = x$, and one usually writes $2 < 3$ instead of $(2, 3) \in <$.

A relation $\equiv$ on a set X is ***reflexive*** if $x \equiv x$ for all $x \in X$; it is ***symmetric*** if, for all $x, y \in X$, $x \equiv y$ implies $y \equiv x$; it is ***transitive*** if, for all $x, y, z \in X$, $x \equiv y$ and $y \equiv z$ imply $x \equiv z$. A relation $\equiv$ on a set X is an ***equivalence relation*** if it is reflexive, symmetric, and transitive.

If $\equiv$ is an equivalence relation on a set X and if $x \in X$, then the ***equivalence class*** of x is

$$[x] = \{y \in X: y \equiv x\} \subset X.$$

Proposition II.1. *Let $\equiv$ be an equivalence relation on a set X. If $x, a \in X$, then $[x] = [a]$ if and only if $x \equiv a$.*

Proof. If $[x] = [a]$, then $x \in [x]$, by reflexivity, and so $x \in [a] = [x]$; that is, $x \equiv a$.

Conversely, if $x \equiv a$, then $a \equiv x$, by symmetry. If $y \in [x]$, then $y \equiv x$. By transitivity, $y \equiv a$, $y \in [a]$, and $[x] \subset [a]$. For the reverse inclusion, if $z \in [a]$, then $z \equiv a$. By transitivity, $z \equiv x$, so that $z \in [x]$ and $[a] \subset [x]$, as desired. ∎

A ***partition*** of a nonempty set X is a family of nonempty subsets $\{S_i: i \in I\}$ such that $X = \bigcup_{i \in I} S_i$ and the subsets are ***pairwise disjoint***: if $i \neq j$, then $S_i \cap S_j = \varnothing$.

Proposition II.2. *If $\equiv$ is an equivalence relation on a nonempty set X, then the family of all equivalence classes is a partition of X.*

Proof. For each $x \in X$, reflexivity gives $x \in [x]$; this shows that the equivalence classes are nonempty and that $X = \bigcup_{x \in X} [x]$. To check pairwise disjointness, assume that $[x] \cap [y] \neq \varnothing$. Therefore, there exists an element $z \in [x] \cap [y]$; that is, $z \equiv x$ and $z \equiv y$. By the first proposition, $[z] = [x]$ and $[z] = [y]$, so that $[x] = [y]$. ■

Proposition II.3. *If $\{S_i: i \in I\}$ is a partition of a nonempty set X, then there is an equivalence relation on X whose equivalence classes are the S_i.*

Proof. If $x, y \in X$, define $x \equiv y$ to mean that there exists S_i containing both x and y. It is plain that $\equiv$ is reflexive and symmetric. To prove transitivity, assume that $x \equiv y$ and $y \equiv z$; that is, $x, y \in S_i$ and $y, z \in S_j$. Since $y \in S_i \cap S_j$, pairwise disjointness gives $S_i = S_j$; hence $x, z \in S_i$ and $x \equiv z$.

If $x \in X$, then $x \in S_i$ for some i. If $y \in S_i$, then $y, x \in S_i$ and $y \equiv x$; that is, $S_i \subset [x]$. For the reverse inclusion, if $z \in [x]$, then $z \equiv x$, and so $z, x \in S_i$; that is, $[x] \subset S_i$. ■

Proposition II.1 signals the importance of equivalence relations. If $\equiv$ is an equivalence relation on a set X and if E is the family of equivalence classes, then $x \equiv y$ in X if and only if $[x] = [y]$ in E; equivalence of elements in X becomes equality of elements in E. The construction of the new set E thus identifies equivalent elements.

For example, the fractions $\frac{1}{2}$ and $\frac{2}{4}$ are called equal if the numerators and denominators satisfy "cross multiplication." In reality, one defines a relation $\equiv$ on $X = \{(a, b) \in \mathbb{Z} \times \mathbb{Z}: b \neq 0\}$ by $(a, b) \equiv (c, d)$ if $ad = bc$, and a straightforward calculation shows that $\equiv$ is an equivalence relation on X. The equivalence class containing (a, b) is denoted by a/b, and the set of all ***rational numbers*** $\mathbb{Q}$ is defined as the family of all such equivalence classes. In particular, $(1, 2)$ and $(2, 4)$ are identified in $\mathbb{Q}$, because $(1, 2) \equiv (2, 4)$, and so $\frac{1}{2} = \frac{2}{4}$.

APPENDIX III

Functions

If X and Y are sets, a ***relation*** from X to Y is a subset f of $X \times Y$ (if $X = Y$, one also says that f is a relation on X). A ***function*** from X to Y, denoted by $f: X \to Y$, is a relation f from X to Y such that for each $x \in X$, there exists a unique $y \in Y$ with $(x, y) \in f$. If $x \in X$, then the unique element y in the definition is denoted by $f(x)$, and it is called the ***value*** of f at x or the ***image*** of x under f. With this notation, the relation f consists of all $(x, f(x)) \in X \times Y$; that is, a function *is* what is usually called its *graph*.

The set X is called the ***domain*** of f and the set Y is called the ***target*** of f. One defines two functions f and g to be ***equal*** if they have the same domain X, the same target Y, and $f(x) = g(x)$ for all $x \in X$ (this says that their graphs are the same subset of $X \times Y$).

In practice, one thinks of a function f as something dynamic: f assigns a value $f(x)$ in Y to each element x in X. For example, the squaring function $f: \mathbb{R} \to \mathbb{R}$ is the parabola consisting of all $(x, x^2) \in \mathbb{R} \times \mathbb{R}$, but one usually thinks of f as assigning x^2 to x; indeed, we often use a footed arrow to denote the value of $f$on a typical element x in the domain: for example, $f: x \mapsto x^2$. Most elementary texts define a function as "a rule which assigns, to each x in X, a unique value $f(x)$ in Y." The idea is correct, but not good enough. For example, consider the functions $f, g: \mathbb{R} \to \mathbb{R}$ defined as follows: $f(x) = (x + 1)^2$; $g(x) = x^2 + 2x + 1$. Are f and g different functions? They are different "rules" in the sense that the procedures involved in computing each of them are different. However, the definition of equality given above shows that $f = g$.

If X is a nonempty set, then a ***sequence*** in X is a function $f: \mathbb{P} \to X$, where $\mathbb{P}$ is the set of positive integers. Usually, one writes x_n instead of $f(n)$ and one describes f by displaying its values: $x_1, x_2, x_3, \ldots$. It follows that two sequences $x_1, x_2, x_3, \ldots$ and $y_1, y_2, y_3, \ldots$ are equal if and only if $x_n = y_n$ for all $n \geq 1$.

The uniqueness of values in the definition of function deserves more comment: it says that a function is "single valued" or, as we prefer to say, it is ***well defined***. For example, if x is a nonnegative real number, then $f(x) = \sqrt{x}$ is usually defined so that $f(x) \geq 0$; with no such restriction, it would not be a function (if $\sqrt{4} = 2$ and $\sqrt{4} = -2$, then a unique value has not been assigned to 4). When attempting to define a function, one must take care that it is well defined lest one define only a relation.

The diagonal $\{(x, x) \in X \times X: x \in X\}$ is a function $X \to X$; it is called the ***identity function*** on X, it is denoted by 1_X, and $1_X: x \mapsto x$ for all $x \in X$. If X is a subset of Y, then the ***inclusion function*** $i: X \to Y$ (often denoted by $i: X \hookrightarrow Y$) is defined by $x \mapsto x$ for all $x \in X$. The only difference between 1_X and i is that they have different targets, but if X is a proper subset of Y, this is sufficient to guarantee that $1_X \neq i$.

If $f: X \to Y$ and $g: Y \to Z$ are functions, then their ***composite*** $g \circ f: X \to Z$ is the function defined by $x \mapsto g(f(x))$. If $h: Z \to W$ is a function, then the associativity formula $h \circ (g \circ f) = (h \circ g) \circ f$ holds, for both are functions $X \to W$ with $x \mapsto h(g(f(x)))$ for all $x \in X$. If $f: X \to A$ and Y is a subset of X with inclusion $i: Y \hookrightarrow X$, then the ***restriction*** $f|Y: Y \to A$ is the composite $f \circ i$; for each $y \in Y$, one has $(f|Y)(y) = f(y)$.

A function $f: X \to Y$ is ***injective*** (or *one-to-one*) if distinct elements of X have distinct values; equivalently, if $f(x) = f(x')$, then $x = x'$ (this is the converse of f being well defined: if $x = x'$, then $f(x) = f(x')$). A function $f: X \to Y$ is ***surjective*** (or *onto*) if each element of Y is a value; that is, for each $y \in Y$, there exists $x \in X$ with $y = f(x)$. (If one did not insist, in the definition of equality of functions, that targets must be the same, then every function would be surjective!) A function is a ***bijection*** (or *one-to-one correspondence*) if it is both injective and surjective. A function $f: X \to Y$ is a bijection if and only if it has an ***inverse function***: there is a function $g: Y \to X$ with $g \circ f = 1_X$ and $f \circ g = 1_Y$ (g is usually denoted by f^{-1}). One must have both composites identity functions, for $g \circ f = 1_X$ implies only that f is injective and g is surjective.

APPENDIX IV

Zorn's Lemma

A relation $\leq$ on a set X is ***antisymmetric*** if $x \leq y$ and $y \leq x$ imply $x = y$, for all $x, y \in X$. A relation $\leq$ on a *nonempty* set X is called a ***partial order*** if it is reflexive, antisymmetric, and transitive. The best example of a partially ordered set is a family of subsets of a set, where $\leq$ means $\subset$.

A partial order on X is a ***simple order*** (or *total order*) if, for each $x, y \in X$, either $x \leq y$ or $y \leq x$.

If S is a nonempty subset of a partially ordered set X, then an ***upper bound*** of S is an element $x \in X$ (not necessarily in S) with $s \leq x$ for all $s \in S$. Finally, a ***maximal element*** in a partially ordered set X is an element $m \in X$ which is smaller than nothing else: if $x \in X$ and $m \leq x$, then $x = m$. For example, if X is the partially ordered set consisting of all the proper subsets of a set A under inclusion, then a maximal element is the complement of a point. Thus, a partially ordered set can have many maximal elements. On the other hand, a partially ordered set may have no maximal elements at all. For example, there are no maximal elements in the set of real numbers $\mathbb{R}$ regarded as a partially ordered set under ordinary inequality (indeed, $\mathbb{R}$ is a simply ordered set).

Zorn's Lemma. *If X is a partially ordered set in which every simply ordered subset has an upper bound, then X has a maximal element.*

Remember that partially ordered sets are, by definition, nonempty.

Zorn's lemma is equivalent to a much more intuitive statement: the ***Axiom of Choice***, which states that the cartesian product of nonempty sets is itself nonempty. (It is easy to prove (by induction) that a cartesian product of a finite number of nonempty sets is nonempty, and so the Axiom of Choice need be invoked only when there are infinitely many factors.) We regard both of these statements as axioms of Mathematics, and we will not be ashamed to use either of them when convenient.

There is a third statement, equivalent to these, which is also useful. A partially ordered set X is ***well-ordered*** if every nonempty subset contains a smallest element; that is, if $S \subset X$ and $S \neq \varnothing$, then there is $s_0 \in S$ with $s_0 \leq s$ for all $s \in S$. (Well-ordered sets must be simply ordered, for every two-element subset has a smallest element.) The set of natural numbers $\mathbb{N} = \{n \in \mathbb{Z}: n \geq 0\}$ is well-ordered, but the set $\mathbb{Z}$ of all integers is not well-ordered.

Well-Ordering Principle. Given a nonempty set X, there exists a partial order $\leq$ on X which is a well-ordering.

Although $\mathbb{Z}$ is not well-ordered under the usual ordering, we can well-order it: $0, 1, -1, 2, -2, \ldots$. Here is an example of a well-ordered set in which an element has infinitely many predecessors: define X to be the following subset of $\mathbb{R}$ with the usual notion of $\leq$:

$$X = \{1 - 1/n: n > 0\} \cup \{2 - 1/n: n > 0\} \cup \{3 - 1/n: n > 0\} \cup \cdots.$$

APPENDIX V

Countability

It is well known that certain "paradoxes" arise if one is not careful about the foundations of Set Theory. We now sketch some features of the foundations we accept. Its primitive undefined terms are ***class***, ***element***, and ***membership***, denoted by $\in$: if X is a member of a class Y, one writes $X \in Y$. A "set" is a special kind of class, described below. Every usage of "set" in the preceding appendices can be replaced by the word "class." In particular, one may speak of cartesian products of classes, so that functions from one class to another are defined; functions between classes may or may not be injective or surjective.

Classes X and Y are called ***equipotent***, denoted by $|X| = |Y|$, if there exists a bijection $f: X \to Y$. It is easy to see that equipotence is an equivalence relation. Define a relation $|X| \leq |Y|$ to mean that there is an injection $f: X \to Y$. It is easy to see that this relation is reflexive and transitive, and the ***Cantor–Schroeder–Bernstein theorem*** shows that it is antisymmetric: $|X| \leq |Y|$ and $|Y| \leq |X|$ imply $|X| = |Y|$; thus, $\leq$ is a partial order.

The foundations allow one to define the *cardinal number* of a set; some classes have a cardinal number and they are called ***sets***; some classes do not have a cardinal (they are too big) and they are called ***proper classes***. For example, the class of all abelian groups is a proper class. The notion of functor in Chapter 10 thus involves a function defined on a proper class.

A set X is ***finite*** if it is empty or if there is a positive integer n and a bijection $f: \{1, 2, \ldots, n\} \to X$ (in this case, we write $|X| = n$); otherwise, X is ***infinite***. There is an elementary, but useful, result here, sometimes called the *pigeonhole principle*.

Theorem. *If X and Y are finite sets with $|X| = |Y|$, then a function $f: X \to Y$ is injective if and only if it is surjective; in either case, therefore, f is a bijection.*

A set X is called ***countable*** if it is finite or if there is a bijection $f: \mathbb{N} \to X$, where $\mathbb{N} = \{0, 1, 2, \dots\}$ is the set of natural numbers; otherwise X is ***uncountable***. The sets $\mathbb{Z}$ and $\mathbb{Q}$ are countable, but the set $\mathbb{R}$ is uncountable.

If X is any infinite set, then the set Y of all its subsets is an uncountable set; moreover, if Z is any set with at least two elements, then the set of all functions $X \to Z$ is uncountable. In particular, the family of all sequences in Z is uncountable.

Here are some facts about countable sets.

1. If X is any countable set, then the family Y of all its *finite* subsets is a countable set.
2. Every subset of a countable set is countable.
3. If X is a countable set and if $f: X \to Y$ is a surjection, then Y is countable.
4. If X and Y are both countable, then so is $X \times Y$.
5. If A_1 and A_2 are countable subsets of a set X, then $A_1 \cup A_2$ is countable. More generally, if $\{A_n: n \geq 0\}$ is a countable family of countable subsets of X, then $\bigcup_n A_n$ is countable.

APPENDIX VI

Commutative Rings

It is assumed that the reader has seen an introduction to commutative rings before beginning this book, and so this appendix is intended only as a reminder of perhaps half-forgotten ideas.

Let R be a commutative ring. In contrast to some authors, we assume that R must contain an element $1 \neq 0$ which, under multiplication, behaves like the number 1: for all $r \in R$, $1r = r$. If R' is a second commutative ring, then a (ring) ***homomorphism*** is a function $f: R \to R'$ such that $f(1) = 1$, $f(r + s) = f(r) + f(s)$, and $f(rs) = f(r)f(s)$. An ***ideal*** I in R is an additive subgroup of R ($0 \in I$ and $a, b \in I$ implies $a - b \in I$) such that $a \in I$ and $r \in R$ imply $ra \in I$. If $f: R \to R'$ is a homomorphism, then its ***kernel*** $= \{r \in R: f(r) = 0\}$ is an ideal. Here is another important example of an ideal: if $a_1, \ldots, a_n \in R$, then the set of all their ***linear combinations*** is called the ideal ***generated*** by $a_1, \ldots, a_n$; it is denoted by $(a_1, \ldots, a_n)$:

$$(a_1, \ldots, a_n) = \{r_1 a_1 + \cdots + r_n a_n: r_i \in R \text{ for all } i\}.$$

In the special case $n = 1$, the ideal generated by $a \in R$, namely,

$$(a) = \{ra: r \in R\},$$

is called the ***principal ideal*** generated by a. If $R = \mathbb{Z}[x]$, the ring of all polynomials with coefficients in $\mathbb{Z}$, one can show that the ideal $I = (x, 2)$, consisting of all polynomials with coefficients in $\mathbb{Z}$ having even constant term, is not a principal ideal.

The following result, which merely states that long division is possible, can be proved by the reader. Recall that the ***zero polynomial*** is the polynomial all of whose coefficients are 0; if $f(x)$ is not the zero polynomial, then we write $f(x) \neq 0$.

Division Algorithm.

(i) *If* $a, b \in \mathbb{Z}$ *with* $a \neq 0$, *then there exist unique* $q, r \in \mathbb{Z}$ *with* $0 \leq r < |a|$ *and*

$$b = qa + r.$$

(ii) *If* k *is a field and* $a(x), b(x) \in k[x]$ *are polynomials with* $a(x) \neq 0$, *then there exist unique* $q(x), r(x) \in k[x]$ *with either* $r(x) = 0$ *or degree* $r(x) <$ *degree* $a(x)$ *and*

$$b(x) = q(x)a(x) + r(x).$$

One calls q (or $q(x)$) the ***quotient*** and r (or $r(x)$) the ***remainder***.

A domain R is a ***principal ideal domain*** (or PID) if every ideal in R is a principal ideal. Our example above shows that $\mathbb{Z}[x]$ is not a principal ideal domain.

Theorem VI.1.
(i) $\mathbb{Z}$ *is a principal ideal domain.*
(ii) *If* k *is a field, then* $k[x]$ *is a principal ideal domain.*

Proof. (i) Let I be an ideal in $\mathbb{Z}$. If I consists of 0 alone, then I is principal, generated by 0. If $I \neq 0$, then it contains nonzero elements; indeed, it contains positive elements (if $a \in I$, then $-a = (-1)a \in I$ also). If a is the smallest positive integer in I, then $(a) \subset I$. For the reverse inclusion, let $b \in I$. By the division algorithm, there are $q, r \in \mathbb{Z}$ with $0 \leq r < a$ and $b = qa + r$. Hence, $r = b - qa \in I$. If $r \neq 0$, then we contradict a being the smallest element in I. Therefore, $r = 0$ and $b = qa \in (a)$; hence $I = (a)$.

(ii) The proof for $k[x]$ is virtually the same as for $\mathbb{Z}$. If I is a nonzero ideal in $k[x]$, then a generator of I is any polynomial $a(x) \in I$ whose degree is smallest among degrees of polynomials in I. ∎

If I is an ideal in R and we forget the multiplication in R, then I is a subgroup of the abelian additive group R. The ***quotient ring*** R/I is the quotient group R/I (whose elements are all cosets $r + I$) made into a commutative ring by defining multiplication:

$$(r + I)(s + I) = rs + I.$$

Let us show that this definition does not depend on the choice of coset representative. By Lemma 2.8 in the text, $r + I = r' + I$ if and only if $r - r' \in I$. If $s + I = s' + I$, then

$$\begin{aligned} rs - r's' &= rs - r's + r's - r's' \\ &= (r - r')s + r'(s - s'). \end{aligned}$$

Since $r - r', s - s' \in I$, we have $rs - r's' \in I$, and so $rs + I = r's' + I$.

The ***natural map*** $\nu: R \to R/I$ is the function defined by $\nu(r) = r + I$; the definition of multiplication in the quotient ring shows that ν is a (surjective) ring homomorphism with kernel I. If $f: R \to R'$ is a ring homomorphism, then the first isomorphism theorem (Theorem 2.23) applies to the additive groups if one forgets multiplication: if I is the kernel of f, then I is a subgroup of R, im f is a subgroup of R', and there is an isomorphism $\varphi: R/I \to \text{im } f$ given by $r + I \mapsto f(r)$. If one now remembers the multiplication, then it is easy to check that I is an ideal in R, im f is a subring of R', and the group isomorphism φ is also a ring isomorphism. The analogue of the correspondence theorem (Theorem 2.27) is valid: If I is an ideal in R, then there is a bijection between all intermediate ideas J with $I \subset J \subset R$ and all ideals of R/I.

If $a, b \in R$, then a ***divides*** b in R (or b is a ***multiple*** of a), denoted by $a|b$, if there is $r \in R$ with $ar = b$. Note that a divides b if and only if $b \in (a)$. If $a_1, \ldots, a_n \in R$, then a ***common divisor*** is an element $c \in R$ which divides each a_i; a ***greatest common divisor*** (gcd), denoted by $d = (a_1, \ldots, a_n)$, is a common divisor which is divisible by every common divisor.

Theorem VI.2. *If R is a PID, then every set of elements $a_1, \ldots, a_n \in R$ has a* gcd d; *moreover, d is a linear combination of $a_1, \ldots, a_n$.*

Proof. The set of all linear combinations of $a_1, \ldots, a_n$ is an ideal I in R. Since R is a PID, I is a principal ideal; let d be a generator. As any element of I, d is a linear combination of $a_1, \ldots, a_n$. It follows that any common divisor c of the a_i also divides d. But d itself is a common divisor, for each a_i lies in (d). ■

We have just shown that $(a_1, \ldots, a_n) = (d)$, where d is a gcd of $a_1, \ldots, a_n$.

An element $u \in R$ is a ***unit*** if $u|1$; that is, there exists $v \in R$ with $uv = 1$. Two elements $a, b \in R$ are ***associates*** if there is a unit u with $a = bu$. An element in R is ***irreducible*** if it is not a unit and its only factors are units and associates. In $\mathbb{Z}$, the irreducibles have the form $\pm p$, where p is a prime; if k is a field, a polynomial $p(x) \in k[x]$ is irreducible if it is not constant and there is no factorization $p(x) = f(x)g(x)$ with degree $f(x) <$ degree $p(x)$ and degree $g(x) <$ degree $p(x)$.

Theorem VI.3. *If R is a PID and $a_1, \ldots, a_n \in R$, then any two* gcd's *of $a_1, \ldots, a_n$ are associates.*

Proof. We may assume that $d \neq 0$ (otherwise $a_i = 0$ for all i). If d and d' are gcd's, then each divides the other: there are $u, v \in R$ with $d = d'u$ and $d' = dv$. Hence, $d = d'u = dvu$, and the cancellation law gives $1 = vu$. Therefore, u is a unit and d and d' are associates. ■

If $R = \mathbb{Z}$, then the only units are 1 and -1, so that two gcd's of $a_1, \ldots, a_n$ differ only in sign. If $R = k[x]$, where k is a field, then the only units are nonzero constants; thus, only one gcd $d(x)$ of $f_1(x), \ldots, f_n(x)$ is ***monic***; that is, it has leading coefficient 1 (i.e., the coefficient of the highest power of x is 1). Thus, if not all $a_i(x) = 0$, any gcd of $a_1(x), \ldots, a_n(x)$ in $k[x]$ is an associate of a monic polynomial.

In light of Theorem VI.3, we may speak of *the* gcd of elements in a PID, and we change the definition, in the special cases of $\mathbb{Z}$ and of $k[x]$, so that the gcd is either positive or monic. The gcd of two elements a and b is denoted by (a, b); we say that a and b are ***relatively prime*** if $(a, b) = 1$.

The ***euclidean algorithm*** shows how to compute gcd's in $\mathbb{Z}$ and in $k[x]$, but we shall not need this here.

Two integers a and b are relatively prime if and only if there are integers s and t with $sa + tb = 1$. In other words, a and b are relatively prime if and only if $[a \ \ b]$ is the first row of a matrix having determinant 1 (let the second row be $[-t \ \ s]$). This result can be generalized. Call a matrix A ***unimodular*** if $\det A = 1$.

Theorem VI.4. *A set $a_1, \ldots, a_n$ of integers is relatively prime if and only if there is a unimodular $n \times n$ matrix A with entries in $\mathbb{Z}$ whose first row is $[a_1 \ldots a_n]$.*

Proof. We prove the theorem by induction on $n \geq 2$, the base step being our observation above. For the inductive step, let d be the gcd of $a_1, \ldots, a_{n-1}$, and let $a_i = db_i$ for $i \leq n - 1$. The integers $b_1, \ldots, b_{n-1}$ are relatively prime (there is no common divisor $c > 1$), and so there exists an $(n-1) \times (n-1)$ unimodular matrix B with entries in $\mathbb{Z}$ whose first row is $[b_1, \ldots, b_{n-1}]$. Now d and a_n are relatively prime (lest there be a common divisor of $a_1, \ldots, a_n$), so there are integers s and t with $sd + ta_n = 1$. If C denotes the lower $n - 2$ rows of B, define A to be the $n \times n$ matrix

$$A = \begin{bmatrix} db_1 & \cdots & db_{n-1} & a_n \\ & C & & 0 \\ -tb_1 & \cdots & -tb_{n-1} & s \end{bmatrix};$$

note that the first row of A is $[a_1 \ \cdots \ a_n]$. Expanding down the last column,

$$\det A = (-1)^{n+1} a_n \det \begin{bmatrix} C \\ -tb \end{bmatrix} + (-1)^{2n} s \det \begin{bmatrix} db \\ C \end{bmatrix}.$$

Now $\det \begin{bmatrix} db \\ C \end{bmatrix} = d \det B = d$ and $\det \begin{bmatrix} C \\ -tb \end{bmatrix} = -t \det \begin{bmatrix} C \\ b \end{bmatrix} =$ $(-t)(-1)^{n+1}(-1)^{n-2} = t$ (because $\begin{bmatrix} C \\ b \end{bmatrix}$ is obtained from B by successively interchanging the top row b with each of the $n - 2$ rows below it). Hence, $\det A = (-1)^{2n+2} ta_n + sd = ta_n + sd = 1$, so that A is unimodular.

For the converse, let A be a unimodular $n \times n$ matrix with entries in $\mathbb{Z}$ whose first row is $[a_1 \ \cdots \ a_n]$. Evaluating the determinant by expanding across the top row gives $1 = \det A$ as a $\mathbb{Z}$-linear combination of $a_1, \ldots, a_n$, and this shows that $a_1, \ldots, a_n$ are relatively prime. ■

Theorem VI.5. *Let R be a* PID *and let $p \in R$ be irreducible. If $I = (p)$ is the principal ideal generated by p, then R/I is a field.*

Proof. If $f + I \in R/I$ is not the zero element, then $f \notin I = (p)$; that is, p does not divide f. Since p is irreducible, its only divisors are units and associates. Therefore, $d = \gcd(p, f)$ is either a unit or an associate of p. But if d were an associate of p, then p would divide f, a contradiction. Hence, $d = 1$, and so there are elements $a, b \in R$ with $1 = ap + bf$. Thus, $bf - 1 \in (p) = I$, so that $(b + I)(f + I) = 1 + I$; that is, $f + I$ is a unit in R/I, and so R/I is a field. ■

Corollary VI.6.

(i) *If p is a prime, then $\mathbb{Z}_p = \mathbb{Z}/(p)$ is a field.*
(ii) *If k is a field and $p(x) \in k[x]$ is an irreducible polynomial, then $k[x]/(p(x))$ is a field.*

When we say that an element a is a product of irreducibles we allow the possibility that a itself is irreducible (there is only one factor).

A domain R is a ***unique factorization domain*** (or **UFD**) if:

(i) every $r \in R$ that is neither 0 nor a unit is a product of irreducibles; and
(ii) if $p_1 \dots p_m = q_1 \dots q_n$, where the p_i and q_j are irreducible, then there is a bijection between the sets of factors (so $m = n$) such that corresponding factors are associates.

We are going to prove that every principal ideal domain R is a unique factorization domain, and our first task is to show that if $r \in R$ is neither 0 nor a unit, then it is a product of irreducibles.

Lemma VI.7. *If R is a PID, then there is no infinite strictly increasing sequence of ideals*

$$I_1 < I_2 < \cdots < I_n < I_{n+1} < \cdots.$$

Proof. If such a sequence exists, then it is easy to check that $J = \bigcup_{n \geq 1} I_n$ is an ideal. Since R is a PID, there is $a \in J$ with $J = (a)$. Now a got into J by being in some I_n. Hence

$$J = (a) \leq I_n < I_{n+1} \leq J,$$

a contradiction. ■

Lemma VI.8. *If R is a PID and $a \in R$ is neither* 0 *nor a unit, then a is a product of irreducibles.*

Proof. If $r \in R$ has a factorization $r = bc$, where neither b nor c is a unit, then we say that b is a *proper factor* of a. It is easy to check, using the hypothesis that R is a domain, that if b is a proper factor of a, then $(a) < (b)$.

Call an element $a \in R$ *good* if it is a product of irreducibles; otherwise, a is *bad*. If a and b are good, then so is their product ab. Thus, if a is bad, then it factors (for irreducibles are good), and at least one of its proper factors is bad. Suppose there is a bad element $a = a_0$ which is neither 0 nor a unit. Assume inductively that there exists $a_0, a_1, \ldots, a_n$ such that each a_{i+1} is a proper bad factor of a_i. Since a_n is bad, it has a proper bad factor a_{n+1}. By induction, there exist such a_n for all $n \geq 0$. There is thus an infinite strictly increasing sequence of ideals $(a_0) < (a_1) < \cdots < (a_n) < (a_{n+1}) < \cdots$, contradicting the previous lemma. Therefore, every $a \in R$ that is neither 0 nor a unit is good; that is, a is a product of irreducibles. ∎

Theorem VI.9 (Euclid's Lemma). *Let R be a PID and let $p \in R$ be irreducible. If $a, b \in R$ and $p|ab$, then $p|a$ or $p|b$.*

Proof. If $p|a$, we are done. Otherwise, p does not divide a, and so the gcd $d = (p, a) = 1$ (as we saw in the proof of Theorem VI.5). By Theorem VI.2, there are elements $s, t \in R$ with $sp + ta = 1$. Therefore, $b = spb + t(ab)$. Since $p|ab$, it follows that $p|b$. ∎

Theorem VI.10 (Fundamental Theorem of Arithmetic). *Every principal ideal domain R is a unique factorization domain.*

Proof. By Lemma VI.8, every $a \in R$ that is neither 0 nor a unit is a product of irreducibles. We have only to verify the uniqueness of such a factorization.

If $p_1 \cdots p_m = q_1 \cdots q_n$, where the p_i and the q_j are irreducible, then $p_1 | q_1 \cdots q_n$. By Euclid's lemma, p_1 divides some q_j. Since q_j is irreducible, p_1 and q_j are associates: there is a unit $u \in R$ with $q_j = up_1$. As R is a domain, we may cancel p_1 to obtain

$$(up_2)p_3 \cdots p_m = \prod_{l \neq j} q_l,$$

and the proof is completed by an induction on $\max\{m, n\}$. ∎

A standard proof of Corollary VI.6 uses the notion of prime ideal, which makes Euclid's lemma into a definition. An ideal I in a commutative ring R is called a ***prime ideal*** if $I \neq R$ and $ab \in I$ implies $a \in I$ or $b \in I$. Note that the ideal 0 is prime if and only if R is a domain.

Theorem VI.11. *A nonzero ideal I in a PID R is prime if and only if $I = (p)$, where p is irreducible.*

Proof. Since R is a PID, there is an element d with $I = (d)$. Assume that I is prime. If d is not irreducible, then $d = ab$, where neither a nor b is a unit. By hypothesis, either $a \in I$ or $b \in I$. But if, say, $a \in I$, then $a = rd$ for some $r \in R$, and hence $d = ab = rdb$. Therefore $1 = rb$, contradicting b not being a unit.

Conversely, assume that p is irreducible and that $ab \in (p)$. Thus, $p|ab$, so that Euclid's lemma gives $p|a$ or $p|b$; that is, $a \in (p)$ or $b \in (p)$. ∎

Theorem VI.12. *An ideal I in a commutative ring R is prime if and only if R/I is a domain.*

Proof. Recall that the zero element of R/I is $0 + I = I$. Assume that I is prime. If $I = (a + I)(b + I) = ab + I$, then $ab \in I$, hence $a \in I$ or $b \in I$; that is, $a + I = I$ or $b + I = I$. Hence, R/I is a domain. Conversely, assume that R/I is a domain and that $ab \in I$. Thus, $I = ab + I = (a + I)(b + I)$. By hypothesis, one of the factors must be zero; that is, $a + I = I$ or $b + I = I$. Hence, $a \in I$ or $b \in I$, so that I is prime. ■

An ideal I in a commutative ring R is a ***maximal ideal*** if $I \neq R$ and there is no ideal J with $I \subsetneqq J \subsetneqq R$. Note that the ideal 0 is maximal if and only if R is a field, for if $a \in R$ is not zero, then the ideal (a) must be all of R. Hence $1 \in (a)$, so there is $r \in R$ with $1 = ra$; that is, a is a unit.

Theorem VI.13. *An ideal I in a commutative ring R is maximal if and only if R/I is a field.*

Proof. If I is maximal, then the correspondence theorem for rings shows that R/I has no proper ideals. Hence the zero ideal is maximal and R/I is a field. Conversely, if there is an ideal J with $I \subsetneqq J \subsetneqq R$, then the correspondence theorem shows that J/I is a proper nonzero ideal in R/I, contradicting the hypothesis tht R/I is a field (the ideal 0 is maximal). ■

Since every field is a domain, if follows that every maximal ideal in a commutative ring is a prime ideal. The converse is not always true; for example, the ideal $I = (x)$ in $\mathbb{Z}[x]$ is prime (for $\mathbb{Z}[x]/I \cong \mathbb{Z}$ is a domain), but it is not maximal because $\mathbb{Z}$ is not a field (or, because $(x) \subsetneqq (x, 2) \subsetneqq \mathbb{Z}[x]$).

Theorem VI.14. *If R is a PID, then every nonzero prime ideal I is maximal.*

Proof. Assume that J is an ideal with $I \subsetneqq J$. Since R is a PID, there are elements $a, b \in R$ with $I = (a)$ and $J = (b)$. Now $I \subsetneqq J$ gives $a \in J = (b)$, so there is $r \in R$ with $a = rb$. But I is prime, so that $b \in I$ or $r \in I$. If $b \in I$, then $J = (b) \subset I$, contradicting $I \subsetneqq J$. Therefore, $r \in I$, so there is $s \in R$ with $r = sa$. Hence, $a = rb = sab$; as R is a domain, $1 = sb \in (b) = J$. Therefore, $J = R$ and I is maximal. ■

One can now give a second proof of Theorem VI.5. If R is a PID and $p \in R$ is irreducible, then $I = (p)$ is a nonzero prime ideal, by Theorem VI.11. By Theorem VI.14, I is a maximal ideal, and so Theorem VI.13 shows that R/I is a field.

Lemma VI.15. *Let k be a field and let $p(x) \in k[x]$ be irreducible. Then there is a field K containing a subfield isomorphic to k and a root of $p(x)$.*

Proof. Let $I = (p(x))$ and let $K = k[x]/I$; since $p(x)$ is irreducible, Theorem VI.5 shows that K is a field. It is easy to check that the family of all cosets of the form $a + I$, where $a \in k$, is a subfield of K isomorphic to k. Let $p(x) = \sum a_i x^i$. We claim that the element $x + I \in K$ is a root of $p(x)$.

$$\begin{aligned} p(x + I) &= \sum a_i(x + I)^i \\ &= \sum a_i(x^i + I) \\ &= \sum (a_i x^i + I) \\ &= (\sum a_i x^i) + I \\ &= p(x) + I = I. \end{aligned}$$

The result follows, for $I = 0 + I$ is the zero element of K. ∎

Theorem VI.16. *If k is a field and $f(x) \in k[x]$, then there is a field F containing k over which $f(x)$ is a product of linear factors; that is, F contains all the roots of $f(x)$.*

Proof. The proof is by induction on n, the degree of $f(x)$. If $n = 1$, then $f(x)$ is linear and we may set $F = k$. If $n > 1$, there is a factorization $f(x) = p(x)g(x)$ in $k[x]$, where $p(x)$ is irreducible (perhaps $g(x) = 1$). By Lemma VI.15, there is a field K containing k and a root β of $p(x)$. There is thus a factorization $p(x) = (x - \beta)h(x)$ in $K[x]$. Since degree $h(x)g(x) < n$, the inductive hypothesis gives a field F containing K over which $h(x)g(x)$, hence $f(x) = (x - \beta)h(x)g(x)$, is a product of linear factors. ∎

If k is a field and $f(x) = a_n x^n + a_{n-1}x^{n-1} + \cdots + a_0 \in k[x]$, then its ***derivative*** is

$$f'(x) = na_n x^{n-1} + (n - 1)a_{n-1}x^{n-2} + \cdots + a_1.$$

It is easy to check that the usual formulas of Calculus hold for derivatives of sums and products of polynomials over arbitrary fields: $(f(x) + g(x))' = f'(x) + g'(x)$; $(f(x)g(x))' = f(x)g'(x) + f'(x)g(x)$.

Lemma VI.17. *Let k be a field, let $f(x) \in k[x]$, and let F be a field containing k which contains all the roots of $f(x)$. Then $f(x)$ has no repeated roots if and only if $f(x)$ and $f'(x)$ are relatively prime.*

Proof. If $f(x)$ has a repeated root, then $f(x) = (x - \beta)^2 g(x)$ in $F[x]$. Hence, $f'(x) = 2(x - \beta)g(x) + (x - \beta)^2 g'(x)$, so that $x - \beta$ is a common divisor of $f(x)$ and $f'(x)$ and they are not relatively prime. Conversely, assume that $x - \beta$ is a common divisor of $f(x)$ and $f'(x)$: say, $f(x) = (x - \beta)g(x)$ and $f'(x) = (x - \beta)h(x)$. By the product formula for derivatives, $f'(x) = (x - \beta)g'(x) + g(x)$, so that $(x - \beta)g'(x) + g(x) = (x - \beta)h(x)$. Therefore, $x - \beta$ divides $g(x)$, $(x - \beta)^2$ divides $f(x)$, and $f(x)$ has a repeated root. ∎

Lemma VI.18. *If k is a field of characteristic $p > 0$, then for all $a, b \in k$ and for all $n \geq 1$,*

$$(a + b)^{p^n} = a^{p^n} + b^{p^n}.$$

Proof. The proof is by induction on n. If $n = 1$, expand $(a + b)^p$ by the binomial theorem. Since p is prime, it divides all the middle binomial coefficients, and hence each of them is 0 mod p. The inductive step is easy. ■

Theorem VI.19 (Galois). *For every prime p and every $n \geq 1$, there exists a field having exactly p^n elements.*

Proof. Let $q = p^n$ and let $f(x) = x^q - x \in \mathbb{Z}_p[x]$. By Theorem VI.16, there is a field F containing $\mathbb{Z}_p$ (so that F has characteristic p) and all the roots of $f(x)$. Define K to be the set of all the roots of $f(x)$ in F. Since $f(x)$ has degree q, it follows that $|K| \leq q$, with equality if $f(x)$ has no repeated roots. Now $f'(x) = qx^{q-1} - 1 = -1$, because F has characteristic p and $q = p^n$. Therefore, $f(x)$ and $f'(x)$ are relatively prime, and Lemma VI.17 shows that $f(x)$ has no repeated roots.

We now show that K is a subfield of F. Let a and b be roots of $f(x)$, so that $a^q = a$ and $b^q = b$. Lemma VI.18 gives $(a - b)^q = a^q - b^q = a - b$, so that $a - b \in K$; moreover, $(ab)^q = a^q b^q = ab$, so that $ab \in K$. Since $1 \in K$, it follows that K is a subring of F. Finally, if $a \neq 0$, then $a^q = a$ implies that $a^{q-1} = 1$, so that $a^{-1} = a^{q-2} \in K$ (because K is a subring). Therefore, K is a field. ■

It is curious that the uniqueness of the finite fields was not established for more than 60 years after their discovery.

Theorem VI.20 (E.H. Moore, 1893). *Any two fields having exactly p^n elements are isomorphic.*

Proof. Let K be a field with exactly $q = p^n$ elements. Since the multiplicative group $K^\times$ of nonzero elements of K has order $q - 1$, Lagrange's theorem gives $a^{q-1} = 1$ for all nonzero $a \in K$. Therefore, every element of K is a root of $f(x) = x^q - x$, so that K is a splitting field of $f(x)$. The result follows from the general fact that any two splitting fields of a given polynomial are isomorphic (the reader may prove this by induction on the degree of $f(x)$, using Lemma 5.5 of the text). ■

The following proof is the polynomial version of the fact that every congruence class $[a] \in \mathbb{Z}_m$ is equal to $[r]$, where r is the remainder after dividing a by m; moreover, any two "remainders," that is, any distinct r and r' between 0 and $r - 1$ are not congruent.

Theorem VI.21. *Let k be a field and let $p(x) \in k[x]$ be an irreducible polyno-*

mial of degree n. If k is a subfield of a field E and if $\alpha \in E$ is a root of $p(x)$, then $[k(\alpha):k] = n$. Indeed, every element in $k(\alpha)$ has a unique expression of the form

$$(*) \qquad b_0 + b_1\alpha + \cdots + b_{n-1}\alpha^{n-1},$$

where $b_i \in k$ for all i.

Proof. The map $\varphi: k[x] \to E$, defined by $f(x) \mapsto f(\alpha)$, is a ring homomorphism with im $\varphi = k(\alpha)$ and ker $\varphi = I \neq 0$ (for $p(x) \in I$). By Theorem VI.11, $I = (q(x))$, where $q(x)$ is irreducible. Hence $q(x)|p(x)$, so that $q(x) = cp(x)$ for some nonzero $c \in k$; that is, $I = (p(x))$. We claim that $X = \{1, \alpha, \ldots, \alpha^{n-1}\}$ is a basis of $k(\alpha)$ viewed as a vector space over k. If so, then $[k(\alpha):k] = n$, as desired. If there are b_i, not all 0, with $\sum_{i=0}^{n-1} b_i\alpha^i = 0$, then α is a root of $g(x) = \sum_{i=0}^{n-1} b_i x^i$, a nonzero polynomial of degree smaller than n. But $g(x) \in$ ker $\varphi = (p(x))$, so that $p(x)|g(x)$ and the degree of $g(x)$ is at least n, a contradiction. Thus, X is linearly independent. To see that X spans $k(\alpha)$, we show, by induction on $m \geq n$, that $\alpha^m \in W = \langle 1, \alpha, \ldots, \alpha^{n-1} \rangle$, the subspace over k spanned by X. If $p(x) = \sum_{i=0}^{n} c_i x^i$, then $0 = p(\alpha) = \sum_{i=0}^{n} c_i\alpha^i$, so that $\alpha^n = -\sum_{i=0}^{n-1} c_i\alpha^i \in W$. For the inductive step, assume that $\alpha^m = \sum_{i=0}^{n-1} d_i\alpha^i$, where $d_i \in k$. Then $\alpha^{m+1} = \alpha\alpha^m = \alpha\sum_{i=0}^{n-1} d_i\alpha^i = \sum_{i=0}^{n-2} d_i\alpha^{i+1} + d_{n-1}\alpha^n \in W$. ■

Bibliography

Albert, A.A., *Introduction to Algebraic Theories*, University of Chicago Press, 1941.
Alperin, J., *Local Representation Theory*, Cambridge University Press, 1986.
Artin, E., *Galois Theory*, Notre Dame, 1955.
———, *Geometric Algebra*, Interscience, 1957.
Aschbacher, M., *Finite Group Theory*, Cambridge University Press, 1986.
———, *Sporadic Groups*, Cambridge University Press, 1994.
Babakhanian, A., *Cohomological Methods in Group Theory*, Dekker, 1972.
Baumslag, G., *Reviews on Infinite Groups*, Amer. Math. Soc., 1974.
Benson, D., *Representations and Cohomology*, 2 volumes, Cambridge University Press, 1991.
Biggs, N.L., *Discrete Mathematics*, Oxford University Press, 1989.
Biggs, N.L., and White, A.T., *Permutation Groups and Combinatorial Structures*, Cambridge University Press, 1979.
Birkhoff, G., and Mac Lane, S., *A Survey of Modern Algebra*, Macmillan, 1977.
Borel, A., *Linear Algebraic Groups*, Benjamin, 1969.
Brown, K., *Cohomology of Groups*, Springer-Verlag, 1982.
Burnside, W., *The Theory of Groups of Finite Order*, Cambridge University Press, 1911.
Cameron, P.J., *Parallelisms of Complete Designs*, Cambridge University Press, 1976.
Carmichael, R., *An Introduction to the Theory of Groups*, Ginn, 1937.
Cartan, H., and Eilenberg, S., *Homological Algebra*, Princeton University Press, 1956.
Carter, R.W., *Simple Groups of Lie Type*, Wiley, 1972.
———, *Finite Groups of Lie Type: Conjugacy Classes and Complex Characters*, Wiley, 1985.
Chandler, B., and Magnus, W., *The History of Combinatorial Group Theory: A Case Study in the History of Ideas*, Springer-Verlag, 1982.
Cohn, P.M., *Universal Algebra*, Harper & Row, 1965.
———, *Algebra*, 2nd edn., vol. I, Wiley, 1982.
Conway, J.H., Curtis, R.T., Norton, S.P., Parker, R.A., and Wilson, R.A., *Atlas of Finite Groups*, Oxford University Press, 1985.
Conway, J.H., and Sloane, N.J.A., *Sphere Packings, Lattices, and Groups*, Springer-Verlag, 1993.

Coxeter, H.S.M., *Projective Geometry*, Springer-Verlag, 1987.
Coxeter, H.S.M., and Moser, W.O., *Generators and Relations for Discrete Groups*, Springer-Verlag, 1965.
Curtis, C., and Reiner, I., *Representation Theory of Finite Groups and Associative Algebras*, Wiley, 1962.
Davis, M., *Computability and Unsolvability*, McGraw-Hill, 1958.
Dicks, W., and Dunwoody, M.J., *Groups Acting on Graphs*, Cambridge University Press, 1989.
Dickson, L.E., *Linear Groups*, Leipzig, 1901; Dover, 1958.
Dieudonné, J., *Sur les Groupes Classiques*, Hermann, 1958.
Dixon, J.D., *Problems in Group Theory*, Blaisdell, 1967.
Dixon, J.D., du Sautoy, M.P.F., Mann, A., and Segal, D., *Analytic Pro-p-Groups*, Cambridge University Press, 1991.
Doerk, K., and Hawkes, T., *Finite Soluble Groups*, de Gruyter, 1992.
Dornhoff, L., *Group Representation Theory*, 2 volumes, Marcel Dekker, 1971.
Epstein, D.B.A., Cannon, J.W., Holt, D.F., Levy, S.V.F., Paterson, M.S., and Thurston, W.P., *Word Processing in Groups*, Jones and Bartlett, 1992.
Evens, L., *The Cohomology of Groups*, Oxford University Press, 1991.
Feit, W., *Characters of Finite Groups*, Benjamin, 1967.
———, *The Representation Theory of Finite Groups*, North-Holland, 1982.
Fuchs, L., *Infinite Abelian Groups*, I and II, Academic Press, 1970 and 1973.
Gersten, S.M., editor, *Essays in Group Theory*, Springer-Verlag, 1987.
Gorenstein, D., *Finite Groups*, Harper & Row, 1968.
———, *Reviews on Infinite Groups*, Amer. Math. Soc., 1974.
———, *Finite Simple Groups*, Plenum, 1982.
———, *The Classification of Finite Simple Groups*; Volume 1: *Groups of Non-characteristic 2 type*, Plenum, 1983.
Griffith, P.A., *Infinite Abelian Group Theory*, University of Chicago Press, 1970.
Gruenberg, K.W., *Cohomological Topics in Group Theory*, Springer SLN 143, 1970.
Gruenberg, K.W., and Weir, A.J., *Linear Geometry*, Springer-Verlag, 1977.
Hall, M., Jr., *The Theory of Groups*, Macmillan, 1959.
Hall, P., *Collected Works*, Oxford University Press, 1987.
Higgins, P.J., *Notes on Categories and Groupoids*, van Nostrand–Reinhold, 1971.
Higman, G., *Existentially Closed Groups*, Oxford University Press, 1988.
Humphreys, J.E., *Reflection Groups and Coxeter Groups*, Cambridge University Press, 1990.
Huppert, B., *Endliche Gruppen*, Springer-Verlag, 1967.
Huppert, B., and Blackburn, N., *Finite Groups*, II and III, Springer-Verlag, 1981 and 1982.
Isaacs, I.M., *Character Theory of Finite Groups*, Academic Press, 1976.
Jacobson, N., *Basic Algebra*, I and II, Freeman, 1974 and 1979.
James, G., and Liebeck, M., *Representations and Characters of Groups*, Cambridge, 1993.
Johnson, D.L., *Topics in the Theory of Group Presentations*, Cambridge University Press, 1980.
Jordan, C., *Traité des Substitutions et des Équations Algébriques*, Gauthier-Villars, 1870.
Kaplansky, I., *Infinite Abelian Groups*, University of Michigan Press, 1969.
———, *Fields and Rings*, University of Chicago Press, 1972.
Karpilovsky, G., *The Schur Multiplier*, Oxford University Press, 1987.
Kegel, O.H., and Wehrfritz, B.A.F., *Locally Finite Groups*, North-Holland, 1973.
Khukhro, E.I., *Nilpotent Groups and their Automorphisms*, de Gruyter, 1993.
Kurosh, A.G., *The Theory of Groups*, 2 volumes, Chelsea, 1956.

Lyndon, R., and Schupp, P., *Combinatorial Group Theory*, Springer-Verlag, 1977.
Mac Lane, S., *Homology*, Springer-Verlag, 1963.
Magnus, W., Karrass, A., and Solitar, D., *Combinatorial Group Theory*, Wiley, 1966.
Massey, W.S., *Algebraic Topology: An Introduction*, Harcourt, Brace, and World, 1967.
Miller, C.F., III, *On Group-Theoretic Decision Problems and Their Classification*, Princeton University Press, 1971.
Miller, G.A., Blichtfeld, H.F., and Dickson, L.E., *Theory and Applications of Finite Groups*, Wiley, 1916.
Milnor, J., *Introduction to Algebraic K-Theory*, Princeton, 1971.
Neumann, H., *Varieties of Groups*, Springer-Verlag, 1967.
Passman, D.S., *Permutation Groups*, Benjamin, 1968.
Powell, M.B., and Higman, G. (eds), *Finite Simple Groups*, Academic Press, 1971.
Puttaswamaiah, B.M., and Dixon, J.D., *Modular Representations of Finite Groups*, Academic Press, 1977.
Robinson, D.J.S., *Finiteness Conditions and Generalized Soluble Groups*, 2 volumes, Springer-Verlag, 1972.
———, *A Course in the Theory of Groups*, Springer-Verlag, 1982.
Rotman, J.J., *An Introduction to Homological Algebra*, Academic Press, 1979.
———, *An Introduction to Algebraic Topology*, Springer-Verlag, 1988.
———, *Galois Theory*, Springer-Verlag, 1990.
Schenkman, E., *Group Theory*, van Nostrand, 1965.
Scott, W.R., *Group Theory*, Prentice-Hall, 1964.
Serre, J.-P., *Linear Representations of Finite Groups*, Springer-Verlag, 1977.
———, *Trees*, Springer-Verlag, 1980.
Specht, W., *Gruppentheorie*, Springer-Verlag, 1956.
Speiser, A., *Die Theorie der Gruppen von Endlicher Ordnung*, Springer-Verlag, 1927.
Stammbach, U., *Homology in Group Theory*, Springer SLN 359, 1980.
Suzuki, M., *Group Theory*, I and II, Springer-Verlag, 1982 and 1986.
van der Waerden, B.L., *Modern Algebra*, Ungar, 1948.
Vaughan-Lee, M., *The Restricted Burnside Problem*, Second Edition, Oxford University Press, 1993.
Wehrfritz, B.A.F., *Infinite Linear Groups*, Springer-Verlag, 1973.
Weiss, E., *Cohomology of Groups*, Academic Press, 1969.
Weyl, H., *Symmetry*, Princeton University Press, 1952.
Wielandt, H., *Finite Permutation Groups*, Academic Press, 1964.
Wussing, H., *Die Genesis des abstrakten Gruppenbegriffes*, VEB Deutscher Verlag der Wissenschaften, 1969 (English translation, *Genesis of the Abstract Group Concept*, MIT Press, 1984).
Zassenhaus, H., *The Theory of Groups*, Chelsea, 1956.
Zieschang, H., Vogt, E., and Coldewey, H.-D., *Surfaces and Planar Discontinuous Groups*, Springer SLN 835, 1980.

Notation

Set Theory and Algebra

$A \subset B$ or $B \supset A$	A is a subset of B
$\varnothing$	empty set
$A \times B$	cartesian product
$f\colon A \to B$	f is a function from A to B
1_X	the identity function $X \to X$
$i\colon A \hookrightarrow B$	i is the inclusion map from the subset $A \subset B$
$f\colon a \mapsto f(a)$	f is a function whose value on the element a in its domain is $f(a)$
$f\|A$	the restriction of f to A; that is, if $f\colon B \to C$ and $A \subset B$, then $f\|A = f \circ i$, where $i\colon A \hookrightarrow B$
$\|X\|$	the number of elements in a set X
$\binom{m}{i}$	binomial coefficient $m!/i!(m-i)!$
$U(R)$	the group of units in the ring R
$K^\times$	if K is a field, $K^\times = K - \{0\}$; the multiplicative group of $K = U(K)$
$G^\#$	if G is a group, $G^\# = G - \{1\}$
$[x]$	equivalence class containing an element x; in particular, the congruence class of an integer $x \in \mathbb{Z}_m$
$A = [a_{ij}]$	matrix whose entry in row i and column j is a_{ij}
δ_{ij}	(***Kronecker delta***) if $i \neq j$, then $\delta_{ij} = 0$; if $i = j$, then $\delta_{ij} = 1$.
E	identity matrix $[\delta_{ij}]$
$\alpha \simeq \beta$	α and β are homotopic paths

$(A, \mathscr{L}_*(A), \alpha)$	affine space
$P^n(K)$	projective space over K of dimension n

Groups and Subgroups

$A \le B$	A is a subgroup of B
$A < B$	A is a proper subgroup of B
$\langle X \rangle$	subgroup generated by a subset X
$A \vee B$	subgroup generated by subgroups A and B
AB	$\{ab \mid a \in A \text{ and } b \in B\}$
$A \lhd B$	A is a normal subgroup of B
G/H	the family of all left cosets of H in G (it is a group when $H \lhd G$)
H char G	H is a characteristic subgroup of G
$A \times B$	direct product of groups
$\prod A_i$	direct product of groups
$A \oplus B$	direct sum of abelian groups
$\sum A_i$	direct sum of abelian groups
$A * B$	free product of groups
$A *_\theta B$	amalgam
$A \rtimes B$ or $A \rtimes_\theta B$	semidirect product
$A \wr B$ or $A \wr_r B$	wreath product or regular wreath product
a^G	conjugacy class of element $a \in G$
$[A, B]$	subgroup generated by all commutators $[a, b] = aba^{-1}b^{-1}$
G'	commutator subgroup $= [G, G]$
$Z(G)$	center of G
$C_G(a)$	centralizer of element a in G
$C_G(H)$	centralizer of subgroup $H \le G$
$N_G(H)$	normalizer of subgroup $H \le G$
$\mathrm{Aut}(G)$	automorphism group of group G
$\mathrm{Inn}(G)$	all inner automorphisms of G
$\mathcal{O}(x)$ or Gx	orbit of element x in a G-set
G_x	stabilizer of element x in a G-set
$\mathrm{Hol}(G)$	holomorph of G
$\mathrm{Hom}(G, B)$	all homomorphisms from G to B
G^*	character group $= \mathrm{Hom}(G, \mathbb{C}^\times)$ ($\cong \mathrm{Hom}(G, \mathbb{Q}/\mathbb{Z})$ when G is finite)
$U(R)$	group of units in a ring R
$\Phi(G)$	Frattini subgroup
dG	maximal divisible subgroup of abelian group G
tG	torsion subgroup of abelian group G
$(X \mid \Delta)$	presentation with generators X and relations Δ

$(G; X'|\Delta')$ given a presentation of a group G, this is a presentation with additional generators X' and additional relations Δ'

Assume that K is an abelian group and that $\theta: Q \to K$ is a homomorphism (when θ is trivial, one omits it from the notation).

$B^2(Q, K, \theta)$ coboundaries
$Z^2(Q, K, \theta)$ factor sets (cocycles)
$H^2(Q, K, \theta) = Z^2(Q, K, \theta)/B^2(Q, K, \theta)$
$M(Q) = H^2(Q, \mathbb{C}^\times)$ = Schur multiplier
$\text{Ext}(Q, K) = H^2(Q, K)$ when both Q and K are abelian and θ is trivial
$\text{Der}(Q, K, \theta)$ all derivations
$\text{PDer}(Q, K, \theta)$ all principal derivations
$H^1(Q, K, \theta) = \text{Der}(Q, K, \theta)/\text{PDer}(Q, K, \theta)$

Names of Groups

A_n alternating group
A_∞ infinite alternating group
$\text{Aff}(n, K)$ affine group
$\text{Aut}(V)$ all affine isomorphisms of vector space V
$\text{Aut}(X, \mathcal{B})$ all automorphisms of Steiner system $(X, \mathcal{B})$
B binary tetrahedral group
B_m braid group
$\mathbb{C}$ complex numbers
$\mathbb{C}^\times$ multiplicative group of nonzero complex numbers
D_{2n} dihedral group of order $2n$
D_∞ infinite dihedral group
$\text{GF}(q)$ finite field having q elements
$\text{GL}(n, K)$ general linear group of all nonsingular $n \times n$ matrices
$\Gamma\text{L}(n, K)$ all nonsingular semilinear transformations
$\text{LF}(K)$ linear fractional transformations over K
M_i Mathieu group $i = 10, 11, 12, 22, 23, 24$
$\text{O}(n, K)$ orthogonal group
$\text{PGL}(n, K)$ projective linear group
$\text{PSL}(2, \mathbb{Z})$ modular group
$\text{P}\Gamma\text{L}(n, K)$ collineation group
$\mathbf{Q}$ quaternions of order 8
$\mathbf{Q}_n$ generalized quaternions of order 2^n
$\mathbb{Q}$ rational numbers
$\mathbb{R}$ real numbers
S_n symmetric group on n letters
S_X symmetric group on a set X
$\text{SL}(n, K)$ all $n \times n$ matrices of determinant 1

$\mathrm{Sp}(2m, K)$	symplectic group
T	nonabelian group of order $12 = \mathbb{Z}_3 \rtimes \mathbb{Z}_4$
$\mathbf{T}$	circle group = multiplicative group $\{z \in \mathbb{C} \mid \lvert z \rvert = 1\}$
$\mathrm{UT}(n, K)$	unitriangular group
$\mathbf{V}$	4-group
$\mathbb{Z}$	integers
$\mathbb{Z}_n$	integers mod n
$\mathrm{Z}(n, K)$	all scalar $n \times n$ matrices
$\mathrm{SZ}(n, K)$	all scalar $n \times n$ matrices of determinant 1

Index

Graduate Texts in Mathematics

continued from page ii

121 LANG. Cyclotomic Fields I and II. Combined 2nd ed.
122 REMMERT. Theory of Complex Functions. *Readings in Mathematics*
123 EBBINGHAUS/HERMES et al. Numbers. *Readings in Mathematics*
124 DUBROVIN/FOMENKO/NOVIKOV. Modern Geometry—Methods and Applications. Part III.
125 BERENSTEIN/GAY. Complex Variables: An Introduction.
126 BOREL. Linear Algebraic Groups.
127 MASSEY. A Basic Course in Algebraic Topology.
128 RAUCH. Partial Differential Equations.
129 FULTON/HARRIS. Representation Theory: A First Course. *Readings in Mathematics*
130 DODSON/POSTON. Tensor Geometry.
131 LAM. A First Course in Noncommutative Rings.
132 BEARDON. Iteration of Rational Functions.
133 HARRIS. Algebraic Geometry: A First Course.
134 ROMAN. Coding and Information Theory.
135 ROMAN. Advanced Linear Algebra.
136 ADKINS/WEINTRAUB. Algebra: An Approach via Module Theory.
137 AXLER/BOURDON/RAMEY. Harmonic Function Theory.
138 COHEN. A Course in Computational Algebraic Number Theory.
139 BREDON. Topology and Geometry.
140 AUBIN. Optima and Equilibria. An Introduction to Nonlinear Analysis.
141 BECKER/WEISPFENNING/KREDEL. Gröbner Bases. A Computational Approach to Commutative Algebra.
142 LANG. Real and Functional Analysis. 3rd ed.
143 DOOB. Measure Theory.
144 DENNIS/FARB. Noncommutative Algebra.
145 VICK. Homology Theory. An Introduction to Algebraic Topology. 2nd ed.
146 BRIDGES. Computability: A Mathematical Sketchbook.
147 ROSENBERG. Algebraic K-Theory and Its Applications.
148 ROTMAN. An Introduction to the Theory of Groups. 4th ed.
149 RATCLIFFE. Foundations of Hyperbolic Manifolds.
150 EISENBUD. Commutative Algebra with a View Toward Algebraic Geometry.
151 SILVERMAN. Advanced Topics in the Arithmetic of Elliptic Curves.
152 ZIEGLER. Lectures on Polytopes.
153 FULTON. Algebraic Topology: A First Course.
154 BROWN/PEARCY. An Introduction to Analysis.
155 KASSEL. Quantum Groups.
156 KECHRIS. Classical Descriptive Set Theory.